Making Hard Decisions
with DecisionTools®

www.duxbury.com

Quality and innovation

Duxbury Titles of Related Interest

Albright, Winston & Zappe, *Data Analysis & Decision Making with Microsoft Excel*
Albright, Winston & Zappe, *Managerial Statistics*
Bell & Schleifer, *Decision Making Under Uncertainty*
Bell & Schleifer, *Risk Management*
Berk & Carey, *Data Analysis with Microsoft Excel*
Brightman, *Data Analysis in Plain English*
Canavos & Miller, *Introduction to Modern Business Statistics,* 2nd ed.
Carlson, *Cases in Managerial Data Analysis*
Clemen, *Making Hard Decisions: An Introduction to Decision Analysis,* 2nd ed.
Davis, *Business Research for Decision Making,* 5th ed.
Derr, *Statistical Consulting: A Guide to Effective Communication*
Farnum, *Modern Statistical Quality Control and Improvement*
Hildebrand & Ott, *Statistical Thinking for Managers,* 4th ed.
Johnson, *Applied Multivariate Methods for Data Analysts*
Keller & Warrack, *Statistics for Management and Economics,* 5th ed.
Kenett & Zacks, *Modern Industrial Statistics: Design of Quality and Reliability*
Kirkwood, *Strategic Decision Making: Multiobjective Decision Analysis with Spreadsheets*
Kuehl, *Design of Experiments,* 2nd ed.
Larsen, Marx & Cooil, *Statistics for Applied Problem Solving and Decision Making*
Lohr, *Sampling: Design and Analysis*
Lunneborg, *Data Analysis by Resampling*
MathSoft Inc., *S-Plus for Windows: Student Edition*
Middleton, *Data Analysis Using Microsoft Excel*
Minh, *Applied Probability Models*
Newton & Harvill, *StatConcepts: A Visual Tour of Statistical Ideas*
SAS Institute Inc., *JMP-IN: Statistical Discovery Software*
Savage, *INSIGHT: Business Analysis Software for Microsoft Excel*
Schaeffer, Mendenhall & Ott, *Elementary Survey Sampling,* 5th ed.
Schleifer & Bell, *Data Analysis, Regression and Forecasting*
Schleifer & Bell, *Decision Making Under Certainty*
Schrage, *Optimization Modeling Using LINDO,* 5th ed.
Shapiro, *Modeling the Supply Chain*
Weiers, *Introduction to Business Statistics,* 3rd ed.
Winston, *Simulation Modeling Using @Risk*
Winston & Albright, *Practical Management Science,* 2nd ed.
Winston, *Operations Research: Applications and Algorithms,* 3rd ed.

To order copies, contact your local bookstore or call 1-800-354-9706. For more information, contact Duxbury at 511 Forest Lodge Road, Pacific Grove, CA, 93950, or go to: **www.duxbury.com**

Making Hard Decisions
with
DecisionTools®

Robert T. Clemen

Fuqua School of Business
Duke University

Terence Reilly

Gillette Professor
Babson College

Australia • Canada • Mexico • Singapore • Spain • United Kingdom • United States

Sponsoring Editor: *Curt Hinrichs*
Marketing Team: *Tom Ziolkowski, Samantha Cabaluna*
Editorial Assistant: *Emily Davidson*
Production Editor: *Keith Faivre*
Production Service: *Scratchgravel Publishing Services*
Manuscript Editor: *Susan Pendleton*
Permissions Editor: *Mary Kay Hancharick*

Interior Design: *Wendy LaChance, Gregory Draus*
Cover Design: *Roy R. Neuhaus*
Cover Illustration: *Rik Olson*
Art Editor: *Scratchgravel Publishing Services*
Print Buyer: *Vena Dyer*
Typesetting: *Scratchgravel Publishing Services*
Printing and Binding: *R.R. Donnelley & Sons, Crawfordsville*

For more information about this or any other Duxbury products, contact:
DUXBURY
511 Forest Lodge Road
Pacific Grove, CA 93950 USA
www.duxbury.com
1-800-423-0563 (Thomson Learning Academic Resource Center)

For permission to use material from this work, contact us by
Web: www.thomsonrights.com
fax: 1-800-730-2215
phone: 1-800-730-2214

Printed in United States of America

10 9

Library of Congress Cataloging-in-Publication Data

Clemen, Robert T. (Robert Taylor), [date]–
    Making hard decisions with DecisionTools / Robert T. Clemen, Terence Reilly.—2nd
  rev. ed.
      p.   cm.
    Earlier ed. published in 1996 under title: Making hard decisions.
    Includes bibliographical references (p.) and index.
    ISBN 0-534-36597-3 (alk. paper)
      1. Decision making. 2. Decision making—Computer programs. 3. DecisionTools. I.
  Reilly, Terence. II. Clemen, Robert T. (Robert Taylor), [date]– Making hard decisions. III.
  Title.

HD30.23 .C577 20001
658.4'03—dc21   00-031451

To Margaret and Cynde

Brief Contents

Contents

Chapter 4   Making Choices   111

Chapter 5   Sensitivity Analysis   174

Chapter 6 Creativity and Decision Making 217

Chapter 9   Theoretical Probability Models   352

Preface

This book provides a one-semester overview of decision analysis for advanced undergraduate and master's degree students. The inspiration to write it has come from many sources, but perhaps most important was a desire to give students access to up-to-date information on modern decision analysis techniques at a level that could be easily understood by those without a strong mathematical background. At some points in the book, the student should be familiar with basic statistical concepts normally covered in an undergraduate applied statistics course. In particular, some familiarity with probability and probability distributions would be helpful in Chapters 7 through 12. Chapter 10 provides a decision-analysis view of data analysis, including regression, and familiarity with such statistical procedures would be an advantage when covering this topic. Algebra is used liberally throughout the book. Calculus concepts are used in a few instances as an explanatory tool. Be assured, however, that the material can be thoroughly understood, and the problems can be worked, without any knowledge of calculus.

The objective of decision analysis is to help a decision maker think hard about the specific problem at hand, including the overall structure of the problem as well as his or her preferences and beliefs. Decision analysis provides both an overall paradigm and a set of tools with which a decision maker can construct and analyze a model of a decision situation. Above all else, students need to understand that the purpose of studying decision-analysis techniques is to be able to represent real-world problems using models that can be analyzed to gain insight and understanding. It is through that insight and understanding—the hoped-for result of the modeling process—that decisions can be improved.

New in this Version: Palisade's DecisionTools

This is not a new edition of *Making Hard Decisions*. It includes virtually all of the material that is in the original version of the second edition. What is different, though, is that this version focuses on the use of an electronic spreadsheet as a platform for modeling and analysis. Spreadsheets are both widely available and powerful tools for decision making, and in the future managers will need to be able to use this flexible tool effectively.

The flexibility of electronic spreadsheets makes them ideal as general-purpose tools for decision analysts. At the same time, though, the analyst needs specialized tools for building and analyzing decision models. This version of *Making Hard Decisions* integrates Palisade Corporation's DecisionTools suite of software. DecisionTools is designed specifically to work with and enhance the capabilities of Microsoft Excel for use by decision analysts.

DecisionTools consists of five programs (PrecisionTree, TopRank, @RISK, BestFit, and RISKView), each designed to help with different aspects of modeling and solving decision problems. PrecisionTree is a versatile program that solves both decision trees and influence diagrams; TopRank performs sensitivity analysis on spreadsheet models; @RISK is a Monte Carlo simulation program; BestFit and RISKView are specialized programs designed to help choose the best probability distribution for modeling an uncertainty; PrecisionTree, TopRank, and @RISK are all spreadsheet add-ins for Microsoft Excel. When one of these programs is opened, its functions are displayed in an Excel toolbar. BestFit and RISKview can operate either within @RISK or as stand-alone programs.

The DecisionTools software can be a powerful ally for the analyst. The software can help create a model as well as analyze it in many different ways. This assumes, of course, that you have learned how to use the software! Thus, we have included at the ends of appropriate chapters instructions for using the programs that correspond to the chapter topic. The instructions provide step-by-step guides through the important features of the programs. Interspersed throughout the instructions you will find explanations of the steps along with tips on interpreting the output. The book's endsheets show where the various programs, features, and analytical techniques are covered. Once you know the software, you will find the problems are easier to work and even fun.

Guidelines for Students

Along with instructions to use the DecisionTools software, this version of *Making Hard Decisions* covers most of the concepts we consider important for a basic understanding of decision analysis. Although the text is meant to be an elementary introduction to decision analysis, this does not mean that the material is itself elementary. In fact, the more we teach decision analysis, the more we realize that the technical level of the math is low, while the level of the analysis is high. Students must be willing to think clearly and analytically about the problems and issues that arise in decision situations. Good decision analysis requires clear thinking; sloppy thinking results in worthless analysis.

Of course, some topics are more demanding than others. The more difficult sections are labeled as "optional." Our faith in students and readers compels us to say that anyone who can handle the "nonoptional" material can, with a bit more effort and thought, also handle the optional material. Thus the label is perhaps best thought of as a warning regarding the upcoming topic. On the other hand, if you do decide to skip the optional material, no harm will be done.

In general, we believe that really serious learning happens when problems are tackled on one's own. We have included a wide variety of exercises, questions, problems, and case studies. The exercises are relatively easy drills of the material. The questions and problems often require thinking beyond the material in the text. Some concepts are presented and dealt with only in the problems. Do not shy away from the problems! You can learn a lot by working through them.

Many case studies are included in *Making Hard Decisions*. A few of the many successful applications of decision analysis show up as case studies in the book. In addition, many issues are explored in the case studies in the context of current events. For example, the AIDS case at the end of Chapter 7 demonstrates how probability techniques can be used to interpret the results of medical tests. In addition to the real-world cases, the book contains many hypothetical cases and examples, as well as fictional historical accounts, all of which have been made as realistic as possible.

Some cases and problems are realistic in the sense that not every bit of information is given. In these cases, appropriate assumptions are required. On one hand, this may cause some frustration. On the other hand, incomplete information is typical in the real world. Being able to work with problems that are "messy" in this way is an important skill.

Finally, many of the cases and problems involve controversial issues. For example, the material on AIDS (Chapter 7) or medical ethics (Chapter 15) may evoke strong emotional responses from some readers. In writing a book like this, there are two choices: We can avoid the hard social problems that might offend some readers, or we can face these problems that need careful thought and discussion. The text adopts the second approach because we believe these issues require society's attention. Moreover, even though decision analysis may not provide the answers to these problems, it does provide a useful framework for thinking about the difficult decisions that our society must make.

A Word to Instructors

Many instructors will want to supplement *Making Hard Decisions* with their own material. In fact, topics that we cover in our own courses are not included here. But, in the process of writing the book and obtaining comments from colleagues, it has become apparent that decision-making courses take on many different forms. Some instructors prefer to emphasize behavioral aspects, while others prefer analytical tools. Other dimensions have to do with competition, negotiation, and group decision making. *Making Hard Decisions* does not aim to cover everything for everyone. Instead, we have tried to cover the central concepts and tools of modern decision analysis with adequate references (and occasionally cases or problems) so that instructors can introduce their own special material where appropriate. For example, in Chapters 8 and 14 we discuss judgmental aspects of probability assessment and decision making, and an instructor can introduce more behavioral material at these points. Likewise, Chapter 15 delves into the additive utility function for decision making. Some instructors may wish to present goal programming or the analytic hierarchy process here.

Regarding the DecisionTools software, we wrote the instructions to be a self-contained tutorial. Although the tutorial approach works well, we also believe that it must be supplemented by guidance from the course instructor. One possible way to supplement the instructions is to walk the students through the instructions in a computer lab. This will allow the instructor to answer questions as they arise and will allow students to learn the software in a more controlled environment. No new material need be prepared for the computer-lab session, and in the text the students have a written copy of the instructions for later reference.

Keeping Up with Changes

The world changes quickly, and decision analysis is changing with it. The good news is that the Internet, and especially the World Wide Web (WWW), can help us keep abreast of new developments. We encourage both students and instructors to visit the WWW site of the Decision Analysis Society at http://www.fuqua.duke.edu/faculty/daweb/. This organization provides focus for decision analysts worldwide and many others with interests in all aspects of decision making. And on the Section's web page, you will find links to many related sites.

While you are keeping up with changes, we hope that you will help us do the same. Regarding the software or instructions in using the software, please send your comments to Terence Reilly at reilly@babson.edu. You may also send regular mail to Terence Reilly, Mathematics Division, Babson College, Babson Park, MA 02457 or to Palisade Corporation, 31 Decker Road, Newfield, NY 14687. You may also contact Palisade at http://www.palisade.com.

For all other non-software matters, please send comments to Robert Clemen at clemen@mail.duke.edu. You may also send regular mail to Robert Clemen, Fuqua School of Business, Duke University, Durham, NC 27708. Please send information about (hopefully the few) mistakes or typos that you may find in the book, innovative ways to teach decision analysis, new case studies, or interesting applications of decision analysis.

Acknowledgments

It is a pleasure to acknowledge the help we have had with the preparation of this text. First mention goes to our students, who craved the notes from which the text has grown. For resource support, thanks to the Lundquist College of Business at the University of Oregon, Decision Research of Eugene, Oregon, Applied Decision Analysis, Inc., of Menlo Park, California, the Fuqua School of Business of Duke University, Babson College, and the Board of Research at Babson College for financial support.

A number of individuals have provided comments on portions of the book at various stages. Thanks to Elaine Allen, Deborah Amaral, Sam Bodily, Adam Borison, Cathy Barnes, George Benson, Dave Braden, Bill Burns, Peter Farquhar, Ken Gaver, Andy Golub, Gordon Hazen, Max Henrion, Don Keefer, Ralph Keeney, Robin Keller, Craig Kirkwood, Don Kleinmuntz, Irv LaValle, George MacKenzie, Allan Murphy, Bob Nau, Roger Pfaffenberger, Steve Powell, Gordon Pritchett, H.V. Ravinder, Gerald Rose, Sam Roy, Rakesh Sarin, Ross Shachter, Jim Smith, Bob Winkler, and Wayne Winston. Special thanks to Deborah Amaral for guidance in writing the Municipal Solid Waste case in Chapter 9; to Dave Braden for outstanding feedback as he and his students used manuscript versions of the first edition; to Susan Brodt for guidance and suggestions for rewriting the creativity material in Chapter 6; and to Kevin McCardle for allowing the use of numerous problems from his statistics course. Vera Gilliland and Sam McLafferty of Palisade Corporation have been very helpful. Thanks also to all of the editors who have worked closely with us on this and previous editions over the years: Patty Adams, Marcia Cole, Mary Douglas, Anne and Greg Draus, Keith Faivre, Curt Hinrichs, and Michael Payne.

Finally, we sincerely thank our families and loved ones for their understanding of the times we were gone and the hours we have spent on this text.

Robert T. Clemen
Terence Reilly

CHAPTER 1

Introduction to Decision Analysis

H ave you ever had a difficult decision to make? If so, did you wish for a straight-forward way to keep all of the different issues clear? Did you end up making the decision based on your intuition or on a "hunch" that seemed correct? At one time or another, all of us have wished that a hard decision was easy to make. The sad fact is that hard decisions are just that—hard. As individuals we run into such diffi-cult decisions frequently. Business executives and governmental policy makers struggle with hard problems all the time. For example, consider the following prob-lem faced by the Oregon Department of Agriculture (ODA) in 1985.

GYPSY MOTHS AND THE ODA

In the winter of 1985, the ODA grappled with the problem of gypsy moth infestation in Lane County in western Oregon. Forest industry representatives argued strongly for an aggressive eradication campaign using potent chemical insecticides. The ODA instead proposed a plan that involved spraying most of the affected area with BT (*Bacillus thuringiensis*), a bacterial insecticide known to be (1) target-specific (that is, it does little damage to organisms other than moths), (2) ecologically safe, and (3) reasonably effective. As well as using BT, the ODA proposed spraying three smaller areas near the city of Eugene with the chemical spray Orthene. Although Orthene was registered as an acceptable insecticide for home garden use, there was some doubt as to its ultimate ecological effects as well as its danger to humans. Forestry officials argued that the chemical insecticide was more potent than BT and was necessary to ensure eradication in the most heavily infested areas. Environmentalists argued that the potential danger from the chemical spray was too great to warrant its use. Some individuals argued that spraying would not help because the infestation already was so advanced that no program would be successful. Others argued that an aggressive spray program could solve the problem once and for all, but only if done immediately. Clearly, in making its final decision the ODA would have to deal with many issues.

The ODA has an extremely complex problem on its hands. Before deciding exactly what course of action to take, the agency needs to consider many issues, including the values of different constituent groups and the uncertainties involving the effectiveness and risks of the pesticides under consideration. The ODA must consider these issues carefully and in a balanced way—but how? There is no escaping the problem: This hard decision requires hard thinking.

Decision analysis provides structure and guidance for thinking systematically about hard decisions. With decision analysis, a decision maker can take action with confidence gained through a clear understanding of the problem. Along with a conceptual framework for thinking about hard problems, decision analysis provides analytical tools that can make the required hard thinking easier.

Why Are Decisions Hard?

What makes decisions hard? Certainly different problems may involve different and often special difficulties. For example, the ODA's problem requires it to think about the interests of various groups as well as to consider only limited information on the possible effects of the sprays. Although every decision may have its own special problems, there are four basic sources of difficulty. A decision-analysis approach can help a decision maker with all four.

First, a decision can be hard simply because of its complexity. In the case of the gypsy moths, the ODA must consider many different individual issues: the uncertainty surrounding the different sprays, the values held by different community groups, the different possible courses of action, the economic impact of any pest-control program, and so on. Simply keeping all of the issues in mind at one time is nearly impossible. Decision analysis provides effective methods for organizing a complex problem into a structure that can be analyzed. In particular, elements of a decision's structure include the possible courses of action, the possible outcomes that could result, the likelihood of those outcomes, and eventual consequences (e.g., costs and benefits) to be derived from the different outcomes. Structuring tools that we will consider include decision trees and influence diagrams as well as procedures for analyzing these structures to find solutions and for answering "what if" questions.

Second, a decision can be difficult because of the inherent uncertainty in the situation. In the gypsy moth case, the major uncertainties are the effectiveness of the different sprays in reducing the moth population and their potential for detrimental ecological and health effects. In some decisions the main issue is uncertainty. For example, imagine a firm trying to decide whether to introduce a new product. The size of the market, the market price, eventual competition, and manufacturing and distribution costs all may be uncertain to some extent, and all have some impact on the firm's eventual payoff. Yet the decision must be made without knowing for sure what these uncertain values will be. A decision-analysis approach can help in identifying important sources of uncertainty and representing that uncertainty in a systematic and useful way.

Third, a decision maker may be interested in working toward multiple objectives, but progress in one direction may impede progress in others. In such a case, a decision maker must trade off benefits in one area against costs in another. In the gypsy moth example, important trade-offs must be made: Are the potential economic benefits to be gained from spraying Orthene worth the potential ecological damage and health risk? In investment decisions a trade-off that we usually must make is between expected return and riskiness. Decision analysis again provides both a framework and specific tools for dealing with multiple objectives.

Fourth, and finally, a problem may be difficult if different perspectives lead to different conclusions. Or, even from a single perspective, slight changes in certain inputs may lead to different choices. This source of difficulty is particularly pertinent when more than one person is involved in making the decision. Different individuals may look at the problem from different perspectives, or they may disagree on the uncertainty or value of the various outcomes. The use of the decision-analysis framework and tools can help sort through and resolve these differences whether the decision maker is an individual or a group of stakeholders with diverse opinions.

Why Study Decision Analysis?

The obvious reason for studying decision analysis is that carefully applying its techniques can lead to better decisions. But what is a good decision? A simple answer might be that it is the one that gives the best outcome. This answer, however, confuses the idea of a lucky outcome with a good decision. Suppose that you are interested in investing an inheritance. After carefully considering all the options available and consulting with investment specialists and financial planners, you decide to invest in stocks. If you purchased a portfolio of stocks in 1982, the investment most likely turned out to be a good one, because stock values increased dramatically during the 1980s. On the other hand, if your stock purchase had been in early 1929, the stock market crash and the following depression would have decreased the value of your portfolio drastically.

Was the investment decision a good one? It certainly could have been if it was made after careful consideration of the available information and thorough deliberation about the goals and possible outcomes. Was the outcome a good one? For the 1929 investor, the answer is no. This example illustrates the difference between a good decision and a lucky outcome: You can make a good decision but still have an unlucky outcome. Of course, you may prefer to have lucky outcomes rather than make good decisions! Although decision analysis cannot improve your luck, it can help you to understand better the problems you face and thus make better decisions. That understanding must include the structure of the problem as well as the uncertainty and trade-offs inherent in the alternatives and outcomes. You may then improve your chances of enjoying a better outcome; more important, you will be less likely to experience unpleasant surprises in the form of unlucky outcomes that were either unforeseen or not fully understood. In other words, you will be making a decision with your eyes open.

The preceding discussion suggests that decision analysis allows people to make effective decisions more consistently. This idea itself warrants discussion. Decision analysis is intended to help people deal with *difficult* decisions. It is a "prescriptive approach designed for normally intelligent people who want to think hard and systematically about some important real problems" (Keeney and Raiffa 1976, p. vii).

This prescriptive view is the most appropriate way to think about decision analysis. It gets across the idea that although we are not perfect decision makers, we can do better through more structure and guidance. We will see that decision analysis is not an idealized theory designed for superrational and omniscient beings. Nor does it describe how people actually make decisions. In fact, ample experimental evidence from psychology shows that people generally do not process information and make decisions in ways that are consistent with the decision-analysis approach. (If they did, then there would be no need for decision analysis; why spend a lot of time studying decision analysis if it suggests that you do what you already do?) Instead, using some fundamental principles, and informed by what we know about human frailties in judgment and decision making, decision analysis offers guidance to normal people working on hard decisions.

Although decision analysis provides structure and guidance for systematic thinking in difficult situations, it does not claim to recommend an alternative that must be blindly accepted. Indeed, after the hard thinking that decision analysis fosters, there should be no need for blind acceptance; the decision maker should understand the situation thoroughly. Instead of providing solutions, decision analysis is perhaps best thought of as simply an information source, providing insight about the situation, uncertainty, objectives, and trade-offs, and possibly yielding a recommended course of action. Thus, decision analysis does not usurp the decision maker's job. According to another author,

> The basic presumption of decision analysis is not at all to replace the decision
> maker's intuition, to relieve him or her of the obligations in facing the problem,
> or to be, worst of all, a competitor to the decision maker's personal style of
> analysis, but to complement, augment, and generally work alongside the decision
> maker in exemplifying the nature of the problem. Ultimately, it is of most value
> if the decision maker has actually learned something about the problem and his
> or her own decision-making attitude through the exercise (Bunn 1984, p. 8).

We have been discussing decision analysis as if it were always used to help an individual make a decision. Indeed, this is what it is designed for, but its techniques have many other uses. For example, one might use decision-analysis methods to solve complicated inference problems (that is, answering questions such as "What conclusions can be drawn from the available evidence?"). Structuring a decision problem may be useful for understanding its precise nature, for generating alternative courses of action, and for identifying important objectives and trade-offs. Understanding trade-offs can be crucial for making progress in negotiation settings. Finally, decision analysis can be used to justify why a previously chosen action was appropriate.

Subjective Judgments and Decision Making

Personal judgments about uncertainty and values are important inputs for decision analysis. It will become clear through this text that discovering and developing these judgments involves thinking hard and systematically about important aspects of a decision.

Managers and policy makers frequently complain that analytical procedures from management science and operations research ignore subjective judgments. Such procedures often purport to generate "optimal" actions on the basis of purely objective inputs. But the decision-analysis approach allows the inclusion of subjective judgments. In fact, decision analysis *requires* personal judgments; they are important ingredients for making good decisions.

At the same time, it is important to realize that human beings are imperfect information processors. Personal insights about uncertainty and preferences can be both limited and misleading, even while the individual making the judgments may demonstrate an amazing overconfidence. An awareness of human cognitive limitations is critical in developing the necessary judgmental inputs, and a decision maker who ignores these problems can magnify rather than adjust for human frailties. Much current psychological research has a direct bearing on the practice of decision-analysis techniques. In the chapters that follow, many of the results from this research will be discussed and related to decision-analysis techniques. The spirit of the discussion is that understanding the problems people face and carefully applying decision-analysis techniques can lead to better judgments and improved decisions.

The Decision-Analysis Process

Figure 1.1 shows a flowchart for the decision-analysis process. The first step is for the decision maker to identify the decision situation and to understand his or her objectives in that situation. Although we usually do not have trouble finding decisions to make or problems to solve, we do sometimes have trouble identifying the exact problem, and thus we sometimes treat the wrong problem. Such a mistake has been called an "error of the third kind." Careful identification of the decision at hand is always important. For example, perhaps a surface problem hides the real issue. For example, in the gypsy moth case, is the decision which insecticide to use to control the insects, or is it how to mollify a vocal and ecologically minded minority?

Understanding one's objectives in a decision situation is also an important first step and involves some introspection. What is important? What are the objectives? Minimizing cost? Maximizing profit or market share? What about minimizing risks? Does risk mean the chance of a monetary loss, or does it refer to conditions potentially damaging to health and the environment? Getting a clear understanding of the crucial objectives in a decision situation must be done before much more can be accomplished. In the next step, knowledge of objectives can help in identifying

Figure 1.1

A decision-analysis process flowchart.

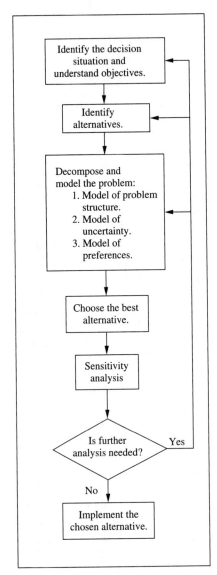

alternatives, and beyond that the objectives indicate how outcomes must be measured and what kinds of uncertainties should be considered in the analysis.

Many authors argue that the first thing to do is to identify the problem and then to figure out the appropriate objectives to be used in addressing the problem. But Keeney (1992) argues the opposite; it is far better, he claims, to spend a lot of effort understanding one's central values and objectives, and then looking for ways—decision opportunities—to achieve those objectives. The debate notwithstanding, the fact is that decisions come in many forms. Sometimes we are lucky enough to shape our decision-making future in the way Keeney suggests, and other times we find ourselves in difficult situations that we may not have anticipated. In either case, establishing the precise nature of the decision situation (which we will later call the *decision context*) goes hand in hand with identifying and understanding one's objectives in that situation.

With the decision situation and pertinent objectives established, we turn to the discovery and creation of alternatives. Often a careful examination and analysis of objectives can reveal alternatives that were not obvious at the outset. This is an important benefit of a decision-analysis approach. In addition, research in the area of creativity has led to a number of techniques that can improve the chance of finding new alternatives.

The next two steps, which might be called "modeling and solution," form the heart of most textbooks on decision analysis, including this one. Much of this book will focus on decomposing problems to understand their structures and measure uncertainty and value; indeed, decomposition is the key to decision analysis. The approach is to "divide and conquer." The first level of decomposition calls for structuring the problem in smaller and more manageable pieces. Subsequent decomposition by the decision maker may entail careful consideration of elements of uncertainty in different parts of the problem or careful thought about different aspects of the objectives.

The idea of *modeling* is critical in decision analysis, as it is in most quantitative or analytical approaches to problems. As indicated in Figure 1.1, we will use models in several ways. We will use influence diagrams or decision trees to create a representation or model of the decision problem. Probability will be used to build models of the uncertainty inherent in the problem. Hierarchical and network models will be used to understand the relationships among multiple objectives, and we will assess utility functions in order to model the way in which decision makers value different outcomes and trade off competing objectives. These models are mathematical and graphical in nature, allowing one to find insights that may not be apparent on the surface. Of course, a key advantage from a decision-making perspective is that the mathematical representation of a decision can be subjected to analysis, which can indicate a "preferred" alternative.

Decision analysis is typically an iterative process. Once a model has been built, *sensitivity analysis* is performed. Such analysis answers "what if" questions: "If we make a slight change in one or more aspects of the model, does the optimal decision change?" If so, the decision is said to be sensitive to these small changes, and the decision maker may wish to reconsider more carefully those aspects to which the decision is sensitive. Virtually any part of a decision is fair game for sensitivity analysis. The arrows in Figure 1.1 show that the decision maker may return even to the identification of the problem. It may be necessary to refine the definition of objectives or include objectives that were not previously included in the model. New alternatives may be identified, the model structure may change, and the models of uncertainty and preferences may need to be refined. The term *decision-analysis cycle* best describes the overall process, which may go through several iterations before a satisfactory solution is found.

In this iterative process, the decision maker's perception of the problem changes, beliefs about the likelihood of various uncertain eventualities may develop and change, and preferences for outcomes not previously considered may mature as more time is spent in reflection. Decision analysis not only provides a structured way to think about decisions, but also more fundamentally provides a structure within which a decision maker can develop beliefs and feelings, those subjective judgments that are critical for a good solution.

Requisite Decision Models

Phillips (1982, 1984) has introduced the term *requisite decision modeling*. This marvelous term captures the essence of the modeling process in decision analysis. In Phillips's words, "a model can be considered requisite only when no new intuitions emerge about the problem" (1984, p. 37), or when it contains everything that is essential for solving the problem. That is, a model is requisite when the decision maker's thoughts about the problem, beliefs regarding uncertainty, and preferences are fully developed. For example, consider a first-time mutual-fund investor who finds high, overall long-term returns appealing. Imagine, though, that in the process of researching the funds the investor begins to understand and become wary of highly volatile stocks and mutual funds. For this investor, a decision model that selected a fund by maximizing the average return in the long run would not be requisite. A requisite model would have to incorporate a trade-off between long-term returns and volatility.

A careful decision maker may cycle through the process shown in Figure 1.1 several times as the analysis is refined. Sensitivity analysis at appropriate times can help the decision maker choose the next modeling steps to take in developing a requisite model. Successful decision analysts artistically use sensitivity analysis to manage the iterative development of a decision model. An important goal of this book is that you begin to acquire this artistic ability through familiarity and practice with the concepts and tools of decision analysis.

Where Is Decision Analysis Used?

Decision analysis is widely used in business and government decision making. Perusing the literature reveals applications that include managing research-and-development programs, negotiating for oil and gas leases, forecasting sales for new products, understanding the world oil market, deciding whether to launch a new product or new venture, and developing ways to respond to environmental risks, to name a few. And some of the largest firms make use of decision analysis, including General Motors, Chevron, and Eli Lilly. A particularly important arena for decision-analysis applications has been in public utilities, especially electric power generation. In part this is because the problems utilities face (e.g., site selection, power-generation methods, waste cleanup and storage, pollution control) are particularly appropriate for treatment with decision-analysis techniques; they involve long time frames and hence a high degree of uncertainty. In addition, multiple objectives must be considered when a decision affects many different stakeholder groups.

In the literature, many of the reported applications relate to public-policy problems and relatively few to commercial decisions, partly because public-policy problems are of interest to such a wide audience. It is perhaps more closely related to the fact that commercial applications often are proprietary; a good decision analysis can create a competitive advantage for a firm, which may not appreciate having its advantage revealed in the open literature. Important public-policy applications have included regulation in the energy (especially nuclear) industry and standard setting in a variety of different situations ranging from regulations for air and water pollution to standards for safety features on new cars.

Another important area of application for decision analysis has been in medicine. Decision analysis has helped doctors make specific diagnoses and individuals to understand the risks of different treatments. Institutional-level studies have been done, such as studying the optimal inventory or usage of blood in a blood bank or the decision of a firm regarding different kinds of medical insurance to provide its employees. On a grander scale, studies have examined policies such as widespread testing for various forms of cancer or the impact on society of different treatment recommendations.

This discussion is by no means exhaustive; the intent is only to give you a feel for the breadth of possible applications of decision analysis and a glimpse at some of the things that have been done. Many other applications are described in cases and examples throughout the book; by the time you have finished, you should have a good understanding of how decision analysis can be (and is) used in many different arenas. And if you feel the need for more, articles by Ulvila and Brown (1982) and Corner and Kirkwood (1991) describe many different applications.

Where Does the Software Fit In?

Included with the text is a CD containing Palisade's DecisionTools suite, which is a set of computer programs designed to help you complete the modeling and solution phase of the decision process. The suite consists of five programs (PrecisionTree, RISKview, BestFit, TopRank, and @RISK), each intended for different steps in the decision process. As you work your way through the text learning the different steps, we introduce the programs that will help you complete each step. We supply detailed instructions on how to use the program and how to interpret the output at the end of certain chapters. Table 1.1 shows where in the decision process each of the five programs is used and the chapter where the instructions appear.

One of the best aspects about the DecisionTools suite is that the programs work together as one program within Excel. When either PrecisionTree, TopRank, or @RISK are opened, their functions are added directly to Excel's toolbar. This allows us to model and solve complex decision problems within an ordinary spreadsheet. These programs are designed to extend the capability of the spreadsheet to handle the types of models used in decision making. If you are already familiar with Excel, then you will not need to learn a completely different system to use PrecisionTree, TopRank, and @RISK.

The programs RISKview and BestFit can be run either as independent programs or as components of @RISK. They are specialized programs specifically designed to help the decision maker determine how best to model an uncertainty. The output

DecisionTools Program	Where It Is Used in the Decision Process	Where in Text
PrecisionTree	Structuring the decision	Chapter 3
	Solving the decision	Chapter 4
	Sensitivity analysis	Chapter 5
	Value of information	Chapter 12
	Modeling preferences	Chapter 13
TopRank	Sensitivity analysis	Chapter 5
RISKview	Modeling uncertainty	Chapters 8 and 9
BestFit	Using data to model uncertainty	Chapter 10
@RISK	Simulation modeling	Chapter 11

from these two programs is used as input to the decision model. Although these programs do not operate within Excel, they do export their output to Excel to be used by the other programs in the suite. As you become more familiar with the theory and the DecisionTools suite, you will see how the five programs are interrelated and form an organized whole.

There is also a strong interconnection between the DecisionTools programs and Excel. These programs do more than use the spreadsheet as an analytical engine. For example, you can link your decision tree to a spreadsheet model. The links will be dynamic; changes made in the spreadsheet are immediately reflected in the decision tree. These dynamic links will pull input values from the tree to the spreadsheet, calculate an output value, and send the output back to the tree.

Linking is one of many ways that Excel and the DecisionTools programs work together. In future chapters we will see other connections. We will also see how flexible and useful electronic spreadsheets can be when constructing decision models. Much of what we will learn about using Excel will extend beyond decision analysis. It is hard to overemphasize the power of modern spreadsheets; these programs can do virtually anything that requires calculations. Spreadsheets have become one of the most versatile and powerful quantitative tools available to business managers. Virtually all managers now have personal computers on their desks with the ability to run these sophisticated spreadsheet programs, which suggests that aspiring managers would be well advised to become proficient in the use of this flexible tool.

The software can help you learn and understand the concepts presented in the chapter. Reading about the concepts and examples will provide a theoretical understanding. The programs will test the understanding when you apply the theory to actual problems. Because the program will carry out your instructions exactly as you input them, it will reflect how well you understand the theory. You will find that your understanding of the concepts greatly increases because the programs force you to think carefully throughout the construction and analysis of the model.

Where Are We Going from Here?

This book is divided into three main sections. The first is titled "Modeling Decisions," and it introduces influence diagrams and decision trees as methods for building models of decision problems. The process is sometimes called *structuring* because it specifies the elements of the decision and how the elements are interrelated (Chapters 2 and 3). We also introduce ways to organize a decision maker's values into hierarchies and networks; doing so is useful when multiple objectives must be considered. We will find out how to analyze our decision models (Chapter 4) and how to conduct sensitivity analysis (Chapter 5). In Chapter 6 we discuss creativity and decision making.

The second section is "Modeling Uncertainty." Here we delve into the use of probability for modeling uncertainty in decision problems. First we review basic probability concepts (Chapter 7). Because subjective judgments play a central role in decision analysis, subjective assessments of uncertainty are the topic of Chapter 8. Other ways to use probability include theoretical probability models (Chapter 9), data-based models (Chapter 10), and simulation (Chapter 11). Chapter 12 closes the section with a discussion of information and how to value it in the context of a probability model of uncertainty within a decision problem.

"Modeling Preferences" is the final section. Here we turn to the development of a mathematical representation of a decision maker's preferences, including the identification of desirable objectives and trade-offs between conflicting objectives. A fundamental issue that we often must confront is how to trade off riskiness and expected value. Typically, if we want to increase our chances at a better outcome, we must accept a simultaneous risk of loss. Chapters 13 and 14 delve into the problem of modeling a decision maker's attitude toward risk. Chapters 15 and 16 complete the section with a treatment of other conflicting objectives. In these chapters we will complete the discussion of multiple objectives begun in Section 1, showing how to construct a mathematical model that reflects subjective judgments of relative importance among competing objectives.

By the end of the book, you will have learned all of the basic techniques and concepts that are central to the practice of modern decision analysis. This does not mean that your hard decisions will suddenly become easy! But with the decision-analysis framework, and with tools for modeling decisions, uncertainty, and preferences, you will be able to approach your hard decisions systematically. The understanding and insight gained from such an approach will give you confidence in your actions and allow for better decisions in difficult situations. That is what the book is about—an approach that will help you to make hard decisions.

SUMMARY

The purpose of decision analysis is to help a decision maker think systematically about complex problems and to improve the quality of the resulting decisions. In this regard, it is important to distinguish between a good decision and a lucky outcome. A good decision is one that is made on the basis of a thorough understanding of the problem and careful thought regarding the important issues. Outcomes, on the other hand, may be lucky or unlucky, regardless of decision quality.

In general, decision analysis consists of a framework and a tool kit for dealing with difficult decisions. The incorporation of subjective judgments is an important aspect of decision analysis, and to a great extent mature judgments develop as the decision maker reflects on the decision at hand and develops a working model of the problem. The overall strategy is to decompose a complicated problem into smaller chunks that can be more readily analyzed and understood. These smaller pieces can then can be brought together to create an overall representation of the decision situation. Finally, the decision-analysis cycle provides the framework within which a decision maker can construct a requisite decision model, one that contains the essential elements of the problem and from which the decision maker can take action.

QUESTIONS AND PROBLEMS

1.1 Give an example of a good decision that you made in the face of some uncertainty. Was the outcome lucky or unlucky? Can you give an example of a poorly made decision whose outcome was lucky?

1.2 Explain how modeling is used in decision analysis. What is meant by the term "requisite decision model"?

1.3 What role do subjective judgments play in decision analysis?

1.4 At a dinner party, an acquaintance asks whether you have read anything interesting lately, and you mention that you have begun to read a text on decision analysis. Your friend asks what decision analysis is and why anyone would want to read a book about it, let alone write one! How would you answer?

1.5 Your friend in Question 1.4, upon hearing your answer, is delighted! "This is marvelous," she exclaims. "I have this very difficult choice to make at work. I'll tell you the facts, and you can tell me what I should do!" Explain to her why you cannot do the analysis for her.

1.6 Give an example in which a decision was complicated because of difficult preference trade-offs. Give one that was complicated by uncertainty.

1.7 In the gypsy moth example, what are some of the issues that you would consider in making this decision? What are the alternative courses of action? What issues involve uncertainty, and how could you get information to help resolve that uncertainty? What are the values held by opposing groups? How might your decision trade off these values?

1.8 Can you think of some different alternatives that the ODA might consider for controlling the gypsy moths?

1.9 Describe a decision that you have had to make recently that was difficult. What were the major issues? What were your alternatives? Did you have to deal with uncertainty? Were there important trade-offs to make?

1.10 "Socially responsible investing" first became fashionable in the 1980s. Such investing involves consideration of the kinds of businesses that a firm engages in and selection of investments that are as consistent as possible with the investor's sense of ethical and moral business activity. What trade-offs must the socially responsible investor make? How are these trade-offs more complicated than those that we normally consider in making investment decisions?

1.11 Many decisions are simple, preprogrammed, or already solved. For example, retailers do not have to think long to decide how to deal with a new customer. Some operations-research models provide "ready-made" decisions, such as finding an optimal inventory level using an order-quantity formula or determining an optimal production mix using linear programming. Contrast these decisions with unstructured or strategic decisions, such as choosing a career or locating a nuclear power plant. What kinds of decisions are appropriate for a decision-analysis approach? Comment on the statement, "Decision making is what you do when you don't know what to do." (For more discussion, see Howard 1980.)

1.12 The argument was made that beliefs and preferences can change as we explore and learn. This even holds for learning about decision analysis! For example, what was your impression of this book before reading the first chapter? Have your beliefs about the value of decision analysis changed? How might this affect your decision about reading more of the book?

CASE STUDIES

DR. JOYCELYN ELDERS AND THE WAR ON DRUGS

After the Nancy Reagan slogan, "Just Say No," and 12 years of Republican administration efforts to fight illegal drug use and trafficking, on December 7, 1993, then Surgeon General Dr. Joycelyn Elders made a startling statement. In response to a reporter's question, she indicated that, based on the experiences of other countries, the crime rate in the United States might actually decrease if drugs were legalized. She conceded that she did not know all of the ramifications and suggested that perhaps some studies should be done.

The nation and especially the Clinton administration were shocked to hear this statement. What heresy after all the efforts to control illegal drugs! Of course, the White House immediately went on the defensive, making sure that everyone understood that President Clinton was not in favor of legalizing drugs. And Dr. Elders had to clarify her statement; it was her personal opinion, not a statement of administration policy.

Questions

1 What decision situation did Dr. Elders identify? What specific values would be implied by choosing to study the legalization of drugs?

2 From a decision-making perspective, which makes more sense: Nancy Reagan's "Just Say No" policy or Elders's suggestion that the issue of legalization be studied? Why?

3 Consider Elders's decision to suggest studying the legalization of drugs. Was her decision to respond to the reporter the way she did a good decision with a bad outcome? Or was it a bad decision in the first place?

4   Why was Elders's suggestion a political hot potato for Clinton's administration? What, if any, are the implications for decision analysis in political situations?

LLOYD BENTSEN FOR VICE PRESIDENT?

In the summer of 1988, Michael Dukakis was the Democratic Party's presidential nominee. The son of Greek immigrants, his political career had flourished as governor of Massachusetts, where he had demonstrated excellent administrative and fiscal skills. He chose Lloyd Bentsen, U.S. Senator from Texas, as his running mate. In an analysis of Dukakis's choice, E. J. Dionne of *The New York Times* (July 13, 1988) made the following points:

1   The main job of the vice presidential nominee is to carry his or her home state. Could Bentsen carry Texas? The Republican presidential nominee was George Bush, whose own adopted state was Texas. Many people thought that Texas would be very difficult for Dukakis to win, even with Bentsen's help. If Dukakis could win Texas's 29 electoral votes, however, the gamble would pay off dramatically, depriving Bush of one of the largest states that he might have taken for granted.

2   Bentsen was a conservative Democrat. Jesse Jackson had run a strong race and had assembled a strong following of liberal voters. Would the Jackson supporters be disappointed in Dukakis's choice? Or would they ultimately come back to the fold and be faithful to the Democratic Party?

3   Bentsen's ties with big business were unusual for a Democratic nominee. Would Democratic voters accept him? The other side of this gamble was that Bentsen was one of the best fund raisers around and might be able to eliminate or even reverse the Republicans' traditional financial advantage. Even if some of the more liberal voters were disenchanted, Bentsen could appeal to a more business-oriented constituency.

4   The safer choice for a running mate would have been Senator John Glenn from Ohio. The polls suggested that with Glenn as his running mate, Dukakis would have no trouble winning Ohio and its 23 electoral votes.

Questions

1   Why is choosing a running mate a hard decision?

2   What objectives do you think a presidential nominee should consider in making the choice?

3   What elements of risk are involved?

4   The title of Dionne's article was "Bentsen: Bold Choice or Risky Gamble?" In what sense was Dukakis's decision a "bold choice," and in what sense was it a "risky gamble"?

DUPONT AND CHLOROFLUOROCARBONS

Chlorofluorocarbons (CFCs) are chemicals used as refrigerants in air conditioners and other cooling appliances, propellants in aerosol sprays, and in a variety of other applications. Scientific evidence has been accumulating for some time that CFCs released into the atmosphere can destroy ozone molecules in the ozone layer 15 miles above the earth's surface. This layer shields the earth from dangerous ultraviolet radiation. A large hole in the ozone layer above Antarctica has been found and attributed to CFCs, and a 1988 report by 100 scientists concluded that the ozone shield above the mid-Northern Hemisphere had shrunk by as much as 3% since 1969. Moreover, depletion of the ozone layer appears to be irreversible. Further destruction of the ozone layer could lead to crop failures, damage to marine ecology, and possibly dramatic changes in global weather patterns.

Environmentalists estimate that approximately 30% of the CFCs released into the atmosphere come from aerosols. In 1978, the U.S. government banned their use as aerosol propellants, but many foreign governments still permit them.

Some $2.5 billion of CFCs are sold each year, and DuPont Chemical Corporation is responsible for 25% of that amount. In early 1988, DuPont announced that the company would gradually phase out its production of CFCs and that replacements would be developed. Already DuPont claims to have a CFC substitute for automobile air conditioners, although the new substance is more expensive.

Questions

Imagine that you are a DuPont executive charged with making the decision regarding continued production of CFCs.

1   What issues would you take into account?

2   What major sources of uncertainty do you face?

3   What corporate objectives would be important for you to consider? Do you think that DuPont's corporate objectives and the way the company views the problem might have evolved since the mid–1970s when CFCs were just beginning to become an issue?

Sources: "A Gaping Hole in the Sky," *Newsweek,* July 11, 1988, pp. 21–23; A. M. Louis (1988), "DuPont to Ban Products That Harm Ozone," *San Francisco Chronicle,* March 25, p. 1.

REFERENCES

The decision-analysis view is distinctly *prescriptive.* That is, decision analysis is interested in helping people make better decisions; in contrast, a *descriptive* view of decision making focuses on how people actually make decisions. Keeney and Raiffa (1976) explain the prescriptive view as well as anyone. For an excellent summary of the descriptive approach, see Hogarth (1987). Bell, Raiffa, and Tversky (1988) provide many readings on these topics.

A fundamental element of the prescriptive approach is discerning and accepting the difference between a good decision and a lucky outcome. This issue has been discussed by many authors, both academics and practitioners. An excellent recent reference is Vlek et al. (1984).

Many other books and articles describe the decision-analysis process, and each seems to have its own twist. This chapter has drawn heavily from Ron Howard's thoughts; his 1988 article summarizes his approach. Other books worth consulting include Behn and Vaupel (1982), Bunn (1984), Holloway (1979), Keeney (1992), Lindley (1985), Raiffa (1968), Samson (1988), and von Winterfeldt and Edwards (1986).

Phillips's (1982, 1984) idea of a requisite decision model is a fundamental concept that we will use throughout the text. For a related view, see Watson and Buede (1987).

Behn, R. D., and J. D. Vaupel (1982) *Quick Analysis for Busy Decision Makers.* New York: Basic Books.

Bell, D., H. Raiffa, and A. Tversky (1988) *Decision Making: Descriptive, Normative, and Prescriptive Interactions.* Cambridge, MA: Cambridge University Press.

Bunn, D. (1984) *Applied Decision Analysis.* New York: McGraw-Hill.

Corner, J. L., and C. W. Kirkwood (1991) "Decision Analysis Applications in the Operations Research Literature, 1970–1989." *Operations Research,* 39, 206–219.

Hogarth, R. (1987) *Judgement and Choice,* 2nd ed. New York: Wiley.

Holloway, C. A. (1979) *Decision Making under Uncertainty: Models and Choices.* Englewood Cliffs, NJ: Prentice-Hall.

Howard, R. A. (1980) "An Assessment of Decision Analysis." *Operations Research,* 28, 4–27.

Howard, R. A. (1988) "Decision Analysis: Practice and Promise," *Management Science,* 34, 679–695.

Keeney, R. (1992) *Value-Focused Thinking.* Cambridge, MA: Harvard University Press.

Keeney, R., and H. Raiffa (1976) *Decisions with Multiple Objectives.* New York: Wiley.

Lindley, D. V. (1985) *Making Decisions,* 2nd ed. New York: Wiley.

Phillips, L. D. (1982) "Requisite Decision Modelling." *Journal of the Operational Research Society,* 33, 303–312.

Phillips, L. D. (1984) "A Theory of Requisite Decision Models." *Acta Psychologica,* 56, 29–48.

Raiffa, H. (1968) *Decision Analysis.* Reading, MA: Addison-Wesley.

Samson, D. (1988) *Managerial Decision Analysis.* Homewood, IL: Irwin.

Ulvila, J. W., and R. V. Brown (1982) "Decision Analysis Comes of Age." *Harvard Business Review,* September–October 1982, 130–141.

Vlek, C., W. Edwards, I. Kiss, G. Majone, and M. Toda (1984) "What Constitutes a Good Decision?" *Acta Psychologica,* 56, 5–27.

von Winterfeldt, D., and W. Edwards (1986) *Decision Analysis and Behavioral Research.* Cambridge: Cambridge University Press.

Watson, S., and D. Buede (1987) *Decision Synthesis.* Cambridge: Cambridge University Press.

EPILOGUE

What did the ODA decide? Its directors decided to use only BT on all 227,000 acres, which were sprayed on three separate occasions in late spring and early summer 1985. At the time, this was the largest gypsy moth–control program ever attempted in Oregon. In 1986, 190,000 acres were sprayed, also with BT. Most of the areas sprayed the second year had not been treated the first year because ODA had found later that the gypsy moth infestation was more widespread than first thought. In the summer of 1986, gypsy moth traps throughout the area indicated that the population was almost completely controlled. In the spring of 1987, the ODA used BT to spray only 7500 acres in 10 isolated pockets of gypsy moth populations on the fringes of the previously sprayed areas. By 1988, the spray program was reduced to a few isolated areas near Eugene, and officials agreed that the gypsy moth population was under control.

1

Modeling Decisions

This first section is about modeling decisions. Chapter 2 presents a short discussion on the elements of a decision. Through a series of simple examples, the basic elements are illustrated: values and objectives, decisions to be made, upcoming uncertain events, and consequences. The focus is on identifying the basic elements. This skill is necessary for modeling decisions as described in Chapters 3, 4, and 5.

In Chapter 3, we learn how to create graphical structures for decision models. First we consider values and objectives, discussing in depth how multiple objectives can be organized in hierarchies and networks that can provide insight and help to generate creative alternatives. We also develop both influence diagrams and decision trees as graphical modeling tools for representing the basic structure of decisions. An influence diagram is particularly useful for developing the structure of a complex decision problem because it allows many aspects of a problem to be displayed in a compact and intuitive form. A decision-tree representation provides an alternative picture of a decision in which more of the details can be displayed. Both graphical techniques can be used to represent single-objective decisions, but we show how they can be used in multiple-objective situations as well. We end Chapter 3 with a discussion of measurement, presenting concepts and techniques that can be used to ensure that we can adequately measure achievement of our objectives, whether those objectives are straightforward (e.g., maximizing dollars or saving time) or more difficult to quantify (e.g., minimizing environmental damage).

Chapters 4 and 5 present the basic tools available to the decision maker for analyzing a decision model. Chapter 4 shows how to solve decision trees and influence diagrams. The basic concept presented is *expected value*. When we are concerned with monetary outcomes, we call this *expected monetary value* and abbreviate it as EMV. In analyzing a decision, EMV is calculated for each of the available alternatives. In many decision situations it is reasonable to choose the alternative with the highest EMV. In addition to the EMV criterion, Chapter 4 also looks briefly at the idea of risk analysis and the uses of a stochastic-dominance criterion for making decisions. Finally, we show how expected value and risk analysis can be used in multiple-objective decisions.

In Chapter 5 we learn how to use sensitivity-analysis tools in concert with EMV calculations in the iterative decision-structuring and analysis process. After an initial basic model is built, sensitivity analysis can tell which of the input variables really matter in the decision and deserve more attention in the model. Thus, with Chapter 5 we bring the discussion of modeling decisions full circle, showing how structuring and analysis are intertwined in the decision-analysis process.

Finally, Chapter 6 delves into issues relating to creativity and decision making. One of the critical aspects of constructing a model of a decision is the determination of viable alternatives. When searching for alternative actions in a decision situation, though, we are subject to a variety of creative blocks that hamper our search for new and different possibilities. Chapter 6 describes these blocks to creativity, discusses creativity from a psychological perspective, and shows how a careful understanding of one's objectives can aid the search for creative alternatives. Several creativity-enhancing techniques are described.

CHAPTER 2

Elements of
Decision Problems

Given a complicated problem, how should one begin? A critical first step is to identify the elements of the situation. We will classify the various elements into (1) values and objectives, (2) decisions to make, (3) uncertain events, and (4) consequences. In this chapter, we will discuss briefly these four basic elements and illustrate them in a series of examples.

Values and Objectives

Imagine a farmer whose trees are laden with fruit that is nearly ripe. Even without an obvious problem to solve or decision to make, we can consider the farmer's objectives. Certainly one objective is to harvest the fruit successfully. This may be important because the fruit can then be sold, providing money to keep the farm operating and a profit that can be spent for the welfare of the family. The farmer may have other underlying objectives as well, such as maximizing the use of organic farming methods.

Before we can even talk about making decisions, we have to understand *values* and *objectives*. "Values" is an overused term that can be somewhat ambiguous; here we use it in a general sense to refer to things that matter to you. For example, you may want to learn how to sail and take a trip around the world. Or you may have an objective of learning how to speak Japanese. A scientist may be interested in resolving a specific scientific question. An investor may want to make a lot of money or gain a controlling interest in a company. A manager, like our farmer with the orchard, may want to earn a profit.

An *objective* is a specific thing that you want to achieve. All of the examples in the previous paragraph refer to specific objectives. As you can tell from the examples, some objectives are related. The farmer may want to earn a profit because it will provide the means to purchase food for the family or to take a trip. The scientist may want to find an answer to an important question in order to gain prestige in the scientific community; that prestige may in turn lead to a higher salary and more research support at a better university.

An individual's objectives taken together make up his or her values. They define what is important to that person in making a decision. We can make an even broader statement: A person's values are the reason for making decisions in the first place! If we did not care about anything, there would not be a reason to make decisions at all, because we would not care how things turned out. Moreover, we would not be able to choose from among different alternatives. Without objectives, it would not be possible to tell which alternative would be the best choice.

Making Money: A Special Objective

In modern western society, most adults work for a living, and if you ask them why, they will all include in their answers something about the importance of making money. It would appear that making money is an important objective, but a few simple questions (Why is money important? What would you do if you had a million dollars?) quickly reveal that money is important because it helps us do things that we want to do. For many people, money is important because it allows us to eat, afford housing and clothing, travel, engage in activities with friends, and generally live comfortably. Many people spend money on insurance because they have an objective of avoiding risks. For very few individuals is money important in and of itself. Unlike King Midas, most of us do not want to earn money simply to have it; money is important because it provides the means by which we can work toward more basic objectives.

Money's role as a trading mechanism in our economy puts it in a special role. Although it is typically not one of our basic objectives, it can serve as a proxy objective in many situations. For example, imagine a young couple who wants to take a vacation. They will probably have to save money for some period of time before achieving this goal, and they will face many choices regarding just how to go about saving their money. In many of these decisions, the main concern will be how much money they will have when they are ready to take their holiday. If they are considering investing their money in a mutual fund, say, they will have to balance the volatility of the fund's value against the amount they can expect to earn over the long run, because most investment decisions require a trade-off between risk and return.

For corporations, money is often a primary objective, and achievement of the objective is measured in terms of increase in the shareholders' wealth through dividends and increased company value. The shareholders themselves can, of course, use their wealth for their own welfare however they want. Because the shareholders have the opportunity to trade their wealth to achieve specific objectives, the company need not be concerned with those objectives but can focus on making its shareholders as wealthy as possible.

Although making money is indeed a special objective, it is important to realize that many situations require a trade-off between making money and some other objective. In many cases, one can *price out* the value of different objectives. When you purchase a car, how much more would you pay to have air conditioning? How much more to get the color of your choice? These questions may be difficult to answer, but we all make related decisions all the time as we decide whether a product or service is worth the price that is asked. In other cases, though, it may not be reasonable to convert everything to dollars. For example, consider the ethical problems faced by a hospital that performs organ transplants. Wealthy individuals can pay more for their operations, and often are willing to do so in order to move up in the queue. The additional money may permit the hospital to purchase new equipment or perform more transplants for needy individuals. But moving the wealthy patient up in the queue will delay surgery for other patients, perhaps with fatal consequences. What if the other patients include young children? Pricing out the lives and risks to the other patients seems like a cold-hearted way to make this decision; in this case, the hospital will probably be better off thinking in terms of its fundamental objectives and how to accomplish them with or without the wealthy patient's fee.

Values and the Current Decision Context

Suppose you have carefully thought about all of your objectives. Among other things you want to do what you can to reduce homelessness in your community, learn to identify birds, send your children to college, and retire at age 55. Having spent the morning figuring out your objectives, you have become hungry and are ready for a good meal. Your decision is where to go for lunch, and it is obvious that the large-scale, overall objectives that you have spent all morning thinking about will not be much help.

You can still think hard about your objectives, though, as you consider your decision. It is just that different objectives are appropriate for this particular decision. Do you want a lot to eat or a little? Do you want to save money? Are you interested in a particular type of ethnic food, or would you like to try a new restaurant? If you are going out with friends, what about their preferences? What about a picnic instead of a restaurant meal?

Each specific decision situation calls for specific objectives. We call the setting in which the decision occurs the *decision context*. In one case, a decision context might be deciding where to go for lunch, in which case the appropriate objectives involve satisfying hunger, spending time with friends, and so on. In another case, the context might be what to choose for a career, which would call for consideration of more global objectives. What do you want to accomplish in your life?

Values and decision context go hand in hand. On one hand, it is worthwhile to think about your objectives in advance to be prepared for decisions when they arise or so that you can identify new decision opportunities that you might not have thought about before. On the other hand, every decision situation involves a specific context, and that context determines what objectives need to be considered. The idea of a requisite model comes into play here. A requisite decision model includes all of the objectives that matter, and only those that matter, in the decision context at hand. Without all of the appropriate objectives considered, you will be left with the gnawing concern that "something is missing" (which would be true), and considering superfluous or inappropriate objectives can distract you from the truly important issues. When the decision context is specified and appropriate objectives aligned with the context, the decision maker knows what the situation is and exactly why he or she cares about making a decision in that situation.

Finding realistic examples in which individuals or companies use their objectives in decision making is easy. In the following example, the Boeing Company found itself needing to acquire a new supercomputer.

BOEING'S SUPERCOMPUTER

As a large-scale manufacturer of sophisticated aircraft, Boeing needs computing power for tasks ranging from accounting and word processing to computer-aided design, inventory control and tracking, and manufacturing support. When the company's engineering department needed to expand its high-power computing capacity by purchasing a supercomputer, the managers faced a huge task of assembling and evaluating massive amounts of information. There were systems requirements and legal issues to consider, as well as price and a variety of management issues. (*Source:* D. Barnhart, (1993) "Decision Analysis Software Helps Boeing Select Supercomputer." *OR/MS Today,* April, 62–63.)

Boeing's decision context is acquiring supercomputing capacity for its engineering needs. Even though the company undoubtedly has global objectives related to aircraft production, maximizing shareholder wealth, and providing good working conditions for its employees, in the current decision context the appropriate objectives are specific to the company's computing requirements.

Organizing all of Boeing's objectives in this decision context is complex because of the many different computer users involved and their needs. With careful thought, though, management was able to specify five main objectives: minimize costs, maximize performance, satisfy user needs, satisfy organizational needs, and satisfy management issues. Each of these main objectives can be further broken down into different aspects, as shown in Figure 2.1.

Figure 2.1
Objectives for Boeing's super- computer.

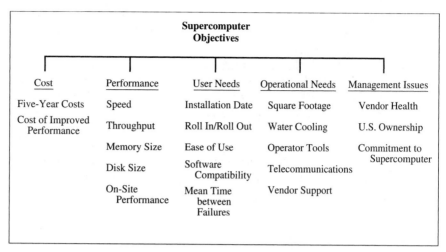

		Supercomputer Objectives		
Cost	Performance	User Needs	Operational Needs	Management Issues
Five-Year Costs	Speed	Installation Date	Square Footage	Vendor Health
Cost of Improved Performance	Throughput	Roll In/Roll Out	Water Cooling	U.S. Ownership
	Memory Size	Ease of Use	Operator Tools	Commitment to Supercomputer
	Disk Size	Software Compatibility	Telecommunications	
	On-Site Performance	Mean Time between Failures	Vendor Support	

Decisions to Make

With the decision context understood and values well in hand, the decision maker can begin to identify specific elements of a decision. Consider our farmer whose fruit crop will need to be harvested soon. If the weather report forecasts mild weather, the farmer has nothing to worry about, but if the forecast is for freezing weather, it might be appropriate to spend some money on protective measures that will save the crop. In such a situation, the farmer has a decision to make, and that decision is whether or not to take protective action. This is a decision that must be made with the available information.

Many situations have as the central issue a decision that must be made right away. There would always be at least two alternatives; if there were no alternatives, then it would not be a matter of making a decision! In the case of the farmer, the alternatives are to take protective action or to leave matters as they are. Of course, there may be a wide variety of alternatives. For example, the farmer may have several strategies for saving the crop, and it may be possible to implement one or more.

Another possibility may be to wait and obtain more information. For instance, if the noon weather report suggests the possibility of freezing weather depending on exactly where a weather system travels, then it may be reasonable to wait and listen to the evening report to get better information. Such a strategy, however, may entail a cost. The farmer may have to pay his hired help overtime if the decision to protect the crop is made late in the evening. Some measures may take time to set up; if the farmer waits, there may not be enough time to implement some of these procedures.

Other possible alternatives are taking out insurance or hedging. For example, the farmer might be willing to pay the harvesting crew a small amount to be available at night if quick action is needed. Insurance policies also may be available to protect against crop loss (although these typically are not available at the last minute). Any of these alternatives might give the farmer more flexibility but would probably cost something up front.

Identifying the immediate decision to make is a critical step in understanding a difficult decision situation. Moreover, no model of the decision situation can be built without knowing exactly what the decision problem at hand is. In identifying the central decision, it is important also to think about possible alternatives. Some decisions will have specific alternatives (protect the crop or not), while others may involve choosing a specific value out of a range of possible values (deciding on an amount to bid for a company you want to acquire). Other than the obvious alternative courses of action, a decision maker should always consider the possibilities of doing nothing, of waiting to obtain more information, or of somehow hedging against possible losses.

Sequential Decisions

In many cases, there simply is no single decision to make, but several sequential decisions. The orchard example will demonstrate this. Suppose that several weeks of the growing season remain. Each day the farmer will get a new weather forecast, and each time there is a forecast of adverse weather, it will be necessary to decide once again whether to protect the crop.

The example shows clearly that the farmer has a number of decisions to make, and the decisions are ordered sequentially. If the harvest is tomorrow, then the decision is fairly easy, but if several days or weeks remain, then the farmer really has to think about the upcoming decisions. For example, it might be worthwhile to adopt a policy whereby the amount spent on protection is less than the value of the crop. One good way to do this would be not to protect during the early part of the growing season; instead, wait until the harvest is closer, and then protect whenever the weather forecast warrants such action. In other words, "If we're going to lose the crop, let's lose it early."

It is important to recognize that in many situations one decision leads eventually to another in a sequence. The orchard example is a special case because the decisions are almost identical from one day to the next: Take protective action or not. In many cases, however, the decisions are radically different. For example, a manufacturer considering a new product might first decide whether or not to introduce it. If the decision is to go ahead, the next decision might be whether to produce it or subcontract the production. Once the production decision is made, there may be marketing decisions about distribution, promotion, and pricing.

When a decision situation is complicated by sequential decisions, a decision maker will want to consider them when making the immediate decision. Furthermore, a future decision may depend on exactly what happened before. For this reason, we refer to these kinds of problems as *dynamic* decision situations. In identifying elements of a decision situation, we want to know not only what specific decisions are to be made, but the sequence in which they will arise. Figure 2.2 shows graphically a sequence of decisions, represented by squares, mapped along a time line.

Figure 2.2

Sequential decisions.
A decision maker
needs to consider
decisions to be made
now and later.

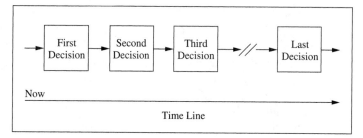

Uncertain Events

In Chapter 1 we saw that decision problems can be complicated because of uncertainty about what the future holds. Many important decisions have to be made without knowing exactly what will happen in the future or exactly what the ultimate outcome will be from a decision made today. A classic example is that of investing in the stock market. An investor may be in a position to buy some stock, but in which company? Some share prices will go up and others down, but it is difficult to tell exactly what will happen. Moreover, the market as a whole may move up or down, depending on economic forces. The best the investor can do is think very carefully about the chances associated with each different security's prices as well as the market as a whole.

The possible things that can happen in the resolution of an uncertain event are called *outcomes.* In the orchard example above, the key uncertain event is the weather, with outcomes of crop damage or no crop damage. With some uncertain events, such as with the orchard, there are only a few possible outcomes. In other cases, such as the stock market, the outcome is a value within some range. That is, next year's price of the security bought today for $50 per share may be anywhere between, say, $0 and $100. (It certainly could never be worth less than zero, but the upper limit is not so well defined: Different individuals might consider different upper limits for the same stock.) The point is that the outcome of the uncertain event that we call "next year's stock price" comes from a range of possible values and may fall anywhere within that range.

Many different uncertain events might be considered in a decision situation, but only some are relevant. How can you tell which ones are relevant? The answer is straightforward; the outcome of the event must have some impact on at least one of your objectives. That is, it should matter to you what actually comes to pass. Although this seems like common sense, in a complex decision situation it can be all too easy to concentrate on uncertain events that we can get information about rather than those that really have an impact in terms of our objectives. One of the best examples comes from risk analysis of nuclear power plants; engineers can make judgments about the chance that a power-plant accident will release radioactive material into the atmosphere, but what may really matter is how local residents react to siting the plant in their neighborhood and to subsequent accidents if they occur.

Of course, a decision situation often involves more than one uncertain event. The larger the number of uncertain but relevant events in a given situation, the more complicated the decision. Moreover, some uncertain events may depend on others. For example, the price of the specific stock purchased may be more likely to go up if the economy as a whole continues to grow or if the overall stock market increases in value. Thus there may be interdependencies among the uncertain events that a decision maker must consider.

How do uncertain events relate to the decisions in Figure 2.2? They must be dovetailed with the time sequence of the decisions to be made; it is important to know at each decision exactly what information is available and what remains unknown. At the current time ("Now" on the time line), all of the uncertain events are just that; their outcomes are unknown, although the decision maker can look into the future and specify which uncertainties will be resolved prior to each upcoming decision. For example, in the dynamic orchard decision, on any given day the farmer knows what the weather has been in the past but not what the weather will be in the future.

Sometimes an uncertain event that is resolved before a decision provides information relevant for future decisions. Consider the stock market problem. If the investor is considering investing in a company that is involved in a lawsuit, one alternative might be to wait until the lawsuit is resolved. Note that the sequence of decisions is (1) wait or buy now, and (2) if waiting, then buy or do not buy after the lawsuit. The decision to buy or not may depend crucially on the outcome of the lawsuit that occurs between the two decisions.

What if there are many uncertain events that occur between decisions? There may be a natural order to the uncertain events, or there may not. If there is, then specifying that order during modeling of the decision problem may help the decision maker. But the order of events between decisions is not nearly as crucial as the dovetailing of decisions and events to clarify what events are unknown and what information is available for each decision in the process. It is the time sequence of the decisions that matters, along with the information available at each decision. In Figure 2.3, uncertain events, represented by circles, are dovetailed with a sequence of decisions. An arrow from a group of uncertain events to a decision indicates that the outcomes of those events are known at the time the decision is made. Of course, the decision maker is like the proverbial elephant and never forgets what has happened. For upcoming decisions, he or she should be able to recall (possibly with the aid of notes and documents) everything that happened (decisions and event outcomes) up to that point.

Figure 2.3

Dovetailing uncertain events and sequential decisions.

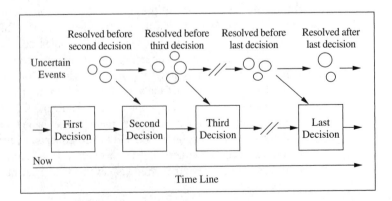

Consequences

After the last decision has been made and the last uncertain event has been resolved, the decision maker's fate is finally determined. It may be a matter of profit or loss as in the case of the farmer. It may be a matter of increase in value of the investor's portfolio. In some cases the final consequence may be a "net value" figure that accounts for both cash outflows and inflows during the time sequence of the decisions. This might happen in the case of the manufacturer deciding about a new product; certain costs must be incurred (development, raw materials, advertising) before any revenue is obtained.

If the decision context requires consideration of multiple objectives, the consequence is what happens with respect to each of the objectives. For example, consider the consequence of a general's decision to storm a hill. The consequence might be good because the army succeeds in taking the hill (a specific objective), but it may be bad at the same time if many lives are lost.

In our graphical scheme, we must think about the consequence at the end of the time line after all decisions are made and all uncertain events are resolved. For example, the consequence for the farmer after deciding whether to protect and then experiencing the weather might be a profit of $15,000, a loss of $3400, or some other dollar amount. For the general it might be "gain the hill, 10 men killed, 20 wounded" or "don't gain the hill, two men killed, five wounded." Thus, the end of the time line is when the decision maker finds out the results. Looking forward from the current time and current decision, the end of the time line is called the *planning horizon*. Figure 2.4 shows how the consequence fits into our graphical scheme.

What is an appropriate planning horizon? For the farmer, the answer is relatively easy; the appropriate planning horizon is at the time of the harvest. But for the general, this question is not so simple. Is the appropriate horizon the end of the next day when he will know whether his men were able to take the hill? Or is it at the end of the war? Or is it sometime in between—say, the end of next month? For the investor, how far ahead should the planning horizon be? A week? A month? Several years? For individuals planning for retirement, the planning horizon may be years in the future. For speculators making trades on the floor of a commodity exchange, the planning horizon may be only minutes into the future.

Figure 2.4

Including the consequence.

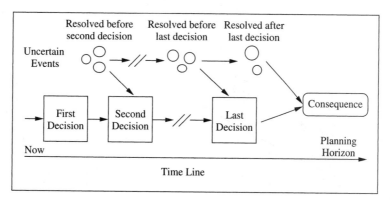

Thus, one of the fundamental issues with which a decision maker must come to grips is how far into the future to look. It is always possible to look farther ahead; there will always be more decisions to make, and earlier decisions may have some effect on the availability of later alternatives. Even death is not an obvious planning horizon because the decision maker may be concerned with effects on future generations; environmental policy decisions provide perfect examples. At some point the decision maker has to stop and say, "My planning horizon is there. It's not worthwhile for me to think beyond that point in time." For the purpose of constructing a requisite model, the idea is to choose a planning horizon such that the events and decisions that would follow after are not essential parts of the immediate decision problem. To put it another way, choose a planning horizon that is consistent with your decision context and the relevant objectives.

Once the dimensions of the consequences and the planning horizon have been determined, the next step is to figure out how to value the consequences. As mentioned, in many cases it will be possible to work in terms of monetary values. That is, the only relevant objective in the decision context is to make money, so all that matters at the end is profit, cost, or total wealth. Or it may be possible to price out nonmonetary objectives as discussed above. For example, a manager might be considering whether to build and run a day care center for the benefit of the employees. One objective might be to enhance the goodwill between the company and the workforce. Enhanced goodwill would in turn have certain effects on the operations of the company, including reduced absenteeism, improved ability to recruit, and a better image in the community. Some of these, such as the reduced absenteeism and improved recruiting, could easily be translated into dollars. The image may be more difficult to translate, but the manager might assess its value subjectively by estimating how much money it would cost in terms of public relations work to improve the firm's image by the same amount.

In some cases, however, it will be difficult to determine exactly how the different objectives should be traded off. In the hospital case discussed earlier, how should the administrator trade off the risks to patients who would be displaced in the queue versus the fee paid by a wealthy patient? How many lives should the general be willing to sacrifice in order to gain the hill? How much damage to the environment are we willing to accept in order to increase the U.S. supply of domestic oil? How much in the way of health risks are we willing to accept in order to have blemish-free fruits and vegetables? Many decisions, especially governmental policy decisions, are complicated by trade-offs like these. Even personal decisions, such as taking a job or purchasing a home, require a decision maker to think hard about the trade-offs involved.

The Time Value of Money: A Special Kind of Trade-Off

One of the most common consequences in personal and business decisions is a stream of cash flows. For example, an investor may spend money on a project (an initial cash outflow) in order to obtain revenue in the future (cash inflows) over a period of years. In such a case, there is a special kind of trade-off: spending dollars today to obtain dollars tomorrow. If a dollar today were worth the same as a dollar next year, there would be no problem. However, this is not the case. A dollar today can be invested in a savings account or other interest-bearing security; at the end of a year, one dollar invested now would be worth one dollar plus the interest paid.

Trade-offs between current and future dollars (and between future dollars at different points in time) refer to the fact that the value of a dollar depends on when it is available to the decision maker. Because of this, we often refer to the "time value of money." Fortunately, there is a straightforward way to collapse a stream of cash flows into a single number. This number is called the *present value,* or value in present dollars, of the stream of cash flows.

Suppose, for example, you have $100 in your pocket. If you put that money into a savings account that earns 10% per year, paid annually, then you would have $100 × 1.1 = $110 at the end of the year. At the end of two years, the balance in the account would be $110 plus another 10%, or $110 × 1.1 = $121. In fact, you can see that the amount you have is just the original $100 multiplied by 1.1 twice: $121 = $100 × 1.1 × 1.1 = $100 × 1.1^2. If you keep the money in the account for five years, say, then the interest compounds for five years. The account balance would be $100 × 1.1^5 = $161.05.

We are going to use this idea of interest rates to work backward. Suppose, for example, that someone promises that you can have $110 next year. What is this worth to you right now? If you have available some sort of investment like a savings account that pays 10% per year, then you would have to invest $100 in order to get $110 next year. Thus, the present value of the $110 that arrives next year is just $110/1.1 = $100. Similarly, the present value of $121 dollars promised at the end of two years is $121/(1.1^2) = $100.

In general, we will talk about the present value of an amount x that will be received at the end of n time periods. Of course, we must know the appropriate interest rate. Let r denote the interest rate per time period in decimal form; that is, if the interest rate is 10%, then $r = 0.10$. With this notation, the formula for calculating present value (PV) is

$$PV(x, n, r) = \frac{x}{(1 + r)^n}$$

The denominator in this formula is a number greater than 1. Thus, dividing x by $(1 + r)^n$ will give a present value that is less than x. For this reason, we often say that we "discount" x back to the present. You can see that if you had the discounted amount now and could invest it at the interest rate r, then after n time periods (days, months, years, and so on) the value of the investment would be the discounted amount times $(1 + r)^n$, which is simply x.

Keeping the interest rate consistent with the time periods is important. For example, a savings account may pay 10% "compounded monthly." Thus, a year is really 12 time periods, and so $n = 12$. The monthly interest rate is 10%/12, or 0.8333%. Thus, the value of $100 deposited in the account and left for a year would be $100 × (1.00833)^{12} = $110.47. Notice that compounding helps because the interest itself earns interest during each time period. Thus, if you have a choice among savings accounts that have the same interest rate, the one that compounds more frequently will end up having a higher eventual payoff.

We can now talk about the present value of a stream of cash flows. Suppose that a friend is involved in a business deal and offers to let you in on it. For $425 paid to him now, he says, you can have $110.00 next year, $121.00 the following year, $133.10 the third year, and $146.41 at the end of Year 4. This is a great deal, he says, because your payments will total $510.51.

What is the present value of the stream of payments? (You probably can guess already!) Let us suppose you put your money into a savings account at 10%, compounded annually. Then we would calculate the present value of the stream of cash flows as the sum of the present values of the individual cash flows:

$$PV = \frac{110.0}{1.1} + \frac{121.00}{(1.1)^2} + \frac{133.10}{(1.1)^3} + \frac{146.41}{(1.1)^4}$$

$$= \$100 + \$100 + \$100 + \$100 = \$400$$

Thus, the deal is not so great. You would be paying $425 for a stream of cash flows that has a present value of only $400. The *net present value* (NPV) of the cash flows is the present value of the cash flows ($400) minus the cost of the deal ($425), or –$25; you would be better off keeping your $425 and investing it in the savings account.

The formula for calculating NPV for a stream of cash flows $x_0, \ldots, x_n$ over n periods at interest rate r is

$$NPV = \frac{x_0}{(1+r)^0} + \frac{x_1}{(1+r)^1} + \cdots + \frac{x_n}{(1+r)^n}$$

$$= \sum_{i=0}^{n} \frac{x_i}{(1+r)^i}$$

In general, we can have both outflows (negative numbers) and inflows. In the example, we could include the cash outflow of $425 as a negative number in calculating NPV:

$$NPV = \frac{-425.00}{(1.1)^0} + \frac{110.00}{(1.1)^1} + \frac{121.00}{(1.1)^2} + \frac{133.10}{(1.1)^3} + \frac{146.41}{(1.1)^4}$$

$$= -\$425 + \$400$$

$$= -\$25$$

[Recall that raising any number to the zero power is equal to 1, and so $(1.1)^0 = 1$.] Clearly, we could deal with any stream of cash flows. There could be one big inflow and then a bunch of outflows (such as with a loan), or there could be a large outflow (buying a machine), then some inflows (revenue), another outflow (maintenance costs), and so on. When NPV is calculated, it reveals the value of the stream of cash flows. A negative NPV for a project indicates that the money would be better invested to earn interest rate r.

We began our discussion by talking about trade-offs. You can see how calculating present values establishes trade-offs between dollars at one point in time and dollars at another. That is, you would be indifferent between receiving $1 now or $1(1 + r)$ at the end of the next time period. More generally, $1 now is worth $1(1 + r)^n$ at the end of n time periods. NPV works by using these trade-off rates to discount all the cash flows back to the present.

Knowing the interest rate is the key in using present-value analysis. What is the appropriate interest rate? In general, it is the interest rate that you could get for investing your money in the next best opportunity. Often we use the interest rate from a savings account, a certificate of deposit, or short-term (money market) securities. For a corporation, the appropriate interest rate to use might be the interest rate they would have to pay in order to raise money by issuing bonds. Often the interest rate is called the *hurdle rate,* indicating that an acceptable investment must earn more than this rate.

We have talked about the elements of decision problems: objectives, decisions to make, uncertain events, and consequences. The discussion of the time value of money showed how a consequence that is a stream of cash flows can be valued through the trade-offs implicit in interest rates. Now it is time to put all of this together and try it out in an example. Imagine the problems that an oil company might face in putting together a plan for dealing with a major oil spill. Here are managers in the fictitious "Larkin Oil" struggling with this situation.

LARKIN OIL

Pat Mills was restless. The Oil Spill Contingency Plan Committee was supposed to come up with a concrete proposal for the top management of Larkin Oil, Inc. The committee had lots of time; the CEO had asked for recommendations within three months. This was their first meeting.

Over the past hour, Sandy Wilton and Marty Kelso had argued about exactly what level of resources should be committed to planning for a major oil spill in the company's main shipping terminal bay.

"Look," said Sandy, "We've been over this so many times. When, and if, an oil spill actually occurs, we will have to move fast to clean up the oil. To do that, we have to have equipment ready to go."

"But having equipment on standby like that means tying up a lot of capital," Chris Brown replied. As a member of the financial staff, Chris was sensitive to committing capital for equipment that would be idle all the time and might actually have to be replaced before it was ever used. "We'd be better off keeping extensive records, maybe just a long list of equipment that would be useful in a major cleanup. We need to know where it is, what it's capable of, what its condition is, and how to transport it."

"Come to think of it, our list will also have to include information on transportation equipment and strategies," Leslie Taylor added.

Pat finally stirred. "You know what bothers me? We're talking about these alternatives, and the fact that we need to do thus and so in order to accomplish such and such. We're getting the cart before the horse. We just don't have our hands on the problem yet. I say we go back to basics. First, how could an oil spill happen?"

"Easy," said Sandy. "Most likely something would happen at the pipeline terminal. Something goes wrong with a coupling, or someone just doesn't pay attention while loading oil on the ship. The other possibility is that a tanker's hull fails for some reason, probably from running aground because of weather."

"Weather may not be the problem," suggested Leslie. "What about incompetence? What if the pilot gets drunk?"

Marty Kelso always was able to imagine the unusual scenarios. "And what about the possibility of sabotage? What if a terrorist decides to wreak environmental havoc?"

"OK," said Pat, "In terms of the actual cleanup, the more likely terminal spill would require a different kind of response than the less likely event of a hull failure. In planning for a terminal accident, we need to think about having some equipment at the terminal. Given the higher probability of such an accident, we should probably spend some money on cleanup equipment that would be right there and available."

"I suppose so," conceded Chris. "At least we would be spending our money on the right kind of thing."

"You know, there's another problem that we're not really thinking about," Leslie offered. "An oil spill at the terminal can be easily contained with relatively little environmental damage. On the other hand, if we ever have a hull failure, we have to act fast. If we don't, and mind you, we may not be able to because of the weather, Larkin Oil will have a terrible time trying to clean up the public relations as well as the beaches. And think about the difference in the PR problem if the spill is due to incompetence on the part of a pilot rather than weather or sabotage."

"Even if we act fast, a huge spill could still be nearly impossible to contain," Pat pointed out. "So what's the upshot? Sounds to me like we need someone who could make a decision immediately about how to respond. We need to recover as much oil as possible, minimize environmental damage, and manage the public relations problem."

"And do this all efficiently," growled Chris Brown. "We still have to do it without having tied up all of the company's assets for years waiting for something to happen."

The committee at Larkin Oil has a huge problem on its hands. The effects of its work now and the policy that is eventually implemented for coping with future accidents will substantially affect the company resources and possibly the environment. We cannot solve the problem entirely, but we can apply the principles discussed so far in the chapter. Let us look at the basic elements of the decision situation.

First, what is the committee's decision context, and what are Larkin's objectives? The context is making recommendations regarding plans for possible future oil spills, and the immediate decision is what policy to adopt for dealing with oil spills. Exactly what alternatives are available is not clear. The company's objectives are well stated by Pat Mills and Chris Brown at the end of the example: (1) recover as much oil as possible, (2) minimize environmental damage, (3) minimize damage to Larkin's public image, and (4) minimize cost. Recovering as much oil as possible is perhaps best viewed as a means to minimize environmental damage as well as the impact on Larkin's image. It also appears that a fundamental issue is how much of the company's resources should be committed to standby status waiting for an accident to occur. In general, the more resources committed, the faster the company could respond and the less damage would be done. Having these objectives out on the table immediately and understanding the inherent trade-offs will help the committee organize their efforts as they explore potential policy recommendations.

Is this a sequential decision problem? Based on Pat's last statement, the immediate decision must anticipate future decisions about responses to specific accident situations. Thus, in figuring out an appropriate policy to adopt now, they must think about possible appropriate future decisions and what resources must be available at the time so that the appropriate action can be taken.

The scenario is essentially about uncertain events. Of course, the main uncertain event is whether an oil spill will ever occur. From Chris Brown's point of view, an important issue might be how long the cleanup equipment sits idle, requiring periodic maintenance, until an accident occurs. Also important are events such as the kind of spill, the location, the weather, the cause, and the extent of the damage. At the present time, imagining the first accident, all of these are unknowns, but if and when a decision must be made, some information will be available (location, current weather, cause), while other factors—weather conditions for the cleanup, extent of the eventual damage, and total cleanup cost—probably will not be known.

What is an appropriate planning horizon for Larkin? No indication is given in the case, but the committee members may want to consider this. How far into the future should they look? How long will their policy recommendations be active? They may wish to specify that at some future date (say three years from the present) another committee be charged with reviewing and updating the policy in light of scientific and technological advances.

The problem also involves fundamental issues about how the different consequences are valued. As indicated, the fundamental trade-off is whether to save money by committing fewer resources or to provide better protection against future possible accidents. In other words, just how much is insurance against damage worth to Larkin Oil? In talking about consequences, the committee can imagine some possible ones and the overall "cost" (in generic terms) to the company: (1) committing substantial resources and never needing them; (2) committing a lot of resources and using them effectively to contain a major spill; (3) committing few resources and never needing them (the best possible outcome); and (4) committing few resources and not being able to clean up a spill effectively (the worst possible outcome).

Just considering the dollars spent, there is a time-value-of-money problem that Chris Brown eventually will want the committee to address. To some extent, dollars can be spent for protection now instead of later on. Alternative financing schemes can be considered to pay for the equipment required. Different strategies for acquiring and maintaining equipment may have different streams of cash flows. Calculating the present value of these different strategies for providing protection may be an important aspect of the decision.

Finally, the committee members also need to think about exactly how to allocate resources in terms of the other objectives stated by Pat Mills. They need to recover oil, minimize environmental damage, and handle public relations problems. Of course, recovering oil and minimizing environmental damage are linked to some extent. Overall, though, the more resources committed to one of these objectives, the less available they are to satisfy the others. The committee may want to specify some guidelines for resource allocation in its recommendations, but for the most part this allocation will be made at the time of future decisions that are in turn made in response to specific accidents.

Can we put all of this together? Figure 2.5 shows the sequence of decisions and uncertain events. This is only a rough picture, intended to capture the elements discussed

Figure 2.5

*A graphical represen-
tation of Larkin Oil's
situation.*

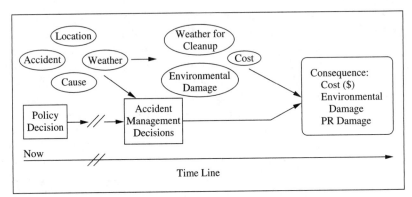

here, a first step toward the development of a requisite decision model. Different decision makers most likely would have different representations of the situation, although
most would probably agree on the essential elements of the values, decisions, uncertain
events, and consequences.

S U M M A R Y

Hard decisions often have many different aspects. The basic elements of decision situations include values and objectives, decisions to be made, uncertain events, and
consequences. This chapter discussed identification of the immediate decision at
hand as well as subsequent decisions. We found that uncertain future events must be
dovetailed with the sequence of decisions, showing exactly what is known before
each decision is made and what uncertainties still remain. We discussed valuing consequences in some depth, emphasizing the specification of a planning horizon and
the identification of relevant trade-offs. The discussion about the time value of
money showed how interest rates imply a special kind of trade-off between cash
flows at different points in time. Finally, the Larkin Oil example served to illustrate
the identification of the basic elements of a major (and messy) decision problem.

Q U E S T I O N S   A N D   P R O B L E M S

2.1   Suppose you are in the market for a new car, the primary use for which would be com-
muting to work, shopping, running errands, and visiting friends.
a   What are your objectives in this situation? What are some different alternatives?
b   Suppose you broaden the decision context. Instead of deciding on a car for commut-
ing purposes, you are interested in having transportation for getting around your community. In this new decision context, how would you describe your objectives? What
are some alternatives that you might not have considered in the narrower decision
context?

c How might you broaden the decision context further? (There are many ways to do this!) In this broader context, what new objectives must you consider? What new alternatives are available?

d Does your planning horizon change when you broaden the decision context in question b? Question c?

2.2 Explain in your own words why it is important in some situations to consider future decisions as well as the immediate decision at hand. Can you give an example from your own experience of an occasion in which you had to make a decision while explicitly anticipating a subsequent decision? How did the immediate decision affect the subsequent one?

2.3 Sometimes broadening the decision context can change the planning horizon. For example, many companies face specific technical problems. Framed in a narrow decision context, the question is how to solve the specific problem, and a reasonable solution may be to hire a consultant. On the other hand, if the decision context is broadened to include solving related problems as well as the current one, the company might want to develop in-house expertise by hiring one or more permanent employees or training an existing employee in the required skills. What is the planning horizon in each case, and why does it change with the broader context? What objectives must be considered in the broader context that can be ignored in the narrower one?

2.4 Explain in your own words why it is important to keep track of what information is known and what events are still uncertain for each decision.

2.5 What alternatives other than specific protection strategies might Larkin Oil consider (for example, insurance)?

2.6 Imagine the difficulties of an employer whose decision context is choosing a new employee from a set of applicants whom he will interview. What do you think the employer's objectives should be? Identify the employer's specific decisions to make and uncertainties, and describe the relevant uncertain events. How does the problem change if the employer has to decide whether to make an offer on the spot after each interview?

2.7 Identify the basic elements of a real-estate investor's decision situation. What are the investor's objectives? Is the situation dynamic (that is, are there sequential decisions)? What are some of the uncertainties that the investor faces? What are the crucial trade-offs? What role does the time value of money play for this investor?

2.8 Describe a decision problem that you have faced recently (or with which you are currently struggling). Describe the decision context and your objectives. What were the specific decisions that you faced, and what were the relevant uncertainties? Describe the possible consequences.

2.9 Calculate the net present value of a business deal that costs $2500 today and will return $1500 at the end of this year and $1700 at the end of the following year. Use an interest rate of 13%.

2.10 Find the net present value of a project that has cash flows of −$12,000 in Year 1, +$5000 in Years 2 and 3, −$2000 in Year 4, and +$6000 in Years 5 and 6. Use an interest rate of 12%. Find the interest rate that gives a net present value of zero.

2.11 A friend asks you for a loan of $1000 and offers to pay you back at the rate of $90 per month for 12 months.

    a Using an annual interest rate of 10%, find the net present value (to you) of loaning your friend the money. Repeat, using an interest rate of 20%.

    b Find an interest rate that gives a net present value of 0. The interest rate for which NPV = 0 is often called the *internal rate of return.*

2.12 Terry Martinez is considering taking out a loan to purchase a desk. The furniture store manager rarely finances purchases, but will for Terry "as a special favor." The rate will be 10% per year, and because the desk costs $600, the interest will come to $60 for a one-year loan. Thus, the total price is $660, and Terry can pay it off in 12 installments of $55 each.

    a Use the interest rate of 10% per year to calculate the net present value of the loan. (Remember to convert to a monthly interest rate.) Based on this interest rate, should Terry accept the terms of the loan?

    b Look at this problem from the store manager's perspective. Using the interest rate of 10%, what is the net present value of the loan to the manager?

    c What is the net present value of the loan to the manager if an interest rate of 18% is used? What does this imply for the real rate of interest that Terry is being charged for the loan?

This kind of financing arrangement was widely practiced at one time, and you can see why from your answers to (c). By law, lenders in the United States now must clearly state the actual annual percentage rate in the loan contract.

2.13 Lynn Rasmussen is deciding what sports car to purchase. In reflecting about the situation, it becomes obvious that after a few years Lynn may elect to trade in the sports car for a new one, although the circumstances that might lead to this choice are uncertain. Should trading in the car count as an uncertain event or a future decision? What are the implications for building a requisite model of the current car-purchase decision if trading in the car is treated as an uncertain event? As a decision?

CASE STUDIES

THE VALUE OF PATIENCE

Robin Briggs, a wealthy private investor, had been approached by Union Finance Company on the previous day. It seemed that Union Finance was interested in loaning money to one of its larger clients, but the client's demands were such that Union could not manage the whole thing. Specifically, the client wanted to obtain a loan for $385,000, offering to repay Union Finance $100,000 per year over seven years.

Union Finance made Briggs the following proposition. Since it was bringing Briggs business, its directors argued, they felt that it was only fair for Briggs to put up a proportionately larger share of the money. If Briggs would put up 60% of the money ($231,000), then Union would put up the remaining 40% ($154,000). They would split the payments evenly, each getting $50,000 at the end of each year for the next seven years.

Questions

1   Union Finance can usually earn 18% on its money. Using this interest rate, what is the net present value of the client's offer to Union?

2   Robin Briggs does not have access to the same investments as Union. In fact, the best available alternative is to invest in a security earning 10% over the next seven years. Using this interest rate, what is Briggs's net present value of the offer made by Union? Should Briggs accept the offer?

3   What is the net present value of the deal to Union if Briggs participates as proposed?

4   The title of this case study is "The Value of Patience." Which of these two investors is more patient? Why? How is this difference exploited by them in coming to an agreement?

EARLY BIRD, INC.

The directors of Early Bird, Inc., were considering whether to begin a sales promotion for their line of specialty coffees earlier than originally planned. "I think we should go ahead with the price cuts," Tracy Brandon said. "After all, it couldn't hurt! At the very worst, we'll sell some coffee cheap for a little longer than we had planned, and on the other side we could beat New Morning to the punch."

"That's really the question, isn't it?" replied Jack Santorini. "If New Morning really is planning their own promotion, and we start our promotion now, we would beat them to the punch. On the other hand, we might provoke a price war. And you know what a price war with that company means. We spend a lot of money fighting with each other. There's no real winner. We both just end up with less profit."

Janice Wheeler, the finance VP for Early Bird, piped up, "The consumer wins in a price war. They get to buy things cheaper for a while. We ought to be able to make something out of that."

Ira Press, CEO for Early Bird, looked at the VP thoughtfully. "You've shown good horse sense in situations like these, Janice. How do you see it?"

Janice hesitated. She didn't like being put on the spot like this. "You all know what the projections are for the six-week promotion as planned. The marketing group tells us to expect sales of 10 million dollars. The objective is to gain at least two percentage points of market share, but our actual gain could be anywhere from nothing to three points. Profits during the promotion are expected to be down by 10 percent, but after the promotion ends, our increased market share should result in more sales and more profits."

Tracy broke in. "That's assuming New Morning doesn't come back with their own promotion in reaction to ours. And you know what our report is from Pete. He says that he figures New Morning is up to something."

"Yes, Pete did say that. But you have to remember that Pete works for our advertising agent. His incentive is to sell advertising. And if he thinks he can talk us into spending more money, he will. Furthermore, you know, he isn't always right. Last time he told us that New Morning was going to start a major campaign, he had the dates right, but it was for a different product line altogether."

Ira wouldn't let Janice off the hook. "But Janice, if New Morning does react to our promotion, would we be better off starting it early?"

Janice thought for a bit. If she were working at New Morning and saw an unexpected promotion begin, how would she react? Would she want to cut prices to match the competition? Would she try to stick with the original plans? Finally she said, "Look, we have to believe that New Morning also has some horse sense. They would not want to get involved in a price war if they could avoid it. At the same time, they aren't going to let us walk away with the market. I think that if we move early, there's about a 30 percent chance that they will react immediately, and we'll be in a price war before we know it."

"We don't have to react to their reaction, you know," replied Ira.

"You mean," asked Jack, "we have another meeting like this to decide what to do if they do react?"

"Right."

"So," Janice said, "I guess our immediate options are to start our promotion early or to start it later as planned. If we start it now, we risk a strong reaction from New Morning. If they do react, then we can decide at that point whether we want to cut our prices further."

Jack spoke up. "But if New Morning reacts strongly and we don't, we would probably end up just spending our money for nothing. We would gain no market share at all. We might even lose some market share. If we were to cut prices further, it might hurt profits, but at least we would be able to preserve what market share gains we had made before New Morning's initial reaction."

At this point, several people began to argue among themselves. Sensing that no resolution was immediately forthcoming, Ira adjourned the meeting, asking everyone to sleep on the problem and to call him with any suggestions or insights they had.

Questions

1   Based on the information in the case, what are Early Bird's objectives in this situation? Are there any other objectives that you think they should consider?

2   Given your answer to the previous question, what do you think Early Bird's planning horizon should be?

3   Identify the basic elements (values, decisions, uncertain events, consequences) of Early Bird's decision problem.

4   Construct a diagram like Figure 2.5 showing these elements.

REFERENCES

Identifying the elements of decision situations is implicit in a decision-analysis approach, although most textbooks do not explicitly discuss this initial step in decision modeling. The references listed at the end of Chapter 1 are all appropriate for discussions of values, objectives, decisions, uncertain events, and consequences.

The idea of understanding one's values as a prerequisite for good decision making is Ralph Keeney's thesis in his book *Value-Focused Thinking* (1992). A good summary is Keeney (1994). In the conventional approach, espoused by most authors on decision analysis, one finds oneself in a situation that demands a decision, identifies available alternatives, evaluates those alternatives, and chooses the best of those alternatives. Keeney argues persuasively that keeping one's values clearly in mind provides the ability to proactively find new decision opportunities and creative alternatives. Of course, the first step, and sometimes a difficult one, is understanding one's values, which we will explore in depth in Chapter 3.

Dynamic decision situations can be very complicated, and many articles and books have been written on the topic. A basic-level textbook that includes dynamic decision analysis is Buchanan (1982). DeGroot (1970) covers many dynamic decision problems at a somewhat more sophisticated level. Murphy et al. (1985) discuss the orchardist's dynamic decision problem in detail.

The time value of money is a standard topic in finance courses, and more complete discussions of net present value, internal rate of return (the implied interest rate in a sequence of cash flows), and related topics can be found in most basic financial management textbooks. Two good ones are Brigham (1985) and Schall and Haley (1986).

Brigham, E. F. (1985) *Financial Management: Theory and Practice,* 4th ed. Hinsdale, IL: Dryden.

Buchanan, J. T. (1982) *Discrete and Dynamic Decision Analysis.* New York: Wiley.

DeGroot, M. H. (1970) *Optimal Statistical Decisions.* New York: McGraw-Hill.

Keeney, R. L. (1992) *Value-Focused Thinking.* Cambridge, MA: Harvard University Press.

Keeney, R. L. (1994) "Creativity in Decision Making with Value-Focused Thinking." *Sloan Management Review,* Summer, 33–41.

Murphy, A. H., R. W. Katz, R. L. Winkler, and W.-R. Hsu (1985) "Repetitive Decision Making and the Value of Forecasts in the Cost-Loss Ratio Situation: A Dynamic Model." *Monthly Weather Review,* 113, 801–813.

Schall, L. D., and C. W. Haley (1986) *Introduction to Financial Management,* 4th ed. New York: McGraw-Hill.

EPILOGUE

On March 24, 1989, the Exxon *Valdez* tanker ran aground on a reef in Prince William Sound after leaving the Valdez, Alaska, pipeline terminal. Over 11 million gallons of oil spilled into Prince William Sound, the largest spill in the United States. In the aftermath, it was revealed that Aleyeska, the consortium of oil companies responsible for constructing and managing the pipeline, had instituted an oil spill contingency plan that was inadequate to the task of cleaning up a spill of such magnitude. As a result of the inadequate plan and the adverse weather immediately after the spill, little oil was recovered. Hundreds of miles of environmentally delicate shoreline were contaminated. Major fisheries were damaged, leading to specific economic harm to individuals who relied on fishing for a livelihood. In addition, the spill proved an embarrassment for all of the major oil companies and sparked new interest in environmental issues, especially upcoming leases for offshore oil drilling. Even though the risk of a major oil spill was very small, in retrospect one might conclude that the oil companies would have been better off with a much more carefully thought out contingency plan and more resources invested in it. (*Source:* "Dead Otters and Silent Ducks," *Newsweek,* April 24, 1989, p. 70.)

CHAPTER 3

Structuring Decisions

H aving identified the elements of a decision problem, how should one begin the modeling process? Creating a decision model requires three fundamental steps. First is identifying and structuring the values and objectives. Structuring values requires identifying those issues that matter to the decision maker, as discussed in Chapter 2. Simply listing objectives, however, is not enough; we also must separate the values into fundamental objectives and means objectives, and we must specify ways to measure accomplishment of the objectives.

The second step is structuring the elements of the decision situation into a logical framework. To do this we have two tools: influence diagrams and decision trees. These two approaches have different advantages for modeling difficult decisions. Both approaches are valuable and, in fact, complement one another nicely. Used in conjunction with a carefully developed value structure, we have a complete model of the decision that shows all of the decision elements: relevant objectives, decisions to make, uncertainties, and consequences.

The final step is the refinement and precise definition of all of the elements of the decision model. For example, we must be absolutely clear on the precise decisions that are to be made and the available alternatives, exactly what the uncertain events are, and how to measure the consequences in terms of the objectives that have been specified. Although many consequences are easily measured on a natural scale (for example, NPV can be measured in dollars), nonquantitative objectives such as increasing health or minimizing environmental impact are more problematic. We will discuss ways to create formal scales to measure achievement of such objectives.

Structuring Values

Our first step is to structure values. In Chapter 2 we discussed the notion of objectives. In many cases, a single objective drives the decision; a manager might want to maximize profits next year, say, or an investor might want to maximize the financial return of an investment portfolio. Often, though, there are multiple objectives that conflict; for example, the manager might want to maximize profits but at the same time minimize the chance of losing money. The investor might want to maximize the portfolio's return but minimize the volatility of the portfolio's value.

If a decision involves a single objective, that objective is often easily identified. Careful thought may be required, however, to define the objective in just the right way. For example, you might want to calculate NPV over three years, using a particular interest rate. The discussion of value structuring that follows can help in the identification and clarification of the objective in a single-objective decision situation.

Even though many pages in this book are devoted to the analysis of single-objective decisions, for many decisions the real problem lies in balancing multiple conflicting objectives. The first step in dealing with such a situation is to understand just what the objectives are. Specifying objectives is not always a simple matter, as we will see in the following example.

Suppose you are an employer with an opening for a summer intern in your marketing department. Under the supervision of a senior employee, the intern would assist in the development of a market survey relating to a line of your company's consumer products.

HIRING A SUMMER INTERN

Many businesses hire students for short-term assignments. Such jobs often are called *internships,* and the employee—or intern—gets a chance to see what a particular kind of job and a specific company are like. Likewise, the company gets to try out a new employee without making a long-term commitment.

In this example, the fictional PeachTree Consumer Products has an opening for a summer intern. Working under the supervision of a senior employee in the marketing group, the intern would focus primarily on the development of a market survey for certain of the company's products. The problem is how to find an appropriate individual to fill this slot. Where should the company go to locate good candidates, which ones should be interviewed, and on the basis of what criteria should a particular candidate be chosen?

Imagine that you are the manager charged with finding an appropriate intern for PeachTree. Your first step is to create a long list of all the things that matter to you in this decision context. What objectives would you want to accomplish in filling this position? Certainly you would want the market survey to be done well. You might want to use the summer as a trial period for the intern, with an eye toward a permanent job for the individual if the internship worked out. You might want to establish or cement a relationship with a college or university placement service. Table 3.1 shows a list of objectives (in no special order) that an employer might write down.

How would you go about generating a list like Table 3.1? Keeney (1994) gives some ideas. For example, think about some possible alternatives and ask what is good or bad about them. Or think about what you would like if you could have anything. Table 3.2 gives eight suggestions for generating your list of objectives.

Table 3.1
Objectives for hiring summer intern.

Maximize quality of market survey.
Sell more consumer products.
Build market share.
Identify new market niches for company's products.
Minimize cost of survey design.
Try out prospective permanent employee.
Establish relationship with local college.
Provide assistance for senior employee.
Free up an employee to be trained for new assignment.
Learn updated techniques from intern:
    Self
    Supervisor
    Market research department
    Entire company
Expose intern to real-world business experience.
Maximize profit.
Improve company's working environment by bringing in new and youthful energy.
Provide financial assistance for college student.

Table 3.2
Techniques for identifying objectives.

1. **Develop a wish list.** What do you want? What do you value? What should you want?

2. **Identify alternatives.** What is a perfect alternative, a terrible alternative, some reasonable alternative? What is good or bad about each?

3. **Consider problems and shortcomings.** What is wrong or right with your organization? What needs fixing?

4. **Predict consequences.** What has occurred that was good or bad? What might occur that you care about?

5. **Identify goals, constraints, and guidelines.** What are your aspirations? What limitations are placed on you?

6. **Consider different perspectives.** What would your competitor or your constituency be concerned about? At some time in the future, what would concern you?

7. **Determine strategic objectives.** What are your ultimate objectives? What are your values that are absolutely fundamental?

8. **Determine generic objectives.** What objectives do you have for your customers, your employees, your shareholders, yourself? What environmental, social, economic, or health and safety objectives are important?

Source: Keeney, R. L. (1994) "Creativity in Decision Making with Value-Focused Thinking," *Sloan Management Review,* Summer, 33–41. Reprinted by permission.

Once you have a list of objectives, what do you do? Structuring the objectives means organizing them so that they describe in detail what you want to achieve and can be incorporated in an appropriate way into your decision model. We start by separating the list into items that pertain to different kinds of objectives. In the summer-intern example, objectives can be sorted into several categories:

- Business performance (sell more products, maximize profit, increase market share, identify market niches)
- Improve the work environment (bring in new energy, assist senior employee)
- Improve the quality and efficiency of marketing activities (maximize survey quality, minimize survey cost)
- Personnel and corporate development (learn updated techniques, free up employee for new assignment, try out prospective employee)
- Community service (financial aid, expose intern to real world, relationship with local college)

Of course, there are other ways to organize these objectives; the idea is to create categories that reflect the company's overall objectives.

Before continuing with the value structuring, we must make sure that the objectives are appropriate for the decision context. Recall that the decision context is hiring a summer intern for the marketing department. This is a relatively narrow context for which some of the listed objectives are not especially relevant. For example, selling more consumer products and maximizing profit, although indeed important objectives, are too broad to be essential in the current decision context. Although hiring the best individual should have a positive impact on overall company performance, more crucial in the specific context of hiring the best intern are the objectives of enhancing marketing activities, personnel development, community service, and enhancing the work environment. These are the areas that hiring an intern may directly affect.

Fundamental and Means Objectives

With a set of objectives that is consistent with the decision context, the next step is to separate *means* from *fundamental* objectives. This is a critical step, because here we indicate those objectives that are important because they help achieve other objectives and those that are important simply because they reflect what we really want to accomplish. For example, working fewer hours may appear to be an important objective, but it may be important only because it would allow an individual to spend more time with his or her family or to pursue other activities that represent fundamental interests, things that are important simply because they are important. Thus, "minimize hours worked" is a means objective, whereas "maximize time with family" is a fundamental objective.

Fundamental objectives are organized into *hierarchies*. The upper levels in a hierarchy represent more general objectives, and the lower levels explain or describe important elements of the more general levels. For example, in the context of defining vehicle regulations, a higher-level fundamental objective might be "Maximize Safety," below which one might find "Minimize Loss of Life," "Minimize Serious Injuries," and "Minimize Minor Injuries." The three lower-level objectives are fundamental objectives that explain what is meant by the higher-level objective "Maximize Safety." The three lower-level objectives are also fundamental; each one describes a specific aspect of safety, and as such each one is inherently important. This hierarchy could be expanded by including another level. For example, we might include the objectives "Minimize Loss of Child Lives" and "Minimize Loss of Adult Lives" as aspects of the loss-of-life objective and similarly distinguish between serious injuries to children and adults. Figure 3.1 displays the hierarchy.

Means objectives, on the other hand, are organized into *networks*. In the vehicle-safety example, some means objectives might be "Minimize Accidents" and "Maximize Use of Vehicle-Safety Features." Both of these are important because they help to maximize safety. Beyond these two means objectives might be other means objectives such as "Maximize Driving Quality," "Maintain Vehicles Properly," and "Maximize Purchase of Safety Features on Vehicles." Figure 3.2 shows a means-objectives network that includes still more means objectives. A key difference between this network and the fundamental-objectives hierarchy in Figure 3.1 is that means objectives can be connected to several objectives, indicating that they help accomplish these objectives. For example, "Have Reasonable Traffic Laws" affects both "Maximize Driving Quality" and "Maintain Vehicles Properly."

Structuring the fundamental-objectives hierarchy is crucial for developing a multiple-objective decision model. As we will see, the lowest-level fundamental objectives will be the basis on which various consequences will be measured. Distinguishing means and fundamental objectives is important at this stage of the game primarily so that the decision maker is certain that the appropriate objectives—fundamental, not means—are specified in the decision model. But the means network has other uses as well. We will see in the last portion of the chapter that an easily measured means objective can sometimes substitute for a fundamental objective that is

Figure 3.1

A fundamental-objectives hierarchy.
Source: Keeney
(1992, p. 70).

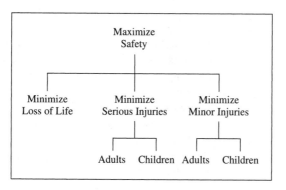

Figure 3.2

*A means-objectives
network.*
Source: Keeney,
(1992, p. 70).

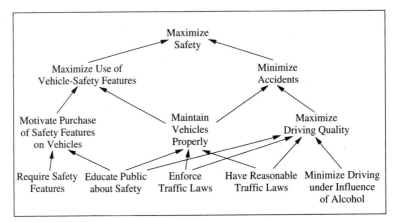

more difficult to measure. And in Chapter 6 we will see how the means-objectives network provides an important basis for generating creative new alternatives.

How do we first separate means and fundamental objectives and then construct
the fundamental-objectives hierarchy and the means-objectives network? A number
of guiding questions are used to accomplish these tasks.

The first question to ask regarding any objective is, *"Why Is That Important?"*
Known as the WITI test, this question does two things: distinguishes between means
and fundamental objectives and reveals connections among the objectives. If the answer to the question is, "This objective is important because it helps accomplish X,"
then you know that the original objective is a means objective and that it has an impact on X. Moreover, a decision maker can continue by asking, "Why is X important?" By continuing to ask why the next objective is important, we can trace out the
connections from one means objective to the next until we arrive at an objective for
which the answer is, "This objective is important just because it is important. It is
one of the fundamental reasons why I care about this decision." In this case, we have
identified a fundamental objective.

As an example, look again at Figure 3.2. We might ask, for example, "Why is it
important to maintain vehicles properly?" The answer is that doing so helps to minimize accidents and maximize the use of vehicle-safety features. Asking why minimizing accidents is important reveals that it helps maximize safety. The same is true
if we ask why maximizing use of safety features is important. Finally, why is safety
important? Maximizing safety is fundamentally important; it is what we care about
in the context of establishing regulations regarding vehicle use. The answers to the
questions trace out the connections among these four objectives and appropriately
identify "Maximize Safety" as a fundamental objective.

The WITI test is useful for moving from means objectives to fundamental objectives. What about going the other way? The obvious question to ask is, "How can this
objective be achieved?" For example, in the vehicle-regulation context we would ask,
"How can we maximize safety?" The answer might give any of the upstream means

objectives that appear in Figure 3.2. Sequentially asking "How can this objective be achieved?" can help to identify means objectives and establish links among them.

What about constructing the fundamental-objectives hierarchy? Starting at the top of the hierarchy, the question to ask is, "What do you mean by that?" In our vehicle example, we would ask, "What does maximize safety mean?" The answer is that we mean minimizing lives lost, serious injuries, and minor injuries. In turn we could ask, "What do you mean by minimizing lives lost?" The answer might be minimizing child lives lost and adult lives lost; that is, it might be useful in this decision context to consider safety issues for children and adults separately, perhaps because different kinds of regulations would apply to these two groups.

Finally, we can work upward in the fundamental-objectives hierarchy, starting at a lower-level objective. Ask the question, "Of what more general objective is this an aspect?" For example, if we have identified saving adult lives as a fundamental objective—it is a fundamental reason we care about vehicle regulations—then we might ask, "Is there a more general objective of which saving adult lives is an aspect?" The answer would be the more general objective of saving lives, and asking the question again with respect to saving lives would lead us to the overall fundamental objective of maximizing safety.

Figure 3.3 summarizes these four techniques for organizing means and fundamental objectives. It is important to realize that one might ask these questions in any order, mixing up the sequence, jumping from the means network to the fundamental-objectives hierarchy and back again. Be creative and relaxed in thinking about your values!

Let us look again at PeachTree's summer-intern decision. Figure 3.4 shows both a fundamental-objectives hierarchy and a means network with appropriate connections between them. The means objectives are shown in italics. Note that some objectives have been added, especially criteria for the intern, such as ability to work with the senior employee, ability to demonstrate new techniques to the staff, and a high level of energy. In the decision context, choosing the best intern for the summer

Figure 3.3

How to construct mean-objectives networks and fundamental-objectives hierarchies.

	Fundamental Objectives	Means Objectives
To Move:	*Downward in the Hierarchy:*	*Away from Fundamental Objectives:*
Ask:	"What do you mean by that?"	"How could you achieve this?"
To Move:	*Upward in the Hierarchy:*	*Toward Fundamental Objectives:*
Ask:	"Of what more general objective is this an aspect?"	"Why is that important?" (WITI)

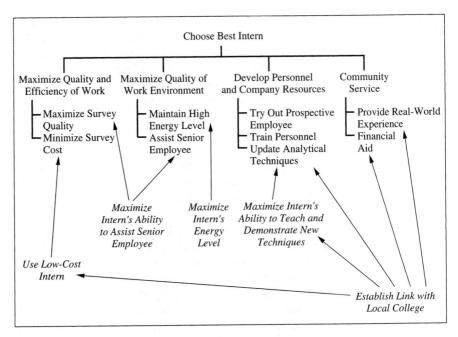

position, these criteria help define what "best" means in terms that relate directly to the company's fundamental objectives.

Insights can be gleaned from Figure 3.4. First, the means objectives give some guidance about what kind of intern to hire; up-to-date technical skills, good "people skills" for working with the senior employee, an ability (and willingness) to demonstrate new techniques for the firm, and a high energy level. In addition, establishing a link with the local college is a very important step. Although this is a means objective and hence not important in and of itself, it has an impact on many other objectives, both means and fundamental.

The fundamental-objectives hierarchy and the means-objectives network can provide a lot of insight even at this initial level. The fundamental objectives tell you why you care about the decision situation and what criteria you should be looking at in evaluating options. For the summer-intern situation, the company cares about the four main-level fundamental objectives, and the lower-level objectives provide more detail. Having sorted out the means objectives, we can rest assured that we will be able to evaluate candidates (and perhaps even develop a strategy for finding good candidates) whose qualities are consistent with the company's concerns. Finally, as we mentioned above, the means network can suggest creative new alternatives. For example, a great strategy would be to become acquainted with professors or career counselors at the local college and to explain to them exactly what the company is looking for in a summer intern.

Getting the Decision Context Right

Recall that the context for PeachTree's decision has been to hire the best intern. What would happen if we were to broaden the context? Suppose we were to set the context as enhancing the company's marketing activities. First, we would want to consider far more options than just hiring an intern. The broader context also suggests looking for permanent new hires or training current employees in new methods. One of the results would be that the means objectives might change; some of the means objectives might broaden from optimizing characteristics of the intern to optimizing characteristics of the marketing group as a whole. "Maximize Intern's Energy Level" might become "maximize marketing group's energy level," which might suggest means objectives such as hiring new high-energy employees or sending employees to a workshop or retreat. You can see that as we broaden the decision context, the objectives change in character somewhat. The more the context is broadened, the greater the change. If we were to go all the way to a *strategic*—broadest possible—context of "maximize profit" or "build market share," for example, then many of the fundamental objectives in Figure 3.4 would become means objectives, and alternatives affecting all parts of the company would have to be considered.

At this point you may be wondering how you know when you have identified the appropriate decision context and its corresponding fundamental-objectives hierarchy and means network. As in Chapter 2, we can invoke the notion of a requisite model to ensure that all appropriate but no superfluous objectives have been included, given the decision context. The real question, though, is the decision context itself. How do you know how broad or narrow to make the context? This question is absolutely fundamental, and unfortunately there is no simple answer. As a decision maker, you must choose a context that fits three criteria. The first is straightforward; ask whether the context you have set really captures the situation at hand. Are you addressing the right problem? For example, searching for a job of the same type as your present one but with a different company is the wrong decision context if your real problem is that you do not enjoy the kind of work required in that job; you should broaden the context to consider different kinds of jobs, careers, or lifestyles. On the other hand, if you really love what you do but are dissatisfied with your current job for reasons related to that particular position or your firm (low salary, poor working conditions, conflicts with fellow workers, and so on), then looking for another similar job with another firm is just the right context.

The second criterion might be called *decision ownership*. Within organizations especially, the broader the decision context, the higher up the organizational ladder are the authority to make the decision and the responsibility for its consequences. Do you have the authority to make decisions within the specified context (or will you be reporting the results of your analysis to someone with that authority)? If you conclude that you do not have this authority, then look for a narrower context that matches the authority you do have.

Feasibility is the final issue; in the specified context, will you be able to do the necessary study and analysis in the time allotted with available resources? Broader contexts often require more careful thought and more extensive analysis; addressing a broad decision context with inadequate time and resources can easily lead to dissatisfaction with the decision process (even though good consequences may result from lucky outcomes). It would be better in such a situation to narrow the context in some way until the task is manageable.

Like most aspects of decision analysis, setting the context and structuring objectives may not be a once-and-for-all matter. After initially specifying objectives, you may find yourself refining the context and modifying the objectives. Refining the context several times and iterating through the corresponding sets of objectives are not signs of poor decision making; instead, they indicate that the decision situation is being taken seriously, and that many different possibilities and perspectives are being considered.

Structuring Decisions: Influence Diagrams

With the fundamental objectives specified, structured, and sorted out from the means objectives, we can turn now to the process of structuring the various decision elements—decisions and alternatives, uncertain events and outcomes, and consequences. We begin with influence diagrams, which can provide simple graphical representations of decision situations. Different decision elements show up in the influence diagram as different shapes. These shapes are then linked with arrows in specific ways to show the relationships among the elements.

In an influence diagram, rectangles represent decisions, ovals represent chance events, and diamonds represent the final consequence or payoff node. A rectangle with rounded corners is used to represent a mathematical calculation or a constant value; these rounded rectangles will have a variety of uses, but the most important is to represent intermediate consequences. The four shapes are generally referred to as *nodes:* decision nodes, chance nodes, payoff nodes, and consequence or calculation nodes. Nodes are put together in a *graph,* connected by arrows, or *arcs.* We call a node at the beginning of an arc a *predecessor* and a node at the end of an arc a *successor.*

Consider a venture capitalist's situation in deciding whether to invest in a new business. For the moment, let us assume that the capitalist has only one objective in this context—to make money (not an unreasonable objective for a person in this line of work). The entrepreneur seeking the investment has impeccable qualifications and has generally done an excellent job of identifying the market, assembling a skilled management and production team, and constructing a suitable business plan. In fact, it is clear that the entrepreneur will be able to obtain financial backing from some source whether the venture capitalist decides to invest or not. The only problem is that the proposed project is extremely risky—more so than most new ventures. Thus, the venture capitalist must decide whether to invest in this highly risky undertaking. If she invests, she may be able to get in on the ground floor of a very successful business. On the other hand, the operation may fail altogether. Clearly, the dilemma is whether the chance of getting in on the ground floor of something big is worth the risk of losing the investment entirely. If she does not invest in this project, she may leave her capital in the stock market or invest in other less risky ventures. Her investment situation appears as an influence diagram in Figure 3.5.

Note that both "Invest?" and "Venture Succeeds or Fails" are predecessors of the final consequence "Return on Investment." The implication is that the consequence depends on both the decision and the chance event. In general, consequences depend on what happens or what is decided in the nodes that are predecessors of the consequence node. Moreover, as soon as the decision is made *and* the uncertain event is resolved, the consequence is determined; there is no uncertainty about the consequence at this point. Note also that no arc points from the chance node to the decision node. The absence of an arc indicates that when the decision is made, the venture capitalist does not know whether the project will succeed. She may have some feeling for the chance of success, and this information would be included in the influence diagram as probabilities of possible levels of success or failure. Thus, the influence diagram as drawn captures the decision maker's current state of knowledge about the situation.

Also note that no arc points from the decision to the uncertain event. The absence of this arrow has an important and subtle meaning. The uncertainty node is about the success of the venture. The absence of the arc from "Invest?" to "Venture Succeeds or Fails" means that the venture's chances for success are not affected by the capitalist's decision. In other words, the capitalist need not concern herself with her impact on the venture.

It is possible to imagine situations in which the capitalist may consider different levels of investment as well as managerial involvement. For example, she may be willing to invest $100,000 and leave the entrepreneur alone. But if she invests $500,000, she may wish to be more active in running the company. If she believes her involvement would improve the company's chance of success, then it would be appropriate to include an arrow from the decision node to the chance node; her investment decision—the level of investment and the concomitant level of involvement—would be relevant for determining the company's chance of success.

Figure 3.5

Influence diagram of venture capitalist's decision.

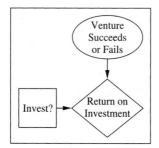

In our simple and stylized example, however, we are assuming that her choice simply is whether to invest and that she has no impact on the company's chance of success.

Influence Diagrams and the Fundamental-Objectives Hierarchy

Suppose the venture capitalist actually has multiple objectives. For example, she might wish to focus on a particular industry, such as personal computers, obtaining satisfaction by participating in the growth of this industry. Thus, in addition to the objective of making money, she would have an objective of investing in the personal-computer industry.

Figure 3.6 shows a simple two-level objectives hierarchy and the corresponding influence diagram for the venture capitalist's decision. You can see in this figure how the objectives hierarchy is reflected in the pattern of consequence nodes in the influence diagram; two consequence nodes labeled "Invest in Computer Industry" and "Return on Investment" represent the lower-level objectives and in turn are connected to the "Overall Satisfaction" consequence node. This structure indicates that in some situations the venture capitalist may have to make a serious trade-off between these two objectives, especially when comparing a computer-oriented business startup with a noncomputer business that has more potential to make money.

Figure 3.6

The venture capitalist's decision with two objectives.

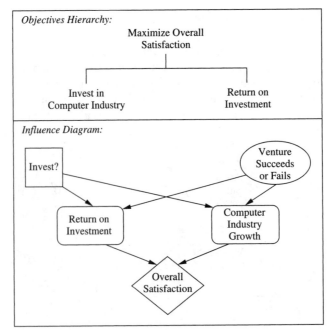

Figure 3.7

Multiple objectives in selecting a bomb-detection system.

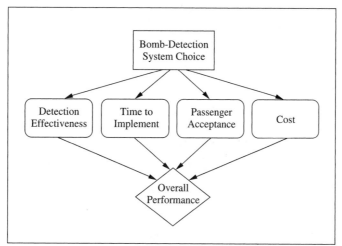

The rounded rectangles for "Computer Industry" and "Return on Investment" are appropriate because these consequences are known after the decision is made and the venture's level of success is determined. The diamond for "Overall Satisfaction" indicates that it is the final consequence. Once the two individual consequences are known, then its value can be determined.

Figure 3.7 shows the influence diagram for another multiple-objective decision. In this situation, the Federal Aviation Administration (FAA) must choose from among a number of bomb-detection systems for commercial air carriers (Ulvila and Brown, 1982). In making the choice, the agency must try to accomplish several objectives. First, it would like the chosen system to be as effective as possible at detecting various types of explosives. The second objective is to implement the system as quickly as possible. The third is to maximize passenger acceptance, and the fourth is to minimize cost. To make the decision and solve the influence diagram, the FAA would have to score each candidate system on how well it accomplishes each objective. The measurements of time and cost would naturally be made in terms of days and dollars, respectively. Measuring detection effectiveness and passenger acceptance might require experiments or surveys and the development of an appropriate measuring device. The "Overall Performance" node would contain a formula that aggregates the individual scores, incorporating the appropriate trade-offs among the four objectives. Assessing the trade-off rates and constructing the formula to calculate the overall score is demonstrated in an example in Chapter 4 and is discussed thoroughly in Chapters 15 and 16.

Using Arcs to Represent Relationships

The rules for using arcs to represent relationships among the nodes are shown in Figure 3.8. In general, an arc can represent either *relevance* or *sequence*. The context of the arrow indicates the meaning. For example, an arrow pointing into a chance

Figure 3.8

Representing influence with arrows. Arrows into chance and consequence nodes represent relevance, and arrows into decision nodes represent sequence.

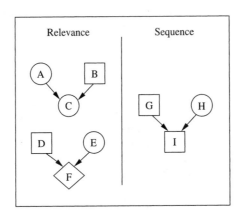

node designates relevance, indicating that the predecessor is relevant for assessing the chances associated with the uncertain event. In Figure 3.8 the arrow from Event A to Event C means that the chances (probabilities) associated with C may be different for different outcomes of A. Likewise, an arrow pointing from a decision node to a chance node means that the specific chosen decision alternative is relevant for assessing the chances associated with the succeeding uncertain event. For instance, the chance that a person will become a millionaire depends to some extent on the choice of a career. In Figure 3.8 the choice taken in Decision B is relevant for assessing the chances associated with Event C's possible outcomes.

Relevance arcs can also point into consequence or calculation nodes, indicating that the consequence or calculation depends on the specific outcome of the predecessor node. In Figure 3.8, consequence F depends on both Decision D and Event E. Relevance arcs in Figure 3.6 point into the "Computer Industry Growth" and "Return on Investment" nodes; the decision made and the success of the venture are relevant for determining these two consequences. Likewise, relevance arcs point from the two individual consequence nodes into the "Overall Satisfaction" node.

When the decision maker has a choice to make, that choice would normally be made on the basis of information available at the time. What information is available? Everything that happens before the decision is made. Arrows that point to decisions represent information available at the time of the decision and hence represent *sequence*. Such an arrow indicates that the decision is made knowing the outcome of the predecessor node. An arrow from a chance node to a decision means that, from the decision maker's point of view, all uncertainty associated with a chance event is resolved and the outcome known when the decision is made. Thus, information is available to the decision maker regarding the event's outcome. This is the case with Event H and Decision I in Figure 3.8; the decision maker is waiting to learn the outcome of H before making Decision I. An arrow from one decision to another decision simply means that the first decision is made before the second, such as Decisions G and I in Figure 3.8. Thus, the sequential ordering of decisions is shown in an influence diagram by the path of arcs through the decision nodes.

The nature of the arc—relevance or sequence—can be ascertained by the context of the arc within the diagram. To reduce the confusion of overabundant notation, all arcs have the same appearance in this book. For our purposes, the rule for determining the nature of the arcs is simple; an arc pointing to a decision represents sequence, and all others represent relevance.

Properly constructed influence diagrams have no *cycles;* regardless of the starting point, there is no path following the arrows that leads back to the starting point. For example, if there is an arrow from A to B, there is no path, however tortuous, that leads back to A from B. Imagine an insect traveling from node to node in the influence diagram, always following the direction of the arrows. In a diagram without cycles, once the insect leaves a particular node, it has no way to get back to that node.

Some Basic Influence Diagrams

In this section, several basic influence diagrams are described. Understanding exactly how these diagrams work will provide a basis for understanding more complex diagrams.

The Basic Risky Decision

This is the most elementary decision under uncertainty that a person can face. The venture-capital example above is a basic risky decision; there is one decision to make and one uncertain event.

Many decision situations can be reduced to a basic risky decision. For example, imagine that you have $2000 to invest, with the objective of earning as high a return on your investment as possible. Two opportunities exist, investing in a friend's business or keeping the money in a savings account with a fixed interest rate. If you invest in the business, your return depends on the success of the business, which you figure could be wildly successful, earning you $3000 beyond your initial investment (and hence leaving you with a total of $5000), or a total flop, in which case you will lose all your money and have nothing. On the other hand, if you put the money into a savings account, you will earn $200 in interest (leaving you with a total of $2200) regardless of your friend's business.

The influence diagram for this problem is shown in Figure 3.9. This figure also graphically shows details underlying the decision, chance, and consequence nodes. The decision node includes the choice of investing in either the business or the savings account. The chance node represents the uncertainty associated with the business and shows the two possible outcomes. The consequence node includes information on the dollar return for different decisions (business investment versus savings) and the outcome of the chance event. This table shows clearly that if you invest in the business, your return depends on what the business does. However, if you put your money into savings, your return is the same regardless of what happens with the business.

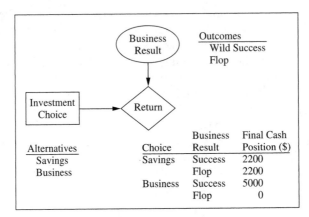

You can see that the essential question in the basic risky decision is whether the potential gain in the risky choice (the business investment) is worth the risk that must be taken. The decision maker must, of course, make the choice by comparing the risky and riskless alternatives. Variations of the basic risky choice exist. For example, instead of having just two possible outcomes for the chance event, the model could include a range of possible returns, a much more realistic scenario. The structure of the influence diagram for this range-of-risk dilemma, though, would look just the same as the influence diagram in Figure 3.9; the difference lies in the details of the chance event, which are not shown explicitly in the structure of the diagram.

Imperfect Information

Another basic kind of influence diagram reflects the possibility of obtaining imperfect information about some uncertain event that will affect the eventual payoff. This might be a forecast, an estimate or diagnosis from an acknowledged expert, or information from a computer model. In the investment example, you might subscribe to a service that publishes investment advice, although such services can never predict market conditions perfectly.

Imagine a manufacturing-plant manager who faces a string of defective products and must decide what action to take. The manager's fundamental objectives are to solve this problem with as little cost as possible and to avoid letting the production schedule slip. A maintenance engineer has been dispatched to do a preliminary inspection on Machine 3, which is suspected to be the source of the problem. The preliminary check will provide some indication as to whether Machine 3 truly is the culprit, but only a thorough and expensive series of tests—not possible at the moment—will reveal the truth. The manager has two alternatives. First, a replacement for Machine 3 is available and could be brought in at a certain cost. If Machine 3 *is* the problem, then work can proceed and the production schedule will not fall behind. If Machine 3 is not the source of the defects, the problem will still exist, and the workers will have to change to another product while the problem is tracked down. Second, the workers could be changed immediately to the other product. This action would certainly cause the production schedule for the current product to fall behind but would avoid the risk (and cost) of unnecessarily replacing Machine 3.

Without the engineer's report, this problem would be another basic risky decision; the manager would have to decide whether to take the chance of replacing Machine 3 based on personal knowledge regarding the chance that Machine 3 is the source of the defective products. However, the manager is able to wait for the engineer's preliminary report before taking action. Figure 3.10 shows an influence diagram for the manager's decision problem, with the preliminary report shown as an example of imperfect information. The diagram shows that the consequences depend on the choice made (replace Machine 3 or change products) and whether Machine 3 actually turns out to be defective. There is no arrow from "Engineer's Report" to the consequence nodes because the report does not have a direct effect on the consequence.

The arrow from "Engineer's Report" to "Manager's Decision" is a sequence arc; the manager will hear from the engineer before deciding. Thus, the engineer's preliminary report is information available at the time of the decision, and this influence diagram represents the situation while the manager is waiting to hear from the engineer. Analyzing the influence diagram will tell the manager how to interpret this information; the appropriate action will depend not only on the engineer's report but also on the extent to which the manager believes the engineer to be correct. The manager's assessment of the engineer's accuracy is reflected in the chances associated with the "Engineer's Report" node. Note that a relevance arc points from "Machine 3 OK?" to "Engineer's Report," indicating that Machine 3's state is relevant for assessing the chances associated with the engineer's report. For example, if the manager believes the engineer is very good at diagnosing the situation, then when Machine 3 really is OK, the chances should be near 100% that the engineer will say so. Likewise, if Machine 3 is causing the defective products, the engineer should be very likely to indicate 3 is the problem. On the other hand, if the manager does not think the engineer is very good at diagnosing the problem—because of lack of fa-

Figure 3.10

Influence diagram for manufacturing-plant manager's imperfect information.

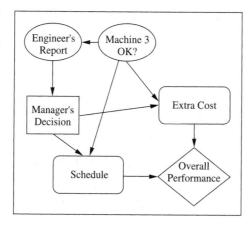

miliarity with this particular piece of equipment, say—then there might be a substantial chance that the engineer makes a mistake.

Weather forecasting provides another example of imperfect information. Suppose you live in Miami. A hurricane near the Bahama Islands threatens to cause severe damage; as a result, the authorities recommend that everyone evacuate. Although evacuation is costly, you would be safe. On the other hand, staying is risky. You could be injured or even killed if the storm comes ashore within 10 miles of your home. If the hurricane's path changes, however, you would be safe without having incurred the cost of evacuating. Clearly, two fundamental objectives are to maximize your safety and to minimize your costs.

Undoubtedly, you would pay close attention to the weather forecasters who would predict the course of the storm. These weather forecasters are not perfect predictors, however. They can provide some information about the storm, but they may not perfectly predict its path because not everything is known about hurricanes.

Figure 3.11 shows the influence diagram for the evacuation decision. The relevance arc from "Hurricane Path" to "Forecast" means that the actual weather situation is relevant for assessing the uncertainty associated with the forecast. If the hurricane is actually going to hit Miami, then the forecaster (we hope) is more likely to predict a hit rather than a miss. Conversely, if the hurricane really will miss Miami, the forecaster should be likely to predict a miss. In either case, though, the forecast may be incorrect because the course of a hurricane is not fully predictable. In this situation, although the forecast actually precedes the hurricane's landfall, it is relatively straightforward to think about the forecaster's tendency to make a mistake conditioned on what direction the hurricane goes. (The modeling choice is up to you, though! If you would feel more confident in assessing the chance of the hurricane hitting Miami by conditioning on the forecast—that is, have the arrow pointing the other way—then by all means do so!)

The consequence node in Figure 3.11 encompasses both objectives of minimizing cost and maximizing safety. An alternative representation might explicitly in-

Figure 3.11

Influence diagram for the evacuation decision.

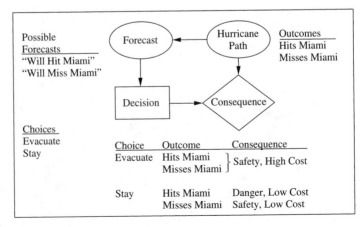

clude both consequences as separate nodes as in Figure 3.10. Moreover, these two objectives are somewhat vaguely defined, as they might be in an initial specification of the decision. A more complete specification would define these objectives carefully, giving levels of cost (probably in dollars) and a scale for the level of danger. In addition, uncertainty about the possible outcomes—ranging from no injury to death—could be included in the influence diagram. You will get a chance in Problem 3.14 to modify and improve on Figure 3.11.

As with the manufacturing example, the influence diagram in Figure 3.11 is a snapshot of your situation as you wait to hear from the forecaster. The sequence arc from "Forecast" to the decision node indicates that the decision is made knowing the imperfect weather forecast. You might imagine yourself waiting for the 6 p.m. weather report on the television, and as you wait, you consider what the forecaster might say and what you would do in each case. The sequence of events, then, is that the decision maker hears the forecast, decides what to do, and then the hurricane either hits Miami or misses. As with the manufacturing example, analyzing this model will result in a strategy that recommends a particular decision for each of the possible statements the forecaster might make.

Sequential Decisions

The hurricane-evacuation decision above can be thought of as part of a larger picture. Suppose you are waiting anxiously for the forecast as the hurricane is bearing down. Do you wait for the forecast or leave immediately? If you wait for the forecast, what you decide to do may depend on that forecast. In this situation, you face a sequential decision situation as diagrammed in Figure 3.12.

The order of the events is implied by the arcs. Because there is no arc from "Forecast" to "Wait for Forecast" but there is one to "Evacuate," it is clear that the sequence is first to decide whether to wait or leave immediately. If you wait, the forecast is revealed, and finally you decide, based on the forecast, whether to evacuate.

In an influence diagram sequential decisions are strung together via sequence arcs, in much the same way that we did in Chapter 2. (In fact, now you can see that the figures in Chapter 2 use essentially the same graphics as influence diagrams!) For another example, let us take the farmer's decision from Chapter 2

Figure 3.12

A sequential version of the evacuation decision.

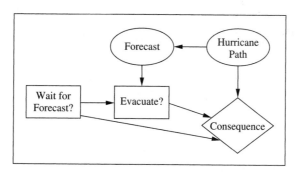

about protecting his trees against adverse weather. Recall that the farmer's decision replayed itself each day; based on the next day's weather forecast, should the fruit crop be protected? Let us assume that the farmer's fundamental objective is to maximize the NPV of the investment, including the costs of protection. Figure 3.13 shows that the influence diagram essentially is a series of imperfect-information diagrams strung together. Between decisions (to protect or not) the farmer observes the weather and obtains the forecast for the next day. The arcs from one decision to the next show the time sequence.

The arrows among the weather and forecast nodes from day to day indicate that the observed weather and the forecast both have an effect. That is, yesterday's weather is relevant for assessing the chance of adverse weather today. Not shown explicitly in the influence diagram are arcs from forecast and weather nodes before the previous day. Of course, the decision maker observed the weather and the forecasts for each prior day. These arcs are not included in the influence diagram but are implied by the arcs that connect the decision nodes into a time sequence. The missing arcs are sometimes called *no-forgetting* arcs to indicate that the decision maker would not forget the outcomes of those previous events. Unless the no-forgetting arcs are critical in understanding the situation, it is best to exclude them because they tend to complicate the diagram.

Finally, although we indicated that the farmer has a single objective, that of maximizing NPV, Figure 3.13 represents the decision as a multiple-objective one, the objectives being to maximize the cash inflow (and hence minimize outflows or costs) each day. The individual cash flows, of course, are used to calculate the farmer's NPV. As indicated in Chapter 2, the interest rate defines the trade-off between earlier and later cash flows.

Figure 3.13

Influence diagram for farmer's sequential decision problem.

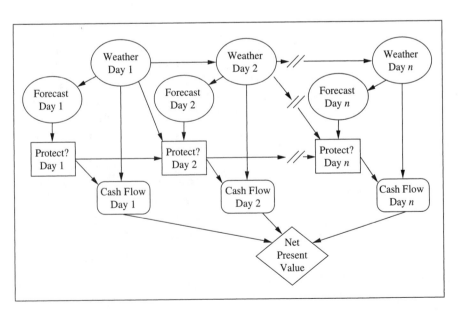

Intermediate Calculations

In some cases it is convenient to include an additional node that simply aggregates results from certain predecessor nodes. Suppose, for example, that a firm is considering introducing a new product. The firm's fundamental objective is the profit level of the enterprise, and so we label the consequence node "Profit." At a very basic level, both cost and revenue may be uncertain, and so a first version of the influence diagram might look like the one shown in Figure 3.14.

On reflection, the firm's chief executive officer (CEO) realizes that substantial uncertainty exists for both variable and fixed costs. On the revenue side, there is uncertainty about the number of units sold, and a pricing decision will have to be made. These considerations lead the CEO to consider a somewhat more complicated influence diagram, which is shown in Figure 3.15.

Figure 3.15 is a perfectly adequate influence diagram. Another representation is shown in Figure 3.16. Intermediate nodes have been included in Figure 3.16 to calculate cost on one hand and revenue on the other; we will call these *calculation nodes,* because they calculate cost and revenue given the predecessors. (In many discussions of influence diagrams, the term *deterministic node* is used to denote a node that represents an intermediate calculation or a constant, and in graphical representation in the influence diagram such a node is shown as a circle with a double outline. The use of the rounded rectangle, the same as the consequence node, is consistent with the

Figure 3.14

Simple influence diagram for new product decision.

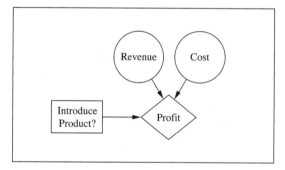

Figure 3.15

New product decision with additional detail.

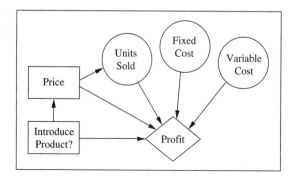

representation in the computer program PrecisionTree and the discussion of these nodes in its documentation.)

Calculation nodes behave just like consequence nodes; given the inputs from the predecessor nodes, the value of a calculation node can be found immediately. No uncertainty exists after the conditioning variables—decisions, chance events, or other calculation nodes—are known. Of course, there is no uncertainty only in a conditional sense; the decision maker can look forward in time and know what the calculation node will be for any possible combination of the conditioning variables. Before the conditioning variables are known, though, the value that the node will eventually have is uncertain.

In general, calculation nodes are useful for emphasizing the structure of an influence diagram. Whenever a node has a lot of predecessors, it may be appropriate to include one or more intermediate calculations to define the relationships among the predecessors more precisely. In Figure 3.16, the calculation of cost and revenue is represented explicitly, as is the calculation of profit from cost and revenue. The pricing decision is clearly related to the revenue side, uncertainty about fixed and variable costs are clearly on the cost side, and uncertainty about sales is related to both.

Another example is shown in Figure 3.17. In this situation, a firm is considering building a new manufacturing plant that may create some incremental pollution. The profitability of the plant depends on many things, of course, but highlighted in Figure 3.17 are the impacts of other pollution sources. The calculation node "Regional Pollution Level" uses information on the number of cars and local industry growth to determine a pollution-level index. The pollution level in turn has an impact on the chances that the new plant will be licensed and that new regulations (either more or less strict) will be imposed.

With the basic understanding of influence diagrams provided above, you should be able to look at any influence diagram (including any that you find in this book) and understand what it means. Understanding an influence diagram is an important decision-analysis skill. On the other hand, actually creating an influence diagram from scratch is considerably more difficult and takes much practice. The following optional section gives an example of the construction process for an influence dia-

Figure 3.16

New product decision with calculation nodes for intermediate calculation of cost and revenue.

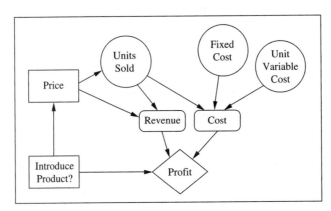

Figure 3.17

Using a calculation node to determine pollution level.

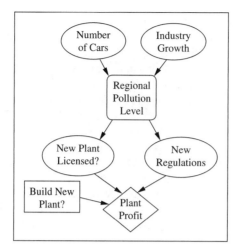

gram and discusses some common mistakes. If you wish to become proficient in constructing influence diagrams, the next section is highly recommended. Working through the reading and exercises, however, is just one possibility; in fact, practice with an influence-diagram program (like PrecisionTree) is the best way to develop skill in constructing influence diagrams. At the end of this chapter, there are instructions on how to construct the influence diagram for the basic risky decision using PrecisionTree.

Constructing an Influence Diagram (Optional)

There is no set strategy for creating an influence diagram. Because the task is to structure a decision that may be complicated, the best approach may be to put together a simple version of the diagram first and add details as necessary until the diagram captures all of the relevant aspects of the problem. In this section, we will demonstrate the construction of an influence diagram for the classic toxic-chemical problem.

TOXIC CHEMICALS AND THE EPA

The Environmental Protection Agency (EPA) often must decide whether to permit the use of an economically beneficial chemical that may be carcinogenic (cancer-causing). Furthermore, the decision often must be made without perfect information about either the long-term benefits or health hazards. Alternative courses of action are to permit the use of the chemical, restrict its use, or to ban it altogether. Tests can be run to learn something about the carcinogenic potential of the material, and survey data can give an indication of the extent to which people are exposed when they do use the chemical. These pieces of information are both important in making the decision. For example, if the chemical is only mildly toxic and the exposure rate is minimal, then restricted use may be reasonable. On the other hand, if the chemical is only mildly toxic but the exposure rate is high, then banning its use may be imperative. (*Source:* Ronald A. Howard & James E. Matheson. 1981. Influence diagrams. In R. Howard & J. Matheson, Eds., *Readings on the Principles and Applications of Decision Analysis, Vol. II,* pp. 719–762. Menlo Park, CA: Strategic Decisions Group.)

The first step should be to identify the decision context and the objectives. In this case, the context is choosing an allowed level of use, and the fundamental objectives are to maximize the economic benefits from the chemicals and at the same time to minimize the risk of cancer. These two objectives feed into an overall consequence node ("Net Value") that aggregates "Economic Value" and "Cancer Cost" as shown in Figure 3.18.

Now let us think about what affects "Economic Value" and "Cancer Cost" other than the usage decision. Both the uncertain carcinogenic character of the chemical and the exposure rate have an effect on the cancer cost that could occur, thus yielding the diagram shown in Figure 3.19. Because "Carcinogenic Potential" and "Exposure Rate" jointly determine the level of risk that is inherent in the chemical, their effects are aggregated in an intermediate calculation node labeled "Cancer Risk." Different values of the predecessor nodes will determine the overall level of "Cancer Risk."

Note that no arrow runs from "Usage Decision" to "Exposure Rate," even though such an arrow might appear to make sense. "Exposure Rate" refers to the extent of contact when the chemical is actually used and would be measured in terms of an amount of contact per unit of time (e.g., grams of dust inhaled per hour). The rate is unknown, and the usage decision does not influence our beliefs concerning the likelihood of various possible rates when the chemical is used.

Figure 3.18

Beginning the toxic-chemical influence diagram.

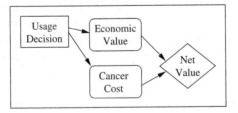

Figure 3.19

Intermediate influence diagram for the toxic-chemical decision.

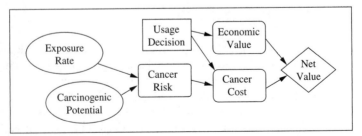

The influence diagram remains incomplete, however, because we have not incorporated the test for carcinogenicity or the survey on exposure. Presumably, results from both the test (called a *bioassay*) and the survey would be available to EPA at the time the usage decision is made. Furthermore, it should be clear that the actual degrees of carcinogenic potential and exposure will influence the test and survey results, and thus "Carcinogenic Potential" and "Exposure Rate" are connected to "Test" and "Survey," respectively, in Figure 3.20. Note that "Test" and "Survey" each represent imperfect information; each one provides some information regarding carcinogenicity or exposure. These two nodes are connected to the decision node. These are sequence arcs, indicating that the information is available when the decision is made. This completes the influence diagram.

This example demonstrates the usefulness of influence diagrams for structuring decisions. The toxic-chemicals problem is relatively complex, and yet its influence diagram is compact and, more important, understandable. Of course, the more complicated the problem, the larger the influence diagram. Nevertheless, influence diagrams are useful for creating easily understood overviews of decision situations.

Some Common Mistakes

First, an easily made mistake in understanding and constructing influence diagrams is to interpret them as flowcharts, which depict the sequential nature of a particular process where each node represents an event or activity. For example, Figure 1.1 is a flowchart of a decision-analysis system, displaying the different things a decision analyst does at each stage of the process.

Even though they look a little like flowcharts, influence diagrams are very different. An influence diagram is a snapshot of the decision situation at a particular time, one that must account for all the decision elements that play a part in the immediate decision. Putting a chance node in an influence diagram means that although the decision maker is not sure exactly what will happen, he or she has some idea of how likely the different possible outcomes are. For example, in the toxic-chemical problem, the carcinogenic potential of the chemical is unknown, and in fact will never be known for sure. That uncertainty, however, can be modeled using probabilities for different

Figure 3.20

Complete influence diagram for the toxic-chemical decision.

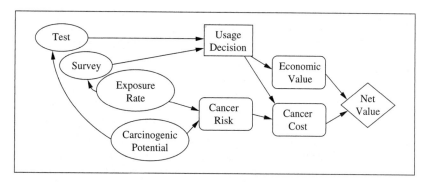

levels of carcinogenic potential. Likewise, at the time the influence diagram is created, the results of the test are not known. The uncertainty surrounding the test results also can be modeled using probabilities. The informational arrow from "Test" to "Usage Decision," however, means that the decision maker will learn the results of the test before the decision must be made.

The metaphor of a picture of the decision that accounts for all of the decision elements also encompasses the possibility of upcoming decisions that must be considered. For example, a legislator deciding how to vote on a given issue may consider upcoming votes. The outcome of the current issue might affect the legislator's future voting decisions. Thus, at the time of the immediate decision, the decision maker foresees future decisions and models those decisions with the knowledge on hand.

A second common mistake, one related to the perception of an influence diagram as a flowchart, is building influence diagrams with many chance nodes having arrows pointing to the primary decision node. The intention usually is to represent the uncertainty in the decision environment. The problem is that the arrows into the decision node are sequence arcs and indicate that the decision maker is waiting to learn the outcome of these uncertain events, which may not be the case. The solution is to think carefully when constructing the influence diagram. Recall that only sequence arcs are used to point to a decision node. Thus, an arrow into a decision node means that the decision maker will have a specific bit of information when making the decision; something will be known for sure, with no residual uncertainty. Before drawing an arrow into a decision node, ask whether the decision maker is waiting for the event to occur *and* will learn the information before the decision is made. If so, the arrow is appropriate. If not, don't draw the arrow!

So how should you include information that the decision maker has about the uncertainty in the decision situation? The answer is simple. Recall that the influence diagram is a snapshot of the decision maker's understanding of the decision situation at a particular point in time. When you create a chance node and connect it appropriately in the diagram, you are explicitly representing the decision maker's uncertainty about that event and showing how that uncertainty relates to other elements of the decision situation.

A third mistake is the inclusion of cycles (circular paths among the nodes). As indicated previously, a properly constructed influence diagram contains no cycles. Cycles are occasionally included in an attempt to denote feedback among the chance and decision nodes. Although this might be appropriate in the case of a flowchart, it is inappropriate in an influence diagram. Again, think about the diagram as a picture of the decision that accounts for all of the decision elements at an instant in time. There is no opportunity for feedback at a single point in time, and hence there can be no cycles.

Influence diagrams provide a graphical representation of a decision's structure, a snapshot of the decision environment at one point in time. All of the details (alternatives, outcomes, consequences) are present in tables that are contained in the nodes, but usually this information is suppressed in favor of a representation that shows off the decision's structure.

Multiple Representations and Requisite Models

Even though your influence diagram may be technically correct in the sense that it contains no mistakes, how do you know whether it is the "correct" one for your decision situation? This question presupposes that a unique correct diagram exists, but for most decision situations, there are many ways in which an influence diagram can appropriately represent a decision. Consider the decision modeled in Figures 3.14, 3.15, and 3.16; these figures represent three possible approaches. With respect to uncertainty in a decision problem, several sources of uncertainty may underlie a single chance node. For example, in Figure 3.16, units sold may be uncertain because the CEO is uncertain about the timing and degree of competitive reactions, the nature of consumer tastes, the size of the potential market, the effectiveness of advertising, and so on. In many cases, and certainly for a first-pass representation, the simpler model may be more appropriate. In other cases, more detail may be necessary to capture all of the essential elements of a situation. In the farmer's problem, for example, a faithful representation of the situation may require consideration of the sequence of decisions rather than looking at each decision as being independent and separate from the others. Thus, different individuals may create different influence diagrams for the same decision problem, depending on how they view the problem. The real issue is determining whether a diagram is appropriate. Does it capture and accurately reflect the elements of the decision problem that the decision maker thinks are important?

How can you tell whether your influence diagram is an appropriate one? The representation that is the most appropriate is the one that is *requisite* for the decision maker along the lines of our discussion in Chapter 1. That is, a requisite model contains everything that the decision maker considers important in making the decision. Identifying all of the essential elements may be a matter of working through the problem several times, refining the model on each pass. The only way to get to a requisite decision model is to continue working on the decision until all of the important concerns are fully incorporated. Sensitivity analysis (Chapter 5) will be a great help in determining which elements are important.

Structuring Decisions: Decision Trees

Influence diagrams are excellent for displaying a decision's basic structure, but they hide many of the details. To display more of the details, we can use a *decision tree*. As with influence diagrams, squares represent decisions to be made, while circles represent chance events. The branches emanating from a square correspond to the choices available to the decision maker, and the branches from a circle represent the possible outcomes of a chance event. The third decision element, the consequence, is specified at the ends of the branches.

Again consider the venture-capital decision (Figure 3.5). Figure 3.21 shows the decision tree for this problem. The decision tree flows from left to right, and so the

Figure 3.21

Decision-tree representation of venture-capital decision.

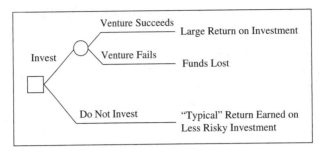

immediate decision is represented by the square at the left side. The two branches represent the two alternatives, invest or not. If the venture capitalist invests in the project, the next issue is whether the venture succeeds or fails. If the venture succeeds, the capitalist earns a large return. However, if the venture fails, then the amount invested in the project will be lost. If the capitalist decides not to invest in this particular risky project, then she would earn a more typical return on another less risky project. These outcomes are shown at the ends of the branches at the right.

The interpretation of decision trees requires explanation. First, the options represented by branches from a decision node must be such that the decision maker can choose only one option. For example, in the venture-capital decision, the decision maker can either invest or not, but not both. In some instances, combination strategies are possible. If the capitalist were considering two separate projects (A and B), for instance, it may be possible to invest in Firm A, Firm B, both, or neither. In this case, each of the four separate alternatives would be modeled explicitly, yielding four branches from the decision node.

Second, each chance node must have branches that correspond to a set of *mutually exclusive* and *collectively exhaustive* outcomes. *Mutually exclusive* means that only one of the outcomes can happen. In the venture-capital decision, the project can either succeed or fail, but not both. *Collectively exhaustive* means that no other possibilities exist; one of the specified outcomes has to occur. Putting these two specifications together means that when the uncertainty is resolved, one and only one of the outcomes occurs.

Third, a decision tree represents all of the possible paths that the decision maker might follow through time, including all possible decision alternatives and outcomes of chance events. Three such paths exist for the venture capitalist, corresponding to the three branches at the right-hand side of the tree. In a complicated decision situation with many sequential decisions or sources of uncertainty, the number of potential paths may be very large.

Finally, it is sometimes useful to think of the nodes as occurring in a time sequence. Beginning on the left side of the tree, the first thing to happen is typically a decision, followed by other decisions or chance events in chronological order. In the venture-capital problem, the capitalist decides first whether to invest, and the second step is whether the project succeeds or fails.

As with influence diagrams, the dovetailing of decisions and chance events is critical. Placing a chance event before a decision means that the decision is made conditional on the specific chance outcome having occurred. Conversely, if a chance node is to the right of a decision node, the decision must be made in anticipation of the chance event. The sequence of decisions is shown in a decision tree by order in the tree from left to right. If chance events have a logical time sequence between decisions, they may be appropriately ordered. If no natural sequence exists, then the order in which they appear in the decision tree is not critical, although the order used does suggest the conditioning sequence for modeling uncertainty. For example, it may be easier to think about the chances of a stock price increasing given that the Dow Jones average increases rather than the other way around.

Decision Trees and the Objectives Hierarchy

Including multiple objectives in a decision tree is straightforward; at the end of each branch, simply list all of the relevant consequences. An easy way to do this systematically is with a *consequence matrix* such as Figure 3.22, which shows the FAA's bomb-detection decision in decision-tree form. Each column of the matrix represents a fundamental objective, and each row represents an alternative, in this case a candidate detection system. Evaluating the alternatives requires "filling in the boxes" in the matrix; each alternative must be measured on every objective. Thus every detection system must be evaluated in terms of detection effectiveness, implementation time, passenger acceptance, and cost.

Figure 3.23 shows a decision-tree representation of the hurricane example. The initial "Forecast" branch at the left indicates that the evacuation decision would be made conditional on the forecast made—recall the imperfect-information decision situation shown in the influence diagram in Figure 3.11. This figure demonstrates that consequences must be considered for every possible endpoint at the right side of the decision tree, regardless of whether those endpoints represent a decision alternative or an uncertain outcome. In addition, Figure 3.23 shows clearly the nature of the risk that the decision to stay entails, and that the decision maker must make a fundamental trade-off between the sure safety of evacuating and the cost of doing so. Finally, the extent of the risk may depend strongly on what the forecast turns out to be!

Figure 3.22

Decision-tree representation of FAA's multiple-objective bomb-detection decision.

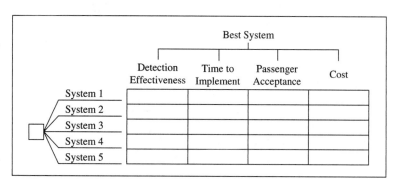

Figure 3.23

*Decision-tree
representation of
evacuation decision.*

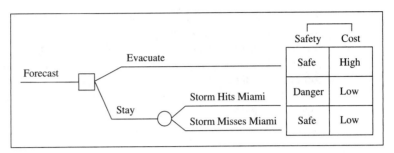

Some Basic Decision Trees

In this section we will look at some basic decision-tree forms. Many correspond to the basic influence diagrams discussed above.

The Basic Risky Decision

Just as the venture-capital decision was the prototypical basic risky decision in our discussion of influence diagrams, so it is here as well. The capitalist's dilemma is whether the potential for large gains in the proposed project is worth the additional risk. If she judges that it is not, then she should not invest in the project.

Figure 3.24 shows the decision-tree representation of the investment decision given earlier in influence-diagram form in Figure 3.9. In the decision tree you can see how the sequence of events unfolds. Beginning at the left side of the tree, the choice is made whether to invest in the business or savings. If the business is chosen, then the outcome of the chance event (wild success or a flop) occurs, and the consequence—the final cash position—is determined. As before, the essential question is whether the chance of wild success and ending up with $5000 is worth the risk of losing everything, especially in comparison to the savings account that results in a bank balance of $2200 for sure.

For another example, consider a politician's decision. The politician's fundamental objectives are to have a career that provides leadership for the country and representation for her constituency, and she can do so to varying degrees by serving in Congress. The politician might have the options of (1) running for reelection to her

Figure 3.24

*The investor's basic
risky decision.*

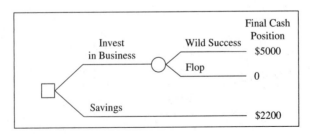

U.S. House of Representatives seat, in which case reelection is virtually assured, or (2) running for a Senate seat. If the choice is to pursue the Senate seat, there is a chance of losing, in which case she could return to her old job as a lawyer (the worst possible outcome). On the other hand, winning the Senate race would be the best possible outcome in terms of her objective of providing leadership and representation. Figure 3.25 diagrams the decision. The dilemma in the basic risky decision arises because the riskless alternative results in an outcome that, in terms of desirability, falls between the outcomes for the risky alternatives. (If this were not the case, there would be no problem deciding!) The decision maker's task is to figure out whether the chance of "winning" in the risky alternative is great enough relative to the chance of "losing" to make the risky alternative more valuable than the riskless alternative. The more valuable the riskless alternative, the greater the chance of winning must be for the risky alternative to be preferred.

A variation of the basic risky decision might be called the *double-risk decision dilemma.* Here the problem is deciding between two risky prospects. On one hand, you are "damned if you do and damned if you don't" in the sense that you could lose either way. On the other hand, you could win either way. For example, the political candidate may face the decision represented by the decision tree in Figure 3.26, in which she may enter either of two races with the possibility of losing either one.

In our discussion of the basic risky decision and influence diagrams, we briefly mentioned the *range-of-risk dilemma,* in which the outcome of the chance event can take on any value within a range of possible values. For example, imagine an individual who has sued for damages of $450,000 because of an injury. The insurance company has offered to settle for $100,000. The plaintiff must decide whether to accept the settlement or go to court; the decision tree is shown as Figure 3.27. The crescent shape

Figure 3.25

The politician's basic risky decision.

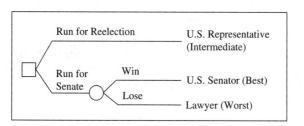

Figure 3.26

Double-risk decision dilemma.

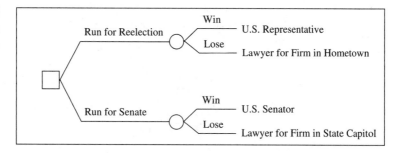

Figure 3.27

A range-of-risk decision dilemma.

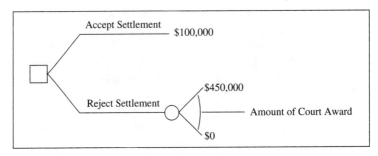

indicates that the uncertain event—the court award—may result in any value between the extremes of zero and $450,000, the amount claimed in the lawsuit.

Imperfect Information

Representing imperfect information with decision trees is a matter of showing that the decision maker is waiting for information prior to making a decision. For example, the evacuation decision problem is shown in Figure 3.28. Here is a decision tree that begins with a chance event, the forecast. The chronological sequence is clear; the forecast arrives, then the evacuation decision is made, and finally the hurricane either hits or misses Miami.

Sequential Decisions

As we did in the discussion of influence diagrams, we can modify the imperfect-information decision tree to reflect a sequential decision situation in which the first choice is whether to wait for the forecast or evacuate now. Figure 3.29 shows this decision tree.

At this point, you can imagine that representing a sequential decision problem with a decision tree may be very difficult if there are many decisions and chance events because the number of branches can increase dramatically under such conditions. Although full-blown decision trees work poorly for this kind of problem, it is possible to use a *schematic* approach to depict the tree.

Figure 3.28

Evacuation decision represented by decision tree.

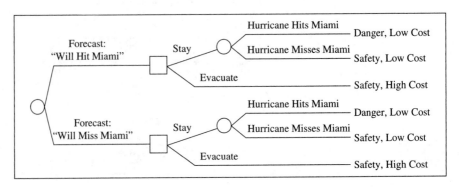

Figure 3.29

Sequential version of evacuation decision in decision-tree form.

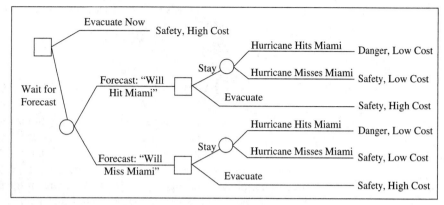

Figure 3.30 shows a schematic version of the farmer's sequential decision problem. This is the decision-tree version of Figure 3.13. Even though each decision and chance event has only two branches, we are using the crescent shape to avoid having the tree explode into a bushy mess. With only the six nodes shown, there would be 2^6, or 64, branches.

We can string together the crescent shapes sequentially in Figure 3.30 because, regardless of the outcome or decision at any point, the same events and decisions follow in the rest of the tree. This ability is useful in many kinds of situations. For example, Figure 3.31 shows a decision in which the immediate decision is whether to invest in an entrepreneurial venture to market a new product or invest in the stock

Figure 3.30

Schematic version of farmer's sequential decision: decision-tree form.

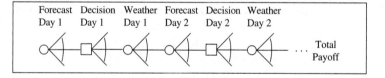

Figure 3.31

An investment decision in schematic form.

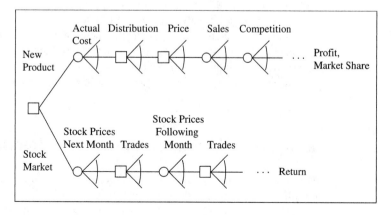

market. Each alternative leads to its own set of decisions and chance events, and each set can be represented in schematic form.

Decision Trees and Influence Diagrams Compared

It is time to step back and compare decision trees with influence diagrams. The discussion and examples have shown that, on the surface at least, decision trees display considerably more information than do influence diagrams. It should also be obvious, however, that decision trees get "messy" much faster than do influence diagrams as decision problems become more complicated. One of the most complicated decision trees we constructed was for the sequential decision in Figure 3.31, and it really does not show all of the intricate details contained in the influence-diagram version of the same problem. The level of complexity of the representation is not a small issue. When it comes time to present the results of a decision analysis to upper-level managers, their understanding of the graphical presentation is crucial. Influence diagrams are superior in this regard; they are especially easy for people to understand regardless of mathematical training.

Should you use decision trees or influence diagrams? Both are worthwhile, and they complement each other well. Influence diagrams are particularly valuable for the structuring phase of problem solving and for representing large problems. Decision trees display the details of a problem. The ultimate decision made should not depend on the representation, because influence diagrams and decision trees are *isomorphic*; any properly built influence diagram can be converted into a decision tree, and vice versa, although the conversion may not be easy. One strategy is to start by using an influence diagram to help understand the major elements of the situation and then convert to a decision tree to fill in details.

Influence diagrams and decision trees provide two approaches for modeling a decision. Because the two approaches have different advantages, one may be more appropriate than the other, depending on the modeling requirements of the particular situation. For example, if it is important to communicate the overall structure of a model to other people, an influence diagram may be more appropriate. Careful reflection and sensitivity analysis on specific probability and value inputs may work better in the context of a decision tree. Using both approaches together may prove useful; the goal, after all, is to make sure that the model accurately represents the decision situation. Because the two approaches have different strengths, they should be viewed as complementary techniques rather than as competitors in the decision-modeling process.

Decision Details: Defining Elements of the Decision

With the overall structure of the decision understood, the next step is to make sure that all elements of the decision model are clearly defined. Beginning efforts to structure decisions usually include some rather loose specifications. For example, when the EPA considers regulating the use of a potentially cancer-causing substance, it would have a fundamental objective of minimizing the social cost of the cancers. (See, for example, Figure 3.20 and the related discussion.) But how will cancer costs be measured? In incremental lives lost? Incremental cases of cancer, both treatable and fatal? In making its decision, the EPA would also consider the rate at which people are exposed to the toxin while the chemical is in use. What are possible levels of exposure? How will we measure exposure? Are we talking about the number of people exposed to the chemical per day or per hour? Does exposure consist of breathing dust particles, ingesting some critical quantity, or skin contact? Are we concerned about contact over a period of time? Exactly how will we know if an individual has had a high or low level of exposure? The decision maker must give unequivocal answers to these questions before the decision model can be used to resolve the EPA's real-world policy problem.

Much of the difficulty in decision making arises when different people have different ideas regarding some aspect of the decision. The solution is to refine the conceptualizations of events and variables associated with the decision enough so that it can be made. How do we know when we have refined enough? The *clarity test* (Howard 1988) provides a simple and understandable answer. Imagine a clairvoyant who has access to all future information: newspapers, instrument readings, technical reports, and so on. Would the clairvoyant be able to determine unequivocally what the outcome would be for any event in the influence diagram? No interpretation or judgment should be required of the clairvoyant. Another approach is to imagine that, in the future, perfect information will be available regarding all aspects of the decision. Would it be possible to tell exactly what happened at every node, again with no interpretation or judgment? The decision model passes the clarity test when these questions are answered affirmatively. At this point, the problem should be specified clearly enough so that the various people involved in the decision are thinking about the decision elements in exactly the same way. There should be no misunderstandings regarding the definitions of the basic decision elements.

The clarity test is aptly named. It requires absolutely clear definitions of the events and variables. In the case of the EPA considering toxic substances, saying that the exposure rate can be either high or low fails the clarity test; what does "high" mean in this case? On the other hand, suppose exposure is defined as high if the average skin contact per person-day of use exceeds an average of 10 milligrams of material per second over 10 consecutive minutes. This definition passes the clarity test. An accurate test could indicate precisely whether the level of exposure exceeded the threshold.

Although Howard originally defined the clarity test in terms of only chance nodes, it can be applied to all elements of the decision model. Once the problem is structured and the decision tree or influence diagram built, consider each node. Is the definition of each chance event clear enough so that an outside observer would know exactly what happened? Are the decision alternatives clear enough so that someone else would know exactly what each one entails? Are consequences clearly defined and measurable? All of the action with regard to the clarity test takes place within the tables in an influence diagram, along the individual branches of a decision tree, or in the tree's consequence matrix. These are the places where the critical decision details are specified. Only after every element of the decision model passes the clarity test is it appropriate to consider solving the influence diagram or decision tree, which is the topic of Chapter 4.

The next two sections explore some specific aspects of decision details that must be included in a decision model. In the first section we look at how chances can be specified by means of probabilities and, when money is an objective, how cash flows can be included in a decision tree. These are rather straightforward matters in many of the decisions we make. However, when we have multiple fundamental objectives, defining ways to measure achievement of each objective can be difficult; it is easy to measure costs, savings, or cash flows in dollars or pounds sterling, but how does one measure damage to an ecosystem? Developing such measurement scales is an important aspect of attaining clarity in a decision model and is the topic of the second section.

More Decision Details: Cash Flows and Probabilities

Many decision situations, especially business decisions, involve some chance events, one or more decisions to make, and a fundamental objective that can be measured in monetary terms (maximize profit, minimize cost, and so on). In these situations, once the decisions and chance events are defined clearly enough to pass the clarity test, the last step is to specify the final details: specific chances associated with the uncertain events and the cash flows that may occur at different times. What are the chances that a particular outcome will occur? What does it cost to take a given action? Are there specific cash flows that occur at different times, depending on an alternative chosen or an event's outcome?

Specifying the chances for the different outcomes at a chance event requires us to use probabilities. Although probability is the topic of Section 2 of the book, we will use probability in Chapters 4 and 5 as we develop some basic analytical techniques. For now, in order to specify probabilities for outcomes, you need to keep in mind only a few basic rules. First, probabilities must fall between 0 and 1 (or equivalently between 0% and 100%). There is no such thing as a 110% chance that some event will occur. Second, recall that the outcomes associated with a chance event must be such that they are mutually exclusive and collectively exhaustive; only one outcome can occur (you can only go down one path), but one of the set must occur (you must go down some path). The implication is that the probability assigned to any given chance outcome (branch) must be between 0 and 1, and for any given chance node, the probabilities for its outcomes must add up to 1.

Indicating cash flows at particular points in the decision model is straightforward. For each decision alternative or chance outcome, indicate the associated cash flow, either as part of the information in the corresponding influence-diagram node or on the appropriate branch in the decision tree. For example, in the toxic-chemical example, there are certainly economic costs associated with different possible regulatory actions. In the new-product decision (Figure 3.16), different cash inflows are associated with different quantities sold, and different outflows are associated with different costs. All of

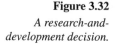

Figure 3.32

*A research-and-
development decision.*

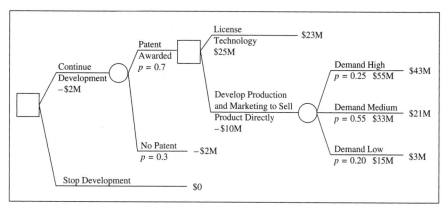

these cash flows must be combined (possibly using net present value if the timing of the cash flows is substantially different) at the end of each branch in order to show exactly what the overall consequence is for a specific path through the decision model.

Figure 3.32 shows a decision tree with cash flows and probabilities fully specified. This is a research-and-development decision. The decision maker is a company that must decide whether to spend $2 million to continue with a particular research project. The success of the project (as measured by obtaining a patent) is not assured, and at this point the decision maker judges only a 70% chance of getting the patent. If the patent is awarded, the company can either license the patent for an estimated $25 million or invest an additional $10 million to create a production and marketing system to sell the product directly. If the company chooses the latter, it faces uncertainty of demand and associated profit from sales.

You can see in Figure 3.32 that the probabilities at each chance node add up to 1. Also, the dollar values at the ends of the branches are the net values. For example, if the company continues development, obtains a patent, decides to sell the product directly, and enjoys a high level of demand, the net amount is $43 million = (−2) + (−10) + 55 million. Also, note that cash flows can occur anywhere in the tree, either as the result of a specific choice made or because of a particular chance outcome.

Defining Measurement Scales for Fundamental Objectives

Many of our examples so far (and many more to come!) have revolved around relatively simple situations in which the decision maker has only one easily measured fundamental objective, such as maximizing profit, as measured in dollars. But the world is not always so accommodating. We often have multiple objectives, and some of those objectives are not easily measured on a single, natural numerical scale. What sort of measure should we use when we have fundamental objectives like maximizing our level of physical fitness, enhancing a company's work environment, or improving the quality of a theatrical production? The answer, not surprisingly, relates back to the ideas embodied in the clarity test; we must find unambiguous ways to measure achievement of the fundamental objectives.

Before going on to the nuts and bolts of developing unambiguous scales, let us review briefly why the measurement of fundamental objectives is crucial in the decision process. The fundamental objectives represent the reasons why the decision maker cares about the decision and, more importantly, how the available alternatives should be evaluated. If a fundamental objective is to build market share, then it makes sense explicitly to estimate how much market share will change as part of the consequence of choosing a particular alternative. The change in market share could turn out to be good or bad depending on the choices made (e.g., bringing a new product to the market) and the outcome of uncertain events (such as whether a competitor launches an extensive promotional campaign). The fact that market share is something which the decision maker cares about, though, indicates that it must be measured.

It is impossible to overemphasize the importance of tying evaluation directly to the fundamental objectives. Too often decisions are based on the wrong measurements because inadequate thought is given to the fundamental objectives in the first place, or certain measurements are easy to make or are made out of habit, or the experts making the measurements have different objectives than the decision maker. An example is trying to persuade the public that high-tech endeavors like nuclear power plants or genetically engineered plants for agricultural use are not risky because few fatalities are expected; the fact is that the public appears to care about many other aspects of these activities as well as potential fatalities! (For example, laypeople are very concerned with technological innovations that may have unknown long-term side effects, and they are also concerned with having little personal control over the risks that they may face because of such innovations.) In complex decision situations there may be many objectives that must be considered. The fundamental-objectives hierarchy indicates explicitly what must be accounted for in evaluating potential consequences.

The fundamental-objectives hierarchy starts at the top with an overall objective, and lower levels in the hierarchy describe important aspects of the more general objectives. Ideally, each of the lowest-level fundamental objectives in the hierarchy would be measured. Thus, one would start at the top and trace down as far as possible through the hierarchy. Reconsider the summer-intern example, in which PeachTree Consumer Products is looking for a summer employee to help with the development of a market survey. Figure 3.4 shows the fundamental-objectives hierarchy (as well as the means network). Starting at the top of this hierarchy ("Choose Best Intern"), we would go through "Maximize Quality and Efficiency of Work" and arrive at "Maximize Survey Quality" and "Minimize Survey Cost." Both of the latter require measurements to know how well they are achieved as a result of hiring any particular individual. Similarly, the other branches of the hierarchy lead to fundamental objectives that must be considered. Each of these objectives will be measured on a suitable scale, and that scale is called the objective's *attribute scale* or simply *attribute*.

As mentioned, many objectives have natural attribute scales: hours, dollars, percentage of market. Table 3.3 shows some common objectives with natural attributes.

Table 3.3

Some common objectives and their natural attributes.

Objective	Attribute
Maximize profit ⎤ Maximize revenue ⎟ Maximize savings ⎟ Minimize cost ⎦	Money (for example, dollars)
Maximize market share ⎤ Maximize rate of return ⎦	Percentage
Maximize proximity	Miles, minutes
Maximize fuel efficiency	Miles per gallon
Maximize time with friends, family	Days, hours
Minimize hypertension	Inches Hg (blood pressure)

In the intern decision, "Minimize Survey Cost" would be easily measured in terms of dollars. How much must the company spend to complete the survey? In the context of the decision situation, the relevant components of cost are salary, fringe benefits, and payroll taxes. Additional costs to complete the survey may arise if the project remains unfinished when the intern returns to school or if a substantial part of the proj-ect must be reworked. (Both of the latter may be important uncertain elements of the de-cision situation.) Still, for all possible combinations of alternative chosen and uncertain outcomes, it would be possible, with a suitable definition of cost, to determine how much money the company would spend to complete the survey.

While "Minimize Survey Cost" has a natural attribute scale, "Maximize Survey Quality" certainly does not. How can we measure achievement toward this objective? When there is no natural scale, two other possibilities exist. One is to use a different scale as a proxy. Of course, the proxy should be closely related to the original objective. For example, we might take a cue from the means-objectives network in Figure 3.4; if we could measure the intern's abilities in survey design and analysis, that might serve as a reasonable proxy for survey quality. One possibility would be to use the intern's grade point average in market research and statistics courses. Another possibility would be to ask one of the intern's instructors to provide a rating of the intern's abilities. (Of course, this latter suggestion gives the instructor the same problem that we had in the first place: how to measure the student's ability when there is no natural scale!)

The second possibility is to construct an attribute scale for measuring achieve-ment of the objective. In the case of survey quality, we might be able to think of a number of levels in general terms. The best level might be described as follows:

> **Best survey quality:** State-of-the-art survey. No apparent crucial issues left un-addressed. Has characteristics of the best survey projects presented at profes-sional conferences.

On the other hand, the worst level might be:

> **Worst survey quality:** Many issues left unanswered in designing survey. Members of the staff are aware of advances in survey design that could have been incorporated but were not. Not a presentable project.

Table 3.4

A constructed scale for survey quality.

RANK:

Best. State-of-the-art survey. No apparent substantive issues left unaddressed. Has characteristics of the best survey projects presented at professional conferences.

Better. Excellent survey but not perfect. Methodological techniques were appropriate for the project and similar to previous projects, but in some cases more up-to-date techniques are available. One substantive issue that could have been handled better. Similar to most of the survey projects presented at professional conferences.

Satisfactory. Satisfactory survey. Methodological techniques were appropriate, but superior methods exist and should have been used. Two or three unresolved substantive issues. Project could be presented at a professional conference but has characteristics that would make it less appealing than most presentations.

Worse. Although the survey results will be useful temporarily, a follow-up study must be done to refine the methodology and address substantive issues that were ignored. Occasionally similar projects are presented at conferences, but they are poorly received.

Worst. Unsatisfactory. Survey must be repeated to obtain useful results. Members of the staff are aware of advances in survey design that could have been incorporated but were not. Many substantive issues left unanswered. Not a presentable project.

We could identify and describe fully a number of meaningful levels that relate to survey quality. Table 3.4 shows five possible levels in order from best to worst. You can see that the detailed descriptions define what is meant by quality of the survey and how to determine whether the survey was well done. According to these defined levels, quality is judged by the extent to which the statistical and methodological techniques were up to date, whether any of the company still has unresolved questions about its consumer products, and a judgmental comparison with similar survey projects presented at professional meetings.

Constructing scales can range from straightforward to complex. The key to constructing a good scale is to identify meaningful levels, including best, worst, and intermediate, and then describe those levels in a way that fully reflects the objective under consideration. The descriptions of the levels must be elaborate enough to facilitate the measurement of the consequences. In thinking about possible results of specific choices made and particular uncertain outcomes, it should be easy to use the constructed attribute scale to specify the corresponding consequences.

The scale in Table 3.4 actually shows two complementary ways to describe a level. First is in terms of specific aspects of survey quality, in this case the methodology and the extent to which the survey successfully addressed the company's concerns about its line of products. The second way is to use a comparison approach; in this case, we compare the survey project overall with other survey projects that have been presented at professional meetings. There is nothing inherently important about the survey's presentability at a conference, but making the comparison can help to measure the level of quality relative to other publicly accessible projects.

Note also from Table 3.4 that we could have extended the fundamental-objectives hierarchy to include "Methodology" and "Address Company's Issues"

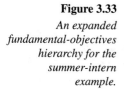

Figure 3.33

*An expanded
fundamental-objectives
hierarchy for the
summer-intern
example.*

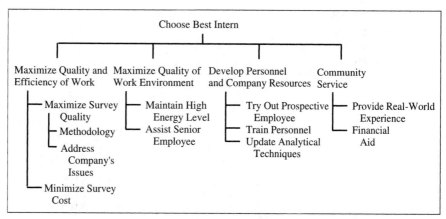

as branches under the "Maximize Survey Quality" branch, as shown in Figure 3.33. How much detail is included in the hierarchy is a matter of choice, and here the principle of a requisite model comes into play. As long as the scale for "Maximize Survey Quality" can adequately capture the company's concerns regarding this objective, then there is no need to use more detailed objectives in measuring quality. If, on the other hand, there are many different aspects of quality that are likely to vary separately depending on choices and chance outcomes, then it may be worthwhile to create a more detailed model of the objective by extending the hierarchy and developing attribute scales for the subobjectives.

Developing the ability to construct meaningful attribute scales requires practice. In addition, it is helpful to see examples of scales that have been used in various situations. We have already seen one such scale in Table 3.4 relating to the summer-intern example. Tables 3.5 and 3.6 show two other constructed attribute scales for biological impacts and public attitudes, respectively, both in the context of selecting a site for a nuclear power generator.

Using PrecisionTree for Structuring Decisions

Decision analysis has benefited greatly from innovations in computers and computer software. Not only have the innovations led to increased computing power allowing for fast and easy analysis of complex decisions, but they have also given rise to user-friendly graphical interfaces that have simplified and enhanced the structuring process. Palisade's DecisionTools, included with this book, consists of five interrelated programs. By acquainting you with the features of DecisionTools, we hope to show you how useful and fun these programs can be. Throughout the text, as we introduce specific decision-analysis concepts, we will also present you with the corresponding DecisionTools software and a step-by-step guide on its use. In this chapter, we explain how to use the program PrecisionTree to construct decision trees and influence diagrams.

Table 3.5

A constructed attribute scale for biological impact.

RANK:

Best • Complete loss of 1.0 square mile of land that is entirely in agricultural use or is entirely urbanized; no loss of any "native" biological communities.

• Complete loss of 1.0 square mile of primarily (75%) agricultural habitat with loss of 25% of second-growth forest; no measurable loss of wetlands or endangered-species habitat.

• Complete loss of 1.0 square mile of land that is 50% farmed and 50% disturbed in some other way (e.g., logged or new second growth); no measurable loss of wetlands or endangered-species habitat.

• Complete loss of 1.0 square mile of recently disturbed (for example, logged, plowed) habitat plus disturbance to surrounding previously disturbed habitat within 1.0 mile of site border; or 15% loss of wetlands or endangered-species habitat.

• Complete loss of 1.0 square mile of land that is 50% farmed (or otherwise disturbed) and 50% mature second-growth forest or other undisturbed community; 15% loss of wetlands or endangered-species habitat.

• Complete loss of 1.0 square mile of land that is primarily (75%) undisturbed mature "desert" community; 15% loss of wetlands or endangered-species habitat.

• Complete loss of 1.0 square mile of mature second-growth (but not virgin) forest community; or 50% loss of big game and upland game birds; or 50% loss of wetlands and endangered-species habitat.

• Complete loss of 1.0 square mile of mature community or 90% loss of productive wetlands and endangered-species habitat.

Worst • Complete loss of 1.0 square mile of mature virgin forest and/or wetlands and/or endangered-species habitat.

Source: Adapted from Keeney (1992, p. 105).

Table 3.6

A constructed attribute scale for public attitudes.

RANK:

Best • *Support.* No groups are opposed to the facility, and at least one group has organized support for the facility.

• *Neutrality.* All groups are indifferent or uninterested.

• *Controversy.* One or more groups have organized opposition, although no groups have action-oriented opposition (for example, letterwriting, protests, lawsuits). Other groups may either be neutral or support the facility.

• *Action-oriented* opposition. Exactly one group has action-oriented opposition. The other groups have organized support, indifference, or organized opposition.

Worst • *Strong action-oriented opposition.* Two or more groups have action-oriented opposition.

Source: Adapted from Keeney (1992, p. 102).

Decision trees and influence diagrams are precise mathematical models of a decision situation that provide a visual representation that is easily communicated and grasped. With the PrecisionTree component of DecisionTools you will be able to construct and solve diagrams and trees quickly and accurately. Features such as pop-up dialog boxes and one-click deletion or insertion of nodes and branches greatly facilitate the structuring process. Visual cues make it easy to distinguish node types: Red circles represent chance nodes, green squares are decision nodes, blue triangles are payoff nodes, and blue rounded rectangles are calculation nodes. Let's put PrecisionTree to work by creating a decision tree for the research-and-development decision (Figure 3.32) and an influence diagram for the basic risky decision (Figure 3.9).

In the instructions below and in subsequent chapters, items in *italics* are words shown on your computer screen. Items in **bold** indicate either the information that you type in or an object that you click with the mouse. The boxes you see below highlight actions you take as the user, with explanatory text between the boxes. Several steps may be described in any given box, so be sure to read and follow the instructions carefully.

Constructing a Decision Tree for the Research-and-Development Decision

In this chapter we will concentrate on PrecisionTree's graphical features, which are specifically designed to help construct decision trees and influence diagrams. Figure 3.34 shows the decision tree for the research-and-development decision that you will generate using the PrecisionTree software.

STEP 1

1.1   Start by opening both the Excel and PrecisionTree programs and enabling the macros if prompted.[1] Figure 3.35 shows the two new toolbars that appear at the top of your screen after opening PrecisionTree.

1.2   To access the on-line help, pull down the PrecisionTree menu and choose **Help,** then **Contents.** Use the on-line help when you have a question concerning the operation of PrecisionTree or any of the other DecisionTools programs. Before proceeding, close the on-line help by choosing **Exit** in the pull-down menu under **File** in the Help window.

1.3   To create your decision tree, return to Excel and click on the **New Tree** button, the first button from the left on the PrecisionTree toolbar. No changes occur until the next step, where you indicate the cell to start the tree.

1.4   Click on the spreadsheet at the location where the tree will start. For this example, choose cell **A1.** The tree's root appears in A1 along with a single end node (blue triangle).

---

[1]To run an add-in within Excel it is necessary to have the "Ignore other applications" option turned off. Choose *Tools* on the menu bar, then *Options,* and click on the *General* tab in the resulting *Options* dialog box. Be sure that the box by *Ignore other applications* is not checked.

Figure 3.34

Decision tree for the research-and-development problem created in PrecisionTree.

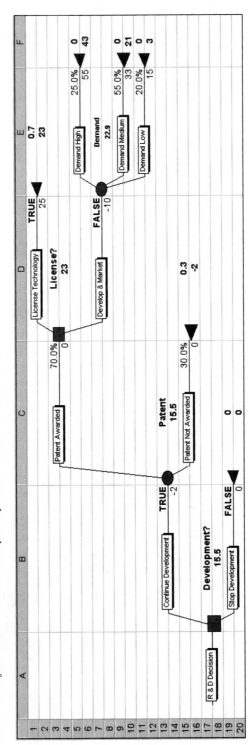

Figure 3.35

PrecisionTree and DecisionTools toolbars that are added to Excel's toolbar.

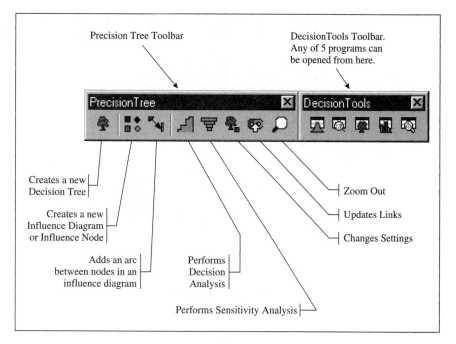

Precision Tree Toolbar

DecisionTools Toolbar. Any of 5 programs can be opened from here.

Creates a new Decision Tree

Creates a new Influence Diagram or Influence Node

Adds an arc between nodes in an influence diagram

Performs Decision Analysis

Performs Sensitivity Analysis

Zoom Out

Updates Links

Changes Settings

1.5   Name the tree by clicking directly on the label with the generic heading, **tree #1,** which brings up the *Tree Settings* dialog box. Note that a pointing hand appears when the cursor is in position to access a dialog box.

1.6   Change the tree name from *tree #1* to **R & D Decision.** Click **OK.**

Two numbers show up at the end of the decision tree, a 1 in cell B1 and a 0 in B2. The 1 represents the probability of reaching this end node and the 0 represents the value attained upon reaching this node. We'll return in the next chapter for a more complete discussion of the values and probabilities. For now, let's focus our attention on structuring the decision.

STEP 2   The next step is to add the "Development?" decision node.

2.1   To create this node, click on the **end node** (blue triangle). The *Node Settings* dialog box pops up.

2.2   Click on the **decision node** button (green square, second from the left) and change the name from *Decision* to **Development?** Your node settings dialog box should now look like Figure 3.36.

2.3   Leave the number of branches at 2 because there are two alternatives: to continue or suspend developing the research project. Click **OK.**

2.4   Rename the branches by clicking on their labels and replacing the word *branch* with **Continue Development** on one and **Stop Development** on the other.

Figure 3.36

Node Settings dialog box for decision trees.

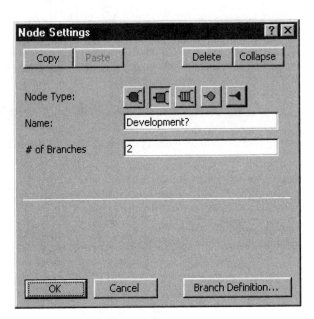

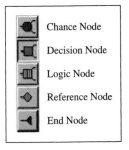

Each alternative or branch of the decision tree can have an associated value, often representing monetary gains or losses. Gains are expressed as positive values, and losses are expressed as negative values. If there is no gain or loss, then the value is zero. In PrecisionTree you enter values by simply typing in the appropriate number below the branch.

2.5   Enter **–2** in the spreadsheet cell below the *Continue Development* branch because it costs $2 million to continue developing the research project.

2.6   Enter **0** in the cell below the *Stop Development* branch because there is no cost in discontinuing the project.

Working with a spreadsheet gives you the option of entering formulas or numbers in the cells. For example, instead of entering a number directly into one of PrecisionTree's value cells, you might refer to a cell that calculates a net present value. The flexibility of referring to other spreadsheet calculations will be useful in later chapters when we model more complex decisions.

STEP 3   If you decide to stop development, no further modeling is necessary and the tree ends for this alternative. On the other hand, if the decision is to continue development, the future uncertainty regarding the patent requires further modeling. The uncertainty is modeled as the "Patent" chance node shown in Figure 3.34.

3.1   To add this chance node, click on the **end node** that follows the *Continue Development* branch.

3.2   In the *Node Settings* box that appears, choose the **chance node** button (red circle, first from the left) and change the name from *Chance* to **Patent.**

3.3   As before, there are two branches, this time because our concern is whether the patent will or will not be awarded. Click **OK.**

3.4   Change the names of the branches to **Patent Awarded** and **Patent Not Awarded** by clicking on the labels and typing in the new names.

Each branch (outcome) of a chance node has both an associated value and a probability. For PrecisionTree, the probabilities are positioned above the branch and the values below the branch. Looking at the diagram you are constructing, you see that PrecisionTree has placed 50.0% above each branch, and has given each branch a value of zero. These are default values and need to be changed to agree with the judgments of the decision maker, as shown in Figure 3.34.

3.5   Click in the spreadsheet cell above *Patent Awarded* and type **0.70** to indicate that there is a 70% chance of the patent being awarded.

3.6   Select the cell above *Patent Not Awarded* and enter either **0.30** or **= 1 – C1,** where C1 is the cell that contains 0.70. In general, it is better to use cell references than numbers when constructing a model so that any changes will automatically be propagated throughout the model. In this case, using the cell reference = 1 – C1 will guarantee that the probabilities add to one even if the patent-award probability changes.

3.7   The values remain at zero because there are no direct gains or losses that occur at the time of either outcome.

STEP 4   Once you know the patent outcome, you must decide whether to license the technology or develop and market the product directly. Model this by adding a decision node following the "Patent Awarded" branch (see Figure 3.34).

4.1   Click the **end node** (blue triangle) of the *Patent Awarded* branch.

4.2   Choose **decision node** (green square) in the *Node Settings* dialog box.

4.3   Name the decision **License?**

4.4   Confirm that the node will have two branches.

4.5   Click **OK.**

4.6   Rename the two new branches **License Technology** and **Develop & Market.**

> 4.7   PrecisionTree defaults to a value of zero for each branch. To change the values, highlight the cell below the *License Technology* branch and type in **25,** because the value of licensing the technology given that the patent has been awarded is $25 million.
>
> 4.8   Similarly, place **–10** in the cell below the *Develop & Market* branch, because a $10 million investment in production and marketing is needed, again assuming we have the patent.

PrecisionTree calculates the end-node values by summing the values along the path that lead to that end node. For example, the end value of the "License Technology" branch in Figure 3.34 is 23 because the values along the path that lead to "License Technology" are –2, 0, and 25, which sum to 23. In Chapter 4 we explain how to input your own specific formula for calculating the end-node values.

> **STEP 5**   Developing the product in-house requires us to model market demand, the final step in the structuring process.
>
> 5.1   Click on the **end node** of the *Develop & Market* branch.
>
> 5.2   Select **chance** as the node type.
>
> 5.3   Enter the name **Demand.**
>
> 5.4   Change *# of Branches* from 2 to **3,** because we have three outcomes for this uncertainty.
>
> 5.5   Click **OK.**
>
> 5.6   Retitle the branches **Demand High, Demand Medium,** and **Demand Low.**
>
> 5.7   Enter the probabilities **0.25, 0.55, 0.20** above and the values **55, 33, 15** below the respective branches *Demand High, Demand Medium,* and *Demand Low.*

Congratulations! You have just completed structuring the research-and-development decision tree.

We have not discussed all of the options in the *Node Settings* box (Figure 3.36). Several are common spreadsheet or word processing functions that are useful in decision-tree construction. You can use the *Copy* and *Paste* buttons for duplicating nodes. When a node is copied, the entire subtree following it is also copied, making it easy to replicate entire portions of the tree. Similarly, *Delete* removes not only the chosen node, but all downstream nodes as well. The *Collapse* button hides all the downstream details but does not delete them; a boldface plus sign next to a node indicates that the subtree that follows has been collapsed.

There are also two additional node types: logic and reference nodes. The logic node (Figure 3.36, purple square, third from the left) is a decision node that determines the chosen alternative by applying a user-specified logic formula to each option (see the PrecisionTree user's manual for more details). The reference node (gray

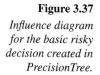

Figure 3.37

Influence diagram for the basic risky decision created in PrecisionTree.

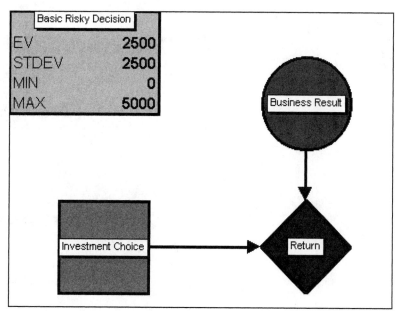

diamond, fourth from the left) allows you to repeat a portion of the tree that has already been constructed without manually reconstructing that portion. Reference nodes are especially useful when constructing a large and complex decision tree because they allow you to prune the tree graphically while retaining all the details. See Chapter 4 for an example that uses reference nodes.

Constructing an Influence Diagram for the Basic Risky Decision

PrecisionTree provides the ability to structure decisions using influence diagrams. Follow the step-by-step instructions below to create an influence diagram for the basic risky decision (Figure 3.37). Our starting point below assumes that PrecisionTree is open and your Excel worksheet is empty. If not, see Step 1 (1.1) above.

STEP 6

6.1   Start by clicking on the **New Influence Diagram/Node** icon (Figure 3.35, PrecisionTree toolbar, second button from the left).

6.2   Move the cursor, which has changed into crosshairs, to the spreadsheet. Although an influence diagram may be started by clicking inside any cell, for this example start the diagram by clicking on cell **B10.**

Figure 3.38

Influence Node
Settings *dialog box for
influence diagrams.*

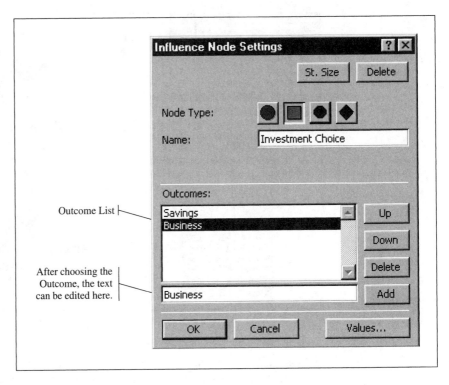

The spreadsheet is altered in three ways:

1. A decision node appears.

2. A dialog box titled *Influence Node Settings* opens, allowing the selection and definition of nodes (Figure 3.38).

3. A gray box labeled *Diagram #1* appears, displaying four summary statistics: expected value, standard deviation, minimum, and maximum. The #N/A appearing for each statistic reflects that we have yet to add values and probabilities to the model.

The first node to add is the decision whether to place $2000 in a savings account or in a friend's business.

6.3   In the *Influence Node Settings* dialog box (Figure 3.38), enter **Investment Choice** as the node name.

6.4   Under the heading *Outcomes* in the lower half of this box is a list of possible outcomes. Naming the outcomes is analogous to naming the branches of a decision tree. To enter outcome names, first click on **Outcome #1.**

6.5   Move the cursor down to where the text can be edited.

6.6   Delete *Outcome #1* from this line by backspacing, and type in the new outcome name **Savings.**

6.7   Change the name of *Outcome #2* by moving the cursor back up to the *Outcomes* list and clicking on **Outcome #2.**

6.8   Return to the editing box, delete *Outcome #2* by backspacing, and replace it with the new outcome name **Business.**

6.9   When you are finished, your *Influence Node Settings* dialog box should look like Figure 3.38. Click **OK.**

Note that the *Up, Down, Delete,* and *Add* buttons in the *Influence Node Settings* dialog box affect only the *Outcomes* list and are not used when renaming the outcomes from the list.

6.10   To name the diagram, click on the generic name **Diagram #1** that straddles cells A1 and B1, which opens the *Influence Diagrams Settings* dialog box. In the *Diagram Name* text box, type **Basic Risky Decision.**

6.11   Click **OK.**

STEP 7   Now let's add the "Business Result" chance node.

7.1   Click on the **New Influence Diagram/Node** button on the PrecisionTree toolbar.

7.2   Click in cell **E2.**

7.3   To make this a chance node, click on the **chance node** icon (red circle, first from the left) in the *Influence Node Settings* dialog box.

7.4   Following the same procedure as before, name the node **Business Result** and the two outcomes **Wild Success** and **Flop.**

7.5   Click **OK.**

Note that when creating an influence diagram, you may create the nodes or name the outcomes in any order.

Influence-diagram nodes can be modified in two ways. Clicking on the node itself rather than the name allows you to edit the graphics, such as resizing the node or dragging it to a new location. The attributes of the node (such as node type, outcome names, and numerical information) are edited in the *Node Settings* dialog box, which you access by clicking directly on the node name (when the cursor changes into a figure of a hand with the index finger extended).

STEP 8   The last node to be added to our diagram is the payoff node. PrecisionTree allows only one payoff node in an influence diagram. Creating a payoff node is similar to creating the other types of nodes except that naming the node is the only available option.

8.1   Start by clicking the **New Influence Diagram/Node** button.

8.2   Click on cell **E10.**

8.3   Click on the **payoff node** icon (blue diamond, fourth from the left) in the *Influence Node Settings* dialog box, enter the name **Return,** and click **OK.**

This creates the third and final node of our diagram.

The next thing is to add arcs. Arcs in PrecisionTree are somewhat more elaborate than what we've described in the text, and so a brief discussion is in order before proceeding. PrecisionTree and the text differ on terminology: What we refer to as relevance arcs, PrecisionTree calls value arcs, and what we know as sequence arcs, PrecisionTree calls timing arcs. We are able to tell by context whether the arc is a relevance (value) or sequence (timing) arc, but the program cannot. Hence, for each arc that you create in PrecisionTree, you must indicate whether it is a value arc, a timing arc, or both. This means that PrecisionTree forces you to think carefully about the type of influence for each arc as you construct your influence diagram. Let's examine these arc types and learn how to choose the right characteristics to represent the relationship between two nodes.

The value arc option is used when the possible outcomes or any of the numerical details (probabilities or numerical values associated with outcomes) in the successor node are influenced by the outcomes of the predecessor node. Ask yourself if knowing the outcomes of the predecessor has an effect on the outcomes, probabilities, or values of the successor node. If you answer yes, then use a value arc. Conversely, if none of the outcomes has an effect, a value arc is not indicated. (Recall that this is the same test for relevance that we used in the text.) Let's demonstrate with some specific examples. The arc from "Investment Choice" to "Return" is a value arc if any of the investment alternatives ("Savings" or "Business") affect the returns on your investment. Because they clearly do, you would make this a value arc. Another example from the hurricane problem is the arc connecting the chance node "Hurricane Path" with the chance node "Forecast." This is also a value arc; we presume that the weather forecast really does bear some relationship to the actual weather and hence that the probabilities for the forecast are related to the path the hurricane takes.

Use a timing arc when the predecessor occurs chronologically prior to the successor. An example, again from the hurricane problem, is the arc from the chance node "Forecast" to the decision node "Decision." It is a timing arc because the decision maker knows the forecast before deciding whether or not to evacuate.

Use both the value and timing options when the arc satisfies both conditions. For example, consider the arc from the "Investment Choice" node to the "Return" node. To calculate the return, you need to know that the investment decision has been made (timing), and you need to know which alternative was chosen (value). In fact, any arc that terminates at the payoff node must be both value and timing, and so the arc from "Business Result" to "Return" also has both characteristics.

Arcs in influence diagrams are more than mere arrows; they actually define mathematical relationships between the nodes they connect. It is necessary to think carefully about each arc so that it correctly captures the relationship between the two nodes. PrecisionTree not only forces you to decide the arc type, but it also supplies feedback on the effect each arc has when values are added to the influence diagram.

STEP 9

9.1   Click on the **New Influence Arc** button on the PrecisionTree toolbar (Figure 3.35, third button from the left).

9.2   Place the cursor, which has become crosshairs, in the predecessor node *Investment Choice*.

9.3   Hold the mouse button down and drag the cursor into the successor node *Return*. Be sure that the arc originates and terminates well inside the boundaries of each node.

When you release the mouse button, an arc appears and the "Influence Arc" dialog box (Figure 3.39) opens. Figure 3.39 shows that both value and timing have been pre-chosen by PrecisionTree. (There is a third influence type shown in Figure 3.39—structure arcs—that will not be discussed. See the PrecisionTree manual or on-line help about structure arcs.)

9.4   Click **OK** in the *Influence Arc* dialog box that pops up.

9.5   To add the second arc, again click on the **New Influence Arc** button and create an arc from the chance node *Business Result* to *Return* as described above.

STEP 10   Now that our decision has been structured, we can add the probabilities and values. We begin by adding the values and probabilities to the chance node.

10.1   Click on the node name **Business Result** to bring up the *Influence Node Settings* dialog box.

Figure 3.39

Influence Arc *dialog box for the arc from the "Investment Choice" node to the "Return" node.*

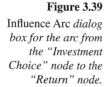

10.2 Click on the **Values...** button in the lower right-hand corner to bring up the *Influence Value Editor* box (Figure 3.40). This box is configured with the names of the two outcomes or branches on the left, a column for entering values in the middle, and a column for entering probabilities on the right.

10.3 Enter the values and probabilities shown in Figure 3.40, hitting the tab key after entering each number, including the last entry. (Be sure to enter a zero if an outcome has no value associated with it, as in the *Value when Skipped* cell.)

10.4 When all the numbers are entered, click **OK.**

STEP 11

11.1 To add the values to the decision node, click on the **Investment Choice** and choose the **Values...** button.

11.2 Enter **0** for the *Value when Skipped,* **2000** for the *Savings,* and **2000** for the *Business* alternatives. Be sure to hit the tab or enter key to confirm each entry, including the last entry.

11.3 When finished, click **OK.**

The payoff relationship is defined in the *Influence Value Editor* box (Figure 3.41). The basic idea is to choose a row, type an equals sign (=) into the value cell, and define a formula that reflects the decisions and outcomes of that row. For example, reading from right to left in the first row, we see that we invested in the savings account, and the business was (or would have been) a wild success. Because we chose the certain return guaranteed by the savings account, the value of our investment is $2000 plus 10%, or $2200. In this case, the value does not depend on whether the business was a success or a flop. Hence, the payoff formula includes only the investment choice and not the business result. In the third row, however, we hit the jackpot with our investment in the business when it becomes a wild success. Thus, the formula for this case will include both the investment choice and the business result.

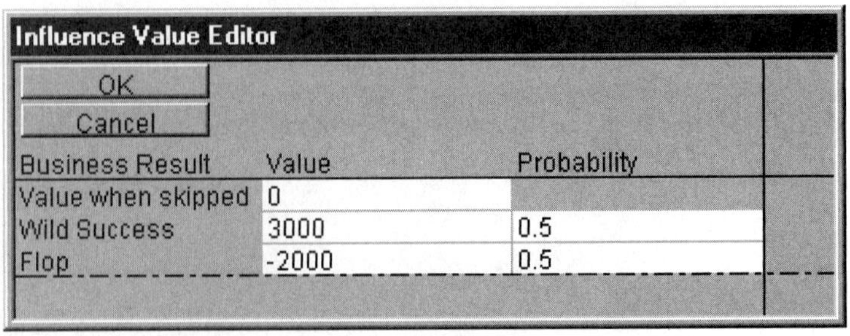

Figure 3.40
Influence Value Editor
for the "Business
Result" chance node.

Influence Value Editor		
OK		
Cancel		
Business Result	Value	Probability
Value when skipped	0	
Wild Success	3000	0.5
Flop	-2000	0.5

Figure 3.41

*Influence Value Editor
for the payoff node.*
This mini-spreadsheet
specifies how
PrecisionTree calcu-
lates the values for the
payoff node.

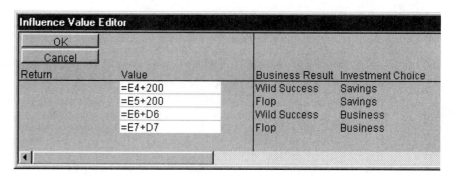

STEP 12   Here are the steps for defining the payoff formulas:

12.1   Open the *Influence Value Editor* box by clicking on the payoff node name **Return** and clicking the **Values...** button.

12.2   Type an equal sign (=) into the first value cell.

12.3   Move the cursor to the word **Savings** to the right of the value cell and below the *Investment Choice* heading, and click. "E4" appears in the value cell next to the equal sign. E4 references the $2000 value we assigned to savings in Step 11. PrecisionTree will substitute this value into the formula.

12.4   To complete this cell, after = E4, type **+200,** the amount of interest earned with the savings account.

Any Excel formula can be placed in these cells, so instead of = E4 + 200, you could equivalently have used =E4 + 0.10*E4, or you could specify the interest rate on a cell on the original spreadsheet and refer to that cell rather than type in 0.10.

12.5   In the next row down, type another = into the value cell.

12.6   Click on **Savings** to the cell's right and two rows below the heading *Investment Choice.* "E5" appears in the cell. Finish by typing **+200.**

12.7   In the third value cell, type **=,** click on **Business** (to access the $2000 investment), type **+,** and click on **Wild Success** (to access the $3000 earnings) to the cell's right and three rows below the heading *Business Result.*

12.8   Using Figure 3.41 as a guide, enter the final set of values, and click **OK.** Note that you use the cells to the right and in the same row as the value cell you are defining.

The summary statistics box should now display the expected value ($2500), standard deviation ($2500), minimum ($0), and maximum ($5000).

Congratulations! You have just completed the influence diagram for the basic risky decision.

This concludes our instructions for building an influence diagram. There are some additional useful influence-diagram structuring tools available in Precision-Tree that we have not covered. For example, we mentioned that calculation nodes are helpful in emphasizing a diagram's structure. These are available in PrecisionTree and are found in the *Node Settings* dialog box as the blue rounded hexagon, third from the left. In addition, PrecisionTree can also convert an influence diagram into a decision tree (see Exercise 3.26), which can help when checking the accuracy of an influence diagram.

SUMMARY

This chapter has discussed the general process of structuring decision problems. It is impossible to overemphasize the importance of the structuring step, because it is here that one really understands the problem and all of its different aspects. We began with the process of structuring values, emphasizing the importance of identifying underlying fundamental objectives and separating those from means objectives. Fundamental objectives, structured in a hierarchy, are those things that the decision maker wants to accomplish, and means objectives, structured in a network, describe ways to accomplish the fundamental objectives.

With objectives specified, we can begin to structure a decision's specific elements. A decision maker may use both influence diagrams and decision trees as tools in the process of modeling decisions. Influence diagrams provide compact representations of decision problems while suppressing many of the details, and thus they are ideal for obtaining overviews, especially for complex problems. Influence diagrams are especially appropriate for communicating decision structure because they are easily understood by individuals with little technical background. On the other hand, decision trees display all of the minute details. Being able to see the details can be an advantage, but in complex decisions trees may be too large and "bushy" to be of much use in communicating with others.

The clarity test is used to ensure that the problem is defined well enough so that everyone can agree on the definitions of the basic decision elements, and we also discussed the specification of probabilities and cash flows at different points in the problem. We also discussed the notion of attribute scales for measuring the extent to which fundamental objectives are accomplished, and we showed how scales can be constructed to measure achievement of those objectives that do not have natural measures. Finally, we introduced PrecisionTree for structuring decisions with both decision trees and influence diagrams.

EXERCISES

3.1   Describe in your own words the difference between a means objective and a fundamental objective. Why do we focus on coming up with attribute scales that measure accomplishment of fundamental objectives, but not means objectives? What good does it do to know what your means objectives are?

3.2 What are your fundamental objectives in the context of renting an apartment while attending college? What are your means objectives? Create a fundamental-objectives hierarchy and a means-objectives network.

3.3 In the context of renting an apartment (Exercise 3.2), some of the objectives may have natural attribute scales. Examples are minimizing rent ($) or minimizing the distance to campus (kilometers or city blocks). But other attributes, such as ambiance, amount of light, or neighbors, have no natural scales. Construct an attribute scale with at least five different levels, ranked from best to worst, for some aspect of an apartment that is important to you but has no natural scale.

3.4 Before making an unsecured loan to an individual a bank orders a report on the applicant's credit history. To justify making the loan, the bank must find the applicant's credit record to be satisfactory. Describe the bank's decision. What are the bank's objectives? What risk does the bank face? What role does the credit report play? Draw an influence diagram of this situation. (*Hint:* Your influence diagram should include chance nodes for a credit report and for eventual default.) Finally, be sure to specify everything (decisions, chance events, objectives) in your model clearly enough to pass the clarity test.

3.5 When a movie producer decides whether to produce a major motion picture, the main question is how much revenue the movie will generate. Draw a decision tree of this situation, assuming that there is only one fundamental objective, to maximize revenue. What must be included in revenue to be sure that the clarity test is passed?

3.6 You have met an acquaintance for lunch, and he is worried about an upcoming meeting with his boss and some executives from his firm's headquarters. He has to outline the costs and benefits of some alternative investment strategies. He knows about both decision trees and influence diagrams but cannot decide which presentation to use. In your own words, explain to him the advantages and disadvantages of each.

3.7 Reframe your answer to Exercise 3.6 in terms of objectives and alternatives. That is, what are appropriate fundamental objectives to consider in the context of choosing how to present the investment information? How do decision trees and influence diagrams compare in terms of these objectives?

3.8 Draw the politician's decision in Figure 3.25 as an influence diagram. Include the tables showing decision alternatives, chance-event outcomes, and consequences.

3.9 A dapper young decision maker has just purchased a new suit for $200. On the way out the door, the decision maker considers taking an umbrella. With the umbrella on hand, the suit will be protected in the event of rain. Without the umbrella, the suit will be ruined if it rains. On the other hand, if it does not rain, carrying the umbrella is an unnecessary inconvenience.

a Draw a decision tree of this situation.

b Draw an influence diagram of the situation.

c Before deciding, the decision maker considers listening to the weather forecast on the radio. Draw an influence diagram that takes into account the weather forecast.

3.10 When patients suffered from hemorrhagic fever, M*A*S*H doctors replaced lost sodium by administering a saline solution intravenously. However, headquarters (HQ) sent a treatment change disallowing the saline solution. With a patient in shock and near death from a disastrously low sodium level, B. J. Hunnicut wanted to administer a low-sodium-concentration saline solution as a last-ditch attempt to save the patient. Colonel Potter looked at B. J. and Hawkeye and summed up the situation. "O.K., let's get this straight. If we go by the new directive from HQ and don't administer saline to replace the sodium, our boy will die for sure. If we try B. J.'s idea, then he may survive, and we'll know how to treat the next two patients who are getting worse. If we try it and he doesn't make it, we're in trouble with HQ and may get court-martialed. I say we have no choice. Let's try it." (*Source:* "Mr. and Mrs. Who." Written by Ronny Graham, directed by Burt Metcalfe, 1980.)

Structure the doctors' decision. What are their objectives? What risks do they face? Draw a decision tree for their decision.

3.11 Here is an example that provides a comparison between influence diagrams and decision trees.

    a Suppose you are planning a party, and your objective is to have an enjoyable party for all the guests. An outdoor barbecue would be the best, but only if the sun shines; rain would make the barbecue terrible. On the other hand, you could plan an indoor party. This would be a good party, not as nice as an outdoor barbecue in the sunshine but better than a barbecue in the rain. Of course, it is always possible to forego the party altogether! Construct an influence diagram and a decision tree for this problem.

    b You will, naturally, consult the weather forecast, which will tell you that the weather will be either "sunny" or "rainy." The forecast is not perfect, however. If the forecast is "sunny," then sunshine is more likely than rain, but there still is a small chance that it will rain. A forecast of "rainy" implies that rain is likely, but the sun may still shine. Now draw an influence diagram for the decision, including the weather forecast. (There should be four nodes in your diagram, including one for the forecast, which will be available at the time you decide what kind of party to have, and one for the actual weather. Which direction should the arrow point between these two nodes? Why?) Now draw a decision tree for this problem. Recall that the events and decisions in a decision tree should be in chronological order.

3.12 The clarity test is an important issue in Exercise 3.11. The weather obviously can be somewhere between full sunshine and rain. Should you include an outcome like "cloudy"? Would it affect your satisfaction with an outdoor barbecue? How will you define rain? The National Weather Service uses the following definition: Rain has occurred if "measurable precipitation" (more than 0.004 inch) has occurred at the official rain gauge. Would this definition be suitable for your purposes? Define a set of weather outcomes that is appropriate relative to your objective of having an enjoyable party.

3.13 Draw the machine-replacement decision (Figure 3.10) as a decision tree.

QUESTIONS AND PROBLEMS

3.14 Modify the influence diagram in Figure 3.11 (the hurricane-forecast example) so that it contains nodes for each of the two objectives (maximize safety and minimize cost). Cost has a natural attribute scale, but how can you define safety? Construct an attribute scale that you could use to measure the degree of danger you might encounter during a hurricane.

3.15 Decision analysis can be used on itself! What do you want to accomplish in studying decision analysis? Why is decision analysis important to you? In short, what are your fundamental objectives in studying decision analysis? What are appropriate means objectives? Is your course designed in a way that is consistent with your objectives? If not, how could the course be modified to achieve your objectives?

3.16 In the spring of 1987 Gary Hart, the leading Democratic presidential candidate, told the news media that he was more than willing to have his private life scrutinized carefully. A few weeks later, the *Miami Herald* reported that a woman, Donna Rice, had been seen entering his Washington townhouse on a Friday evening but not leaving until Saturday evening. The result was a typical political scandal, with Hart contending that Rice had left Friday evening by a back door that the reporter on the scene was not watching. The result was that Hart's credibility as a candidate was severely damaged, thus reducing his chance of winning both the Democratic nomination and the election. The decision he had to make was whether to continue the campaign or to drop out. Compounding the issue was a heavy debt burden that was left over from his unsuccessful 1984 presidential bid.

Using both an influence diagram and a decision tree, structure Hart's decision. What is the main source of uncertainty that he faces? Are there conflicting objectives, and if so, what are they? What do you think he should have done? (He decided to drop out of the race. However, he eventually reentered, only to drop out again because of poor showings in the primary elections.)

3.17 When an amateur astronomer considers purchasing or building a telescope to view deep-sky objects (galaxies, clusters, nebulae, etc.), the three primary considerations are minimizing cost, having a stable mounting device, and maximizing the *aperture* (diameter of the main lens or mirror). The aperture is crucial because a larger aperture gathers more light. With more light, more detail can be seen in the image, and what the astronomer wants to do is to see the image as clearly as possible. As an example, many small telescopes have lens or mirrors up to 8 inches in diameter. Larger amateur telescopes use concave mirrors ranging from 10 to 16 inches in diameter. Some amateurs grind their own mirrors as large as 40 inches.

Saving money is important, of course, because the less spent on the telescope, the more can be spent on accessories (eyepieces, star charts, computer-based astronomy programs, warm clothing, flashlights, and so on) to make viewing as easy and comfortable as possible. Money might also be spent on an observatory to house a large telescope or on trips away from the city (to avoid the light pollution of city skies and thus to see images more clearly).

Finally, a third issue is the way the telescope is mounted. First, the mount should be very stable, keeping the telescope perfectly still. Any vibrations will show up dramatically in the highly magnified image, thus reducing the quality of the image and the detail that can be seen. The mount should also allow for easy and smooth movement of the telescope to view any part of the sky. Finally, if the astronomer wants to use the telescope to take photographs of the sky (astrophotos), it is important that the mount includes some sort of tracking device to keep the telescope pointing at the same point in the sky as the earth rotates beneath it.

Based on this description, what are the amateur astronomer's fundamental objectives in choosing a telescope? What are the means objectives? Structure these objectives into a fundamental-objectives hierarchy and a means-objectives network. (*Hint:* If you feel the need for more information, look in your library for recent issues of *Astronomy* magazine or *Sky and Telescope,* two publications for amateur astronomers.)

3.18 Consider the following situations that involve multiple objectives:

a Suppose you want to go out for dinner. What are your fundamental objectives? Create a fundamental-objectives hierarchy.

b Suppose you are trying to decide where to go for a trip over spring break. What are your fundamental objectives? What are your means objectives?

c You are about to become a parent (surprise!), and you have to choose a name for your child. What are important objectives to consider in choosing a name?

d Think of any other situation where choices involve multiple objectives. Create a fundamental-objectives hierarchy and a means-objectives network.

3.19 Thinking about fundamental objectives and means objectives is relatively easy when the decision context is narrow (buying a telescope, renting an apartment, choosing a restaurant for dinner). But when you start thinking about your strategic objectives—objectives in the context of what you choose to do with your life or your career—the process becomes more difficult. Spend some time thinking about your fundamental strategic objectives. What do you want to accomplish in your life or in your career? Why are these objectives important? Try to create a fundamental-objectives hierarchy and a means-objectives network for yourself.

If you succeed in this problem, you will have achieved a deeper level of self-knowledge than most people have, regardless of whether they use decision analysis. That knowledge can be of great help to you in making important decisions, but you should revisit your fundamental objectives from time to time; they might change!

3.20 Occasionally a decision is sensitive to the way it is structured. The following problem shows that leaving out an important part of the problem can affect the way we view the situation.

a Imagine that a close friend has been diagnosed with heart disease. The physician recommends bypass surgery. The surgery should solve the problem. When asked about the risks, the physician replies that a few individuals die during the operation, but most recover and the surgery is a complete success. Thus, your friend can (most likely) anticipate a longer and healthier life after the surgery. Without surgery, your friend will have a shorter and gradually deteriorating life. Assuming that your friend's objective is to maximize the quality of her life, diagram this decision with both an influence diagram and a decision tree.

b Suppose now that your friend obtains a second opinion. The second physician suggests that there is a third possible outcome: Complications from surgery can develop which will require long and painful treatment. If this happens, the eventual outcome can be either a full recovery, partial recovery (restricted to a wheelchair until death), or death within a few months. How does this change the decision tree and influence diagram that you created in part a? Draw the decision tree and influence diagram that represent the situation after hearing from both physicians. Given this new structure, does surgery look more or less positive than it did in part a? [For more discussion of this problem, see von Winterfeldt and Edwards (1986, pp. 8–14).]

c Construct an attribute scale for the patient's quality of life. Be sure to include levels that relate to all of the possible outcomes from surgery.

3.21 Create an influence diagram and a decision tree for the difficult decision problem that you described in Problem 1.9. What are your objectives? Construct attribute scales if necessary. Be sure that all aspects of your decision model pass the clarity test.

3.22   To be, or not to be, that is the question:
Whether 'tis nobler in the mind to suffer
The slings and arrows of outrageous fortune
Or to take arms against a sea of troubles,
And by opposing end them. To die—to sleep—
No more; and by a sleep to say we end
The heartache, and the thousand natural shocks
That flesh is heir to. 'Tis a consummation
Devoutly to be wished. To die—to sleep.
To sleep—perchance to dream: ay, there's the rub!
For in that sleep of death what dreams may come
When we have shuffled off this mortal coil,
Must give us pause. There's the respect
That makes calamity of so long life.
For who would bear the whips and scorns of time,
the oppressor's wrong, the proud man's contumely,
The pangs of despised love, the law's delay,
The insolence of office, and the spurns
That patient merit of the unworthy takes,
When he himself might his quietus make
With a bare bodkin? Who would these fardels bear,
To grunt and sweat under a weary life,
But that the dread of something after death—
The undiscovered country, from whose bourn
No traveller returns—puzzles the will,
And makes us rather bear those ills we have
Than fly to others that we know not of?

—Hamlet, Act III, Scene 1

Describe Hamlet's decision. What are his choices? What risk does he perceive? Construct a decision tree for Hamlet.

3.23   On July 3, 1988, the USS *Vincennes* was engaged in combat in the Persian Gulf. On the radar screen a blip appeared that signified an incoming aircraft. After repeatedly asking the aircraft to identify itself with no success, it appeared that the aircraft might be a hostile Iranian F-14 attacking the *Vincennes*. Captain Will Rogers had little time to make his decision. Should he issue the command to launch a missile and destroy the plane? Or should he wait for positive identification? If he waited too long and the plane was indeed hostile, then it might be impossible to avert the attack and danger to his crew.

Captain Rogers issued the command, and the aircraft was destroyed. It was reported to be an Iranian Airbus airliner carrying 290 people. There were no survivors.

What are Captain Rogers's fundamental objectives? What risks does he face? Draw a decision tree representing his decision.

3.24 Reconsider the research-and-development decision in Figure 3.32. If you decide to continue the project, you will have to come up with the $2 million this year (Year 1). Then there will be a year of waiting (Year 2) before you know if the patent is granted. If you decide to license the technology, you would receive the $25 million distributed as $5 million per year beginning in Year 3. On the other hand, if you decide to sell the product directly, you will have to invest $5 million in each of Years 3 and 4 (to make up the total investment of $10 million). Your net proceeds from selling the product, then, would be evenly distributed over Years 5 through 9.

Assuming an interest rate of 15%, calculate the NPV at the end of each branch of the decision tree.

3.25 When you purchase a car, you may consider buying a brand-new car or a used one. A fundamental trade-off in this case is whether you pay repair bills (uncertain at the time you buy the car) or make loan payments that are certain.

Consider two cars, a new one that costs $15,000 and a used one with 75,000 miles for $5500. Let us assume that your current car's value and available cash amount to $5500, so you could purchase the used car outright or make a down payment of $5500 on the new car. Your credit union is willing to give you a five-year, 10% loan on the $9500 difference if you buy the new car; this loan will require monthly payments of $201.85 per month for five years. Maintenance costs are expected to be $100 for the first year and $300 per year for the second and third years.

After taking the used car to your mechanic for an evaluation, you learn the following. First, the car needs some minor repairs within the next few months, including a new battery, work on the suspension and steering mechanism, and replacement of the belt that drives the water pump. Your mechanic has estimated that these repairs will cost $150.00. Considering the amount you drive, the tires will last another year but will have to be replaced next year for about $200. Beyond that, the mechanic warns you that the cooling system (radiator and hoses) may need to be repaired or replaced this year or next and that the brake system may need work. These and other repairs that an older car may require could lead you to pay anywhere from $500 to $2500 in each of the next three years. If you are lucky, the repair bills will be low or will come later. But you could end up paying a lot of money when you least expect it.

Draw a decision tree for this problem. To simplify it, look at the situation on a yearly basis for three years. If you buy the new car, you can anticipate cash outflows of $12 \times \$201.85 = \$2422.20$ plus maintenance costs. For the used car, some of the repair costs are known (immediate repairs this year, tires next year), but we must model the uncertainty associated with the rest. In addition to the known repairs, assume that in each year there is a 20% chance that these uncertain repairs will be $500, a 20% chance they will be $2500, and a 60% chance they will be $1500. (*Hint:* You need 3 chance nodes: one for each year!)

To even the comparison of the two cars, we must also consider their values after three years. If you buy the new car, it will be worth approximately $8000, and you will still owe $4374. Thus, its net salvage value will be $3626. On the other hand, you would own the used car free and clear (assuming you can keep up with the repair bills!), and it would be worth approximately $2000.

Include all of the probabilities and cash flows (outflows until the last branch, then an inflow to represent the car's salvage value) in your decision tree. Calculate the net values at the ends of the branches.

3.26 PrecisionTree will convert any influence diagram into the corresponding decision tree with the click of a button. This provides an excellent opportunity to explore the meaning of arrows in an influence diagram because you can easily see the effect of adding or deleting an arrow in the corresponding decision tree. Because we are only concerned with the structural effect of arrows, we will not input numerical values.

a Construct the influence diagram in Exercise 3.11(a) in PrecisionTree. Convert the influence diagram to a decision tree by clicking on the influence diagram's name (straddling cells A1 and B1) and clicking on **Convert To Tree** in the influence diagram settings dialog box.

b Add an arrow from the "Weather" chance node to the "Party" decision node and convert the influence diagram into a decision tree. Why would the decision tree change in this way?

c Construct the influence diagram in Exercise 3.11(b) in PrecisionTree. Convert the influence diagram to a decision tree. How would the decision tree change if the arrow started at the "Party" decision node and went into the "Forecast" node? What if, in addition, the arrow started at the "Forecast" chance node and went into the "Weather" chance node?

CASE STUDIES

COLD FUSION

On March 23, 1989, Stanley Pons and Martin Fleischmann announced in a press conference at the University of Utah that they had succeeded in creating a small-scale nuclear fusion reaction in a simple apparatus at room temperature. They called the process "cold fusion." Although many details were missing from their description of the experiment, their claim inspired thoughts of a cheap and limitless energy supply, the raw material for which would be ocean water. The entire structure of the world economy potentially would change.

For a variety of reasons, Pons and Fleischmann were reluctant to reveal all of the details of their experiment. If their process really were producing energy from a fusion reaction, and any commercial potential existed, then they could become quite wealthy. The state of Utah also considered the economic possibilities and even went so far as to approve $5 million to support cold-fusion research. Congressman Wayne Owens from Utah introduced a bill in the U.S. House of Representatives requesting $100 million to develop a national cold-fusion research center at the University of Utah campus.

But were the results correct? Experimentalists around the world attempted to replicate Pons and Fleischmann's results. Some reported success, while many others did not. A team at Texas A&M claimed to have detected neutrons, the telltale sign of fusion. Other teams detected excess heat as had Pons and Fleischmann. Many experiments failed to confirm a fusion reaction, however, and several physicists claimed that the Utah pair simply had made mistakes in their measurements.

Questions

1   Consider the problem that a member of the U.S. Congress would have in deciding whether to vote for Congressman Owens's bill. What alternatives are available? What are the key uncertainties? What objectives might the Congress member consider? Structure the decision problem using an influence diagram and a decision tree.

2   A key part of the experimental apparatus was a core of palladium, a rare metal. Consider a speculator who is thinking of investing in palladium in response to the announcement. Structure the investor's decision. How does it compare to the decision in Question 1?

Sources: "Fusion in a Bottle: Miracle or Mistake," *Business Week,* May 8, 1989, pp. 100–110; "The Race for Fusion," *Newsweek,* May 8, 1989, pp. 49–54.

PRESCRIBED FIRE

Using fire in forest management sounds contradictory. Prescribed fire, however, is an important tool for foresters, and a recent article describes how decision analysis is used to decide when, where, and what to burn. In one example, a number of areas in the Tahoe National Forest in California had been logged and were being prepared for replanting. Preparation included prescribed burning, and two possible treatments were available: burning the slash as it lay on the ground, or "yarding of unmerchantable material" (YUM) prior to burning. The latter treatment involves using heavy equipment to pile the slash. YUM reduces the difficulty of controlling the burn but costs an additional $100 per acre. In deciding between the two treatments, two uncertainties were considered critical. The first was how the fire would behave under each scenario. For example, the fire could be fully successful, problems could arise which could be controlled eventually, or the fire could escape, entailing considerable losses. Second, if problems developed, they could result in high, low, or medium costs.

Questions

1   What do you think the U.S. Forest Service's objectives should be in this decision? In the article, only one objective was considered, minimizing cost (including costs associated with an escaped fire and the damage it might do). Do you think this is a reasonable criterion for the Forest Service to use? Why or why not?

2   Develop an influence diagram and a decision tree for this situation. What roles do the two diagrams play in helping to understand and communicate the structure of this decision?

Source: D. Cohan, S. Haas, D. Radloff, and R. Yancik (1984) "Using Fire in Forest Management: Decision Making under Uncertainty." *Interfaces,* 14, 8–19.

THE SS *KUNIANG*

In the early 1980s, New England Electric System (NEES) was deciding how much to bid for the salvage rights to a grounded ship, the SS *Kuniang*. If the bid were successful, the ship could be repaired and fitted out to haul coal for its power-generation stations. The value of doing so, however, depended on the outcome of a Coast Guard judgment about the salvage value of the ship. The Coast Guard's judgment involved an obscure law regarding domestic shipping in coastal waters. If the judgment indicated a low salvage value, then NEES would be able to use the ship for its shipping needs. If the judgment were high, the ship would be considered ineligible for domestic shipping use unless a considerable amount of money was spent in fitting her with fancy equipment. The Coast Guard's judgment would not be known until after the winning bid was chosen, so there was considerable risk associated with actually buying the ship as a result of submitting the winning bid. If the bid failed, the alternatives included purchasing a new ship for $18 million or a tug barge combination for $15 million. One of the major issues was that the higher the bid, the more likely that NEES would win. NEES judged that a bid of $3 million would definitely not win, whereas a bid of $10 million definitely would win. Any bid in between was possible.

Questions

1   Draw an influence diagram and a decision tree for NEES's decision.

2   What roles do the two diagrams play in helping to understand and communicate the structure of this decision? Do you think one representation is more appropriate than the other? Why?

Source: David E. Bell (1984) "Bidding for the SS *Kuniang*." *Interfaces*, 14, 17–23.

REFERENCES

Decision structuring as a topic of discussion and research is relatively new. Traditionally the focus has been on modeling uncertainty and preferences and solution procedures for specific kinds of problems. Recent discussions of structuring include von Winterfeldt and Edwards (1986, Chapter 2), Humphreys and Wisudha (1987), and Keller and Ho (1989).

The process of identifying and structuring one's objectives comes from Keeney's (1992) *Value-Focused Thinking*. Although the idea of specifying one's objectives clearly as part of the decision process has been accepted for years, Keeney has made this part of decision structuring very explicit. Value-focused thinking captures the ultimate in common sense; if you know what you want to accomplish, you will be able to make choices that help you accomplish those things. Thus, Keeney advocates focusing on values and objectives first, before considering your alternatives. For a more compact description of value-focused thinking, see Keeney (1994).

Relatively speaking, influence diagrams are brand-new on the decision-analysis circuit. Developed by Strategic Decisions Group as a consulting aid in the late seventies, they first appeared in the decision-analysis literature in Howard and Matheson (1984). Bodily (1985) presents an overview of influence diagrams. For more technical details, consult Shachter (1986, 1988) and Oliver and Smith (1989).

The idea of representing a decision with a network has spawned a variety of different approaches beyond influence diagrams. Two in particular are valuation networks (Shenoy, 1992) and sequential decision diagrams (Covaliu and Oliver, 1995). A recent overview of influence diagrams and related network representations of decisions can be found in Matzkevich and Abramson (1995).

Decision trees, on the other hand, have been part of the decision-analysis tool kit since the discipline's inception. The textbooks by Holloway (1979) and Raiffa (1968) provide extensive modeling using decision trees. This chapter's discussion of basic decision trees draws heavily from Behn and Vaupel's (1982) typology of decisions.

The clarity test is another consulting aid invented by Ron Howard and his associates. It is discussed in Howard (1988).

Behn, R. D., and J. D. Vaupel (1982) *Quick Analysis for Busy Decision Makers.* New York: Basic Books.

Bodily, S. E. (1985) *Modern Decision Making.* New York: McGraw-Hill.

Covaliu, Z., and R. Oliver (1995) "Representation and Solution of Decision Problems Using Sequential Decision Diagrams." *Management Science,* 41, in press.

Holloway, C. A. (1979) *Decision Making under Uncertainty: Models and Choices.* Englewood Cliffs, NJ: Prentice Hall.

Howard, R. A. (1988) "Decision Analysis: Practice and Promise." *Management Science,* 34, 679–695.

Howard, R. A., and J. E. Matheson (1984) "Influence Diagrams." In R. Howard and J. Matheson (eds.) *The Principles and Applications of Decision Analysis, Vol. II,* pp. 719–762. Palo Alto, CA: Strategic Decisions Group.

Humphreys, P., and A. Wisudha (1987) "Methods and Tools for Structuring and Analyzing Decision Problems: A Catalogue and Review." Technical Report 87-1. London: Decision Analysis Unit, London School of Economics and Political Science.

Keeney, R. L. (1992) *Value-Focused Thinking.* Cambridge, MA: Harvard University Press.

Keeney, R. L. (1994) "Creativity in Decision Making with Value-Focused Thinking." *Sloan Management Review,* Summer, 33–41.

Keller, L. R., and J. L. Ho (1989) "Decision Problem Structuring." In A. P. Sage (ed.) *Concise Encyclopedia of Information Processing in Systems and Organizations.* Oxford, England: Pergamon Press.

Matzkevich, I., and B. Abramson (1995) "Decision-Analytic Networks in Artificial Intelligence." *Management Science,* 41, 1–22.

Oliver, R. M., and J. Q. Smith (1989) *Influence Diagrams, Belief Nets and Decision Analysis* (Proceedings of an International Conference 1988, Berkeley). New York: Wiley.

Raiffa, H. (1968) *Decision Analysis.* Reading, MA: Addison-Wesley.

Shachter, R. (1986) "Evaluating Influence Diagrams." *Operations Research,* 34, 871–882.

Shachter, R. (1988) "Probabilistic Inference and Influence Diagrams." *Operations Research,* 36, 589–604.

Shenoy, P. (1992) "Valuation-Based Systems for Bayesian Decision Analysis." *Operations Research,* 40, 463–484.

Ulvila, J., and R. B. Brown (1982) "Decision Analysis Comes of Age." *Harvard Business Review,* Sept–Oct, 130–141.

von Winterfeldt, D., and W. Edwards (1986) *Decision Analysis and Behavioral Research.* Cambridge: Cambridge University Press.

EPILOGUE

Toxic Chemicals   The trade-off between economic value and cancer cost can be very complicated and lead to difficult decisions, especially when a widely used substance is found to be carcinogenic. Imposing an immediate ban can have extensive economic consequences. Asbestos is an excellent example of the problem. This material has been in use since Roman times and was used extensively after World War II. However, pioneering research by Dr. Irving Selikoff of the Mt. Sinai School of Medicine showed that breathing asbestos particles can cause lung cancer. This caused the EPA to list it as a hazardous air pollutant in 1972. In 1978, the EPA imposed further restrictions and banned spray-on asbestos insulation. Finally, in the summer of 1989 the EPA announced a plan that would result in an almost total ban of the substance by the year 1996. (*Sources:* "U.S. Orders Virtual Ban on Asbestos." *Los Angeles Times,* July 7, 1989; "Asbestos Widely Used Until Researcher's Warning," *The Associated Press,* July 7, 1989.)

Cold Fusion   At a conference in Santa Fe, New Mexico, at the end of May 1989, Pons and Fleischmann's results were discussed by scientists from around the world. After many careful attempts by the best experimentalists in the world, no consensus was reached. Many researchers reported observing excess heat, while others observed neutrons. Many had observed nothing. With no agreement, research continued.

Over the next year, many labs attempted to replicate Pons and Fleischmann's experiments. The most thorough attempts were made at Caltech and MIT, and both failed to find evidence for a fusion reaction. In what appeared to be the death blow, exactly one year later the journal *Nature* published an article reporting work by Michael Salamon of the University of Utah. Using the electrolytic cells of Pons and Fleischmann, and observing them for several weeks, still no evidence of fusion was observed. In the same issue, *Nature* editor David Lindley wrote an editorial that essentially was an epitaph for cold fusion. In addition, two books by scientist John Huizenga (*Cold Fusion: The Scientific Fiasco of the Century.* Rochester, NY: University of Rochester Press, 1992) and journalist Gary Taubes (*Bad Science: The Short Life and Weird Times of Cold Fusion.* New York: Random House, 1993) have attempted to close the door definitively on cold fusion.

Surprisingly, though, the controversy continues. Although the top-level scientific journals no longer publish their articles, cold-fusion experimenters from around the world continue to hold conferences to report their results, and evidence is growing that some unusual and poorly understood phenomenon is occurring and can be reproduced in carefully controlled laboratory conditions. EPRI (the Electric Power Research Institute) has provided funding for cold-fusion research for several years. In its May/June 1994 cover story, *Technology Review* summarized the collected evidence relating to cold fusion and possible explanations—none consistent with conventional physical theory—of the phenomenon. Undoubtedly, research will continue for some time. Eventually the experimental effects will be confirmed and explained, or the entire enterprise will be debunked for good! (*Sources:* David Lindley (1990) "The Embarrassment of Cold Fusion." *Nature,* 344, 375–376; Robert Pool (1989) "Cold Fusion: End of Act I." *Science,* 244; 1039–1040; Edmund Storms (1994) "Warming Up to Cold Fusion." *Technology Review,* May/June, 20–29.)

CHAPTER 4

Making Choices

I n this chapter, we will learn how to use the details in a structured problem to find a preferred alternative. "Using the details" typically means analysis: making calculations, creating graphs, and examining the results so as to gain insight into the decision. We will see that the kinds of calculations we make are essentially the same in solving decision trees and influence diagrams. We also introduce risk profiles and dominance considerations, ways to make decisions without doing all those calculations.

We begin by studying the analysis of decision models that involve only one objective or attribute. Although most of the examples we give use money as the attribute, it could be anything that can be measured as discussed in Chapter 3. After discussing calculation of expected values and the use of risk profiles for single-attribute decisions, we turn to decisions with multiple attributes and present some simple analytical approaches. The chapter concludes with a discussion of software for doing decision-analysis calculations on personal computers.

Our main example for this chapter is from the famous Texaco-Pennzoil court case.

TEXACO VERSUS PENNZOIL

In early 1984, Pennzoil and Getty Oil agreed to the terms of a merger. But before any formal documents could be signed, Texaco offered Getty a substantially better price, and Gordon Getty, who controlled most of the Getty stock, reneged on the Pennzoil deal and sold to Texaco. Naturally, Pennzoil felt as if it had been dealt with unfairly and immediately filed a lawsuit against Texaco alleging that Texaco had interfered illegally in the Pennzoil-Getty negotiations. Pennzoil won the case; in late 1985, it was awarded $11.1 billion, the largest judgment ever in the United States at that time. A Texas appeals court reduced the judgment by $2 billion, but interest and penalties drove the total back up to $10.3 billion. James Kinnear, Texaco's chief executive officer, had said that Texaco would file for bankruptcy if Pennzoil obtained court permission to secure the judgment by filing liens against Texaco's assets. Furthermore, Kinnear had promised to fight the case all the way to the U.S. Supreme Court if necessary, arguing in part that Pennzoil had not followed Security and Exchange Commission regulations in its negotiations with Getty. In April 1987, just before Pennzoil began to file the liens, Texaco offered to pay Pennzoil $2 billion to settle the entire case. Hugh Liedtke, chairman of Pennzoil, indicated that his advisors were telling him that a settlement between $3 and $5 billion would be fair.

What do you think Liedtke (pronounced "lid-key") should do? Should he accept the offer of $2 billion, or should he refuse and make a firm counteroffer? If he refuses the sure $2 billion, then he faces a risky situation. Texaco might agree to pay $5 billion, a reasonable amount in Liedtke's mind. If he counteroffered $5 billion as a settlement amount, perhaps Texaco would counter with $3 billion or simply pursue further appeals. Figure 4.1 is a decision tree that shows a simplified version of Liedtke's problem.

The decision tree in Figure 4.1 is simplified in a number of ways. First, we assume that Liedtke has only one fundamental objective: maximizing the amount of the settlement. No other objectives need be considered. Also, Liedtke has a more varied set of decision alternatives than those shown. He could counteroffer a variety of possible values in the initial decision, and in the second decision, he could counteroffer some amount between $3 and $5 billion. Likewise, Texaco's counteroffer, if it

Figure 4.1

Hugh Liedtke's decision in the Texaco-Pennzoil affair.

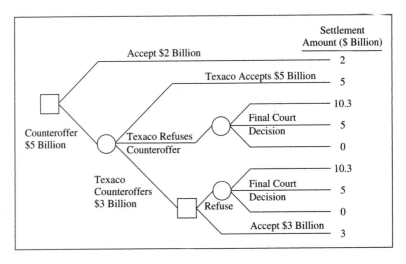

makes one, need not be exactly $3 billion. The outcome of the final court decision could be anything between zero and the current judgment of $10.3 billion. Finally, we have not included in our model of the decision anything regarding Texaco's option of filing for bankruptcy.

Why all of the simplifications? A straightforward answer (which just happens to have some validity) is that for our purposes in this chapter we need a relatively simple decision tree to work with. But this is just a pedagogical reason. If we were to try to analyze Liedtke's problem in all of its glory, how much detail should be included? As you now realize, all of the relevant information should be included, and the model should be constructed in a way that makes it easy to analyze. Does our representation accomplish this? Let us consider the following points.

1 *Liedtke's objective.* Certainly maximizing the amount of the settlement is a valid objective. The question is whether other objectives, such as minimizing attorney fees or improving Pennzoil's public image, might also be important. Although Liedtke may have other objectives, the fact that the settlement can range all the way from zero to $10.3 billion suggests that this objective will swamp any other concerns.

2 *Liedtke's initial counteroffer.* The counteroffer of $5 billion could be replaced by an offer for another amount, and then the decision tree reanalyzed. Different amounts may change the chance of Texaco accepting the counteroffer. At any rate, other possible counteroffers are easily dealt with.

3 *Liedtke's second counteroffer.* Other possible offers could be built into the tree, leading to a Texaco decision to accept, reject, or counter. The reason for leaving these out reflects an impression from the media accounts (especially *Fortune,* May 11, 1987, pp. 50–58) that Kinnear and Liedtke were extremely tough negotiators and that further negotiations were highly unlikely.

4 *Texaco's counteroffer.* The $3 billion counteroffer could be replaced by a fan representing a range of possible counteroffers. It would be necessary to find a "break-even" point, above which Liedtke would accept the offer and below which he would refuse. Another approach would be to replace the $3 billion value with other values, recomputing the tree each time. Thus, we have a variety of ways to deal with this issue.

5 *The final court decision.* We could include more branches, representing additional possible outcomes, or we could replace the three branches with a fan representing a range of possible outcomes. For a first-cut approximation, the possible outcomes we have chosen do a reasonably good job of capturing the uncertainty inherent in the court outcome.

6 *Texaco's bankruptcy option.* A detail left out of the case is that Texaco's net worth is much more than the $10.3 billion judgment. Thus, even if Texaco does file for bankruptcy, Pennzoil probably would still be able to collect. In reality, negotiations can continue even if Texaco has filed for bankruptcy; the purpose of filing is to protect the company from creditors seizing assets while the company proposes a financial reorganization plan. In fact, this is exactly what Texaco needs

to do in order to figure out a way to deal with Pennzoil's claims. In terms of Liedtke's options, however, whether Texaco files for bankruptcy appears to have no impact.

The purpose of this digression has been to explore the extent to which our structure for Liedtke's problem is requisite in the sense of Chapter 1. The points above suggest that the main issues in the problem have been represented in the problem. While it may be necessary to rework the analysis with slightly different numbers or structure later, the structure in Figure 4.1 should be adequate for a first analysis. The objective is to develop a representation of the problem that captures the essential features of the problem so that the ensuing analysis will provide the decision maker with insight and understanding.

One small detail remains before we can solve the decision tree. We need to specify the chances associated with Texaco's possible reactions to the $5 billion counteroffer, and we also need to assess the chances of the various court awards. The probabilities that we assign to the outcome branches in the tree should reflect Liedtke's beliefs about the uncertain events that he faces. For this reason, any numbers that we include to represent these beliefs should be based on what Liedtke has to say about the matter or on information from individuals whose judgments in this matter he would trust. For our purposes, imagine overhearing a conversation between Liedtke and his advisors. Here are some of the issues they might raise:

- Given the tough negotiating stance of the two executives, it could be an even chance (50%) that Texaco will refuse to negotiate further. If Texaco does not refuse, then what? What are the chances that Texaco would accept a $5 billion counteroffer? How likely is this outcome compared to the $3 billion counteroffer from Texaco? Liedtke and his advisors might figure that a counteroffer of $3 billion from Texaco is about twice as likely as Texaco accepting the $5 billion. Thus, because there is already a 50% chance of refusal, there must be a 33% chance of a Texaco counteroffer and a 17% chance of Texaco accepting $5 billion.

- What are the probabilities associated with the final court decision? In the *Fortune* article cited above, Liedtke is said to admit that Texaco could win its case, leaving Pennzoil with nothing but lawyer bills. Thus, there is a significant possibility that the outcome would be zero. Given the strength of Pennzoil's case so far, there is also a good chance that the court will uphold the judgment as it stands. Finally, the possibility exists that the judgment could be reduced somewhat (to $5 billion in our model). Let us assume that Liedtke and his advisors agree that there is a 20% chance that the court will award the entire $10.3 billion and a slightly larger, or 30%, chance that the award will be zero. Thus, there must be a 50% chance of an award of $5 billion.

Figure 4.2 shows the decision tree with these chances included. The chances have been written in terms of probabilities rather than percentages.

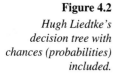

Figure 4.2

Hugh Liedtke's decision tree with chances (probabilities) included.

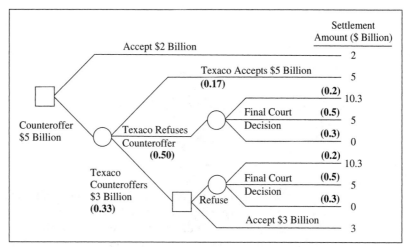

Decision Trees and Expected Monetary Value

One way to choose among risky alternatives is to pick the alternative with the highest *expected value* (EV). When the decision's consequences involve only money, we can calculate the *expected monetary value* (EMV). Finding EMVs when using decision trees is called "folding back the tree" for reasons that will become obvious. (The procedure is called "rolling back" in some texts.) We start at the endpoints of the branches on the far right-hand side and move to the left, (1) calculating expected values (to be defined momentarily) when we encounter a chance node, or (2) choosing the branch with the highest value or expected value when we encounter a decision node. These instructions sound rather cryptic. It is much easier to understand the procedure through a few examples. We will start with a simple example, the double-risk dilemma shown in Figure 4.3.

Recall that a double-risk dilemma is a matter of choosing between two risky alternatives. The situation is one in which you have a ticket that will let you participate in a game of chance (a lottery) that will pay off $10 with a 45% chance, and nothing with a 55% chance. Your friend has a ticket to a different lottery that has a 20% chance of paying $25 and an 80% chance of paying nothing. Your friend has offered to let you have his ticket if you will give him your ticket plus one dollar. Should you agree to the trade and play to win $25, or should you keep your ticket and have a better chance of winning $10?

Figure 4.3 displays your decision situation. In particular, notice that the dollar consequences at the ends of the branches are the net values as discussed in Chapter 3. Thus, if you trade tickets and win, you will have gained a net amount of $24, having paid one dollar to your friend.

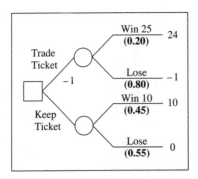

To solve the decision tree using EMV, begin by calculating the expected value of keeping the ticket and playing for $10. This expected value is simply the weighted average of the possible outcomes of the lottery, the weights being the chances with which the outcomes occur. The calculations are

$$\text{EMV(Keep Ticket)} = 0.45(10) + 0.55(0)$$
$$= \$4.5$$

One interpretation of this EMV is that playing this lottery many times would yield an average of approximately $4.50 per game. Calculating EMV for trading tickets gives

$$\text{EMV(Trade Ticket)} = 0.20(24) + 0.80(-1)$$
$$= \$4$$

Now we can replace the chance nodes in the decision tree with their expected values, as shown in Figure 4.4. Finally, choosing between trading and keeping the ticket amounts to choosing the branch with the highest expected value. The double slash through the "Trade Ticket" branch indicates that this branch would not be chosen.

This simple example is only a warm-up exercise. Now let us see how the solution procedure works when we have a more complicated decision problem. Consider Hugh Liedtke's situation as diagrammed in Figure 4.2. Our strategy, as indicated, will be to work from the right-hand side of the tree. First, we will calculate the expected value of the final court decision. The second step will be to decide whether it is better for Liedtke to accept a $3 billion counteroffer from Texaco or to refuse and take a chance on the final court decision. We will do this by comparing the expected value of the judgment with the sure $3 billion. The third step will be to calculate the

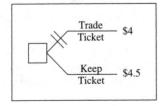

expected value of making the $5 billion counteroffer, and finally we will compare this expected value with the sure $2 billion that Texaco is offering now.

The expected value of the court decision is the weighted average of the possible outcomes:

$$EMV(\text{Court Decision}) = [P(\text{Award} = 10.3) \times 10.3] + [P(\text{Award} = 5) \times 5]$$
$$+ [P(\text{Award} = 0) \times 0]$$
$$= [0.2 \times 10.3] + [0.5 \times 5] + [0.3 \times 0]$$
$$= 4.56$$

We replace both uncertainty nodes representing the court decision with this expected value, as in Figure 4.5. Now, comparing the two alternatives of accepting and refusing Texaco's $3 billion counteroffer, it is obvious that the expected value of $4.56 billion is greater than the certain value of $3 billion, and hence the slash through the "Accept $3 Billion" branch.

To continue folding back the decision tree, we replace the decision node with the preferred alternative. The decision tree as it stands after this replacement is shown in Figure 4.6. The third step is to calculate the expected value of the alternative "Counteroffer $5 Billion." This expected value is

$$EMV(\text{Counteroffer \$5 Billion}) = [P(\text{Texaco Accepts}) \times 5]$$
$$+ [P(\text{Texaco Refuses}) \times 4.56]$$
$$+ [P(\text{Texaco Counteroffers}) \times 4.56]$$
$$= [0.17 \times 5] + [0.50 \times 4.56] + [0.33 \times 4.56]$$
$$= 4.63$$

Replacing the chance node with its expected value results in the decision tree shown in Figure 4.7. Comparing the values of the two branches, it is clear that the expected value of $4.63 billion is preferred to the $2 billion offer from Texaco. According to

Figure 4.5

Hugh Liedtke's decision tree after calculating expected value of court decision.

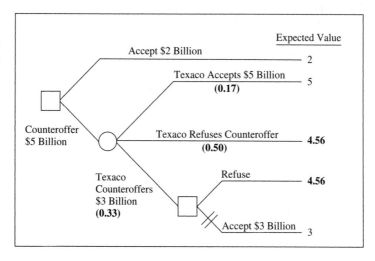

Figure 4.6

Hugh Liedtke's decision tree after decision node replaced with expected value of preferred alternative.

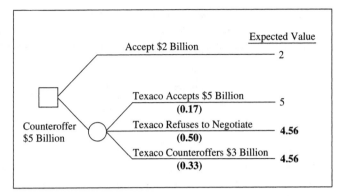

Figure 4.7

Hugh Liedtke's decision tree after original tree completely folded back.

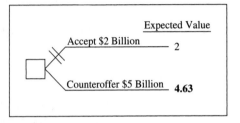

this solution, which implies that decisions should be made by comparing expected values, Liedtke should turn down Texaco's offer but counteroffer a settlement of $5 billion. If Texaco turns down the $5 billion and makes another counteroffer of $3 billion, Liedtke should refuse the $3 billion and take his chances in court.

We went through this decision in gory detail so that you could see clearly the steps involved. In fact, in solving a decision tree, we usually do not redraw the tree at each step, but simply indicate on the original tree what the expected values are at each of the chance nodes and which alternative is preferred at each decision node. The solved decision tree for Liedtke would look like the tree shown in Figure 4.8, which shows all of the details of the solution. Expected values for the chance nodes are placed above the nodes. The 4.56 above the decision node indicates that if Liedtke gets to this decision point, he should refuse Texaco's offer and take his chances in court for an expected value of $4.56 billion. The decision tree also shows that his best current choice is to make the $5 billion counteroffer with an expected payoff of $4.63 billion.

The decision tree shows clearly what Liedtke should do if Texaco counteroffers $3 billion: He should refuse. This is the idea of a contingent strategy. If a particular course of events occurs (Texaco's counteroffer), then there is a specific course of action to take (refuse the counteroffer). Moreover, in deciding whether to accept Texaco's current $2 billion offer, Liedtke must know what he will do in the event that Texaco returns with a counteroffer of $3 billion. This is why the decision tree is solved backward. In order to make a good decision at the current time, we have to know what the appropriate contingent strategies are in the future.

Figure 4.8

*Hugh Liedtke's solved
decision tree.*

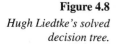

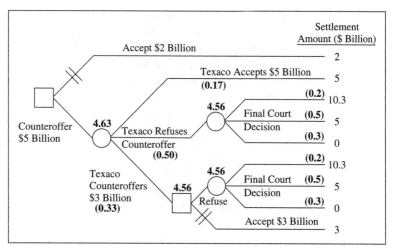

Solving Influence Diagrams: Overview

Solving decision trees is straightforward, and EMVs for small trees can be calculated by hand relatively easily. The procedure for solving an influence diagram, though, is somewhat more complicated. Fortunately, computer programs such as PrecisionTree are available to do the calculations. In this short section we give an overview of the issues involved in solving an influence diagram. For interested readers, the following optional section goes through a complete solution of the influence diagram of the Texaco-Pennzoil decision.

While influence diagrams appear on the surface to be rather simple, much of the complexity is hidden. Our first step is to take a close look at how an influence diagram translates information into an internal representation. An influence diagram "thinks" about a decision in terms of a symmetric expansion of the decision tree from one node to the next.

Figure 4.9

Umbrella problem.

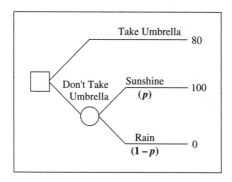

For example, suppose we have the basic decision tree shown in Figure 4.9, which represents the "umbrella problem" (see Exercise 3.9). The issue is whether or not to take your umbrella. If you do not take the umbrella, and it rains, your good clothes (and probably your day) are ruined, and the consequence is zero (units of satisfaction). However, if you do not take the umbrella and the sun shines, this is the best of all possible consequences with a value of 100. If you decide to take your umbrella, your clothes will not get spoiled. However, it is a bit of a nuisance to carry the umbrella around all day. Your consequence is 80, between the other two values.

If we were to represent this problem with an influence diagram, it would look like the diagram in Figure 4.10. Note that it does not matter whether the sun shines or not if you take the umbrella. If we were to reconstruct exactly how the influence diagram "thinks" about the umbrella problem in terms of a decision tree, the representation would be that shown in Figure 4.11. Note that the uncertainty node on the "Take Umbrella" branch is an unnecessary node. The payoff is the same regardless of the weather. In a decision-tree model, we can take advantage of this fact by not even drawing the unnecessary node. Influence diagrams, however, use the symmetric decision tree, even though this may require unnecessary nodes (and hence unnecessary calculations).

With an understanding of the influence diagram's internal representation, we can talk about how to solve an influence diagram. The procedure essentially solves the symmetric decision tree, although the terminology is somewhat different. Nodes are *reduced;* reduction amounts to calculating expected values for chance nodes and choosing the largest expected value at decision nodes, just as we did with the decision tree. Moreover, also parallel with the decision-tree procedure, as nodes are reduced, they are removed from the diagram. Thus, solving the influence diagram in Figure 4.10 would require first reducing the "Weather" node (calculating the expected values) and then reducing the "Take Umbrella?" node by choosing the largest expected value.

Figure 4.10

Influence diagram of the umbrella problem.

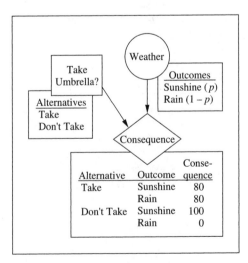

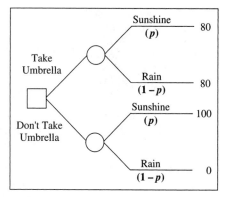

Figure 4.11

How the influence diagram "thinks" about the umbrella problem.

Solving Influence Diagrams: The Details (Optional)

Consider the Texaco-Pennzoil case in influence-diagram form, as shown in Figure 4.12. This diagram shows the tables of alternatives, outcomes (with probabilities), and consequences that are contained in the nodes. The consequence table in this case is too complicated to put into Figure 4.12. We will work with it later in great detail, but if you want to see it now, it is displayed in Table 4.1.

Figure 4.12 needs explanation. The initial decision is whether to accept Texaco's offer of $2 billion. Within this decision node a table shows that the available alternatives are to accept the offer or make a counteroffer. Likewise, under the "Pennzoil Reaction" node is a table that lists "Accept 3" and "Refuse" as alternatives. The chance node "Texaco Reaction" contains a table showing the probabilities of Texaco accepting a counteroffer of $5 billion, making an offer of $3 billion, or refusing to

Figure 4.12

Influence diagram for Liedtke's decision.

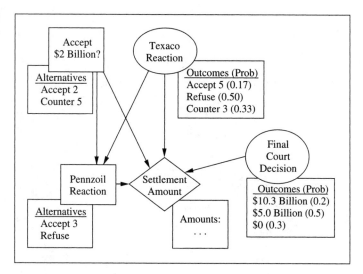

Table 4.1

Consequence table for the influence diagram of Liedtke's decision.

Accept $2 Billion?	Texaco Reaction ($ Billion)	Pennzoil Reaction ($ Billion)	Final Court Decision ($ Billion)	Settlement Amount ($ Billion)
Accept 2	Accept 5	Accept 3	10.3	2.0
			5	2.0
			0	2.0
		Refuse	10.3	2.0
			5	2.0
			0	2.0
	Offer 3	Accept 3	10.3	2.0
			5	2.0
			0	2.0
		Refuse	10.3	2.0
			5	2.0
			0	2.0
	Refuse	Accept 3	10.3	2.0
			5	2.0
			0	2.0
		Refuse	10.3	2.0
			5	2.0
			0	2.0
Offer 5	Accept 5	Accept 3	10.3	5.0
			5	5.0
			0	5.0
		Refuse	10.3	5.0
			5	5.0
			0	5.0
	Offer 3	Accept 3	10.3	3.0
			5	3.0
			0	3.0
		Refuse	10.3	10.3
			5	5.0
			0	0.0
	Refuse	Accept 3	10.3	10.3
			5	5.0
			0	0.0
		Refuse	10.3	10.3
			5	5.0
			0	0.0

negotiate. Finally, the "Final Court Decision" node has a table with its outcomes and associated probabilities.

The thoughtful reader should have an immediate reaction to this. After all, whether Texaco reacts depends on whether Liedtke makes his $5 billion counteroffer in the first place! Shouldn't there be an arrow from the decision node "Accept $2 Billion" to the "Texaco Reaction" node? The answer is yes, there could be such an arrow, but it is unnecessary and would only complicate matters. The reason is that, as with the umbrella example above, the influence diagram "thinks" in terms of a symmetric expansion of the decision tree. Figure 4.13 shows a portion of the tree that deals with Liedtke's initial decision and Texaco's reaction. An arrow in Figure 4.12 from "Accept $2 Billion" to "Texaco Reaction" would indicate that the decision made (accepting or rejecting the $2 billion) would affect the chances associated with Texaco's reaction to a counteroffer. But the uncertainty about Texaco's response to a $5 billion counteroffer does not depend on whether Liedtke accepts the $2 billion. Essentially, the influence diagram is equivalent to a decision tree that is symmetric.

For similar reasons, there are no arrows between "Final Court Decision" and the other three nodes. If some combination of decisions comes to pass so that Pennzoil and Texaco agree to a settlement, it does not matter what the court decision would be. The influence diagram implicitly includes the "Final Court Decision" node with the agreed-upon settlement regardless of the "phantom" court outcome.

How is all of this finally resolved in the influence-diagram representation? Everything is handled in the consequence node. This node contains a table that gives Liedtke's settlement for every possible combination of decisions and outcomes. That table (Table 4.1) shows that the settlement is $2 billion if Liedtke accepts the current offer, regardless of the other outcomes. It also shows that if Liedtke counteroffers $5 billion and Texaco accepts, then the settlement is $5 billion regardless of the court decision or Pennzoil's reaction (neither of which have any impact if Texaco accepts the $5 billion). The table also shows the details of the

Figure 4.13

How the influence diagram "thinks" about the Texaco-Pennzoil case.

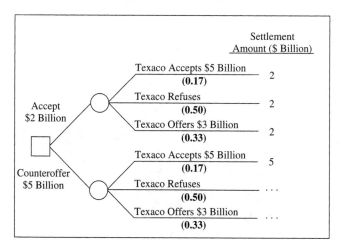

court outcomes if either Texaco refuses to negotiate after Liedtke's counteroffer or if Liedtke refuses a Texaco counteroffer. And so on. The table shows exactly what the payoff is to Pennzoil under all possible combinations. The column headings in Table 4.1 represent nodes that are predecessors of the value node. In this case, both decision nodes and both chance nodes are included because all are predecessors of the value node. We can now discuss how the algorithm for solving an influence diagram proceeds. Take the Texaco-Pennzoil diagram as drawn in Figure 4.12. As mentioned above, our strategy will be to *reduce nodes* one at a time. The order of reduction is reminiscent of our solution in the case of the decision tree. The first node reduced is "Final Court Decision," resulting in the diagram in Figure 4.14. In this first step, expected values are calculated using the "Final Court Decision" probabilities, which yields Table 4.2. All combinations of decisions and possible outcomes of Texaco's reaction are shown. For example, if Liedtke counteroffers $5 billion and Texaco refuses to negotiate, the expected value of $4.56 billion is listed regardless of the decision in the "Pennzoil Reaction" node (because that decision is meaningless if Texaco initially refuses to negotiate). If Liedtke accepts the $2 billion offer, the expected value is listed as $2 billion, regardless of other outcomes. (Of course, there is nothing uncertain about this outcome; the value that we know will happen is the expected value.) If Liedtke offers 5, Texaco offers 3, and finally Liedtke refuses to continue negotiating, then the expected value is given as 4.56. And so on.

The next step is to reduce the "Pennzoil Reaction" node. The resulting influence diagram is shown in Figure 4.15. Now the table in the consequence node (Table 4.3) reflects the decision that Liedtke should choose the alternative with the highest expected value (refuse to negotiate) if Texaco makes the counteroffer of $3 billion. Thus, the table now says that, if Liedtke offers $5 billion and Texaco either refuses to negotiate or counters with $3 billion, the expected value is $4.56 billion. If Texaco accepts the $5 billion counteroffer, the expected value is $5 billion, and if Liedtke accepts the current offer, the expected value is $2 billion. (Again, there is nothing uncertain about these values; the expected value in these cases is just the value that we know will occur.)

Figure 4.14

First step in solving the influence diagram.

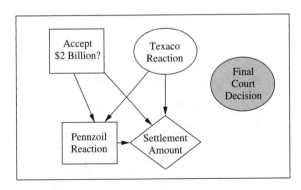

The third step is to reduce the "Texaco Reaction" node, as shown in Figure 4.16. As with the first step, this involves taking the table of consequences (now expected values) within the "Settlement Amount" node and calculating expected values again. The resulting table has only two entries (Table 4.4). The expected value of Liedtke accepting $2 billion is just $2 billion, and the expected value of countering with $5 billion is $4.63 billion.

The fourth and final step is simply to figure out which decision is optimal in the "Accept $2 Billion?" node and to record the result. This final step is shown in Figure 4.17. The table associated with the decision node indicates that Liedtke's optimal choice is to counteroffer $5 billion. The payoff table now contains only one value, $4.63 billion, the expected value of the optimal decision.

Reviewing the procedure, you should be able to see that it followed basically the same steps that we followed in folding back the decision tree.

Table 4.2

Table for Liedtke's decision after reducing the "Final Court Decision" node.

Accept $2 Billion?	Texaco Reaction	Pennzoil Reaction	Expected Value ($ Billion)
Accept 2	Accept 5	Accept 3	2
		Refuse	2
	Offer 3	Accept 3	2
		Refuse	2
	Refuse	Accept 3	2
		Refuse	2
Offer 5	Accept 5	Accept 3	5
		Refuse	5
	Offer 3	Accept 3	3
		Refuse	4.56
	Refuse	Accept 3	4.56
		Refuse	4.56

Figure 4.15

Second step in solving the influence diagram.

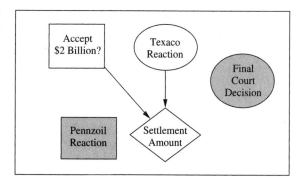

Table 4.3

Table for Liedtke's decision after reducing "Final Court Decision" and "Pennzoil Reaction" nodes.

Accept $2 Billion?	Texaco Reaction	Expected Value ($ Billion)
Accept 2	Accept 5	2.00
	Offer 3	2.00
	Refuse	2.00
Offer 5	Accept 5	5.00
	Offer 3	4.56
	Refuse	4.56

Table 4.4

Table for Liedtke's decision after reducing "Final Court Decision," "Pennzoil Reaction," and "Texaco Reaction" nodes.

Accept $2 Billion?	Expected Value ($ Billion)
Accept 2	2.00
Offer 5	4.63

Figure 4.16

Third step in solving the influence diagram.

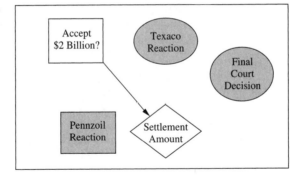

Figure 4.17

Final step in solving the influence diagram.

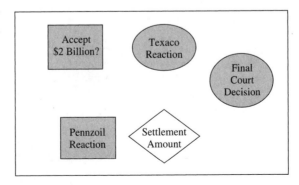

Solving Influence Diagrams: An Algorithm (Optional)

The example above should provide some insight into how influence diagrams are solved. Fortunately, you will not typically have to solve influence diagrams by hand; computer programs are available to accomplish this. It is worthwhile, however, to spend a few moments describing the procedure that is used to solve influence diagrams. A set procedure for solving a problem is called an *algorithm*. You have already learned the algorithm for solving a decision tree (the folding-back procedure). Now let us look at an algorithm for solving influence diagrams.

1 First, we simply clean up the influence diagram to make sure it is ready for solution. Check to make sure the influence diagram has only one consequence node (or a series of consequence nodes that feed into one "super" consequence node) and that there are no cycles. If your diagram does not pass this test, you must fix it before it can be solved. In addition, if any nodes other than the consequence node have arrows into them but not out of them, they can be eliminated. Such nodes are called *barren nodes* and have no effect on the decision that would be made. Replace any intermediate-calculation nodes with chance nodes. (This includes any consequence nodes that feed into a "super" consequence node representing a higher-level objective in the objectives hierarchy.) For each possible combination of the predecessor node outcomes, such a node has only one outcome that happens with probability 1.

2 Next, look for any chance nodes that (a) directly precede the consequence node and (b) do *not* directly precede any other node. Any such chance node found should be reduced by calculating expected values. The consequence node then inherits the predecessors of the reduced nodes. (That is, any arrows that went into the node you just reduced should be redrawn to go into the consequence node.)

This step is just like calculating expected values for chance nodes at the far right-hand side of a decision tree. You can see how this step was implemented in the Texaco-Pennzoil example. In the original diagram, Figure 4.12, the "Final Court Decision" node is the only chance node that directly precedes the consequence node *and* does not precede any decision node. Thus it is reduced by the expected-value procedure, resulting in Table 4.2. The consequence node does not inherit any new direct predecessors as a result of this step because "Final Court Decision" has no direct predecessors.

3 Next, look for a decision node that (a) directly precedes the consequence node and (b) has as predecessors all of the other direct predecessors of the consequence node. *If you do not find any such decision node, go directly to Step 5.* If you find such a decision node, you can reduce it by choosing the optimum value. When decision nodes are reduced, the consequence node does not inherit any new predecessors. This step may create some barren nodes, which can be eliminated from the diagram.

This step is like folding a decision tree back through a decision node at the far right-hand side of the tree. In the Texaco-Pennzoil problem, this step was implemented when we reduced "Pennzoil Reaction." In Figure 4.14, this node satisfies the criteria for reduction because it directly precedes the consequence node, and the other nodes that directly precede the consequence node also precede "Pennzoil Reaction." In reducing this node, we choose the option for "Pennzoil Reaction" that gives the highest expected value, and as a result we obtain Table 4.3. No barren nodes are created in this step.

4 Return to Step 2 and continue until the influence diagram is completely solved (all nodes reduced). This is just like working through a decision tree until all of the nodes have been processed from right to left.

5 You arrived at this step after reducing all possible chance nodes (if any) and then not finding any decision nodes to reduce. How could this happen? Consider the influence diagram of the hurricane problem in Figure 3.12. None of the chance nodes satisfy the criteria for reduction, and the decision node also cannot be reduced. In this case, one of the arrows between chance nodes must be reversed. This is a procedure that requires probability manipulations through the use of Bayes' theorem (Chapter 7). We will not go into the details of the calculations here because most of the simple influence diagrams that you might be tempted to solve by hand will not require arrow reversals.

Finding an arrow to reverse is a delicate process. First, find the correct chance node. The criteria are that (a) it directly precedes the consequence node and (b) it does not directly precede any decision node. Call the selected node A. Now look at the arrows out of node A. Find an arrow from A to chance node B (call it A → B) such that there is no other way to get from A to B by following arrows. The arrow A → B can be reversed using Bayes' theorem. Afterward, both nodes inherit each other's direct predecessors *and* keep their own direct predecessors.

After reversing an arrow, return to Step 2 and continue until the influence diagram is solved. (More arrows may need to be reversed before a node can be reduced, but that only means that you may come back to Step 5 one or more times in succession.)

This description of the influence-diagram solution algorithm is based on the complete (and highly technical) description given in Shachter (1986). The intent is not to present a "cookbook" for solving an influence diagram because, as indicated, virtually all but the simplest influence diagrams will be solved by computer. The description of the algorithm, however, is meant to show the parallels between the influence-diagram and decision-tree solution procedures.

Risk Profiles

The idea of expected value is appealing, and comparing two alternatives on the basis of their EMVs is straightforward. For example, Liedtke's expected values are $2 billion and $4.63 billion for his two immediate alternatives. But you might have noticed that these two numbers are not exactly perfect indicators of what might happen. In particular, suppose that Liedtke decides to counteroffer $5 billion: He might end up with $10.3 billion, $5 billion, or nothing, given our simplification of the situation. Moreover, the interpretation of EMV as the average amount that would be obtained by "playing the game" a large number of times is not appropriate here. The "game" in this case amounts to suing Texaco—not a game that Pennzoil will play many times!

That Liedtke could come away from his dealings with Texaco with nothing indicates that choosing to counteroffer is a somewhat risky alternative. In later chapters we will look at the idea of risk in more detail. For now, however, we can intuitively grasp the relative riskiness of alternatives by studying their *risk profiles.*

A risk profile is a graph that shows the chances associated with possible consequences. Each risk profile is associated with a *strategy,* a particular immediate alternative, as well as specific alternatives in future decisions. For example, the risk profile for the "Accept $2 Billion" alternative is shown in Figure 4.18. There is a 100% chance that Liedtke will end up with $2 billion. The risk profile for the strategy "Counteroffer $5 Billion; Refuse Texaco Counteroffer" is somewhat more complicated and is shown in Figure 4.19. There is a 58.5% chance that the eventual settlement is $5 billion, a 16.6% chance of $10.3 billion, and a 24.9% chance of nothing. These numbers are easily calculated. For example, take the $5 billion amount. This can happen in three different ways. There is a 17% chance that it happens because Texaco accepts. There is a 25% chance that it happens because Texaco refuses and the judge awards $5 billion. (That is, there is a 50% chance that Texaco refuses times a 50% chance that the award is $5 billion.) Finally, the chances are 16.5% that the settlement is $5 billion because Texaco counteroffers $3 billion, Liedtke refuses and goes to court, and the judge awards $5 billion. That is, 16.5% equals 33% times 50%. Adding up, we get the chance of $5 billion = 17% + 25% + 16.5% = 58.5%.

In constructing a risk profile, we collapse a decision tree by multiplying out the probabilities on sequential chance branches. At a decision node, only one branch is taken; in the case of "Counteroffer $5 Billion; Refuse Texaco Counteroffer," we use only the indicated alternative for the second decision, and so this decision node need not be included in the collapsing process. You can think about the process as one in which nodes are gradually removed from the tree in much the same sense as we did

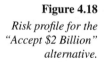

Figure 4.18

Risk profile for the "Accept $2 Billion" alternative.

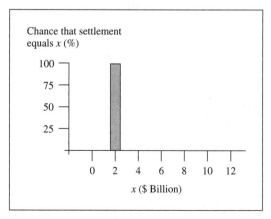

Chance that settlement equals x (%)

x ($ Billion)

Figure 4.19

Risk profile for the "Counteroffer $5 Billion; Refuse Texaco Counteroffer" strategy.

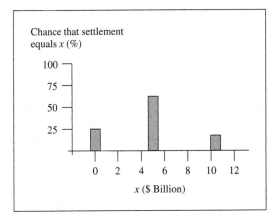

with the folding-back procedure, except that in this case we keep track of the possible outcomes and their probabilities. Figures 4.20, 4.21, and 4.22 show the progression of collapsing the decision tree in order to construct the risk profile for the "Counteroffer $5 Billion; Refuse Texaco Counteroffer" strategy.

By looking at the risk profiles, the decision maker can tell a lot about the riskiness of the alternatives. In some cases a decision maker can choose among alternatives on the basis of their risk profiles. Comparing Figures 4.18 and 4.19, it is clear that the worst possible consequence for "Counteroffer $5 Billion; Refuse Texaco Counteroffer" is less than the value for "Accept $2 billion." On the other hand, the largest amount ($10.3 billion) is much better than $2 billion. Hugh Liedtke has to decide whether the risk of perhaps coming away empty-handed is worth the possibility of getting more than $2 billion. This is clearly a case of a basic risky decision, as we can see from the collapsed decision tree in Figure 4.22.

Risk profiles can be calculated for strategies that might not have appeared as optimal in an expected-value analysis. For example, Figure 4.23 shows the risk profile for

Figure 4.20

First step in collapsing the decision tree to make a risk profile for "Counteroffer $5 Billion; Refuse Texaco Counteroffer" strategy. The decision node has been removed to leave only the outcomes associated with the "Refuse" branch.

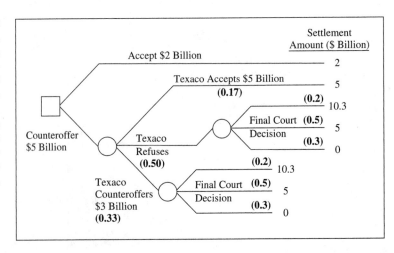

Figure 4.21

Second step in collapsing the decision tree to make a risk profile. The three chance nodes have been collapsed into one chance node. The probabilities on the branches are the product of the probabilities from sequential branches in Figure 4.20.

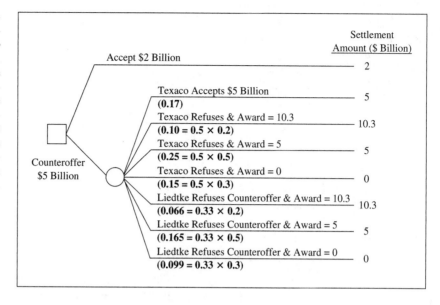

Figure 4.22

Third step in collapsing the decision tree to make a risk profile. The seven branches from the chance node in Figure 4.21 have been combined into three branches.

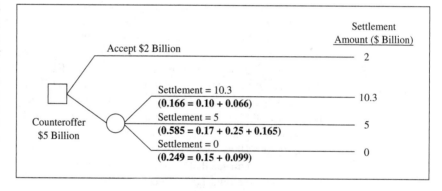

"Counteroffer $5 Billion; Accept $3 Billion," which we ruled out on the basis of EMV. Comparing Figures 4.23 and 4.19 indicates that this strategy yields a smaller chance of getting nothing, but also less chance of a $10.3 billion judgment. Compensating for this is the greater chance of getting something in the middle: $3 or $5 billion.

Although risk profiles can in principle be used as an alternative to EMV to check every possible strategy, for complex decisions it can be tedious to study many risk profiles. Thus, a compromise is to look at strategies only for the first one or two decisions, on the assumption that future decisions would be made using a decision rule such as maximizing expected value, which is itself a kind of strategy. (This is the approach used by many decision-analysis computer programs, PrecisionTree included.) Thus, in the Texaco-Pennzoil example, one might compare only the "Accept $2 Billion" and "Counteroffer $5 Billion; Refuse Texaco Counteroffer" strategies.

Figure 4.23

Risk profile for the "Counteroffer $5 Billion; Accept $3 Billion" strategy.

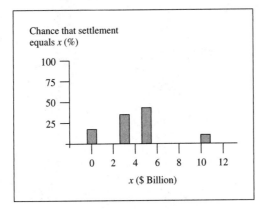

Cumulative Risk Profiles

We also can present the risk profile in cumulative form. Figure 4.24 shows the *cumulative risk profile* for "Counteroffer 5 Billion; Refuse Texaco Counteroffer." In this format, the vertical axis is the chance that the payoff is less than or equal to the corresponding value on the horizontal axis. This is only a matter of translating the information contained in the risk profile in Figure 4.19. There is no chance that the settlement will be less than zero. At zero, the chance jumps up to 24.9%, because there is a substantial chance that the court award will be zero. The graph continues at 24.9% across to $5 billion. (For example, there is a 24.9% chance that the settlement is less than or equal to $3.5 billion; that is, there is the 24.9% chance that the settlement is zero, and that is less than $3.5 billion.) At $5 billion, the line jumps up to 83.4% (which is 24.9% + 58.5%), because there is an 83.4% chance that the settlement is less than or equal to $5 billion. Finally, at $10.3 billion, the cumulative graph jumps up to 100%: The chance is 100% that the settlement is less than or equal to $10.3 billion.

Thus, you can see that creating a cumulative risk profile is just a matter of adding up, or accumulating, the chances of the individual payoffs. For any specific value

Figure 4.24

Cumulative risk profile for "Counteroffer $5 Billion; Refuse Texaco Counteroffer."

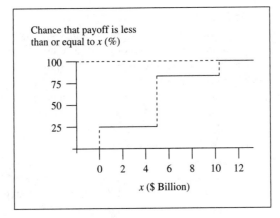

along the horizontal axis, we can read off the chance that the payoff will be less than or equal to that specific value. Later in this chapter, we show how to generate risk profiles and cumulative risk profiles in PrecisionTree. Cumulative risk profiles will be very helpful in the next section in our discussion of dominance.

Dominance: An Alternative to EMV

Comparing expected values of different risky prospects is useful, but in many cases EMV inadequately captures the nature of the risks that must be compared. With risk profiles, however, we can make a more comprehensive comparison of the risks. But how can we choose one risk profile over another? Unfortunately, there is no clear answer that can be used in all situations. By using the idea of *dominance,* though, we can identify those profiles (and their associated strategies) that can be ignored. Such strategies are said to be *dominated,* because we can show logically, according to some rules relating to cumulative risk profiles, that there are better risks (strategies) available.

Suppose we modify Liedtke's decision as shown in Figure 4.2 so that $2.5 billion is the minimum amount that he believes he could get in a court award. This decision is diagrammed in Figure 4.25. Now what should he do? It is rather obvious. Because he believes that he could do no worse than $2.5 billion if he makes a counteroffer, he should clearly shun Texaco's offer of 2 billion. This kind of dominance is called *deterministic dominance,* signifying that the dominating alternative pays off at least as much as the one that is dominated.

We can show deterministic dominance in terms of the cumulative risk profiles displayed in Figure 4.26. The cumulative risk profile for "Accept $2 Billion" goes from zero to 100% at $2 billion, because the settlement for this alternative is bound to be $2 billion. The risk profile for "Counteroffer $5 Billion; Refuse Texaco

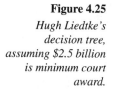

Figure 4.25

Hugh Liedtke's decision tree, assuming $2.5 billion is minimum court award.

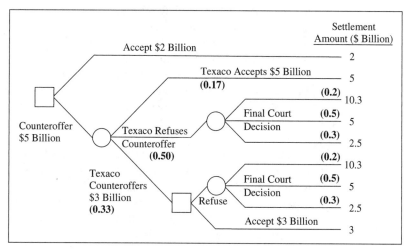

Figure 4.26

Cumulative risk profiles for alternatives in Figure 4.25.

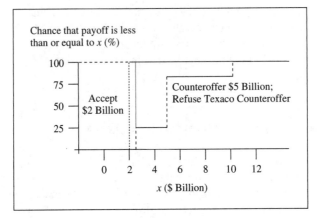

Counteroffer" starts at $2.5 billion but does not reach 100% until $10.3 billion. Deterministic dominance can be detected in the risk profiles by comparing the value where one cumulative risk profile reaches 100% with the value where another risk profile begins. If there is a value x such that the chance of the payoff being less than or equal to x is 100% in alternative B, and the chance of the payoff being *less than x* is 0% in Alternative A, then A deterministically dominates B. Graphically, continue the vertical line where alternative A first leaves 0% (the vertical line at $2.5 billion for "Counteroffer $5 Billion"). If that vertical line corresponds to 100% for the other cumulative risk profile, then A dominates B. Thus, even if the minimum court award had been $2 billion instead of $2.5 billion, "Counteroffer $5 Billion" still would have dominated "Accept $2 Billion."

The following example shows a similar kind of dominance. Suppose that Liedtke is choosing between two different law firms to represent Pennzoil. He considers both law firms to be about the same in terms of their abilities to deal with the case, but one charges less in the event that the case goes to court. The full decision tree for this problem appears in Figure 4.27. Which choice is preferred? Again, it's rather obvious; the settlement amounts for choosing Firm A are the same as the corresponding amounts for choosing Firm B, except that Pennzoil gets more with Firm A if the case results in a damage award in the final court decision. Choosing Firm A is like choosing Firm B and possibly getting a bonus as well. Firm A is said to display *stochastic dominance* over Firm B. Many texts also use the term *probabilistic dominance* to indicate the same thing. (Strictly speaking, this is first-order stochastic dominance. Higher-order stochastic dominance comes into play when we consider preferences regarding risk.)

The cumulative risk profiles corresponding to Firms A and B (and assuming that Liedtke refuses a Texaco counteroffer) are displayed in Figure 4.28. The two cumulative risk profiles almost coincide; the only difference is that Firm A's profile is slightly to the right of Firm B's at $5 and $10 billion, which represents the possibility of Pennzoil having to pay less in fees. Stochastic dominance is represented in the cumulative risk profiles by the fact that the two profiles do not cross and that there is some space between them. That is, if two cumulative risk profiles are such that no part

Figure 4.27

A decision tree comparing two law firms. Firm A charges less than Firm B if Pennzoil is awarded damages in court.

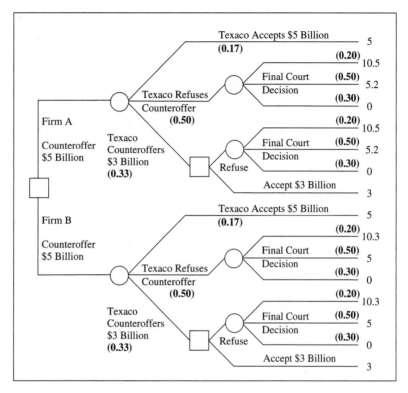

Figure 4.28

Cumulative risk profiles for two law firms in Figure 4.27. Firm A stochastically dominates Firm B.

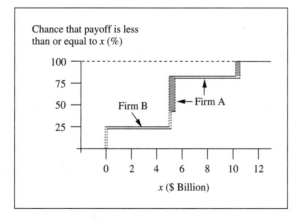

of Profile A lies to the left of B, and at least some part of it lies to the right of B, then the strategy corresponding to Profile A stochastically dominates the strategy for Profile B.

The next example demonstrates stochastic dominance in a slightly different form. Instead of the consequences, the pattern of the probability numbers makes the preferred alternative apparent. Suppose Liedtke's choice is between two law firms that he considers to be of different abilities. The decision tree is shown in Figure 4.29. Carefully examine the probabilities in the branches for the final court decision. Which law firm is preferred? This is a somewhat more subtle situation than the preceding one. The essence of the problem is that for Firm C, the larger outcome values have higher probabilities. The settlement with Firm C is not bound to be at least as great or greater than that with Firm D, but with Firm C the settlement is more likely to be greater. Think of Firm C as being a better gamble if the situation comes down to a court decision. Situations like this are characterized by two alternatives that offer the same possible consequences, but the dominating alternative is more likely to bring a better consequence.

Figure 4.30 shows the cumulative risk profiles for the two law firms in this example. As in the last example, the two profiles nearly coincide, although space is found between the two profiles because of the different probabilities associated with the court award. Because Firm C either coincides with or lies to the right of Firm D, we can conclude that Firm C stochastically dominates Firm D.

Stochastic dominance can show up in a decision problem in several ways. One way is in terms of the consequences (as in Figure 4.27), and another is in terms of the probabilities (as in Figure 4.29). Sometimes stochastic dominance may emerge as a mixture of the two; both slightly better payoffs and slightly better probabilities may lead to one alternative dominating another.

Figure 4.29

Decision tree comparing two law firms. Firm C has a better chance of winning a damage award in court than does Firm D.

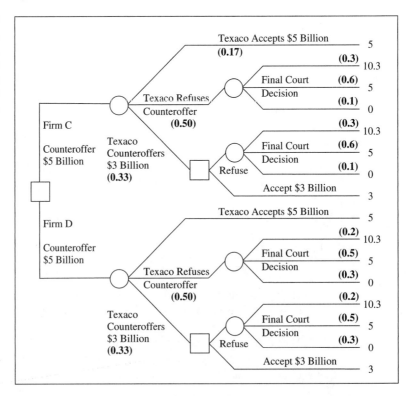

Figure 4.30

Cumulative risk profiles for law firms in Figure 4.29.
Firm C stochastically dominates Firm D.

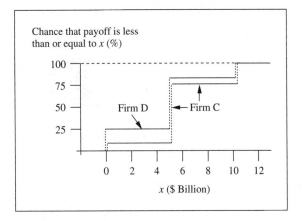

What is the relationship between stochastic dominance and expected value? It turns out that if one alternative dominates another, then the dominating alternative must have the higher expected value. This is a property of dominant alternatives that can be proven mathematically. To get a feeling for why it is true, think about the cumulative risk profiles, and imagine the EMV for a dominated Alternative B. If Alternative A dominates B, then its cumulative risk profile must lie at least partly to the right of the profile for B. Because of this, the EMV for A must also lie to the right of, and hence be greater than, the EMV for B.

Although this discussion of dominance has been fairly brief, one should not conclude that dominance is not important. Indeed, screening alternatives on the basis of dominance begins implicitly in the structuring phase of decision analysis, and, as alternatives are considered, they usually are at least informally compared to other alternatives. Screening alternatives formally on the basis of dominance is an important decision-analysis tool. If an alternative can be eliminated early in the selection process on that basis, considerable cost can be saved in large-scale problems. For example, suppose that the decision is where to build a new electric power plant. Analysis of proposed alternatives can be exceedingly expensive. If a potential site can be eliminated in an early phase of the analysis on the grounds that another dominates it, then that site need not undergo full analysis.

Making Decisions with Multiple Objectives

So far we have learned how to analyze a single-objective decision; in the Texaco-Pennzoil example, we have focused on Liedtke's objective of maximizing the settlement amount. How would we deal with a decision that involves multiple objectives? In this section, we learn how to extend the concepts of expected value and risk profiles to multiple-objective situations. In contrast to the grandiose Texaco-Pennzoil example, consider the following down-to-earth example of a young person deciding which of two summer jobs to take.

THE SUMMER JOB

Sam Chu was in a quandary. With two job offers in hand, the choice he should make was far from obvious. The first alternative was a job as an assistant at a local small business; the job would pay minimum wage ($5.25 per hour), it would require 25 to 35 hours per week, and the hours would be primarily during the week, leaving the weekends free. The job would last for three months, but the exact amount of work, and hence the amount Sam could earn, was uncertain. On the other hand, the free weekends could be spent with friends.

The second alternative was to work as a member of a trail-maintenance crew for a conservation organization. This job would require 10 weeks of hard work, 40 hours per week at $6.50 per hour, in a national forest in a neighboring state. The job would involve extensive camping and backpacking. Members of the maintenance crew would come from a large geographic area and spend the entire 10 weeks together, including weekends. Although Sam had no doubt about the earnings this job would provide, the real uncertainty was what the staff and other members of the crew would be like. Would new friendships develop? The nature of the crew and the leaders could make for 10 weeks of a wonderful time, 10 weeks of misery, or anything in between.

From the description, it appears that Sam has two objectives in this context: earning money and having fun this summer. Both are reasonable, and the two jobs clearly differ in these two dimensions; they offer different possibilities for the amount of money earned and the quality of summer fun.

The amount of money to be earned has a natural scale (dollars), and like most of us Sam prefers more money to less. The objective of having fun has no natural scale, though. Thus, a first step is to create such a scale. After considering the possibilities, Sam has created the scale in Table 4.5 to represent different levels of summer fun in the context of choosing a summer job. Although living in town and living in a forest camp pose two very different scenarios, the scale has been constructed in such a way that it can be applied to either job (as well as to any other prospect that might arise). The levels are numbered so that the higher numbers are more preferred.

Table 4.5

A constructed scale for summer fun.

5	(Best) A large, congenial group. Many new friendships made. Work is enjoyable, and time passes quickly.
4	A small but congenial group of friends. The work is interesting, and time off work is spent with a few friends in enjoyable pursuits.
3	No new friends are made. Leisure hours are spent with a few friends doing typical activities. Pay is viewed as fair for the work done.
2	Work is difficult. Coworkers complain about the low pay and poor conditions. On some weekends it is possible to spend time with a few friends, but other weekends are boring.
1	(Worst) Work is extremely difficult, and working conditions are poor. Time off work is generally boring because outside activities are limited or no friends are available.

With the constructed scale for summer fun, we can represent Sam's decision with the influence diagram and decision tree shown in Figures 4.31 and 4.32, respectively. The influence diagram shows the uncertainty about fun and amount of work, and that these have an impact on their corresponding consequences. The tree reflects Sam's belief that summer fun with the in-town job will amount to Level 3 in the constructed scale, but there is considerable uncertainty about how much fun the forest job will be. This uncertainty has been translated into probabilities based on Sam's uncertainty; how to make such judgments is the topic of Chapter 8. Likewise, the decision tree reflects uncertainty about the amount of work available at the in-town job.

Figure 4.31

Influence diagram for summer-job example.

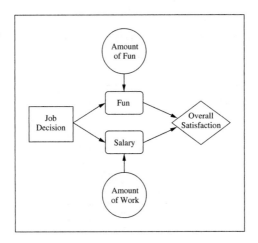

Figure 4.32

Decision tree for summer-job example.

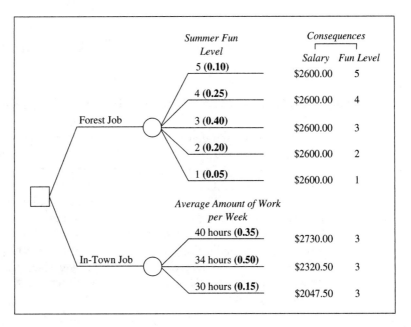

Analysis: One Objective at a Time

One way to approach the analysis of a multiple-objective decision is to calculate the expected value or create the risk profile for each individual objective. In the summer-job example, it is easy enough to do these things for salary. For the forest job, in which there is no uncertainty about salary, the expected value is $2600, and the risk profile is a single bar at $2600, as in Figure 4.33. For the in-town job, the expected salary is

$$E(Salary) = 0.35(\$2730.00) + 0.50(\$2320.50) + 0.15(\$2047.50)$$
$$= \$2422.88$$

The risk profile for salary at the in-town job is also shown in Figure 4.33.

Figure 4.33

Risk profiles for salary in the summer-job example.

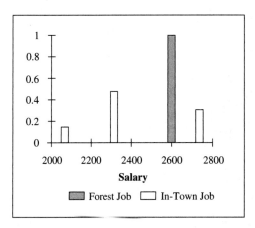

Subjective Ratings for Constructed Attribute Scales

For the summer-fun constructed attribute scale, risk profiles can be created and compared (Figure 4.34), but expected-value calculations are not meaningful because no meaningful numerical measurements are attached to the specific levels in the scale. The levels are indeed ordered, but the ordering is limited in what it means. The labels do not mean, for example, that going from Level 2 to Level 3 would give Sam the same increase in satisfaction as going from Level 4 to Level 5. Thus, before we can do any meaningful analysis, Sam must *rate* the different levels in the scale, indicating how much each level is worth (to Sam) relative to the other levels. This is a subjective judgment on Sam's part. Different people with different preferences would be expected to give different ratings for the possible levels of summer fun.

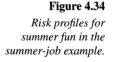

Figure 4.34

Risk profiles for summer fun in the summer-job example.

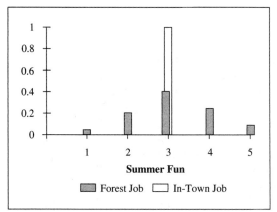

To make the necessary ratings, we begin by setting the endpoints of the scale. Let the best possible level (Level 5 in the summer-job example) have a value of 100 and the worst possible level (Level 1) a value of 0. Now all Sam must do is indicate how the intermediate levels rate on this scale from 0 to 100 points. For example, Level 4 might be worth 90 points, Level 3, 60 points, and Level 2, 25 points. Sam's assessments indicate that going from Level 3 to Level 4, with an increase of 30 points, is three times as good as going from Level 4 to Level 5 with an increase of only 10 points. Note that there is no inherent reason for the values of the levels to be evenly spaced; in fact, it might be surprising to find perfectly even spacing.

This same procedure can be used to create meaningful measurements for any constructed scale. The best level is assigned 100 points, the worst 0 points, and the decision maker must then assign rating points between 0 and 100 to the intermediate levels. A scale like this assigns more points to the preferred consequences, and the rating points for intermediate levels should reflect the decision maker's relative preferences for those levels.

With Sam's assessments, we can now calculate and compare the expected values for the amount of fun in the two jobs. For the in-town job, this is trivial because there is no uncertainty; the expected value is 60 points. For the forest job, the expected value is

$$E(\text{Fun Points}) = 0.10(100) + 0.25(90) + 0.40(60) + 0.20(25) + 0.05(0)$$
$$= 61.5$$

With individual expected values and risk profiles, alternatives can be compared. In doing so, we can hope for a clear winner, an alternative that dominates all other alternatives on all attributes. Unfortunately, comparing the forest and in-town jobs does not produce a clear winner. The forest job appears to be better on salary, having no risk and a higher expected value. Considering summer fun, the news is mixed. The in-town job has less risk but a lower expected value. It is obvious that going from one job to the other involves trading risks. Would Sam prefer a slightly higher salary for sure and take a risk on how much fun the summer will be? Or would the in-town job be better, playing it safe with the amount of fun and taking a risk on how much money will be earned?

Assessing Trade-Off Weights

The summer-job decision requires Sam to make an explicit trade-off between the objectives of maximizing fun and maximizing salary. How can Sam make this trade-off? Although this seems like a formidable task, a simple thought experiment is possible that will help Sam to understand the relative value of salary and fun.

In order to make the comparison between salary and fun, it is helpful to measure these two on similar scales, and the most convenient arrangement is to put salary on the same 0 to 100 scale that we used for summer fun. As before, the best ($2730) and worst ($2047.50) take values of 100 and 0, respectively. To get the values for the intermediate salaries ($2320.50 and $2600), a simple approach is to calculate them proportionately. Thus, we find that $2320.50 is 40% of the way from $2047.50 to $2730, and so it gets a value of 40 on the converted scale. (That is, [$2320.50 – $2047.50]/[$2730 – $2047.50] = 0.40) Likewise, $2600 is 81% of the way from $2047.50 to $2730, and so it gets a value of 81. (In Chapter 15, we will call this approach *proportional scoring*.) With the ratings for salary and summer fun, we now can create a new consequence matrix, giving the decision tree in Figure 4.35.

Now the trade-off question can be addressed in a straightforward way. The question is how Sam would trade points on the salary scale for points on the fun scale. To do this we introduce the idea of *weights*. What we want to do is assign weights to salary and fun to reflect their relative importance to Sam. Call the weights k_s and k_f, where the subscripts s and f stand for salary and fun, respectively. We will use the weights to calculate a weighted average of the two ratings for any given consequence in order to get an overall score. For example, suppose that $k_s = 0.70$ and

Figure 4.35

Decision tree with ratings for consequences.

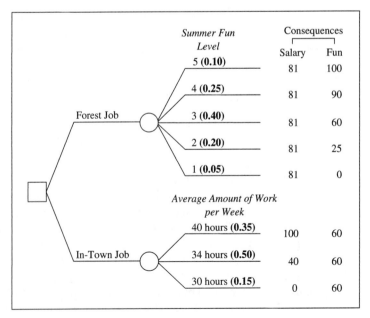

Summer Fun Level	Consequences	
	Salary	Fun
5 (**0.10**)	81	100
4 (**0.25**)	81	90
3 (**0.40**)	81	60
2 (**0.20**)	81	25
1 (**0.05**)	81	0

Forest Job

Average Amount of Work per Week		
40 hours (**0.35**)	100	60
34 hours (**0.50**)	40	60
30 hours (**0.15**)	0	60

In-Town Job

$k_f = 0.30$, reflecting a judgment that salary is a little more than twice as important as fun. The overall score (U) for the forest job with fun at Level 3 would be

$$U(\text{Salary: } 81, \text{ Fun: } 60) = 0.70(81) + 0.30(60)$$
$$= 74.7$$

It is up to Sam to make an appropriate judgment about the relative importance of the two attributes. Although details on making this judgment are in Chapter 15, one important issue in making this judgment bears discussion here. Sam *must* take into consideration the ranges of the two attributes. Strictly speaking, the two weights should reflect the relative value of going from best to worst on each scale. That is, if Sam thinks that improving salary from \$2047.50 to \$2730 is three times as important as improving fun from Level 1 to Level 5, this judgment would imply weights $k_s = 0.75$ and $k_f = 0.25$.

Paying attention to the ranges of the attributes in assigning weights is crucial. Too often we are tempted to assign weights on the basis of vague claims that Attribute A (or its underlying objective) is worth three times as much as Attribute B. Suppose you are buying a car, though. If you are looking at cars that all cost about the same amount but their features differ widely, why should price play a role in your decision? It should have a low weight in the overall score. In the Texaco-Pennzoil case, we argued that we could legitimately consider only the objective of maximizing the settlement amount because its range was so wide; any other objectives would be overwhelmed by the importance of moving from worst to best on this one. In an overall score, the weight for settlement amount would be near 1, and the weight for any other attribute would be near zero.

Suppose that, after carefully considering the possible salary and summer-fun outcomes, Sam has come up with weights of 0.6 for salary and 0.4 for fun, reflecting a judgment that the range of possible salaries is 1.5 times as important as the range of possible summer-fun ratings. With these weights, we can collapse the consequence matrix in Figure 4.35 to get Figure 4.36. For example, if Sam chooses the forest job and the level of fun turns out to be Level 4, the overall score is $0.6(81) + 0.4(90) = 84.6$. The other endpoint values in Figure 4.36 can be found in the same way.

In these last two sections we have discussed some straightforward ways to make subjective ratings and trade-off assessments. These topics are treated more completely in Chapters 13, 15, and 16. For now you can rest assured that the techniques described here are fully compatible with those described in later chapters.

Analysis: Expected Values and Risk Profiles for Two Objectives

The decision tree in Figure 4.36 is now ready for analysis. The first thing we can do is fold back the tree to calculate expected values. Using the overall scores from Figure 4.36, the expected values are:

Figure 4.36

Decision tree with overall scores for summer-job example. Weights used are $k_s = 0.60$ and $k_f = 0.40$. For example, consider the forest job that has an outcome of Level 4 on the fun scale. The rating for salary is 81, and the rating for fun is 90. Thus, the overall score is $0.60(81) + 0.40(90) = 84.6$.

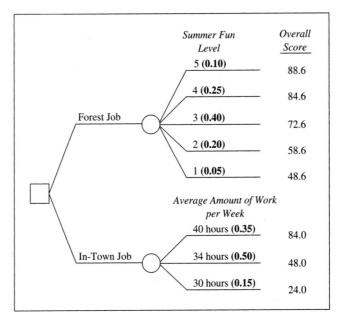

$$E(\text{Score for Forest Job}) = 0.10(88.6) + 0.25(84.6)$$
$$+ 0.40(72.6) + 0.20(58.6) + 0.05(48.6)$$
$$= 73.2$$
$$E(\text{Score for In-Town Job}) = 0.35(84) + 0.50(48) + 0.15(24)$$
$$= 57$$

Can we also create risk profiles for the two alternatives? We can; the risk profiles would represent the uncertainty associated with the overall weighted score Sam will get from either job. To the extent that this weighted score is meaningful to Sam as a measure of overall satisfaction, the risk profiles will represent the uncertainty associated with Sam's overall satisfaction. Figures 4.37 and 4.38 show the risk profiles and cumulative risk profiles for the two jobs. Figure 4.38 shows that, given the ratings and the trade-off between fun and salary, the forest job stochastically dominates the in-town job in terms of the overall score. Thus, the decision may be clear for Sam at this point; given Sam's assessed probabilities, ratings, and the trade-off, the forest job is a better risk. (Before making the commitment, though, Sam may want to do some *sensitivity analysis,* the topic of Chapter 5; small changes in some of those subjective judgments might result in a less clear choice between the two.)

Two final caveats are in order regarding the risk profiles of the overall score. First, it is important to understand that the overall score is something of an artificial outcome; it is an amalgamation in this case of two rating scales. As indicated above, Figures 4.37 and 4.38 only make sense to the extent that Sam is willing to interpret them as representing the uncertainty in the overall satisfaction from the two jobs.

Second, the stochastic dominance displayed by the forest job in Figure 4.38 is a relatively weak result; it relies heavily on Sam's assessed trade-off between the two attributes. A stronger result—one in which Sam could have confidence that the forest job is preferred regardless of his trade-off—requires that the forest job stochastically dominate the in-town job on each individual attribute. (Technically, however, even individual stochastic dominance is not quite enough; the risk profiles for the attributes must be combined into a single two-dimensional risk profile, or *bivariate probability distribution,* for each attribute. Then these two-dimensional risk profiles must be compared in much the same way we did with the single-attribute risk profiles. The good news is that as long as amount of work and amount of fun are *independent* (no arrow between these two chance nodes in the influence diagram in Figure 4.31), then finding that the same job stochastically dominates the other on each attribute guarantees that the same relationship holds in terms of the technically correct two-dimensional risk profile. Independence and stochastic dominance for multiple attributes will be discussed in Chapter 7.)

Figure 4.37

Risk profiles for summer jobs.

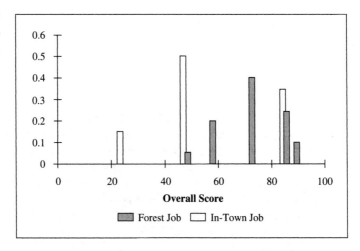

Figure 4.38

Cumulative risk profiles for summer jobs. The forest job stochastically dominates the in-town job.

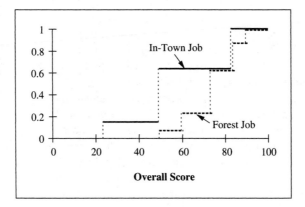

Decision Analysis Using PrecisionTree

In this section, we will first learn how to analyze decision trees and influence diagrams, then we will learn how to model multiple-objective decisions in a spreadsheet. Although this chapter concludes our discussion of PrecisionTree's basic capabilities, we will revisit PrecisionTree in later chapters to introduce other features, such as sensitivity analysis (Chapter 5), simulation (Chapter 11), and utility curves (Chapter 14).

A major advantage of using a program like PrecisionTree is the ease with which it performs an analysis. With the click of a button, any tree or influence diagram can be analyzed, various calculations performed, and risk profiles generated. Because it is so easy to run an analysis, however, there can be a temptation to build a quick model, analyze it, and move on. As you learn the steps for running an analysis using PrecisionTree, keep in mind that the insights you are working toward come from careful modeling, then analysis, and perhaps iterating through the model-and-analysis cycle several times. We encourage you to "play" with your model, searching for different insights using a variety of approaches. Take advantage of the time and effort you've saved with the automated analysis by investing it in building a requisite model.

Decision Trees

Figure 4.39 shows the Texaco-Pennzoil decision tree. You can see the structure we developed early in the chapter: The first decision is whether to accept the $2 billion offer or to make a $5 billion counteroffer. Then there is a chance node indicating Kinnear's response, and so on through the tree. PrecisionTree automatically calculates expected values through the tree. The expected value for each chance node appears in the cell below the node's name. Likewise, at each decision node, the largest branch value (for the preferred alternative) can be seen in the cell below the decision node's name. The word "TRUE" identifies the preferred alternative for a decision, with all other alternatives labeled "FALSE" for that node. (You can define other decision criteria besides maximizing the expected value. See the on-line help manual, Chapter 5: Settings Command, for instructions.)

Before analyzing the Texaco-Pennzoil decision, we have to create the decision tree. You have two choices at this point. You may wish to build the tree from scratch, using the skills you learned in the last chapter. If so, be sure that your tree looks just like the one in Figure 4.39 before proceeding.

Alternatively, you may open the existing spreadsheet on the Palisade CD located at Examples\Chapter 4\TexPenDT.xls. If you use this spreadsheet, you will see that it is not complete; we want you to practice your tree-construction skills at least a little! In particular, you should notice two things about the tree at first glance. First, you will have to modify the values and probabilities from the default values supplied by the software. Make sure your expected values match those in Figure 4.39 before proceeding. Second, you will notice that the chance node "Final Court Decision" is

Figure 4.39

Texaco-Pennzoil decision tree.

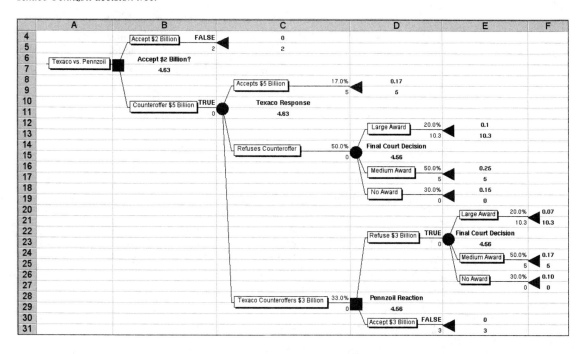

missing following the "Refuse $3 Billion" branch. There are three possibilities for completing this part of the tree:

ALTERNATIVE 1: You can create the chance node using the techniques you learned in Chapter 3.

ALTERNATIVE 2: You can copy and paste the chance node by following these steps:

A2.1 Click on the **Final Court Decision** node (on the circle itself when the cursor changes to a hand).

A2.2 In the *Node Settings* dialog box click the **Copy** button.

A2.3 Now click on the end node (blue triangle) of the *Refuse $3 Billion* branch, and in the *Node Settings* dialog box click **Paste.**

PrecisionTree will re-create the "Final Court Decision" at the end of the branch.

ALTERNATIVE 3: The third possibility is to use the existing "Final Court Decision" node as a "reference node." In this case, you are instructing PrecisionTree to refer back to the "Final Court Decision" chance node (and all of its subsequent structure) at the end of the "Refuses Counteroffer" branch. To do this:

A3.1   Click on the end node of the *Refuse $3 Billion* branch and select the **Reference Node** option (gray diamond, fourth from the left).

A3.2   In the name box, type **Final Court Decision.**

A3.3   Click in the entry box that is situated to the right of the option button labeled *node of this tree.* Move the cursor outside the dialog box to the spreadsheet and click in the cell below the name *Final Court Decision* (Figure 4.39, **cell D15**). (Be sure to point to cell D15, which contains the value of the chance node, not cell D14, which contains the name of the node.)

A3.4   Click **OK.**

The dotted line that runs between the reference node (gray diamond at the end of the "Refuse $3 Billion" branch) and the "Final Court Decision" chance node indicates that the tree will be analyzed as if there was an identical "Final Court Decision" chance node at the position of the reference node. Reference nodes are useful as a way to graphically prune a tree without leaving out any of the mathematical details.

With the decision tree structured and the appropriate numbers entered, it takes only two clicks of the mouse to run an analysis.

STEP 1

1.1   Click on the **Decision Analysis** button (fourth button from the left on the PrecisionTree toolbar).

1.2   In the *Decision Analysis* dialog box that appears (Figure 4.40), choose the **Analyze All Choices** option located under the *Initial Decision* heading in the lower right-hand corner.

1.3   Click **OK.**

At this point, PrecisionTree creates a new workbook with several worksheets. The *Statistics Report* contains seven statistics on each of the two alternatives: "Accept $2 Billion" and "Counteroffer $5 Billion." (If we had not chosen *Analyze All Options,* only the optimal choice, "Counteroffer $5 Billion," would be reported.) The *Policy Suggestion Report* (Figure 4.41) clarifies the optimal strategy by trimming away all suboptimal choices at the decision nodes. A look at Figure 4.41 reveals that Liedtke should refuse the $2 billion offer and counter with an offer of $5 billion, and if Kinnear counters by offering $3 billion, then Liedtke should again refuse and go to court.

The remaining three worksheets all convey the same information, each in a different type of graph. For example, Figure 4.42 illustrates the *Cumulative Profile* (which we have called the "cumulative risk profile") for the two alternatives. (Again, choosing *Analyze All Options* forces PrecisionTree to include profiles for all of the

Figure 4.40

Decision Analysis dialog box.

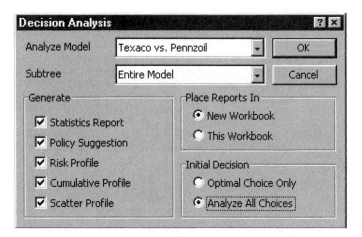

Figure 4.41

Policy Suggestion for Texaco-Pennzoil problem.

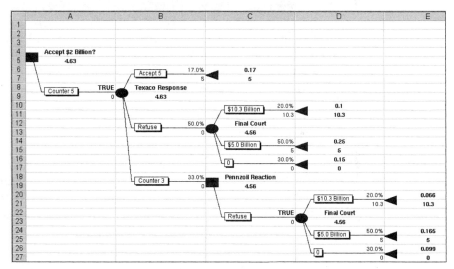

Figure 4.42

Cumulative risk profiles for the "Accept $2 Billion" and the "Counteroffer $5 Billion" alternatives.

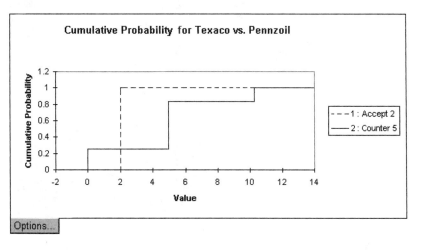

initial options.) At the bottom left of the graph is an *Option* button that lets you modify the graph's characteristics, such as the minimum and maximum values on the two axes.

The *Decision Analysis* dialog box (Figure 4.40) offers several choices regarding which model to analyze or whether to analyze all or a portion of the tree. Because it is possible to analyze any currently open tree or influence diagram, be sure to select the right model, especially if you have more than one on the same spreadsheet. Specify the tree or portion thereof that you wish to analyze using the pull-down list to the right of *Analyze Model.*

Influence Diagrams

PrecisionTree analyzes influence diagrams as readily as decision trees, which we demonstrate next with the Texaco-Pennzoil influence diagram. Because the numerical information of an influence diagram is contained in hidden tables, it is important that you carefully check that your diagram accurately models the decision before performing an analysis. One simple check is to verify that the values and probabilities are correctly entered. We will also see how PrecisionTree converts an influence diagram into a decision tree for a more thorough check. After you are satisfied that your influence diagram accurately models the decision, running an analysis is as simple as clicking two buttons.

We need to create the Texaco versus Pennzoil influence diagram, as shown in Figure 4.43, before analyzing it. One option would be to construct the influence diagram from scratch using the skills you learned in the last chapter. If so, be sure that the summary statistics box displays the correct expected value ($4.63 billion), standard deviation ($3.28 billion), minimum ($0 billion), and maximum ($10 billion).

Alternatively, you may open the existing spreadsheet on the Palisade CD located at Examples\Chapter 4\TexPenID.xls. Again, to encourage you to practice your

Figure 4.43

Influence diagram for Texaco versus Pennzoil.

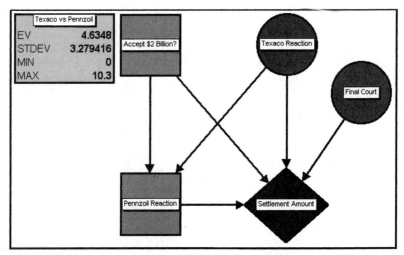

influence-diagram construction skills, this spreadsheet contains a partially completed model. Step 2 below describes how to complete this model by adding the values and probabilities, Step 3 describes how to convert the influence diagram into a decision tree, and Step 4 describes how to analyze the influence diagram.

> **STEP 2**   As a general rule, numbers are added to a node of an influence diagram by clicking on the name of the node, clicking the *Values* button in the *Influence Node Settings* dialog box, and entering the appropriate numbers in the *Influence Value Editor* dialog box. Remember to hit the *Return* button after each value is specified. (Refer to Chapter 3 for a more detailed review if necessary.)
>
> 2.1   Enter the numerical values and probabilities for the nodes "Accept $2 Billion?," "Texaco Response," and "Final Court Decision" using Figure 4.39 as a guide. For example, for the values of the "Accept $2 Billion?" node, enter 2 if accepted and enter 0 if $5 billion is counteroffered. Similarly, enter 5 if Texaco's response is to accept the $5 billion counteroffer and 0 if they refuse or counteroffer $3 billion. Enter 0 for *value if skipped* for these three nodes.

Adding values to the other two nodes ("Pennzoil Reaction" and "Settlement Amount") is slightly more complex:

> 2.2   Click on the decision node name **Pennzoil Reaction** and click on the **Values** button. A spreadsheet pops up titled *Influence Value Editor,* in which we use the entries on the right to specify the values in the *Value* column (Figure 4.44). Essentially, the *Influence Value Editor* is a consequence table similar to Table 4.1 except the columns are read from right to left.

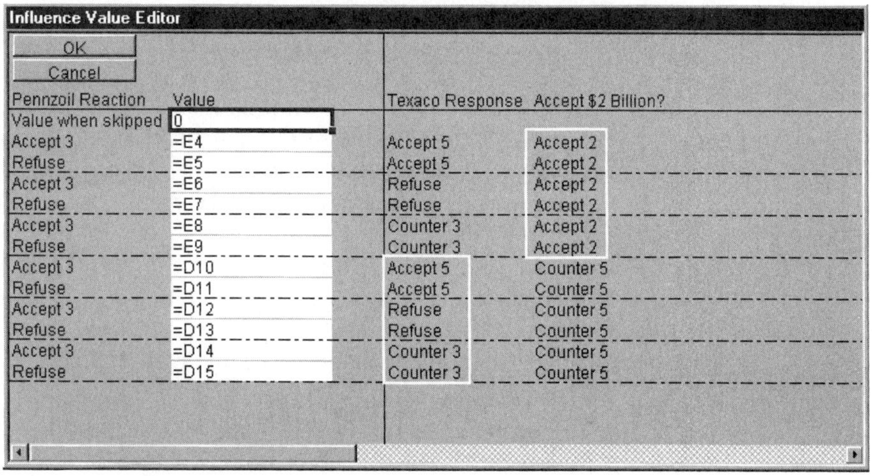

Figure 4.44

Influence Value Editor for "Pennzoil Reaction" decision node.

Influence Value Editor			
OK			
Cancel			
Pennzoil Reaction	Value	Texaco Response	Accept $2 Billion?
Value when skipped	0		
Accept 3	=E4	Accept 5	Accept 2
Refuse	=E5	Accept 5	Accept 2
Accept 3	=E6	Refuse	Accept 2
Refuse	=E7	Refuse	Accept 2
Accept 3	=E8	Counter 3	Accept 2
Refuse	=E9	Counter 3	Accept 2
Accept 3	=D10	Accept 5	Counter 5
Refuse	=D11	Accept 5	Counter 5
Accept 3	=D12	Refuse	Counter 5
Refuse	=D13	Refuse	Counter 5
Accept 3	=D14	Counter 3	Counter 5
Refuse	=D15	Counter 3	Counter 5

2.3   Start by typing **0** in the first row (*Value when skipped*) and hit **Enter.**

2.4   Type the equals sign (=) in the second row. For this row, the alternative we are entering is found by reading the rightmost entry—accept the $2 billion offer. Hence its value is 2, which is accessed by clicking on the first **Accept 2** in the column below the *Accept $2 Billion* heading. Row 2 should now read: =E4. Hit **Enter.**

2.5   The next five entries also involve accepting the $2 billion, so repeat Step 2.4 five times, and for each row click on the corresponding **Accept 2** in the rightmost column. Alternatively, because this is an Excel spreadsheet, you can use the fill-down command. As a guide in completing this node, we have outlined the outcomes that you are to reference with boxes in Figure 4.44.

2.6   Continuing onto rows 7 –12, the rightmost alternative is to counteroffer $5 billion. Since Pennzoil has counteroffered, the values are now determined by Texaco's response. Specify the appropriate value by clicking on the corresponding row under the *Texaco Response* heading.

2.7   Click **OK** when your *Influence Value Editor* matches Figure 4.44.

2.8   Following the same procedure, open the *Influence Value Editor* for **Settlement Amount** and enter the appropriate values as you read from left to right. Figure 4.45 highlights the cells to click on.

When finished entering the values, the summary statistics box should display the correct expected value ($4.63 billion), standard deviation ($3.28 billion), minimum ($0 billion), and maximum ($10 billion).

We can also convert the influence diagram into a decision tree.

STEP 3

3.1   Click on the name of the influence diagram **Texaco versus Pennzoil.**

3.2   In the *Influence Diagram Settings* dialog box, click on **Convert to Tree.**

A new spreadsheet is added to your workbook containing the converted tree. Comparing the converted tree to Figure 4.39 clearly shows the effect of assuming symmetry in an influence diagram, as discussed on page 120. For example, the decision tree in Figure 4.39 stops after the "Accept 2" branch, whereas the converted tree has all subsequent chance and decision nodes following the "Accept 2" branch. These extra branches are due to the symmetry assumption and explain why the payoff table for the influence diagram required so many entries. PrecisionTree has the ability to incorporate asymmetry into influence diagrams via structure arcs. Although we do not cover this feature of PrecisionTree, you can learn about structure arcs from the user's manual and on-line help.

We are now ready to analyze the influence diagram.

Figure 4.45

Settlement Amount Values dialog box for Texaco-Pennzoil model.

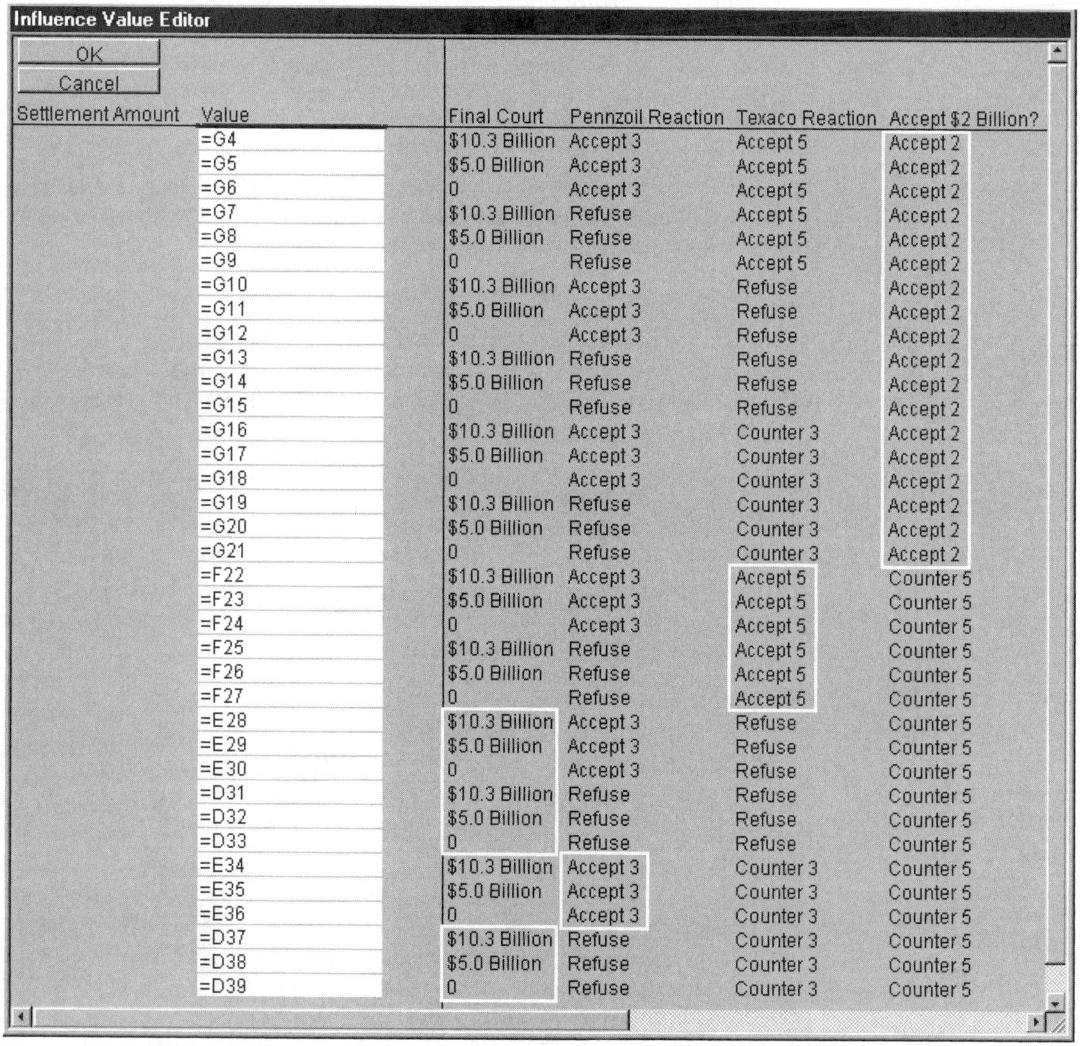

Influence Value Editor					
OK					
Cancel					
Settlement Amount	Value	Final Court	Pennzoil Reaction	Texaco Reaction	Accept $2 Billion?
	=G4	$10.3 Billion	Accept 3	Accept 5	Accept 2
	=G5	$5.0 Billion	Accept 3	Accept 5	Accept 2
	=G6	0	Accept 3	Accept 5	Accept 2
	=G7	$10.3 Billion	Refuse	Accept 5	Accept 2
	=G8	$5.0 Billion	Refuse	Accept 5	Accept 2
	=G9	0	Refuse	Accept 5	Accept 2
	=G10	$10.3 Billion	Accept 3	Refuse	Accept 2
	=G11	$5.0 Billion	Accept 3	Refuse	Accept 2
	=G12	0	Accept 3	Refuse	Accept 2
	=G13	$10.3 Billion	Refuse	Refuse	Accept 2
	=G14	$5.0 Billion	Refuse	Refuse	Accept 2
	=G15	0	Refuse	Refuse	Accept 2
	=G16	$10.3 Billion	Accept 3	Counter 3	Accept 2
	=G17	$5.0 Billion	Accept 3	Counter 3	Accept 2
	=G18	0	Accept 3	Counter 3	Accept 2
	=G19	$10.3 Billion	Refuse	Counter 3	Accept 2
	=G20	$5.0 Billion	Refuse	Counter 3	Accept 2
	=G21	0	Refuse	Counter 3	Accept 2
	=F22	$10.3 Billion	Accept 3	Accept 5	Counter 5
	=F23	$5.0 Billion	Accept 3	Accept 5	Counter 5
	=F24	0	Accept 3	Accept 5	Counter 5
	=F25	$10.3 Billion	Refuse	Accept 5	Counter 5
	=F26	$5.0 Billion	Refuse	Accept 5	Counter 5
	=F27	0	Refuse	Accept 5	Counter 5
	=E28	$10.3 Billion	Accept 3	Refuse	Counter 5
	=E29	$5.0 Billion	Accept 3	Refuse	Counter 5
	=E30	0	Accept 3	Refuse	Counter 5
	=D31	$10.3 Billion	Refuse	Refuse	Counter 5
	=D32	$5.0 Billion	Refuse	Refuse	Counter 5
	=D33	0	Refuse	Refuse	Counter 5
	=E34	$10.3 Billion	Accept 3	Counter 3	Counter 5
	=E35	$5.0 Billion	Accept 3	Counter 3	Counter 5
	=E36	0	Accept 3	Counter 3	Counter 5
	=D37	$10.3 Billion	Refuse	Counter 3	Counter 5
	=D38	$5.0 Billion	Refuse	Counter 3	Counter 5
	=D39	0	Refuse	Counter 3	Counter 5

STEP 4   The procedure for analyzing an influence diagram is the same as analyzing a decision tree.

4.1   Click on the **Decision Analysis** button (fourth button from the left on the PrecisionTree toolbar) and click **OK.**

The *Analyze All Choices* option that we instructed you to choose for decision trees is not available for influence diagrams. Thus, PrecisionTree's output for influence diagrams reports only on the optimal alternative via one set of statistics and one risk profile. The output for influence diagrams is interpreted the same as for decision trees.

Multiple-Attribute Models

This section presents two methods for modeling multiple-attribute decisions. Both methods take advantage of the fact that PrecisionTree runs in a spreadsheet and hence provides easy access to side calculations. We will explore these two methods using the summer-job example for illustration.

Method 1

The first method uses the fact that the branch value is entered into a spreadsheet cell, which makes it possible to use a specific formula in the cell to calculate the weighted scores. The formula we use is $U(s, f) = k_s*s + k_f*f$, where k_s and k_f are the weights and s and f are the scaled values for "Salary" and "Fun," respectively.

STEP 5

5.1 Build the decision tree, as shown in Figure 4.46, using the given probabilities. Leave the values (the numbers below the branches) temporarily at zero. Alternatively, the base tree can be found in Palisade's CD, Examples\Chapter 4\Method1.xls.

STEP 6

6.1 Create the weights table by typing **0.6** in cell B3 and **=1-B3** in cell C3.

STEP 7

7.1 Create the consequence table by entering the appropriate scaled salary and scaled fun scores corresponding to the branch values in columns F and G, as shown in Figure 4.46.

7.2 Compute the weighted scores for the top branch by clicking in cell **E5** and type **=$B$3*F5+$C$3*G5**. A score of 88.6 should appear in E5.

7.3 Click on cell **E5** and copy (either Ctrl-C or Edit-Copy).

7.4 Click into cell **E7** and paste. A score of 84.6 should appear in E7. Continue pasting the formula into the cells corresponding to each of the branches. You can streamline the process by copying the formula once, then holding down the control key while highlighting each cell into which you want to paste the formula. Now choose paste, and the formula is inserted into each highlighted cell. The weighted scores should be the same as those in Figure 4.46.

Figure 4.46

Modeling multiattribute decisions using the spreadsheet for calculations.

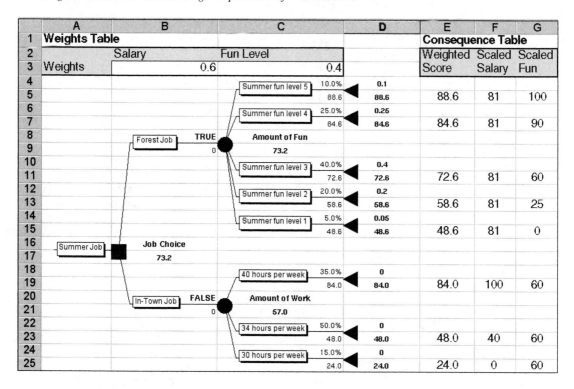

STEP 8

8.1 We now place the weighted scores into the decision tree. Click in cell **C5** and type **=E5**.

8.2 Copy the contents of **C5** and paste into the cells below each branch of the decision tree. At this point, your tree should be identical to the one in Figure 4.46.

Solving the tree and examining the risk profiles demonstrates that the "Forest Job" stochastically dominates the "In-Town Job."

Method 2

The second method is more sophisticated and eliminates the need to enter a separate formula for each outcome. Instead, a link is established between the decision tree and an Excel table along which input values pass from the tree to the table. The end-point value is calculated for the given input values, and passed back to the corresponding end node. Specifically, for the summer-job example, we will import the salary and fun levels from the tree, compute the weighted score, and export it back into the tree. The table acts as a template for calculating end-node values. Let's try it out.

STEP 9

9.1   Construct the "Summer Job" decision tree as shown in Figure 4.47 and name the tree **Linked Tree.** The base tree can also be found on the Palisade CD, Examples\Chapter 4\LinkSummer.xls.

STEP 10   Next, we construct the tables as shown at the top of Figure 4.47. The table for computing the end-node values has one column for each alternative. When we are finished, this table will import the specific values from each branch and then export the overall weighted score to each end node.

Figure 4.47

A decision tree linked to a table that computes the end-node values.

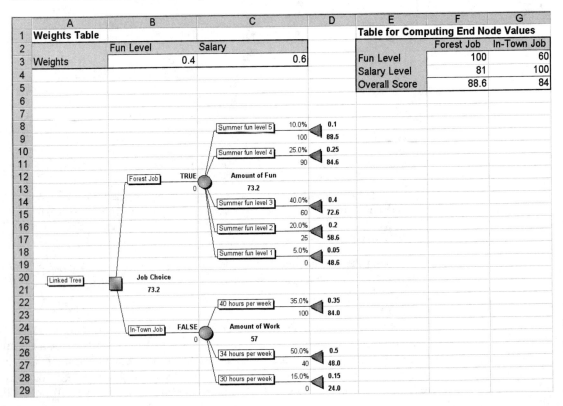

10.1   Enter **0.40** in B3 and **= 1 − B3** in C3. These are the weights for "Fun" and "Salary," respectively.

10.2   Enter the values **100, 81, 60,** and **100,** respectively, in cells F3, F4, G3, and G4 as shown in Figure 4.47.

10.3   Enter the overall score formula $U(s, f) = k_f * f + k_s * s$ in cell F5 as **= F3*$B$3+F4*$C$3**.

10.4   Enter the corresponding formula in C5 as **= G3*$B$3+G4*$C$3** (or use the fill-right command). This table will be used to compute the values for each node. Currently, the end-node values are the scaled fun or salary levels, such as 100 for the top branch of the "Forest Job" alternative. These values need to be changed to reflect the overall score.

STEP 11   Now that we have built the tree and the table, we will link them together. We first link each of the end nodes to the formulas for the overall score. This tells PrecisionTree what formula to use when calculating the end-node values.

11.1   Click directly on the tree's name: **Linked Tree.**

11.2   Under the *Payoff Calculation* heading in the *Tree Settings* dialog box, choose the **Link to Spreadsheet Model** option button and choose the **Automatically Update Link** box.

11.3   Click **OK.** The nodes have turned white, indicating that the links can now be established.

11.4   One by one, click on each end node (the white triangle) of the "Forest Job" alternative, choose the option button **Cell** under *Link Payoffs Values From,* and either type **F5** in the text box or click on **F5** in the spreadsheet. After you click **OK,** two changes occur: The end node turns blue, indicating the link has been established, and the end-node value changes to 88.6, which is the "Forest Job" overall score when $s = 81$ and $f = 100$. Be sure to change all five end nodes.

11.5   Notice that each end node has a value of 88.6, which is correct only for the top branch. Type **90** into F3. Now all the end-node values are 84.6. Because we have not yet linked cell F3 to the chance node *Amount of Fun,* PrecisionTree does not know to input the different fun level scores from the tree into the formula. We link the chance nodes to the formula in Step 12 below.

11.6   Now move to the "In-Town" alternative. One by one, click on each end node (white triangles) of the "In-Town" alternative, choose the option button **Cell** under *Link Payoffs Values From,* and either type **G5** in the text box or click on **G5** in the spreadsheet. This changes the end-node value to 84.0, which is the "In-Town" overall score when $s = 100$ and $f = 60$. Be sure to change all three end nodes. Again, all end-node values are 84 because we have not linked the chance node *Amount of Work* to the formula.

> **STEP 12**   We have connected the table output to the end-node values. Now
> we need to link the two chance nodes with the table so that the spreadsheet
> calculates the overall scores at the end nodes using the appropriate inputs.
>
> 12.1   To link the *Amount of Fun* chance node to the *Fun Level* in the table, click
> on the **Amount of Fun** node (the circle, not the name) to access the *Node
> Settings* dialog box.
>
> 12.2   Under the heading *Link Branch Value To,* click the **Cell** option button,
> click in the text box to the right, type **F3,** and click **OK.** Thus, for each
> branch, PrecisionTree calculates the end-node value by first sending the
> branch value to the table, then calculating the overall score, and finally
> sending that value back to the tree.

The end-node values for the "Forest Job" have been recalculated and now match
Figure 4.47. Color has returned to the "Amount of Fun" chance node.

> **STEP 13**
>
> 13.1   Finally, to link the *Amount of Work* chance node, click directly on the
> **Amount of Work** node.
>
> 13.2   Click the option button **Cell,** click in the text box to the right, and type
> **G4.**
>
> 13.3   Click **OK.** Your worksheet should match Figure 4.47.

This completes the construction of the linked tree. Analysis of this tree would
proceed as described above. The linked-tree method becomes more advantageous as
the decision tree grows in size and complexity. Also, it often happens that a decision
is first modeled in a spreadsheet, and later developed into a decision tree. It may be
natural in such a case to use the linked-tree method where the end-node values are
computed using the existing spreadsheet calculations.

S U M M A R Y   This chapter has demonstrated a variety of ways to use quantitative tools to make
choices in uncertain situations. We first looked at the solution process for decision trees
using expected value [or expected monetary value (EMV) when consequences are dol-
lars]. This is the most straightforward way to analyze a decision model; the algorithm
for solving a decision tree is easy to apply, and expected values are easy to calculate.

We also explored the process of solving influence diagrams using expected values. To understand the solution process for influence diagrams, we had to look at
their internal structures. In a sense, we had to fill in certain gaps left from Chapter 3
about how influence diagrams work. The solution procedure works out easily once
we understand how the problem's numerical details are represented internally. The

procedure for reducing nodes involves calculating expected values in a way that parallels the solution of a decision tree.

Risk profiles can be used to compare the riskiness of strategies and give comprehensive views of risks faced by a decision maker. Thus, risk profiles provide additional information to the decision maker trying to gain insight into the decision situation and the available alternatives. We also showed how cumulative risk profiles can be used to identify dominated alternatives.

The chapter ended with a set of detailed instructions on how to solve both decision trees and influence diagrams in PrecisionTree. We also described two different ways to model multiple-objective decisions in PrecisionTree. Both methods made use of the fact that PrecisionTree works within a spreadsheet, allowing us to input formulas that combine multiple-objective scores into a single score. The first method required that we define a separate formula for each end branch of the decision tree. The second method linked the end branches to one formula whose inputs came from the branch values of the tree.

EXERCISES

4.1   Is it possible to solve a decision-tree version of a problem and an equivalent influence-diagram version and come up with different answers? If so, explain. If not, why not?

4.2   Explain in your own words what it means when one alternative stochastically dominates another.

4.3   The analysis of the Texaco-Pennzoil example shows that the EMV of counteroffering with $5 billion far exceeds $2 billion. Why might Liedtke want to accept the $2 billion anyway? If you were Liedtke, what is the smallest offer from Texaco that you would accept?

4.4   Solve the decision tree in Figure 4.48.

Figure 4.48

Generic decision tree for Exercise 4.4.

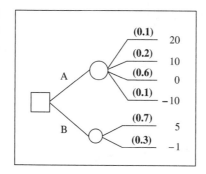

4.5   Draw and solve the influence diagram that corresponds to the decision tree in Figure 4.48.

4.6   Solve the decision tree in Figure 4.49. What principle discussed in Chapter 4 is illustrated by this decision tree?

Figure 4.49

Generic decision tree for Exercise 4.6.

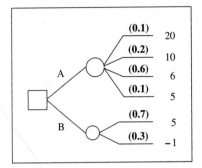

4.7   Which alternative is preferred in Figure 4.50? Do you have to do any calculations? Explain.

Figure 4.50

Generic decision tree for Exercise 4.7.

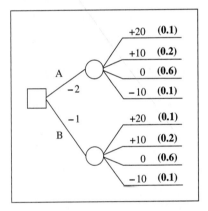

4.8   Solve the decision tree in Figure 4.51.

Figure 4.51

Generic decision tree for Exercise 4.8.

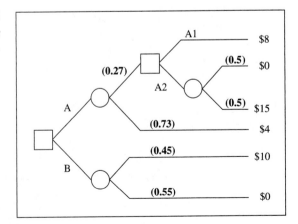

4.9 Create risk profiles and cumulative risk profiles for all possible strategies in Figure 4.51. Is one strategy stochastically dominant? Explain.

4.10 Draw and solve the influence diagram that corresponds to the decision tree in Figure 4.51.

4.11 Explain why deterministic dominance is a special case of stochastic dominance.

4.12 Explain in your own words why it is important to consider the ranges of the consequences in determining a trade-off weight.

4.13 Solve the influence diagram for the umbrella problem shown in Figure 4.10.

QUESTIONS AND PROBLEMS

4.14 A real-estate investor has the opportunity to purchase an apartment complex. The apartment complex costs $400,000 and is expected to generate net revenue (net after all operating and finance costs) of $6000 per month. Of course, the revenue could vary because the occupancy rate is uncertain. Considering the uncertainty, the revenue could vary from a low of -$1000 to a high of $10,000 per month. Assume that the investor's objective is to maximize the value of the investment at the end of 10 years.

    a Do you think the investor should buy the apartment complex or invest the $400,000 in a 10-year certificate of deposit earning 9.5%? Why?

    b The city council is currently considering an application to rezone a nearby empty parcel of land. The owner of that land wants to build a small electronics-assembly plant. The proposed plant does not really conflict with the city's overall land use plan, but it may have a substantial long-term negative effect on the value of the nearby residential district in which the apartment complex is located. Because the city council currently is divided on the issue and will not make a decision until next month, the real-estate investor is thinking about waiting until the city council makes its decision.

    If the investor waits, what could happen? What are the trade-offs that the investor has to make in deciding whether to wait or to purchase the complex now?

    c Suppose the investor could pay the seller $1000 in earnest money now, specifying in the purchase agreement that if the council's decision is to approve the rezoning, the investor can forfeit the $1000 and forego the purchase. Draw and solve a decision tree showing the investor's three options. Examine the alternatives for dominance. If you were the investor, which alternative would you choose? Why?

4.15 A stock market investor has $500 to spend and is considering purchasing an option contract on 1000 shares of Apricot Computer. The shares themselves are currently selling for $28.50 per share. Apricot is involved in a lawsuit, the outcome of which will be known within a month. If the outcome is in Apricot's favor, analysts expect Apricot's stock price to increase by $5 per share. If the outcome is unfavorable, then the price is expected to drop by $2.75 per share. The option costs $500, and owning the option would allow the investor to purchase 1000 shares of Apricot stock for $30 per share. Thus, if the investor buys the option and Apricot prevails in the lawsuit, the investor would make an immediate profit. Aside from purchasing the option, the investor could (1) do nothing and earn about 8% on his money, or (2) purchase $500 worth of Apricot shares.

a Construct cumulative risk profiles for the three alternatives, assuming Apricot has a 25% chance of winning the lawsuit. Can you draw any conclusions?

b If the investor believes that Apricot stands a 25% chance of winning the lawsuit, should he purchase the option? What if he believes the chance is only 10%? How large does the probability have to be for the option to be worthwhile?

4.16 Johnson Marketing is interested in producing and selling an innovative new food processor. The decision they face is the typical "make or buy" decision often faced by manufacturers. On one hand, Johnson could produce the processor itself, subcontracting different sub-assemblies, such as the motor or the housing. Cost estimates in this case are as follows:

Alternative: Make Food Processor	
Cost per Unit ($)	Chance (%)
35.00	25
42.50	25
45.00	37
49.00	13

The company also could have the entire machine made by a subcontractor. The subcontractor, however, faces similar uncertainties regarding the costs and has provided Johnson Marketing with the following schedule of costs and chances:

Alternative: Buy Food Processor	
Cost per Unit ($)	Chance (%)
37.00	10
43.00	40
46.00	30
50.00	20

If Johnson Marketing wants to minimize its expected cost of production in this case, should it make or buy? Construct cumulative risk profiles to support your recommendation. (*Hint:* Use care when interpreting the graph!)

4.17 Analyze the difficult decision situation that you identified in Problem 1.9 and structured in Problem 3.21. Be sure to examine alternatives for dominance. Does your analysis suggest any new alternatives?

4.18 Stacy Ennis eats lunch at a local restaurant two or three times a week. In selecting a restaurant on a typical workday, Stacy uses three criteria. First is to minimize the amount of travel time, which means that close-by restaurants are preferred on this attribute. The next objective is to minimize cost, and Stacy can make a judgment of the average lunch cost at most of the restaurants that would be considered. Finally, variety comes into play. On any given day, Stacy would like to go someplace different from where she has been in the past week.

Today is Monday, her first day back from a two-week vacation, and Stacy is considering the following six restaurants:

	Distance (Walking Time)	Average Price ($)
Sam's Pizza	10	3.50
Sy's Sandwiches	9	2.85
Bubba's Italian Barbecue	7	6.50
Blue China Cafe	2	5.00
The Eating Place	2	7.50
The Excel-Soaring Restaurant	5	9.00

a If Stacy considers distance, price, and variety to be equally important (given the range of alternatives available), where should she go today for lunch? (*Hints:* Don't forget to convert both distance and price to similar scales, such as a scale from 0 to 100. Also, recall that Stacy has just returned from vacation; what does this imply for how the restaurants compare on the variety objective?)

b Given your answer to part a, where should Stacy go on Thursday?

4.19 The national coffee store Farbucks needs to decide in August how many holiday-edition insulated coffee mugs to order. Because the mugs are dated, those that are unsold by January 15 are considered a loss. These premium mugs sell for $23.95 and cost $6.75 each. Farbucks is uncertain of the demand. They believe that there is a 25% chance that they will sell 10,000 mugs, a 50% chance that they will sell 15,000, and a 25% chance that they will sell 20,000.

a Build a linked-tree in PrecisionTree to determine if they should order 12,000, 15,000, or 18,000 mugs. Be sure that your model does not allow Farbucks to sell more mugs than it ordered. You can use the IF( ) command in Excel. If demand is less than the order quantity, then the amount sold is the demand. Otherwise, the amount sold is the order quantity. (See Excel's help or function wizard for guidance.)

b Now, assume that any unsold mugs are discounted and sold for $5.00. How does this affect the decision?

C A S E   S T U D I E S

GPC'S NEW PRODUCT DECISION

The executives of the General Products Company (GPC) have to decide which of three products to introduce, A, B, or C. Product C is essentially a risk-free proposition, from which the company will obtain a net profit of $1 million. Product B is considerably more risky. Sales may be high, with resulting net profit of $8 million, medium with net profit of $4 million, or low, in which case the company just breaks even. The probabilities for these outcomes are

P(Sales High for B) = 0.38

P(Sales Medium for B) = 0.12

P(Sales Low for B) = 0.50

Product A poses something of a difficulty; a problem with the production system has not yet been solved. The engineering division has indicated its confidence in solving the problem, but there is a slight (5%) chance that devising a workable solution may take a long time. In this event, there will be a delay in introducing the product, and that delay will result in lower sales and profits. Another issue is the price for Product A. The options are to introduce it at either high or low price; the price would not be set until just before the product is to be introduced. Both of these issues have an impact on the ultimate net profit.

Finally, once the product is introduced, sales can be either high or low. If the company decides to set a low price, then low sales are just as likely as high sales. If the company sets a high price, the likelihood of low sales depends on whether the product was delayed by the production problem. If there was no delay and the company sets a high price, the probability is 0.4 that sales will be high. However, if there is a delay and the price is set high, the probability is only 0.3 that sales will be high. The following table shows the possible net profit figures (in millions) for Product A:

	Price	High Sales ($ Million)	Low Sales ($ Million)
Time delay	High	5.0	(0.5)
	Low	3.5	1.0
No delay	High	8.0	0.0
	Low	4.5	1.5

Questions

1   Draw an influence diagram for GPC's problem. Specify the possible outcomes and the probability distributions for each chance node. Specify the possible alternatives for each decision node. Write out the complete table for the consequence node. (If possible, use a computer program for creating influence diagrams.)

2   Draw a complete decision tree for GPC. Solve the decision tree. What should GPC do? (If possible, do this problem using a computer program for creating and solving decision trees.)

3   Create cumulative risk profiles for each of the three products. Plot all three profiles on one graph. Can you draw any conclusions?

4   One of the executives of GPC is considerably less optimistic about Product B and assesses the probability of medium sales as 0.3 and the probability of low sales as

0.4. Based on expected value, what decision would this executive make? Should this executive argue about the probabilities? Why or why not? (*Hint:* Don't forget that probabilities have to add up to 1!)

5   Comment on the specification of chance outcomes and decision alternatives. Would this specification pass the clarity test? If not, what changes in the problem must be made in order to pass the clarity test?

SOUTHERN ELECTRONICS, PART I

Steve Sheffler is president, CEO, and majority stockholder of Southern Electronics, a small firm in the town of Silicon Mountain. Steve faces a major decision: Two firms, Big Red Business Machines and Banana Computer, are bidding for Southern Electronics.

Steve founded Southern 15 years ago, and the company has been extremely successful in developing progressive computer components. Steve is ready to sell the company (as long as the price is right!) so that he can pursue other interests. Last month, Big Red offered Steve $5 million and 100,000 shares of Big Red stock (currently trading at $50 per share and not expected to change substantially in the future). Until yesterday, Big Red's offer sounded good to Steve, and he had planned on accepting it this week. But a lawyer from Banana Computer called last week and indicated that Banana was interested in acquiring Southern Electronics. In discussions this past week, Steve has learned that Banana is developing a new computer, code-named EYF, that, if successful, will revolutionize the industry. Southern Electronics could play an important role in the development of the machine.

In their discussions, several important points have surfaced. First, Banana has said that it believes the probability that the EYF will succeed is 0.6, and that if it does, the value of Banana's stock will increase from the current value of $30 per share. Although the future price is uncertain, Banana judges that, conditional on the EYF's success, the expected price of the stock is $50 per share. If the EYF is not successful, the price will probably decrease slightly. Banana judges that if the EYF fails, Banana's share price will be between $20 and $30, with an expected price of $25.

Yesterday Steve discussed this information with his financial analyst, who is an expert regarding the electronics industry and whose counsel Steve trusts completely. The analyst pointed out that Banana has an incentive to be very optimistic about the EYF project. "Being realistic, though," said the analyst, "the probability that the EYF succeeds is only 0.4, and if it does succeed, the expected price of the stock would be only $40 per share. On the other hand, I agree with Banana's assessment for the share price if the EYF fails."

Negotiations today have proceeded to the point where Banana has made a final offer to Steve of $5 million and 150,000 shares of Banana stock. The company's representative has stated quite clearly that Banana cannot pay any more than this in a straight transaction. Furthermore, the representative claims, it is not clear why Steve will not accept the offer because it appears to them to be more valuable than the Big Red offer.

Questions

1   In terms of expected value, what is the least that Steve should accept from Banana? (This amount is called his *reservation price*.)

2   Steve obviously has two choices, to accept the Big Red offer or to accept the Banana offer. Draw an influence diagram representing Steve's decision. (If possible, do this problem using a computer program for structuring influence diagrams.)

3   Draw and solve a complete decision tree representing Steve's decision. (If possible, do this problem using a computer program for creating and solving decision trees.)

4   Why is it that Steve cannot accept the Banana offer as it stands?

SOUTHERN ELECTRONICS, PART II

Steve is well aware of the difference between his probabilities and Banana's, and he realizes that because of this difference, it may be possible to design a contract that benefits both parties. In particular, he is thinking about put options for the stock. A put option gives the owner of the option the right to sell an asset at a specific price. (For example, if you own a put option on 100 shares of General Motors (GM) with an exercise price of $75, you could sell 100 shares of GM for $75 per share before the expiration date of the option. This would be useful if the stock price fell below $75.) Steve reasons that if he could get Banana to include a put option on the stock with an exercise price of $30, then he would be protected if the EYF failed.

Steve proposes the following deal: He will sell Southern Electronics to Banana for $530,000 plus 280,000 shares of Banana stock and a put option that will allow him to sell the 280,000 shares back to Banana for $30 per share any time within the next year (during which time it will become known whether the EYF succeeds or fails).

Questions

1   Calculate Steve's expected value for this deal. Ignore tax effects and the time value of money.

2   The cost to Banana of their original offer was simply

$$\$5,000,000 + 150,000(\$30) = \$9,500,000$$

Show that the expected cost to Banana of Steve's proposed deal is less than $9.5 million, and hence in Banana's favor. Again, ignore tax effects and the time value of money.

STRENLAR

Fred Wallace scratched his head. By this time tomorrow he had to have an answer for Joan Sharkey, his former boss at Plastics International (PI). The decision was difficult to make. It involved how he would spend the next 10 years of his life.

Four years ago, when Fred was working at PI, he had come up with an idea for a revolutionary new polymer. A little study—combined with intuition, hunches, and educated guesses—had convinced him that the new material would be extremely strong for its weight. Although it would undoubtedly cost more than conventional materials, Fred discovered that a variety of potential uses existed in the aerospace, automobile manufacturing, robotics, and sporting goods industries.

When he explained his idea to his supervisors at PI, they had patiently told him that they were not interested in pursuing risky new projects. His appeared to be even riskier than most because, at the time, many of the details had not been fully worked out. Furthermore, they pointed out that efficient production would require the development of a new manufacturing process. Sure, if that process proved successful, the new polymer could be a big hit. But without that process the company simply could not provide the resources Fred would need to develop his idea into a marketable product.

Fred did not give up. He began to work at home on his idea, consuming most of his evenings and weekends. His intuition and guesses had proven correct, and after some time he had worked out a small-scale manufacturing process. With this process, he had been able to turn out small batches of his miracle polymer, which he dubbed Strenlar. At this point he quietly began to assemble some capital. He invested $100,000 of his own, managed to borrow another $200,000, and quit his job at PI to devote his time to Strenlar.

That was 15 months ago. In the intervening time he had made substantial progress. The product was refined, and several customers eagerly awaited the first production run. A few problems remained to be solved in the manufacturing process, but Fred was 80% sure that these bugs could be worked out satisfactorily. He was eager to start making profits himself; his capital was running dangerously low. When he became anxious, he tried to soothe his fears by recalling his estimate of the project's potential. His best guess was that sales would be approximately $35 million over 10 years, and that he would net some $8 million after costs.

Two weeks ago, Joan Sharkey at PI had surprised him with a telephone call and had offered to take Fred to lunch. With some apprehension, Fred accepted the offer. He had always regretted having to leave PI, and was eager to hear how his friends were doing. After some pleasantries, Joan came to the point.

"Fred, we're all impressed with your ability to develop Strenlar on your own. I guess we made a mistake in turning down your offer to develop it at PI. But we're interested in helping you out now, and we can certainly make it worth your while. If you will grant PI exclusive rights to Strenlar, we'll hire you back at, say $40,000 a year, and we'll give you a 2.5 percent royalty on Strenlar sales. What do you say?"

Fred didn't know whether to laugh or become angry. "Joan, my immediate reaction is to throw my glass of water in your face! I went out on a limb to develop the product, and now you want to capitalize on my work. There's no way I'm going to sell out to PI at this point!"

The meal proceeded, with Joan sweetening the offer gradually, and Fred obstinately refusing. After he got back to his office, Fred felt confused. It would be nice to work at PI again, he thought. At least the future would be secure. But there would never be the potential for the high income that was possible with Strenlar. Of course, he thought grimly, there was still the chance that the Strenlar project could fail altogether.

At the end of the week, Joan called him again. PI was willing to go either of two ways. The company could hire him for $50,000 plus a 6% royalty on Strenlar gross sales. Alternatively, PI could pay him a lump sum of $500,000 now plus options to purchase up to 70,000 shares of PI stock at the current price of $40 any time within the next three years. No matter which offer Fred accepted, PI would pay off Fred's creditors and take over the project immediately. After completing development of the manufacturing process, PI would have exclusive rights to Strenlar. Furthermore, it turned out that PI was deadly serious about this game. If Fred refused both of these offers, PI would file a lawsuit claiming rights to Strenlar on the grounds that Fred had improperly used PI's resources in the development of the product.

Consultation with his attorney just made him feel worse. After reviewing Fred's old contract with PI, the attorney told him that there was a 60% chance that he would win the case. If he won the case, PI would have to pay his court costs. If he lost, his legal fees would amount to about $20,000.

Fred's accountant helped him estimate the value of the stock options. First, the exercise date seemed to pose no problem; unless the remaining bugs could not be worked out, Strenlar should be on the market within 18 months. If PI were to acquire the Strenlar project and the project succeeded, PI's stock would go up to approximately $52. On the other hand, if the project failed, the stock price probably would fall slightly to $39.

As Fred thought about all of the problems he faced, he was quite disturbed. On one hand, he yearned for the comradery he had enjoyed at PI four years ago. He also realized that he might not be cut out to be an entrepreneur. He reacted unpleasantly to the risk he currently faced. His physician had warned him that he may be developing hypertension and had tried to persuade him to relax more. Fred knew that his health was important to him, but he had to believe that he would be able to weather the tension of getting Strenlar onto the market. He could always relax later, right? He sighed as he picked up a pencil and pad of paper to see if he could figure out what he should tell Joan Sharkey.

Question

1   Do a complete analysis of Fred's decision. Your analysis should include at least structuring the problem with an influence diagram, drawing and solving a decision tree, creating risk profiles, and checking for stochastic dominance. What do you think Fred should do? Why? (*Hint:* This case will require you to make certain assumptions in order to do a complete analysis. State clearly any assumptions you make, and be careful that the assumptions you make are both reasonable and con-

sistent with the information given in the case. You may want to analyze your decision model under different sets of assumptions. Do not forget to consider issues such as the time value of money, riskiness of the alternatives, and so on.)

JOB OFFERS

Robin Pinelli is considering three job offers. In trying to decide which to accept, Robin has concluded that three objectives are important in this decision. First, of course, is to maximize disposable income—the amount left after paying for housing, utilities, taxes, and other necessities. Second, Robin likes cold weather and enjoys winter sports. The third objective relates to the quality of the community. Being single, Robin would like to live in a city with a lot of activities and a large population of single professionals.

Developing attributes for these three objectives turns out to be relatively straightforward. Disposable income can be measured directly by calculating monthly take-home pay minus average monthly rent (being careful to include utilities) for an appropriate apartment. The second attribute is annual snowfall. For the third attribute, Robin has located a magazine survey of large cities that scores those cities as places for single professionals to live. Although the survey is not perfect from Robin's point of view, it does capture the main elements of her concern about the quality of the singles community and available activities. Also, all three of the cities under consideration are included in the survey.

Here are descriptions of the three job offers:

1   MPR Manufacturing in Flagstaff, Arizona. Disposable income estimate: $1600 per month. Snowfall range: 150 to 320 cm per year. Magazine score: 50 (out of 100).

2   Madison Publishing in St. Paul, Minnesota. Disposable income estimate: $1300 to $1500 per month. (This uncertainty here is because Robin knows there is a wide variety in apartment rental prices and will not know what is appropriate and available until spending some time in the city.) Snowfall range: 100 to 400 cm per year. Magazine score: 75.

3   Pandemonium Pizza in San Francisco, California. Disposable income estimate: $1200 per month. Snowfall range: negligible. Magazine score: 95.

Robin has created the decision tree in Figure 4.52 to represent the situation. The uncertainty about snowfall and disposable income are represented by the chance nodes as Robin has included them in the tree. The ratings in the consequence matrix are such that the worst consequence has a rating of zero points and the best has 100.

Questions

1   Verify that the ratings in the consequence matrix are proportional scores (that is, that they were calculated the same way we calculated the ratings for salary in the summer-fun example in the chapter).

Figure 4.52

Robin Pinelli's decision tree.

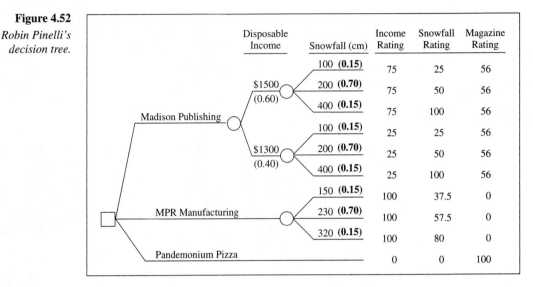

2   Comment on Robin's choice of annual snowfall as a measure for the cold-weather–winter-sports attribute. Is this a good measure? Why or why not?

3   After considering the situation, Robin concludes that the quality of the city is most important, the amount of snowfall is next, and the third is income. (Income is important, but the variation between $1200 and $1600 is not enough to make much difference to Robin.) Furthermore, Robin concludes that the weight for the magazine rating in the consequence matrix should be 1.5 times the weight for the snowfall rating and three times as much as the weight for the income rating. Use this information to calculate the weights for the three attributes and to calculate overall scores for all of the end branches in the decision tree.

4   Analyze the decision tree using expected values. Calculate expected values for the three measures as well as for the overall score.

5   Do a risk-profile analysis of the three cities. Create risk profiles for each of the three attributes as well as for the overall score. Does any additional insight arise from this analysis?

6   What do you think Robin should do? Why?

SS *KUNIANG*, PART II

This case asks you to find the optimal amount for NEES to bid for the SS *Kuniang* (page 107). Before doing so, though, you need additional details. Regarding the Coast Guard's (CG) salvage judgment, NEES believes that the following probabilities are an appropriate representation of its uncertainty about the salvage-value judgment:

$$P(\text{CG judgment} = \$9 \text{ million}) = 0.185$$
$$P(\text{CG judgment} = \$4 \text{ million}) = 0.630$$
$$P(\text{CG judgment} = \$1.5 \text{ million}) = 0.185$$

The obscure-but-relevant law required that NEES pay an amount (including both the winning bid and refitting cost) at least 1.5 times the salvage value for the ship in order to use it for domestic shipping. For example, if NEES bid $3.5 million and won, followed by a CG judgment of $4 million, then NEES would have to invest at least $2.5 million more: $3.5 + $2.5 = $6 = 4×1.5. Thus, assuming NEES submits the winning bid, the total investment amount required is either the bid or 1.5 times the CG judgment, whichever is greater.

As for the probability of submitting the highest bid, recall that winning is a function of the size of the bid; a bid of $3 million is sure to lose, and a bid of $10 million is sure to win. For this problem, we can model the probability of winning (P) as a linear function of the bid: $P = (\text{Bid} - \$3 \text{ million})/(\$7 \text{ million})$.

Finally, NEES's values of $18 million for the new ship and $15 million for the tug-barge alternatives are adjusted to reflect differences in age, maintenance, operating costs, and so on. The two alternatives provide equivalent hauling capacity. Thus, at $15 million, the tug-barge combination appears to be the better choice.

Questions

1 Reasonable bids may fall anywhere between $3 and $10 million. Some bids, though, have greater expected values and some less. Describe a strategy you can use to find the optimal bid, assuming that NEES's objective is to minimize the cost of acquiring additional shipping capacity. (*Hint:* This question just asks you to describe an approach to finding the optimal bid.)

2 Use your structure of the problem (or one supplied by the instructor), along with the details supplied above, to find the optimal bid.

REFERENCES

The solution of decision trees as presented in this chapter is commonly found in text-books on decision analysis, management science, and statistics. The decision-analysis texts listed at the end of Chapter 1 can provide more guidance in the solution of decision trees if needed. In contrast, the material presented here on the solution of influence diagrams is relatively new. For additional basic instruction in the construction and analysis of decisions using influence diagrams, the user's manual for PrecisionTree and other influence-diagram programs can be helpful.

The solution algorithm presented here is based on Shachter (1986). The fact that this algorithm deals with a decision problem in a way that corresponds to solving a symmetric decision tree means that the practical upper limit for the size of an influence diagram that can be solved using the algorithm is relatively small. Recent work has explored a variety of ways to exploit asymmetry in decision models and to solve influence diagrams and related representations more efficiently (Call and Miller 1990; Covaliu and Oliver 1995; Smith et al. 1993; Shenoy 1993).

An early and quite readable article on risk profiles is that by Hertz (1964). We have developed them as a way to examine the riskiness of alternatives in a heuristic way and also as a basis for examining alternatives in terms of deterministic and stochastic dominance. Stochastic dominance itself is an important topic in probability. Bunn (1984) gives a good introduction to stochastic dominance. Whitmore and Findlay (1978) and Levy (1992) provide thorough reviews of stochastic dominance.

Our discussion of assigning rating points and trade-off rates is necessarily brief in Chapter 4. These topics are covered in depth in Chapters 13 to 16. In the meantime, interested readers can get more information from Keeney (1992) and Keeney and Raiffa (1976).

Bodily, S. E. (1985) *Modern Decision Making.* New York: McGraw-Hill.

Bunn, D. (1984) *Applied Decision Analysis.* New York: McGraw-Hill.

Call, H., and W. Miller (1990) "A Comparison of Approaches and Implementations for Automating Decision Analysis." *Reliability Engineering and System Safety,* 30, 115–162.

Covaliu, Z., and R. Oliver (1995) "Representation and Solution of Decision Problems Using Sequential Decision Diagrams." *Management Science,* 41.

Hertz, D. B. (1964) "Risk Analysis in Capital Investment." *Harvard Business Review.* Reprinted in *Harvard Business Review,* September–October, 1979, 169–181.

Keeney, R. L. (1992) *Value-Focused Thinking.* Cambridge, MA: Harvard University Press.

Keeney, R., and H. Raiffa (1976) *Decisions with Multiple Objectives.* New York: Wiley.

Levy, H. (1992) "Stochastic Dominance and Expected Utility: Survey and Analysis." *Management Science,* 38, 555–593.

Shachter, R. (1986) "Evaluating Influence Diagrams." *Operations Research,* 34, 871–882.

Shenoy, P. (1993) "Valuation Network Representation and Solution of Asymmetric Decision Problems." Working paper, School of Business, University of Kansas, Lawrence.

Smith, J., S. Holtzman, and J. Matheson (1993) "Structuring Conditional Relationships in Influence Diagrams." *Operations Research,* 41, 280–297.

Whitmore, G. A., and M. C. Findlay (1978) *Stochastic Dominance.* Lexington, MA: Heath.

EPILOGUE

What happened with Texaco and Pennzoil? You may recall that in April of 1987 Texaco offered a $2 billion settlement. Hugh Liedtke turned down the offer. Within days of that decision, and only one day before Pennzoil began to file liens on Texaco's assets, Texaco filed for protection from creditors under Chapter 11 of the federal bankruptcy code, fulfilling its earlier promise. In the summer of 1987, Pennzoil submitted a financial reorganization plan on Texaco's behalf. Under their proposal, Pennzoil would receive approximately $4.1 billion, and the Texaco shareholders would be able to vote on the plan. Finally, just before Christmas 1987, the two companies agreed on a $3 billion settlement as part of Texaco's financial reorganization.

CHAPTER 5

Sensitivity Analysis

The idea of sensitivity analysis is central to the structuring and solving of decision models using decision-analysis techniques. In this chapter we will discuss sensitivity-analysis issues, think about how sensitivity analysis relates to the overall decision-modeling strategy, and introduce a variety of graphical sensitivity-analysis techniques.

The main example for this chapter is a hypothetical one in which the owner of a small airline considers expanding his fleet.

EAGLE AIRLINES

Dick Carothers, president of Eagle Airlines, had been considering expanding his operation, and now the opportunity was available. An acquaintance had put him in contact with the president of a small airline in the Midwest that was selling an airplane. Many aspects of the situation needed to be thought about, however, and Carothers was having a hard time sorting them out.

Eagle Airlines owned and operated three twin-engine aircraft. With this equipment, Eagle provided both charter flights and scheduled commuter service among several communities in the eastern United States. Scheduled flights constituted approximately 50% of Eagle's flights, averaging only 90 minutes of flying time and a distance of some 300 miles. The remaining 50% of flights were chartered. The mixture of charter flights and short scheduled flights had proved profitable, and

Carothers felt that he had found a niche for his company. He was aching to increase the level of service, especially in the area of charter flights, but this was impossible without more aircraft.

A Piper Seneca was for sale at a price of $95,000, and Carothers figured that he could buy it for between $85,000 and $90,000. This twin-engine airplane had been maintained according to FAA regulations. In particular, the engines were almost new, with only 150 hours of operation since a major overhaul. Furthermore, having been used by another small commercial charter service, the Seneca contained all of the navigation and communication equipment that Eagle required. There were seats for five passengers and the pilot, plus room for baggage. Typical airspeed was approximately 175 nautical miles per hour (knots), or 200 statute miles per hour (mph). Operating cost was approximately $245 per hour, including fuel, maintenance, and pilot salary. Annual fixed costs included insurance ($20,000) and finance charges. Carothers figured that he would have to borrow some 40% of the money required, and he knew that the interest rate would be two percentage points above the prime rate (currently 9.5% but subject to change). Based on his experience at Eagle, Carothers knew that he could arrange charters for $300 to $350 per hour or charge a rate of approximately $100 per person per hour on scheduled flights. He could expect on average that the scheduled flights would be half full. He hoped to be able to fly the plane for up to 1000 hours per year, but realized that 800 might be more realistic. In the past his business had been approximately 50% charter flights, but he wanted to increase that percentage if possible.

The owner of the Seneca has told Carothers that he would either sell the airplane outright or sell Carothers an option to purchase it within a year at a specified price. (The current owner would continue to operate the plane during the year.) Although the two had not agreed on a price for the option, the discussions had led Carothers to believe that the option would cost between $2500 and $4000. Of course, he could always invest his cash in the money market and expect to earn about 8%.

As Carothers pondered this information, he realized that many of the numbers he was using were estimates. Furthermore, some were within his control (for example, the amount financed and prices charged) while others, such as the cost of insurance or the operating cost, were not. How much difference did these numbers make? What about the option? Was it worth considering? Last, but not least, did he really want to expand the fleet? Or was there something else that he should consider?

Sensitivity Analysis: A Modeling Approach

Sensitivity analysis answers the question, "What makes a difference in this decision?" Returning to the idea of requisite decision models discussed in Chapter 1, you may recall that such a model is one whose form and content are just sufficient to solve a particular problem. That is, the issues that are addressed in a requisite decision model are the ones that matter, and those issues left out are the ones that do not matter. Determining what matters and what does not requires incorporating sensitivity analysis throughout the modeling process.

No "optimal" sensitivity-analysis procedure exists for decision analysis. To a great extent, model building is an art. Because sensitivity analysis is an integral part of the modeling process, its use as part of the process also is an art. Thus, in this chapter we will discuss the philosophy of model building and how sensitivity analysis helps with model development. Several sensitivity-analysis tools are available, and we will see how they work in the context of the Eagle Airlines example.

Problem Identification and Structure

The flowchart of the decision-analysis process in Figure 1.1 shows that sensitivity analysis can lead the decision maker to reconsider the very nature of the problem. The question that we ask in performing sensitivity analysis at this level is, "Are we solving the right problem?" The answer does not require quantitative analysis, but it does demand careful thought and introspection about the appropriate decision context. Why is this an important sensitivity-analysis concern? The answer is quite simple: Answering a different question, addressing a different problem, or satisfying different objectives can lead to a very different decision.

Solving the wrong problem sometimes is called an "error of the third kind." The terminology contrasts this kind of a mistake with Type I and Type II errors in statistics, where incorrect conclusions are drawn regarding a particular question. An error of the third kind, or Type III error, implies that the wrong question was asked; in terms of decision analysis, the implication is that an inappropriate decision context was used, and hence the wrong problem was solved.

Examples of Type III errors abound; we all can think of times when a symptom was treated instead of a cause. Consider lung disease. Researchers and physicians have developed expensive medical treatments for lung disease, the objective apparently being to reduce the suffering of lung-disease patients. If the fundamental objective is to reduce suffering from lung disease in general, however, these treatments are not as effective as antismoking campaigns. We can, in fact, broaden the context further. Is the objective really to reduce patient suffering? Or is it to reduce discomfort in general, including patient suffering as well as the discomfort of nonsmokers exposed to second-hand smoke? Considering the broader problem suggests an entirely different range of options.

For another example, think about a farmer who considers using expensive sprays in the context of deciding how to control pests and disease in an orchard. To a great extent, the presence of pests and disease in orchards result from the practice of monoculture—that is, growing a lot of one crop rather than a little each of many crops. A monoculture does not promote a balanced ecological system in which diseases and pests are kept under control naturally. Viewed from this broader perspective, the farmer might want to consider new agricultural practices rather than relying exclusively on sprays. Admittedly a long-term project, this requires the development of efficient methods for growing, harvesting, and distributing crops that are grown on a smaller scale.

How can one avoid a Type III error? The best solution is simply to keep asking whether the problem on the surface is the real problem. Is the decision context properly specified? What exactly is the "unscratched itch" that the decision maker feels? In the case of Eagle Airlines, Carothers appears to be eager to expand operations by acquiring more aircraft. Could he "scratch his itch" by expanding in a different direction? In particular, even though he, like many pilots, may be dedicated to the idea of flying for a living, it might be wise to consider the possibility of helping his customers to communicate more effectively at long distance. To some extent, efficient communication channels such as those provided by computer links and facsimile service, coupled with an air cargo network, can greatly reduce the need for travel. Pursuing ideas such as these might satisfy Carothers's urge to expand while providing a more diversified base of operations. So the real question may be how to satisfy Carothers's desires for expansion rather than simply how to acquire more airplanes.

We also can talk about sensitivity analysis in the context of problem structuring. Problem 3.20 gave an example in a medical context in which a decision might be sensitive to the structure. In this situation, the issue is the inclusion of a more complete description of outcomes; coronary bypass surgery can lead to complications that require long and painful treatment. Inclusion of this outcome in a decision tree might make surgery appear considerably less appealing. Von Winterfeldt and Edwards (1986) describe a problem involving the setting of standards for pollution from oil wells in the North Sea. This could have been structured as a standard regulatory problem: Different possible standards and enforcement policies made up the alternatives, and the objective was to minimize pollution while maintaining efficient oil production. The problem, however, was perhaps more appropriately structured as a competitive situation in which the players were the regulatory agency, the industry, and the potential victims of pollution. This is an example of how a decision situation might be represented in a variety of different ways. Sensitivity analysis can aid the resolution of the problem of multiple representations by helping to identify the appropriate perspective on the problem as well as by identifying the specific issues that matter to the decision maker.

Is problem structuring an issue in the Eagle Airlines case? In this case, the alternatives are to purchase the airplane, the option, or neither. Although Carothers might consider a variety of fundamental objectives, such as company growth or increased influence in the community, in the context of deciding whether to purchase the Seneca, it seems reasonable for him to focus on one objective: maximize profit. Carothers could assess the probabilities associated with the various unknown quantities such as operating costs, amount of business, and so on. Thus, it appears that a straightforward decision tree or influence diagram may do the trick.

Figure 5.1 shows an initial influence diagram for Eagle Airlines. Note that the diagram consists entirely of decision nodes and rounded rectangles. "Profit" is obviously the consequence node, and "Finance Cost," "Total Cost," and "Revenue" are intermediate-calculation nodes. All of the other rounded rectangles ("Interest Rate," "Price," "Insurance," "Operating Cost," "Hours Flown," "Capacity of Scheduled Flights," "Proportion of Chartered Flights") represent inputs to the calculations, and for now we represent these inputs as being constant. (Thus, in Figure 5.1 you can see the different

Figure 5.1

Influence diagram representing the Eagle Airlines decision.

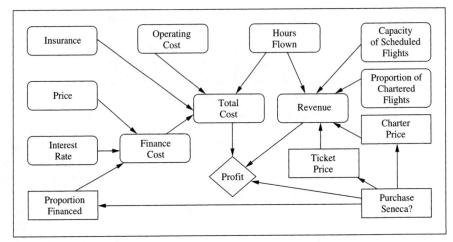

roles—constants and intermediate calculations—that rounded rectangles can play. Although these different roles may seem confusing, the basic idea is the same in each case; for any variable represented by a rounded rectangle, as soon as you know what its inputs are, you can calculate the value of the variable. In the case of the constants, there are no inputs, and so there is no calculation to do!)

Table 5.1 provides a description of the input and decision variables. This table also includes estimates (base values) and reasonable upper and lower bounds. The upper and lower bounds represent Carothers's ideas about how high and how low each of these variables might be. He might specify upper and lower bounds as absolute extremes, beyond which he is absolutely sure that the variable cannot fall. Another approach would be to specify the bounds such that he would be "very surprised" (a 1-in-10 chance, say) that the variable would fall outside the bounds.

The "Base Value" column in Table 5.1 indicates Carothers's initial guess regarding the 10 input variables. We can use these to make an estimate of annual profit (ignoring taxes for simplicity). The annual profit would be the total annual revenue minus the total annual cost:

$$
\begin{aligned}
\text{Total Revenue} &= \text{Revenue from Charters} + \text{Revenue from Scheduled Flights} \\
&= (\text{Charter Proportion} \times \text{Hours Flown} \times \text{Charter Price}) \\
&\quad + [(1 - \text{Charter Proportion}) \times \text{Hours Flown} \times \text{Ticket Price} \\
&\quad \times \text{Number of Passenger Seats} \times \text{Capacity of Scheduled Flights}] \\
&= (0.5 \times 800 \times \$325) + (0.5 \times 800 \times \$100 \times 5 \times 0.5) \\
&= \$230,000
\end{aligned}
$$

$$
\begin{aligned}
\text{Total Cost} &= (\text{Hours Flown} \times \text{Operating Cost}) + \text{Insurance} + \text{Finance Cost} \\
&= (\text{Hours Flown} \times \text{Operating Cost}) + \text{Insurance} \\
&\quad + (\text{Price} \times \text{Proportion Financed} \times \text{Interest Rate}) \\
&= (800 \times \$245) + \$20,000 + (\$87,500 \times 0.4 \times 11.5\%) \\
&= \$220,025
\end{aligned}
$$

Table 5.1

Input variables and ranges of possible values for Eagle Airlines aircraft-purchase decision.

Variable	Base Value	Lower Bound	Upper Bound
Hours Flown	800	500	1000
Charter Price/Hour	$325	$300	$350
Ticket Price/Hour	$100	$95	$108
Capacity of Scheduled Flights	50%	40%	60%
Proportion of Chartered Flights	0.50	0.45	0.70
Operating Cost/Hour	$245	$230	$260
Insurance	$20,000	$18,000	$25,000
Proportion Financed	0.40	0.30	0.50
Interest Rate	11.5%	10.5%	13%
Purchase Price	$87,500	$85,000	$90,000

Thus, using the base values, Carothers's annual profit is estimated to be $230,000 – $220,025 = $9975. This represents a return of approximately 19% on his investment of $52,500 (60% of the purchase price).

One-Way Sensitivity Analysis

The sensitivity-analysis question in the Eagle airlines case is, what variables really make a difference in terms of the decision at hand? For example, do different possible interest rates really matter? Does it matter that we can set the ticket price? If Hours Flown changes by some amount, how much impact is there on Profit? We can begin to address questions like these with one-way sensitivity analysis.

Let us consider Hours Flown. From Table 5.1, we see that Carothers is not at all sure what Hours Flown might turn out to be, and that it can vary from 500 to 1000 hours. What does this imply for Profit? The simplest way to answer this question is with a one-way sensitivity graph as in Figure 5.2. The upward-sloping line in Figure 5.2 shows profit as Hours Flown varies from 500 to 1000; to create this line, we have substituted different values for Hours Flown into the calculations detailed above. The horizontal line represents the amount of money ($4200) that Carothers could earn from the money market. The point where these lines cross is the threshold at which the two alternatives each yield the same profit ($4200), which occurs when Hours Flown equals 664. The heavy line indicates the maximum profit Carothers could obtain at different values of Hours Flown, and the different segments of this line are associated with different strategies (buy the Seneca versus invest in the money market). The fact that Carothers believes that Hours Flown could be above or below 664 suggests that this is a crucial variable and that he may need to think more carefully about the uncertainty associated with it.

Figure 5.2

One-way sensitivity analysis of hours flown.

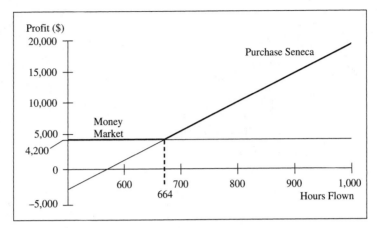

Tornado Diagrams

A *tornado diagram* allows us to compare one-way sensitivity analysis for many input variables at once. Suppose we take each input variable in Table 5.1 and "wiggle" that variable between its high and low values to determine how much change is induced in Profit. Figure 5.3 graphically shows how annual profit varies as the input variables are independently wiggled between the high and low values. For instance, with everything else held at the base value, setting Capacity of Scheduled Flights at 0.4 instead of 0.5 implies a loss of $10,025. That is, plug all the base values into the revenue equation above, except use 0.4 for Capacity of Scheduled Flights:

Total Revenue = Revenue from Charters + Revenue from Scheduled Flights
 = (Charter Proportion × Hours Flown × Charter Price)
 + [(1 − Charter Proportion) × Hours Flown × Ticket Price
 × Number of Passenger Seats × Capacity on Scheduled Flights]
 = (0.5 × 800 × $325) + (0.5 × 800 × $100 × 5 × 0.4)
 = $210,000

Nothing in the cost equation changes, and so cost still is estimated as $220,025. The estimated loss is just the difference between cost and revenue: $210,000 − $220,025 = −$10,025. This is plotted on the graph as the left end of the bar labeled Capacity of Scheduled Flights. On the other hand, setting Capacity of Scheduled Flights at the high end of its range, 0.6, leads to a profit of $29,975. (Again, plug all of the same values into the revenue equation, but use 0.6 for capacity.) Thus, the right end of the capacity bar is at $29,975.

We follow this same procedure for each input variable. The length of the bar for any given variable represents the extent to which annual profit is sensitive to this variable. The graph is laid out so that the most sensitive variable—the one with the

Figure 5.3

Tornado diagram for the Eagle Airlines case. The bars represent the range for the annual profit when the specified quantity is varied from one end of its range to the other, keeping all other variables at their base values.

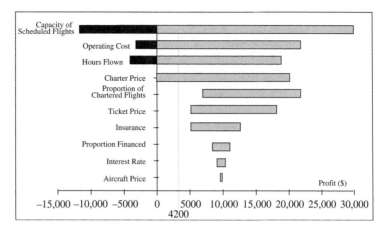

longest bar—is at the top, and the least sensitive is at the bottom. With the bars arranged in this order, it is easy to see why the graph is called a tornado diagram. Later in this chapter we will learn how to generate tornado diagrams in PrecisionTree.

The vertical line at $4200 represents what Carothers could make on his investment if he left his $52,500 in the money market account earning 8%. If he does not think he can earn more than $4200, he should not purchase the Seneca.

Interesting insights can be gleaned from Figure 5.3. For example, Carothers's uncertainty regarding Capacity of Scheduled Flights is extremely important. On the other hand, the annual profit is very insensitive to Aircraft Price. What can we do with information like this? The tornado diagram tells us which variables we need to consider more closely and which ones we can leave at their base values. In this case, annual profit is insensitive to Proportion Financed, Interest Rate, and Aircraft Price, so in further analyzing this decision we simply can leave these variables at their base values. And yet Capacity of Scheduled Flights, Operating Cost, Hours Flown, and Charter Price all have substantial effects on the annual profit; the bars for these four variables cross the critical $4200 line. Proportion of Chartered Flights, Ticket Price, and Insurance each has a substantial effect on the profit, but the bars for all of these variables lie entirely above the $4200 line. In a first pass, these variables might be left at their base values, and the analyst might perform another sensitivity analysis at a later stage.

Dominance Considerations

In our discussion of making decisions in Chapter 4, we learned that alternatives can be screened on the basis of deterministic and stochastic dominance, and inferior alternatives can be eliminated. Identifying dominant alternatives can be viewed as a version of sensitivity analysis for use early in an analysis. In sensitivity-analysis terms, analyzing alternatives for dominance amounts to asking whether there is any way that one alternative could end up being better than a second. If not, then the first alternative is dominated by the second and can be ignored.

In the case of Eagle Airlines, an immediate question is whether purchasing the option is a dominated alternative. Why would Carothers want to buy the option? There are two possibilities. First, it would allow him to lock in a favorable price on a suitable aircraft while he tried to gather more information. Having constructed a tornado diagram for the problem, we can explore the potential value of purchasing the option by considering the amount of information that we might obtain and the potential impact of this information. A second typical motivation for purchasing an option is to wait and see whether the economic climate for the venture becomes more favorable. In this case, if the commuter/charter air-travel market deteriorates, then Carothers has only lost the cost of the option. (Some individuals also purchase options to lock in a price while they raise the required funds. Carothers, however, appears to have the required capital and credit.)

It is conceivable that Carothers could obtain more accurate estimates of certain input variables. Considering the tornado diagram, he would most like to obtain information about the more critical variables. Some information regarding market variables (Capacity of Scheduled Flights, Hours Flown, and Charter Ratio) might be obtainable through consumer-intentions surveys, but it would be far from perfect as well as costly. The best way to obtain such information would be to purchase or lease an aircraft for a year and try it—but then he might as well buy the Seneca!

What about Operating Cost and Insurance? The main source of uncertainty for Operating Cost is fuel cost, and this is tied to the price of oil, which can fluctuate dramatically. Increases in Insurance are tied to changes in risk as viewed by the insurance companies. Rates have risen dramatically over the years, and stability is not expected. The upshot of this discussion is that good information regarding many of the input variables probably is not available. As a result, if Carothers is interested in acquiring the option in order to have the chance to gather information, he might discover that he is unable to find what he needs.

What about the second motivation, waiting to see whether the climate improves? The question here is whether any uncertainty will be resolved during the term of the option, and whether or not the result would be favorable to Eagle Airlines. In general, considerable uncertainty regarding all of the market variables will remain regardless of how long Carothers waits. Market conditions can fluctuate, oil prices can jump around, and insurance rates can change. On the other hand, if some event is anticipated, such as settlement of a major lawsuit or the creation of new regulations, then the option could protect Carothers until this uncertainty is resolved. (Notice that, even in this case, the option provides Carothers with an opportunity to collect information—all he must do is wait until the uncertain situation is resolved.) But Carothers does not appear to be awaiting the resolution of some major uncertainty. Thus, if his motivation for purchasing the option is to wait to see whether the climate improves, it is not clear whether he would be less uncertain about the economic climate when the option expires.

What are the implications of this discussion? It is fairly clear that, unless an inexpensive information-gathering strategy presents itself, purchasing the option probably is a dominated alternative. For the purposes of the following analysis, we will assume that Carothers has concluded that no such information-gathering strategy exists, and that purchasing the option is unattractive. Thus, we can reduce his alternatives to (1) buying the airplane outright and (2) investing in the money market.

Two-Way Sensitivity Analysis

The tornado-diagram analysis provides considerable insights, although these are limited to what happens when only one variable changes at a time. Suppose we wanted to explore the impact of several variables at one time? This is a difficult problem, but a graphical technique is available for studying the interaction of two variables.

Suppose, for example, that we want to consider the joint impact of changes in the two most critical variables, Operating Cost and Capacity of Scheduled Flights. Imagine a rectangular space (Figure 5.4) that represents all of the possible values that these two variables could take. Now, let us find those values of Operating Cost and Capacity for which the annual profit would be less than $4200. If this is to be the case, then we must have total revenues minus total costs less than $4200 or total revenues less than total costs plus $4200:

(Charter Proportion × Hours Flown × Charter Price) + [(1 − Charter Proportion)

    × Hours Flown × Ticket Price × Number of Seats

    × Capacity of Scheduled Flights]

  < (Hours Flown × Operating Cost) + Insurance

    + (Price × Percent Financed × Interest Rate) + 4200

Inserting the base values for all but the two variables of interest, we obtain

Figure 5.4

Two-way sensitivity graph for Eagle Airlines. The Line AB represents the points for which profit would be $4200.

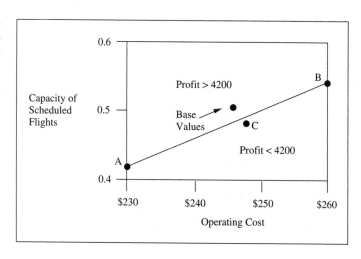

$$(0.5 \times 800 \times 325) + [0.5 \times 800 \times 100 \times 5 \times \text{Capacity}]$$
$$< (800 \times \text{Operating Cost}) + 20{,}000 + (87{,}500 \times 0.4 \times 0.115) + 4200$$

which reduces to

$$130{,}000 + (200{,}000 \times \text{Capacity}) < (800 \times \text{Operating Cost}) + 28{,}225$$

Now solve this inequality for Capacity in terms of Operating Cost to get

$$\text{Capacity} < 0.004 \times \text{Operating Cost} - 0.509$$

This inequality defines the region in which purchasing the airplane would lead to a profit of less than $4200. When the "<" sign is replaced with an equality, we have the points for which profit equals $4200, or the line of points where the venture just breaks even relative to investing in the money market. To plot this line, notice that we only need two points. The simplest way to come up with these is to plug in the extreme values for Operating Cost and calculate the corresponding values for Capacity. Doing this gives the break-even points A (Capacity = 0.411 when Operating Cost = 230) and B (Capacity = 0.531 when Operating Cost = 260). These points define Line AB in Figure 5.4. The area below the line (Capacity < 0.004 × Operating Cost − 0.509) represents the region when the profit would be less than $4200. The area above the line represents the region in which the profit would be greater than $4200.

What insight can Carothers gain from Figure 5.4? The point labeled "Base Values" shows that when we plug in the base values for the capacity and operating-cost variables, we get an estimated profit that is greater than $4200, and so the project looks promising. But Carothers might be wondering how likely it is that the two variables might work together to lead to a profit of less than $4200. For example, suppose that Operating Cost was slightly more than the base value, say $248, and that Capacity was just slightly less than the base value, say 48%. Taken individually, these two values do not seem to cause a problem. That is, substituting either one into the profit calculations, while keeping the other at its base value, leads to profit that is still greater than $4200. When we consider these two values jointly (Point C in Figure 5.4), however, they lead to a situation in which it would have been better not to buy the airplane. If Carothers thinks that the two variables might be likely to fall in the "Profit < $4200" region, then he may wish to forego the purchase. But such a situation would indicate that he really needs to model his uncertainty about these variables using probability methods. In the next section we will see how two-way sensitivity analysis can be used in conjunction with probabilities.

Sensitivity to Probabilities

The next step in our analysis will be to model the uncertainty surrounding the critical variables identified by our analysis of the tornado diagram. The four critical variables were (1) Capacity of Scheduled Flights, (2) Operating Cost, (3) Hours

Flown, and (4) Charter Price. We only need to think about uncertainty for the first three, because charter price is a decision variable set by Carothers. For the purposes of the example here, let us assume that, in an initial attempt to model the uncertainty, Carothers chooses two values for each variable, one representing an optimistic and one a pessimistic scenario. The influence diagram is shown in Figure 5.5 and shows changes in the model based on the sensitivity analysis so far. Operating Cost, Hours Flown, and Capacity of Scheduled Flights have been changed to chance nodes. The remaining input variables (Interest Rate, Proportion Financed, Price, Insurance, Ticket Price, and Proportion of Chartered Flights) have been left at their base values and hence continue to be represented by rounded rectangles (constants). The decision tree in Figure 5.6 shows the pessimistic and optimistic values for the three uncertain variables.

Now that we have simplified the problem somewhat, we can include considerations regarding the interdependence of the remaining chance variables. In Figures 5.5 and 5.6, the probability distribution for Hours Flown is judged to depend on the Capacity of Scheduled Flights: If Capacity is low, then this may actually result in some flights being canceled and thus fewer total hours. Thus, a relevance arc leads from "Capacity of Scheduled Flights" to "Hours Flown" in the influence diagram, and in the decision tree the value for $r = P(\text{Low Hours} \mid \text{Low Capacity})$ may not be the same as the value for $s = P(\text{Low Hours} \mid \text{High Capacity})$. In fact, our argument suggests that r will be greater than s. On the other hand, Operating Cost is judged to be independent of the other variables.

The next thing to do is to assess some values for probabilities p, q, r, and s. Let us suppose that Carothers is comfortable with an assessment that $p = 0.5$, or that

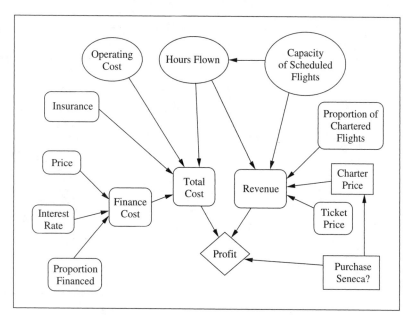

Figure 5.5

Influence diagram of Eagle Airlines decision. Note that only three variables are considered to be uncertain, and that Hours Flown and Capacity are considered to be probabilistically dependent.

Figure 5.6

Decision tree for Eagle Airlines with uncertainty for three variables. Profit is calculated with all other variables held at their base values.

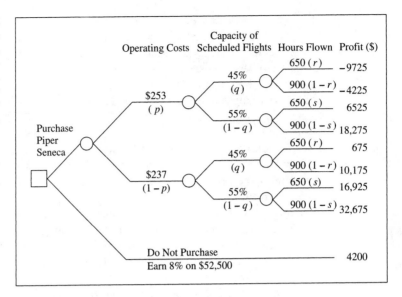

Operating Cost is just as likely to be high ($253) as low ($237). Furthermore, suppose that Carothers feels that a reasonable way to represent the dependence between Hours and Capacity is to let s be 80% of r. That is, if Capacity is high (55%), then the probability that Hours = 650 is only 80% of the probability that Hours = 650 when Capacity is low. With these two specifications, we now have only two unspecified probabilities left to consider, q and r. Figure 5.7 shows the modified decision tree with $p = 0.5$ and $s = 0.8r$.

Figure 5.7

Eagle Airlines' decision tree with probabilities substituted for p and s. This decision tree is now ready for a two-way sensitivity analysis on q and r.

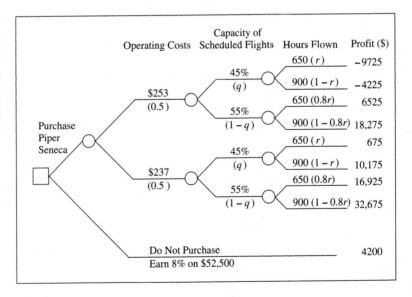

We now can create a two-way sensitivity graph for q and r. As with the two-way sensitivity analysis above, the graph will show regions for which the expected value of purchasing the Seneca is greater than investing in the money market.

To create the graph, we first write out the expected value of purchasing the airplane in terms of q and r, including the specifications that $p = 0.5$ and $s = 0.8r$. This equation comes from solving the decision tree:

$$\text{EMV(Purchase)} = 0.5\{q[-9725r - 4225(1 - r)] + (1 - q)[6525(0.8r)$$
$$+ 18{,}275(1 - 0.8r)]\} + 0.5\{q[675r + 10{,}175(1 - r)]$$
$$+ (1 - q)[16{,}925(0.8r) + 32{,}675(1 - 0.8r)]\}$$

After algebraic reduction, this expression becomes

$$\text{EMV(Purchase)} = q(3500r - 22{,}500) - 11{,}000r + 25{,}475$$

We would want to purchase the airplane if EMV(Purchase) > 4200. Thus, we can solve the following inequality for q in terms of r:

$$q(3500r - 22{,}500) - 11{,}000r + 25{,}475 > 4200$$
$$25{,}475 - 4200 - 11{,}000r > q(22{,}500 - 3500r)$$

This inequality reduces to

$$\frac{21{,}275 - 11{,}000r}{22{,}500 - 3500r} > q$$

Using this inequality, we can create a two-way sensitivity graph for Eagle Airlines (Figure 5.8). The curve separating the two regions represents the values of q and r for which EMV(Purchase) = \$4200. It was plotted by plugging values for r between 0 and 1 into the inequality above. For these values of q and r, Carothers should be indifferent (in terms of EMV) between buying the airplane and not. The area below the line contains points where $q < (21{,}275 - 11{,}000r)/(22{,}500 - 3500r)$; for these (q, r) points, EMV(Purchase) > \$4200. The graph makes sense because q and r are probabilities of the pessimistic scenarios—low Capacity and low number of Hours Flown. If Carothers thinks that the pessimistic scenarios are likely (q and r close to 1), then he would not want to buy the airplane.

The importance of Figure 5.8 is that Carothers may not have especially firm ideas of what the probabilities q and r should be. Suppose, for example, that in the process of coming up with the probabilities he feels that q could be between 0.4 and 0.5 and that r could be between 0.5 and 0.65. These probabilities are represented by the points inside Rectangle A in Figure 5.8. All of these points fall within the "Purchase Seneca" region, and so the conclusion is that the Seneca should be purchased. The decision is not sensitive to the assessment of the probabilities. If, on the other hand, Carothers thinks that reasonable values of q and r fall in Rectangle B, then the optimal choice is not clear. (No wonder the decision is a hard one!) In this situation, he could reflect on the chances associated with Capacity and Hours Flown

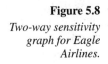

Figure 5.8

Two-way sensitivity graph for Eagle Airlines.

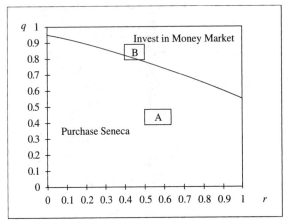

and try to refine his model of the uncertainty. Decision-analysis tools for modeling uncertainty more carefully are discussed in Chapters 7 through 12.

The value of the two-way sensitivity graph is to provide guidance in determining how much effort is needed to model uncertainty in a decision problem. Looking at it another way, the graph can reveal whether the decision is sensitive to the uncertainty in the problem and to the modeling of that uncertainty.

Performing a sensitivity analysis by hand involves manipulating and solving algebraic equations and inequalities. Who wants to go through all this when you can have a computer run your sensitivity analyses! Two of the programs (PrecisionTree and @RISK) in Palisade's DecisionTools have built-in sensitivity-analysis features, and one of the programs (TopRank) is devoted entirely to sensitivity analysis. These programs approach sensitivity analysis by recalculating the spreadsheet many times for different input values. They monitor the output values for each set of input values and create various tables and graphs of the output. Each of the programs will generate tornado diagrams, and one-way and two-way sensitivity graphs. From these tables and graphs it will be easy to identify the impact of each variable and determine the most influential variables. Instructions on how to use PrecisionTree and TopRank are given at the end of this chapter.

Two-Way Sensitivity Analysis for Three Alternatives (Optional)

As we have analyzed it here, the Eagle Airlines case involves only two alternatives, and so the sensitivity graph has two regions. What happens when there are more than two alternatives? The graph may contain a region for each alternative. Let us consider a stock market–investment problem.

INVESTING IN THE STOCK MARKET

An investor has funds available to invest in one of three choices: a high-risk stock, a low-risk stock, or a savings account that pays a sure $500. If he invests in the stocks, he must pay a brokerage fee of $200.

His payoff for the two stocks depends in part on what happens to the market as a whole. If the market goes up (as measured, say, by the Standard and Poor's 500 Index increasing 8% over the next 12 months), he can expect to earn $1700 from the hish-risk stock and $1200 from the low-risk stock. Finally, if the stock market goes down (as indicated by the index decreasing by 3% or more), he will lose $800 with the high-risk stock but still gain $100 with the low-risk stock. If the market stays at roughly the same level, his payoffs for the high- and low-risk stocks will be $300 and $400, respectively.

The decision tree is given in Figure 5.9, with unspecified probabilities $t = $ P(market up) and $v = $ P(market same). Of course, P(market down) $= 1 - t - v$ because the probabilities must sum to 1.

To construct the graph, we must compare the alternatives two at a time. First we have to realize that $t + v$ must be less than or equal to 1. Thus, the graph (see Figure 5.10) is a triangle rather than a rectangle because all of the points above a line from $(t = 1, v = 0)$ to $(t = 0, v = 1)$ are not feasible. To find the strategy regions, begin by finding the area where the savings account would be preferred to the low-risk stock, or

$$\text{EMV(Savings Account)} > \text{EMV(Low-Risk Stock)}$$

$$500 > t(1000) + v(200) - (1 - t - v)\,100$$

Solving for v in terms of t, we get

$$v < 2 - \frac{11t}{3}$$

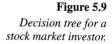

Figure 5.9

Decision tree for a stock market investor.

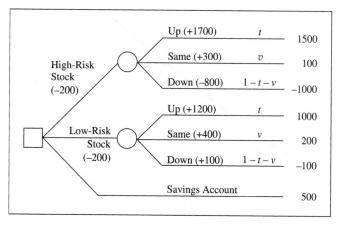

Figure 5.10

Beginning the analysis of the stock market problem. Note that $t + v$ must be less than or equal to 1, and so the only feasible points are within the large triangular region.

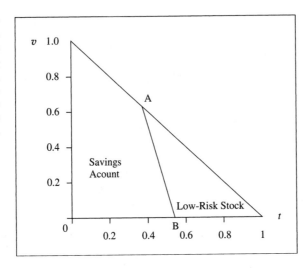

Figure 5.10 shows the regions for the savings account and the low-risk stock divided by Line AB.

Now let us find the regions for the high- and low-risk stocks. Begin by setting up the inequality

EMV(Low-Risk Stock) > EMV(High-Risk Stock)

$$t(1000) + v(200) - (1 - t - v)100 > t(1500) + v(100) - (1 - t - v)1000$$

This reduces to

$$v < \frac{9}{8} - \frac{7t}{4}$$

Using this inequality, we can add another line to our graph (Figure 5.11). Now Line CDE separates the graph into regions in which EMV(Low-Risk Stock) is greater or less than EMV(High-Risk Stock).

From Figure 5.11 we can tell what the optimal strategy is in all but one portion of the graph. For example, in ADEG, we know that the high-risk stock is preferred to the low-risk stock and that the low-risk stock is preferred to the savings account. Thus, the high-risk stock would be preferred overall. Likewise, in HFBDC the savings account would be preferred, and in DBE the low-risk stock would be preferred. But in CDA, all we know is that the low-risk stock is worse than the other two, but we do not know whether to choose the savings account or the high-risk stock.

If the decision maker is sure that the probabilities t and v do not fall into the region CDA, then the sensitivity analysis could stop here. If some question remains (or even if we feel compelled to finish the job), then we can complete the graph by comparing EMV(Savings Account) with EMV(High-Risk Stock):

EMV(Savings Account) > EMV(High-Risk Stock)

$$500 > t(1500) + v(100) - (1 - t - v)1000$$

Figure 5.11

Second stage in analysis of the stock market problem. A second inequality has been incorporated. The optimal strategy is clear now for all regions except CDA.

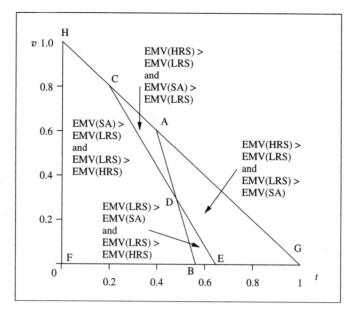

Figure 5.12

Completed two-way sensitivity graph for the stock market problem. Line ID has split region CDA.

This inequality reduces to

$$v < \frac{15}{11} - \frac{25t}{11}$$

Incorporating this result into the graph allows us to see that region CDA actually is split between the high-risk stock and the savings account as indicated by Line ID in Figure 5.12.

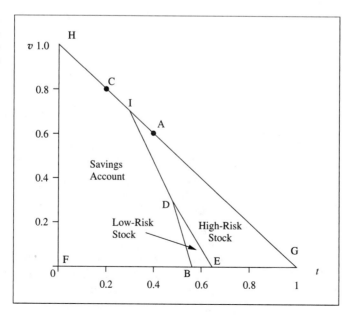

With the analysis completed, the investor now can think about probabilities t and v. As in the Eagle Airlines case, it should be possible to tell whether the optimal investment decision is sensitive to these probabilities and whether additional effort should be spent modeling the uncertainty about the stock market.

Sensitivity Analysis in Action

Is sensitivity analysis ever used in the real world? Indeed it is. This fundamental approach to modeling is the source of important insights and understanding in many real-world problems. The following example comes from medical decision making, showing how sensitivity-analysis graphs can improve decisions in an area where hard decisions are made even harder by the stakes involved.

HEART DISEASE IN INFANTS

Macartney, Douglas, and Spiegelhalter used decision analysis to study alternative treatments of infants who suffered from a disease known as coarctation of the aorta. Difficult to detect, the fundamental uncertainty is whether the disease is present at all. Three alternative treatments exist if an infant is suspected of having the disease. The first is to do nothing, which may be appropriate if the disease is not present. The second is to operate. The third alternative is to catheterize the heart in an attempt to confirm the diagnosis, although it does not always yield a perfect diagnosis. Moreover, catheterizing the heart of a sick infant is itself a dangerous undertaking and may lead to death. The difficulty of the problem is obvious; with all of the uncertainty and the risk of death from operating or catheterization, what is the appropriate treatment?

Source: F. Macartney, J. Douglas, and D. Spiegelhalter (1984) "To Catheterise or Not to Catheterise?" *British Heart Journal*, 51, 330–338.

In their analysis Macartney et al. created a two-way sensitivity graph (Figure 5.13) showing the sensitivity of the decision to two probabilities. The two probabilities are (1) the disease is present, which is along the horizontal axis, and (2) the mortality rate for cardiac catheterization, which is along the vertical axis. The mortality rate also could be interpreted as the physician's judgment regarding the chance that the infant would die as a result of catheterization.

The graph shows three regions, reflecting the three available alternatives. The location of the three regions makes good sense. If the physician believes that the chances are low that the disease is present and that the risk of catheterizing the infant is high, then the appropriate response is to do nothing. On the other hand, if the risk of catheterization is high relative to the chance that the disease is present, then operating without catheterizing is the prescribed treatment. Catheterization is recommended only for situations with relatively low risk from the procedure.

Figure 5.13

Two-way sensitivity analysis for the heart disease treatment decision.
Source: Macartney et al. (1984).

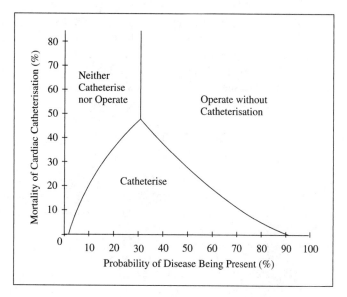

Sensitivity Analysis Using TopRank and PrecisionTree

In this section, we demonstrate how to perform sensitivity analysis in two ways, first with a program called TopRank and second with PrecisionTree. Whereas Precision-Tree's sensitivity analysis is designed specifically to analyze decision trees, TopRank is a general sensitivity-analysis program designed to analyze spreadsheet models. Both programs make running a sensitivity analysis easy. With only a few keystrokes, you can systematically recalculate your model and create a variety of tables and graphs. First, we will describe how to perform a sensitivity analysis on a spreadsheet model of the Eagle Airlines example using TopRank, and then on the decision-tree model using PrecisionTree.

TopRank

With TopRank, you can either run a fully automated analysis or design a sensitivity analysis to meet your specific needs. The automated feature is very convenient, taking only two clicks of the mouse to analyze a spreadsheet model. The downside, however, is that you must accept the default settings, such as varying each input value plus and minus 10%, which may not be suitable for every input variable. Let's begin by running a fully automated sensitivity analysis, and then later we will learn how to modify the settings. We will not cover all of TopRank's features, but you can use the on-line manual to learn about TopRank's capabilities.

TopRank works by replacing input values with *Vary* functions. *Vary* functions determine how the input values are varied, or wiggled, during the sensitivity analysis. For example, in the automated analysis, an input value of 90 is replaced by

AutoVary(90, –10, 10), indicating that the base value 90 will be varied from –10% to +10% of base; i.e., from 81 to 99. The default number of steps is preset at 4, which means that the spreadsheet will be recalculated 4 times at equally spaced values; i.e., once at 81, once at 87, once at 93, and finally at 99. If your model has a total of 10 input variables, then the automated sensitivity analysis will recalculate the spreadsheet 40 times: 4 steps for each of the 10 input variables.

TopRank—Automated Analysis

STEP 1

1.1   Open both Excel and TopRank, enabling any macros if prompted. A TopRank toolbar is added to the Excel toolbar and is explained in Figure 5.14.

1.2   At any time you may access TopRank's on-line help by clicking on the **Show TopRank Window** button (second from the right), and choosing **Help**, then **Contents** in the pull-down menu.

1.3   Build the spreadsheet model shown in Figure 5.15. The equations in cells B16 to B19 are shown in cells C16 to C19. Alternatively, open the spreadsheet on the Palisade CD: Examples\Chapter5\EagleAir.xls.

You are now only two clicks away from completing a sensitivity analysis!

STEP 2

2.1   To set "Profit" as the output variable, highlight cell **B19** and click on the **Add Output** button on the TopRank toolbar (fourth button from the left).

Figure 5.14

TopRank's toolbar and its functions.

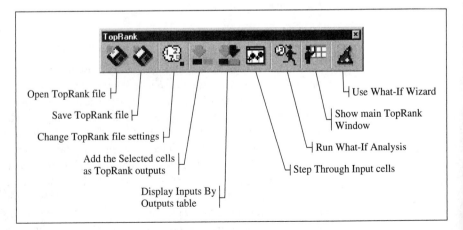

Figure 5.15

Spreadsheet model of Eagle Airlines' purchase decision. The formulas for cells B16 through B19 are displayed in C16 to C19.

	A	B	C	D	E	F
4		Base Value		Lower Bound		Upper Bound
5	Hours Flown	800		500		1000
6	Capacity	50%		40%		60%
7	Ticket Price	$ 100	$	95	$	108
8	Charter Price	$ 325	$	300	$	350
9	Charter Proportion	50%		45%		70%
10	Operating Cost	$ 245	$	230	$	260
11	Insurance	$ 20,000	$	18,000	$	25,000
12	Price	$ 87,500	$	85,000	$	90,000
13	Interest Rate	11.5%		10.5%		13.0%
14	Proportion Financed	40%		30%		50%
15						
16	Finance Cost	$ 4,025	=B12*B13*B14			
17	Revenue	$ 230,000	=(B5*B8*B9)+((1-B9)*B5*B6*5*B7)			
18	Total Cost	$ 220,025	=(B5*B10)+B11+B16			
19	Profit	$ 9,975	=B17-B18			

This action identifies B19 ("Profit") as the output variable and replaces all the cell values that B19 refers to with *AutoVary* functions. Because each cell from B5 to B14 is used to calculate B19, each has had its value replaced by *AutoVary* functions. For example, the value of 800 for "Hours Flown" in cell B5 has been replaced by *RiskAutoVary*(800,−10,10). The −10 and 10 mean that this cell will be varied from a low of −10% of 800, or 720, to a high of 10% of 800, or 880. Notice that the lower bound of 720 and the upper bound of 880 differ from the bounds of 500 and 1000 given by Carothers. This is the price you pay for automation. Later, we show how to modify TopRank so that it uses the correct boundary values.

Note that TopRank searches the spreadsheet for any numerical value that affects the output cell and replaces it with an *AutoVary* function. TopRank only replaces values and not cell references, so the contents of cells B16, B17, and B18 are left as they are even though they do affect B19.

STEP 3   We now perform the sensitivity analysis.

3.1   Click on the **Run What-If Analysis** button on the TopRank toolbar (Figure 5.14, third button from the right).

This action recalculates "Profit" for the 40 different input values and displays the output in two windows: *Results* and *Detail by Input*. The *Results* window (Figure 5.16) summarizes the analysis, and the *Detail by Input* window lists the specific input values used in calculating "Profit" along with the corresponding percentage change.

Reviewing the *Results* window in Figure 5.16, we see that TopRank's sensitivity analysis ranks "Operating Cost" as the most important profitability factor, followed

Figure 5.16

Results window of What-If Analysis for Eagle Airlines.

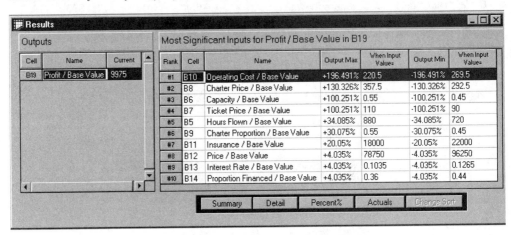

by "Charter Price," then "Capacity," and so on. TopRank calculates the percentage change in the output using the formula

$$100 * \frac{(\text{Profit Calculated at New Input } - \text{ Profit at Base})}{(\text{Profit at Base})}\%$$

The numerator measures the amount that the output variable has changed for the given input value, and this is converted to a percentage of the base profit.

Returning to the results in Figure 5.16, we see that when "Operating Cost" is at the minimum value of $220.50, then the percentage change in "Profit" increases by 196%. Conversely, raise "Operating Cost" to the maximum value of $269.50, and "Profit" drops by the same 196%. You are not limited to viewing the sensitivity-analysis results in terms of percentage change. You may, for example, prefer to view the output in the original units, which in this case is dollars. There are a number of other options as well:

3.2 Click the **Actuals** button that is along the bottom of the *Results* window and dollar values replace the percentage change values in the *Output Max* and *Output Min* columns. The dollar value that corresponds to the 196% increase is $29,575. Note that the *Actuals* button also changes the output values in the *Detail by Input* window to dollar values.

3.3 To return to the original window, click the **Percent%** button next to the *Actuals* button.

3.4 Click the **Detail** button and an upward pointing blue arrow marks those variables that cause increases above 20%, while a downward pointing red arrow marks those that cause decreases below 20%. Scrolling down you can see that all 40 input values are shown along with the corresponding percentage changes and dollar values.

3.5   Click the **Summary** button and only the minimum and maximum output values are shown.

3.6   Click the **Graph** button above the *Results* window and a dialog box pops up, in which three graph options are presented: *tornado graph, sensitivity graph*, and *spider graph*.

3.7   Click the **tornado graph,** then click **OK**, and a tornado diagram is produced, as shown in Figure 5.17.

3.8   Highlight **Hours Flown** in the *Results* window (cell B5, ranked #5), then successively click the **Graph** button, the **sensitivity** option, and **OK**. A graph is produced with "Hours Flown" along the *x* axis and "Profit" along the *y* axis (Figure 5.18). This graph shows how "Profit" values change as "Hours Flown" moves from –10% of base to +10% of base. Because TopRank uses the currently highlighted variable as the *x* axis variable, be sure you select the desired variable in the *Results* window before clicking the graph button.

3.9   Click the **Graph** button, the **spider** option, and **OK.** A spider graph is produced, which is a compilation of all the sensitivity graphs (Figure 5.19). Its name and its look derive from the fact that all the curves intersect at the same point (base value) and radiate out, like the legs of a spider. Notice that the *x* axis for the spider graph is percent change. Because the input variables typically have different units and scales, TopRank uses percent change as a common scale so that the variables can be plotted in a single graph.

Figure 5.17

Tornado graph for the automated sensitivity analysis on the Eagle Airlines model.

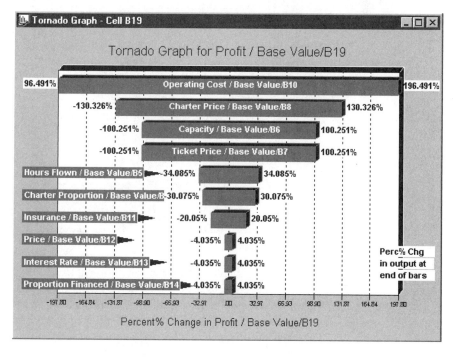

Figure 5.18

Sensitivity graph of "Hours Flown" for the automated sensitivity analysis on the Eagle Airlines model.

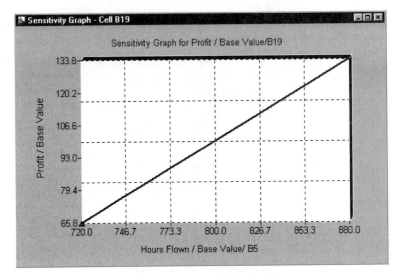

Figure 5.19

Spider graph for the automated sensitivity analysis on the Eagle Airlines model.

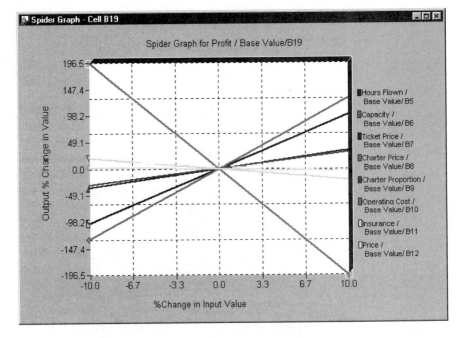

Remember that the results here are different from those previously presented, because the automated analysis has varied each input plus and minus 10%, which is not consistent with the range of values given in Table 5.1.

You can edit these graphs in either TopRank or Excel. To edit, place the cursor on the graph and click the right mouse button. A small pop-up window appears with two

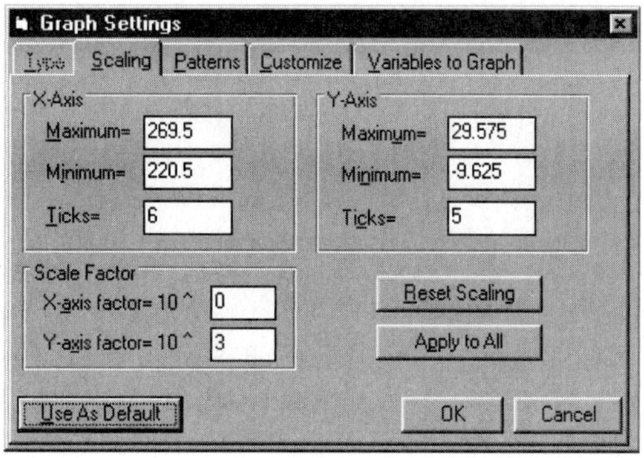

Figure 5.20

Graph Settings dialog box for modifying the sensitivity graphs.

options: *Format* for editing in TopRank, or *Graph in Excel* for exporting the graph and graphing data to Excel. Choosing *Format* opens the *Graph Settings* window shown in Figure 5.20. The four tabs along the top of the *Graph Settings* window allow you to make various alterations to the graph:

- The *Scaling* tab allows you to change the scale (max and min) of the *x* and *y* axes.

- The *Patterns* tab allows you to change the patterns and colors of the graph and text.

- The *Customize* tab allows you to modify the title of the graph and the axes.

- The *Variables to Graph* tab allows you to select the variables you wish to graph. For example, you can unclutter a tornado or spider graph by eliminating variables.

STEP 4

4.1   Close all the graph windows. To return to the spreadsheet, you can either minimize the TopRank window or you can click on the **Hide** button, second from the right. We do not want to close TopRank because in the next section we learn how to modify the settings.

TopRank—Custom Analysis

Now that you have seen how to run an automated analysis, let's learn how to modify the settings to customize the analysis. You can either modify the global settings of all the input variables using the *What-If Wizard* or you can modify the individual settings of each variable using the *Step Through Input* command. For example, you would use the *What-If Wizard* if you wanted to vary each input by plus and minus 15%, and you would use the *Step Through Input* command if you wanted to vary each variable its own specific amount. For a demonstration, we will use the *Step Through Input* command.

> **STEP 5**
>
> 5.1   For the following, we assume that you have a working model of the spreadsheet shown in Figure 5.15 and that "Profit" (B19) has been chosen as the output cell. If not, follow Steps 1 and 2 in the previous section.
>
> 5.2   Click **Step Through Input Cells** on the TopRank toolbar (sixth button from the left). Figure 5.21 shows the resulting *Step By Input* dialog box with "Hours Flown" as the chosen variable.
>
> 5.3   If you do not have "Hours Flown" chosen, then select it from the list in the pull-down menu accessed to the right of *Cell/Name*.

You have three options under the *Type of Range* heading for defining the bounds of each input variable: You can define the percentage change from the base value, the actual change from the base value, or the actual bounds themselves. To set the bounds of 500 and 1000 for "Hours Flown," it is easiest to choose the *Actual Min and Max* option, which lets us directly input the values 500 and 1000, as shown in Figure 5.21. Alternatively, you can choose *+/- Percent%* and then set the minimum equal to −37.5 because 500 is 37.5% below the base value of 800, and set the maximum equal to 25 because 1000 is 25% above 800. Likewise, choosing *+/- Actual Values* implies setting the minimum equal to −300 and the maximum equal to 200.

> 5.4   Choose the **Actual Min and Max** option, type **500** in the text box to the right of *Minimum*, and type **1000** in the text box to the right of *Maximum*.

This automatically changes the formula from *RiskAutoVary*(800, −10, 10) to *RiskVary*(800, 500, 1000, 2).

> **STEP 6**
>
> 6.1   Move on to the next input variable by either clicking on the forward button in the lower left corner or by unfurling the pull-down menu at the far right of *Cell/Name*.

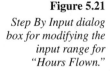

Figure 5.21

Step By Input dialog box for modifying the input range for "Hours Flown."

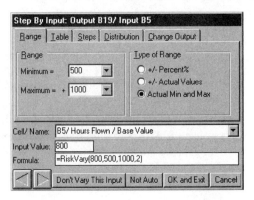

> 6.2 For each input, set the lower and upper bounds, using Figure 5.15 as a guide. You can choose the *Type of Range* options that you wish, but we recommend choosing the **Actual Min and Max** for each of the 10 inputs. (*Note:* Do not type in commas or percent signs. See explanation below.)
>
> 6.3 Click **OK and Exit** when finished with all 10 inputs.

Sometimes a #VALUE! unexpectedly appears in a cell of the worksheet. This indicates that you entered a value that TopRank does not understand. For example, TopRank does not always interpret percentage signs the way Excel does. If you choose *Actual Min and Max* and enter "40%," this will be interpreted as 40.0 and not as 0.40. Another common misunderstanding occurs if you type a comma in the number, such as 85,000. TopRank interprets this as two numbers, with 85 as the first number and 000 as the second. Finally, make sure that the lower/upper-bound number is smaller/larger than the base value. To fix such problems, just click in the cell and make the necessary adjustment, or return to the *Step Through Input* window and modify the cell from there.

> ## STEP 7
>
> 7.1 Click on **Run What-If Analysis**.
>
> 7.2 Verify your work by clicking the **Graph** button at the top of the *Results* window and compare your tornado diagram to Figure 5.3. They should be the same.
>
> 7.3 Exit TopRank by closing all the windows.

You can further customize the sensitivity analysis using the other tabs along the top of the *Step By Input* dialog box shown in Figure 5.21. For example,

- The *Steps* tab allows you to configure the number of steps (equally spaced values) for each input. For example, one input might have 4 steps and another might have 12.

- The *Table* tab allows you to specify the exact values at which the spreadsheet is recalculated.

- The *Distribution* tab is an advanced feature that allows you to choose the step values using the percentiles of a distribution.

You can find detailed explanations of these and related features in the TopRank online manual.

PrecisionTree

In this section, we perform a sensitivity analysis on the Eagle Airlines decision tree (Figure 5.7) using PrecisionTree. Running a sensitivity analysis in PrecisionTree is similar to constructing a customized sensitivity analysis in TopRank: First, the range is specified for each variable, and then the analysis is run. It is natural here to build a linked tree because we have already set up a spreadsheet that calculates profit. Thus, we link the end nodes to the profit formula and link the uncertainties ("Operating Costs," "Capacity," and "Hours Flown") in the tree to the corresponding inputs of the profit formula. To complete Step 8, you may find it necessary to review the section of Chapter 4 on linked trees.

STEP 8

8.1   Open Excel and PrecisionTree, enabling any macros.

8.2   Either construct the linked tree as shown in Figure 5.22, linking the tree to the profit model in Figure 5.15, or open the workbook on the Palisade CD: Examples\Chapter5\EagleTree.xls.

Figure 5.22

Eagle Airlines' decision tree linked to spreadsheet model.

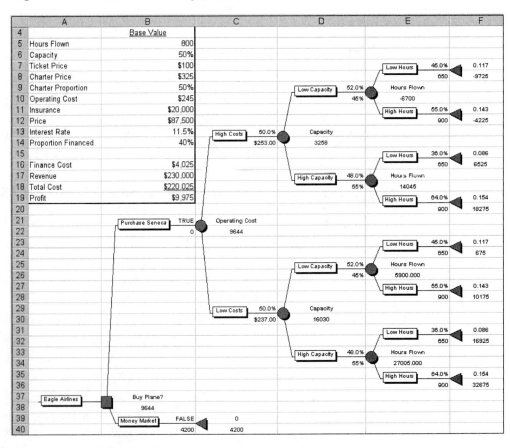

We will perform a sensitivity analysis on probabilities, so we need to control the input values carefully in order to ensure that the probabilities sum to 1. For example, as the probability of "High Costs" varies between 0 and 1, we need to be sure that the probabilities of "High Costs" and "Low Costs" will always add to 1.

The necessary formulas have been entered in the Excel file on the Palisade CD. For example, instead of entering 50% for the probability of "Low Costs," it is calculated in cell C29 as 1 minus the probability of "High Costs" (cell C13), thereby ensuring the two probabilities will always add to 1. Similar formulas were used for the probability of "High Capacity" (D17) and the probability of "High Hours" (E11). Additionally, we guaranteed that the probabilities of the two capacity chance nodes are always the same. We did this by inputting =D9 into cell D25, and =D17 into cell D33. In a similar manner, we forced the "Hours Flown" chance nodes to be the same. Finally, as shown in Figure 5.7, we used the formula =0.8*E7 in cell E15 to model the dependence between "Hours" and "Capacity." You will need to enter these formulas if you are building your model from scratch.

Step 9

9.1   Click the **Sensitivity Analysis** button on the PrecisionTree toolbar (fifth button from the left), and the *Sensitivity Analysis* dialog box appears (Figure 5.23). Your box will appear different because we have not yet indicated which cells to vary.

9.2   Click in the *Cell to Analyze* text box, delete the current entry, move the pointer to the spreadsheet, and click on cell **B38**. This identifies the EMV value of $9,644 as the output variable.

9.3   Click in the *Name* text box and type **Expected Profit**.

9.4   Reading down, you'll notice that PrecisionTree will perform either a one-way or two-way analysis. Choose **One-Way** for now.

We also recommend that you choose *Update Display* the first few times you run an analysis to see how the spreadsheet values change during the analysis. Now we specify which variables to vary and by how much.

9.5   Click in the *Cell* text box under the *Input Editor* heading (see Figure 5.23). Move the pointer to the decision tree and click on the probability **50.0%** above the "High Costs" branch of "Operating Cost." This places $C$13 in the *Cell* text box.

9.6   Click in the *Name* text box directly below and type **Prob High OC**.

9.7   For the *Min, Max, Base,* and *Steps* inputs, click **Suggest Values**, then change the *Min* to **0.0** and the *Max* to **1.0**.

9.8   Click **Add**. You have specified the first variable for the sensitivity analysis. Specifically, the probability for high operating cost will vary between zero and one.

Figure 5.23

PrecisionTree's Sensitivity Analysis dialog box.

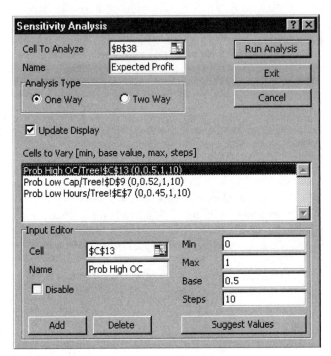

9.9 Following the procedures in Steps 9.5 to 9.8, add the probability of low capacity in cell D9 and the probability of low hours in E7, as shown in Figure 5.23.

9.10 When all three input variables have been entered, click **Run Analysis.**

The output of PrecisionTree's one-way sensitivity analysis is a new workbook that contains three sheets of graphs and supporting data from the sensitivity analysis. One worksheet contains a tornado diagram, the second contains a spider graph, and the third has sensitivity graphs.

Figure 5.24 shows two sensitivity graphs. Figure 5.24a shows the effect on expected profit while the probability of low capacity varies from 0 to 1, and Figure 5.24b shows the effect on expected profit for each of the two alternatives as this probability varies from 0 to 1. Because we are maximizing expected profit, PrecisionTree chooses the alternative with the highest expected profit in Figure 5.24a, and therefore the graph never drops below the 4200 mark.

These two graphs show that expected profit decreases as the probability of low capacity increases, and that purchasing the airplane has the higher expected profit as long as this probability is less than approximately 0.8. If the probability of low capacity is greater than the crossover point, then the money market alternative has the

Figure 5.24

Sensitivity graphs from the Eagle Airlines decision tree.
(a) Expected profit for purchasing the plane is higher than the money market account until the probablility of low capacity reaches 0.8.
(b) The expected profit graphed for both the alternatives of purchasing the plane and the money market account.

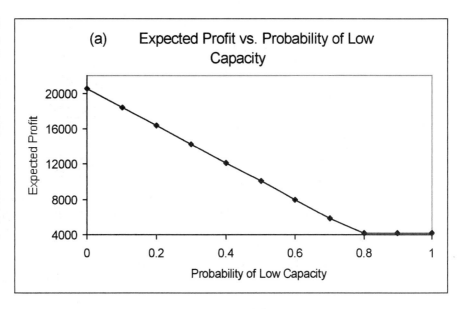

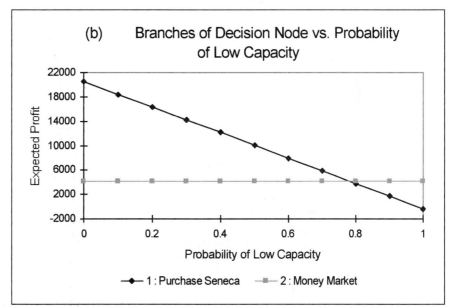

higher expected value. A more accurate estimate of the crossover point can be found by examining the data that accompany the graph. The data are hidden under the graphs on the left side of the window and can be viewed by moving the graphs.

Running a two-way sensitivity analysis is simple once the variables and their ranges have been input into the sensitivity-analysis window.

STEP 10

10.1   To perform a two-way sensitivity analysis in PrecisionTree, return to the spreadsheet and click the **Sensitivity Analysis** button in PrecisionTree's toolbar.

10.2   Choose the **Two Way** option in the *Sensitivity Analysis* dialog box.

10.3   Click **Run Analysis.**

The result of running a two-way analysis is again a new workbook with a sheet for every pair of selected input variables. Each sheet contains both a two-dimensional and three-dimensional sensitivity graph similar to those shown in Figure 5.25. Figure 5.25a is essentially the same as Figure 5.8, where each point represents the two input values at which the decision tree was evaluated. Figure 5.25b is the three-dimensional counterpart of Figure 5.25a, where the height is computed as the maximum expected profit over both alternatives. Hence, the height never drops below the 4200 mark. Sometimes it is easier to interpret three-dimensional graphs by viewing them from different vantage points. You can rotate these graphs by double-clicking on a corner of the graph, then hold the mouse button down, and drag the corner.

Sensitivity Analysis: A Built-In Irony

There is a strange irony in sensitivity analysis and decision making. We begin by structuring a decision problem, part of which involves identifying several alternatives. Then some alternatives are eliminated on the grounds of dominance. The remaining ones are difficult to choose from. Being difficult to choose from, they lead us to unveil our array of decision-analysis tools. But also being difficult to choose from, they probably are not too different in expected value; and if so, then it does not matter much which alternative one chooses, does it? For the analyst who wants to be quite sure of making the best possible choice, this realization can be terribly frustrating; almost by definition, hard decisions are sensitive to our assessments. For those who are interested in modeling to improve decision making, the thought is comforting; the better the model, the better the decision, but only a small degree of improvement may be available from rigorous and exquisite modeling of each minute detail. Adequate modeling is all that is necessary. The best way to view sensitivity analysis is as a source of guidance in modeling a decision problem. It provides the guidance for each successive iteration through the decision-analysis cycle. You can see now how the cycle is composed of modeling steps, followed by sensitivity analysis, followed by more modeling, and so on. The ultimate objective of this cycle of modeling and analysis is to arrive eventually at a requisite decision model and to analyze it just enough to understand clearly which alternative should be chosen. By the time the decision maker reaches this point, all important issues will be included in the decision model, and the choice should be clear.

Figure 5.25

*Two-way sensitivity
graphs from the Eagle
Airlines decision tree.*
(a) Two-dimensional
grid indicating the al-
ternative with the high-
est expected value for
different probabilities
of low capacity and
different probabilities
of low hours given low
capacity for the Eagle
Airlines example.
(b) Three-dimensional
graph of expected
profit.

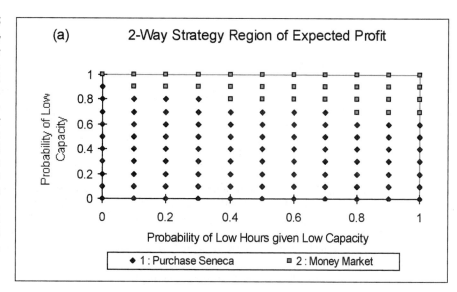

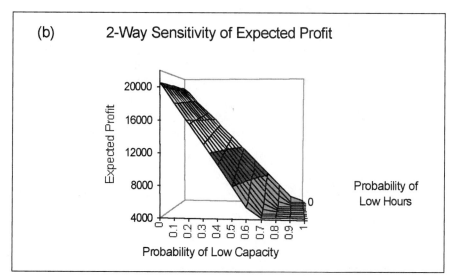

SUMMARY

This chapter has presented an approach and several tools for performing sensitivity analysis. We have considered sensitivity analysis in terms of identifying and structuring problems, dominance among alternatives, and probability assessment. Tornado diagrams and one- and two-way sensitivity graphs were developed, and we discussed ways to perform sensitivity analysis using computers. The purpose of sensitivity analysis in the decision-analysis cycle is to provide guidance for the development of a requisite decision model.

EXERCISES

5.1   What is the fundamental question that sensitivity analysis answers?

5.2   Some friends of yours have been considering purchasing a new home. They currently live 20 miles from town on a two-acre tract. The family consists of the mother, father, and two small children. The parents also are considering having more children, and they realize that as the children grow, they may become more involved in activities in town. As it is, most of the family's outings take place in town. Describe the role that sensitivity analysis could play in your friends' decision. What variables could be subjected to sensitivity analysis?

5.3   Over dinner, your father mentions that he is considering retiring from real-estate sales. He has found a small retail business for sale, which he is considering acquiring and running. There are so many issues to think about, however, that he has a difficult time keeping them all straight. After hearing about your decision-analysis course, he asks you whether you have learned anything that might help him in his decision. What kinds of issues are important in deciding whether to buy a retail business? Describe how he might use sensitivity analysis to explore the importance of these issues.

5.4   When purchasing a home, one occasionally hears about the possibility of "renting with an option to buy." This arrangement can take various forms, but a common one is that the renter simply pays rent and may purchase the house at an agreed-upon price. Rental payments typically are not applied toward purchase. The owner is not permitted to sell the house to another buyer unless the renter/option holder waives the right to purchase. The duration of the option may or may not be specified.

Suppose that a buyer is considering whether to purchase a house outright or rent it with an option to buy. Under what circumstances would renting with an option be a dominated alternative? Under what circumstances would it definitely not be dominated?

5.5   What role does sensitivity analysis play in the development of a requisite decision model?

5.6   Explain why the lines separating the three regions in Figure 5.12 all intersect at Point D.

QUESTIONS AND PROBLEMS

5.7   *Cost-to-loss ratio problem.* Consider the decision problem shown in Figure 5.26. This basic decision tree often is called a cost-to-loss ratio problem and is characterized as a decision situation in which the question is whether to take some protective action in the face of possible adverse circumstances. For example, the umbrella problem (Figure 4.9) is a cost-to-loss ratio problem. Taking the umbrella incurs a fixed cost and protects against possible adverse weather. A farmer may face a cost-to-loss ratio problem if there is a threat of freezing weather that could damage a fruit crop. Steps can be taken to protect the orchard, but they are costly. If no steps are taken, the air temperature may or may not become cold enough to damage the crop.

Figure 5.26

Cost-to-loss ratio problem.

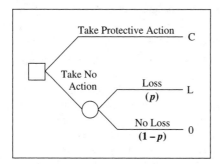

Sensitivity analysis is easily performed for the cost-to-loss ratio problem. How large can the probability p become before "Take Protective Action" becomes the optimal (minimum expected cost) alternative? Given your answer, what kind of information does the decision maker need in order to make the decision? (*Hint:* This is an algebra problem. If that makes you uncomfortable, substitute numerical values for C, L, and p.)

5.8 *The cost-to-loss ratio problem continued.* The cost-to-loss ratio problem as shown in Figure 5.26 may be considered a simplified version of the actual situation. The protective action that may be taken may not provide perfect protection. Suppose that, even with protective action, damage D will be sustained with probability q. Thus, the decision tree appears as Figure 5.27. Explain how sensitivity analysis could be used to determine whether it is important to include the upper chance node with probability q and damage D.

Figure 5.27

More general version of the cost-to-loss problem.

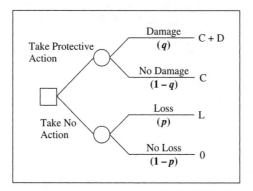

5.9 An orange grower in Florida faces a dilemma. The weather forecast is for cold weather, and there is a 50% chance that the temperature tonight will be cold enough to freeze and destroy his entire crop, which is worth some $50,000. He can take two possible actions to try to alleviate his loss if the temperature drops. First, he could set burners in the orchard; this would cost $5000, but he could still expect to incur damage of approximately $15,000 to $20,000. Second, he could set up sprinklers to spray the trees. If the temperature drops, the water would freeze on the fruit and provide some insulation. This method is cheaper ($2000), but less effective. With the sprinklers he could expect to incur as much as $25,000 to $30,000 of the loss with no protective action.

Compare the grower's expected values for the three alternatives he has, considering the various possible loss scenarios for the burners and the sprinklers. Which alternative would you suggest the grower take? Why?

5.10 An important application of sensitivity analysis occurs in problems involving multiple attributes. Many decision makers experience difficulty in assessing trade-off weights. A sensitivity analysis of the trade-off weight, though, can reveal whether a decision maker must make a more precise judgment. Reconsider the summer-job example described and analyzed in Chapter 4 (pages 138–145). In the analysis, we used trade-off weights of $k_s = 0.60$ for salary and $k_f = 0.40$ for fun (see Figure 4.36).

Suppose Sam Chu is uncomfortable with the precise assessment that $k_s = 0.60$. Sam does believe, though, that k_s could range from 0.50 up to 0.75. (Recall that k_s and k_f add up to 1, so by implication, k_f can range from 0.50 to 0.25, depending on the value of k_s.) Perform a sensitivity analysis on the expected overall score for the two jobs by varying k_s over this range. Is the forest job preferred for all values of k_s between 0.50 and 0.75?

5.11 A friend of yours can invest in a multiyear project. The cost is $14,000. Annual cash flows are estimated to be $5000 per year for six years but could vary between $2500 and $7000. Your friend estimates that the cost of capital (interest rate) is 11%, but it could be as low as 9.5% and as high as 12%. The basis of the decision to invest will be whether the project has a positive net present value. Construct a tornado diagram for this problem. On the basis of the tornado diagram, advise your friend regarding either (1) whether to invest or (2) what to do next in the analysis.

5.12 Reconsider Hugh Liedtke's decision as diagrammed in Figure 4.2. Note that three strategies are possible: (1) accept $2 billion, (2) counteroffer $5 billion and then accept $3 billion if Texaco counteroffers, and (3) counteroffer $5 billion and then refuse $3 billion if Texaco counteroffers. Suppose that Liedtke is unsure about the probabilities associated with the final court outcome. Let $p = P(10.3)$ and $q = P(5)$ so that $1 - p - q = P(0)$. Create a two-way sensitivity graph that shows optimal strategies for Liedtke for possible values of p and q. (*Hint:* What is the constraint on $p + q$?) If Liedtke thinks that p must be at least 0.15 and q must be more than 0.35, can he make a decision without further probability assessment?

CASE STUDIES

THE *STARS AND STRIPES*

In 1987, the United States won the prestigious America's Cup sailing race, winning the trophy from Australia. The race normally is run every four years, but in 1988 New Zealand invoked an obscure provision in the race charter and challenged the U.S. team to a match. Furthermore, New Zealand proposed to race with the largest boat permitted rather than a standard 12-meter craft. The new yacht, dubbed the *New Zealand,* was 133 feet long, designed and built using space-age material, and equipped with state-of-the-art computer equipment to monitor performance.

Not to be outdone, the U.S. team countered by designing and building a catamaran, the *Stars and Stripes*. A conventional sailboat like the *New Zealand* drags a heavy keel through the water to maintain stability. A catamaran relies on two long, narrow hulls for stability and thus can be considerably lighter and faster. Furthermore, the *Stars and Stripes* was outfitted with a rigid sail designed like an airplane wing. With slots and flaps controlled by wires, the sail could be adjusted precisely for optimum airflow.

New Zealand counterattacked with a lawsuit claiming that the vague deed that established the competition implied that the match was to be between similar boats. But the New York Supreme Court ruled that the race should go on and that protests should be filed afterward. The race began on September 7, 1988. The *Stars and Stripes* won easily, by 18 minutes in the first race and 21 minutes in the second.

Questions

1 Designing world-class racing sailboats involves thousands of decisions about shape, size, materials, and countless other details. What are some objectives that might be reasonable in designing such a sailboat?

2 What are some specific design decisions that must be made (for example, the shape of the sail)?

3 How can sensitivity analysis be used to decide which design decisions are more important than others?

DUMOND INTERNATIONAL, PART I

"So that's the simplified version of the decision tree based on what appear to be the critical issues," Nancy Milnor concluded. "Calculating expected values, it looks as though we should introduce the new product. Now, I know that we don't all agree on the numbers in the tree, so why don't we play around with them a little bit. I've got the data in the computer here. I can make any changes you want and see what effect they have on the expected value."

Nancy had just completed her presentation to the board of directors of DuMond International, which manufactured agricultural fertilizers and pesticides. The decision the board faced was whether to go ahead with a new pesticide product to replace an old one or whether to continue to rely on the current product, which had been around for years and was a good seller. The problem with the current product was that evidence was beginning to surface that showed that the chemical's use could create substantial health risks, and there even was some talk of banning the product. The new product still required more development, and the question was whether all of the development issues could be resolved in time to meet the scheduled introduction date. And once the product was introduced, there was always the question of how well it would be received. The decision tree (Figure 5.28) that Nancy had presented to the board captured these concerns.

Figure 5.28

*DuMond's new
product decision.*

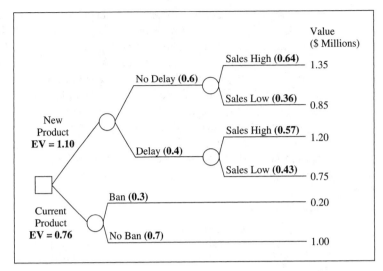

The boardroom was beginning to get warm. Nancy sat back and relaxed as she listened to the comments.

"Well, I'll start," said John Dilts. "I don't have much trouble with the numbers in the top half of the tree. But you have the chance of banning the current product pinned at 30 percent. That's high. Personally, I don't think there's more than a 10 percent chance of an out-and-out ban."

"Yeah, and even if there were, the current product ought to be worth $300,000 at least," added Pete Lillovich. "With a smaller chance of a ban and a higher value, surely we're better off with the old product!"

"Well, I don't know about you two," said Marla Jenkins. "I think we have a pretty good handle on what's going on with the current product. But I'd like to play the new product a little more conservatively. I know that the values at the ends of the branches on the top half of the tree are accounting's best guesses based on a complete analysis, but maybe they should all be reduced by $100,000 just to play it safe. And maybe we should just set the probability of high sales equal to 50 percent regardless of the delay."

Steven Kellogg had been involved in the preliminary development of the new product more than anyone else. He piped up, "And the delay is actually more likely than no delay. I'd just reverse those probabilities so that there's a 60 percent chance of a delay. But I wouldn't make any changes on the lower part of the tree. I agree with Marla that we have a good idea about the performance of the current product and the prospects for a ban."

"I don't think it matters," countered Lillovich. "The changes John and I suggest make the current product look better than it does in Nancy's analysis. Marla's and Steven's changes make the new product look worse. Either way, the effect is the same."

Nancy had kept track of the comments and suggested changes. She sat down at the computer and started to enter the changes. After a few moments, she grinned and turned to the board. "In spite of your changes," she said, "I believe I can persuade you that DuMond should go with the new product."

Question

1   Explain why Nancy believes that DuMond should go with the new product.

STRENLAR, PART II

Question

1   The Strenlar case study at the end of Chapter 4 (page 167) required substantial modeling. Use sensitivity analysis to refine your model. In particular, you might consider (1) the interest rate used to calculate net present value, (2) legal fees, (3) the eventual price of PI's stock, (4) Strenlar's gross sales, (5) Fred's profits if Strenlar is successful (remembering that profits are linked to sales), (6) the probability of Strenlar being successful, and (7) the probability of winning the lawsuit. Do you think that Fred's decision is sensitive to any of these variables? Try wiggling one variable at a time away from its base value (the value given in the case) while holding everything else at base value. How much can you wiggle the variable before the decision changes? At the end of your analysis, discuss your results and the implications for Fred's decision model. If he were to refine his model, what refinements should he make?

FACILITIES INVESTMENT AND EXPANSION

Spetzler and Zamora describe a decision analysis concerning whether a major U.S. corporation should make a $20 million investment in a new plant with the possibility of a $5 million expansion later. The product was a "brightener," and there was some chance that the process also could generate significant quantities of a valuable by-product. Unfortunately, the primary chemical reaction was difficult to control. The exact yields were uncertain and were subject to the amounts of impurities in the raw material. Other substantial uncertainties surrounded the decision, such as raw material costs, inflation effects, federal regulatory intervention, the development of a full-scale production process based on the pilot project, and so on.

At an early stage of the analysis, the focus was on the numerous uncertain variables that could affect the value of the investment. Table 5.2 shows the effect of 18 different variables on the project's present value.

Question

1   Which variables definitely should be kept in the model? What additional information is needed in order to decide whether any of the other variables can be eliminated?

Source: C. S. Spetzler and R. M. Zamora (1984) "Decision Analysis of a Facilities Investment and Expansion Decision." In R. A. Howard and J. E. Matheson (eds.) *The Principles and Applications of Decision Analysis.* Menlo Park, CA: Strategic Decisions Group.

Table 5.2

Results of deterministic sensitivity analysis on 18 different variables in facilities investment decision.

Variable		Base Case	Tested Range From	Tested Range To	PV Range ($ Million) From	PV Range ($ Million) To	Change in PV ($ Millions)
1	By-Product Production	36.00 lb/ton	0.00 lb/ton	80.00 lb/ton	−30	35	65
2	Market Price of Brightener in 1980	$0.27/lb	$0.15/lb	$0.35/lb	−12	45	57
3	Raw Material Cost Growth	5.00%/yr	0.00%/yr	8.00%/yr	−7	40	47
4	Raw Material Costs	$7.00/ton	$2.00/ton	$18.00/ton	−9	35	44
5	Impurities in Raw Material	4.00 lb/ton	2.00 lb/ton	6.00 lb/ton	−10	30	40
6	Cost Multiplier on Investment	90.00%	70.00%	125.00%	−12	25	37
7	Brightener Price Growth after 1980	4.00%/yr	2.00%/yr	6.00%/yr	−5	28	33
8	Cost Multiplier on Operations Expenses	110.00%	80.00%	150.00%	3	32	29
9	Cost Multiplier on Maintenance Expenses	100.00%	70.00%	120.00%	2	30	28
10	By-Product Price Growth	$0.03/yr	$0.01/yr	$0.06/yr	3	30	27
11	Water Reclamation Costs	$0.03/gal	$0.02/gal	$0.04/gal	4	31	27
12	By-Product Price in 1970	$0.50/lb	$0.40/lb	$0.60/lb	3	29	26
13	Plant Efficiency	75.00%	50.00%	110.00%	0	26	26
14	Brightener Production	45.00 lb/ton	40.00 lb/ton	50.00 lb/ton	4	30	26
15	Market Price of Brightener in 1974	$0.25/lb	$0.15/lb	$0.30/lb	4	28	24
16	Government Regulation Costs	$0.02/lb	$0.01/lb	$0.03/lb	3	25	22
17	Market Price of Other By-Products	$0.10/lb	$0.07/lb	$0.12/lb	4	24	20
18	Other By-Products Produced	84.00 lb/ton	70.00 lb/ton	96.00 lb/ton	4	24	20

Source: Spetzler and Zamora (1984).

JOB OFFERS, PART II

Questions

1 Reconsider Robin Pinelli's dilemma in choosing from among three job offers (page 169). Suppose that Robin is unwilling to give a precise probability for disposable income with the Madison Publishing job. Conduct a sensitivity analysis on the

expected value of the Madison Publishing job assuming that the probability of disposable income being $1500 could range anywhere from zero to 1. Does the optimal choice—the job with the highest expected overall score—depend on the value of this probability? [*Hint:* Remember that probabilities must add up to 1, so P(Disposable Income = $1300) must equal 1 – P(Disposable Income = $1500).]

2   Suppose Robin is unable to come up with an appropriate set of trade-off weights. Assuming that any combination of weights is possible as long as all three are positive and add up to 1, conduct a sensitivity analysis on the weights. Create a graph that shows the regions for which the different job offers are optimal (have the highest expected overall score). (*Hint:* Given that the three weights must sum to 1, this problem reduces to a two-way sensitivity analysis like the stock market example on pages 189–192.)

REFERENCES

Sensitivity analysis is one of the more recent additions to the decision analyst's bag of tricks. As more complicated decision problems have been tackled, it has become obvious that sensitivity analysis plays a central role in guiding the analysis and interpreting the results. Recent overviews of sensitivity analysis can be found in Samson (1988), von Winterfeldt and Edwards (1986), and Watson and Buede (1987). In particular, Watson and Buede use real-world examples to show how sensitivity analysis is a central part of the decision-analysis modeling strategy. Phillips (1982) describes an application in which sensitivity analysis played a central part in obtaining consensus among the members of a board of directors.

Howard (1988) presents tornado diagrams and gives them their name. This approach to deterministic sensitivity analysis, along with other sensitivity-analysis tools, is discussed by McNamee and Celona (1987). Another sensitivity tool worth mentioning is the spiderplot (Eschenbach, 1992). The spiderplot is similar to the tornado diagram in that it allows simultaneous comparison of the impact of several different variables on the consequence or expected value.

Eschenbach, T. G. (1992) "Spiderplots versus Tornado Diagrams for Sensitivity Analysis." *Interfaces,* 22(6), 40–46.

Howard, R. A. (1988) "Decision Analysis: Practice and Promise." *Management Science,* 34, 679–695.

McNamee, P., and J. Celona (1987) *Decision Analysis for the Professional with Supertree.* Redwood City, CA: Scientific Press.

Phillips, L. D. (1982) "Requisite Decision Modelling." *Journal of the Operational Research Society,* 33, 303–312.

Samson, D. (1988) *Managerial Decision Analysis.* Homewood, IL: Irwin.

von Winterfeldt, D., and W. Edwards (1986) *Decision Analysis and Behavioral Research.* Cambridge: Cambridge University Press.

Watson, S., and D. Buede (1987) *Decision Synthesis.* Cambridge: Cambridge University Press.

E P I L O G U E   After the America's Cup race, New Zealand filed its lawsuit against the U.S. team, claiming that the deed of gift that established the race called for a "fair match." The New York Yacht Club, which had held the trophy from 1851 until 1983, even filed an affidavit with the court supporting New Zealand's contention. In April 1989, the court awarded the trophy to New Zealand. But in September 1989, the New York State Supreme Court issued a 4–1 decision that the cup should go back to the United States. (*Sources:* "The Cup Turneth Over," *Time,* April 10, 1989, p. 42; *Raleigh News and Observer,* September 20, 1989.)

CHAPTER 6

Creativity and Decision Making

The majority of businessmen are incapable of original thinking, because they are unable to escape from the tyranny of reason. Their imaginations are blocked. (David Ogilvy, *Confessions of an Advertising Man*)

H ow ironic that we have spent so much effort in discussing and demonstrating the systematic analysis of decisions only to find this quote. Decision analysis *is* reason exemplified, is it not? If we relentlessly practice the form of reason that we have learned as decision analysis, will we find ourselves trapped by that reason, unable to be creative?

The emphatic answer is no. From a decision-making perspective, we need not be stuck with the alternatives that present themselves to us; in fact, good decision making includes active creation of new and useful alternatives. Moreover, the very process of decision analysis—especially the specification of objectives—provides an excellent basis for developing creative new alternatives.

Although it may seem unusual, a chapter on creativity in a decision-making text makes good sense. Everyone is frustrated occasionally by an inability to think creatively and thus can use a hand in being more creative. Perhaps not so clear, however, is the increasing need for creative and innovative solutions. In *Thriving on Chaos,* Tom Peters (1988) depicts the modern business climate as one in which conditions change rapidly. Modern managers must do more than simply cope with radical transformations: They must be on the attack. To be successful, a manager must learn to view new situations as opportunities for beneficial change rather than as problems to overcome somehow without rocking the boat too much. Indeed, Peters argues that the core paradox a manager faces is building an organization that is stable in its ability to innovate rapidly and flexibly. Solving problems creatively must become part of a manager's and a firm's essence. And corporate America appears to take this message seriously; Bazerman (1994) reports that a majority of large U.S. corporations engaged in formal creativity training during the early 1980s.

This chapter presents a short overview of creativity. The literature on creativity is large and growing; interested readers will have no problem locating additional material. We frame the discussion first by defining creativity in decision making and looking at some of the psychological theories that have been developed to help us understand the creative process. What follows is a discussion of the many different ways in which creativity can be blocked. Finally, we discuss ways to enhance the creative process, especially the process of generating alternatives in a decision-making situation. One important technique promised back in Chapter 3 is the use of the means-objectives network. Other techniques use fundamental objectives, and still others (many developed by decision analysts) can be used to break through various blocks to find creative new alternatives.

What Is Creativity?

One thing is certain regarding the definition of creativity—it is much easier to identify creative acts than it is to define the term itself (Barron 1965). We readily recognize creative acts, and we often use adjectives like novel, insightful, clever, unique, different, or imaginative. But coming up with a coherent and useful definition of the term *creativity* is not easy.

Many different scholars have attempted to define creativity. All definitions include some aspect of novelty. But there is also an element of effectiveness that must be met. Slinging buckets of mud at customers as they arrive at a used-car lot is indeed a novel greeting but may not be very effective in selling cars. But offering coupons for a cosmetic mud pack, an evening at the local mud-wrestling arena, or a therapeutic and relaxing mud bath might be very effective as well as novel. And what about creating a television advertisement in which the car lot's owner offers customers the opportunity to dunk him in a mud bath set up at the lot specifically for this purpose? ("My name is mud, because I ordered too many cars! So come on down and make the name stick! Throw me into the pit! And after you do that, check out some of these great deals. . . ." Well, it would be no worse than many similar ads.)

For our purposes in this chapter, we are particularly concerned with the development of creative alternatives in decision problems. To be sure, creativity arises in many different situations; a novel and elegant proof of a mathematical theorem, an artist's creativity in painting or music, and a storyteller's clever retelling of an old tale are a few examples. When we think of creativity in decision making, though, we will be looking for new alternatives with elements that achieve fundamental objectives in ways previously unseen. Thus, a creative alternative has both elements of novelty and effectiveness, where effectiveness is thought of in terms of satisfying objectives of a decision maker, a group of individuals, or even the diverse objectives held by different stakeholders in a negotiation.

Theories of Creativity

Why do creative thoughts seem to come more readily to some people than to others? Or in certain kinds of situations? Many scholars have tried to understand the creative process, and in this section we review some of the psychological bases of creativity.

Perhaps the most basic approach relates creativity to Maslow's (1954) concept of self-actualization. For example, Davis (1986) describes self-actualization as, among other things, being able to perceive reality accurately and compare cultures objectively, having a degree of genuine spontaneity, and being able to look at things in a fresh, naive, and simple way. Davis claims that these and other qualities help people, even those without special talent, to act creatively, and he reviews some recent psychological evidence to support this proposition. Thus, Davis's position is good news for many of us. Self-actualization, happy lives, and creativity all seem to go hand in hand and to some extent can be developed by anyone. One need not have the special talent of an Einstein, Mozart, or Alexander the Great to reap the creative rewards that follow from self-actualization.

Others have attempted to delve more deeply into the process of creative thought itself. Psychoanalytic theories (Kris 1952, Kubie 1958, Rugg 1963) generally maintain that creative productivity is the result of preconscious mental activity. These theories suggest that our brain is processing information at a level that is not accessible to our conscious thoughts. Behavioristic theories (Maltzman 1960, Skinner 1972) argue that our behavior, including creative behavior, is simply a conglomerate of responses to environmental stimuli. Appropriate rewards (stimuli) can lead to more creative behavior.

A cognitive approach suggests that creativity stems from a capacity for making unusual and new mental associations of concepts (Campbell 1960, Mednick 1962, Staats 1968). Campbell proposes that creative thought is just one manifestation of a general process by which people acquire new knowledge and thereby learn about the world. This process includes as the first step the production of "variations," a result of mentally associating elements of a problem in new ways. People who are more creative are better at generating a wider range of variations as they think about the problems they face. Having a broader range of life experiences and working in the right kind of environment can facilitate the production of variations. Finally, some people simply are better at recognizing and seizing appropriate creative solutions as they arise; the ability to come up with creative solutions is not very helpful if one ignores those solutions later.

Before going on, try the following problem. Do not read further until you have devoted some genuine effort to finding a solution.

CHAINS OF THOUGHT

You have four three-link chain segments as shown in Figure 6.1. A jeweler has offered to connect the segments to make a complete circle but to do so must open and then resolder some of the links. Opening and closing a link costs $50. When you

Figure 6.1.

Chains of thought.
Connect the segments
by opening and closing
only three links.

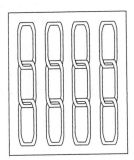

point out that you have only $150, the jeweler says the job can be done for that amount. How can the jeweler connect the segments by opening and closing only three links? [*Source:* Adapted from Winkelgren (1974).]

Most people see the obvious solution of opening one link on each of the four segments and connecting to the adjacent segment, but this approach requires opening and closing four links. The creative and efficient solution, however, opens all three links of one segment and uses those three to connect the remaining three segments. This is an example of making a new association of the elements of the problem. Rather than thinking about the four segments as the elements of the problem, the creative solution considers open links as one kind of element, connected segments of closed links as another kind of element, and associates these by specifying that an open link must be used to connect two segments.

Phases of the Creative Process

A number of authors have identified phases of the individual creative thought process. For example, Wallas (1926) identified preparation, incubation, illumination, and verification.

1 Preparation   In this first stage, the individual learns about the problem. This includes understanding the elements of the problem and how they relate to each other. It may include looking at the problem from different perspectives or asking other people what they know or think about the problem. From a decision-making point of view, this stage is very similar to problem structuring as delineated in Chapters 2 and 3. Spending effort understanding fundamental objectives, decisions that must be made (along with the immediately available set of alternatives), uncertainties inherent in the situation, and how these elements relate to each other prepares the decision maker for creative identification of new alternatives.

2 Incubation   In the second stage, the prepared decision maker explores, directly or indirectly, a multitude of different paths toward new alternatives. We might also use the terms *production* or *generation* of alternatives. The decision maker may do many things that seem to have a low chance of generating a new alternative, such as eliminating assumptions or adopting an entirely different perspective. Apparently frivolous activities may evoke the idea of the decision maker "playing" with the decision.

Many authors have included in this phase unconscious processing of information known about the decision. The literature on creativity contains several well-known and oft-quoted examples of this unconscious process, including Kekule's dream that led to the discovery of the chemical structure of the benzene ring and Poincaré's discovery of the fuchsian family of groups and functions in mathematics. More recently, Gay Balfour of Colorado made the U.S. national news (*Associated Press,* January 24, 1992) by literally dreaming up a way to control prairie dog populations on ranches; he sucks them up with a giant vacuum machine created from a septic-system cleaning truck. (He hastens to explain that sucking the critters out of their holes causes them no harm at all; even though they may be slightly confused, shortly after landing in the holding area they begin burrowing in the soft dirt sucked up with them!) Balfour's new technique has reportedly been quite successful in the southwestern United States.

One explanation of unconscious incubation as a valid element of the creative process has been suggested by researchers in artificial intelligence. The explanation is based on a "blackboard" model of memory in the human brain. When the brain is in the process of doing other things—when a problem is incubating—parts of the blackboard are erased and new items put up. Every so often, the new information just happens to be pertinent to the original problem, and the juxtaposition of the new and old information suggests a creative solution; in other words, the process of coming up with a new and unusual association can result simply from the way the brain works. An attractive feature of this theory is that it explains why incubation works only a small percentage of the time. Too bad it works so infrequently!

Beyond the literature's examples and the above speculation about how the brain works, however, there is little hard evidence that unconscious processes are at work searching for creative solutions (Baron 1988). Still, it is a romantic thought, and if it provides an excuse for many of us to relax a little, perhaps the reduced stress and refocus of our attention is enough to help us be more creative when we are trying to be.

3 Illumination   This is the instant of becoming aware of a new candidate solution to a problem, that flash of insight when all the pieces come together, either spontaneously (Aha!) or as the result of careful study and work. Wallas described illumination as a separate stage, but you can see that illumination is better characterized as the culmination of the incubation stage.

4 Verification   In the final step the decision maker must verify that the candidate solution does in fact have merit. (How many times have you thought you had the answer to a difficult problem, only to realize later—sometimes moments, sometimes much later—that your "dream solution" turned out to be just that: an impossible dream?) The verification stage requires the careful thinker to turn back to the hard logic of the problem at hand to evaluate the quality of the candidate solution. In our decision-making context, this means looking very carefully at a newly invented alternative in terms of whether it satisfies the constraints of the problem and how well it performs relative to the fundamental objectives.

Although there are many ways to think about the creative thought process, the cognitive approach described above, including the stages of creativity, can help us to frame the following discussion. We turn now to ways in which our creativity can be hindered, and we follow with suggestions about how to reduce or eliminate such blocks and thereby increase our creativity.

Blocks to Creativity

With a clear understanding of the creative process, we are now in a position to discuss ways in which that process can be derailed, albeit inadvertently. This section describes three kinds of creativity blocks, drawing heavily from work by Adams (1979), Baron (1988), Bazerman (1994), Hogarth (1987), and Kleindorfer, Kunreuther, and Schoemaker (1993). All of these blocks interfere with the creativity process by hindering the generation and recognition of new and unusual solutions to a problem or alternatives in a decision situation.

Framing and Perceptual Blocks

These blocks arise because of the ways in which we tend to perceive, define, and examine the problems and decisions that we face. To get a feel for these blocks, consider the following two problems. As with the earlier chain puzzle, give serious effort to solving these problems before reading on.

THE MONK AND THE MOUNTAIN

At dawn one day, a monk begins to walk along a path from his home to the top of a mountain. Never straying from the path, he takes his time, traveling at various speeds, stopping to rest here and there, and arrives at the top of the mountain as the sun sets. He meditates at the top of the mountain overnight and for the next full day. At dawn the following morning, he begins to make his way back down the mountain along the same path, again relaxing and taking his time, and arrives home in the afternoon. Prove that there is a spot along the path that the monk occupies at the same time of day going up and coming down. [*Source:* Adapted from Hogarth (1987, p. 161).]

MAKING CIGARS

Lonesome Molly loves to smoke cigars, and she has learned to make one out of five cigar butts. Suppose she collects 25 butts. How many cigars can she make? [*Source:* Adapted from Bartlett (1978).]

Did you solve either one? The monk problem is difficult to solve if you try to maintain the frame in which it is cast: that the same monk travels up and down on two different days. Change the frame, though, and imagine identical monks taking the uphill and downhill journeys on the same day, each one starting at dawn. Now it is easy to see that the monks will meet somewhere along the path at some time during the day. At that instant, they are at the same spot in the path; hence in the original problem there must be a point along the path that the monk in the problem occupies at the same time going up and coming down.

Lonesome Molly can obviously make at least five cigars out of the 25 butts that she finds. But you were right to suspect that the obvious answer is not correct! The answer lies in thinking not about the gross requirement of butts to make a cigar but to frame the problem in terms of net usage. Molly indeed requires five butts to make a cigar. For each one she makes (and smokes), however, she has one butt left. The net consumption per cigar is four butts. So if she has 25 butts to start with, she can make six cigars and have one butt left over: She makes five cigars out of the original 25 butts and smokes those five cigars, which yields five butts from which she can make a sixth cigar. After she smokes the last one, she has one butt left (and needs to find only four more for her next smoke).

Here are some specific blocks relating to framing and perception that hinder our creative potential:

1 Stereotyping   Suppose you are a personnel manager, and an individual with long hair and no necktie applies for a job as an engineer. Imagine your reaction. What would you think about the person? A typical mental strategy that most people use is to fit observations (people, things, events, and so on) into a standard category or stereotype. Much of the time this strategy works well because the categories available are rich enough to represent most observations adequately. But when new phenomena present themselves, stereotyping and associated preconceived notions can interfere with good judgment.

2 Tacit Assumptions   Consider the classic nine-dot puzzle. Lay out nine dots in a square, three dots per row (Figure 6.2), and then, without lifting your pencil, draw four straight lines that cross all nine dots. Try it before you read on. The epilogue to this chapter gives the standard solution as well as many surprising solutions that Adams (1979) has collected from creative readers.

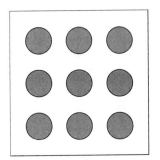

Figure 6.2

Nine-dot puzzle.
Connect the dots using
four straight lines
without lifting your
pencil.
[*Source:* From
*Conceptual
Blockbusting* by
James L. Adams.
© 1986 by James L.
Adams. Reprinted by
permission of Perseus
Books Publishers, a
member of Perseus
Books, L.L.C.]

The nine-dot puzzle is a nice example, but what does this block have to do with decision making? People often look at problems with tacitly imposed constraints, which are sometimes appropriate and sometimes not. Suppose you believe you need more warehouse space for your business, so you have your real-estate agent look for warehouses of a specific size to rent. The size, however, may be an inappropriate constraint. Perhaps so much space is not necessary or may be divided among several smaller warehouses. Perhaps some characteristic of your product would permit a smaller warehouse to be modified in a clever way to provide adequate storage.

3 Inability to Understand a Problem at Different Levels   This block can be manifested in different ways. First is the familiar issue of isolating the precise decision context that requires attention. Suppose you are a national sales manager for a line of boots. Sales in the Rocky Mountain states are down. Knowing your regional salesperson, you suspect that the problem is motivational. The "obvious" solution is to threaten or cajole the salesperson into better sales. But is the problem just what you think? Could it be a marketing problem—for example, competition with a regional brand that has been developed specifically for the area? What about a distribution problem? Perhaps it is difficult for the one warehouse in the region to supply the area's special needs. Perhaps customers in the region finally are getting tired of the same old style that has been the company's cash cow for many years. Even if the problem does lie with the salesperson, other possibilities exist, such as personal problems or personality conflicts with local business owners.

Another manifestation is focusing too much on detail and not being able to reframe the decision in a broader context, a problem commonly called "not seeing the forest for the trees." Many decisions require attention to a large amount of detailed information. For example, consider the issues involved in deciding whether to attempt a takeover of another firm, or where to site a new manufacturing plant. The sheer volume of information to be processed can keep the decision maker from seeing new and promising alternatives.

4 Inability to See the Problem from Another Person's Perspective   Where the previous block relates to seeing the problem itself in different ways, this one relates to seeing the problem through someone else's eyes and with their values. When a decision involves multiple stakeholders, it is always important to understand the values, interests, and objectives of other parties. Really creative solutions incorporate and satisfy as many competing objectives as possible, and an inability to understand others' values can interfere with the development of such solutions. For example, finding a meaningful way to achieve peace in the Middle East requires the parties to consider the interests of both Israelis and Palestinians, as well as other nations in the region.

Value-Based Blocks

Blocks in this category relate to the values we hold. In many cases our values and objectives can interfere with our ability to seek or identify truly creative alternatives in a decision situation.

1 Fear of Taking a Risk   To get a feel for this block, try the following game at a party with a lot of friends. Each person is assigned to be a particular kind of barnyard animal: cow, donkey, chicken, goat, sheep, or whatever else you designate. The more people, the better. After everyone has been assigned to be an animal, the organizer counts to three. On the count of three, each person looks directly at his or her nearest neighbor and makes the sound of his or her animal as loudly as possible. For obvious reasons, this is called the Barnyard Game (Adams 1979). Almost all participants feel some reluctance to play because they risk appearing silly in front of their friends.

There is nothing inherently wrong with being afraid to take a risk. In fact, the idea of risk aversion is a basic concept in decision making under uncertainty; we have seen, for example, that the basic risky decision as described in Chapter 3 requires the decision maker to determine whether the risk of a loss (relative to a sure thing) is justified by a possible but uncertain gain. It may be counterproductive, though, not to offer a creative alternative for consideration in a decision problem because you risk others thinking your idea is impossible, too "far out," or downright silly. What are the consequences of presenting a far-out idea that turns out to be unacceptable? The worst that might happen is that the idea is immediately determined to be infeasible. (Making far-out suggestions can have a more subtle value. Outsiders often have a difficult time understanding exactly what the problem is. Presenting far-out ideas for action is a sure way to get a clear statement of the problem, couched in an explicit and often supercilious explanation of why the idea will not work. Although this technique cannot be used in every situation, when it works the result is a better understanding of the decision situation.)

2 Status Quo Bias   Decision making automatically means that the decision maker is considering at least one alternative that is different from the status quo. As indicated in the opening of this chapter, the ability to deal with change is becoming increasingly important for managers and decision makers. Studies show, however, that many people have a built-in bias toward the status quo. The stronger that bias, the more difficulty one may have coming up with creative problem solutions and alternatives.

3 Reality versus Fantasy   An individual may place a lot of value on being realistic and a low value on fantasizing. Creative people must be able to control their imagination, and they need complete access to it. Many exercises are available for developing an enhanced imagination and the ability to fantasize. Richard de Mille's *Put Your Mother on the Ceiling* (1976) has many imagination games. Although designed primarily for children, going through one of these games as an exercise in using fantasy can provide a remarkable experience for anyone. An excerpt from one of these games is reproduced in *Breathing* (see below). For the best effect, have a friend read this to you, pausing at the slash marks, while you sit quietly with your eyes closed.

4 Judgment and Criticism　This block arises from applying one's values too soon in the creative process. Rather than letting ideas flow freely, some individuals tend to find fault with ideas as they arise. Such fault finding can discourage the creation of new ideas and can prevent ideas—one's own or someone else's—from maturing and gathering enough detail to become usable. Making a habit of judging one's own thoughts inevitably sacrifices some creative potential.

Breathing

Let us imagine that we have a goldfish in front of us. / Have the fish swim around. / Have the fish swim into your mouth. / Take a deep breath and have the fish go down into your lungs, into your chest. / Have the fish swim around in there. / Let out your breath and have the fish swim out into the room again.

Now breathe in a lot of tiny goldfish. / Have them swim around in your chest. / Breathe them all out again.

Let's see what kinds of things you can breathe in and out of your chest. / Breathe in a lot of rose petals. / Breathe them out again. / Breathe in a lot of water. / Have it gurgling in your chest. / Breathe it out again. / Breathe in a lot of dry leaves. / Have them blowing around in your chest. / Breathe them out again. / Breathe in a lot of raindrops. / Have them pattering in your chest. / Breathe them out again. / Breathe in a lot of sand. / Have it blowing around in your chest. / Breathe it out again. / Breathe in a lot of little firecrackers. / Have them all popping in your chest. / Breathe out the smoke and bits of them that are left. / Breathe in a lot of little lions. / Have them roaring in your chest. / Breathe them out again.

Breathe in some fire. / Have it burning and crackling in your chest. / Breathe it out again. / Breathe in some logs of wood. / Set fire to them in your chest. / Have them roaring as they burn up. / Breathe out the smoke and ashes. . . .

Be a fish. / Be in the ocean. / Breathe the water of the ocean, in and out. / How do you like that? / Be a bird. / Be high in the air. / Breathe the cold air, in and out. / How do you like that? / Be a camel. / Be on the desert. / Breathe the hot wind of the desert, in and out. / How does that feel? / Be an old-fashioned steam locomotive. / Breathe out steam and smoke all over everything. / How is that? / Be a stone. / Don't breathe. / How do you like that? / Be a boy (girl). / Breathe the air of this room in and out. / How do you like that?

Source: de Mille, Richard (1976). *Put Your Mother on the Ceiling.* New York: Viking Penguin. Reprinted by permission of the author.

Cultural and Environmental Blocks

All decisions are made in some sort of social and cultural environment. The blocks that we describe here represent ways in which that environment may hinder the production and recognition of creative alternatives in decision situations.

1 Taboos   This type of block has to do with what is "proper behavior" or "acceptable" in a cultural sense; taboos may exist for no apparently good reason. The following problem (adapted from Adams 1979) demonstrates this block. As before, think about this problem before reading further.

PING-PONG BALL IN A PIPE

You are in a room with six other people. The room is entirely barren except for a steel pipe embedded solidly in the concrete floor and extending 25 centimeters above the floor. The upper end of the pipe is open, and a Ping-Pong ball is at the bottom of the pipe as in Figure 6.3. Your job is to get the Ping-Pong ball out of the pipe without damaging the ball, the pipe, or the concrete. You have only the following items available that you can use to extricate the ball:

10 feet of cotton string

5 ounces of dry cereal

A wire coat hanger

A steel file

A chisel

A hammer with a wooden handle

A monkey wrench

A light bulb

Often people come up with solutions like fashioning a long set of tweezers out of the coat hanger or smashing the hammer handle and using the splinters to retrieve the ball. But occasionally a few intrepid individuals cross a cultural borderline and realize that the simplest way to get the ball out is to have people in the group urinate in the pipe! Of course, urinating in such a public situation is somewhat taboo in our Western culture, effective though it may be for solving the problem.

For a more realistic example, suppose one of your co-workers has a new baby and wishes to bring the child to work so that she can continue to nurse the child. Certain taboos are involved here, including nursing in public and having a child in the workplace during "serious" work time. Should the taboos be violated? A creative alternative would find a way to accommodate the mother without grossly violating the taboos.

Figure 6.3

Ping-Pong ball in a pipe. Extract the Ping-Pong ball from the steel pipe without damaging the concrete, pipe, or ball.

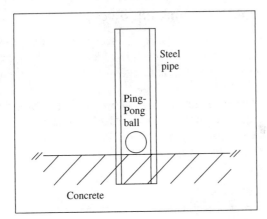

Steel pipe

Ping-Pong ball

Concrete

2 Strength of Tradition As we mentioned previously, individuals can resist change because of a bias toward the status quo. There is a cultural counterpart to this block; in many cases, the sociocultural environment in which a decision maker operates places a high value on maintaining tradition. Adopting change can be difficult in such a situation, which in turn can hinder the production of creative suggestions in the first place. For example, the musical *Fiddler on the Roof* describes the tradition-bound culture of Russian Jews in the early twentieth century and a father's difficulty in dealing with his daughters' new ways of finding husbands.

3 Reason and Logic versus Humor, Fantasy, and Artistic Thinking There is a clear block against using feelings, intuitions, and emotions in business problem solving. Certainly valuable insights and understanding come from analytical treatments of any given problem; indeed these skills are important in decision making, and a course in decision analysis offers to teach such skills. However, valuable cues and ideas can also arise by admitting and examining feelings, intuitions, and emotions. For example, doing so can help understand the values of others who may have a stake in a decision. In the example above regarding bringing a child to work, reluctance to allow a child in the workplace may be due to the values of the workers who feel that the playfulness, fantasy, and humor that children represent should not displace reason and logic at work.

In a decision-making course much of the emphasis is on the development of analytical thinking. Unfortunately, little effort is put into more artistically oriented thinking skills such as using imagery, being playful, storytelling, or expressing and appreciating feelings. Such activities tend to be culturally blocked because of the stress placed on analysis. From the discussion in this chapter, it would appear that artistic thinking can play an important role in the development of creative alternatives. The best possible arrangement is for an individual to be "mentally ambidextrous," or good at switching between analytical and artistic thinking styles. This enhances creative development of potential alternatives without sacrificing subsequent careful analysis.

Organizational Issues

Without a doubt, different organizations have different characteristics or cultures, and organizational culture can have a strong influence on decision making. Many of the issues that we have already discussed can be a part of an organization's decision-making culture. For example, an organization may have a culture that in subtle ways promotes criticism and judging of ideas, stereotyping, or being risk-averse. Humor, playfulness, or artistic thinking may be frowned upon, or change may be resisted in order to preserve company traditions. For all of the reasons discussed above, such characteristics can reduce the creative potential of individuals in the organization.

By their very nature, organizations can impede creative thought. As Adams (1979, p. 143) points out, "the natural tendency of organizations to routinize, decrease uncertainty, increase predictability, and centralize functions and controls is certainly at odds with creativity." Other features of organizations also can hinder creativity. Examples include excessive formal procedures (red tape) or lack of cooperation and trust among co-workers. Hierarchical organizational structures can hinder creativity, which in turn can be exacerbated by supervisors who tend to be autocratic.

Teresa Amabile has studied creativity and organizations for over twenty years. Her work has led to a detailed model of individual creativity in the organizational context (Amabile 1988). First, individual creativity requires three ingredients: expertise in the domain, skill in creative thinking, and intrinsic motivation to do the task well. In other words, we need someone who is good at what he or she does, who likes to do it just because it is interesting and fun, and who has some skill in creative thinking, perhaps along the lines of the creativity-enhancing techniques we discuss later in this chapter.

Amabile's work shows how the organizational environment can influence individual creativity. In particular, she warns that expecting detailed and critical evaluation, being closely watched, focusing on tangible rewards, competing with other people, and having limited choices and resources for doing the job all can hinder one's creativity. When she compared high- and low-creativity scenarios in organizations, though, the results indicated that a delicate balance must be maintained. For example, workers need clear overall goals, but at the same time they need latitude in how to achieve those goals. Likewise, evaluation is good as long as it is focused on the work itself (as opposed to the person) and provides informative and constructive help. Such evaluation ideally involves peers as well as supervisors. Although a focus on tangible rewards can be detrimental, knowing that one's successful creative efforts will be recognized is important. A sense of urgency can create a challenging atmosphere, particularly if individuals understand the importance of the problem on which they are working. If the challenge is viewed as artificial, however, such as competing with another division in the company or having an arbitrary deadline, the effect can be to decrease creativity. Thus, although creativity is essentially an individual phenomenon, managers can have a significant impact on creativity in their organizations through goal setting, evaluation, recognition and rewards, and creating pressure that reflects a genuine need for a creative solution.

Finally, even though managers can help individuals in their organizations be more creative, one can develop a "blind spot" because of a long-term association with a particular firm; it becomes difficult to see things in a new light simply because certain procedures have been followed or perspectives adopted for a long time. The German word *betriebsblind* for this situation literally means "company-blind." One of the important roles that consultants serve is bringing a new perspective to the client's situation.

Value-Focused Thinking for Creating Alternatives

Keeney (1992, Chapters 7 and 8) describes a number of different ways in which fundamental and means objectives can be used as a basis for creating new alternatives for decision alternatives. In this section we review some of these techniques.

Fundamental Objectives

The most basic techniques use the fundamental objectives directly. For example, take one fundamental objective and, ignoring the rest, invent a (possibly hypothetical) alternative that is as good as it could be on that one objective. Do this for each fundamental objective one at a time, and keep track of all of the alternatives you come up with. Now go back and consider pairs of objectives; what are good alternatives that balance these two objectives? After doing this for various combinations of objectives, look at the alternatives you have listed. Could any of them be modified so that they would be feasible or perhaps satisfy the remaining objectives better? Can any of the alternatives be combined?

A related approach is to consider all of the fundamental objectives at once and imagine what an alternative would look like that is perfect in all dimensions; call this the *ideal* alternative. Most likely it is impossible, but what makes it impossible? If the answer is constraints, perhaps some of those constraints can be removed or relaxed.

Still another possibility is to go in the opposite direction. Find a good alternative and think of ways to improve it. The fact that the alternative is a good one in the first place can reduce the pressure of finding a better one. In searching for a better one, examine the alternative carefully in terms of the objectives: On which objectives does it perform poorly? Can it be improved in these dimensions? For example, Keeney (1992) describes an analysis of possible sites for a hydroelectric power plant. One of the potential sites was very attractive economically but had a large environmental impact, which made it substantially less desirable. On further study, however, the design of the facility at the site was modified to reduce the environmental impact while maintaining the economic advantage.

Means Objectives

We mentioned back in Chapter 3 that the means objectives can provide a particularly fruitful hunting ground for new alternatives. The reason for this is simply that the means objectives provide guidance on what to do to accomplish the fundamental objectives. In complicated problems with many fundamental objectives and many related means objectives, this approach can generate many possible courses of action. For example, consider the following decision situation.

TRANSPORTATION OF NUCLEAR WASTE

One of the problems with the use of fission reactors for generating electricity is that the reactors generate substantial amounts of radioactive waste that can be highly toxic for extremely long periods of time. Thus, management of the waste is necessary, and one possibility is to place it in a storage facility of some sort. Transporting the waste is itself hazardous, though. In describing the problem, Keeney (1992) notes that the decision situation includes the selection of a type of storage cask in which the material will be shipped, followed by the selection of a transportation route and a choice as to how many casks to ship at once. The uncertainties include whether an accident occurs, the amount of radiation released, and whether an efficient evacuation plan exists when and if an accident occurs.

Means objectives are associated with each of the decisions and uncertainties. For example, a means objective is to select the best possible cask, and that might include designing a special kind of cask out of a particular material with appropriate size and wall thickness specifications. Selecting a transportation route that travels through a sparsely populated area is a means objective to reduce potential exposure in the case of an accident. In selecting the number of casks to ship at once, one would want to balance the chance of smaller accidents with more frequent but smaller shipments against the chance of a larger accident with larger and less frequent shipments.

In examining the uncertainties, obvious means objectives come to mind. For example, an important means objective is to reduce the chance of an accident, which in turn suggests strict rules for nuclear-waste transportation (slow speeds, driving during daylight hours, special licensing of drivers, additional maintenance of roads along the route, and so on). Reducing the amount of radiation released in an accident and increasing the chance of an efficient evacuation plan being in place suggest the development of special emergency teams and procedures at all points along the transportation route (Keeney 1992, pp. 205–207).

In another example, we can use a means network directly. Take the automotive-safety example that we discussed in Chapter 3; the problem is to find ways to maximize safety of automotive travel, and according to the fundamental objectives (Figure 3.1), maximizing safety means minimizing minor injuries, serious injuries, and deaths of both adults and children. The means-objectives network is reproduced here as Figure 6.4. You can see that each node in the network suggests particular alternatives. The objective of maximizing use of vehicle-safety features suggests that currently available features might be mandated, or incentives might be provided to

Figure 6.4

*A means-objectives
network for improving
automotive safety.
Source:* Keeney
(1992, p. 70).

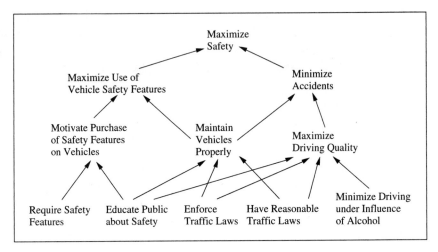

manufacturers. Likewise, incentives (e.g., a reduced insurance premium) might be given to automobile drivers for using safety features. Farther down in the network we find the objective of educating the public. We already have widespread driver-education programs for new drivers, but what about providing incentives for adult drivers to take periodic refresher courses? To combat drunken driving in particular, many states in the United States have implemented a combination of tougher traffic laws and public-education programs.

You can see from these examples how useful the means objectives can be for identifying new alternatives. Moreover, using a means-objectives network can ensure that as many aspects of the decision problem as possible are covered; the decision maker can see exactly what the new alternatives help achieve and can perhaps further develop the alternatives to attain a level of balance among the fundamental and means objectives. For example, in examining a set of safety proposals, matching the proposals against means and fundamental objectives might reveal that safety of children has not been explicitly addressed or that the set of proposals currently under consideration do nothing to encourage proper vehicle maintenance.

The Decision Context

Finally, it is always possible to broaden the decision context as part of the search for new ideas. We mentioned this possibility back in Chapter 3 in discussing decision structuring. As argued above, part of the creative process requires that the decision maker look at a problem from as many different perspectives as possible, and considering a broader context is guaranteed to reveal a different view of a decision situation. Take the automotive-safety example, and suppose the issue is broadened from automotive safety to transportation safety in general. At a microlevel in a city, this might imply a comprehensive education program for local cyclists, pedestrians, and motorists. At a county or state level, one might consider incentives to reduce urban sprawl and increase the use of alternative transportation methods.

For another example, reconsider the nuclear-waste transportation problem. Deciding how to transport the waste may arise from a prior decision to place a storage site in a designated location. But suppose we broaden the decision context to consider a variety of storage options. Doing so suggests possibilities like storing waste at the nuclear power plants themselves, which in turn suggests finding ways to reduce waste production or to store it more efficiently at these sites. Another possibility is to store the waste in smaller but more numerous waste facilities rather than in fewer and larger sites. Although some of these alternatives may not be reasonable to pursue, broadening the decision context can help uncover new ways to address the decision at hand and may suggest solutions of a more general nature.

Other Creativity Techniques

Fluent and Flexible Thinking

Fluency and flexibility of thinking are important in enhancing creativity. Fluency is the ability to come up with many new ideas quickly. Flexibility, on the other hand, stimulates variety among these new ideas. An individual who can write down many ideas quickly, regardless of what they may be, would be a fluent thinker. The flexible thinker might have a shorter list of ideas, but the ideas would tend to cover a broader range of possibilities. An individual who is both fluent and flexible can write down many different ideas quickly. One useful analogy compares thinking with digging holes. Fluent thinking is seen as the ability to dig one hole very deep and very quickly by taking a lot of dirt from one place. Flexible thinking, however, is more like the ability to dig many smaller holes in many different places.

A common exercise that is used to demonstrate these ideas is to think of new uses for common red construction bricks. The goal is to come up with marketing ideas that a brickyard owner could pursue to get out of financial difficulties. A list with many uses that are variations on a common theme indicates fluency. For example, many of the uses may be ways to use bricks to build things, or uses that take advantage of a brick's weight (for example, as a paperweight or ballast for hot-air balloons). A list with a lot of variety in different attributes of the bricks indicates flexibility. For example, flexible thinking would use many different attributes of a brick: strength, weight, color, texture, hardness, shape, and so on. Understanding the difference between the two different kinds of thinking is the first step toward developing enhanced thinking skills in either fluency or flexibility or even both.

Idea Checklists

One classic technique for enhancing creativity uses checklists that cover many potential sources of creative solutions to problems. Most of us use lists in a rather natural way. The yellow pages provide a simple and ubiquitous list that can provide many ideas for solutions to specific problems. (Are any plumbers available on Sunday? Any do-it-yourself stores close by?) Mail-order and retail catalogs are examples of lists that we may use to solve a gift-giving problem. Many authors have written general-purpose lists for problem solving. The best known is Osborn's (1963) 73 Idea-Spurring Questions, reproduced below. Of course, Osborn's list can be extended with other descriptive verbs such as multiply, squeeze, lighten, propel, and flatten.

Another creativity-enhancing technique is to write down attributes of a problem, list alternative options under each attribute, and then consider various combinations and permutations of the alternatives. This use of lists was suggested by Koberg and Bagnall (1974), who dubbed the technique *morphological forced connections* to emphasize the combination of morphological attributes in design problems. The technique is not limited to designing new products. In fact, an early variant of this technique was used to think about objectives in a decision situation. Dole et al. (1968 a,b) describe an application by the National Aeronautics and Space Administration (NASA) for determining the scientific objectives of space-exploration missions. Action phrases (such as "measure tidal deformations in . . . ") were combined with target features (such as "the interior of . . . ") and target subjects (Jupiter, for example) to create a possible scientific objective. The candidate objective then was considered to determine whether it was a valid scientific objective. If so, it was included

---

Osborn's 73 Idea-Spurring Questions

Put to other uses? New ways to use as is? Other uses if modified?

Adapt? What else is like this? What other idea does this suggest? Does the past offer a parallel? What could I copy? Whom could I emulate?

Modify? New twist? Change meaning, color, motion, sound, odor, form, shape? Other changes?

Magnify? What to add? More time? Greater frequency? Stronger? Higher? Longer? Thicker? Extra value? Plus ingredient? Duplicate? Multiply? Exaggerate?

Minify? What to subtract? Smaller? Condensed? Miniature? Lower? Shorter? Lighter? Omit? Streamline? Split up? Understate?

Substitute? Who else instead? What else instead? Other ingredient? Other material? Other process? Other power? Other place? Other approach? Other tone of voice?

Rearrange? Interchange components? Other pattern? Other layout? Other sequence? Transpose cause and effect? Change pace? Change schedule?

Reverse? Transpose positive and negative? How about opposites? Turn it backward? Turn it upside down? Reverse roles? Change shoes? Turn tables? Turn other cheek?

Combine? How about a blend, an alloy, an assortment, an ensemble? Combine units? Combine purposes? Combine appeals? Combine ideas?

Source: From A. F. Osborn, *Applied Imagination,* 3rd edition. Copyright © 1963. All rights reserved. Adapted by permission of Allyn & Bacon.

in the list. For example, "Establish the structure of the interior of the sun" was an objective, but "Determine the characteristic circulation patterns in the photosphere of the space environment" was not.

The value of morphological forced connections is not so much to find all possible combinations as much as to provide a framework within which all imaginable combinations can be screened easily to determine the most appropriate candidates. In the NASA example, candidate objectives were generated readily and then screened for validity. Another example, reported by Howard (1988), comes from an advertising claim by a fast food hamburger chain that, with its custom service, one can order 1024 different kinds of hamburgers; that is, a customer could order a burger with or without each of 10 possible ingredients. Not all of the possible combinations are reasonable, however; for example, one combination is a "burger" that consists only of lettuce, or only of ketchup. One is the "nullburger": nothing at all. Most individuals would agree that many of these unusual combinations must be screened out. In fact, a burger may not be a burger without the beef patty and a bun.

A more serious example also comes from Howard (1988), who suggests the *strategy-generation table.* Figure 6.5 shows a typical strategy-generation table for an energy conglomerate that is considering possible expansion. Dividend payout and dividend-to-equity ratio also are important attributes for the conglomerate to consider in strategic planning. For the most part, the table is self-explanatory. An overall strategy is one that has individual elements in each column. As with the hamburger, however, not all combinations make sense. For example, because of cash constraints it is not likely that the conglomerate could pursue aggressive expansion in each of its five existing businesses, acquire another firm, and also have a high dividend payout. The strategy outlined in Figure 6.5 is one that might be described by executives as a "service business" strategy. Other feasible strategies could be assembled using the strategy-generation table. The point here is that the morphological forced connections technique facilitates both generation and screening to come up with a list of reasonable alternatives.

Figure 6.5

Strategy-generation table.

Source: Reprinted by permission of Ron Howard, "Decision Analysis: Practice and Promise," *Management Science,* 34, No. 6, June 1988, pp. 679–695. Copyright 1988, The Institute of Management Sciences.

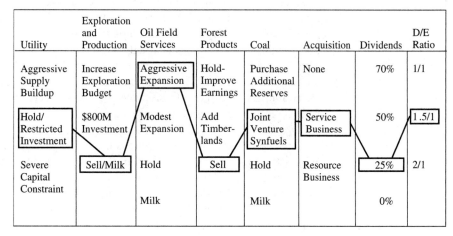

Utility	Exploration and Production	Oil Field Services	Forest Products	Coal	Acquisition	Dividends	D/E Ratio
Aggressive Supply Buildup	Increase Exploration Budget	Aggressive Expansion	Hold-Improve Earnings	Purchase Additional Reserves	None	70%	1/1
Hold/Restricted Investment	$800M Investment	Modest Expansion	Add Timberlands	Joint Venture Synfuels	Service Business	50%	1.5/1
Severe Capital Constraint	Sell/Milk	Hold	Sell	Hold	Resource Business	25%	2/1
		Milk		Milk		0%	

Brainstorming

Brainstorming is another popular way of generating a long list of ideas quickly. To be effective, a brainstorming session should include at least two people, and probably no more than 8 or 10 (it can be difficult to keep up with all of the ideas generated by a group that is very large). The rules for a brainstorming session are simple:

1   No evaluation of any kind is permitted.
2   All participants should think of the wildest ideas possible.
3   The group should be encouraged to come up with as many ideas as possible.
4   Participants should try to build upon or modify ideas of others.

Brainstorming works well for several reasons. The most important is probably the lack of any judgment, which eliminates an important block for many people. That so many ideas are created rapidly reassures those who have little faith in their own creative potential: The enthusiasm of a few individuals tends to be contagious, and a "one-upmanship" game usually develops as participants try to top previous ideas. Practitioners report that the technique tends to generate a first rash of ideas utilizing common solutions. After this initial phase, participants must come up with new concepts. Naturally, the newer concepts are the most valuable result of the brainstorming exercise.

Metaphorical Thinking

Given our earlier discussion of fantasy and imagination, it should come as no surprise that creativity can be enhanced by the use of metaphors. Much of the work on metaphorical thinking comes from William J. J. Gordon (1961, 1969), who founded Synectics, Inc., a consulting corporation that specializes in creative problem solving. Three kinds of metaphors can be used systematically to enhance creative potential: direct analogy, personal analogy, and fantasy analogy.

Direct analogy involves thinking about how others have solved problems similar to the one under consideration. Often the most productive approach is to examine solutions found in nature. For example, wet leaves, which pack together snugly, suggested metaphorically how potato chips might be packaged. The result was Pringles potato chips, which are stacked and sold in a can. The inventor of Velcro was inspired while removing burrs from his dog's fur. Suppose the problem is to design a home security system. What kinds of security systems do plants and animals have? Many animals make a lot of noise when threatened. Perhaps a security system could be designed that could sense intruders and then make noise—by turning on a stereo or TV inside a house—to frighten away intruders.

Personal analogy is closely related to the kinds of games that are played in de Mille's *Put Your Mother on the Ceiling*. The idea is to imagine a variety of personal situations that are pertinent in some way to the problem at hand. Personal analogy can be extremely helpful in computer programming, for example. Imagining the precise steps (and perhaps shortcuts) that would be taken to solve a problem by hand can help in designing a computer program to solve the same kind of problem.

In *fantasy analogy,* the group tries to come up with truly far-fetched, fantastic, and ideal solutions. The classic example arose when a group was trying to design an airtight closure for space suits that would be easily operated. Here is a transcript of part of the session, taken from Gordon (1961):

G: Okay. That's over. Now what we need here is a crazy way to look at this mess. A real insane viewpoint . . . a whole new room with a viewpoint!

T: Let's imagine you could will the suit closed . . . and it would do just as you wanted by wishing . . . [fantasy analogy]

G: Wishing will make it so . . .

F: Ssh, okay. Wish fulfillment. Childhood dream . . . you wish it closed, and invisible microbes, working for you, cross hands across the opening and *pull* it tight . . .

B: A zipper is kind of a mechanical bug [direct analogy]. But not airtight . . . or strong enough . . .

G: How do we build a psychological model of "willing-it-to-be-closed"?

R: What are you talking about?

B: He means that if we could conceive of how "willing-it-to-be-closed" might happen in an actual model—then we . . .

R: There are two days left to produce a working model—and you guys are talking about childhood dreams! Let's make a list of all the ways there are of closing things.

F: I hate lists. It goes back to my childhood and buying groceries . . .

R: F, I can understand your oblique approach when we have time, but now, with this deadline . . . and you still talking about wish fulfillment.

G: All the crappy solutions in the world have been rationalized by deadlines.

T: Trained insects?

D: What?

B: You mean, train insects to close and open on orders? 1–2–3 Open! Hup! 1–2–3 Close!

F: Have two lines of insects, one on each side of the closure—on the order to close they all clasp hands . . . or fingers . . . or claws . . . whatever they have . . . and then closure closes tight . . .

G: I feel like a kind of Coast Guard insect [personal analogy].

D: Don't mind me. Keep talking . . .

G: You know the story . . . worst storm of the winter—vessel on the rocks . . . can't use lifeboats . . . some impatient hero grabs the line in his teeth and swims out . . .

B: I get you. You've got an insect running up and down the closure, manipulating the little latches . . .

G: And I'm looking for a demon to do the closing for me. When I will it to be closed [fantasy analogy], Presto! It's closed!

B: Find the insect—he'd do the closing for you!

R: If you used a spider . . . he could spin a thread . . . and sew it up [direct analogy].

T: Spider makes thread . . . gives it to a flea . . . Little holes in the closure . . . flea runs in and out of the holes closing as he goes . . .

G: Okay. But those insects reflect a low order of power . . . When the Army tests this thing, they'll grab each lip in a vise one inch wide and they'll pull 150 pounds on it . . . Those idiot insects of yours will have to pull steel wires behind them in order . . . They'd have to stitch with steel. *Steel.*

B: I can see one way of doing that. Take the example of that insect pulling a thread up through the holes . . . You could do it mechanically . . . Same insect . . . put holes in like so . . . and twist a spring like this . . . through the holes all the way up to the damn closure . . . twist, twist, twist, . . . Oh, crap! It would take hours! And twist your damn arm off!

G: Don't give up yet. Maybe there's another way of stitching with steel . . .

B: Listen . . . I have a picture of another type of stitching . . . That spring of yours . . . take two of the . . . let's say you had a long demon that forced its way up . . . like this . . .

R: I see what he's driving at . . .

B: If that skinny demon were a wire, I could poke it up to where, if it got a start, it could pull the whole thing together . . . the springs would be pulled together closing the mouth . . . Just push it up . . . push—and it will pull the rubber lips together . . . Imbed the springs in rubber . . . and then you've got it stitched with steel!

Source: Gordon (1961).

Other Techniques

Several other techniques are available to enhance a group's creative potential. Many rely on methods for improving group interaction in general. For example, Nominal Group Technique (NGT) (Delbecq, Van de Ven, and Gustafson 1975) begins with no interaction. Individuals in the group each write down as many ideas as they can on pieces of paper. Then each individual in turn presents one of his or her ideas. The group leader records these ideas on a flipchart or chalkboard. Discussion begins after ideas from each participant are written down. At the end, each individual writes down his or her ranking or rating of the ideas. These are then combined mathematically to arrive at a group decision.

The main advantage of NGT is that the group leader manages the interaction of the group in such a way that certain blocks are avoided and the environment is enhanced. For example, discussion is not permitted until after the ideas are presented, thus creating a more supportive environment.

Other techniques actually use a more adversarial approach. Devil's advocacy and dialectical inquiry are techniques in which individuals take sides in a debate. On the surface, this might appear to hamper creative thought, but when such techniques are used only after ideas have been generated and a healthy creative environment has been established, they can work well. It also helps if all participating members understand what the techniques are meant to do. The main advantage of this kind of approach is that it can help a group of individuals consider a problem from multiple perspectives. Being forced into an alternative viewpoint can lead to new creative ideas. Another advantage is that the group is less likely to overlook basic issues that may be hidden from certain vantage points.

The role of the leader in group discussion techniques is paramount. A good leader sets the tone of the session, and a positive tone promotes an atmosphere that is conducive to healthy discussion and that encourages the free flow of ideas. It is easy to see that a group with such a leader probably will have more success in generating creative ideas and solving problems.

Creating Decision Opportunities

Creativity in decision making can be much more than generating new alternatives. A really creative decision maker is one who creates decision opportunities. Keeney (1992) stresses that an individual, group, or organization which understands its values and objectives clearly is perfectly positioned to look for decision opportunities proactively rather than merely reacting to decision problems served up by life. As you may agree, life does not always generously provide decision situations with many attractive options, but instead often seems to pose difficult decision problems that must be addressed, often under trying circumstances.

In Chapter 9 of *Value-Focused Thinking,* Keeney describes a wide variety of ways to use one's objectives to help identify new decision opportunities. The more obvious ones are to look at fundamental and means objectives and find ways to achieve them; simply choosing to look for ways to achieve ones objective is in itself creating an opportunity. As we discussed above, broadening a decision context can be worthwhile. Simply learning to say no to those who ask for time, money, or assistance can create decision opportunities for those who have a difficult time turning down meritorious requests. And, as Keeney explains, when you just do not know what to do, that is the perfect occasion to create a decision opportunity in which you can tailor the alternatives to match your objectives. The decision maker who searches actively and creatively for new opportunities can look forward to many exciting possibilities and a full life.

SUMMARY

Creativity is important in decision making because the available alternatives determine the boundaries of the decision. We discussed various theories of creativity, including the phases of creative thought. We then introduced many different kinds of creativity blocks that hinder creative efforts. In discussing creativity-enhancing techniques, we began with the use of one's objectives as a basis for developing new ideas and showed how means objectives in particular can provide a fertile ground for generation of new alternatives. Other creativity techniques include becoming proficient in both flexible and fluent thinking, list making, brainstorming, and metaphorical thinking. Group discussion techniques can promote creativity through appropriate management of interaction in the group and can also enhance the creative environment. Finally, we argued that a truly creative decision maker goes beyond the creation of alternatives all the way to the development of new decision opportunities.

QUESTIONS AND PROBLEMS

6.1   In talking about cultural and environmental blocks, we discussed the matter of a woman bringing an infant into the workplace and the possible reactions of others. Consider the situation in which a classmate wants to bring her child to school in order to continue nursing. Are the taboos the same? What are the differences between the two situations?

6.2   Choose a decision that you currently face. What are your objectives in this situation? List your means objectives, and for each means objective list at least one alternative that could help achieve that objective.

6.3   An important objective of most academic programs is to get students to be more creative. How can this be done? What specific exercises can you think of that would help? What kinds of things can an instructor do? Any changes to the curriculum or requirements? Try some of the creativity-enhancing techniques that we discussed. For example, you might try any of the following:

   a   List fundamental and means objectives that you think would be appropriate for your academic program. Use these objectives as a basis for generating ways to make students more creative.

   b   List making, especially morphological forced connections. What are important attributes to list? What are alternatives under those attributes? Can you use Osborn's list? Can you think of extensions to his list?

   c   Brainstorming. To do this, you need to find a small group of people to brainstorm with. The exact size of the group is not important, but you need at least one more besides yourself, and probably no more than 8 or 10 altogether.

   d   Metaphorical thinking. Direct analogy is difficult here. How do other disciplines teach students to be creative? How do animals teach offspring to deal with new situations? Personal analogy is somewhat more straightforward. Imagine that you are the boss. What do you want your employee to be able to do? Imagine that you are a professor. Would you want your students to be creative? How would you encourage it? Can you imagine that you are a problem that needs to be solved? How would you like to be solved? Try fantasy analogy. Try to imagine the most fantastic and ideal creativity training program. Describe it in detail; exercise your imagination! What kinds of resources would be required? How many people? How much time? What kinds of interactions with others? Would there be instructors and students, or just participants, all on the same level? Think of other questions.

   You are not required to use any of these techniques, nor are you limited to them. Your job is to generate some creative ideas any way you can.

6.4   How would you design an organization so that it could, in Tom Peters's words, "thrive on chaos"? What characteristics would such an organization have? What kind of people would you try to hire? How would the role of the managers differ from the traditional view of what a manager does?

6.5   In discussing the perceptual block of stereotyping, we used the example of a person with long hair and no necktie applying for a job as an engineer. Did you imagine this person as male or female? Why? Was there a block involved in your perception? Describe it.

6.6   The point is often made that formal schooling can actually discourage young children from following their natural curiosity. That curiosity is an important element of creativity, and so it may be the case that schools indirectly and inadvertently are causing children to become less creative than they might be. What does this imply for those of us who have attended school for many years? What can you suggest to today's educators as ways to encourage children to be curious and creative?

6.7 Use the means-objectives network in Figure 6.4 to create a list of alternatives for improving automotive safety. Try to create at least one alternative for each objective listed.

6.8 Reconsider the summer-intern decision discussed in Chapter 3. Figure 3.4 shows the fundamental and means objectives. With a group of friends, use these objectives along with brainstorming to generate a list of at least 20 things PeachTree can do to help locate and hire a good summer intern.

6.9 The lists that were discussed focused on the generation of alternatives. However, the same technique can be used to determine objectives in a decision situation. Write a list of questions that you might ask yourself when searching for a job. What do you want to accomplish? For example, do you want to save money? Save time? Improve your lifestyle? What else is important?

6.10 Describe a situation in which unconscious incubation worked for you. Describe one in which it did not. Can you explain why it worked in the first case but not in the second?

6.11 One of the technological problems that we face as a society is the increasing use of plastics in disposable items. Landfills are becoming more and more expensive and difficult to maintain, and land for new ones can only be obtained at premium prices. Furthermore, the plastics that are dumped in the landfills may release dioxins (toxic chemical substances) into the soil.

  Of course, many different kinds of plastic exist. Spend 10 minutes writing down possible ways to recycle one-gallon plastic milk jugs. You can assume that the milk jugs are received rinsed out and reasonably clean, but not sterile. Look at your list. Does it reflect fluent thinking, flexible thinking, or both?

CASE STUDIES

MODULAR OLYMPICS

Seoul, Korea, was the site of the 1988 Summer Olympic Games. However, the *Wall Street Journal* (June 29, 1987) reported that because of political unrest in Korea, there was some concern about having the games there. Would it have been possible to change sites as late as 1987? It seems unlikely because of the expense and planning associated with putting the games on. Wouldn't it be nice to be able to disassemble the Olympics from one site, ship them to a new site, reassemble, and proceed? Modular Olympics.

Question

1 Use whatever creativity-enhancing techniques you can to help think about ways to modify the Olympics to make it possible to change locations more easily than is currently the case.

BURNING GRASS-SEED FIELDS

Grass-seed farmers in the Willamette Valley in western Oregon have been burning their fields since the 1940s. After the grass seed is harvested, the leftover straw is burned to remove it and to sterilize the fields. However, burning large fields of straw creates a lot of pollution. Another burning method is available that uses a propane torch to burn the stubble in a field after the straw has been removed.

In the face of controversy over the negative effects of burning, the grass-seed industry is considering alternative ways to sterilize the fields. The propane method is the most promising, but it first requires removal of the straw.

Question

1   What can be done with the straw? Spend 10 minutes writing down all the possible uses you can think of for straw. If possible, form a group and have a brainstorming session. Or try any other creativity technique that you can think of. Does your list demonstrate flexibility or fluency of ideas?

REFERENCES

It is rather unusual for a book on decision theory to have a chapter on creativity. The "management science" tradition would suggest that there is a set way of attacking a specific problem and that all the decision maker must do is apply the appropriate technique. However, this is too simple for the complex problems that managers face these days. Hence, this chapter is meant to dispel the notion that analytical thinking is all that is required to solve real problems.

Very little literature exists within management science on creativity. Keller and Ho (1989) review and summarize this literature. Some of Keeney and Raiffa's (1976, Chapter 2) discussion of the structuring of objectives is pertinent to creativity in decision making. As indicated in the text, Keeney (1992) devotes many pages to creativity. Kleindorfer, Kunreuther, and Schoemaker (1993) also discuss creativity from a management science perspective.

Some of the literature on creativity, like Adams (1979), comes from an engineering/design/inventing perspective. (In fact a recent creativity book by Adams (1986) contains a chapter on decision analysis!) Most of the literature, however, comes from psychology, as the many references at the beginning of the chapter indicate. Baron (1988), Bazerman (1994), and Hogarth (1987) provide good reviews of the creativity literature from a psychological perspective. Johnson, Parrott, and Stratton (1968) provide some insights on the value of brainstorming.

Adams, J. L. (1979) *Conceptual Blockbusting: A Guide to Better Ideas,* 2nd ed. Stanford, CA: Stanford Alumni Association.

Adams, J. L. (1986) *The Care and Feeding of Ideas.* Stanford, CA: Stanford Alumni Association.

Amabile, T. (1988) "A Model of Creativity and Innovation in Organizations." *Research in Organizational Behavior,* 10, 123–167.

Baron, J. (1988) *Thinking and Deciding*. Cambridge: Cambridge University Press.

Barron, F. (1965) "The Psychology of Creativity." In T. M. Newcomb (ed.) *New Directions in Psychology,* Vol. 2, pp. 1–134. New York: Holt.

Bartlett, S. (1978) "Protocol Analysis in Creative Problem Solving," *Journal of Creative Behavior,* 12, 181–191.

Bazerman, M. H. (1994) *Judgment in Managerial Decision Making,* 3rd ed. New York: Wiley.

Campbell, D. T. (1960) "Blind Variation and Selective Retention in Creative Thought as in Other Knowledge Processes," *Psychological Review,* 67, 380–400.

Davis, G. (1986) *Creativity Is Forever,* 2nd ed. Dubuque, IA: Kendall/Hunt.

Delbecq, A. L., A. H. Van de Ven, and D. H. Gustafson (1975) *Group Techniques for Program Planning.* Glenview, IL: Scott, Foresman.

De Mille, R. (1976) *Put Your Mother on the Ceiling.* Middlesex, England: Penguin.

Dole, S. H., H. G. Campbell, D. Dreyfuss, W. D. Gosch, E. D. Harris, D. E. Lewis, T. M. Parker, J. W. Ranftl, and J. String, Jr. (1968a) Methodologies for Analyzing the Comparative Effectiveness and Costs of Alternate Space Plans. *RM-5656-NASA,* Vol. 1 (Summary). Santa Monica, CA: The Rand Corporation.

Dole, S. H., H. G. Campbell, D. Dreyfuss, W. D. Gosch, E. D. Harris, D. E. Lewis, T. M. Parker, J. W. Ranftl, and J. String, Jr. (1968b) Methodologies for Analyzing the Comparative Effectiveness and Costs of Alternate Space Plans. *RM-5656-NASA,* Vol. 2. Santa Monica, CA: The Rand Corporation.

Gordon, W. J. J. (1961) *Synectics.* New York: Harper & Row.

Gordon, W. J. J. (1969) *The Metaphorical Way of Learning and Knowing.* Cambridge, MA: SES Associates.

Hogarth, R. (1987) *Judgement and Choice,* 2nd ed. Chichester, England: Wiley.

Howard, R. A. (1988) "Decision Analysis: Practice and Promise," *Management Science,* 34, 679–695.

Johnson, D. M., G. L. Parrott, and R. P. Stratton (1968) "Production and Judgment of Solutions to Five Problems," *Journal of Educational Psychology Monograph Supplement,* 59, 1–21.

Keeney, R. (1992) *Value-Focused Thinking*. Cambridge, MA: Harvard University Press.

Keeney, R., and H. Raiffa (1976) *Decisions with Multiple Objectives.* New York: Wiley.

Keller, L. R., and J. L. Ho (1989) "Decision Problem Structuring: Generating Options," *IEEE Transactions on Systems, Man, and Cybernetics,* 18, 715–728.

Kleindorfer, P. R., H. C. Kunreuther, and P. J. H. Schoemaker (1993) *Decision Sciences: An Integrated Perspective.* Cambridge: Cambridge University Press.

Koberg, D., and J. Bagnall (1974) *The Universal Traveler.* Los Altos, CA: William Kaufmann.

Kris, E. (1952) "On Preconscious Mental Processes." In *Psychoanalytic Explorations in Art.* New York: International Universities Press.

Kubie, L. S. (1958) *Neurotic Distortion of the Creative Process.* Lawrence: University of Kansas Press.

Maltzman, I. (1960) "On the Training of Originality." *Psychological Review,* 67, 229–242.

Maslow, A. (1954) *Motivation and Personality.* New York: Harper & Row.

Mednick, S. A. (1962) "The Associative Basis of the Creative Process." *Psychological Review,* 69, 220–232.

Osborn, A. F. (1963) *Applied Imagination,* 3rd ed. New York: Scribner.

Peters, T. (1988) *Thriving on Chaos.* New York: Knopf.

Rugg, H. (1963) *Imagination: An Inquiry into the Sources and Conditions that Stimulate Creativity.* New York: Harper & Row.

Skinner, B. F. (1972) *Cumulative Record: A Selection of Papers,* 3rd ed. Englewood Cliffs, NJ: Prentice Hall.

Staats, A. W. (1968) *Learning, Language, and Cognition.* New York: Holt.

Wallas, G. (1926) *The Art of Thought.* New York: Harcourt, Brace.

Winkelgren, W. A. (1974) *How to Solve Problems.* San Francisco: Freeman.

EPILOGUE

The basic solution to the nine-dot puzzle is shown in Figure 6.6. Many people tacitly assume that the lines may not go beyond the square that is implied by the dots, and so they fail to solve the puzzle. Figure 6.7 shows how to connect nine fat dots with three straight lines, removing the block that the line has to go through the centers of the dots.

It is possible, with enough effort to remove the necessary blocks, to connect the dots with one line. Some solutions include:

• Fold the paper in a clever way so that the dots line up in a row. Then just draw one straight line through all nine dots.

• Roll the paper up and tape it so that you can draw a spiral through all of the dots.

• Cut the paper in strips and tape it together so that the dots are in one row.

• Draw large dots, wad the paper up, and stab it with a pencil. Unfold it and see if the pencil went through all the dots. If not, try again. "Everybody wins!"

Figure 6.6

The standard solution to the nine-dot puzzle.

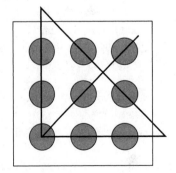

Figure 6.7

An unblocked three-line solution.

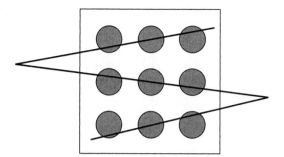

- Fold the paper carefully so that the dots are lying one on top of the other. Now stab your pencil through the nine layers of paper. It helps to sharpen your pencil first.

- Draw very small dots very close together in the square pattern and then draw a fat line through all of them at once. This one is courtesy of Becky Buechel when she was 10 years old.

Source: From *Conceptual Blockbusting* by James L. Adams. © 1986 by James L. Adams. Reprinted by permission of Perseus Books Publishers, a member of Perseus Books, L.L.C.

2

Modeling Uncertainty

A s we have seen, uncertainty is a critical element of many decisions that we face. In the next five chapters we will consider a variety of ways to model uncertainty in decision problems by using probability. We begin with a brief introduction to probability in Chapter 7. This introduction has three objectives: to remind you of probability basics, to show some ways that probability modeling can be useful in decision problems, and to give you a chance to polish your ability to manipulate probabilities. The problems and cases at the end of Chapter 7 are recommended especially to help you accomplish the last goal.

Chapter 8 introduces the topic of subjective probability. In the introduction to this text the claim was made that subjective judgments are critical in a decision-analysis approach. Most of us are comfortable making informal statements that reflect our uncertainty. For example, we use terms such as "there's a chance that such-and-such will happen." In a decision-analysis approach, however, there is a need for more precision. We can use probability to model subjective beliefs about uncertainty, and Chapter 8 presents techniques to do this. Thus, our beliefs and feelings about uncertainty can be translated into probability numbers that can be included in a decision tree or influence diagram.

In many cases the uncertainty that we face has characteristics that make it similar to certain prototypical situations. In these cases, it may be possible to represent the uncertainty with a standard mathematical model and then derive probabilities on the basis of the mathematical model. Chapter 9 presents a variety of theoretical probability models that are useful for representing uncertainty in many standard situations.

Chapter 10 discusses the use of historical data as a basis for developing probability distributions. If data about uncertainty in a decision situation are available, a decision maker would surely want to use them. First, we see how to use data alone to create histograms and continuous distributions. Second, we discuss the use of data to model relationships among variables. Finally, we develop an approach that uses data to update a decision maker's probability beliefs via Bayes' theorem.

It is also possible to "create data" through computer simulation, or by what is known as Monte Carlo simulation. That is, one can construct a model of a complex decision situation and use a computer to simulate the situation many times. By tracking the outcomes, the decision maker can obtain a fair idea of the probabilities associated with various outcomes. Monte Carlo simulation techniques are discussed in Chapter 11.

When faced with uncertainty most decision makers do their best to reduce it. The basic strategy that we follow is to collect information. In this day and age of extensive telecommunications and computer databases, information is anything but scarce, but determining what information is appropriate and then processing it can be costly. How much is information worth to you? Moreover, in a problem with many sources of uncertainty, calculating the value of information can help to guide the decision analysis, thus indicating where the decision maker can best expend resources to reduce uncertainty. Chapter 12 explores the value of information within the decision-analysis framework.

Probability Basics

One of the central principles of decision analysis is that we can represent uncertainty of any kind through the appropriate use of probability. We already have been using probability in straightforward ways to model uncertainty in decision trees and influence diagrams. This chapter presents some of the basic principles for working with probability and probability models. Our objective is to be able to create and analyze a model that represents the uncertainty faced in a decision. The nature of the model created naturally depends on the nature of the uncertainty faced, and the analysis required depends on the exigencies of the decision situation.

We have used the term *chance event* to refer to something about which a decision maker is uncertain. In turn, a chance event has more than one possible *outcome*. When we talk about probabilities, we are concerned with the chances associated with the different possible outcomes. For convenience, we will refer to chance events with boldface letters (e.g., Chance Events **A** and **B**), and to outcomes with lightface letters (e.g., Outcomes A_1, B_j, or C). Thus Chance Event **B**, for example, might represent a particular chance node in an influence diagram or decision tree, and Outcomes B_1, B_2, and B_3 would represents **B**'s possible outcomes.*

After reading the chapter and working the problems and cases, you should be (1) reasonably comfortable with probability concepts, (2) comfortable in the use of probability to model simple uncertain situations, (3) able to interpret probability statements in terms of the uncertainty that they represent, and (4) able to manipulate and analyze the models you create.

---

*The terminology we adopt in this book is slightly unconventional. Most authors define an *outcome space* that includes all possible elemental outcomes that may occur. For example, if the uncertain event is how many orders arrive at a mail-order business in a given day, the outcome space is composed of the integers 0, 1, 2, 3, . . . , and each integer is a possible elemental outcome (or simply an outcome). An *event* is then defined as a set of possible outcomes. Thus, we might speak of the event that the number of orders equals zero or the event that more than 10 orders arrive in a day.

In this book we use the term *outcome* to refer to what can occur as the result of an uncertain event. Such occurrences, which we represent as branches from chance nodes in decision trees, can be either events or elemental outcomes in the conventional terminology. Thus, our usage of *outcome* includes the conventional *event* as well as *outcome*. Why the change? For our purposes it is not necessary to distinguish between elemental outcomes and sets of outcomes. Also, we avoid the potential confusion that can arise by using *uncertain event* to refer to a process and *event* to refer to a result of that process.

A Little Probability Theory

Probabilities must satisfy the following three requirements.

1 Probabilities Must Lie Between 0 and 1   Every probability (p) must be positive, and between 0 and 1, inclusive ($0 \le p \le 1$). This is a sensible requirement. In informal terms it simply means nothing can have more than a 100% chance of occurring or less than a 0% chance.

2 Probabilities Must Add Up   Suppose two outcomes are mutually exclusive (only one can happen, not both). The probability that one or the other occurs is then the sum of the individual probabilities. Mathematically, we write $P(A_1 \text{ or } A_2) = P(A_1) + P(A_2)$ if A_1 and A_2 cannot both happen. For example, consider the stock market. Suppose there is a 30% chance that the market will go up and a 45% chance that it will stay the same (as measured by the Dow Jones average). It cannot do both at once, and so the probability that it will either go up or stay the same must be 75%.

3 Total Probability Must Equal 1   Suppose a set of outcomes is mutually exclusive and collectively exhaustive. This means that one (and only one) of the possible outcomes must occur. The probabilities for this set must sum to 1. Informally, if we have a set of outcomes such that one of them has to occur, then there is a 100% chance that one of them will indeed come to pass.

We have seen this in decision trees; the branches emanating from a chance node must be such that one and only one of the branches occurs, and the probabilities for all the branches must add to 1. Consider the stock market example again. If we say that the market can go up, down, or stay the same, then one of these three outcomes must happen. The probabilities for these outcomes must sum to 1—that is, there is a 100% chance that one of them will occur.

Venn Diagrams

Venn diagrams provide a graphic interpretation of probability. Figure 7.1 shows a simple Venn diagram in which two Outcomes, A_1 and A_2, are displayed. Think of the diagram as a whole representing a chance event $(\mathbf{A})_1$ and areas representing possible outcomes. The circle labeled A_1 thus represents outcome A_1. Because the areas of A_1 and A_2 do not overlap, A_1 and A_2 cannot both occur at the same time; they are mutually exclusive.

We can use Figure 7.1 to interpret the three requirements of probability mentioned above. The first requirement is that a probability must lie between 0 and 1. Certainly an outcome cannot be represented by a negative area. Furthermore, an outcome cannot be represented by an area larger than the entire rectangle. For the second requirement, we see that A_1 and A_2 are mutually exclusive because they do not

Figure 7.1

A Venn diagram.
Outcomes A_1 and A_2
are mutually exclusive
events.

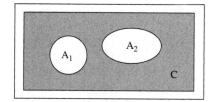

overlap. Thus, the probability of A_1 or A_2 occurring must be just the sum of the probability of A_1 plus the probability of A_2. For the third requirement, label the shaded portion of the rectangle as C. This is what must happen if neither A_1 nor A_2 happens. Because A_1, A_2, and C together make up the whole rectangle, then one of the three must occur. Moreover, only one of them can occur. The upshot is that their probabilities must sum to 1. Alternatively, there is a 100% chance that A_1, A_2, or C will occur.

More Probability Formulas

The following definitions and formulas will make it possible to use probabilities in a wide variety of decision situations.

4 Conditional Probability   Suppose that you are interested in whether a particular stock price will increase. You might use a probability P(Up) to represent your uncertainty. If you find out that the Dow Jones industrial average rose, however, you would want to base your probability on this condition. Figure 7.2 represents the situation with a Venn diagram. Now the entire rectangle actually represents two chance events, what happens with the Dow Jones index and what happens with the stock price. For the Dow Jones event, one possible outcome is that the index goes up, represented by the circle. Likewise, the oval represents the possible outcome that the stock price goes up. Because these two outcomes can happen at once, the two areas overlap in the diagonally shaded area. When both outcomes occur, this is called the *joint outcome* or *intersection*. It also is possible for one to rise while the other does not, which is represented in the diagram by the nonoverlapping portions of the "Dow Jones Up" and "Stock Price Up" areas. And, of course, the gray area surrounding the circle and oval represents the joint outcome that neither the Dow Jones index nor the stock price goes up.

Once we know that the Dow Jones has risen, then the entire rectangle is no longer appropriate. At this point, we can restrict our attention to the "Dow Jones Up" circle. We want to know what the probability is that the stock price will increase given that the Dow Jones average is up, and so we are interested in the probability associated with the area "Stock Price Up" in the restricted space.

Given that we are looking at the restricted space, the conditional probability of "Stock Price Up given Dow Jones Up" would be represented by the proportion of "Dow Jones Up" in the original diagram that is the joint outcome "Stock Price Up" *and* "Dow Jones Up" (the diagonally shaded area). This intuitive approach leads to

Figure 7.2

A Venn-diagram representation of conditional probability.

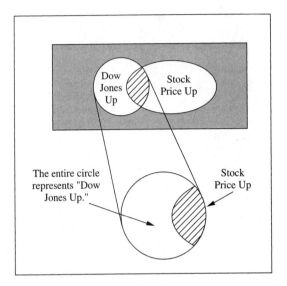

the conclusion that P(Stock Price Up given Dow Jones Up) = P(Stock Price Up and Dow Jones Up)/P(Dow Jones Up).

Mathematically, we write P(A|B) to represent the conditional probability of A given that B occurs. Read it as "Probability of A given B." The definition is

$$P(A \mid B) = \frac{P(A \text{ and } B)}{P(B)}$$

Informally, we are looking only at the occasions when Outcome B occurs, and P(A|B) is the proportion of those times that Outcome A also occurs. The probability P(A and B) is often called a *joint* probability.

5 Independence   The definition of probabilistic independence is as follows:

Chance Events **A** (with Outcomes $A_1, \ldots, A_n$) and **B** (with Outcomes $B_1, \ldots, B_m$) are *independent* if and only if

$$P(A_i \mid B_j) = P(A_i)$$

for all possible Outcomes A_i and B_j.

In words, knowing which of **B**'s outcomes occurred will not help you find probabilities for **A**'s outcomes. Furthermore, if Chance Events **A** and **B** are independent, then we can write

$$P(A_i) = P(A_i \mid B_j) = \frac{P(A_i \text{ and } B_j)}{P(B_j)}$$

From this we can see that $P(A_i \text{ and } B_j) = P(A_i) P(B_j)$. Thus, when two events are independent, we can find the probability of a joint outcome by multiplying the probabilities of the individual outcomes.

Independence between two chance events is shown in influence diagrams by the absence of an arrow between chance nodes. This is fully consistent with the definitions given in Chapter 3. If one event is not relevant in determining the chances associated with another event, there is no arrow between the chance nodes. An arrow from chance node **B** to chance node **A** would mean that the probabilities associated with **A** are conditional probabilities that depend on the outcome of **B**.

As an example of independent chance events, consider the probability of the Dow Jones index increasing and the probability of the Dow Jones increasing *given* that it rains tomorrow. It seems reasonable to conclude that

$$P(\text{Dow Jones Up}) = P(\text{Dow Jones Up} \mid \text{Rain})$$

because knowing about rain does not help to assess the chance of the index going up.

Independent chance events are not to be confused with mutually exclusive outcomes. Two outcomes are mutually exclusive if only one can happen at a time. Clearly, however, independent chance events can have outcomes that happen together. For example, it is perfectly possible for the Dow Jones to increase and for rain to occur tomorrow. Rain and No Rain would constitute mutually exclusive outcomes; only one occurs at a time.

Finally, if two chance events are probabilistically dependent, this does *not* imply a causal relationship. As an example, consider economic indicators. If a leading economic indicator goes up in one quarter, then it is unlikely that a recession will occur in the next quarter; the change in the indicator and the occurrence of a recession are dependent events. But this is not to say that the indicator going up causes the recession not to happen, or vice versa. In some cases there may be a causal chain linking the events, but it may be a very convoluted one. In general, dependence does not imply causality.

Conditional Independence. This is an extension of the idea of independence. Conditional independence is best demonstrated with an influence diagram. In Figure 7.3, Events **A** and **B** are conditionally independent given Event **C**. Note that **C** can be a chance event or a decision, as in Figure 7.3b. The only connection between **A** and **B** goes through **C**; there is no arrow directly from **A** to **B**, or vice versa. Mathematically, we would write

---

Events **A** and **B** are conditionally independent given **C** if and only if

$$P(A_i \mid B_j, C_k) = P(A_i \mid C_k),$$

for all possible Outcomes A_i, B_j, and C_k.

In words, suppose we are interested in Event **A**. If **A** and **B** are conditionally independent given **C**, then learning the outcome of **B** adds no new information regarding **A** if

Figure 7.3

Conditional indepen-dence in an influence diagram. In these in-fluence diagrams, *A* and *B* are conditionally independent given *C*. As shown, the condi-tioning event can be either (a) a chance event or (b) a decision.

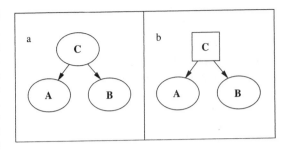

the outcome of **C** already is known. Alternatively, conditional independence means that

$$P(A_i \text{ and } B_j \,|\, C_k) = P(A_i \,|\, C_k) \, P(B_j \,|\, C_k)$$

Conditional independence is the same as normal (unconditional) independence ex-cept that every probability has the same conditions to the right side of the vertical bar. When constructing influence diagrams, identification of conditional indepen-dence can ease the burden of finding probabilities for the chance events.

As an example of conditional independence, consider a situation in which you wonder whether or not a particular firm will introduce a new product. You are ac-quainted with the CEO of the firm and may be able to ask about the new product. But one of your suppliers, who has chatted with the CEO's assistant, reports that the company is on the brink of announcing the new product. Your probability that the company will indeed introduce the product thus would change:

P(Introduce Product | Supplier's Report) ≠ P(Introduce Product)

Thus, these two events are not independent when considered by themselves. Consider, however, the information from the CEO. Given that information, the sup-plier's report might not change your probability:

P(Introduce Product | Supplier's Report, Information from CEO)

= P(Introduce Product | Information from CEO)

Thus, given the information from the CEO, the supplier's report and the event of the product introduction are conditionally independent.

Conditional independence is an important concept when thinking about causal effects and dependence. As mentioned above, dependence between two events does not necessarily mean that one causes the other. For example, more drownings tend to be associated with more ice cream consumption, but it seems unlikely that these two are causally related. The explanation lies in a common cause; both tend to happen during summer, and it might be reasonable to assume that drownings and ice cream consumption are conditionally independent given the season. Another example is that high skirt hemlines tend to occur when the stock market is advancing steadily (a bull market), although no one believes that hemlines cause the stock market to ad-vance, or vice versa. Perhaps both reflect a pervasive feeling of confidence and adventure in our society. If we had an adequate index of "societal adventuresomeness," hemline height and stock market activity might well be conditionally independent given the index.

6 Complements   Let $\overline{B}$ ("Not B" or "B-bar") represent the outcome that is the complement of B. This means that if B does not occur, then $\overline{B}$ must occur. Because probabilities must add to 1 (requirement 3 above),

$$P(\overline{B}) = 1 - P(B)$$

The Venn diagram in Figure 7.4 demonstrates complements. If the area labeled B represents Outcome B, then everything outside of the oval must represent what happens if B does not happen. For another example, the lightly shaded area in Figure 7.2 represents the complement of the Outcome "Dow Jones Up *or* Stock Price Up," the union of the two individual outcomes.

7 Total Probability of an Event   A convenient way to calculate P(A) is with this formula:

$$P(A) = P(A \text{ and } B) + P(A \text{ and } \overline{B})$$
$$= P(A \mid B)\, P(B) + P(A \mid \overline{B})\, P(\overline{B})$$

To understand this formula, examine Figure 7.5. Clearly, Outcome A is composed of those occasions when A and B occur and when A and $\overline{B}$ occur. Because the joint Outcomes "A and B" and "A and $\overline{B}$" are mutually exclusive, the probability of A must be the sum of the probability of "A and B" plus the probability of "A and $\overline{B}$."

As an example, suppose we want to assess the probability that a stock price will increase. We could use its relationship with the Dow Jones index (Figure 7.2) to help make the assessment:

P(Stock Price Up) = P(Stock Price Up | Dow Jones Up) × P(Dow Jones Up)
+ P(Stock Price Up | Dow Jones Not Up)
× P(Dow Jones Not Up)

Figure 7.4

Venn diagram illustrating the idea of an outcome's complement.

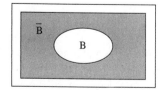

Figure 7.5

Total probability.
Outcome A is made up
of Outcomes "A and
B" and "A and $\overline{B}$."

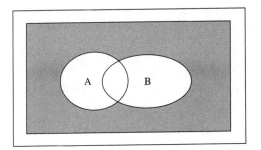

Although it may appear that we have complicated matters by requiring three probabilities instead of one, it may be quite easy to think about (1) the probability of the stock price movement conditional on the change in the Dow Jones index and (2) the probabilities associated with changes in the index.

8 Bayes' Theorem　Because of the symmetry of the definition of conditional probability, we can write

$$P(B \mid A)\, P(A) = P(A \mid B)\, P(B)$$

from which we can derive

$$P(B \mid A) = \frac{P(A \mid B)\, P(B)}{P(A)}$$

Now expanding P(A) with the formula for total probability, we obtain

$$P(B \mid A) = \frac{P(A \mid B)\, P(B)}{P(A \mid B)\, P(B) + P(A \mid \overline{B})\, P(\overline{B})}$$

This formula often is referred to as *Bayes' theorem*. It is extremely useful in decision analysis, especially when using information. We will not bother with an example showing its application here, but we will see numerous applications of Bayes' theorem in the examples at the end of this chapter, as well as in Chapters 9, 10, and 12.

Uncertain Quantities

Many uncertain events have quantitative outcomes. For example, we already have mentioned stock prices and the level of the Dow Jones index. Another example would be tomorrow's maximum temperature. If an event is not quantitative in the first place, we might define a variable that has a quantitative outcome based on the original event. For example, we might be concerned about precipitation. If we consider the amount of precipitation (in centimeters) that falls tomorrow, this uncertain event is quantitative. Moreover, we could define an uncertain quantity X: Let $X = 1$ if precipitation occurs, and $X = 0$ if not.

The set of probabilities associated with all possible outcomes of an uncertain quantity is called its *probability distribution*. For example, consider the probability distribution for the number of raisins in an oatmeal cookie, which we could denote by Y. We might have $P(Y = 0) = 0.02$, $P(Y = 1) = 0.05$, $P(Y = 2) = 0.20$, $P(Y = 3) = 0.40$, and so on. Of course, the probabilities in a probability distribution must add to 1 because the events—numerical outcomes—are mutually exclusive. Uncertain quantities (often called *random variables*) and their probability distributions play a central role in decision analysis.

In general, we will use capital letters to represent uncertain quantities. Thus, we will write $P(X = 3)$ or $P(Y > 0)$, for example, which are read as "the probability that the uncertain quantity X equals 3," and "the probability that the uncertain quantity Y is greater than 0." Occasionally we will need to use a more general form. Lowercase letters will denote outcomes or realizations of an uncertain quantity. An example would be $P(X = x)$, where capital X denotes the uncertain quantity itself and lowercase x represents the actual outcome.

In general, it is helpful to distinguish between *discrete* and *continuous* uncertain quantities. In the next section we will describe in detail discrete uncertain quantities, their probability distributions, and certain characteristics of those distributions. Then we will turn to some continuous quantities and show how their probability distributions and their characteristics are analogous to the discrete case.

Discrete Probability Distributions

The discrete probability distribution case is characterized by an uncertain quantity that can assume a finite or countable number of possible values. We already have seen two examples. The first was the precipitation example; we defined a discrete uncertain quantity that could take only the values 0 and 1. The other example was the number of raisins in an oatmeal cookie. Other examples might be the number of operations a computer performs in any given second or the number of games that will be won by the Chicago Cubs next year. Strictly speaking, future stock prices quoted on the New York Stock Exchange are discrete uncertain quantities because they can take only values that are in eighths: $10\frac{5}{8}$, $11\frac{3}{4}$, $12\frac{1}{2}$, for example.

When we specify a probability distribution for a discrete uncertain quantity, we can express the distribution in several ways. Two approaches are particularly useful. The first is to give the *probability mass function*. This function lists the probabilities for each possible discrete outcome. For example, suppose that you think that no cookie in a batch of oatmeal cookies could have more than five raisins. A possible probability mass function would be:

$P(Y = 0 \text{ raisins}) = 0.02$ $P(Y = 3 \text{ raisins}) = 0.40$

$P(Y = 1 \text{ raisin}) = 0.05$ $P(Y = 4 \text{ raisins}) = 0.22$

$P(Y = 2 \text{ raisins}) = 0.20$ $P(Y = 5 \text{ raisins}) = 0.11$

This mass function can be displayed in graphical form (Figure 7.6). Such a graph often is called a *histogram*.

The second way to express a probability distribution is as a *cumulative distribution function* (CDF). A cumulative distribution gives the probability that an uncertain quantity is less than or equal to a specific value: $P(X \leq x)$. For our example, the CDF is given by

$$P(Y \leq 0 \text{ raisins}) = 0.02 \qquad P(Y \leq 3 \text{ raisins}) = 0.67$$
$$P(Y \leq 1 \text{ raisin}) = 0.07 \qquad P(Y \leq 4 \text{ raisins}) = 0.89$$
$$P(Y \leq 2 \text{ raisins}) = 0.27 \qquad P(Y \leq 5 \text{ raisins}) = 1.00$$

Cumulative probabilities can be graphed; the CDF for the oatmeal cookie is graphed in Figure 7.7.

Note that the graph actually covers all the points along the horizontal axis. That is, we can read from the graph not only $P(Y \leq 3)$, but also $P(Y \leq 4.67)$, for example. In fact, $P(Y \leq 4.67) = P(Y \leq 4) = 0.89$, because it is not possible for a cookie to have a fractional number of raisins (assuming whole raisins, of course).

You may recognize this idea of a probability mass function and a CDF. When we constructed risk profiles and cumulative risk profiles in Chapter 4, we were working with these two representations of probability distributions.

Figure 7.6

A probability mass function displayed as a histogram.

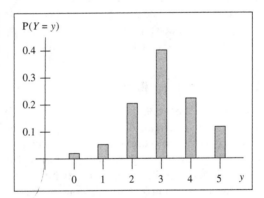

Figure 7.7

Cumulative distribution function (CDF) for number of raisins in an oatmeal cookie.

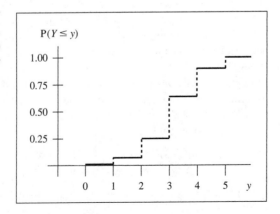

Expected Value

A discrete uncertain quantity's *expected value* is the probability-weighted average of its possible values. That is, if X can take on any value in the set $\{x_1, x_2, \ldots, x_n\}$, then the expected value of X is simply the sum of x_1 through x_n, each weighted by the probability of its occurrence. Mathematically,

$$\text{Expected value of } X = x_1 P(X = x_1) + x_2 P(X = x_2) + \cdots + x_n P(X = x_n)$$

$$= \sum_{i=1}^{n} x_i P(X = x_i).$$

The expected value of X also is referred to as the average or mean of X and is denoted by $E(X)$ or occasionally μ_x (Greek mu).

The expected value can be thought of as the "best guess" for the value of an uncertain quantity or random variable. If it were possible to observe many outcomes of the random variable, the average of those outcomes would be very close to the expected value. We already have encountered expected values in calculating EMVs to solve influence diagrams and decision trees. The expected monetary value is the expected value of a random variable that happens to be the monetary outcome in a decision situation.

Suppose that X is used to calculate some other quantity, say Y. Then it is possible to talk about the expected value of Y, or the expected value of this function of X:

If

$$Y = f(X)$$

then

$$E(Y) = [f(X)]$$

A particularly useful function is a linear function of X, one in which X is multiplied by a constant and has a constant added to it. If Y is a linear function of X, then $E(Y)$ is particularly easy to find:

If

$$Y = a + bX$$

then

$$E(Y) = a + bE(X)$$

That is, plug $E(X)$ into the linear formula to get $E(Y)$. Unfortunately, this does not hold for nonlinear functions (log, square root, and so on); in general, it applies only to linear ones. That is, suppose you have some function $f(X)$ and you want to find the expected value of $E[f(X)]$. In general, you cannot plug $E(X)$ into the function and get the correct answer; $E[f(X)] \neq f[E(X)]$, unless $f(X)$ is a linear function like $a + bX$.

To go one step further, suppose we have several uncertain quantities, $X_1, \ldots, X_k$, and we add them to get uncertain quantity Y. Then the expected value of Y is the sum of the expected values:

If
$$Y = X_1 + X_2 + \cdots + X_k$$
then
$$E(Y) = E(X_1) + E(X_2) + \cdots E(X_k)$$

For instance, if we know the expected amount of precipitation for each of the next seven days, the expected amount of precipitation for the entire week is simply the sum of the seven daily expected values.

Variance and Standard Deviation

Another useful measure of a probability distribution is the *variance*. The variance of uncertain quantity X is denoted by $Var(X)$ or σ_X^2 (Greek sigma) and is calculated mathematically by

---

$$Var(X) = [x_1 - E(X)]^2 P(X = x_1) + [x_2 - E(X)]^2 P(X = x_2)$$
$$+ \cdots + [x_n - E(X)]^2 P(X = x_n)$$
$$= \sum_{i=1}^{n} [x_i - E(X)]^2 P(X = x_i)$$
$$= E[X - E(X)]^2$$

In words, calculate the difference between the expected value and x_i and square that difference. Do this for each possible x_i. Now find the expected value of these squared differences.

As with expected values, we can find variances of functions for X. In particular, the variance of a linear function of X is easily found:

If
$$Y = a + bX$$
then
$$Var(Y) = b^2 Var(X)$$

For example, suppose that a firm will sell an uncertain number (X) of units of a product. The expected value of X is 1000 units, and the variance is 400. If the price is $3 per unit, then the revenue (Y) is equal to $3X$. Because this is a linear function of X, $E(Y) = 3E(X) = \$3000$, and $Var(Y) = 3^2\, Var(X) = 9(400) = 3600$.

We also can talk about the variance of a sum of independent uncertain quantities. That is, *as long as the uncertain quantities are probabilistically independent*—the probability distribution for one does not depend on the others—then the variance of the sum is the sum of the variances:

If
$$Y = X_1 + X_2 + \cdots + X_k$$
then
$$Var(Y) = Var(X_1) + Var(X_2) + \cdots + Var(X_k)$$

So, if our firm sells two products, one for $3 and another for $5, and the variance for these two products is 400 and 750, respectively, then the variance of the firm's revenue is $Var(Y) = 3^2 Var(X_1) + 5^2 Var(X_2) = 9(400) + 25(750) = 22,350$.

The *standard deviation* of X, denoted by σ_x is just the square root of the variance. Because the variance is the expected value of the squared differences, the standard deviation can be thought of as a "best guess" as to how far the outcome of X might lie from $E(X)$. A large standard deviation and variance means that the probability distribution is quite spread out; a large difference between the outcome and the expected value is anticipated. For this reason, the variance and standard deviation of a probability distribution are used as measures of variability. A large variance or standard deviation would indicate a situation in which the outcomes are highly variable.

To illustrate the ideas of expected value, variance, and standard deviation, consider the double-risk dilemma depicted in Figure 7.8. Choices A and B both lead to uncertain dollar outcomes. Given Decision A, for example, there are three possible profit outcomes having probability mass function $P(\text{Profit} = \$20 \mid A) = 0.24$, $P(\text{Profit} = \$35 \mid A) = 0.47$, and $P(\text{Profit} = \$50 \mid A) = 0.29$. Likewise, for B we have $P(\text{Profit} = -\$9 \mid B) = 0.25$, $P(\text{Profit} = \$0 \mid B) = 0.35$, and $P(\text{Profit} = \$95 \mid B = 0.40$. Now we can calculate the expected profits conditional on choosing A or B:

$$E(\text{Profit} \mid A) = 0.24(\$20) + 0.47(\$35) + 0.29(\$50) = \$35.75$$
$$E(\text{Profit} \mid B) = 0.25(-\$9) + 0.35(\$0) + 0.40(\$95) = \$35.75$$

These two uncertain quantities have exactly the same expected values. (Note that the expected profit does not have to be one of the possible outcomes!)

We also can calculate the variances and standard deviations (σ) for A and B:

$$\text{Var(Profit} \mid A) = (20 - 35.75)^2(0.24) + (35 - 35.75)^2(0.47)$$
$$+ (50 - 35.75)^2(0.29)$$
$$= 118.69 \text{ "dollars squared"}$$

$$\sigma_A = \sqrt{118.69} = \$10.89$$

$$\text{Var(Profit} \mid B) = (-9 - 35.75)^2(0.25) + (0 - 35.75)^2(0.35)$$
$$+ (95 - 35.75)^2(0.40)$$
$$= 2352.19 \text{ "dollars squared"}$$

$$\sigma_B = \sqrt{2352.19} = \$48.50$$

The variance and standard deviation of B are much larger than those for A. The outcomes in B are more spread out or more variable than the outcomes for A, which are clustered fairly closely around the expected value.

The example also points out the fact that variance, being a weighted sum of squared terms, is expressed in squared units. In this case, the variance is in "dollars squared" because the original outcomes are in dollars. Taking the square root to find the standard deviation brings us back to the original units, or, in this case, dollars. For this reason, the standard deviation is interpreted more easily as a measure of variability.

You might have noticed that, while the variance and standard deviation can be used to gauge the riskiness of an option, the cumulative probabilities also can be useful in this respect. For example, we have P(Profit $\leq 0 \mid$ A) = 0, but P(Profit $\leq 0 \mid$ B) = 0.6, and B thus looks somewhat riskier than A. At the other extreme, of course, B looks better. Project A cannot produce a profit greater than $50: P(Profit $\leq \$50 \mid$ A) = 1.00. For B, however, P(Profit $\leq \$50 \mid$ B) = 0.60.

Covariance and Correlation for Measuring Dependence (Optional)

We have discussed the notion of probabilistic dependence above and indicated that dependence is defined in terms of conditional distributions. In some cases, though, the use of conditional distributions can be difficult, and another approach to measuring

Figure 7.8

A choice between two uncertain prospects.

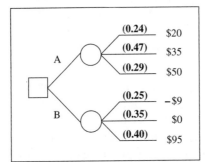

dependence is worthwhile. *Covariance* is a quantity that is closely related to the idea of variance. Covariance and its close relative *correlation* can be used to measure certain kinds of dependence.

The covariance between two uncertain quantities X and Y is calculated mathematically by

$$
\begin{aligned}
\text{Cov}(X,\ Y) &= [x_1 - E(X)][y_1 - E(Y)]P(X = x_1 \text{ and } Y = y_1) \\
&+ [x_1 - E(X)][y_2 - E(Y)]P(X = x_1 \text{ and } Y = y_2) \\
&+ \cdots + [x_1 - E(X)][y_m - E(Y)]P(X = x_1 \text{ and } Y = y_m) \\
&+ [x_2 - E(X)][y_1 - E(Y)]P(X = x_2 \text{ and } Y = y_1) \\
&+ \cdots + [x_2 - E(X)][y_m - E(Y)]P(X = x_2 \text{ and } Y = y_m) \\
&+ \cdots + [x_n - E(X)][y_1 - E(Y)]P(X = x_n \text{ and } Y = y_1) \\
&+ \cdots + [x_n - E(X)][y_m - E(Y)]P(X = x_n \text{ and } Y = y_m) \\[8pt]
&= \sum_{i=1}^{n} \{ [x_i - E(X)][y_1 - E(Y)]P(X = x_i \text{ and } Y = y_1) \\
&+ \cdots + [x_i - E(X)][y_m - E(Y)]P(X = x_i \text{ and } Y = y_m) \} \\[8pt]
&= \sum_{i=1}^{n} \sum_{j=1}^{m} [x_i - E(X)][y_j - E(Y)]P(X = x_i \text{ and } Y = y_j) \\[8pt]
&= E[(X - E(X))(Y - E(Y))]
\end{aligned}
$$

Although this is a complicated formula, with a little interpretation we can get some insight into it. First, it really is similar to the formula for variance. There we calculated an average of squared deviations of X from its expected value $E(X)$. Here, instead of squaring the deviations, we multiply X's deviation times Y's deviation. This is sometimes called a *cross product* because it is a product of deviations for two different quantities (X and Y).

Although it may not be evident on the surface, the covariance can be either positive or negative. Suppose that large values of X tend to occur with large values of Y, and small with small. On the other hand, the probability that high values of X and low values of Y (and vice versa) occur together is low. Now consider what happens with the cross products in the formula. When both X and Y are high, they will both be above their corresponding expected values, making both deviations and their cross product positive. When both quantities are low, they will both be below their expected values. Both deviations will be negative, but the cross product will again be positive. When one is high and one is low (that is, one above its expected value and one below), the cross product will be negative. If large X's tend to go with large Y's, and small with small, then the positive cross products will get more weight (higher probabilities) than negative cross products in the formula, and the overall calculation will yield a positive covariance. On the other hand, if large X's tend to occur with small Y's, and vice versa, then the negative cross products will get more weight, resulting in a negative covariance. Thus, a positive covariance reflects quantities that tend to move in the same direction, and this is called a *positive* relationship. Likewise, a negative covariance indicates that the quantities tend to move in opposite directions, which is called a *negative* relationship.

A simple example will help clarify the idea of covariance and its calculation. Suppose an investor is considering purchasing shares in American Rivets Corporation (ARC). The investor already has shares of Sundance Solar Power (SSP). One of the things the investor would like to accomplish is to stabilize the rate of return of the portfolio; ideally, when the return on one stock goes down, the other would go up. On the other hand, a positive relationship would be bad because the returns would tend to go up and down together, making the overall return on the portfolio vary considerably.

A simple model of the returns on the two stocks involves only two possible outcomes for each one. ARC could have returns of 10% or -5%. SSP, on the other hand, could have returns of 12% or -8%. The probabilities of the possible outcomes are

$$P(ARC = 10\% \text{ and } SSP = 12\%) = 0.35$$

$$P(ARC = 10\% \text{ and } SSP = -8\%) = 0.10$$

$$P(ARC = -5\% \text{ and } SSP = 12\%) = 0.15$$

$$P(ARC = -5\% \text{ and } SSP = -8\%) = 0.40$$

You can see that the two stocks tend to move in the same direction. There is a 75% chance that they are both high or both low. Likewise there is only a 25% chance that one is high and the other low. Thus, we might expect the covariance to be positive.

Calculating the covariance requires first calculating the expected values for each stock. Considering ARC first, we can use the law of total probability:

$$P(ARC = 10\%) = P(ARC = 10\% \text{ and } SSP = 12\%)$$
$$+ P(ARC = 10\% \text{ and } SSP = -8\%)$$
$$= 0.35 + 0.10$$
$$= 0.45$$

Thus, $P(ARC = -5\%) = 1 - P(ARC = 10\%) = 0.55$, and we can calculate

$$E(ARC) = 0.45(10\%) + 0.55(-5\%)$$
$$= 1.75\%$$

Likewise, we can calculate $E(SSP)$:

$$P(SSP = 12\%) = P(ARC = 10\% \text{ and } SSP = 12\%)$$
$$+ P(ARC = -5\% \text{ and } SSP = 12\%)$$
$$= 0.35 + 0.15$$
$$= 0.50$$

$$P(SSP = -8\%) = 1 - P(SSP = 12\%)$$
$$= 0.50$$

$$E(SSP) = 0.50(12\%) + 0.50(-8\%)$$
$$= 2\%$$

Now we can calculate the covariance between ARC and SSP:

Cov(ARC, SSP)

$$= [10\% - 1.75\%][12\% - 2\%]P(ARC = 10\% \text{ and } SSP = 12\%)$$
$$+ [10\% - 1.75\%][-8\% - 2\%]P(ARC = 10\% \text{ and } SSP = -8\%)$$
$$+ [-5\% - 1.75\%][12\% - 2\%]P(ARC = -5\% \text{ and } SSP = 12\%)$$
$$+ [-5\% - 1.75\%][-8\% - 2\%]P(ARC = -5\% \text{ and } SSP = -8\%)$$
$$= 8.25\% \times 10\% \times 0.35$$
$$+ 8.25\% \times (-10\%) \times 0.10$$
$$+ (-6.75\%) \times 10\% \times 0.15$$
$$+ (-6.75\%) \times (-10\%) \times 0.40$$
$$= 37.5(\% \text{ squared})$$

As expected, the covariance is positive. The problem, however, is that the magnitude of the covariance is not very meaningful because it depends on the range of variation in the two quantities. Also, as with the variance, the covariance carries units that are meaningful. In the case of the two stock returns, the units are % squared. But suppose we wanted to calculate the covariance between hemline height and stock market return; the calculation would involve multiplying inches times percentage points of return, and so the units would be % inches. What in the world is a % inch?

To solve these two problems, we often transform the covariance to get a standardized measure of dependence. This standardized measure is called the *correlation* coefficient, and the Greek symbol ρ (rho) is used to represent it. To calculate ρ, divide the covariance of X and Y by the standard deviations of these two uncertain quantities:

$$\rho_{XY} = \frac{\text{Cov}(X, Y)}{\sigma_X \sigma_Y}$$

The correlation ρ has a number of useful properties. First, it ranges between +1 (perfect positive dependence) and −1 (perfect negative dependence). A correlation of zero suggests no relationship, although certain kinds of dependence are possible even though the correlation is zero (see Exercise 7.11). Also, ρ has no units. In the hemline–stock market example, we would divide the covariance (% inches) by the standard deviation of return (%) and the standard deviation of hemline height (inches), and the units would cancel each other out. An implication is that the correlation is useful for comparing the strength of the relationship in one case with the strength of the relationship in another that involves different variables altogether.

To continue the example, we can calculate the correlation between the returns for stocks ARC and SSP. To do this calculation, we must first calculate the standard deviation for each of the two individual stocks. These are $\sigma_{ARC} = 7.46\%$ and $\sigma_{SSP} = 10\%$. Thus,

$$\rho_{ARC,SSP} = \frac{37.5}{7.46 \times 10} = 0.503$$

This correlation of 0.503 gives the investor an indication of the extent to which the returns are related. By comparing the correlations of different pairs of stocks, the investor can try to locate those with lower (or even negative) correlations in order to accomplish the objective of stabilizing the return of the overall portfolio. This is the principle of portfolio diversification.

One final warning is in order before leaving the ideas of covariance and correlation. Although these measures of dependence are widely used, they only provide insight into a certain kind of dependence. That is, as long as the relationship is such that an increase in one variable always suggests an increase (or always a decrease) in the other, then the covariance and correlation will reflect this relationship. If the relationship is more complex, however, such a relationship may not be adequately reflected in the covariance and correlation. For example as X increases up to a certain point, Y might be expected to increase. As X continues to increase, though, Y might be expected to decrease. This kind of *nonmonotonic* relationship may not be reflected by the covariance and correlation.

Continuous Probability Distributions

The discussion above has focused on discrete uncertain quantities. Now we turn briefly to continuous uncertain quantities. In this case, the uncertain quantity can take any value within some range. For example, the temperature tomorrow at O'Hare Airport in Chicago at noon is an uncertain quantity that can be anywhere between, say, −50°F and 120°F. The length of time until some anticipated event (for example, the next major earthquake in California) is a continuous uncertain quantity, as are locations in space (the precise location of the next earthquake), as well as various measurements such as height, weight, and speed (for example, the peak "ground acceleration" in the next major earthquake).

With continuous uncertain quantities, it is not reasonable to speak of the probability that a specific value occurs. In fact, the probability of a particular value occurring is equal to zero: $P(Y = y) = 0$. Intuitively, there are infinitely many possible values, and so the probability of any particular value must be infinitely small. Instead, we typically speak of interval probabilities: $P(a \leq Y \leq b)$. The CDF for a continuous uncertain quantity can be constructed on the basis of such intervals.

The easiest way to understand this process is through a simple example. Let us suppose we are interested in a movie star's age. For a variety of reasons, we may be certain that she is older than 29 and no older than 65. We can translate these into probability statements: $P(Age \leq 29) = 0$, and $P(Age \leq 65) = 1$. Now, suppose that

Figure 7.9

Cumulative distribution function for movie star's age.

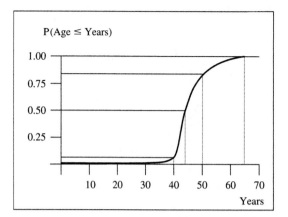

P(Age ≤ Years)

you decide that she most likely is between 40 and 50 years old, and that P(40 < Age ≤ 50) = 0.8. Also suppose you figure that P(Age ≤ 40) = 0.05, and P(Age > 50) = 0.15. Naturally, if you were so inclined, you also could make several other judgments. For example, you might figure that her age is just as likely to be 44 or less as it is to be greater than 44. This would translate into P(Age ≤ 44) = 0.50.

We can take the probabilities from this last example, transform them, and create the following table of cumulative probabilities:

$$P(\text{Age} \le 29) = 0.00$$
$$P(\text{Age} \le 40) = 0.05$$
$$P(\text{Age} \le 44) = 0.50$$
$$P(\text{Age} \le 50) = 0.85$$
$$P(\text{Age} \le 65) = 1.00$$

These probabilities then can be displayed in a graph, with years along the horizontal axis and P(Age ≤ Years) along the vertical axis. The graph, shown in Figure 7.9, allows you to select a number of years on the horizontal axis and then read off the probability that the movie star's age is less than or equal to that number of years. As with the discrete case, this graph represents the cumulative distribution function, or CDF.

As before, the CDF allows us to calculate the probability for any interval. For example, P(40 < Age ≤ 44) = P(Age ≤ 44) − P(Age ≤ 40) = 0.45. If we were to make more assessments, we could refine the curve in Figure 7.9, but it should always slope upward. If it were to slope downward, it would imply that some interval had a negative probability!

Stochastic Dominance Revisited

Chapter 4 introduced the concept of stochastic dominance as it related to cumulative risk profiles for discrete uncertain quantities. The same principles also hold true in the continuous case.

Figure 7.10

CDFs for three invest-ment alternatives. Investment B stochas-tically dominates Investment A.

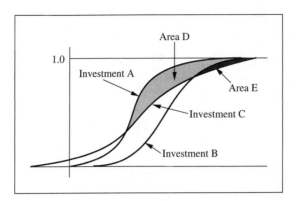

Figure 7.10 shows CDFs for three investment alternatives, each of which is modeled as a continuous uncertain quantity. Investment B stochastically dominates Investment A because the CDF for B lies entirely to the right of the CDF for A. On the other hand, Investment C neither dominates nor is dominated by the other two alternatives. For instance, Area E represents the portion of B's CDF that lies to the left of the CDF for Investment C. Because C's CDF crosses the other two, no decision can be made between C and B (or between C and A) on the basis of stochastic dominance. There is a sense in which B dominates C, however. Imagine an investor who is averse to risk; such an investor would gladly trade any risky prospect for its expected value. B is better than C for such an investor because it has less risk (the distribution is less spread out) and a greater expected value (the distribution is shifted to the right).

Stochastic Dominance and Multiple Attributes (Optional)

Now we are prepared to follow up the discussion of stochastic dominance with multiple attributes in Chapter 4. In the case of multiple attributes, one must consider the joint distribution for all of the attributes together. To develop this, we need to introduce some notation.

First, let $F(X)$ denote the CDF for a variable X. That is, $F(X) = P(X \leq x)$. For example, in the case of the movie star's age, $F(50) = P(\text{Age} \leq 50)$. Now we can write the condition for stochastic dominance in terms of the F's. Considering the investments in Figure 7.10, B dominates A because $F_B(x) \leq F_A(x)$ for all values of x on the horizontal axis. This condition asserts that the CDF for B must lie to the right of the CDF for A.

When there are more attributes, the CDF must encompass all of the attributes. For example, recall the summer-job example from Chapter 4, in which we discussed uncertainty about both salary and summer fun. We would have to look at the CDF for both uncertain quantities. The CDF would be denoted by $F(x_s, x_f) = P(\text{Salary} \leq x_s$ and $\text{Fun} \leq x_f)$. An alternative B (a specific job like the in-town job) dominates alternative A if $F_B(x_s, x_f) \leq F_A(x_s, x_f)$ for all values of x_s and x_f and is strictly less for some x_s and x_f. If we could draw the picture of the graph in three-dimensional space, you could see that this means that the CDF for B must be entirely below the CDF for A and shifted toward larger values for both Salary and Fun.

In Chapter 4, we made the claim that if the uncertain quantities are independent, then stochastic dominance on each of the individual attributes implies overall stochastic dominance. To see this, consider the summer-job example further. Stochastic dominance requires that

$$F_B(x_s, x_f) \leq F_A(x_s, x_f)$$

for all values of x_s and x_f, which is the same as

$$P_B(\text{Salary} \leq x_s \text{ and Fun} \leq x_f) \leq P_A(\text{Salary} \leq x_s \text{ and Fun} \leq x_f)$$

for all values of x_s and x_f. If Salary and Fun are independent for both alternatives A and B, we now know that the joint probabilities in this condition can be rewritten as the product of the individual (or *marginal*) probabilities:

$$P_B(\text{Salary} \leq x_s)P_B(\text{Fun} \leq x_f) \leq P_A(\text{Salary} \leq x_s)P_A(\text{Fun} \leq x_f)$$

Now, suppose that B dominates A individually on each attribute. This means that

$$P_B(\text{Salary} \leq x_s) \leq P_A(\text{Salary} \leq x_s)$$

and

$$P_B(\text{Fun} \leq x_f) \leq P_A(\text{Fun} \leq x_f)$$

If this is true, it is certainly the case that the overall stochastic-dominance condition is met, because the product $P_B(\text{Salary} \leq x_s)$ $P_B(\text{Fun} \leq x_f)$ must be less than or equal to $P_A(\text{Salary} \leq x_s)$ $P_A(\text{Fun} \leq x_f)$.

A final word of caution is in order here. The reasoning above only goes in one direction. That is, *if* the attributes are independent and *if* the individual stochastic-dominance conditions are met, *then* the overall stochastic-dominance condition is also met. That is, we have identified sufficient conditions for overall stochastic dominance. However, it is possible for overall stochastic dominance to exist even though the uncertain quantities are not independent or do not display stochastic dominance in the individual attributes. In other words, in some cases, you might have to go back to the definition of overall stochastic dominance [$F_B(x_1, \ldots, x_n) \leq F_A(x_1, \ldots, x_n)$ for all $x_1, \ldots, x_n$] in order to determine whether B dominates A.

Probability Density Functions

The CDF for a continuous uncertain quantity corresponds closely to the CDF for the discrete case. Is there some representation that corresponds to the probability mass function? The answer is yes, and that representation is called a *probability density function*, which, if we are speaking of uncertain quantity X, would be denoted typically as $f(x)$. The density function $f(x)$ can be built up from the CDF. It is a function in which the area under the curve within a specific interval represents the probability that the uncertain quantity will fall in that interval. For example, the density function $f(\text{Age})$ for the movie star's age might look something like the graph in Figure 7.11. The total area under the curve equals 1 because the uncertain quantity must take on

Figure 7.11

Probability density function for movie star's age.

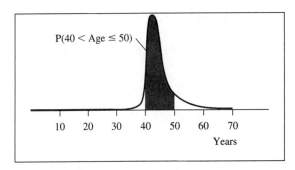

some value. The shaded area in Figure 7.11 corresponds to P(40 < Age ≤ 50) and so this area must be equal to 0.80.

Expected Value, Variance, and Standard Deviation: The Continuous Case

As in the discrete case, a continuous probability distribution can have an expected value, variance, and standard deviation. But the definition is not as easy as it was before because now we do not have probabilities for specific values, only probabilities for intervals. Without going into a lot of detail, these characteristics of a continuous probability distribution are defined by using calculus. The definitions for a continuous uncertain quantity X correspond to the discrete case, except that the summation sign is replaced with an integral sign and the density function is used in place of the probabilities:

$$E(X) = \int_{x^-}^{x^+} x f(x) dx$$

$$\mathrm{Var}(X) = \sigma_X^2 = \int_{x^-}^{x^+} [x - E(X)]^2 f(x) dx$$

where x^- and x^+ represent the lower and upper bounds for the uncertain quantity X. As before, the standard deviation σ_X is the square root of the variance.

The interpretation of these formulas also corresponds closely to the summation in the discrete case. It turns out that integration is really the continuous equivalent of the summing operation. Consider the formula for the expected value, for example. Each possible value x between x^- and x^+ is multiplied by (weighted by) the height of the density function $f(x)$. The integration then adds all of these $xf(x)$ products to find E(X). It is as if we had carved up the density function into a very large number of quite narrow but equally wide intervals (Figure 7.12). The relative likelihood of X falling in these different intervals corresponds to the relative height of the density function in each interval. That is, if the density function is (on average) twice as high in one interval as in another, then X is twice as likely to fall into the first interval as the second.

Figure 7.12

A density function f(x) in narrow intervals. The probability that X falls in interval A is about twice as great as for B because $f(x)$ is about twice as high in A as it is in B.

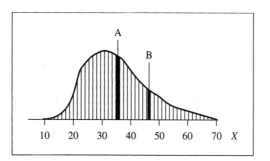

As the intervals become infinitesimally narrow, the height of the interval becomes equal to the value $f(x)$. In the limit, as the width of each interval approaches 0, we take each x, multiply by $f(x)$, add these products (by integration), and—bingo!—we get E(X).

Rest assured—you will not be required to perform any integration (in this book anyway). In general, the integration of density functions to obtain expected values and variances is a difficult task and beyond the technical scope of this textbook. Fortunately, mathematicians have studied many different kinds of probability distributions and have performed the integration for you, providing you with formulas for the expected values and variances. We will encounter several of these in Chapter 9.

What about all of those formulas for the expected value and variance of Y, a function of X, or for linear combinations of random variables? Fortunately, *all of those formulas carry over to the continuous case.* If you know the expected value and variance of several uncertain quantities, you can apply the formulas given above regardless of whether the uncertain quantities are continuous or discrete. For example, if $Y = a + bX$, then E(Y) = $a + b$ E(X). If $Y = aX_1 + bX_2$, and X_1 and X_2 are independent, then Var(Y) = a^2 Var(X_1) + b^2 Var(X_2). It does not matter whether the X's are discrete or continuous.

Covariance and Correlation: The Continuous Case (Optional)

Covariance and correlation also have counterparts when the uncertain quantities are continuous. As with expected value and variance, the definition of covariance uses an integral sign instead of a summation:

$$\text{Cov}(X, Y) = \int_{x^-}^{x^+} \int_{y^-}^{y^+} [x - \text{E}(X)][y - \text{E}(Y)]f(x, y)dx\, dy$$

As before, the correlation ρ_{XY} is calculated by dividing Cov(X,Y) by $\sigma_X\sigma_Y$. The double integral in the formula above replaces the double summation in the previous formula for the covariance of two discrete uncertain quantities. The term $f(x, y)$ refers to the *joint density function* for uncertain quantities X and Y. This joint density function is a natural extension of the density function for a single variable; it can be interpreted as a function that indicates the relative likelihood of different (x, y) pairs occurring, and the probability that X and Y fall into any given region can be calculated from $f(x, y)$.

Examples

The formulas and definitions given above are the essential elements of probability theory. With these few tools we will be able to go quite a long way in the construction of uncertainty models for decision situations. The intent has been to present the tools and concepts so that they are easily grasped. It may be worthwhile to memorize the formulas, but most important is understanding them at an intuitive level. To help you cement the concepts in your mind and to show the formulas in action, we turn now to specific examples.

OIL WILDCATTING

An oil company is considering two sites for an exploratory well. Because of budget constraints, only one well can be drilled. Site 1 is fairly risky, with substantial uncertainty about the amount of oil that might be found. On the other hand, Site 2 is fairly certain to produce a low level of oil. The characteristics of the two sites are as follows:

Site 1: Cost to Drill $100K	
Outcome	**Payoff**
Dry	−100K
Low producer	150K
High producer	500K

If the rock strata underlying Site 1 are characterized by what geologists call a "dome" structure (see Figure 7.13), the chances of finding oil are somewhat greater than if no dome structure exists. The probability of a dome structure is P(Dome) = 0.6. The conditional probabilities of finding oil at Site 1 are given in Table 7.1.

Site 2 is considerably less complicated. A well there is not expected to be a high producer, so the only outcomes considered are a dry hole and a low producer. The cost, outcomes, payoffs, and probabilities are as follows:

Site 2: Cost to Drill $200K		
Outcome	**Payoff**	**Probability**
Dry	−200K	0.2
Low producer	50K	0.8

Figure 7.13

Rock strata forming a dome structure. Oil tends to pool at the top of the dome in an oil-bearing layer if the layer above is impermeable.

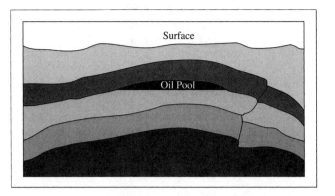

Table 7.1

Conditional probabilities of outcomes at Site 1.

If Dome Structure Exists

Outcome	P(Outcome|Dome)
Dry	0.600
Low	0.250
High	0.150
	1.000

If No Dome Structure Exists

Outcome	P(Outcome No|Dome)
Dry	0.850
Low	0.125
High	0.025
	1.000

The decision tree is shown in Figure 7.14. The problem with it as drawn, however, is that we cannot assign probabilities immediately to the outcomes for Site 1. To find these probabilities, we must use the conditional probabilities and the law of total probability to calculate P(Dry), P(Low), and P(High):

$$P(\text{Dry}) = P(\text{Dry} \mid \text{Dome}) \, P(\text{Dome}) + P(\text{Dry} \mid \text{No Dome}) \, P(\text{No Dome})$$
$$= 0.6(0.6) + 0.85(0.4) = 0.70$$

$$P(\text{Low}) = P(\text{Low} \mid \text{Dome}) \, P(\text{Dome}) + P(\text{Low} \mid \text{No Dome}) \, P(\text{No Dome})$$
$$= 0.25(0.6) + 0.125(0.4) = 0.20$$

Figure 7.14

Decision tree for oil-wildcatting problem.

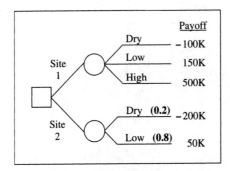

$$P(\text{High}) = P(\text{High} \mid \text{Dome})\, P(\text{Dome}) + P(\text{High} \mid \text{No Dome})\, P(\text{No Dome})$$
$$= 0.15(0.6) + 0.025(0.4) = 0.10$$

Everything works out as it should. The three probabilities are for mutually exclusive and collectively exhaustive outcomes, and they add to 1, just as they should. Folding back the decision tree, we find that Site 1 has the higher EMV (expected payoff):

$$\text{EMV}(\text{Site 1}) = 0.7(-100\text{K}) + 0.2(150\text{K}) + 0.1(500\text{K})$$
$$= 10\text{K}$$

$$\text{EMV}(\text{Site 2}) = 0.2(-200\text{K}) + 0.8(50\text{K})$$
$$= 0$$

We also can calculate the variance and standard deviation of the payoffs for each site:

$$\sigma_1^2 = 0.7(-100 - 10)^2 + 0.2(150 - 10)^2 + 0.1(500 - 10)^2$$
$$= 0.7(-110)^2 + 0.2(140)^2 + 0.1(490)^2$$
$$= 36{,}400\text{K}^2$$

$$\sigma_1 = 190.79\text{K}$$

$$\sigma_2^2 = 0.2(-200 - 0)^2 + 0.8(50 - 0)^2$$
$$= 0.2(-200)^2 + 0.8(50)^2$$
$$= 10{,}000\text{K}^2$$

$$\sigma_2 = 100.00\text{K}$$

If we treat these numbers as measures of variability, then it is clear that Site 1, with its higher variance and standard deviation, is more variable than Site 2. In this sense, Site 1 might be considered to be riskier than Site 2.

Note that we could have drawn the decision tree as in Figure 7.15, with "stacked" probabilities. That is, we could have drawn it with a first chance node indicating whether or not a dome is present, followed by chance nodes for the amount of oil. These chance nodes would include the conditional probabilities from Table 7.1.

We also could have created a probability table as in Table 7.2. The probabilities in the cells of the table are the joint probabilities of both outcomes happening at the same time. Calculating the joint probabilities for the table requires the definition of conditional probability. For example,

$$P(\text{Low and Dome}) = P(\text{Dome}) \, P(\text{Low} \mid \text{Dome})$$
$$= 0.60 \, (0.25)$$
$$= 0.15$$

The probability table is easy to construct and easy to understand. Once the probabilities of the joint outcomes are calculated, the probabilities of the individual outcomes then are found by adding across the rows or down the columns. For example, from Table 7.2, we can tell that $P(\text{Dry}) = 0.36 + 0.34 = 0.70$. (You can remember that these are called *marginal* probabilities because they are found in the margins of the table!)

Suppose that the company drills at Site 1 and the well is a high producer. In light of this evidence, does it seem more likely that a dome structure exists? Can we figure out $P(\text{Dome} \mid \text{High})$? This question is part of a larger problem. Figure 7.16 shows a "probability tree" (a decision tree without decisions) that reflects the information that we have been given in the problem. We know $P(\text{Dome})$ and $P(\text{No Dome})$, and we know the conditional probabilities of the amount of oil given the presence or absence of a dome. Thus, our probability tree has the chance node representing the presence or absence of a dome on the left and the chance node representing the amount of oil on the right. Finding $P(\text{Dome} \mid \text{High})$ is a matter of "flipping the tree" so that the chance node for the amount of oil is on the left and the node for the presence or absence of a dome is on the right, as in Figure 7.17.

Flipping a probability tree is the same as reversing an arrow between two chance nodes in an influence diagram. In Figure 7.18a, the direction of the arrow represents the probabilities as they are given in Table 7.1. Because the arrow points to the "Oil

Figure 7.15

An alternative decision tree for the oil-wildcatting problem.

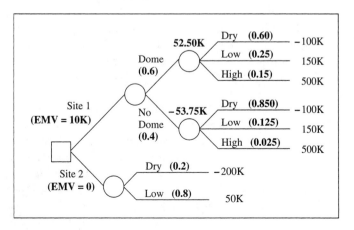

Table 7.2

Calculating the probabilities for Site 1.

	Dome	No Dome	
Dry	0.36	0.34	0.70
Low	0.15	0.05	0.20
High	0.09	0.01	0.10
	0.60	0.40	1.00

Figure 7.16

Probability tree for the uncertainty faced at Site 1.

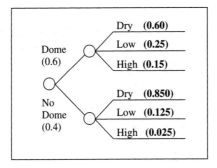

Figure 7.17

Flipping the probability tree in Figure 7.16.

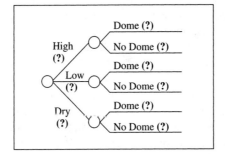

Production" node, the probabilities for the level of oil production are conditioned on whether or not there is a dome. Figure 7.18b shows the arrow reversed. Now the probability of a dome or no dome is conditioned on the amount of oil produced. Changing the direction of the arrow has the same effect as flipping the probability tree.

Simply flipping the tree or turning the arrow around in the influence diagram is not enough. The question marks in Figure 7.17 indicate that we do not yet know what the probabilities are for the flipped tree. The probabilities in the new tree must be consistent with the probabilities in the original version. Consistency means that if we were to calculate P(Dome) and P(No Dome) using the probabilities in the new tree, we would get 0.60 and 0.40, the probabilities we started with in Figure 7.16. How do we ensure consistency?

Figure 7.18

Reversing arrows be-
tween chance nodes in
an influence diagram.

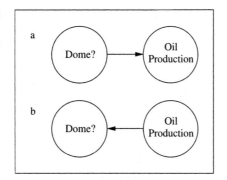

We can find P(High), P(Low), and P(Dry) by using the law of total probability. We did these calculations above for Figure 7.14, and we found that P(High) = 0.10, P(Low) = 0.20, and P(Dry) = 0.70. How about finding the new conditional probabilities? Bayes' theorem provides the answer. For example:

$$P(\text{Dome} \mid \text{High}) = \frac{P(\text{High} \mid \text{Dome})\, P(\text{Dome})}{P(\text{High})}$$

$$= \frac{P(\text{High} \mid \text{Dome})\, P(\text{Dome})}{P(\text{High} \mid \text{Dome})\, P(\text{Dome}) + P(\text{High} \mid \text{No Dome})\, P(\text{No Dome})}$$

$$= \frac{0.15(0.6)}{0.15(0.6) + 0.025(0.4)}$$

$$= \frac{0.15(0.6)}{0.10} = 0.90$$

Probabilities that have the same conditions must add to 1, and so P(No Dome | High) must be equal to 0.10. Likewise, we can calculate the conditional probabilities of a dome or no dome given a dry hole or a low producer. These probabilities are shown in Figure 7.19.

If you did not enjoy using Bayes' theorem directly to calculate the conditional probabilities required in flipping the tree, you may be pleased to learn that the probability table (Table 7.2) has all the information needed. Recall that the entries in the cells inside the table are the joint probabilities. For example, P(High and Dome) = 0.09. We need

Figure 7.19

The flipped probabil-
ity tree with new
probabilities.

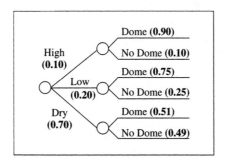

P(Dome | High) = P(High and Dome)/P(High), which is just 0.09/0.10 = 0.90. That is, we take the joint probability from inside the table and divide it by the probability of the outcome on the right side of the vertical bar. This probability is found in the margin of the probability table.

Whether we use Bayes' theorem directly or the probability-table approach to flip the probability tree, the conditional probabilities that we obtain have an interesting interpretation. Recall that we started with P(Dome) = 0.60. After finding that the well was a high producer, we were able to calculate P(Dome | High). This probability sometimes is called a *posterior* probability, indicating that it is the result of revising the original probability after gathering data. In contrast, the probability with which we started, in this case P(Dome) = 0.60, sometimes is called the *prior* probability. One way to think about Bayes' theorem is that it provides a mechanism to update prior probabilities when new information becomes available. [*Source:* The oil-wildcatting example and analysis are adapted from C. A. Holloway (1979) *Decision Making under Uncertainty: Models and Choices,* pp. 195–200. Englewood Cliffs, NJ: Prentice-Hall.]

JOHN HINCKLEY'S TRIAL

In 1982 John Hinckley was on trial, accused of having attempted to kill President Reagan. During Hinckley's trial, Dr. Daniel R. Weinberger told the court that when individuals diagnosed as schizophrenics were given computerized axial tomography (CAT) scans, the scans showed brain atrophy in 30% of the cases compared with only 2% of the scans done on normal people. Hinckley's defense attorney wanted to introduce as evidence Hinckley's CAT scan, which showed brain atrophy. The defense argued that the presence of atrophy strengthened the case that Hinckley suffered from mental illness.

We can use Bayes' theorem easily to analyze this situation. We want to know the probability that Hinckley was schizophrenic given that he had brain atrophy. Approximately 1.5% of people in the United States suffer from schizophrenia. This is the *base rate,* which we can interpret as the prior probability of schizophrenia before we find out about the condition of an individual's brain. Thus, $P(S) = 0.015$, where S means schizophrenia. We also have $P(A \mid S) = 0.30$ (A means atrophy) and $P(A \mid \overline{S}) = 0.02$. We want $P(S \mid A)$. Bayes' theorem provides the mathematical mechanism for flipping the probability from $P(A \mid S)$ to $P(S \mid A)$:

$$P(S \mid A) = \frac{P(A \mid S) \, P(S)}{P(A \mid S) \, P(S) + P(A \mid \overline{S}) \, P(\overline{S})}$$

$$= \frac{0.30(0.015)}{0.30(0.015) + 0.02(0.985)}$$

$$= 0.186$$

Thus, given that his brain showed such atrophy, Hinckley still has less than a 1-in-5 chance of being schizophrenic. Given the situation, this is perhaps a surprisingly low probability. The intuition behind this result is that there are many false-positive tests. If we tested 100,000 individuals, some 1500 of them would be schizophrenic and 98,500 would be normal (or at least not schizophrenic). Of the 1500, only 30%, or approximately 450, would show atrophy. Of the 98,500, 2% (some 1970) would show brain atrophy. If a single individual has atrophy, is he one of the 450 with schizophrenia or one of the 1970 without? Note that

$$0.186 = \frac{450}{450 + 1970}$$

The real question is whether this is good news or bad news for Hinckley. The prosecution might argue that the probability of schizophrenia is too small to make any difference; even in light of the CAT-scan evidence, Hinckley is less than one-fourth as likely to be schizophrenic as not. On the other hand, the defense would counter, 0.186 is much larger than 0.015. Thus, the CAT-scan results indicate that Hinckley was more than 12 times as likely to be schizophrenic as a randomly chosen person on the street.

Now, however, consider what we have done in applying Bayes' theorem. We have used a prior probability of 0.015, which essentially is the probability that a randomly chosen person from the population is schizophrenic. *But Hinckley was not randomly chosen.* In fact, it does not seem reasonable to think of Hinckley, a man accused of attempted assassination, as the typical person on the street.

If 0.015 is not an appropriate prior probability, what is? It may not be obvious what an appropriate prior probability should be, so let us consider a sensitivity-analysis approach and see what different priors would imply. Imagine a juror who, before encountering the CAT-scan evidence, believes that there is only a 10% chance that Hinckley is schizophrenic. For most of us, this would be a fairly strong statement; Hinckley is nine times as likely to be normal as schizophrenic. Now consider the impact of the CAT-scan evidence on this prior probability. We can calculate this juror's posterior probability:

$$P(S \mid A) = \frac{0.30(0.10)}{0.30(0.10) + 0.02(0.90)}$$
$$= 0.63$$

$P(S \mid A) = 0.63$ is a substantial probability. We can do this for a variety of values for the prior probability. Figure 7.20 shows the posterior probability $P(S \mid A)$ graphed as a function of the prior probability that Hinckley was schizophrenic.

As a result of this discussion, it is clear that a juror need not have a very strong prior belief for the CAT-scan evidence to have an overwhelming effect on his or her posterior belief of Hinckley's mental illness. Furthermore, no matter what the juror's prior belief, the CAT-scan result must increase the probability that Hinckley was schizophrenic. [*Source:* A. Barnett, I. Greenberg, and R. Machol (1984) "Hinckley and the Chemical Bath." *Interfaces*, 14, 48–52.]

Figure 7.20

Graph of the posterior probability that Hinckley was schizophrenic plotted against the prior probability.

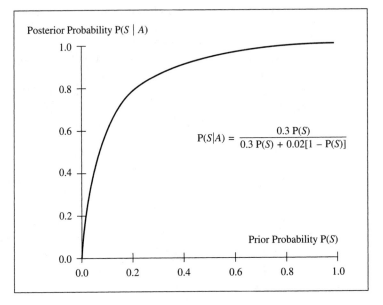

Posterior Probability P(S | A)

$$P(S|A) = \frac{0.3\ P(S)}{0.3\ P(S) + 0.02[1 - P(S)]}$$

Prior Probability P(S)

Decision-Analysis Software and Bayes' Theorem

As we have noted above, flipping conditional probabilities by means of Bayes' theorem can seem like a complicated task. In fact, the rules are straightforward: Calculate the joint probability and divide by the appropriate marginal probability. The problems are being sure that the joint probability is calculated correctly and selecting the appropriate marginal probability. Fortunately, if all of the probabilities are defined (for example, in a probability tree or table), applying Bayes' theorem is something that computers can do much more readily than humans!

PrecisionTree automatically flips probabilities as needed when analyzing a decision model. PrecsionTree allows you to enter probability data into an influence diagram that must be flipped before the model can be analyzed. By converting the influence diagram into its corresponding decision tree, you can see the revised probabilities. You can also easily program Excel to perform Bayes' theorem calculations based on a probability table like that shown in Table 7.2.

SUMMARY

The definitions and formulas at the beginning of the chapter are the building blocks of probability theory. Venn diagrams provide a way to visualize these probability "laws." We also discussed uncertain quantities, their probability distributions, and characteristics of those distributions such as expected value, variance, and standard deviation. The use of probability concepts and the manipulation of probabilities were demonstrated in the oil-wildcatting and Hinckley trial examples.

EXERCISES

7.1   Explain why probability is important in decision analysis.

7.2   Explain in your own words what an uncertain quantity or random variable is. Why is the idea of an uncertain quantity important in decision analysis?

7.3   You are given the following probability table:

		A	$\overline{A}$	
B	B	0.12	0.53	0.65
$\overline{B}$	$\overline{B}$	0.29	0.06	0.35
		0.41	0.59	1.00

Use the probability table to find the following:

P(A and B), P(A and $\overline{B}$), P(A), P(B), P($\overline{B}$), P(B|A), P(A|B), P($\overline{A}$|$\overline{B}$)

7.4   Use the probability table in Exercise 7.3 to find P(A or B), or the outcome where either A occurs or B occurs (or both).

7.5   The Outcome A or B sometimes is called the union of A and B. The union event occurs if either Outcome A or Outcome B (or both) occurs. Suppose that both A and B can happen at the same time (that is, their areas overlap in a Venn diagram). Show that P(A or B) = P(A) + P(B) − P(A and B). Use a Venn diagram to explain. Why is this result consistent with the second requirement that probabilities add up?

7.6   Often it is difficult to distinguish between the probability of an intersection of outcomes (joint probability) and the probability of a conditional outcome (conditional probability). Classify the following as joint probability statements or conditional probability statements. [If in doubt, try to write down the probability statement; for example, P(Crash Landing | Out of Fuel) or P(Dow Jones Up and Stock Price Up).]

a   Eight percent of the students in a class were left-handed and red-haired.

b   Of the left-handed students, 20% had red hair.

c   If the Orioles lose their next game, then the Cubs have a 90% chance of winning the pennant.

d   Fifty-nine percent of the people with a positive test result had the disease.

e   For 78% of patients, the surgery is a success and the cancer never reappears.

f   If the surgery is a success, the cancer is unlikely to reappear.

g   Given the drought, food prices are likely to increase.

h   There is an even chance that a farmer who loses his crop will go bankrupt.

i   If the temperature is high and there is no rain, farmers probably will lose their crops.

j John probably will be arrested because he is trading on insider information.

k John probably will trade on insider information and get caught.

7.7 Calculate the variance and standard deviation of the payoffs for Products B and C in the GPC case at the end of Chapter 4 (pages 163–165).

7.8 $P(A) = 0.42$, $P(B \mid A) = 0.66$, and $P(B \mid \overline{A}) = 0.25$. Find the following:

$P(\overline{A})$, $P(\overline{B} \mid A)$, $P(\overline{B} \mid \overline{A})$, $P(B)$, $P(\overline{B})$, $P(A \mid B)$, $P(\overline{A} \mid B)$, $P(A \mid \overline{B})$, $P(\overline{A} \mid \overline{B})$.

7.9 $P(A) = 0.10$, $P(B \mid A) = 0.39$, and $P(B \mid \overline{A}) = 0.39$. Find the following:

$P(\overline{A})$, $P(\overline{B} \mid A)$, $P(\overline{B} \mid \overline{A})$, $P(B)$, $P(\overline{B})$, $P(A \mid B)$, $P(\overline{A} \mid B)$, $P(A \mid \overline{B})$, $P(\overline{A} \mid \overline{B})$.

How would you describe the relationship between Outcomes A and B?

7.10 Consider the following probabilities:

$$P(X = 2) = 0.3$$
$$P(X = 4) = 0.7$$
$$P(Y = 10 \mid X = 2) = 0.9$$
$$P(Y = 20 \mid X = 2) = 0.1$$
$$P(Y = 10 \mid X = 4) = 0.25$$
$$P(Y = 20 \mid X = 4) = 0.75$$

Calculate the covariance and correlation between X and Y.

7.11 Consider the following joint probability distribution for uncertain quantities X and Y:

$$P(X = -2 \text{ and } Y = 2) = 0.2$$
$$P(X = -1 \text{ and } Y = 1) = 0.2$$
$$P(X = 0 \text{ and } Y = 0) = 0.2$$
$$P(X = 1 \text{ and } Y = 1) = 0.2$$
$$P(X = 2 \text{ and } Y = 2) = 0.2$$

a Calculate the covariance and correlation between X and Y.

b Calculate $P(Y = 2)$, $P(Y = 2 \mid X = -2)$, $P(Y = 2 \mid X = 2)$, $P(Y = 2 \mid X = 0)$.

c Calculate $P(X = -2)$, $P(X = -2 \mid Y = 2)$, $P(X = -2 \mid Y = 0)$.

d Are X and Y dependent or independent? How would you describe the relationship between X and Y?

7.12 Write probability statements relating hemline height, stock market prices, and "adventuresomeness" that correspond to the following description of these relationships:

Another example is that high skirt hemlines tend to occur when the stock market is advancing steadily (a bull market), although no one believes that hemlines cause the stock market to advance, or vice versa. Perhaps both reflect a pervasive feeling of confidence and adventure in our society. If we had an adequate index of "societal adventuresomeness," hemline height and stock market activity might well be conditionally independent given the index.

7.13 Even though we distinguish between continuous and discrete uncertain quantities, in reality everything is discrete if only because of limitations inherent in our measuring devices. For example, we can only measure time or distance to a certain level of precision. Why, then, do we make the distinction between continuous and discrete uncertain quantities? What value is there in using continuous uncertain quantities in a decision model?

7.14 $P(A) = 0.68$, $P(B \mid A) = 0.30$, and $P(B \mid \overline{A}) = 0.02$. Find $P(\overline{A})$, $P(A \text{ and } B)$, and $P(\overline{A} \text{ and } B)$. Use these to construct a probability table. Now use the table to find the following:

$$P(\overline{B} \mid A), \ P(\overline{B} \mid \overline{A}), \ P(B), \ P(\overline{B}), \ P(A \mid B), \ P(\overline{A} \mid B), \ P(A \mid \overline{B}), \ P(\overline{A} \mid \overline{B}).$$

7.15 Julie Myers, a graduating senior in accounting, is preparing for an interview with a Big Eight accounting firm. Before the interview, she sets her chances of eventually getting an offer at 50%. Then, on thinking about her friends who have interviewed and gotten offers from this firm, she realizes that of the people who received offers, 95% had good interviews. On the other hand, of those who did not receive offers, 75% said they had good interviews. If Julie Myers has a good interview, what are her chances of receiving an offer?

7.16 Find the expected value, variance, and standard deviation of X in the following probability distributions:

a $P(X = 1) = 0.05$, $P(X = 2) = 0.45$, $P(X = 3) = 0.30$, $P(X = 4) = 0.20$.

b $P(X = -20) = 0.13$, $P(X = 0) = 0.58$, $P(X = 100) = 0.29$.

c $P(X = 0) = 0.368$, $P(X = 1) = 0.632$.

7.17 If $P(X = 1) = p$ and $P(X = 0) = 1 - p$, show that $E(X) = p$ and $Var(X) = p(1 - p)$.

7.18 If $P(A \mid B) = p$, must $P(A \mid \overline{B}) = 1 - p$? Explain.

7.19 Suppose that a company produces three different products. The sales for each product are independent of the sales for the others. The information for these products is given below:

Product	Price ($)	Expected Unit Sales	Variance of Unit Sales
A	3.50	2000	1000
B	2.00	10,000	6400
C	1.87	8500	1150

a What are the expected revenue and variance of the revenue from Product A alone?

b What are the company's overall expected revenue and variance of its revenue?

7.20 A company owns two different computers, which are in separate buildings and operated entirely separately. Based on past history, Computer 1 is expected to break down 5.0 times a year, with a variance of 6, and costing $200 per breakdown. Computer 2 is expected to break down 3.6 times per year, with a variance of 7, and costing $165 per breakdown. What is the company's expected cost for computer breakdowns and the variance of the breakdown cost? What assumption must you make to find the variance? Is this a reasonable assumption?

7.21   A firm is negotiating with a local club to supply materials for a party. The firm's manager expects to sell 100 large bags of pretzels for $3 each or 300 for $2 each; these two outcomes are judged to be equally likely. The expected number of bags sold is 200 = (100 + 300)/2, and expected price is $2.50 = ($3 + $2)/2. The manager then calculates expected revenue as the expected number sold times the expected price: E(Revenue) = 200($2.50) = $500. What is wrong with the manager's calculation?

7.22   Flip the probability tree shown in Figure 7.21.

7.23   Figure 7.22 shows part of an influence diagram for a chemical that is considered potentially carcinogenic. How would you describe the relationship between the test results and the field results?

7.24   Let CP denote carcinogenic potential, TR test results, and FR field results. Suppose that for Figure 7.22 we have the following probabilities:

$$P(CP\ High) = 0.27 \qquad P(CP\ Low) = 0.73$$

Figure 7.21

A probability tree representing the diagnostic performance of a medical test.

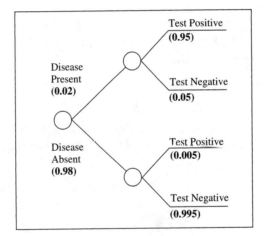

Figure 7.22

An influence diagram for a potentially carcinogenic chemical.

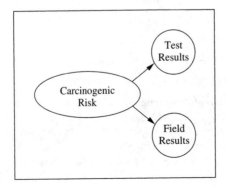

$$P(\text{TR Positive} \mid \text{CP High}) = 0.82 \qquad P(\text{TR Positive} \mid \text{CP Low}) = 0.21$$
$$P(\text{TR Negative} \mid \text{CP High}) = 0.18 \qquad P(\text{TR Negative} \mid \text{CP Low}) = 0.79$$

$$P(\text{FR Positive} \mid \text{CP High}) = 0.95 \qquad P(\text{FR Positive} \mid \text{CP Low}) = 0.17$$
$$P(\text{FR Negative} \mid \text{CP High}) = 0.05 \qquad P(\text{FR Negative} \mid \text{CP Low}) = 0.83$$

Find the following:

$$P(\text{TR Positive and FR Positive} \mid \text{CP High})$$
$$P(\text{TR Positive and FR Negative} \mid \text{CP High})$$
$$P(\text{TR Negative and FR Negative} \mid \text{CP Low})$$
$$P(\text{TR Negative and FR Positive} \mid \text{CP Low})$$

QUESTIONS AND PROBLEMS

7.25 Linda is 31 years old, single, outspoken, and very bright. She majored in philosophy. As a student, she was deeply concerned with issues of discrimination and social justice and also participated in antinuclear demonstrations. Use your judgment to rank the following statements by their probability, using 1 for the most probable statement and 8 for the least probable:

a Linda is a teacher in an elementary school.

b Linda works in a bookstore and takes Yoga classes.

c Linda is active in the feminist movement.

d Linda is a psychiatric social worker.

e Linda is a member of the League of Women Voters.

f Linda is a bank teller.

g Linda is an insurance salesperson.

h Linda is a bank teller and is active in the feminist movement.

[*Source:* Amos Tversky and Daniel Kahneman (1982) "Judgments of and by Representativeness." In D. Kahneman, P. Slovic, and A. Tversky (eds.), *Judgment under Uncertainty: Heuristics and Biases,* pp. 84–98. Cambridge: Cambridge University Press.]

7.26 The description and statements given in Problem 7.25 often elicit responses that are not consistent with probability requirements. If you are like most people, you ranked statement h (Linda is a bank teller and is active in the feminist movement) as more probable than statement f (Linda is a bank teller).

a Explain why you ranked statements h and f as you did.

b Statement h is actually a compound event. That is, for h to be true, Linda must be both a bank teller (Outcome A) and active in the feminist movement (Outcome B). Thus, statement h is represented by the Outcome "A *and* B." Use a Venn diagram to explain why statement h must be less probable than statement f.

c Suppose that you have presented Problem 7.25 to a friend, who ranks statement h as more probable than statement f. Your friend argues as follows: "Well, it's not very likely that Linda is a bank teller in the first place. But if she is a bank teller, then she is very likely to be active in the feminist movement. So h would appear to be more likely than f." How is your friend interpreting statement h? Explain why this is not an appropriate interpretation.

7.27   Suppose you are a contestant on the television game show, *Let's Make a Deal.* You must choose one of three closed doors, labeled A, B, and C, and you will receive whatever is behind the chosen door. Behind two of the doors is a goat, and behind the third is a new car. Like most people, you have no use for a goat, but the car would be very welcome. Suppose you choose Door A. Now the host opens Door B, revealing a goat, and offers to let you switch from Door A to Door C. Do you switch? Why or why not? [*Hint:* What can you assume about the host's behavior? Would it ever be the case that the host would open a door to reveal the car before asking if you want to switch?]

7.28   On the fictitious television game show, "Marginal Analysis for Everyone," the host subjects contestants to unusual tests of mental skill. On one, a contestant may choose one of two identical envelopes labeled A and B, each of which contains an unknown amount of money. The host reveals, though, that one envelope contains twice as much as the other. After choosing A, the host suggests that the contestant might want to switch.

"Switching is clearly advantageous," intones the host persuasively. "Suppose you have amount x in your Envelope A. Then B must contain either $x/2$ (with probability 0.5) or $2x$ (also with probability 0.5). Thus, the expected value of switching is $1.25x$. In fact now that I think about it, I'll only let you switch if you give me a 10% cut of your winnings. What do you say? You'll still be ahead."

"No deal," replies the contestant. "But I'll be happy to switch for free. In fact, I'll even let you choose which envelope I get. I won't even charge you anything!"

What is wrong with the host's analysis?

[*Source:* This problem was suggested by Ross Shachter.]

7.29   Draw an influence diagram for the oil-wildcatting problem. Construct the necessary tables represented by the nodes, including the payoff table, and solve the influence diagram. If possible, use a computer program to do this problem.

7.30   Finding the variance and standard deviation for the payoff from Product A in the GPC case at the end of Chapter 4 (page 163) is somewhat complicated because a pricing decision must be made. How would you calculate the variance and standard deviation for Product A's payoff? (Recall the approach we used in solving decision trees. We were able to "prune" decision branches that were suboptimal.)

7.31   Calculate the variance and standard deviation of the payoffs in the final court decision in the Texaco-Pennzoil case as diagrammed in Figure 4.2.

7.32   In the oil-wildcatting problem, suppose that the company could collect information from a drilling core sample and analyze it to determine whether a dome structure exists at Site 1. A positive result would indicate the presence of a dome, and a negative result would indicate the absence of a dome. The test is not perfect, however. The test is highly accurate for detecting a dome; if there is a dome, then the test shows a positive result 99% of the time. On the other hand, if there is no dome, the probability of a negative result is only 0.85. Thus, P(+ | Dome) = 0.99 and P(− | No Dome) = 0.85. Use these probabilities, the information given in the example, and Bayes' theorem to find the posterior probabilities P(Dome | +) and P(Dome | −). If the test gives a positive result, which site should be selected? Calculate expected values to support your conclusion. If the test result is negative, which site should be chosen? Again, calculate expected values.

7.33 In Problem 7.32, calculate the probability that the test is positive and a dome structure exists [P(+ and Dome)]. Now calculate the probability of a positive result, a dome structure, and a dry hole [P(+ and Dome and Dry)]. Finally, calculate P(Dome | + and Dry).

7.34 Referring to the oil-wildcatting decision diagrammed in Figure 7.15, suppose that the decision maker has not yet assessed P(Dome) for Site 1. Find the value of P(Dome) for which the two sites have the same EMV. If the decision maker believes that P(Dome) is somewhere between 0.55 and 0.65, what action should be taken?

7.35 Again referring to Figure 7.15, suppose the decision maker has not yet assessed P(Dry) for Site 2 or P(Dome) for Site 1. Let P(Dry) = p and P(Dome) = q. Construct a two-way sensitivity analysis graph for this decision problem.

7.36 Refer to Exercises 7.23 and 7.24. Calculate P(FR Positive) and P(FR Positive | TR Positive). [*Hints:* P(FR Positive) is a fairly simple calculation. To find P(FR Positive | TR Positive), first let B = FR Positive, C = TR Positive, A = CP High, and $\overline{A}$ = CP Low. Now expand P(FR Positive | TR Positive) using the law of total probability in this form:

$$P(B \mid C) = P(B \mid A, C) \, P(A \mid C) + P(B \mid \overline{A}, C) \, P(\overline{A} \mid C)$$

Now all of the probabilities on the right-hand side can be calculated using the information in the problem.]

Compare P(FR Positive) and P(FR Positive | TR Positive). Would you say that the test results and field results are independent? Why or why not? Discuss the difference between conditional independence and regular independence.

CASE STUDIES

DECISION ANALYSIS MONTHLY

Peter Finch looked at the numbers for the renewals of subscriptions to *Decision Analysis Monthly* magazine. For both May and June he had figures for the percentage of expiring subscriptions that were gift subscriptions, promotional subscriptions, and from previous subscribers. Furthermore, his data showed what proportion of the expiring subscriptions in each category had been renewed (see Table 7.3).

Finch was confused as he considered these numbers. Robert Calloway, who had assembled the data, had told him that the overall proportion of renewals had dropped from May to June. But the figures showed clearly that the proportion renewed had increased in each category. How could the overall proportion possibly have gone down? Peter got a pencil and pad of paper to check Calloway's figures. He had to report to his boss this afternoon and wanted to be able to tell him whether these figures represented good news or bad news.

Question

1 Do the data represent good news or bad news regarding renewal trends? Why?

Table 7.3

Subscription data for
*Decision Analysis
Monthly.*

May Subscription Data	Expiring Subscriptions, %	Proportion Renewed
Gift Subscriptions	70	0.75
Promotional Subscriptions	20	0.50
Previous Subscribers	10	0.10
Total	100	

June Subscription Data	Expiring Subscriptions, %	Proportion Renewed
Gift Subscriptions	45	0.85
Promotional Subscriptions	10	0.60
Previous Subscribers	45	0.20
Total	100	

SCREENING FOR COLORECTAL CANCER

The fecal occult blood test, widely used both in physicians' offices and at home to screen patients for colon and rectal cancer, examines a patient's stool sample for blood, a condition indicating that the cancer may be present. A recent study funded by the National Cancer Institute found that of 15,000 people tested on an annual basis, 10% were found to have blood in their stools. These 10% underwent further testing, including *colonoscopy,* the insertion of an optical-fiber tube through the rectum in order to inspect the colon and rectum visually for direct indications of cancer. Only 2.5% of those having colonoscopy actually had cancer. Additional information in the study suggests that, of the patients who were tested, approximately 5 out of 1000 tested negative (no blood in the stool) but eventually did develop cancer.

Questions

1   Create a probability table that shows the relationship between blood in a stool sample and colorectal cancer. Calculate P(Cancer | Blood) and P(Cancer | No Blood).

2   The study results have led some medical researchers to agree with the American Cancer Society's long-standing recommendation that all U.S. residents over 50 years of age be tested annually. On the other hand, many researchers claim the costs of such screening, including the cost of follow-up testing on 10% of the population, far exceeds its value. Assume that the test can be performed for as little as $10 per person, that colonoscopy costs $750 on average, and that about 60 million people in the United States are over age 50. What is the expected cost (including follow-up colonoscopy) of implementing a policy of screening everyone over age 50? What is the expected number of people who must undergo colonoscopy? What is the expected number of people who must undergo colonoscopy only to find that they do not have cancer after all?

3   Over 13 years of follow-up study, 0.6% (6 out of 1000) of those who were screened annually with the fecal occult blood test died from colon cancer anyway. Of those who were not screened, 0.9% (9 out 1000) died of colon cancer during the same 13 years. Thus, the screening procedure saves approximately 3 lives per 1000 every 13 years. Use this information, along with your calculations from Questions 1 and 2, to determine the expected cost of saving a life by implementing a policy requiring everyone over 50 to be screened every year.

4   What is your conclusion? Do you think everyone over 50 should be screened? From your personal point of view, informed now by your calculations above, would the saved lives be worth the money spent and the inconvenience, worry, discomfort, and potential complications of subjecting approximately 6 million people each year to colonoscopy even though relatively few of them actually have detectable and curable cancer?

Source: J. S. Mandel, J. H. Bond, T. R. Church, D. C. Snover, G. M. Braley, L. M. Schuman, and F. Ederer (1993) "Reducing Mortality from Colorectal Cancer by Screening for Fecal Occult Blood. (Minnesota Colon Cancer Control Study)." *New England Journal of Medicine,* 328 (19), 1365–1371.

AIDS

Acquired immune deficiency syndrome (AIDS) is the most frightening disease of the late twentieth century. The disease attacks and disables the immune system, leaving the body open to other diseases and infection. It is almost always fatal, although years may pass between infection and the development of the disease. As of December 31, 1993, there were 361,509 confirmed AIDS cases and 220,871 deaths due to AIDS in the United States.

Even more frightening is the process by which the disease travels through the population. AIDS is caused by a virus (human T-lymphotropic virus, Type III, or HTLV-III, although more commonly listed as HIV). The virus is transmitted through blood, semen, and vaginal secretions, and may attack virtually anyone who engages in any of several risky behaviors. The extent of the concern about AIDS among public health officials is reflected in the fact that the U.S. Surgeon General's office mailed brochures on AIDS to 107 million households in May and June 1988, the largest single mailing undertaken by the federal government to that time.

When an individual becomes infected, the body produces a special antibody in an effort to counteract the virus. But it can be as long as 12 weeks before these antibodies appear, and it may be years before any signs or symptoms of AIDS infection appear. During this time the individual may not be aware that he or she is infected and thus can spread the disease inadvertently.

Because of the delayed reaction of the virus, there are many more infected individuals than reported AIDS cases. Epidemiologists estimate that about 1 million people in the United States are infected with HIV and thus are potentially infectious of others. Worldwide, the estimate is that over 13 million people are infected. Because of this and because of the way the disease is transmitted, the best way to avoid AIDS simply is to avoid risky behavior. Do not share drug needles. Use a latex condom during intercourse unless you are certain that your partner is not infected.

To help reduce the rate at which AIDS is spread, a number of local (and often controversial) programs have sprung up to provide condoms to sexually active teenagers and clean needles to intravenous drug users.

If 1 million people in the United States *are* HIV-infected, this amounts to an overall rate of approximately 0.38% in a population of 260 million people. The term *seroprevalence* refers to the incidence of HIV in the population. Seroprevalence rates vary dramatically depending on the particular subpopulation. For example, the overall seroprevalence rate for adult men in the United States is about 1%, but for adult women it is 0.13%. Figures 7.23 and 7.24 show seroprevalence rates for male and female military recruits, respectively, in each of the 50 states plus the District of Columbia and Puerto Rico. For higher risk groups the seroprevalence rates are much higher. For example, nationally about 25.5% of gay and bisexual men are infected, but the rate ranges from 3.9% in Providence, Rhode Island, to 47.4% in Atlanta, Georgia. For intravenous drug users, the seroprevalence rate is 7.5% nationwide, reaching as high as 40.4% in New York City.

The best tests available for AIDS detect the antibodies rather than the virus. Two such tests are generally available and widely used. The first is the *enzyme-linked immunosorbent assay* (ELISA). An individual is considered to have a positive result on the ELISA test only if both of two separate trials yield positive results. The performance of such a diagnostic test can be measured by the probabilities associated with correct diagnosis. The probability that an infected individual tests positive, P(ELISA+ | Infected), is called the *sensitivity* of the test. The probability that an uninfected individual tests negative, P(ELISA– | Not Infected), is called the *specificity* of the test. A negative result for an infected individual is called a *false-negative,* and

Figure 7.23

HIV seroprevalence among male applicants for military service by state of residence, United States, January 1991 through December 1992. Source: Department of Defense.

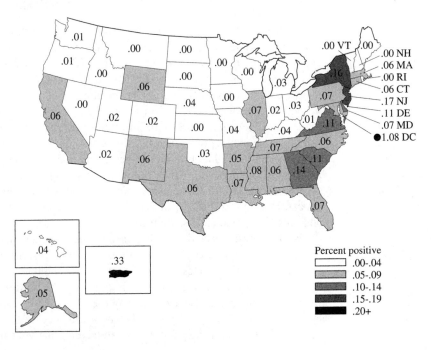

Percent positive
- .00–.04
- .05–.09
- .10–.14
- .15–.19
- .20+

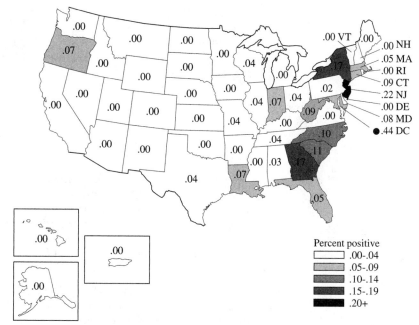

a positive result for someone who is not infected is called a *false-positive*. An ideal test would have 100% sensitivity and 100% specificity, giving correct diagnoses all the time with neither false-negatives nor false-positives. In 1990, the Centers for Disease Control (CDC) reported a study indicating that the sensitivity and specificity of the ELISA test are 0.997 and 0.985, respectively.

The *Western blot* test generally is used to confirm or disconfirm a positive ELISA result. For some time the Western blot was considered to be a perfect test, but this may not be true. The Western blot is a labor-intensive test, the results of which require interpretation by a skilled laboratory technician. The same CDC study indicated sensitivity and specificity of the Western blot to be 0.993 and 0.916, respectively. Table 7.4 summarizes the performance characteristics of the two tests.

Questions

1   Given that the tests are not perfect, it is worthwhile to calculate the probability of being infected, given the various combinations of results on the tests. Calculate the probability of being infected given a positive ELISA test result, P(Inf | ELISA+). Use as your prior probability the P(Inf) = 0.0038, the estimated overall rate of infection in the United States. For P(ELISA + | Inf) and P(ELISA + | Not Inf), use the numbers from Table 7.4.

2   Create a graph like Figure 7.20 that shows P(Inf | ELISA+) as a function of P(Inf), the prior probability or seroprevalence rate. Indicate on your graph the appropriate prior and posterior probabilities for female military recruits in New Jersey, gay men in Providence, Rhode Island, and intravenous drug users in New York City.

Table 7.4

Performance of ELISA and Western blot tests.

Test	Sensitivity P(+ | Inf)	False-Negative Rate P(− | Inf)	Specificity P(− | Not Inf)	False-Positive Rate P(+ | Not Inf)
ELISA	0.997	0.003	0.985	0.015
Western blot	0.993	0.007	0.916	0.084

3   Repeat Questions 1 and 2, but this time find the probability of being infected given a negative ELISA result, P(Inf | ELISA−).

4   Calculate the probability of being infected given a positive ELISA test result and a positive Western blot, P(Inf | ELISA+, WB+) and the probability of being infected given a positive ELISA and negative Western blot, P(Inf | ELISA+, WB−). [*Hint:* *All* of the information in the case regarding the performance of the Western blot assumes a positive ELISA. The calculations required here can be done with Bayes' theorem, using as the prior probability P(Inf | ELISA+), the quantity calculated in Question 1.]

5   Create graphs like those for Questions 2 and 3 that show P(Inf | ELISA+, WB+) and P(Inf | ELISA+, WB−) as functions of the prior probability P(Inf). [*Hint:* Note that this prior probability enters the calculations through P(Inf | ELISA+).]

6   Some public health officials have called for widespread testing for HIV infection. Certainly there is considerable value to society in identifying HIV carriers, although there are costs inherent in incorrect diagnoses. For example, suppose that you were forced to be tested, and both ELISA and Western blot tests gave a positive result. Imagine that a later tissue culture revealed no HIV exposure. Thus, for some time you would have been falsely labeled as an AIDS carrier. On the other hand, suppose that the tests had been falsely negative. Then you may have engaged in risky behavior under the assumption that you had no infection. Discuss the social trade-offs of costs and benefits that are involved in using imperfect screening tests for AIDS.

Sources: This case study was prepared using several publications available from the Centers for Disease Control National AIDS Clearinghouse, P.O. Box 6003, Rockville, MD 20849. The key publications used were "Surgeon General's Report to the American Public on HIV Infection and AIDS" (1994); "Serologic Testing for HIV–1 Antibody—United States, 1988 and 1989"; *Morbidity and Mortality Weekly Report* (1990), 39, 380–383 (abstracted in *Journal of the American Medical Association,* July 11, 1990, 171–173); and "National HIV Serosurveillance Summary: Results Through 1992," HIV/NCID/11-93/036.

DISCRIMINATION AND THE DEATH PENALTY

Is there a relationship between the race of convicted defendants in murder trials and the imposition of the death penalty on such defendants? This question has been debated extensively, with one side claiming that white defendants are given the death sentence much less frequently than nonwhites. This case can help you understand one reason for the debate. Table 7.5 shows information regarding 326 cases.

Questions

1 On the basis of Table 7.5, estimate P(Death Penalty | Defendant White) and P(Death Penalty | Defendant Black). What is your conclusion on the basis of these calculations?

2 Table 7.6 shows the same data disaggregated on the basis of the race of the victim. Use Table 7.6 to estimate the following probabilities:

P(Death Penalty | Defendant White, Victim White)

P(Death Penalty | Defendant Black, Victim White)

P(Death Penalty | Defendant White, Victim Black)

P(Death Penalty | Defendant Black, Victim Black)

Now what is your conclusion?

3 Explain the apparent contradiction between your answers to Questions 1 and 2.

Table 7.5

Death-penalty and racial status for 326 convicted murderers.

Race of Defendant	Death Penalty Imposed		Total Defendants
	Yes	No	
White	19	141	160
Black	17	149	166
Total	36	290	326

Table 7.6

Death-penalty and racial status for 326 convicted murderers, disaggregated by victim's race.

Race of Victim	Race of Defendant	Death Penalty Imposed		Total Defendants
		Yes	No	
White	White	19	132	151
	Black	11	52	63
Total		30	184	214
Black	White	0	9	9
	Black	6	97	103
Total		6	106	112
Total		36	290	326

Source: M. Radelet (1981) "Racial Characteristics and Imposition of the Death Penalty." *American Sociological Review,* 46, 918–927.

REFERENCES

Probability basics appear in a wide variety of books on probability and statistics, presenting the material at various levels of sophistication. Good lower-level introductions can be found in elementary statistics textbooks such as McClave and Benson (1988), Mendenhall et al. (1989), Sincich (1989), and Wonnacott and Wonnacott (1984). Two excellent resources written at higher levels are Olkin, Gleser, and Derman (1980) and Feller (1968).

All decision-analysis textbooks seem to have at least one example about oil wildcatting. True, this is the quintessential decision-analysis problem, and it includes many sources of uncertainty, concerns about attitudes toward risk, and opportunities for gathering information. But many problems have these characteristics. Probably the real reason for the oil-wildcatting scenario is that, in 1960, C. J. Grayson published one of the first applied dissertations using decision theory, and its area of application was oil drilling. Decision theorists ever since have used oil drilling as an example!

Feller, W. (1968) *An Introduction to Probability Theory and Its Applications,* Vol. 1, 3rd ed. New York: Wiley.

Grayson, C. J. (1960) *Decisions under Uncertainty: Drilling Decisions by Oil and Gas Operators.* Cambridge, MA: Division of Research, Harvard Business School.

McClave, J. T., and P. G. Benson (1988) *Statistics for Business and Economics,* 4th ed. San Francisco: Dellen.

Mendenhall, W., J. Reinmuth, and R. Beaver (1989) *Statistics for Management and Economics,* 6th ed. Boston: PWS-KENT.

Olkin, I., L. J. Gleser, and C. Derman (1980) *Probability Models and Applications.* New York: Macmillan.

Sincich, T. (1989) *Business Statistics by Example,* 3rd ed. San Francisco: Dellen.

Wonnacott, T. H., and R. J. Wonnacott (1984) *Introductory Statistics for Business and Economics.* New York: Wiley.

EPILOGUE

John Hinckley   Hinckley's defense attorney was not permitted to introduce the CAT scan of Hinckley's brain. In spite of this, the jury's verdict found Hinckley "not guilty by reason of insanity" on all counts, and he was committed to Saint Elizabeth's Hospital in Washington, D.C. The trial caused a substantial commotion among the public, many people viewing the insanity plea and the resulting verdict as a miscarriage of justice. Because of this, some lawmakers initiated efforts to tighten legal loopholes associated with the insanity plea.

AIDS   New diagnostic tests for AIDS are under continual development as research on this frightening disease continues. For example, in December 1994, the U.S. Food and Drug Administration approved an AIDS diagnostic test that uses saliva instead of blood. This test may be easier to use, and the hope is that more people will be willing to be tested. The ELISA and Western blot tests described in the case study are typical of medical diagnostic tests in general, and the analysis performed shows how to evaluate such tests.

CHAPTER 8

Subjective Probability

A ll of us are used to making judgments regarding uncertainty, and we make them frequently. Often our statements involve informal evaluations of the uncertainties that surround an event. Statements such as "The weather is likely to be sunny today," "I doubt that the Democrats will win the next presidential election," or "The risk of cancer from exposure to cigarette smoke is small" all involve a personal, subjective assessment of uncertainty at a fundamental level. As we have seen, subjective assessments of uncertainty are an important element of decision analysis. A basic tenet of modern decision analysis is that subjective judgments of uncertainty can be made in terms of probability. In this chapter we will explore how to make such judgments and what they imply.

Although most people can cope with uncertainty informally, perhaps, it is not clear that it is worthwhile to develop a more rigorous approach to measure the uncertainty that we feel. Just how important is it to deal with uncertainty in a careful and systematic way? The following vignettes demonstrate the importance of uncertainty assessments in a variety of public-policy situations.

UNCERTAINTY AND PUBLIC POLICY

Fruit Frost   Farmers occasionally must decide whether to protect a crop from potentially damaging frost. The decision must be made on the basis of weather forecasts that often are expressed in terms of probability (U.S. National Weather Service is responsible for providing these forecasts). Protecting a crop can be costly, but less so than the potential damage. Because of such potential losses, care in assessing probabilities is important.

Earthquake Prediction   Geologists are beginning to develop ways to assess the probability of major earthquakes for specific locations. In 1988 the U.S. Geological Survey published a report that estimated a 0.60 probability of a major earthquake (7.5–8 on the Richter scale) occurring in Southern California along the southern portion of the San Andreas Fault within the next 30 years. Such an earthquake could cause catastrophic damage in the Los Angeles metropolitan area.

Environmental Impact Statements   Federal and state regulations governing environmental impact statements typically require assessments of the risks associated with proposed projects. These risk assessments often are based on the probabilities of various hazards occurring. For example, in projects involving pesticides and herbicides, the chances of cancer and other health risks are assessed.

Public Policy and Scientific Research   Often scientists learn of the possible presence of conditions that may require action by the government. But action sometimes must be taken without absolute certainty that a condition exists. For example, scientists in 1988 reported that the earth had begun to warm up because of the "greenhouse effect," a condition presumably resulting from various kinds of pollution and the destruction of tropical forests. James Hansen of NASA expressed his beliefs in probabilistic terms, saying that he was 99% certain that the greenhouse effect was upon us.

Medical Diagnosis   Many physicians in hospital intensive-care units (ICUs) have access to a complex computer system known as APACHE III (Acute Physiology, Age, and Chronic Health Evaluation). Based on information about a patient's medical history, condition, treatment, and lab results, APACHE III evaluates the patient's risk as a probability of dying either in the ICU or later in the hospital.

Some of the above examples include more complicated and more formal probability assessments as well as subjective judgments. For example, the National Weather Service forecasts are based in part on a large-scale computer model of the global atmospheric system. The computer output is just one bit of information used by a forecaster to develop an official forecast that involves his or her subjective judgment regarding the uncertainty in local weather. Some risk assessments are based on cancer studies performed on laboratory animals. The results of such studies must be extrapolated subjectively to real-world conditions to derive potential effects on humans. Because of the high stakes involved in these examples and others, it is important for policy makers to exercise care in assessing the uncertainties they face.

At a reduced scale, personal decisions also involve high stakes and uncertainty. Personal investment decisions and career decisions are two kinds of decisions that typically involve substantial uncertainty. Perhaps even harder to deal with are personal medical decisions in which the outcomes of possible treatments are not known in advance. If you suffer from chronic chest pain, would you undergo elective surgery in an attempt to eliminate the pain? Because of the risks associated with open heart surgery, this decision must be made under a condition of uncertainty. You would want to think carefully about your chances on the operating table, considering not only statistics regarding the operation but also what you know about your own health and the skills of your surgeon and the medical staff.

Probability: A Subjective Interpretation

Many introductory textbooks present probability in terms of long-run frequency. For example, if a die is thrown many times, it would land with the five on top approximately one-sixth of the time; thus, the probability of a five on a given throw of the die is one-sixth. In many cases, however, it does not make sense to think about probabilities as long-run frequencies. For example, in assessing the probability that the California condor will be extinct by the year 2010 or the probability of a major nuclear power plant failure in the next 10 years, thinking in terms of long-run frequencies or averages is not reasonable because we cannot rerun the "experiment" many times to find out what proportion of the times the condor becomes extinct or a power plant fails. We often hear references to the chance that a catastrophic nuclear holocaust will destroy life on the planet. Let us not even consider the idea of a long-run frequency in this case!

Even when a long-run frequency interpretation might seem appropriate, there are times when an event has occurred, but we remain unsure of the final outcome. For example, consider the following:

1 You have flipped a coin that has landed on the floor. Neither you nor anyone else has seen it. What is the probability that it is heads?

2 What is the probability that Oregon beat Stanford in their 1970 football game?

3 What is the probability that the coin that was flipped at the beginning of that game came up heads?

4 What is the probability that Millard Fillmore was President in 1850?

For most of us the answers to these questions are not obvious. In every case the actual event has taken place. But unless you know the answer, you are uncertain.

The point of this discussion is that we can view uncertainty in a way that is different from the traditional long-run frequency approach. In Examples 1 and 3 above, there was a random event (flipping the coin), but the randomness is no longer in the coin. You are uncertain about the outcome because you do not know what the outcome was; the uncertainty is in your mind. In all of the examples, the uncertainty lies in your own brain cells. When we think of uncertainty and probability in this way, we are adopting a subjective interpretation, with a probability representing an individual's *degree of belief* that a particular outcome will occur.

Decision analysis requires numbers for probabilities, not phrases such as "common," "unusual," "toss-up," or "rare." In fact, there is considerable evidence from the cognitive psychologists who study such things that the same phrase has different connotations to different people in different contexts. For example, in one study (Beyth-Marom, 1982), the phrase "there is a non-negligible chance . . ." was given specific probability interpretations by individuals that ranged from below 0.36 to above 0.77. Furthermore, it may be the case that we interpret such phrases differently depending on the context. The phrase "a slight chance that it will rain tomorrow" may carry a very different probability interpretation from the phrase "a slight chance that the space shuttle will explode."

The problems with using verbal representations of uncertainty can be seen in the following financial-accounting policy.

ACCOUNTING FOR CONTINGENT LOSSES

When a company prepares its financial statements, in some instances it must disclose information about possible future losses. These losses are called *contingent* because they may or may not occur contingent on the outcome of some future event (e.g., the outcome of a lawsuit or a subsidiary's default on a loan that the parent company guaranteed). In their *Statement of Financial Accounting Standards No. 5,* "Accounting for Contingencies," the Financial Accounting Standards Board provides guidelines for different accounting treatments of accounting losses, depending on whether the contingent loss is judged to be "probable," "remote," or "reasonably possible." In defining these verbal terms of uncertainty, "probable" is taken to mean that the future event is likely to occur. Likewise, "remote" means that the chance the event will occur is slight. Finally, "reasonably possible" means that the chance of the event occurring is somewhere between slight and likely. [*Source*: Financial Accounting Standards Board (1991). *Original Pronouncements: Accounting Standards. Vol. 1: FASB Statement of Standards.* Homewood, IL: Irwin.]

In this example, verbal terms of uncertainty are defined using other verbal terms; no precise guidance is provided. The wording of the standard moves us from wondering about the meaning of "probable" and "remote" to a concern with "likely to occur" and "slight." How much more straightforward the accountant's job would be if the standard specified the different degrees of risk in terms of quantitative judgments made by a knowledgeable person!

One of the main topics of this chapter is how to assess probabilities—the numbers—that are consistent with one's subjective beliefs. Of the many concepts in decision analysis, the idea of subjective probability is one that seems to give students trouble. Some are uncomfortable assessing a degree of belief, because they think there must be a "correct" answer. There are no correct answers when it comes to subjective judgment; different people have different degrees of belief and hence will assess different probabilities.

If you disagree with a friend about the probability that your favorite team will win a game, do you try to persuade your friend that your probability is better? You might discuss different aspects of the situation, such as which team has the home advantage, which players are injured, and so forth. But even after sharing your information, the two of you still might disagree. Then what? You might place a bet. For many people, betting reflects their personal subjective probabilities. Some people bet on anything even if the outcome is based on some "objectively" random event (flipping a coin, playing cards, and so on). One of the most common bets might be investing—betting—in the stock market. For example, you might be willing to purchase a stock now if you think its value is likely to increase.

We will begin with the assessment of probabilities. We will show how you can view different situations in terms of the bets you might place involving small cash amounts or in terms of hypothetical lotteries. Following the discussion of the assessment of discrete probabilities, we will see how to deal with continuous probability distributions (the fan or crescent shape in the range-of-risk dilemma that was introduced in Chapter 3). Special psychological phenomena are associated with probability assessment, and we will explore the cognitive heuristics that we tend to use in probability assessment. The last two sections discuss procedures for decomposing probability assessments and what it means to be "coherent" in assessing probabilities.

Assessing Discrete Probabilities

There are three basic methods for assessing probabilities. The first is simply to have the decision maker assess the probability directly by asking, "What is your belief regarding the probability that event such and such will occur?" The decision maker may or may not be able to give an answer to a direct question like this and may place little confidence in the answer given.

The second method is to ask about the bets that the decision maker would be willing to place. The idea is to find a specific amount to win or lose such that the decision maker is indifferent about which side of the bet to take. If he or she is indifferent about which side to bet, then the expected value of the bet must be the same regardless of which is taken. Given these conditions, we can then solve for the probability.

As an example, suppose that the Los Angeles Lakers are playing the Detroit Pistons in the NBA finals this year. We are interested in finding the decision maker's probability that the Lakers will win the championship. The decision maker is willing to take either of the following two bets:

Bet 1	Win $X if the Lakers win.
	Lose $Y if the Lakers lose.
Bet 2	Lose $X if the Lakers win.
	Win $Y if the Lakers lose.

Figure 8.1

Decision-tree representation for assessing subjective probability via the betting method. The assessor's problem is to find X and Y so that he or she is indifferent about betting for or against the Lakers.

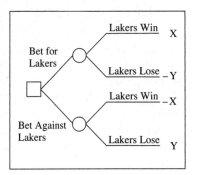

In these bets, X and Y can be thought of as the amounts that each person puts into the "pot." The winner of the bet takes all of the money therein. Bets 1 and 2 are symmetric in the sense that they are opposite sides of the same bet. Figure 8.1 displays the decision tree that the decision maker faces.

If the decision maker is indifferent between Bets 1 and 2, then in his or her mind their expected values must be equal:

$$X \, P(\text{Lakers Win}) - Y[1 - P(\text{Lakers Win})]$$
$$= -X \, P(\text{Lakers Win}) + Y[1 - P(\text{Lakers Win})]$$

which implies that

$$2\{X \, P(\text{Lakers Win}) - Y[1 - P(\text{Lakers Win})]\} = 0$$

We can divide through by 2 and expand the left-hand side to get

$$X \, P(\text{Lakers Win}) - Y + Y \, P(\text{Lakers Win}) = 0$$

Collecting terms gives

$$(X + Y) \, P(\text{Lakers Win}) - Y = 0$$

which reduces to

$$P(\text{Lakers Win}) = \frac{Y}{X + Y}$$

For example, a friend of yours might be willing to take either side of the following bet:

Win $2.50 if the Lakers win.

Lose $3.80 if the Lakers lose.

His subjective probability that the Lakers win, as implied by his betting behavior, is $3.80/(2.50 + 3.80) = 0.603$.

Finding the bet for which a decision maker would be willing to take either side is fairly straightforward. Begin by offering a bet that is highly favorable to one side or the other, and note which side of the bet she would take. Then offer a bet that favors the opposite side, and ask which side of this new bet she would prefer. Continue offering bets that first favor one side and then the other, gradually adjusting the payoffs on each round. By adjusting the bet appropriately, making it more or less attractive depending on the response to the previous bet, the indifference point can be found.

The betting approach to assessing probabilities appears straightforward enough, but it does suffer from a number of problems. First, many people simply do not like the idea of betting (even though most investments can be framed as a bet of some kind). For these people, casting the judgment task as a bet can be distracting. Most people also dislike the prospect of losing money; they are *risk-averse*. Thus, the bets that are considered must be for small enough amounts of money that the issue of risk aversion does not arise, but some people may be risk-averse even for very small amounts. Finally, the betting approach also presumes that the individual making the bet cannot make any other bets on the specific event (or even related events). That is, the individual cannot protect himself or herself from losses by "hedging" one bet with another.

To get around the problems with direct assessment or with the betting approach, a third approach adopts a thought-experiment strategy in which the decision maker compares two lotterylike games, each of which can result in a Prize (A or B). For convenience, set it up so that the decision maker prefers A to B. (Prize A might be a fully paid two-week vacation in Hawaii, and Prize B a coupon for a free beer.) We would ask the decision maker to compare the lottery

Win Prize A if the Lakers win.

Win Prize B if the Lakers lose.

with the lottery

Win Prize A with known probability p.

Win Prize B with probability $1 - p$.

The decision-tree representation is shown in Figure 8.2. The second lottery is called the *reference lottery,* for which the probability mechanism must be well specified. A typical mechanism is drawing a colored ball from an urn in which the proportion of colored balls is known to be p. Another mechanism is to use a "wheel of fortune" with a known area that represents "win"; if the wheel were spun and the pointer landed in the win area, then the decision maker would win Prize A.

Once the mechanism is understood by the decision maker, the trick is to adjust the probability of winning in the reference lottery until the decision maker is indifferent between the two lotteries. Indifference in this case means that the decision maker has no preference between the two lotteries, but slightly changing probability p makes one or the other lottery clearly preferable. If the decision maker is indifferent, then her subjective probability that the Lakers win must be the p that makes her indifferent.

How do we find the p that makes the decision maker indifferent? The basic idea is to start with some p_1 and ask which lottery she prefers. If she prefers the reference lottery, then p_1 must be too high; she perceives that the chance of winning in the reference

Figure 8.2

*Decision-tree repre-
sentation for assessing
subjective probability
with equivalent-lottery
method.* The assessor's
problem is to find a
value of p so that the
two lotteries are
equivalent.

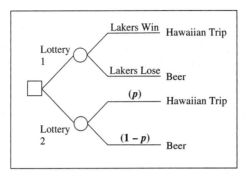

lottery is higher. In this case, choose p_2 less than p_1 and ask her preference again. Continue adjusting the probability in the reference lottery until the indifference point is found. It is important to begin with extremely wide brackets and to converge on the indifference probability slowly. Going slowly allows the decision maker plenty of time to think hard about the assessment, and she probably will be much happier with the final result than she would be if rushed. This is an important point for finding your own probability assessments. Be patient, and hone in on your indifference point gradually.

The wheel of fortune is a particularly useful way to assess probabilities. By changing the setting of the wheel to represent the probability of winning in the reference lottery, it is possible to find the decision maker's indifference point quite easily. Furthermore, the use of the wheel avoids the bias that can occur from using only "even" probabilities (0.1, 0.2, 0.3, and so on). With the wheel, a probability can be any value between 0 and 1. Figure 8.3 shows the probability wheel corresponding to the probability assessments of Texaco's reaction to Pennzoil's $5 billion offer.

The lottery-based approach to probability assessment is not without its own shortcomings. Some people have a difficult time grasping the hypothetical game that they are asked to envision, and as a result they have trouble making assessments. Others dislike the idea of a lottery or carnival-like game. These same people, though, do make trade-offs with their own money whenever they purchase insurance, invest in a small business, or purchase shares of stock in a company. In some cases it may be better to recast the assessment procedure in terms of risks that are similar to the kinds of financial risks an individual might take.

The last step in assessing probabilities is to check for consistency. Many problems will require the decision maker to assess several interrelated probabilities. It is important that these probabilities be consistent among themselves; they should obey the probability laws introduced in Chapter 7. For example, if P(A), P(B | A), and P(A and B) were all assessed, then it should be the case that

$$P(A)\, P(B\,|\,A) = P(A \text{ and } B)$$

If a set of assessed probabilities is found to be inconsistent, then the decision maker should reconsider and modify the assessments as necessary to achieve consistency.

Figure 8.3

The Texaco reaction chance node and the corresponding probability wheel.

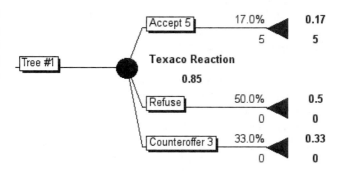

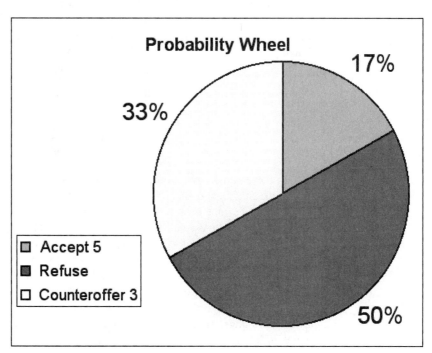

Assessing Continuous Probabilities

The premise of this chapter is that it always is possible to model a decision maker's uncertainty using probabilities. How would this be done in the case of an uncertain but continuous quantity? We already have learned how to assess individual probabilities; we will apply this technique to assess several cumulative probabilities and then use these to plot a rough CDF. We will discuss two strategies for assessing a subjective CDF.

Let us reexamine the example of the movie star's age that we introduced in Chapter 7 (page 266). As you recall, the problem was to derive a probability distribution representing a probability assessor's uncertainty regarding a particular movie star's age. In that example, several probabilities were found, and these were transformed into cumulative probabilities.

A typical cumulative assessment would be to assess $P(\text{Age} \leq a)$, where a is a particular value. For example, consider $P(\text{Age} \leq 46)$. The outcome "Age ≤ 46" is an outcome just like any other, and so a decision maker could assess the probability of this event by using any of the three techniques discussed above. For example, a wheel of fortune might be used as an assessment aid to find the probability p that would make the decision maker indifferent between the two lotteries shown in Figure 8.4.

Using this technique to find a CDF amounts to assessing the cumulative probability for a number of points, plotting them, and drawing a smooth curve through the plotted points. Suppose the following assessments were made:

$$P(\text{Age} \leq 29) = 0.00$$
$$P(\text{Age} \leq 40) = 0.05$$
$$P(\text{Age} \leq 44) = 0.50$$
$$P(\text{Age} \leq 50) = 0.85$$
$$P(\text{Age} \leq 65) = 1.00$$

Plotting these cumulative probabilities would result in the graph that we originally drew in Figure 7.9, and which is reproduced here as Figure 8.5.

The strategy that we have used here is to choose a few values from the horizontal axis (some ages) and then to find the cumulative probabilities that correspond to those ages. This is a perfectly reasonable strategy for assessing a CDF. Another strategy builds up the graph the other way around. That is, we pick a few cumulative probabilities from the vertical axis and find the corresponding ages. For example, suppose we pick probability 0.35. Now we want the number of years $a_{0.35}$ such that $P(\text{Age} \leq a_{0.35}) = 0.35$. The number $a_{0.35}$ is called the 0.35 *fractile* of the distribution. In general, the p fractile of a distribution for X is the value x_p such that $P(X \leq x_p) = p$. We can see from Figure 8.5 that the 0.35 fractile of the distribution is approximately 42 years. We know from the assessments that were

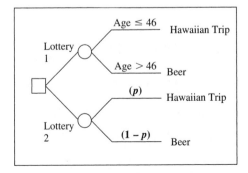

Figure 8.4

Decision-tree representation for assessing $P(\text{Age} \leq 46)$ *in movie star example.*

Figure 8.5

Cumulative distribution function for movie star's age.

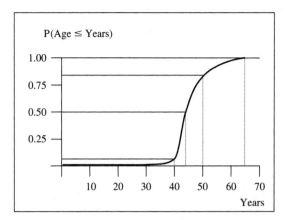

made that the 0.05 fractile is 40 years, the 0.50 fractile is 44 years, and the 0.85 fractile is 50 years.

How could you go about assessing a fractile? Figure 8.6 shows a decision tree that represents the process for assessing the 0.35 fractile. In Lottery B, or the reference lottery, the probability of winning is fixed at 0.35. The assessment task is to adjust the number x in Lottery A until indifference between the lotteries is achieved. Indifference would mean that the probability of winning in Lottery A must be 0.35. Hence, there must be a 0.35 chance that X is less than or equal to the assessed x. By definition, then, x must be the 0.35 fractile of the distribution.

It is important to recognize the difference between Figures 8.4 and 8.6. In Figure 8.4 we adjusted p in the reference lottery to find indifference. To assess the 0.35 fractile in Figure 8.6, we fix the probability in the reference lottery at 0.35, and we adjust x in the upper lottery.

The term *fractile* is a general one, but other similar terms are useful for referring to specific fractiles. The idea of a *median* may be familiar. If we can find an amount such that the uncertain quantity is as likely to be above as below that amount, then we have found the median. The median is defined as the 0.50 fractile; the median for the movie star's age is 44 years. We also can speak of *quartiles*. The first quartile is

Figure 8.6

Decision tree for assessing the 0.35 fractile of a continuous distribution for X. The decision maker's task is to find x in Lottery A that results in indifference between the two lotteries.

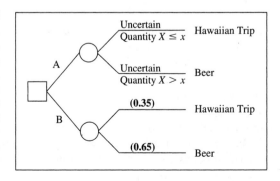

Figure 8.7

Decision tree for assessing the median of the distribution for the movie star's age. The assessment task is to adjust the number of years a in Lottery A to achieve indifference.

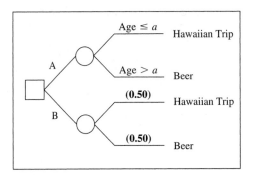

an amount such that P($X \leq$ first quartile) = 0.25, or the 0.25 fractile. In our example, the first quartile appears to be around 42 years. Likewise, the third quartile is defined as the 0.75 fractile. The third quartile of our example is approximately 48 years because P(Age $\leq$ 48) is approximately 0.75. The second quartile is, of course, the median. Fractiles also can be expressed as *percentiles*. For example, the 90th percentile is defined as the 0.90 fractile of a distribution.

As mentioned above, we can exploit this idea of fractiles so as to assess a continuous distribution. In general, the strategy will be to select specific cumulative probabilities and assess the corresponding fractiles. The first step might be to find the uncertain quantity's extreme values. How small or large could this quantity be? Because it often is difficult, and sometimes even misleading, to think in terms of probabilities of 0 or 1, we take the 0.05 and 0.95 fractiles (or the fifth and 95th percentiles). In our movie star example, the 0.05 fractile is a value $a_{0.05}$ such that there is only a 5% chance that the movie star's age would be less than or equal to $a_{0.05}$. Likewise, the 0.95 fractile is the value $a_{0.95}$ such that there is a 95% chance that the age would be less than or equal to $a_{0.95}$. Informally, we might think of these as the smallest and largest values that the uncertain quantity could reasonably assume. Anything beyond these values would be quite surprising.

After assessing the extreme points, the median might be assessed. For example, Figure 8.7 shows the decision tree that corresponds to the assessment for the median in our example. The task would be for the decision maker to find an age a that makes the two lotteries equivalent. The value of a that leaves the decision maker indifferent is the median of the distribution and can be plotted as a point on the decision maker's subjective CDF.

Next, assess the first and third quartiles. These assessments can be made using a lottery setup similar to that in Figure 8.6. Another way to think of the quartiles is that they "split" the probability intervals above and below the median. For example, the first quartile is a value x such that the uncertain quantity is just as likely to fall below x as between x and the median.

To do this, the decision maker must find a point such that the uncertain quantity is equally likely to fall above or below this point. Having assessed the extreme points, the median, and the quartiles, we have five points on the cumulative distribution function. These points can be plotted on a graph, and a smooth curve drawn through the points.

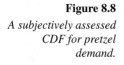

Figure 8.8

A subjectively assessed CDF for pretzel demand.

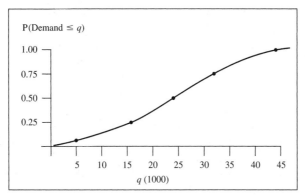

As an example, suppose you have developed a new soft pretzel that you are thinking about marketing through sidewalk kiosks. You are interested in assessing the annual demand for the pretzels as a continuous quantity. You might make the following assessments:

- 0.05 fractile for demand = 5000.
- 0.95 fractile for demand = 45,000.
- Demand is just as likely to be above 23,000 as below or equal to 23,000.
- There is a 0.25 chance that demand will be below 16,000.
- There is a 0.75 chance that demand will be below 31,000.

The last three assessments establish the median to be 23,000, the first quartile to be 16,000, and the third quartile to be 31,000. Plotting the points, we obtain the graph in Figure 8.8. A smooth curve drawn through the five points represents your subjective cumulative probability distribution of demand for the new pretzels.

Later in the chapter, we will learn how to use the program RISKview to draw smooth distribution curves that pass through the subjective assessments. RISKview is relatively simple to master because you need only type in the assessed probabilities and their corresponding values, and the program does the rest. It draws a distribution curve through the points; it calculates the expected value, the standard deviation, and the various other statistics; it provides a simple method for calculating any probability value; and it provides various ways to reshape the distribution, including smoothing out the curve. Because RISKview provides you with both the visual graph and the numerical summaries of the fitted distribution, you can use this information to check the consistency and accuracy of the assessments. For example, you can report the mean or a few of the implied fractiles of the fitted distribution to your expert. If the expert does not agree, then you can easily modify the fitted distribution. Once you are satisfied with the fitted distribution, RISKview will export the formula to Excel, so that it can be used to model your uncertainty. Essentially, RISKview allows you to view distributions from various perspectives so that you may better understand the uncertainty or risk that you are modeling.

Once we assess a continuous distribution, how can we use it? Apparently, our motivation for assessing it in the first place was that we faced uncertainty in the form of a range-of-risk dilemma. In our example, we may be deciding whether to go ahead and market the pretzels. We need the probability distribution to fill out our decision tree, calculate expected values, and find an optimum choice. But how will we calculate an expected value for this subjectively assessed distribution? There are several possibilities, some of which we will explore in later chapters. For example, in the next chapter, we will see how to fit a theoretical distribution to an assessed one; we then can use the mathematical properties of the theoretical distribution. Advanced simulation and numerical integration techniques also are possible. At this point, however, we will content ourselves with some simple and useful approximation techniques.

The easiest way to use a continuous distribution in a decision tree or influence diagram is to approximate it with a discrete distribution. The basic idea is to find a few representative points in the distribution and then to assign those points specific probability values. A particularly simple approach, from Keefer and Bodily (1983), is called the *extended Pearson-Tukey method.* This three-point approximation uses the median and the 0.05 and 0.95 fractiles. Thus, one primary advantage of this method is that we can use assessments that already have been made. In assigning probabilities, the median gets probability 0.63, and the 0.05 and 0.95 fractiles each have probability 0.185. (These probabilities do not appear to admit any obvious interpretation. They are used because the resulting approximation is reasonably accurate for a wide variety of distributions.) For the pretzel-demand example, we can create the three-point discrete approximation as in Figure 8.9. The extended Pearson-Tukey method works best for approximating symmetric distributions. Given its simplicity, however, it also works surprisingly well for asymmetric distributions.

A slightly more complex approximation technique is to find *bracket medians.* Suppose that we consider an interval in which an uncertain quantity could fall: $a \leq X \leq b$. The bracket median is a value m^* between a and b such that $P(a \leq X \leq m^*) = P(m^* \leq X \leq b)$. Figure 8.10 shows how a bracket median relates to the underlying CDF. The bracket median divides the probability of the original interval in half and is associated with a cumulative probability halfway between the cumulative probabilities for a and b.

To use bracket medians, the typical approach is to break the subjective probability distribution into several equally likely intervals and then to assess the bracket median for each interval. In practice, three, four, or five intervals are typically used; the more intervals used, the better the approximation. With five intervals, the assessor would use the extreme points and the 0.20, 0.40, 0.60, and 0.80 fractiles. These would correspond to cumulative probabilities as follows:

$$P(X \leq x_{0.0}) = 0.00$$
$$P(X \leq x_{0.2}) = 0.20$$
$$P(X \leq x_{0.4}) = 0.40$$
$$P(X \leq x_{0.6}) = 0.60$$

Figure 8.9

*Replacing a continu-
ous distribution with a
three-branch discrete
uncertainty node in a
decision tree.*

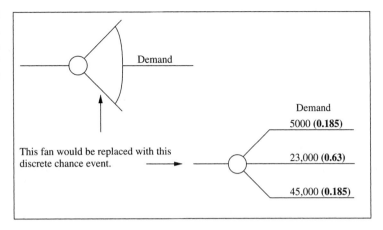

Figure 8.10

*Finding the bracket me-
dian for the interval be-
tween a and b. The cu-
mulative probabilities p
and q correspond to a
and b, respectively.
Bracket median m* is
associated with a cu-
mulative probability
that is halfway between
p and q.*

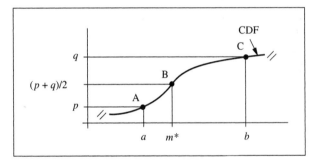

$$P(X \le x_{0.8}) = 0.80$$
$$P(X \le x_{1.0}) = 1.00$$

Because these six points define five intervals, we need five bracket medians $(m_1 - m_5)$ with the five cumulative probabilities

$$P(X \le m_1) = 0.10$$
$$P(X \le m_2) = 0.30$$
$$P(X \le m_3) = 0.50$$
$$P(X \le m_4) = 0.70$$
$$P(X \le m_5) = 0.90$$

These bracket medians can be assessed either by using the general assessment technique for assessing fractiles above or by assessing the value m_i so that X is just as likely
to be between the lower interval bound and m_i as between m_i and the upper bound.

Figure 8.11

Finding bracket medians for the pretzel-demand distribution.

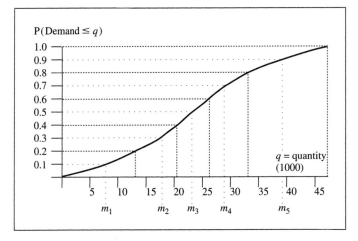

Figure 8.11 shows the determination of five bracket medians for the pretzel-demand distribution. These bracket medians then can be used as "representative points" for their respective brackets. Thus, the discrete approximation to the original distribution would have

$$P(X = m_1 = 8) \ = 0.20$$
$$P(X = m_2 = 18) = 0.20$$
$$P(X = m_3 = 23) = 0.20$$
$$P(X = m_4 = 29) = 0.20$$
$$P(X = m_5 = 39) = 0.20$$

as shown in Figure 8.12.

Both methods that we have described for creating a discrete approximation are straightforward and work reasonably well. The extended Pearson-Tukey method has

Figure 8.12

Replacing a continuous distribution with bracket medians in a decision tree.

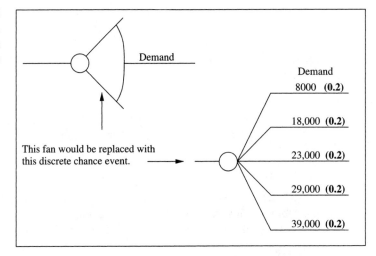

the advantage of requiring no additional assessments. The advantages of the bracket-median approach are its ability to approximate virtually any kind of distribution and the intuitive nature of the bracket-median assessments.

Which method should you use? Both work well, but a general strategy exists that dovetails perfectly with the sensitivity-analysis approach we described in Chapter 5. Suppose that you have constructed a tornado diagram for a decision problem and have identified several variables that are candidates for probabilistic modeling. A simple first step would be to use the extended Pearson-Tukey method with the base value as the median, and the upper and lower values as the 0.95 and 0.05 fractiles, respectively. In many cases, this may be an adequate model of the uncertainty. If not, then it may be worthwhile to use bracket medians to construct a more complete model of the uncertainty. (Additional information on discrete approximations is given in the reference section at the end of the chapter.)

Pitfalls: Heuristics and Biases

The methods presented above make probability assessment sound easy. As you probably realize, however, thinking in terms of probabilities is not easy. It takes considerable practice before one is comfortable making probability assessments. Even then, we tend to use rather primitive cognitive techniques to make our probability assessments. Tversky and Kahneman (1974) have labeled these techniques *heuristics*. In general, heuristics can be thought of as rules of thumb for accomplishing tasks. For example, an inventory-control heuristic might be to keep 10% of total yearly demand for any given item on hand. When placing an order for an item that sells some 200 units per year, one then would check the stock and order enough to have 20 units on hand. Heuristics tend to be simple, are easy to perform, and usually do not give optimal answers.

Heuristics for assessing probabilities operate in basically the same way. They are easy and intuitive ways to deal with uncertain situations, but they tend to result in probability assessments that are biased in different ways depending on the heuristic used. In this section we will look at the various heuristics and the biases they can create. Before we begin, however, consider the case of Tom W.

TOM W.

"Tom W. is of high intelligence, although lacking in true creativity. He has a need for order and clarity, and for neat and tidy systems in which every detail finds its appropriate place. His writing is rather dull and mechanical, occasionally enlivened by somewhat corny puns and by flashes of imagination of the sci-fi type. He has a strong drive for competence. He seems to have little feel and little sympathy for other people and does not enjoy interacting with others. Self-centered, he nonetheless has a deep moral sense."

The preceding personality sketch of Tom W. was written during his senior year in high school by a psychologist on the basis of projective tests. Tom W. is now a graduate student. Please rank the following nine fields of graduate specialization in order of the likelihood that Tom W. is now a graduate student in each of these fields:

a Business administration

b Computer science

c Engineering

d Humanities and education

e Law

f Library science

g Medicine

h Physical and life sciences

i Social science and social work

Write down your rankings before you read on. [*Source:* Kahneman and Tversky (1973).]

Representativeness

If you are like most people, you wrote down your ranks on the basis of how similar the description of Tom W. is to your preconceived notions of the kinds of people in the nine different fields of study. Specifically, Tom W.'s description makes him appear to be a "nerd," and so most people think that he has a relatively high chance of being in engineering or computer science. But judging the probability of membership in a group on the basis of similarity ignores important information. There are many more graduate students in humanities and education and in social science and social work than in computer science or engineering. Information relating to the incidence or *base rate* of occurrence in the different fields is ignored, however, when we make probability judgments on the basis of similarity.

Making such judgments on similarity is one example of a kind of heuristic that Kahneman and Tversky (1973) call *representativeness*. In its most fundamental form, the representativeness heuristic is used to judge the probability that someone or something belongs to a particular category. Using the representativeness heuristic means that the judgment is made by comparing the information known about the person or thing with the stereotypical member of the category. The closer the similarity between the two, the higher the judged probability of membership in the category.

The representativeness heuristic surfaces in many different situations and can lead to a variety of different biases. For example, as in the Tom W. problem, people can be insensitive to base rates or prior probabilities. If one were to consider base rates carefully in the Tom W. problem, humanities and education may well be the most likely category. Insensitivity to sample size is another possible result of the representativeness heuristic. Sometimes termed the *law of small numbers,* people (even scientists!) sometimes draw conclusions from highly representative small samples even though small samples are subject to considerably more statistical error than are large samples. Other situations in which representativeness operates include relying on old and unreliable information to make predictions, and making equally precise predictions regardless of the inherent uncertainty in a situation.

Misunderstanding of random processes is another phenomenon attributed to the representativeness heuristic. One of the most important aspects of this situation for managers relates to changes over time and misunderstanding the extent to which a process can be controlled. For example, Kahneman and Tversky relate their experience with a flight instructor in Israel. The instructor had been in the habit of praising students for good landings and scolding them for poor ones. He observed that after receiving praise for a good landing, a pilot's subsequent landing tended to be worse. Conversely, after a pilot received a scolding, his next landing tended to be better. The instructor concluded that scolding was effective feedback and that praise was not. In fact, this phenomenon is more easily explained by what is known as the statistical phenomenon *regression toward the mean.* If performance or measurements are random, then extreme cases will tend to be followed by less extreme ones. Landing a jet is not an easy task, and the pilot must deal with many different problems and conditions on each landing. It is perfectly reasonable to assume that performance for any pilot will vary from one landing to the next. Regression toward the mean suggests that a good landing probably will be followed by one that is not as good, and that a poor one will most likely be followed by one that is better.

Availability

Another heuristic that people tend to use to make probability judgments is termed *availability.* According to this heuristic, we judge the probability that an event will occur according to the ease with which we can retrieve similar events from memory. As with the representativeness heuristic, availability comes into play in several ways. External events and influences, for example, can have a substantial effect on the availability of similar incidents. Seeing a traffic accident can increase one's estimate of the chance of being in an accident. Being present at a house fire can have more effect on the retrievability of fire incidents than reading about the fire in the newspaper. Furthermore, differential attention by the news media to different kinds of incidents can result in availability bias. Suppose the local newspaper plays up deaths resulting from homicide but plays down traffic deaths. To some extent, the unbalanced reporting can affect readers' judgments of the relative incidence of homicides and traffic fatalities, thus affecting the community's overall perception.

Bias from availability arises in other ways as well. For example, some situations are simply easier to imagine than others. In other cases, it may be difficult to recall contexts in which a particular event occurs. Another situation involves *illusory correlation.* If a pair of events is perceived as happening together frequently, this perception can lead to an incorrect judgment regarding the strength of the relationship between the two events.

Anchoring and Adjusting

This heuristic refers to the notion that in making estimates we often choose an initial anchor and then adjust that anchor based on our knowledge of the specific event in question. An excellent example is sales forecasting. Many people make such forecasts by considering the sales figures for the most recent period and then adjusting those values based on new circumstances. The problem is that the adjustment usually is insufficient.

The anchor-and-adjust heuristic affects the assessment of probability distributions for continuous uncertain quantities more than it affects discrete assessments. We tend to begin by assessing an anchor, say, the mean or median. Then extreme points or fractiles are assessed by adjusting away from the anchor. Because of the tendency to underadjust, most subjectively assessed probability distributions are too narrow, inadequately reflecting the inherent variability in the uncertain quantity. One of the consequences is that we tend to be overconfident, having underestimated the probability of extreme outcomes, often by a substantial amount (Capen 1976). The technique described above for assessing a subjective CDF using the median and quartiles is subject to this kind of overconfidence from anchoring and adjusting. This is one reason why we assess the 0.05 and 0.95 fractiles instead of the 0.00 and 1.00 fractiles. Assessing the 0.05 and 0.95 fractiles essentially admits that there is a remote possibility that the uncertain quantity could fall beyond these assessed points. Also, we assess the 0.05 and 0.95 fractiles before the median to reduce the tendency to anchor on a central value.

Motivational Bias

The cognitive biases described above relate to the ways in which we as human beings process information. But we also must be aware of motivational biases. Incentives often exist that lead people to report probabilities or forecasts that do not entirely reflect their true beliefs. For example, a salesperson asked for a sales forecast may be inclined to forecast low so that he will look good (and perhaps receive a bonus) when he sells more than the amount forecasted. Occasionally incentives can be quite subtle or even operate at a subconscious level. For example, some evidence suggests that weather forecasters, in assessing the probability of precipitation, persistently err on the high side; they tend to overstate the probability of rain. Perhaps they would rather people were prepared for bad weather (and were pleasantly surprised by sunshine) instead of expecting good weather and being unpleasantly surprised. Even though forecasters generally are good probability assessors and strive for accurate forecasts, their assessments may indeed be slightly affected by such implicit incentives.

Heuristics and Biases: Implications

This discussion of heuristics and biases in probability judgments sounds quite pessimistic. If people really are subject to such deficiencies in assessing probabilities, is there any hope? There is indeed. First, some evidence suggests that individuals can learn to become good at assessing probabilities. As mentioned, weather forecasters are good probability assessors; in general, they provide accurate probabilities. For example, on those occasions when a forecaster says the probability of rain is 0.20, rain actually occurs very nearly 20% (or slightly less) of the time. Weather forecasters have three advantages; they have a lot of specialized knowledge about the weather, they make many forecasts, and they receive immediate feedback regarding the outcome. All of these appear to be important in improving probability-assessment performance.

Second, awareness of the heuristics and biases may help individuals make better probability assessments. If nothing else, knowing about some of the effects, you now may be able to recognize them when they occur. For example, you may be able to recognize regression toward the mean, or you may be sensitive to availability effects that result from unbalanced reporting in the news media. Moreover, when you obtain information from other individuals, you should realize that their judgments are subject to these same problems.

Third, the techniques we have discussed for assessing probabilities involve thinking about lotteries and chances in a structured way. These contexts are quite different from the way that most people think about uncertainty. By thinking hard about probabilities using these methods, it may be possible to avoid some heuristic reasoning and attendant biases. At the very least, thinking about lotteries provides a new perspective in the assessment process.

Finally, some problems simply cannot be addressed well in the form in which they are presented. In many cases it is worthwhile to decompose a chance event into other events. The result is that more assessments must be made, although they may be easier. In the next section we will see how decomposition may improve the assessment process.

Decomposition and Probability Assessment

In many cases it is possible to break a probability assessment into smaller and more manageable chunks. This process is known as *decomposition*. There are at least three different scenarios in which decomposition of a probability assessment may be appropriate. In this section, we will discuss these different scenarios.

In the simplest case, decomposition involves thinking about how the event of interest is related to other events. A simple example might involve assessing the probability that a given stock price increases. Instead of considering only the stock itself, we might think about its relationship to the market as a whole. We could assess the probability that the market goes up (as measured by the Dow Jones average, say), and then assess the conditional probabilities that the stock price increases given that the market increases and given that the market does not increase. Finding the probability that the stock price increases is then a matter of using the law of total probability:

P(Stock Price Up) = P(Stock Price Up | Market Up) P(Market Up)

+ P(Stock Price Up | Market Not Up) P(Market Not Up)

The reason for performing the assessment in this way is that it may be more comfortable to assess the conditional probabilities and the probability about the market rather than to assess P(Stock Price Up) directly. In terms of an influence diagram or a probability tree, we are adding a chance node that is relevant to the assessment of the probabilities in which we are interested. Figure 8.13 shows the decomposition of the stock-price assessment.

In the second scenario, it is a matter of thinking about what kinds of uncertain outcomes could eventually lead to the outcome in question. For example, if your car will not start, there are many possible reasons why it will not. The decomposition strategy would be to think about the chances that different possible things could go wrong and the chance that the car will not start given each of these specific underlying problems or some combination of them.

For a more complicated example that we can model, suppose that you are an engineer in a nuclear power plant. Your boss calls you into his office and explains that the Nuclear Regulatory Commission has requested safety information. One item that the commission has requested is an assessment of the probability of an accident resulting in the release of radioactive material into the environment. Your boss knows that you have had a course in decision analysis, and so you are given the job of assessing this probability.

How would you go about this task? Of course, one way is to sit down with a wheel of fortune and think about lotteries. Eventually you would be able to arrive at a probability assessment. Chances are that as you thought about the problem, however, you would realize that many different kinds of situations could lead to an accident. Thus, instead of trying to assess the probability directly, you might construct an influence diagram that includes some of the outcomes that could lead to an accident. Figure 8.14 shows the simple influence diagram that you might draw.

Figure 8.13

Decomposing the probability assessment for stock-price movement.

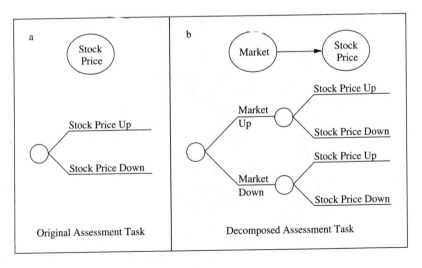

Figure 8.14

Simple influence diagram for assessing the probability of a nuclear power plant accident.

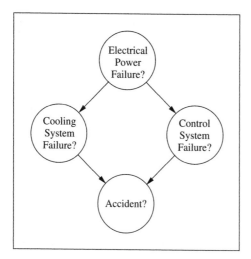

The intuition behind Figure 8.14 is that an accident could result from a failure of the cooling system or of the control system. The cooling system could either spring a leak itself, thus spilling radioactive material, or its pumps could fail, allowing the reactor core to overheat, and thus resulting in a possible accident. The control system also is critical. If the control system fails, it may become impossible to maintain safe operation of the reactor, and thus possibly result in an accident. Furthermore, the control system and the cooling system do not operate independently; they both depend on the electrical power system within the plant. Thus, failure of the electrical system would make failure of both the cooling and the control systems more likely. (Many other relationships also are possible. An influence diagram to assess the probability of an accident in an actual nuclear power plant would be considerably more complicated than Figure 8.14.)

For each of the four chance nodes in Figure 8.14, we have two possible outcomes. Because failures of both the cooling and control systems are relevant to the assessment of an accident, we have four conditional probabilities to assess. Let A denote the outcome of an accident, L the outcome of a cooling system failure, N the outcome of a control system failure, and E the outcome of an electrical system failure. The four conditional probabilities we must assess for Outcome A are $P(A \mid L, N)$, $P(A \mid \overline{L}, N)$, $P(A \mid L, \overline{N})$, and $P(A \mid \overline{L}, \overline{N})$. For the cooling system node, probabilities $P(L \mid E)$ and $P(L \mid \overline{E})$ must be assessed. Likewise, for the control system node $P(N \mid E)$ and $P(N \mid \overline{E})$ must be assessed. Finally, $P(E)$ must be assessed for the electrical system node.

There are nine assessments in all. Again, the reason for decomposing the assessment task into multiple assessments is that you may be more comfortable with the assessments that are required in the decomposed version. For example, you may be able to conclude that $P(A \mid \overline{L}, \overline{N}) = 0$, or that if neither the cooling system nor the control system fails, then the probability of an accident is zero.

Assembling the probabilities in this case again is a matter of using the law of total probability, although it must be used more than once. Start out by using the law of total probability to expand $P(A)$:

$$P(A) = P(A \mid L, N)P(L, N) + P(A \mid \overline{L}, N)P(\overline{L}, N)$$
$$+ P(A \mid L, \overline{N}) P(L, \overline{N}) + P(A \mid \overline{L}, \overline{N})P(\overline{L}, \overline{N})$$

Now the problem is to find $P(L, N)$, $P(\overline{L}, N)$, $P(L, \overline{N})$, and $P(\overline{L}, \overline{N})$. Each in turn can be expanded using the law of total probability. For example, consider $P(L, N)$:

$$P(L, N) = P(L, N \mid E)P(E) + P(L, N \mid \overline{E})P(\overline{E})$$

Now we must find $P(L, N \mid E)$ and $P(L, N \mid \overline{E})$. From the influence diagram, the only connection between the cooling system (L) and the control system (N) is through the electrical system. Thus, cooling and control failures are conditionally independent given the state of the power system. From the definition of conditional independence in Chapter 7, we can write

$$P(L, N \mid E) = P(L \mid E) \, P(N \mid E)$$

and

$$P(L, N \mid \overline{E}) = P(L \mid \overline{E}) \, P(N \mid \overline{E})$$

Thus, by expanding out the probabilities, it is possible to build up the probability $P(A)$ from the nine assessments and their complements.

The third scenario is related to the second. In this case, however, it is not a matter of different possible underlying causes but a matter of thinking through all of the different events that must happen before the outcome in question occurs. For example, in assessing the probability of an explosion at an oil refinery, an engineer would have to consider the chances that perhaps some critical pipe would fail, that all of the different safety measures also would fail at the same time, and that no one would notice the problem before the explosion occurred. Thus, many different individual outcomes would have to occur before the explosion. In contrast, the nuclear power plant example involved alternative paths that could lead to a failure. Of course, the second and third scenarios can be combined. That is, there may be alternative paths to a failure, each requiring that certain individual outcomes occur. This kind of analysis often is called *fault-tree analysis* because it is possible to build a tree showing the relationship of prior outcomes to the outcome in question, which often is the failure of some complicated system.

As you may have noticed in the nuclear power plant example, the probability manipulations can become somewhat complicated. Fortunately, in complicated assessment problems, computer programs can perform the probability manipulations for us. Using such a program allows us to focus on thinking hard about the assessments rather than on the mathematics.

Figure 8.15

Influence diagram for a decision analysis of alternative sites for a nuclear-waste repository.
Source: Merkhofer (1987b), p. 105. Reprinted by permission of Kluwer Academic Publishers.

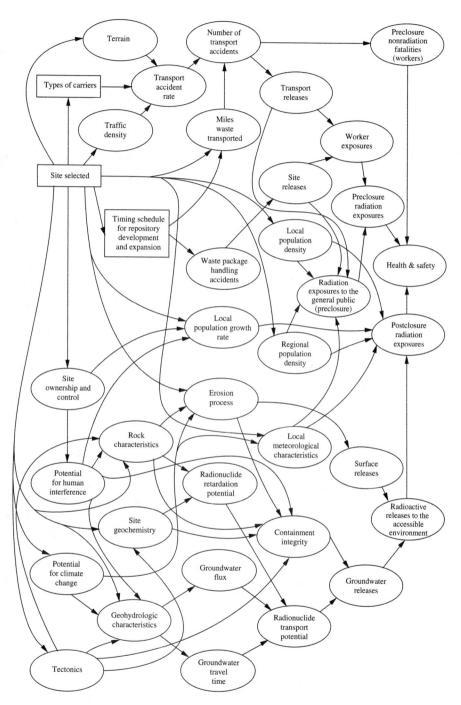

Figure 8.16

Influence diagram (or belief network) for forecasting crude oil prices. Rounded rectangles are used instead of ovals to represent chance nodes. Nodes with bold outlines represent entire submodels. *Source*: Abramson and Finizza (1995).

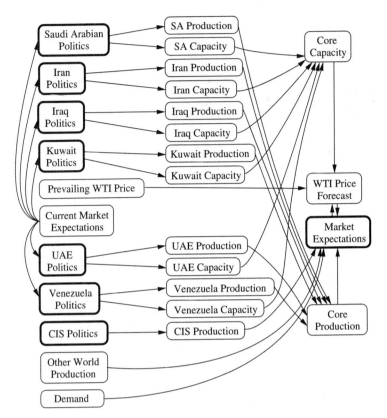

As with many decision-analysis techniques, there may be more than one way to decompose a probability assessment. The whole reason to use decomposition is to make the assessment process easier. The best decomposition to use is the one that is easiest to think about and that gives the clearest view of the uncertainty in the decision problem.

As indicated above, decomposition in decision-analysis assessment of probabilities is important for many reasons. Perhaps most important, though, is that it permits the development of large and complex models of uncertainty. Two examples are given in Figures 8.15, 8.16, and 8.17. The influence diagram in Figure 8.15 was constructed as part of an analysis of alternative sites for a nuclear-waste repository in the United States (Merkhofer 1987b). Figures 8.16 and 8.17 show parts of the influence diagram for a probabilistic model for forecasting crude oil prices (Abramson and Finizza, 1995). In Figures 8.16 and 8.17 the authors have used rounded rectangles to represent chance nodes. In addition, each of the nodes with bold outlines in Figure 8.16 actually represents an entire submodel. For example, Figure 8.17 shows the generic structure for each of the "Politics" nodes. In all, this model of the oil market includes over 150 chance nodes.

Figure 8.17

Details of the "Politics" submodels in Figure 8.16.
Source: Abramson and Finizza (1995).

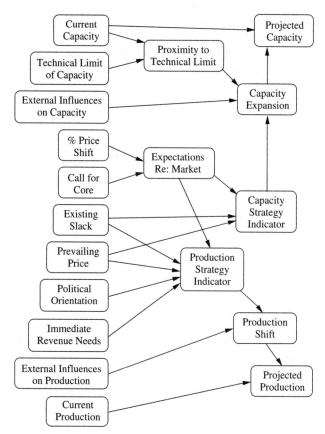

Experts and Probability Assessment: Pulling It All Together

Our discussion thus far has taken a conventional decision-analysis approach; the decision maker makes the probability assessments that are required in the decision model. In practice, though, decisions are often quite complex, and the decision maker must rely on experts to provide information—in the form of probability assessments—regarding crucial uncertainties. For example, all of the probability assessments required in the influence diagrams in Figures 8.15, 8.16, and 8.17 were provided by experts regarding nuclear-waste technology and the world oil market, respectively.

In such complex problems, expert risk assessment plays a major role in the decision-making process. As such, the process by which the expert information was acquired must stand up to professional scrutiny, and thus policy makers who acquire and use expert information must be able to document the assessment process. In general, this is no different from standard scientific principles of data collection; scientists who run experiments or conduct surveys are expected to adhere to standards that ensure the scientific validity of their data. Adherence to these standards is necessary if the conclusions drawn on the basis of the data are to have any validity.

The assessment of expert judgments must also adhere to standards. Not surprisingly, though, the standards for experts are quite different from those for data collection; expert judgments are not as well behaved as data collected in a laboratory. There are important parallels between the two situations, however. For example, the definition of an expert in any given situation is not always without controversy. Thus, the policy maker must be able to document and justify the expert-selection process, just as the data-collecting scientist must be able to document and justify the process by which specific data points were selected. Also, as we have mentioned, experts can be subject to numerous biases. Thus, policy makers must be able to show that the environment in which the judgments were made did as much as possible to reduce or avoid these biases. The counterpart in data collection is that the scientist must be able to show that measurements were taken without bias. If judgments from multiple experts are combined to obtain a single probability distribution, then issues of relative expertise and redundancy among the experts must be taken into account. The corresponding situation in data collection occurs when a scientist combines multiple data sets or uses results from multiple studies to draw conclusions.

Over the past 20 years, as the use of expert information has grown in importance, procedures have been developed for acquiring expert probability assessments. In general, the approach requires the creation of a *protocol* for expert assessment that satisfies the need for professional scrutiny. Thorough discussions of protocol development are found in Merkhofer (1987a) and Morgan and Henrion (1990). Although procedures vary (for example, Morgan and Henrion describe three different approaches), every assessment protocol should include the following steps.

1 Background   The first step is to identify those variables for which expert assessment is needed. Although this sounds obvious, it is an important first step. Relevant scientific literature should be searched to determine the extent of scientific knowledge. The objectives of stakeholders should be examined to be sure that information is being obtained about pertinent concerns. For example, if stake holders in an environmental-management situation care about habitat for endangered species, ecosystem viability, extraction of natural resources for economic purposes, recreation opportunities, and saving wilderness for future generations, then information on all five of these issues must be acquired. Some may require expert assessment, while others may be better addressed by conventional scientific studies. In many cases, experts may be required to make issue-specific probabilistic judgments based on state-of-the-art (but more general) scientific knowledge.

2 Identification and Recruitment of Experts   Identification of appropriate experts can range from straightforward to very difficult. Often an organization can find in-house expertise. In some cases, experts must be recruited externally. Recommendations by peers (for example, through professional associations) can help to justify the selection of specific experts.

3 Motivating Experts   Experts often are leery of the probability-assessment process. Typically they are scientists themselves and prefer to rely on the process of science to generate knowledge. Their opinions may or may not be "correct," and hence they hesitate to express those opinions. The fact remains, though, that a decision must be made with the limited information available, and the choice was made in Step 1 that expert opinion is the appropriate way to obtain that information. Thus, it is important to establish rapport with the experts and to engender their enthusiasm for the project.

4 Structuring and Decomposition   This step might be called *knowledge exploration*. This step identifies specific variables for which judgments are needed and explores the experts' understanding of causal and statistical relationships among the relevant variables. The objective is to develop a general model (expressed, for example, as an influence diagram) that reflects the experts' thinking about the relationships among the variables. The resulting model may be an elaborate decomposition of the original problem, showing which probability distributions must be assessed conditional on other variables. The model thus gives an indication of the order in which the probability assessments must be made.

5 Probability-Assessment Training   Because many experts do not have specific training in probability assessment, it is important to explain the principles of assessment, to provide information on the inherent biases in the process and ways to counteract those biases, and to give the experts an opportunity to practice making probability assessments.

6 Probability Elicitation and Verification   In this step the experts make the required probability assessments, typically under the guidance of an individual trained in the probability-elicitation process. The expert's assessments are checked to be sure they are consistent (probabilities sum to 1, conditional probabilities are consistent with marginal and joint probability assessments, and so on). As part of this process, an expert may provide detailed chains of reasoning for the assessments. Doing so can help to establish a clear rationale for specific aspects of the assessed distributions (e.g., a variable's extreme values or particular dependence relationships). At the same time, encouraging a thorough examination of the expert's knowledge base can help to counteract the biases associated with the psychological heuristics of availability, anchoring, and representativeness. Thus, as output this step produces the required probability assessments and a documentation of the reasoning behind the assessments.

7 Aggregation of Experts' Probability Distributions   If multiple experts have assessed probability distributions, then it may be necessary to aggregate their assessed probability distributions into a single distribution for the use of the decision maker. Another reason for aggregation is that a single distribution may be needed in a probabilistic model of a larger system. In general, two approaches are possible. One is to ask the experts themselves to generate a consensus distribution. Doing so may require considerable sharing of information, clarification of individual definitions, and possibly compromise on the parts of individuals. Unfortunately, the necessary interactions can also lead to biases in the consensus opinion. For this reason, several methods have been proposed to control the interaction among the individual experts.

If the experts are unable to reach a consensus, either because of irreconcilable differences or because convening the group is logistically inconvenient, then one can aggregate their individual distributions using a mathematical formula. Simply averaging the distributions is a straightforward (and intuitively appealing) aggregation approach, but it ignores the relative expertise among the experts as well as the extent to which their information is redundant or dependent.

The seven steps described above give a feel for the process of obtaining expert probability assessments. In a full-blown risk assessment, this process can involve dozens of people and may take several months to complete. The following example describes a risk assessment in which expert climatologists provided probability assessments regarding possible future climate changes at the site of the proposed nuclear-waste repository in Nevada.

CLIMATE CHANGE AT YUCCA MOUNTAIN, NEVADA

The U.S. government has proposed construction of a long-term nuclear-waste repository at Yucca Mountain, about 100 miles northwest of Las Vegas, Nevada. Spent fuel rods from nuclear reactors around the nation would be sealed in large steel casks, shipped to Yucca Mountain, and stored in a large cavern carved out of solid rock about 300 meters below the surface. When full, the repository will be sealed, but the nuclear waste will remain dangerous for millennia. Thus, requirements for licensing the facility include showing that the repository can safely contain the radioactive material for 10,000 years.

One of several risks that the repository faces is that the local climate is expected to change over time as the global climate changes. In broad-brush terms, one might expect that increased human activity will lead to some warming but that natural climate cycles will lead to a global cooling or even another ice age. Climatologists have studied general climate trends over long periods of time and have produced global climate-change forecasts. What does the future hold for Yucca Mountain, Nevada, though? One of the attractions of Yucca Mountain is the dry climate; it experiences an annual average precipitation of about 15 centimeters. If the future climate changes enough, groundwater could enter the repository. How much future precipitation is likely at Yucca Mountain over the next 10,000 years?

To address this question, the Center for Nuclear Waste Repository Analyses (CNWRA) undertook a project to obtain the opinions of expert climatologists. The seven steps described above helped to provide a framework for the project, during which the experts assessed subjective probability distributions for a variety of different climatological variables (average annual rainfall, average temperature, amount of cloud cover, and others) at several different points in time over the next 10,000 years.

Although the exceptionally long forecast horizon suggests that forecasting will be difficult if not impossible, the experts found this to be an interesting project, and they were able to develop carefully thought-out rationales for different scenarios and assess probability distributions that reflected their reasoning. In general, the experts agreed that the Yucca Mountain area would warm up slightly in the short term due to atmospheric buildup of greenhouse gases such as carbon dioxide and water vapor, followed by slight cooling as the earth enters a "mini ice age," but they disagreed regarding the extent and persistence of the global-warming effect. This disagreement shows up, for example, in Figure 8.18a, which shows the medians of the experts' CDFs for average temperature traced over time. Some experts think the effect of global warming will be short-lived, followed quickly by global cooling, while others believe it will last longer and be more pronounced. This disagreement is also seen in their assessed CDFs. Figures 8.18b and c, for example, show the experts' CDFs for change in average temperature (°C) and change in average annual precipitation (mm) 3000 years in the future.

Figure 8.18

Some assessments of climate change at Yucca Mountain. (a) Medians of CDFs for average temperature over time. (b) CDFs for change in average temperature 3000 years in future. (c) CDFs for change in average annual precipitation 3000 years in future. *Source*: DeWispelare et al. (1993). *Expert elicitation of future climate in the Yucca Mountain vicinity*, Technical Report CNWRA 93–016. Prepared for Nuclear Regulatory Commission. San Antonio, TX: Southwest Research Institute.

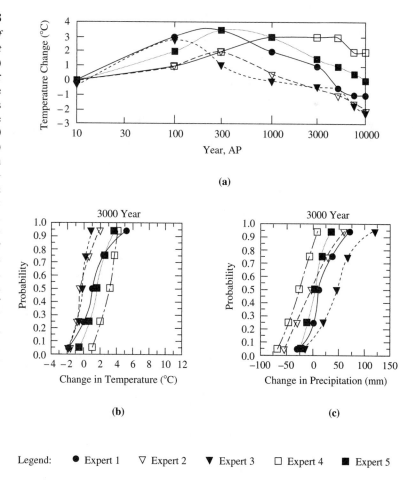

Legend: ● Expert 1   ▽ Expert 2   ▼ Expert 3   □ Expert 4   ■ Expert 5

The Yucca Mountain climate forecast is just one example of a probabilistic risk analysis that assessed expert opinions in the form of probability distributions. Other examples of such analyses involve sales forecasting, research-and-development decisions, analysis of health risks due to ozone, policy making regarding hazardous waste, power plant design analysis, and many others. Risk analyses are performed by both government agencies and private corporations and are becoming more important over time as decision makers, both public and private, face important decisions with substantial uncertainty and imperfect information.

Coherence and the Dutch Book (Optional)

Subjective probabilities must obey the same postulates and laws as so-called objective probabilities. That is, probabilities must be between 0 and 1, probabilities for mutually exclusive and collectively exhaustive sets of events must add to 1, and the probability of any event occurring from among a set of mutually exclusive events must be the sum of the probabilities of the individual events. All of the other properties of probabilities follow from these postulates, and subjective probabilities thus also must obey all of the probability "laws" and formulas that we studied in Chapter 7.

If an individual's subjectively assessed probabilities obey the probability postulates, the person is said to be *coherent*. What benefit is there in being coherent? The mathematician de Finetti proved a famous theorem, the Dutch Book Theorem, which says that if a person is not coherent, then it is possible to set up a Dutch book against him or her. A Dutch book is a series of bets that guarantees your opponent will lose and you will win.

For example, suppose that it is nearly the end of the NBA season, and that the Los Angeles Lakers and the Boston Celtics are playing in the finals. A friend of yours says that the probability is 0.4 that the Lakers will win and 0.5 that the Celtics will win. You look at this statement and point out that his probabilities only add to 0.9. Is there some other possible outcome? He sullenly replies that those are his probabilities, and that there is nothing wrong with them.

Let us find what is wrong. If those really are his probabilities, then he should be willing to agree to the following bets:

> **Bet 1**   He wins $40 if the Lakers lose.
> You win $60 if the Lakers win.

> **Bet 2**   You win $50 if the Celtics win.
> He wins $50 if the Celtics lose.

Note that, according to his stated probabilities, his expected value for each bet is 0. In Bet 1 he has a 0.6 chance of winning $40 and a 0.4 chance of losing $60. His expected value is $0.6(40) - 0.4(60) = 0$. In Bet 2, his expected value is $0.5(50) - 0.5(50) = 0$.

What can happen? If the Lakers win (and hence the Celtics lose), then he pays you $60 in Bet 1 and you pay him $50 in Bet 2; he has a net loss of $10. If the Lakers lose and the Celtics win, you pay him $40 in Bet 1 and he pays you $50 in Bet 2. Again he has a net loss of $10. He is bound to pay you $10 no matter what happens!

He wants to know how you figured this out, so we will show him. First, the bets are determined in the following way. Let p represent his stated probability that the Lakers win and q his probability that the Celtics win. To be coherent, $p + q$ should equal 1. To make up the two bets with zero expected value to him, consider his sides of the bets:

Bet 1   He wins the amount pX with probability $1 - p$
    (that is, if the Lakers lose/Celtics win).
    You win (he loses) the amount $(1 - p) X$ with probability p.

Bet 2   You win the amount $(1 - q) Y$ with probability q
    (that is, if the Celtics win/Lakers lose).
    He wins the amount qY with probability $1 - q$.

As before, we can think of Y and X as the total stakes that are involved in each bet. For example, in Bet 1 you put pX into the pot and he puts in $(1 - p) X$, and whoever wins the bet gets all of the money in the pot. It is easy to verify that his expected value for each bet is zero; for example, in Bet 1 his expected value is

$$(1 - p)(pX) - p[(1 - p) X] = 0$$

Now, think about the equations that give his net gain or loss. If the Lakers win, his net position is L:

$$-(1 - p)X + qY = L$$

The two terms on the left-hand side of the equation are his loss from Bet 1 and his gain from Bet 2. Likewise, if the Celtics win, his position is K:

$$pX - (1 - q)Y = K$$

Now, here's the trick. He has supplied p and q. *You have decided to set L and K each equal to –$10 so that he is guaranteed to pay you $10.* Now it is just a matter of solving for X and Y to find the bets necessary to guarantee that he loses $10. There are two equations and two unknowns (X and Y). What could be easier?

"Aha!" he exclaims, "In my linear algebra course, I learned about solving linear equations simultaneously. No matter what L and K are, you should be able to find an X and Y to defeat me. Unless . . . Oh, I see!" What does he see? Solve the second equation for X to get

$$X = \frac{K + (1 - q)Y}{p}$$

Now substitute this into the first equation and solve for Y. Rearranging yields

$$Y = \frac{L - K + \dfrac{K}{p}}{1 - \dfrac{1 - q}{p}}$$

This equation always can be solved for Y unless the denominator $[1 - (1 - q)/p]$ on the right-hand side happens to equal 0. That will occur only when $p = 1 - q$, or $p + q = 1$. This was exactly the condition he violated in the first place; his probabilities did not sum to 1. If his probabilities had satisfied this condition, you would not be able to find an X and Y that would solve the two equations, and the Dutch book would have been impossible.

This example points out how incoherence can be exploited. In fact, if a person's probabilities do not conform to the probability laws, it is always possible to do this kind of thing, no matter which probability law is violated. The contribution of de Finetti was to prove that it is only possible not to be exploited if subjective probabilities obey the probability postulates and laws. The practical importance of this is not that you can set up Dutch books against incoherent probability assessors—because no one in his or her right mind would agree to such a series of bets—but to provide insight into why subjective probability should work the same way as "long-run frequency" probability. Because no one in his or her right mind would make decisions in a way that could be exploited, coherence is a reasonable condition to guide our assessments of probabilities for decision-making purposes. When we assess probabilities for use in a decision tree or influence diagram, those probabilities should obey all the normal probability properties.

The idea of coherence has an important implication for assessment when a decision analysis involves assessing several probabilities. Once all probabilities have been assessed, it is important to check for coherence. If the assessments are not coherent, the decision maker should be made aware of this and given the chance to adjust the assessments until they are coherent.

Constructing Distributions Using RISKview

This chapter has focused on assessing subjective probabilities. Let's take this one step further and, with the introduction of a program called RISKview, construct a probability distribution based on assessed probabilities. We will use RISKview to fit a piecewise linear distribution to our assessments and then demonstrate various ways to modify the fitted distribution. After having made the desired modifications, we can export the formula to Excel and incorporate it into a decision model. We demonstrate RISKview using the assessed values from the pretzel problem.

Keep in mind that our goal is to construct a distribution, based on our assessments, that accurately captures the underlying uncertainty. RISKview helps us by translating the assessments into a graph of the distribution. The graph provides a visual check of our assessments and highlights properties that are not apparent from a list of numerical values. The pitfalls of probability assessment that we discussed are hard to avoid, and sometimes we do not realize our assessments need to be modified until we see the resulting graph. For example, the graph may reveal that we overstated the probability of certain outcomes at the expense of other outcomes. By carefully scrutinizing the visual representation, we can see whether the distribution represents our beliefs in an appropriate way.

Let's construct a distribution for the monthly quantity demanded in the pretzel problem. The assessments are as follows:

Monthly Demand X	Cumulative Probability P(Monthly Demand $\leq X$)
5,000	0.05
16,000	0.25
23,000	0.50
31,000	0.75
45,000	0.95

We also need a minimum and maximum value for demand. For this example, let's assume the minimum equals 0 and the maximum equals 60,000.

STEP 1

1.1   Open Excel and then @RISK, enabling any macros if requested. RISKview is both a stand-alone program and embedded within @RISK. We demonstrate RISKview within @RISK to highlight the integration between the two programs.

1.2   There are four on-line help options available. Pull down the **@RISK** menu and choose the **Help** command. *"How Do I"* answers frequently asked questions about @RISK, *Online Manual* is an electronic copy (pdf file) of the user's guide, *@RISK Help* is a searchable database of the user's guide, and *PDF Help* lists various facts about the probability distribution functions available in @RISK and RISKview.

STEP 2

2.1   Highlight cell **A1** in Excel.

2.2   Open RISKview by either clicking on the **Define Distributions** button (third from the left in the *@RISK* toolbar) or **right-clicking** the mouse and choosing **@RISK,** then **Define Distributions**. An *@RISK Definition for A1* window is opened, as shown in Figure 8.19.

2.3   At first glance this window may seem complex, but you will appreciate its capabilities as you learn about its features. The current window shows the normal distribution in the center with summary facts about this distribution, such as its mean and standard deviation, listed to the right. When a different distribution is chosen using the dialog boxes to the left of the graph, the center graph and its summary values change accordingly.

Figure 8.19

RISKview's distribution viewing window showing the standard normal distribution.

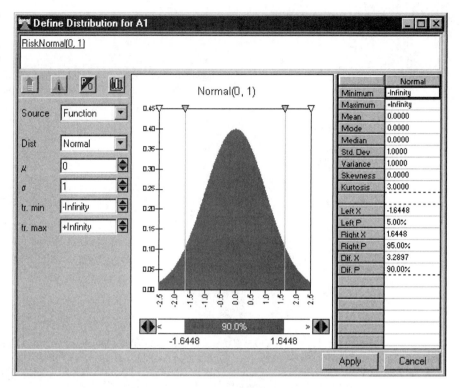

STEP 3

3.1 To construct the distribution for our assessed values, we need to choose the cumulative distribution and enter the values in the area to the left of the graph. Unfurl the pop-up distribution list by clicking on the **black triangle** to the right of the *Dist* name box that currently contains the name *Normal*. Scroll and choose **Cumul**, which is an abbreviation for cumulative distribution.

3.2 Rather than showing a cumulative distribution as you might expect, RISKview shows the probability density function that corresponds to cumulative assessments. In the absence of more information, RISKview assumes that the density is a step function.

3.3 Let's enter the assessed values. Type **0** for the *min* and **60000** for the *max*. If you double-click on the entry, it is highlighted and then deleted when you type in your value.

3.4 To enter the first *x* value, click the cell directly below X and type in the value **5000**.

3.5 Pressing **enter** moves the cursor into the cell to the right and directly below P. Enter the corresponding cumulative probability **0.05**. Pressing **enter** again places the cursor in the second row in the X column directly below 5000.

3.6   Continue repeating Steps 3.4 and 3.5 for successive rows until you have entered all the ordered pairs: **(16000, 0.25)**, **(23000, 0.50)**, **(31000, 0.75)**, and **(45000, 0.95)**. This is shown in Figure 8.20. Be sure to enter the *x* values in ascending order. You have constructed a distribution based on the assessed values!

Now lets see how to modify the appearance of the graph (Step 4) and how to read probabilities directly from the graph (Step 5).

STEP 4

4.1   Click on the **Graph Formatting Options** button (first button to the left and slightly above the graph) or **right-click** the mouse while the cursor is over the graph and choose **Format Graph,** as shown in Figure 8.20.

4.2   A dialog box titled *Graph Format* appears. The options in this dialog box are similar to those in TopRank and PrecisionTree and are explained below. For example, to view the cumulative ascending curve, choose **Cumulative Ascending** under *Type* and click **OK**.

4.3   To view the line graph, click the **Graph Formatting Options** button again, choose the **Style** tab along the top of the dialog box, choose **Line**

Figure 8.20

Probability density graph for a custom cumulative distribution.

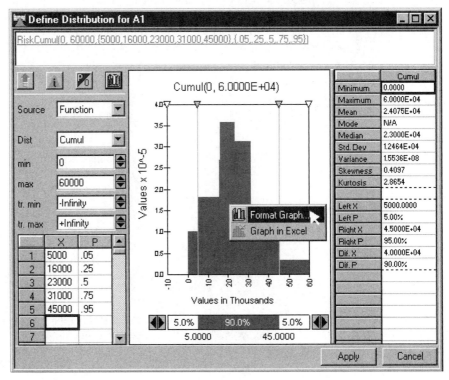

> as the format of the primary curve style, and click **OK**. The graph should
> now be piecewise linear similar to Figure 8.8.
>
> 4.4   You can also export the graph and supporting data to Excel if you wish by
> right-clicking while over the graph and choosing the *Graph in Excel* op-
> tion, as shown in Figure 8.20.

The five tabs along the top of the *Graph Format* dialog box accessed in Step 4.1
allow you to make various alterations to the graph:

- The *Type* tab allows you to view either a density function or a CDF.
- The *Scaling* tab allows you to change the scale (the units and the maximum and
  minimum) of the x and y axes.
- The *Style* tab allows you to change the patterns and colors of the graph.
- The *Titles* tab allows you to label the axis, title the graph, and add a legend.
- The *Delimiters* tab allows you to remove the delimiters, which are the vertical
  lines that extend downward from the gray triangles along the top of the graph.

The two delimiters (gray vertical lines overlaying the graph) are markers that
allow you to determine the cumulative probability of an x value for the distribution.
In Figure 8.20, the leftmost delimiter is at the fifth percentile, as shown by the 5.0%
in the scrollbar below the curve. This delimiter is at the x value of 5000, as shown by
the 5.0000 below the scrollbar (value in thousands). Therefore, there is a 5% probability that the monthly quantity demand for pretzels will be less than or equal to
5000. The 90.0% shown in the scrollbar indicates that there is a 90% probability that
demand will be greater than 5000 but less than 45,000. The rightmost delimiter
shows that there is a 95% probability that the monthly quantity demand for pretzels
will be less than or equal to 45,000.

> ## STEP 5
>
> 5.1   You can move the delimiter bars to find the cumulative probability value
> corresponding to an x value or to find the x value corresponding to a cu-
> mulative probability. Click on the **gray triangle** at the top of a delimiter
> bar and drag the mouse left or right while holding the mouse button
> down. The bar moves accordingly. Note that the smallest increment it will
> move is one unit of x, which is currently in thousands. Find the probabil-
> ity that demand will be between **10,000** and **40,000**. (Answer: 73.8%)
>
> 5.2   Clicking on the scrollbar below the graph also moves the bar, but this
> time one percentile. Find the x values so that there is a **20%** chance that
> demand will be below the first x value and a **20%** chance that demand
> will be above the second x value. (Answer: 13,250 and 34,500, respec-
> tively)
>
> 5.3   You can also type exact values for either the x value or the percentile into
> the spreadsheet cells located to the right of the graph. The *Left X* and *Left*

P refer to the leftmost delimiter and *Right X* and *Right P* refer to the right-most. If you enter the *x* value, RISKview reports the *p*-value, and vice versa. The bottom two cells calculate the difference in the *x* values and *p*-values; these cells cannot be typed in. Enter **.25** for the *Left P* and **.75** for the *Right P*. The corresponding *x* values are 1.6000E+4, or 16,000, and 3.1000E+4, or 31,000, which correspond to our assessed values.

The delimiters help verify that the distribution is consistent with the experts' beliefs. They allow us to report the implied probability values for any *x* value. For example, the above assessments imply that there is a 10% chance that demand will be less than or equal to 7,750. If the expert does not agree with this number, then the distribution should be modified. We explain in Step 7 how to modify the distribution.

 We now export the distribution constructed in RISKview to Excel in order to use it in our model.

STEP 6

6.1   If the formula for the distribution is in the text box above the curve, then you need only click **Apply** to export the formula to Excel. The formula for your distribution is placed in the spreadsheet cell A1. To place the formula in the text box, if it is not there, click on the **Replace Selected Formula with Distribution** button (upward pointing blue arrow situated at the upper left of the curve).

STEP 7   If, after viewing the graph, you believe changes are necessary, you can either enter modified values and probabilities for the original ones or you can manipulate the curve in the Distribution Artist window.

7.1   Click on the **Display List of Outputs and Inputs** button (double-headed red and blue arrow in the @RISK toolbar). A new window opens, titled *@RISK—Model*, in which the inputs and outputs are listed in the left section. You should have one input (cell A1) and no outputs.

7.2   Highlight A1 by clicking directly on **A1** under *Inputs*.

7.3   Either click the **Define Distributions** button in the *@RISK—Model* toolbar or **right-click** the mouse when the cursor is directly over *A1* and choose **Define Distributions,** as shown in Figure 8.21.

7.4   The *@RISK Definition for A1* window opens with the predefined cumulative distribution for monthly quantity demand for pretzels shown. At this point, you can modify the distribution by entering different values and/or probabilities in the *X* and *P* columns.

7.5   Alternatively, RISKview has an option for manipulating the actual curve: Click on the **Insert** heading and choose **Artist Window**. A *Distribution Artist* window is opened in the *@RISK—Model* window, as shown in

Figure 8.21

The @RISK window showing how to modify Inputs and Outputs.

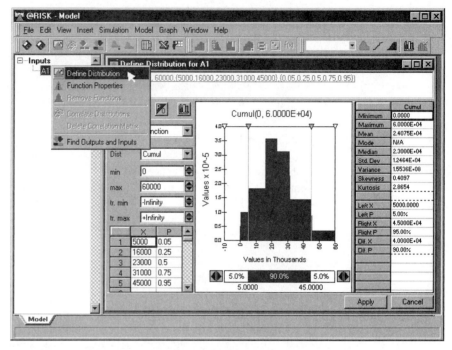

Figure 8.22

The @RISK Distribution Artist window showing the options available in the pop-up menu.

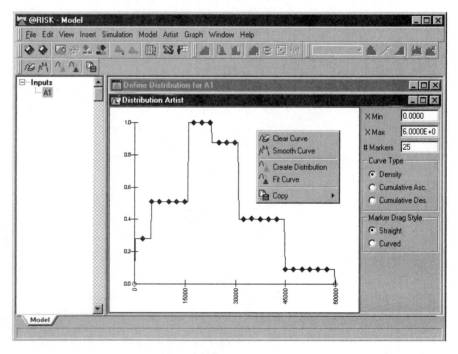

Figure 8.22. The curve from the definition window has been sampled at 250 points and pasted in the *Distribution Artist* window with markers at every tenth point (25 markers).

7.6   The markers allow you to reposition the height of the curve at that point. Try it! Click on any marker and move the cursor up or down while keeping the mouse button depressed.

7.7   To better understand the effects of moving a point, you can toggle between viewing the density curve and the cumulative ascending curve with the buttons below *Curve Type* to the right of the graph.

7.8   While the cursor is in the graph area, **right-click** the mouse. A menu pops up that provides additional options for manipulating the curve and is shown in Figure 8.22. *Clear Curve* deletes the current distribution; *Smooth Curve* smoothes out jagged edges; *Copy* copies the curve or data points; *Create Distribution* places the modified curve into a new definition window; and *Fit Curve* calls up BestFit. BestFit is one of the Palisade programs that searches for a theoretical distribution that "best fits" the assessed values; we will discuss this feature in Chapter 10. Note that the *Copy Graph* option copies only the image of the chart and not the formula; *Create Distribution* copies the formula.

7.9   Smooth the curve a few times to see the effects. When you are ready to export the distribution to the definition window, click **Create Distribution**.

STEP 8

8.1   To replace the cumulative distribution in cell A1 with the newly modified distribution, click **Apply**.

We have not covered every feature that *Distribution Artist* has to offer, and among these is one that allows you to actually draw the distribution curve using the cursor as a stylus. Undeniably fun, drawing your own distribution is nonetheless fraught with obvious hazards. It is very difficult to draw a curve freehand so that the area under the curve accurately represents your probabilities. This is made even more difficult because the mouse provides only limited control of the cursor.

We end this section with a word of caution: RISKview provides the opportunity to modify the distribution in many ways. Be careful that any changes accurately reflect your expert's beliefs about the variable that you are modeling. RISKview makes it easy and tempting to play around with the numbers, but unless you have a good reason and sound data for making changes—DON'T! Distributions are subtle, and small changes can have big effects. Always check the summary statistics when you make changes to see that they match your expert's judgment. Trying to make all your distributions "bell-shaped," although tempting, is at best a siren's song. It is better to keep a steady course and make sure all changes match your expert's judgment.

SUMMARY   Many of the decision problems that we face involve uncertain future events. We may have some feeling for such uncertainties, and we can build models of them using subjective probability-assessment techniques. The basic approach to assessing a probability involves either setting up a bet or comparing lotteries. We also have considered assessment methods for continuous uncertain quantities and found that it is straightforward to assess continuous distributions in terms of cumulative distribution functions. A reasonable and practical way to incorporate a continuous distribution into a decision tree is to use a discrete approximation. We discussed bracket medians and the Pearson-Tukey three-point approximation.

Our discussion also touched on the pitfalls of probability assessment. Individuals tend to use cognitive heuristics to judge probabilities. Heuristics such as representativeness, availability, and anchoring and adjustment can lead to bias in probability assessment. Certain ideas for improving probability assessments were discussed, including decomposition of the assessment task. We also presented a protocol-based approach to obtaining expert probability assessments and showed how this approach was used in a risk assessment of a nuclear-waste repository.

Finally, we discussed the idea of coherence—that is, that subjective probabilities must obey the same probability laws that "long-run frequency" probabilities do. Being coherent in probability assessment means being sure that one cannot be exploited through a Dutch book.

EXERCISES

8.1   Explain in your own words the idea of subjective probability.

8.2   An accounting friend of yours has gone to great lengths to construct a statistical model of bankruptcy. Using the model, the probability that a firm will file for bankruptcy within a year is calculated on the basis of financial ratios. On hearing your explanation of subjective probability, your friend says that subjective probability may be all right for decision analysis, but his model gives objective probabilities on the basis of real data. Explain to him how his model is, to a great extent, based on subjective judgments. Comment also on the subjective judgments that a bank officer would have to make in using your friend's model.

8.3   Explain in your own words the difference between assessing the probability for a discrete event and assessing a probability distribution for a continuous unknown quantity.

8.4   For each of the following phrases write down the probability number that you feel is represented by the phrase. After doing so, check your answers to be sure they are consistent. [For example, is your answer to e less than your answer to j?]

   a   "There is a better than even chance that . . . "

   b   "A possibility exists that . . . "

   c   ". . . has a high likelihood of occurring."

   d   "The probability is very high that . . . "

    e "It is very unlikely that . . . "

    f "There is a slight chance . . . "

    g "The chances are better than even . . . "

    h "There is no probability, no serious probability, that . . . "

    i ". . . is probable."

    j ". . . is unlikely."

    k "There is a good chance that . . . "

    l ". . . is quite unlikely."

    m ". . . is improbable."

    n ". . . has a high probability."

    o "There is a chance that . . . "

    p ". . . is very improbable."

    q ". . . is likely."

    r "Probably . . . "

8.5 Suppose your father asked you to assess the probability that you would pass your decision-analysis course. How might you decompose this probability assessment? Draw an influence diagram to represent the decomposition.

QUESTIONS AND PROBLEMS

8.6 Assess your probability that the following outcomes will occur. Use the equivalent-lottery method as discussed in the chapter. If possible, use a wheel of fortune with an adjustable win area, or a computer program that simulates such a wheel. What issues did you account for in making each assessment?

    a It will rain tomorrow in New York City.

    b You will have been offered a job before you graduate.

    c The women's track team at your college will win the NCAA championship this year.

    d The price of crude oil will be more than $30 per barrel on January 1, 2010.

    e The Dow Jones industrial average will go up tomorrow.

    f Any other uncertain outcome that interests you.

8.7 Consider the following two outcomes:

    a You will get an A in your most difficult course.

    b You will get an A or a B in your easiest course.

Can you assess the probability of these outcomes occurring? What is different about assessing probabilities regarding your own performance as compared to assessing probabilities for outcomes like those in Problem 8.6?

8.8 Describe a decomposition strategy that would be useful for assessing the probabilities in Problem 8.7.

8.9   Many people deal with uncertainty by assessing odds. For example, in horse racing different horses' odds of winning are assessed. Odds of "*a* to *b* for Outcome *E*" means that $P(E) = a/(a + b)$. Odds of "*c* to *d* against Outcome *E*" means that $P(\overline{E}) = c / (c + d)$. For the outcomes in Problem 8.6, assess the odds for that outcome occurring. Convert your assessed odds to probabilities. Do they agree with the probability assessments that you made in Problem 8.6?

8.10   It is said that Napoleon assessed probabilities at the Battle of Waterloo in 1815. His hopes for victory depended on keeping the English and Prussian armies separated. Believing that they had not joined forces on the morning of the fateful battle, he indicated his belief that he had a 90% chance of defeating the English; P(Napoleon Wins) = 0.90. When told later that elements of the Prussian force had joined the English, Napoleon revised his opinion downward on the basis of this information, but his posterior probability was still 60%; P(Napoleon Wins | Prussian and English Join Forces) = 0.60.

Suppose Napoleon were using Bayes' theorem to revise his information. To do so, he would have had to make some judgments about P(Prussian and English Join Forces | Napoleon Wins) and P(Prussian and English Join Forces | Napoleon Loses). In particular, he would have had to judge the ratio of these two probabilities. Based on the prior and posterior probabilities given above, what is that ratio?

8.11   Should you drop your decision-analysis course? Suppose you faced the following problem: If you drop the course, the anticipated salary in your best job offer will depend on your current GPA:

Anticipated Salary | Drop = ($4000 × Current GPA) + $16,000

If you take the course, the anticipated salary in your best job offer will depend on both your current GPA and your overall score (on a scale of 0 to 100) in the course:

Anticipated Salary | Do Not Drop = 0.6($4000 × Current GPA)
+ 0.4($170 × Course Score)
+ $16,000

The problem is that you do not know how well you will do in the course. You can, however, assess a distribution for your score. Assuming that 90–100 is an A, 80–89 is a B, 70–79 a C, 60–69 a D, and 0–59 an F, assess a continuous probability distribution for your numerical score in the course. Use that distribution to decide whether or not to drop the course. Figure 8.23 shows your decision tree.

8.12   Assess these fractiles for the following uncertain quantities: 0.05 fractile, 0.25 fractile (first quartile), 0.50 (median), 0.75 fractile (third quartile), and 0.95 fractile. Plot your assessments to create graphs of your subjective CDFs.

a   The closing Dow Jones industrial average (DJIA) on the last Friday of the current month.

b   The closing DJIA on the last Friday of next year.

c   The exchange rate, in Japanese yen per dollar, at the end of next Monday.

d   The official high temperature at O'Hare International Airport tomorrow.

e   The number of fatalities from airline accidents in the United States next year.

f   The number of casualties from nuclear power plant accidents in the United States over the next 10 years.

g   The value of the next jackpot won in the California state lottery.

Figure 8.23

Decision tree for Question 8.11. Should you drop your decision-analysis course?

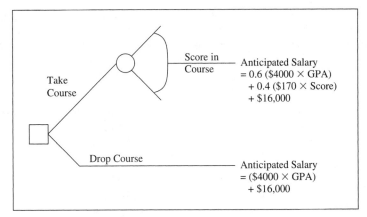

8.13 For each of the following 10 items, assess the 0.05 and 0.95 fractiles. That is, choose upper and lower estimates such that you are 90% sure that the actual value falls between your estimates. Your challenge is to be neither too narrow (i.e., overconfident) nor too wide (underconfident).

	0.05 Fractile (Low)	0.95 Fractile (High)
1. Martin Luther King's age at death	_____	_____
2. Length of the Nile River	_____	_____
3. Number of countries that are members of OPEC	_____	_____
4. Number of books in the Old Testament	_____	_____
5. Diameter of the moon	_____	_____
6. Weight of an empty Boeing 747	_____	_____
7. Year of Wolfgang Amadeus Mozart's birth	_____	_____
8. Gestation period of an Asian elephant	_____	_____
9. Air distance from London to Tokyo	_____	_____
10. Depth of deepest known point in the oceans	_____	_____

If you have done a good job, most of your intervals contain the actual value (given in the back of the book—no peeking!). If three or more of your intervals missed the actual value, though, you have demonstrated overconfidence. Given your results in this question, would you adjust your assessments in Problem 8.12? [*Source:* Adapted from J. E. Russo and P. J. H. Schoemaker (1989) *Decision Traps: The Ten Barriers to Brilliant Decision Making and How to Overcome Them.* New York: Fireside. Reprinted by permission.]

8.14 Forecasters often provide only point forecasts, which are their best guesses as to an upcoming event. For example, an economic forecaster might predict that U.S. gross national product (GNP) will increase at a 3% annual rate over the next three months.

Occasionally a forecaster also will provide an estimate of the degree of confidence in the point forecast to indicate how sure (or unsure) the forecaster is.

In what sense can your answers to Problem 8.12 be interpreted as forecasts? What advantages do subjective probability distributions have over typical point forecasts? What disadvantages? How could a decision maker use probabilistic forecasts such as those in Problem 8.12?

8.15   Choose a course that you are currently taking in which the final exam is worth 100 points. Treating your score on the exam as if it were a continuous uncertain quantity, assess the subjective probability distribution for your score. After you have finished, check your assessed distribution for consistency by:

    a   choosing any two intervals you have judged to have equal probability content, and

    b   determining whether you would be willing to place small even-odds bets that your score would fall in one of the two intervals. (The bet would be called off if the score fell elsewhere.)

    c   After assessing the continuous distribution, construct a three-point approximation to this distribution with the extended Pearson-Tukey method. Use the approximation to estimate your expected exam score.

    d.   Now construct a five-point approximation with bracket medians. Use this approximation to estimate your expected exam score. How does your answer compare with the estimate from part c?

8.16   Compare the discrete-approximation methods by doing the following:

    a   Use the extended Pearson-Tukey method to create three-point discrete approximations for the continuous distributions assessed in Problem 8.12. Use the approximations to estimate the expected values of the uncertain quantities.

    b   Repeat part a, but construct five-point discrete approximations using bracket medians. Compare your estimated expected values from the two methods.

8.17   Assess the probability that the New York Mets will win the World Series (WS) next year. Call this probability p. Now assess the following probabilities: P(Win WS | Win Pennant) and P(Win Pennant). Use these to calculate q = P(Win WS) = P(Win WS | Win Pennant) P(Win Pennant) = P(Win WS and Win Pennant). (To play in the World Series, the team must first win the pennant.)

    a   Does $p = q$? Which of the two assessments are you more confident about? Would you adjust your assessments to make $p = q$? Why or why not?

    b   If you do not like the Mets, do this problem for your favorite major league baseball team.

8.18   Assess the probability that you will be hospitalized for more than one day during the upcoming year. In assessing this probability, decompose the assessment based on whether or not you are in an automobile accident. With this decomposition, you must assess P(Hospitalized | Accident), P(Hospitalized | No Accident), and P(Accident). In what other ways might this assessment be decomposed?

8.19   Choose a firm in which you are particularly interested, perhaps one where you might like to work. Go to the library and read about this firm, about its industry, and about the relationship of this firm and the industry to the economy as a whole. After your reading, assess a subjective CDF for the firm's revenue over the next fiscal year. Discuss the assessment process. In particular, what did you learn in your research that had an impact on the assessment? What kinds of decomposition helped you in your assessment process?

8.20 After observing a long run of red on a roulette wheel, many gamblers believe that black is bound to occur. Such a belief often is called the *gambler's fallacy* because the roulette wheel has no memory. Which probability-assessment heuristic is at work in the gambler's fallacy? Explain.

8.21 Look again at Problems 7.25 and 7.26. These problems involve assessments of the relative likelihood of different statements. When an individual ranks "Linda is a bank teller and is active in the feminist movement" as more probable than the statement "Linda is a bank teller," which of the three probability-assessment heuristics may be at work? Explain.

8.22 Suppose that you are a solid B student. Your grades in your courses have always been B's. In your statistics course, you get a D on the midterm. When your parents express concern, what statistical phenomenon might you invoke to persuade them not to worry about the D?

8.23 When we assess our subjective probabilities, we are building a model of the uncertainty we face. If we face a range-of-risk problem, and we begin to assess a continuous distribution subjectively, then clearly it is possible to perform many assessments, which would make the sketched CDF smoother and smoother. How do we know when to stop making assessments and when to keep going? When is the model of our uncertainty adequate for solving the problem, and when is it inadequate?

8.24 Most of us have a hard time assessing probabilities with a lot of precision. For instance, in assessing the probability of rain tomorrow, even carefully considering the lotteries and trying to adjust a wheel of fortune to find the indifference point, many people would eventually say something like this: "If you set $p = 0.2$, I'd take Lottery A, and if $p = 0.3$, I'd take Lottery B. My indifference point must be somewhere in between these two numbers, but I am not sure where."

How could you deal with this kind of imprecision in a decision analysis? Illustrate how your approach would work using the umbrella problem (Figure 4.9). (*Hint:* The question is *not* how to get more precise assessments. Rather, given that the decision maker refuses to make precise assessments, and you are stuck with imprecise assessments, what kinds of decision-analysis techniques could you apply to help the individual make a decision?)

8.25 It is not necessary to have someone else set up a series of bets against you in order for incoherence to take its toll. It is conceivable that one could inadvertently get oneself into a no-win situation through inattention to certain details and the resulting incoherence, as this problem shows.

Suppose that an executive of a venture-capital investment firm is trying to decide how to allocate his funds among three different projects, each of which requires a $100,000 investment. The projects are such that one of the three will definitely succeed, but it is not possible for more than one to succeed. Looking at each project as an investment, the anticipated payoff is good, but not wonderful. If a project succeeds, the payoff will be a net gain of $150,000. Of course, if the project fails, he loses all of the money invested in that project. Because he feels as though he knows nothing about whether a project will succeed or fail, he assigns a probability of 0.5 that each project will succeed, and he decides to invest in each project.

a According to his assessed probabilities, what is the expected profit for each project?

b What are the possible outcomes of the three investments, and how much will he make in each case?

   c   Do you think he invested wisely? Can you explain why he is in such a predicament?

   d   If you feel you know nothing about some event, is it reasonable to assess equal probabilities for the outcomes? Give an example where this might be reasonable and another where it might not.

8.26   Could you ever set up a Dutch book against a bookie who places bets for a living? Work through the following problem and think about this question.

      Consider a baseball season when the Chicago Cubs and the New York Yankees meet in the World Series. A friend of yours is a Cubs fan, and you are trying to find out how confident he is about the Cubs' chances of winning. He tells you that he would bet on the Cubs at odds of 3:2 *or better* and on the Yankees at odds of 1:2 *or better*. [Odds of *a:b* for an event means the probability of the event is $a/(a + b)$.] This means that he would be happy with either of the following bets:

        Bet 1   He wins $20 if the Cubs win.

                       He loses $30 if the Yankees win.

        Bet 2   He loses $20 if the Cubs win.

                       He wins $40 if the Yankees win.

   a   Explain why he would be happy to accept any modification of these bets as long as the amount he would win increases and the amount he would lose stays the same or is lower. Explain in terms of his expected bet value.

   b   Show mathematically that $0.60 \leq P(\text{Cubs Win}) \leq 0.67$. (*Hint:* If he is willing to accept either bet, what does that imply about his expected value for each bet?)

   c   Would it be possible to set up a Dutch book against this individual? If your answer is yes, what bets would you place, and how much would you be sure to win? If your answer is no, explain why not. (Be careful! Make sure he is willing to accept the bets you propose.)

   d   Most individuals have a hard time assessing subjective probabilities with a very high degree of precision. (See Problem 8.23.) Does this problem shed any light on the issue?

8.27   **Ellsberg Paradox** A barrel contains a mixture of 90 red, blue, and yellow balls. Thirty of the balls are red, and the remaining 60 are a mixture of blue or yellow, but the proportion of blue and yellow is unknown. A single ball will be taken randomly from the barrel.

   a   Suppose you are offered the choice between gambles A and B:

        A: Win $1000 if a red ball is chosen.

        B: Win $1000 if a blue ball is chosen.

     Would you prefer A or B? Why?

   b   Now suppose that you are offered a choice between gambles C and D:

        A: Win $1000 if either a red or a yellow ball is chosen.

        B: Win $1000 if either a blue or a yellow ball is chosen.

     Would you prefer C or D? Why?

   c   Many people prefer A in the first choice and D in the second. Do you think this is inconsistent? Explain.

CASE STUDIES

ASSESSING CANCER RISK—FROM MOUSE TO MAN

Cancer is a poorly understood and frightening disease. Its causes are essentially unknown, and the biological process that creates cancerous cells from healthy tissue remains a mystery. Given this state of ignorance, much research has been conducted into how external conditions relate to cancer. For example, we know that smoking leads to substantially higher incidence of lung cancer in humans. Cancer appears spontaneously, however, and its onset seems to be inherently probabilistic, as shown by the fact that some people smoke all their lives without developing lung cancer.

Some commentators claim that the use of new and untested chemicals is leading to a cancer epidemic. As evidence they point to the increase in cancer deaths over the years. Indeed, cancer deaths have increased, but people generally have longer life spans, and more elderly people now are at risk for the disease. When cancer rates are adjusted for the increased life span, cancer rates have not increased substantially. In fact, when data are examined this way, some cancer rates (liver, stomach, and uterine cancer) are less common now than they were 50 years ago (1986 data of the American Cancer Society). Nevertheless, the public fears cancer greatly. The Delaney Amendment to the Food, Drug, and Cosmetics Act of 1954 outlaws residues in processed foods of chemicals that pose any risk of cancer to animals or humans. One of the results of this fear has been an emphasis in public policy on assessing cancer risks from a variety of chemicals.

Scientifically speaking, the best way to determine cancer risk to humans would be to expose one group of people to the chemical while keeping others away from it. But such experiments would not be ethical. Thus, scientists generally rely on experiments performed on animals, usually mice or rats. The laboratory animals in the experimental group are exposed to high doses of the substance being tested. High doses are required because low doses probably would not have a statistically noticeable effect on the relatively small experimental group. After the animals die, cancers are identified by autopsy. This kind of experiment is called a *bioassay*. Typical cancer bioassays involve 600 animals, require two to three years to complete, and cost several hundred thousand dollars.

When bioassays are used to make cancer risk assessments, two important extrapolations are made. First, there is the extrapolation from high doses to low doses. Second, it is necessary to extrapolate from effects on test species to effects on humans. On the basis of data from laboratory experiments and these extrapolations, assessments are made regarding the incidence of cancer when humans are exposed to the substance.

Questions

1   Clearly, the extrapolations that are made are based on subjective judgments. Because cancer is viewed as being an inherently probabilistic phenomenon, it is

reasonable to view these judgments as probability assessments. What kinds of assessments do you think are necessary to make these extrapolations? What issues must be taken into account? What kind of scientific evidence would help in making the necessary assessments?

2   It can be argued that most cancer risk assessments are weak evidence of potential danger or lack thereof. To be specific, a chemical manufacturer and a regulator might argue different sides of the same study. The manufacturer might claim that the study does not conclusively show that the substance is dangerous, while the regulator might claim that the study does not conclusively demonstrate safety. Situations like this often arise, and decisions must be made with imperfect information. What kind of strategy would you adopt for making these decisions? What trade-offs does your strategy involve?

3   In the case of risk assessment, as with many fields of scientific inquiry, some experiments are better than others for many reasons. For example, some experiments may be more carefully designed or use larger samples. In short, some sources of information are more "credible."

For a simple, hypothetical example, suppose that you ask three "experts" whether a given coin is fair. All three report that the coin is fair; for each one the best estimate of P(Heads) is 0.50. You learn, however, that the first expert flipped the coin 10,000 times and observed heads on 5000 occasions. The second flipped the coin 20 times and observed 10 heads. The third expert did not flip the coin at all, but gave it a thorough physical examination, finding it to be perfectly balanced, as nearly as he could measure.

How should differences in credibility of information be accounted for in assessing probabilities? Would you give the same weight to information from the three examiners of the coin? In the case of putting together information on cancer risk from multiple experiments and expert sources, how might you deal with the information sources' differential credibility?

Source: D. A. Freedman and H. Zeisel (1988) "From Mouse-to-Man: The Quantitative Assessment of Cancer Risks." *Statistical Science,* 3, 3–56. Includes discussion.

BREAST IMPLANTS

The controversy surrounding breast implants is a good example of scientific uncertainty. Yanked from the market by the Food and Drug Administration in 1991 because of some evidence of dangerous side effects, breast implants received a reprieve in 1994 when the *New England Journal of Medicine* published an article that found no evidence of danger. An editorial in the *San Jose Mercury News* opined regarding the controversy:

> The wisdom of letting science settle this question [regarding breast implants] will win the ready assent of everyone. "Scientific certainty" has such a reassuring ring to it.

But the implant case is a sterling example of the limits of science. Regarding long-term harm, science often can't provide definitive answers within the time a decision needs to be made.

And many policy decisions are out of the realm of science. What level of risk is reasonable for a woman who's lost a breast to cancer? What about reasonable risk for a woman who wants a fuller figure? Who decides what's reasonable? Is cosmetic breast enhancement a misuse of medicine to reinforce outdated notions of female beauty? [*Source:* "Uncertain Science," *San Jose Mercury News,* June 17, 1994, p. 10B.]

Questions

1   What kinds of information should a jury consider when deciding whether a plaintiff's claims of damages are reasonable? Anecdotes? The number of plaintiffs filing similar lawsuits? Scientific studies? What does a juror need to know in order to evaluate the quality of a scientific study? Do you think the average juror (or judge for that matter!) in the United States has the ability to critically evaluate the quality of scientific studies?

2   Discuss the questions asked in the last paragraph of the quote above. Do these questions relate to uncertainty (scientific or otherwise) or to values? Given the imperfect information available about breast implants, what role should the manufacturers play in deciding what risks are appropriate for which women? What role should government agencies play? What role should individual consumers play?

THE SPACE SHUTTLE *CHALLENGER*

On January 28, 1986, the space shuttle *Challenger* lifted off from an ice-covered launch pad. Only 72 seconds into the flight, the shuttle exploded, killing all seven astronauts aboard. The United States and the rest of the world saw the accident firsthand as films from NASA were shown repeatedly by the television networks.

Before long the cause of the accident became known. The shuttle's main engines are fueled by liquid hydrogen and oxygen stored in a large tank carried on the shuttle's belly. Two auxiliary rockets that use solid fuel are mounted alongside the main fuel tank and provide additional thrust to accelerate the shuttle away from the launch pad. These boosters use their fuel rapidly and are jettisoned soon after launch.

The solid rocket boosters are manufactured in sections by Morton Thiokol, Inc. (MTI), in Utah. The sections are shipped individually to Kennedy Space Center (KSC) in Florida where they are assembled. The joints between sections of the rocket are sealed by a pair of large rubber O-rings, whose purpose is to contain the hot gases and pressure inside the rocket. In the case of the *Challenger,* one of the joint seals failed. Hot gases blew past the O-rings and eventually burned through the large belly tank, igniting the highly explosive fuel inside. The resulting explosion destroyed the spacecraft.

Before long it also became known that the launch itself was not without controversy. MTI engineers had been aware of the problems with the O-rings for some time, having observed eroded O-rings in the boosters used on previous flights. A special task force was formed in 1985 to try to solve the problem, but ran into organizational problems. One memo regarding the task force began, "Help! The seal task force is constantly being delayed by every possible means." The problem came to a head when, on the evening before the launch, MTI engineers recommended not launching the shuttle because of the anticipated cold temperatures on the launch pad. After a teleconference involving officials at KSC and the Marshal Space Flight Center (MSFC) in Alabama, management officials at MTI reversed their engineers' recommendation and approved the launch.

Questions

1   To a great extent, the engineers were concerned about the performance of the O-ring under anticipated cold weather conditions. The coldest previous flight had been 53°F, and, knowing of the existing problems with the seals, the engineers hesitated to recommend a launch under colder conditions. Technically, the problem was that an O-ring stiffens as it gets colder, thus requiring a longer time to seal a joint. The real problem, however, was that the engineers did not know much about the performance of the O-rings at cold temperatures. Robert K. Lund, vice president of engineering for MTI, testified to the presidential commission investigating the accident, "We just don't know how much further we can go below the 51 or 53 degrees or whatever it was. So we were concerned with the unknown. . . . They [officials at MSFC] said they didn't accept that rationale" (*Report of the Presidential Commission on the Space Shuttle* Challenger *Accident,* p. 94).

The MTI staff felt as if it were in the position of having to prove that the shuttle was unsafe to fly instead of the other way around. Roger Boisjoly, an MTI engineer, testified, "This was a meeting where the determination was to launch, and it was up to us to prove beyond a shadow of a doubt that it was not safe to do so. This is in total reverse to what the position usually is in a preflight conversation or a flight readiness review. It is usually exactly opposite that" (*Report,* p. 93).

NASA solicited information regarding ice on the launch pad from Rockwell International, the shuttle's manufacturer. Rockwell officials told NASA that the ice was an unknown condition. Robert Glaysher, a vice president at Rockwell, testified that he had specifically said to NASA, "Rockwell could not 100 percent assure that it is safe to fly" (*Report,* p. 115). In this case, the presidential commission also found that "NASA appeared to be requiring a contractor to prove that it was not safe to launch, rather than proving it was safe" (*Report,* p. 118).

The issue is how to deal with unknown information. What do you think the policy should be regarding situations in which little or no information is available? Discuss the problems faced by both MTI and NASA. What incentives and pressures might they have faced?

2   Professor Richard Feynman, Nobel Laureate in physics, was a member of the commission. He issued his own statement, published as an appendix to the report, taking

NASA to task for a variety of blunders. Some of his complaints revolved around assessments of the probability of failure.

Failure of the solid rocket boosters. A study of 2900 flights of solid-fuel rockets revealed 121 failures, or approximately 1 in 25. Because of improved technology and special care in the selection of parts and in inspection, Feynman is willing to credit a failure rate of better than 1 in 100 but not as good as 1 in 1000. But in a risk analysis prepared for the Department of Energy (DOE) that related to DOE radioactive material aboard the shuttle, NASA officials used a figure of 1 in 100,000. Feynman writes:

> If the real probability is not so small [as 1 in 100,000], flights would show troubles, near failures, and possibly actual failures with a reasonable number of trials, and standard statistical methods could give a reasonable estimate. In fact, previous NASA experience had shown, on occasion, just such difficulties, near accidents, and accidents, all giving warning that the probability of flight failure was not so very small. (*Report,* p. F–1)

Failure of the liquid fuel engine. In another section of his report, Feynman discussed disparate assessments of the probability of failure of the liquid fuel engine. His own calculations suggested a failure rate of approximately 1 in 500. Engineers at Rocketdyne, the engine manufacturer, estimated the probability to be approximately 1 in 10,000. NASA officials estimated 1 in 100,000. An independent consultant for NASA suggested that a failure rate of 1 or 2 per 100 would be a reasonable estimate.

How is it that these probability estimates could vary so widely? How should a decision maker deal with probability estimates that are so different?

3   To arrive at their overall reliability estimates, NASA officials may have decomposed the assessment, estimated the reliability of many different individual components, and then aggregated their assessments. Suppose that, because of an optimistic viewpoint, each probability assessment had been slightly overoptimistic (that is, a low assessed probability of failure). What effect might this have on the overall reliability estimate?

4   In an editorial in *Space World* magazine, editor Tony Reichhardt commented on the accident:

> One person's safety is another's paranoia. How safe is safe? What is acceptable risk? It's no small question, in life or in the space program. It's entirely understandable that astronauts would come down hard on NASA policies that appear to be reckless with their lives. But unless I'm misreading the testimony [before the commission], at the end of the teleconference that night of January 27, most of the participating engineers believed that it was safe to go ahead and launch. A few argued that it was not safe *enough*. There was an element of risk in the decision, and in many others made prior to *Challenger's* launch, and seven people were killed.
>
> Whether this risk can be eliminated is a question of monumental importance to the space program. Those who have put the blame squarely on NASA launch managers need to think hard about the answer. If no Shuttle takes off until everyone at every level of responsibility is in complete agreement, then it may never be launched again. No single person can be absolutely sure that the whole system will work. On this vehicle, or on some other spacecraft next year

or 30 years from now—even if we ease the financial and scheduling pressures—something will go wrong again. [*Source*: T. Reichhardt (1986) "Acceptable Risk," *Space World*, April, p. 3. Reprinted by permission.]

Comment on Reichhardt's statement. What is an acceptable risk? Does it matter whether we are talking about risks to the general public from cancer or risks to astronauts in the space program? Would your answer change if you were an astronaut? A NASA official? A manufacturer of potentially carcinogenic chemicals? A cancer researcher? How should a policy maker take into account the variety of opinions regarding what constitutes an acceptable risk?

Source: Information for this case was taken from many sources, but by far the most important was the report by the Rogers Commission (1986) *Report of the Presidential Commission on the Space Shuttle Challenger Accident.* Washington, DC: U.S. Government Printing Office.

REFERENCES

The subjective interpretation of probability is one of the distinguishing characteristics of decision theory and decision analysis. This interpretation was presented first by Savage (1954) and has been debated by probabilists and statisticians ever since. Winkler (1972) provides an excellent introduction to subjective probability and Bayesian statistics (so called because of the reliance on Bayes' theorem for inference), as well as extensive references to the literature.

Although we argued that verbal expressions of uncertainty are by nature less precise than numerical probabilities, Wallsten, Budescu, and Zwick (1993) show how to *calibrate* a set of verbal probability labels so that the meaning of each label is precisely understood.

Spetzler and Staël von Holstein (1975) is the standard reference on probability assessment. Winkler (1972) also covers this topic. Wallsten and Budescu (1983) review the field from a psychological perspective.

The construction of an appropriate discrete distribution is an interesting problem that has occupied a number of decision-analysis researchers. Many different approaches exist; the extended Pearson-Tukey and bracket median approaches are two of the more straightforward ones that perform well. Other good ones include the *extended Swanson-Megill* method, in which the distribution's median gets probability 0.4 and the 0.10 and 0.90 fractiles get probability 0.3 each. More complicated (and more precise) methods create discrete distributions with mathematical characteristics that match those of the continuous distribution (e.g., Miller and Rice 1983, Smith 1993); such approaches are difficult to use "by hand" but are easily implemented in computer software. Studies that report on the relative performance of different discrete-approximation methods include Keefer and Bodily (1983), Keefer (1994), and Smith (1993).

The literature on heuristics and biases is extensive. One reference that covers the topic at an introductory level is Tversky and Kahneman (1974). Hogarth (1987) provides a unique overview of this material in the context of decision analysis. Kahneman, Slovic, and Tversky (1982) have collected key research papers in the area. With respect to the issue of overconfidence in particular, Capen (1976) reports an experiment that demonstrates the extent of overconfidence in subjective probability assessments and possible ways to cope with this phenomenon.

Probability decomposition is a topic that has not been heavily researched, although there are many applications. For example, see Bunn (1984) and von Winterfeldt and Edwards (1986) for discussions and examples of fault trees. A recent research paper by Ravinder, Kleinmuntz, and Dyer (1988) discusses when decomposition is worth doing. Fault-tree analysis is widely used in reliability engineering and risk analysis. It is described in Gottfried (1974), Bunn (1984), and Merkhofer (1987b). Covello and Merkhofer (1993) provide additional references regarding fault trees.

As mentioned, the concept of coherence was first introduced by de Finetti (1937). Thinking in terms of coherence as did de Finetti contrasts with the axiomatic approach taken by Savage (1954). The idea of coherence has been developed considerably; examples of ways in which various decision rule or probability assessments are incoherent are given by Bunn (1984), French (1986), and Lindley (1985).

Finally, a topic that arose in Problems 8.23–8.26 deserves mention. This is the matter of vagueness or ambiguity in probability assessments. Problem 8.26 is the classic paradox of Ellsberg (1961), essentially showing that people shy away from risky prospects with vague probabilities. Ongoing research is attempting to understand this phenomenon both descriptively and prescriptively. Einhorn and Hogarth (1985) present a psychological model of the process. Frisch and Baron (1988) define vagueness in terms of the lack of information that could have an impact on a probability assessment. The setup in Problem 8.25 is a simplified version of Nau's (1990) model in which a decision maker places constraints on the size of the bets that would be acceptable at different odds levels.

Abramson, B., and A. J. Finizza (1995) "Probabilistic Forecasts from Probabilistic Models Case Study: The Oil Market." *International Journal of Forecasting,* 11, 63–72.

Beyth-Marom, R. (1982) "How Probable Is Probable? A Numerical Translation of Verbal Probability Expressions." *Journal of Forecasting*, 1, 257–269.

Bunn, D. (1984) *Applied Decision Analysis.* New York: McGraw-Hill.

Capen, E. C. (1976) "The Difficulty of Assessing Uncertainty." *Journal of Petroleum Technology,* August, 843–850. Reprinted in R. Howard and J. Matheson (eds.) (1983) *The Principles and Applications of Decision Analysis.* Menlo Park, CA: Strategic Decisions Group.

Covello, V. T., and M. W. Merkhofer (1993) *Risk Assessment Methods: Approaches for Assessing Health and Environmental Risks.* New York: Plenum.

de Finetti, B. (1937) "La Prévision: Ses Lois Logiques, Ses Sources Subjectives." *Annales de l'Institut Henri Poincaré*, 7, 1–68. Translated by H. E. Kyburg in H. E. Kyburg, Jr., and H. E. Smokler (eds.) (1964) *Studies in Subjective Probability.* New York: Wiley.

Einhorn, H., and R. M. Hogarth (1985) "Ambiguity and Uncertainty in Probabilistic Inference." *Psychological Review*, 92, 433–461.

Ellsberg, D. (1961) "Risk, Ambiguity, and the Savage Axioms." *Quarterly Journal of Economics*, 75, 643–669.

French, S. (1986) *Decision Theory: An Introduction to the Mathematics of Rationality.* London: Wiley.

Frisch, D., and J. Baron (1988) "Ambiguity and Rationality." *Journal of Behavioral Decision Making*, 1, 149–157.

Gottfried, P. (1974) "Qualitative Risk Analysis: FTA and FMEA." *Professional Safety,* October, 48–52.

Hogarth, R. M. (1987) *Judgment and Choice,* 2nd ed. New York: Wiley.

Kahneman, D., P. Slovic, and A. Tversky (eds.) (1982) *Judgment under Uncertainty: Heuristics and Biases.* Cambridge: Cambridge University Press.

Kahneman, D., and A. Tversky (1973) "On the Psychology of Prediction." *Psychological Review*, 80, 237–251.

Keefer, D. L. (1994) "Certainty Equivalents for Three-Point Discrete-Distribution Approximations." *Management Science,* 40, 760–773.

Keefer, D., and S. E. Bodily (1983) "Three-Point Approximations for Continuous Random Variables." *Management Science,* 29, 595–609.

Lindley, D. V. (1985) *Making Decisions*, 2nd ed. New York: Wiley.

Merkhofer, M. W. (1987a) "Quantifying Judgmental Uncertainty: Methodology, Experiences, and Insights." *IEEE Transactions on Systems, Man, and Cybernetics,* 17, 741–752.

Merkhofer, M. W. (1987b) *Decision Science and Social Risk Management.* Dordrect, Holland: Reidel.

Miller, A. C., and T. R. Rice (1983) "Discrete Approximations of Probability Distributions." *Management Science,* 29, 352–362.

Morgan, M. G., and M. Henrion (1990) *Uncertainty: A Guide to Dealing with Uncertainty in Quantitative Risk and Policy Analysis.* Cambridge: Cambridge University Press.

Nau, R. (1990) "Indeterminate Probabilities on Finite Sets." *Annals of Statistics,* 20, 1737–1787.

Ravinder, H. V., D. Kleinmuntz, and J. S. Dyer (1988) "The Reliability of Subjective Probabilities Obtained Through Decomposition." *Management Science*, 34, 186–199.

Savage, I. J. (1954) *The Foundations of Statistics.* New York: Wiley.

Smith, J. E. (1993) "Moment Methods for Decision Analysis." *Management Science*, 39, 340–358.

Spetzler, C. S., and C. A. Staël von Holstein (1975) "Probability Encoding in Decision Analysis." *Management Science,* 22, 340–352.

Tversky, A., and D. Kahneman (1974) "Judgments under Uncertainty: Heuristics and Biases." *Science*, 185, 1124–1131.

von Winterfeldt, D., and W. Edwards (1986) *Decision Analysis and Behavioral Research.* Cambridge: Cambridge University Press.

Wallsten, T. S., and D. V. Budescu (1983) "Encoding Subjective Probabilities: A Psychological and Psychometric Review." *Management Science*, 29, 151–173.

Wallsten, T. S., D. V. Budescu, and R. Zwick (1993) "Comparing the Calibration and Coherence of Numerical and Verbal Probability Judgments." *Management Science*, 39, 176–190.

Winkler, R. L. (1972) *Introduction to Bayesian Inference and Decision.* New York: Holt.

E P I L O G U E   Stanford beat Oregon in their 1970 football contest. The score was 33–10. Jim Plunkett was Stanford's quarterback that season and won the Heisman trophy.

Millard Fillmore was Zachary Taylor's vice president. Taylor died in 1850 while in office. Fillmore succeeded him and was president from 1850 through 1853.

CHAPTER 9

Theoretical Probability Models

The last chapter dealt with the subjective assessment of probabilities as a method of modeling uncertainty in a decision problem. Using subjective probabilities often is all that is necessary, although they occasionally are difficult to come up with, and the nature of the uncertainty in a decision situation can be somewhat complicated. In these cases, we need another approach.

An alternative source for probabilities is to use theoretical probability models and their associated distributions. We can consider the characteristics of the system from which the uncertain event of interest arises, and, if the characteristics correspond to the assumptions that give rise to a standard distribution, we may use the distribution to generate the probabilities. It is important to realize, however, that in this situation a substantial subjective judgment is being made: that the physical system can be represented adequately using the model chosen. In this sense, such probability models would be just as "subjective" as a directly assessed probability distribution.

One approach is to assess a subjective probability distribution and then find a standard distribution that provides a close "fit" to those subjective probabilities. A decision maker might do this simply to make the probability and expected-value calculations easier. Of course, in this case the underlying subjective judgment is that the theoretical distribution adequately fits the assessed judgments.

How important are theoretical probability distributions for decision making? Consider the following applications of theoretical models.

THEORETICAL MODELS APPLIED

Educational Testing Most major educational and intelligence tests generate distributions of scores that can be well represented by the normal distribution, or the familiar bell-shaped curve. Many colleges and universities admit only individuals whose scores are above a stated criterion that corresponds to a specific percentile of the normal distribution of scores.

Market Research In many market research studies, a fundamental issue is whether a potential customer prefers one product to another. The uncertainty involved in these problems often can be modeled using the binomial or closely related distributions.

Quality Control How many defects are acceptable in a finished product? In some products, the occurrence of defects, such as bubbles in glass or blemishes in cloth, can be modeled quite nicely with a Poisson process. Once the uncertainty is modeled, alternative quality-control strategies can be analyzed for their relative costs and benefits.

Predicting Election Outcomes How do the major television networks manage to extrapolate the results of exit polls to predict the outcome of elections? Again, the binomial distribution forms the basis for making probability statements about who wins an election based on a sample of results.

Capacity Planning Do you sometimes feel that you spend too much time standing in lines waiting for service? Service providers are on the other side; their problem is how to provide adequate service when the arrival of customers is uncertain. In many cases, the number of customers arriving within a period of time can be modeled using the Poisson distribution. Moreover, this distribution can be extended to the placement of orders at a manufacturing facility, the breakdown of equipment, and other similar processes.

Environmental Risk Analysis In modeling the level of pollutants in the environment, scientists often use the lognormal distribution, a variant of the normal distribution. With uncertainty about pollutant levels modeled in this way, it is possible to analyze pollution-control policies so as to understand the relative effects of different policies.

You might imagine that many different theoretical probability models exist and have been used in a wide variety of applications. We will be scratching only the surface here, introducing a few of the more common distributions. For discrete probability situations, we will discuss the binomial and Poisson distributions, both of which are commonly encountered and easily used. Continuous distributions that will be discussed are the exponential, normal, and beta distributions. We also have access to over 30 theoretical distributions via Palisade's DecisionTools. RISKview will help us understand the different distribution types through manipulation of interactive graphs. The references at the end of this chapter will direct you to sources on other distributions.

The Binomial Distribution

Perhaps the easiest place to begin is with the *binomial* distribution. Suppose, for example, that you were in a race for mayor of your hometown, and you wanted to find out how you were doing with the voters. You might take a sample, count the number of individuals who indicated a preference for you, and then, based on this information, judge your chances of winning the election. In this situation, each voter interviewed can be either for you or not. This is the kind of situation in which the binomial distribution can play a large part. Of course, it is not limited to the analysis of voter preferences but also can be used in analyses of quality control where an item may or may not be defective, in market research, and in many other situations.

The binomial distribution arises from a situation that has the following characteristics:

1   *Dichotomous outcomes.* Uncertain events occur in a sequence, each one having one of two possible outcomes—success/failure, heads/tails, yes/no, true/false, on/off, and so on.

2   *Constant probability.* Each event, or *trial* has the same probability of success. Call that probability p.

3   *Independence.* The outcome of each trial is independent of the outcomes of the other trials. That is, the probability of success does not depend on the preceding outcomes. (As we noted in Problem 8.20 ignoring this sometimes is called the *gambler's fallacy:* Just because a roulette wheel has come up red five consecutive times does not mean that black is "due" to occur. Given independence from one game to the next, the probability of black still is 0.5, regardless of the previous outcomes.)

Now suppose we look at a sequence of n trials; for example, four tosses of a loaded coin that has $P(\text{Heads}) = 0.8$. How many successes (heads) could there be? Let R denote the uncertain quantity or random variable that is the number of successes in the sequence of trials. Clearly, there can be no more than n successes: $0 \leq R \leq n$. In our example, $0 \leq R \leq 4$, or there will be somewhere between zero and four heads.

What is the probability of four heads? Let H_i denote the event that a head occurs on the *i*th toss. Then four heads can happen only if the following sequence occurs: H_1, H_2, H_3, H_4. What is $P(H_1, H_2, H_3, H_4)$?

$$P(R = 4 \mid n = 4, p = 0.8) = P(H_1, H_2, H_3, H_4)$$

$$= P(H_1)\, P(H_2 \mid H_1)\, P(H_3 \mid H_2, H_1)\, P(H_4 \mid H_3, H_2, H_1)$$

(by using conditional probabilities)

$$= P(H_1)\, P(H_2)\, P(H_3)\, P(H_4)$$

(by the independence property)

$$= p^4 \quad \text{(by the constant probability property)}$$

$$= 0.8^4 \quad [P(\text{Heads}) = 0.8 \text{ for the loaded coin}]$$

$$= 0.41$$

What is the probability of three heads? In this case, three heads can occur in any of four ways:

$$H_1, H_2, H_3, T$$
$$H_1, H_2, T, H_4$$
$$H_1, T, H_3, H_4$$
$$T, H_2, H_3, H_4$$

The probability for any one of these sequences is $p^3(1 - p)$. Because these four sequences are mutually exclusive, the probability of three heads is simply the sum of the individual probabilities, or P(Three Heads in Four Tosses) = $P(R = 3 \mid n = 4, p = 0.8) = 4p^3(1 - p)^1 = 4(0.8^3)(0.2)$.

In general, the probability of obtaining r successes in n trials is given by

$$P_B(R = r \mid n, p) = \frac{n!}{r!(n - r)!} p^r (1 - p)^{n-r} \qquad (9.1)$$

where the subscript B indicates that this is a binomial probability. The term with the factorials is called the *combinatorial term*. It gives the number of ways that a sequence of n trials can have r successes. The second term is just the probability associated with a particular sequence with r successes in n trials. The expected number of successes is simply $E(R) = np$. (If I have n trials, I expect proportion p of them to be successes.) The variance of the number of successes is $\text{Var}(R) = np(1 - p)$.

Binomial probabilities are not difficult to calculate using Formula (9.1), but you do not have to calculate them yourself. Individual probabilities [$P_B(r$ Successes)] and cumulative terms [$P_B(r$ or Fewer Successes)] can be found in the tables in Appendixes A and B at the end of this book. For example, $P_B(R = 2 \mid n = 9, p = 0.12)$ is found in the following way, as illustrated in Figure 9.1. First, we find the value $n = 9$ along the left margin, and then the value for $r = 2$. Now read across that row until you find the column headed by $p = 0.12$. The probability you read is 0.212. Thus, $P_B(R = 2 \mid n = 9, p = 0.12) = 0.212$.

Figure 9.1

Using the binomial table to find $P_B(R = 2 \mid n = 9, p = 0.12)$.

n	r	p	.10	.12	.14
.	.		.	.	.
.	.		.	.	.
.	.		.	.	.
9	0		.387	.316	.257
	1		.387	.388	.377
	2	....	.172	.212	.245
	3		.045	.067	.093
	.		.	.	.
	.		.	.	.

The tables can be used to find binomial probabilities in many different forms. For example, what is $P_B(R > 3 \mid n = 15, p = 0.2)$? To find this probability, all that is necessary is to consider the complement:

$$P_B(R > 3 \mid n = 15, p = 0.2) = 1 - P_B(R \leq 3 \mid n = 15, p = 0.2)$$

$$= 1 - 0.648 \quad \text{(from Appendix B)}$$

$$= 0.352$$

You may have noticed that the values for p only go up to 0.50. What if you need to find a binomial probability with a p larger than 0.50? Let us try it. What is $P_B(R \leq 5 \mid n = 8, p = 0.7)$? To do this, we will look at the flip side of the situation. Getting five or fewer successes in eight trials is equivalent to getting three or more failures in eight trials, where $P(\text{Failure}) = 1 - P(\text{Success})$. Thus, we are arguing that

$$P_B(R \leq 5 \mid n = 8, p = 0.7) = P_B(R' \geq 3 \mid n = 8, p = 0.3)$$

where $R' = n - R$, or the number of failures. Now, to find $P_B(R' \geq 3 \mid n = 8, p = 0.3)$, use the idea of a complement:

$$P_B(R' \geq 3 \mid n = 8, p = 0.3) = 1 - P_B(R' \leq 2 \mid n = 8, p = 0.3)$$

$$= 1 - 0.552$$

$$= 0.448$$

An Example: Melissa Bailey   It is the beginning of winter term at the College of Business at Eastern State. Melissa Bailey says she is eager to study, but she has plans for a ski trip each weekend for 12 weeks. She will go only if there is good weather. The probability of good weather on any weekend during the winter is approximately 0.65. What is the probability that she will be gone for eight or more weekends?

Are all of the requirements satisfied for the binomial distribution? Weekend weather is good or bad, satisfying the dichotomous outcomes property. We will assume that the probabilities are the same from one weekend to the next, and moreover it seems reasonable to assume that the weather on one weekend during the winter is independent of the weather of previous weekends. Given these assumptions, the binomial distribution is an appropriate model for the uncertainty in this problem. Keep in mind that we are building a model of the uncertainty. Although our assumptions may not be exactly true, the binomial distribution should provide a good approximation.

To solve the problem, we must find $P_B(R \geq 8 \mid n = 12, p = 0.65)$. Of course, it is always possible to calculate this probability directly using the formula; however, let us see how to use the tables. The first problem is that the tables only give probabilities for p less than or equal to 0.50. To obtain the probability we want, we look at the problem in terms of bad weather rather than good. Being gone 8 or more weekends (good weather) out of 12 is the same as staying home on 4 or fewer weekends (bad weather). Find $P_B(R' \leq 4 \mid n = 12, p = 0.35)$. Using Appendix B, this is 0.583. Thus, there is more than a 50% chance that Melissa will be home on four or fewer weekends and gone on eight or more weekends.

Another Example: Soft Pretzels   Having just completed your degree in business, you are eager to try your skills as an entrepreneur by marketing a new pretzel that you have developed. You estimate that you should be able to sell them at a competitive price of 50 cents each. The potential market is estimated to be 100,000 pretzels per year. Unfortunately, because of a competing product, you know you will not be able to sell that many. After careful research and thought, you conclude that the following model of the situation captures the relevant aspects of the problem: Your new pretzel might be a hit, in which case it will capture 30% of the market in the first year. On the other hand, it may be a flop, in which case the market share will be only 10%. You judge these outcomes to be equally likely.

Being naturally cautious, you decide that it is worthwhile to bake a few pretzels and test market them. You bake 20, and in a taste test against the competing product, 5 out of 20 people preferred your pretzel. Given these new data, what do you think the chances are that your new pretzel is a hit? The following analysis is one way that you might analyze the situation.

The question we are asking is this: What is P(New Pretzel a Hit | 5 of 20 Preferred New Pretzel)? How can we get a handle on this probability? A problem like this that involves finding a probability given some new evidence almost certainly requires an application of Bayes' theorem. Let us use some notation to make our life simpler. Let "Hit" and "Flop" denote the outcomes that the new pretzel is a hit or a flop, respectively. Let R be the number of tasters (out of 20) who preferred the new pretzel. Now we can write down Bayes' theorem using this notation:

$$\text{P(New Pretzel a Hit} \mid 5 \text{ of 20 Preferred New Pretzel)}$$
$$= \text{P(Hit} \mid R = 5)$$
$$= \frac{\text{P}(R = 5 \mid \text{Hit)} \, \text{P(Hit)}}{\text{P}(R = 5 \mid \text{Hit)} \, \text{P(Hit)} + \text{P}(R = 5 \mid \text{Flop)} \, \text{P(Flop)}}$$

Next we must fill in the appropriate probabilities on the right-hand side of the Bayes' theorem equation. The probabilities P(Hit) and P(Flop) are easy. Based on the judgment (stated above) that these two outcomes are considered to be equally likely, we can say that P(Hit) = P(Flop) = 0.50.

What about $P(R = 5 \mid \text{Hit})$ and $P(R = 5 \mid \text{Flop})$? These are a bit trickier. Consider $P(R = 5 \mid \text{Hit})$. This is the binomial probability that 5 out of 20 people prefer your pretzel, given that 30% ($p = 0.30$) of the entire population of pretzel customers would prefer yours. That is, if your pretzel is a hit, you will capture 30% of the market. How does this idea of 30% across the entire population relate to the chance of 5 out of a sample of 20 preferring your pretzel? Provided that we can view the 20 people in our sample as "randomly selected," we can apply the binomial distribution. ("Randomly selected" means that each member of the population had the same chance of being chosen. That is, each taster has the same chance of preferring your pretzel, thus satisfying the independence and constant probabilities for the binomial distribution.) We have $p = 0.30$ (Hit) and $n = 20$ (the sample size), and so $P(R = 5 \mid \text{Hit}) = P_B(R = 5 \mid n = 20, p = 0.30)$. We can use Formula (9.1) or the table to find $P_B(R = 5 \mid n = 20, p = 0.30) = 0.179$.

The same argument can be made regarding P(R = 5 | Flop). Now the condition is that the pretzels are a flop. This means that only 10% of the population prefer your pretzel over the other. Thus, we have $p = 0.10$ in this case. This gives us P(R = 5 | Flop) = P_B(R = 5 | n = 20, p = 0.10). From the table, this probability is 0.032.

We now have everything we need to do the calculations that are required by Bayes' theorem:

$$P(\text{New Pretzel a Hit} \mid 5 \text{ of } 20 \text{ Preferred New Pretzel})$$
$$= P(\text{Hit} \mid R = 5)$$
$$= \frac{P(R = 5 \mid \text{Hit}) \, P(\text{Hit})}{P(R = 5 \mid \text{Hit}) \, P(\text{Hit}) + P(R = 5 \mid \text{Flop}) \, P(\text{Flop})}$$
$$= \frac{0.179(0.50)}{0.179(0.50) + 0.032(0.50)}$$
$$= 0.848$$

Thus, this evidence (5 out of 20 people preferring your pretzel) is good news. Your posterior probability that your pretzel will be a hit is almost 85%. Of course, we did the analysis on the basis of prior probabilities being P(Hit) = P(Flop) = 0.50. If you had assessed different prior probabilities, your answer would be different, although in any case your probability that the pretzel is a hit increases with the evidence.

The Poisson Distribution

While the binomial distribution is particularly good for representing successes in several trials, the *Poisson* distribution is good for representing occurrences of a particular event over time or space. Suppose, for example, that you are interested in the number of customers who arrive at a bank in one hour. Clearly this is an uncertain quantity; there could be none, one, two, three, and so on. The Poisson distribution also may be appropriate for modeling the uncertainty surrounding the number of machine breakdowns in a factory over some period of time. Other Poisson applications include modeling the uncertain number of blemishes in a bolt of fabric or the number of chocolate chips in a chocolate chip cookie.

The Poisson distribution requires the following:

1   Events can happen at any of a large number of places within the unit of measurement (hour, square yard, and so on), and preferably along a continuum.

2   At any specific point, the probability of an event is small. This simply means that the events do not happen too frequently. For example, we would be interested in a steady flow of customers to a bank, not a run on the bank.

3   Events happen independently of other events. In other words, the probability of an event at any one point is the same regardless of the time (or location) of other events.

4 The average number of events over a unit of measure (time or space) is constant no matter how far or how long the process has gone on.

Let X represent the uncertain number of events in a unit of time or space. Under the conditions given above, the probability that $X = k$ events is given by

$$P_P(X = k) = \frac{e^{-m}m^k}{k!} \qquad (9.2)$$

where the subscript P indicates this is a Poisson probability, e is the constant 2.718 . . . (the base of the natural logarithms), and m is a parameter that characterizes the distribution. In particular, m turns out to be both the expected number of events and the variance of the number of events. In symbols, $E(X) = m$ and $Var(X) = m$.

It is easy to calculate Poisson probabilities using Formula (9.2) and a good calculator. For example,

$$P_P(X = 2 \mid m = 1.5) = \frac{e^{-1.5}(1.5)^2}{2!}$$
$$= 0.251$$

Again, however, tables (Appendixes C and D) are available that give both individual Poisson probabilities as well as cumulative probabilities (the probability of k or fewer events) for different values of m. Using these tables is much like using the binomial tables. Find the value for m across the top and the value for k along the left side. Figure 9.2 illustrates the use of the Poisson table in Appendix C to confirm our answer above.

An Example: Blemishes in Fabric   As a simple example, suppose that you are interested in estimating the number of blemishes in 200 yards of cloth. Based on earlier experience with the cloth manufacturer, you estimate that a blemish occurs (on average) every 27 yards. At a rate of 1 blemish per 27 yards, this amounts to an approximate 7.4 blemishes in the 200 yards of cloth.

Is the Poisson distribution appropriate? Condition 1 is satisfied—we are looking at a continuous 200 yards of fabric. Condition 2 also is satisfied; apparently there are

Figure 9.2

Using the Poisson table to find $P_P(X = 2 \mid m = 1.5)$.

k	m	1.4	1.5	1.6
0		.247	.223	.202
1		.345	.335	.323
2	....	.242	.251	.258
3		.113	.126	.138

only a few blemishes in the 200 yards. Conditions 3 and 4 both should be satisfied unless blemishes are created by some machine malfunction that results in many blemishes occurring together. Thus, the Poison distribution appears to provide an appropriate model of the uncertainty in this problem.

The expected value of 7.4 suggests that we could use a Poisson distribution with $m = 7.4$. Probabilities can be calculated using Formula (9.2). For example, the probability of nine blemishes to the cloth is

$$P_P(X = 9 \mid m = 7.4) = \frac{e^{-7.4}(7.4)^9}{9!}$$
$$= 0.112$$

We also can confirm this answer by looking in Appendix C, although information from Appendix D may be more useful. For example, the probability of 4 or fewer blemishes is $P_P(X \leq 4 \mid m = 7.4) = 0.140$. We also can find the probability of more than 10 blemishes: $P_P(X > 10 \mid m = 7.4) = 1 - P_P(X \leq 10 \mid m = 7.4) = 1.000 - 0.871 = 0.129$. Although it is theoretically possible for there to be a very large number of blemishes, we can see that the probability of more than 18 is extremely low. In fact, it is less than 0.0005 (the probabilities in Appendixes C and D are rounded to three decimal places).

Soft Pretzels, Continued   Let us continue with the problem of the soft pretzels. You introduced the pretzels, and they are doing quite well. You have been distributing the product through several stores as well as through one street vendor. This vendor has been able to sell an average of 20 pretzels per hour. He had tried a different location earlier but had to move; sales there were only some 8 pretzels per hour, which was not enough to support his business. Now you are ready to try a second vendor in an altogether different location. The new location could be a "good" one, meaning an average of 20 pretzels per hour, "bad" with an average of only 10 per hour, or "dismal" with an average of 6 per hour. You have carefully assessed the probabilities of good, bad, or dismal using the assessment techniques from Chapter 8. Your judgment is that the new location is likely (probability 0.7) to be a good one. On the other hand, it could be (probability 0.2) a bad location, and it is just possible (probability 0.1) that the sales rate will be dismal.

After having the new stand open for a week, or long enough to establish a presence in the neighborhood, you decide to run a test. In 30 minutes you sell 7 pretzels. Now what are your probabilities regarding the quality of your new location? That is, what are P(Good | $X = 7$), P(Bad | $X = 7$), and P(Dismal | $X = 7$)?

As in the binomial case, we are interested in finding posterior probabilities given some new evidence. We will use Bayes' theorem to solve the problem:

P(Good | $X = 7$)

$$= \frac{P(X = 7 \mid \text{Good}) \, P(\text{Good})}{P(X = 7 \mid \text{Good}) \, P(\text{Good}) + P(X = 7 \mid \text{Bad}) \, P(\text{Bad}) + P(X = 7 \mid \text{Dismal}) \, P(\text{Dismal})}$$

We have our prior probabilities, P(Good) $= 0.7$, P(Bad) $= 0.2$, and P(Dismal) $= 0.1$. What about the probabilities P($X = 7$ | Good), P($X = 7$ | Bad), and P($X = 7$ | Dismal)?

First, note that we are talking about a 30-minute period. Thus, the expected number of sales is either 10, 5, or 3 per half hour, depending on whether the location is good, bad, or dismal. If the conditions for the Poisson distribution hold, then P(X = 7 | Good) is the Poisson probability of 7 occurrences when $m = 10$, $P_P(X = 7 \mid m = 10)$. From Appendix C, this probability is 0.090. Likewise, P(X = 7 | Bad) is the Poisson probability of 7 occurrences when $m = 5$, or 0.104. Finally, P(X = 7 | Dismal) is 0.022.

Now we can plug these values into Bayes' theorem:

$$P(\text{Good} \mid X = 7)$$

$$= \frac{P(X = 7 \mid \text{Good})\, P(\text{Good})}{P(X = 7 \mid \text{Good})\, P(\text{Good}) + P(X = 7 \mid \text{Bad})\, P(\text{Bad}) + P(X = 7 \mid \text{Dismal})\, P(\text{Dismal})}$$

$$= \frac{0.090(0.7)}{0.090(0.7) + 0.104(0.2) + 0.022(0.1)}$$

$$= 0.733$$

Likewise, we can calculate P(Bad | X = 7) = 0.242 and P(Dismal | X = 7) = 0.025. The posterior probabilities P(Good | X = 7) and P(Bad | X = 7) are not much different from the corresponding prior probabilities, so you might conclude that this information did not tell you much. On the basis of the information, however, the probability of the new location being dismal is now quite small.

The Exponential Distribution

The Poisson and binomial distributions are examples of discrete probability distributions because the outcome can take on only specific "discrete" values. What about continuous uncertain quantities? For example, tomorrow's high temperature could be any value between, say, 0° and 100°F. The per-barrel price of crude oil at the end of the year 2010 might be anywhere between $10 and $100. As discussed in Chapter 7, it is more natural in these cases to speak of the probability that the uncertain quantity falls within some interval. In the weather example we might look at historical weather patterns and determine that on 50% of days during May in Columbus, Ohio, the daily high temperature has been 70° or lower. On the basis of this, we could assess P(High Temperature in Columbus on May 28th ≤ 70°) = 0.50.

In this section, we will look briefly at the *exponential* distribution for a continuous random variable. In fact, the exponential distribution is closely related to the Poisson. If in the Poisson we were considering the number of arrivals within a specified period of time, then the uncertain time between arrivals (T) has an exponential distribution. The two go hand in hand; if the four conditions listed for the Poisson hold, then the time (or space) between events follows an exponential distribution.

For the binomial and Poisson distributions above, we were able to express in Formulas (9.1) and (9.2) the probability of a specific value for the random variable.

The corresponding expression for a continuous random variable is the *density function*. This function shows the relative likelihood for the different values that the uncertain quantity can take. For the exponential, the density function is

$$f_E(t \mid m) = me^{-mt} \qquad\qquad (9.3)$$

where *m* is the same average rate that we used in the Poisson and *t* represents the possible values for the uncertain quantity *T*. An exponential density function with $m = 2$ is illustrated in Figure 9.3. Recall from Chapter 7 that areas under the density function correspond to probabilities. Thus, the area from 0 to *a* in Figure 9.3 represents $P_E(T \le a \mid m = 2)$, where the subscript E indicates that this is a probability from an exponential distribution.

The exponential distribution turns out to be an easy distribution to work with. Probabilities are calculated by using the following formulas:

$$P_E(T \le a \mid m) = 1 - e^{-am}$$

$$P_E(T > a \mid m) = 1 - P(T \le a \mid m) = e^{-am}$$

$$P_E(b < T \le a \mid m) = P(T \le a \mid m) - P(T \le b \mid m) = e^{-bm} - e^{-am}$$

For example, we can calculate that

$$P_E(T > 15 \text{ Min} \mid m = 2 \text{ Arrivals per Hr}) = P_E(T > 0.25 \text{ Hr} \mid m = 2 \text{ Arrivals per Hr})$$

$$= e^{-0.25(2)}$$

$$= 0.607$$

Moreover, we can check this particular probability by using the Poisson table! The outcome "The time until the next arrival is greater than 15 minutes" is equivalent to the outcome "There are no arrivals in the next 15 minutes." Thus,

$$P_E(T > 15 \text{ Min} \mid m = 2 \text{ Arrivals per Hr})$$

$$= P_P(X = 0 \text{ Arrivals in 15 Min} \mid m = 2 \text{ Arrivals per Hr})$$

$$= P_P(X = 0 \text{ Arrivals in 15 Min} \mid m = 0.5 \text{ Arrival per 15 Min})$$

$$= 0.607$$

from Appendix C.

Figure 9.3

Exponential density function with parameter $m = 2$. The shaded area represents $P_E(T \le a \mid m = 2)$.

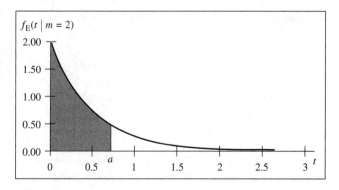

The expected value of an exponential random variable is $E(T) = 1/m$, and the variance is $Var(T) = 1/m^2$. Remember, these are units of time. For example, if $m = 6$ events per hour, then the expected time until the next event is $\frac{1}{6}$ of an hour, or 10 minutes.

Soft Pretzels, Again   The problem at hand is whether a pretzel can be prepared during the time between customers. If it takes 3.5 minutes to cook a pretzel, what is the probability that the next customer will arrive after it is finished, $P(T > 3.5 \text{ Min})$?

This is a difficult calculation because we do not know exactly what the rate m would be for an exponential distribution. But we can expand $P(T > 3.5 \text{ Min})$ by using the total probability formula:

$$P(T > 3.5 \text{ Min}) = P_E(T > 3.5 \text{ Min} \mid m = 20/\text{Hr}) \, P(m = 20)$$
$$+ \, P_E(T > 3.5 \text{ Min} \mid m = 10/\text{Hr}) \, P(m = 10)$$
$$+ \, P_E(T > 3.5 \text{ Min} \mid m = 6/\text{Hr}) \, P(m = 6)$$
$$= P_E(T > 0.0583 \text{ Hr} \mid m = 20/\text{Hr}) \, P(m = 20)$$
$$+ \, P_E(T > 0.0583 \text{ Hr} \mid m = 10/\text{Hr}) \, P(m = 10)$$
$$+ \, P_E(T > 0.0583 \text{ Hr} \mid m = 6/\text{Hr}) \, P(m = 6)$$
$$= e^{-0.0583(20)} \, P(m = 20) + e^{-0.0583(10)} \, P(m = 10)$$
$$+ \, e^{-0.0583(6)} \, P(m = 6)$$
$$= 0.3114 \, P(m = 20) + 0.5580 \, P(m = 10) + 0.7047 \, P(m = 6)$$

Now we can substitute in the posterior probabilities that we calculated above, $P(\text{Good} \mid X = 7) = 0.733$, $P(\text{Bad} \mid X = 7) = 0.242$, and $P(\text{Dismal} \mid X = 7) = 0.025$:

$$P(T > 3.5 \text{ Min}) = 0.3114(0.733) + 0.5580(0.242) + 0.7047(0.025)$$
$$= 0.3809$$

Thus, the probability is 0.3809 that the time between arrivals is greater than 3.5 minutes. Put another way, because $P(T \leq 3.5 \text{ min}) = 1 - 0.3809 = 0.6191$, we can see that the majority of customers will arrive before the next pretzel pops out of the oven.

The Normal Distribution

Another particularly useful continuous distribution is the *normal* distribution, which is the familiar bell-shaped curve. The normal distribution is particularly good for modeling situations in which the uncertain quantity is subject to many different sources of uncertainty or error. For example, in measuring something, errors may be introduced by a wide range of environmental conditions, equipment malfunctions, human error, and so on. Many measured biological phenomena (height, weight, length) often follow a bell-shaped curve that can be represented well with a normal distribution.

We will let Y represent an uncertain quantity that follows a normal distribution. If this is the case, the density function for Y is

$$f_N(y \mid \mu, \sigma) = \frac{1}{\sigma\sqrt{2\pi}} e^{-(y-\mu)^2/2\sigma^2} \qquad (9.4)$$

where μ and σ are parameters of the distribution and y represents the possible values that Y can take. In fact, it turns out that $E(Y) = \mu$ and $Var(Y) = \sigma^2$. Figure 9.4 illustrates a normal density function. As with the exponential, the area under the density function represents the probability that the random variable falls into the corresponding interval. For example, the shaded area in Figure 9.4 represents $P_N(a \le Y \le b \mid \mu, \sigma)$. Strictly speaking, a normal random variable can take values anywhere between plus and minus infinity. But the probabilities associated with values more than three or four standard deviations from the mean are negligible, so we often use the normal to represent values that have a restricted range (for example, weight or height, which can only be positive) as long as the extreme points are several standard deviations from the mean.

In the case of the normal distribution, there are no simple formulas for finding probabilities as there are for the exponential. A simple rule of thumb exists for the normal, however. The probability is approximately 0.68 that a normal random variable is within one standard deviation of the mean μ, and the probability is approximately 0.95 that it is within two standard deviations of the mean. In symbols:

$$P_N(\mu - \sigma \le Y \le \mu + \sigma) \approx 0.68$$
$$P_N(\mu - 2\sigma \le Y \le \mu + 2\sigma) \approx 0.95$$

Figure 9.4

Normal density function. The shaded area represents $P_N(a \le Y \le b \mid \mu, \sigma)$.

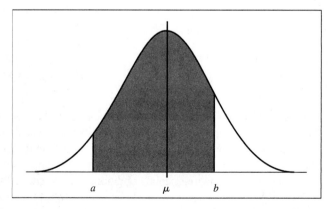

These are handy approximations for normal random variables and sometimes are called the *Empirical Rule* because these probabilities are found to hold (approximately) in many real-world situations.

A table has been provided in Appendix E so that you can find cumulative normal probabilities. But there is a catch! The table is for a *standard normal* distribution, one that has $\mu = 0$ and $\sigma = 1$. To use this table, we must know how to convert from a normal with any mean and standard deviation to a standard normal. The conversion works by subtracting the mean and dividing by the standard deviation. In symbols,

$$P_N(Y \le a \mid \mu, \sigma) = P_N(Z \le \frac{a - \mu}{\sigma} \mid \mu = 0, \sigma = 1)$$

$$= P_N\left(Z \le \frac{a - \mu}{\sigma}\right)$$

In most cases, we will leave off the values $\mu = 0$, $\sigma = 1$ when we talk about the standard normal random variable Z.

For example, if Y has a normal distribution with mean 10 and variance 400, then the probability that Y is less than or equal to 35 is

$$P_N(Y \le 35 \mid \mu = 10, \sigma^2 = 400) = P_N\left(Z \le \frac{35 - 10}{20}\right)$$

$$= P_N(Z \le 1.25)$$

Find this probability by looking in Appendix E as illustrated in Figure 9.5. This table has cumulative probabilities listed for values of z from −3.50 up to 3.49. Find $z = 1.25$, and read off $P(Z \le 1.25) = 0.8944$.

Figure 9.5

Using the normal distribution table to find $P_N(Z \le 1.25)$.

....	z	$P(Z \le z)$	....
....	....	....	....
	1.20	.8849	
....	1.21	.8869	....
	1.22	.8888	
....	1.23	.8907	....
	1.24	.8925	
	1.25	.8944	
....	1.26	.8962	....
	1.27	.8980	
....	1.28	.8997	....
	1.29	.9015	
....	....	....	....

An Investment Example   Consider three alternative investments, A, B, and C. Investing in C yields a sure return of $40. Investment A has an uncertain return (X), which is modeled in this case by a normal distribution with mean $50 and standard deviation $10. Investment B's return (represented by Y) also is modeled with a normal distribution having a mean of $59 and a standard deviation of $20. On the basis of expected values, B is the obvious choice because $59 is greater than $50 or $40. The decision tree is shown in Figure 9.6.

Although it is obvious that the expected payoff for B is greater than the sure $40 for C, we might be interested in the probability that B's payoff will be less than $40. To find this probability, we must convert the value of 40 in our problem to a *standardized* value. Subtract the mean and then divide by the standard deviation to get a standardized random variable Z:

$$P_N(B \le 40 \,|\, \mu = 59, \sigma = 20) = P_N\!\left(Z \le \frac{40 - 59}{20}\right)$$
$$= P_N(Z \le -0.95)$$

From Appendix E we find that

$$P(Z < -0.95) = 0.1711$$

Doing the same thing for Investment A, we want $P_N(A \le \$40 \,|\, \mu = 50, \sigma = 10)$:

$$P_N(A \le 40 \,|\, \mu = 50, \sigma = 10) = P\!\left(Z \le \frac{40 - 50}{10}\right)$$
$$= P(Z \le -1.0)$$
$$= 0.1587$$

Thus, even though Investment A has a lower expected value than does B, A has a smaller probability of having a return of less than $40. Why? The larger variance for B means that the distribution for B is spread out more than is the distribution for A.

Figure 9.6

Decision tree for three alternative investments.

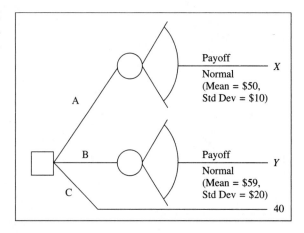

We also might be interested in the probability of a particularly large return, say,

$$P_N(B > 78 \mid \mu = 59, \sigma = 20) = P_N\left(Z > \frac{78 - 59}{20}\right)$$

$$= P_N(Z > 0.95)$$

Because Appendix E shows cumulative probabilities $P(Z \le z)$, we must do extra work in this case. We can find the probability of the complement $P(Z \le 0.95)$ and subtract this from 1:

$$P_N(Z > 0.95) = 1 - P_N(Z \le 0.95)$$

$$= 1 - 0.8289$$

$$= 0.1711$$

The probability $P(Z > 0.95)$ turns out to be the same as $P(Z \le -0.95)$. This example points out the symmetry of the normal distribution. For any z, $P(Z > z) = P(Z \le -z)$. To get our probability $P(Z > 0.95)$, we could have found it directly in the table by looking up $P(Z \le -0.95)$. Figure 9.7 shows this property graphically.

Example: Quality Control   Suppose that you are the manager for a manufacturing plant that produces disk drives for personal computers. One of your machines produces a part that is used in the final assembly. The width of this part is important to the disk drive's operation; if it falls below 3.995 or above 4.005 millimeters (mm), the disk drive will not work properly and must be repaired at a cost of $10.40.

The machine can be set to produce parts with a width of 4 mm, but it is not perfectly accurate. In fact, the actual width of a part is normally distributed with mean 4 mm and a variance that depends on the speed of the machine. If the machine is run at a slower speed, the width of the produced parts has a standard deviation of 0.0019 mm. At the higher speed, however, the machine is less precise, producing parts with a standard deviation of 0.0026 mm.

Figure 9.7

The symmetry of the normal distribution.

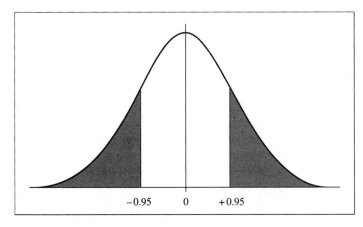

-0.95 0 $+0.95$

Of course, the higher speed means that more parts can be produced in less time, thus reducing the overall cost of the disk drive. In fact, it turns out that the cost of the disk drive when the machine is set at high speed is $20.45. At low speed, the cost is $20.75.

The question that you as plant manager face is whether it would be better to run the machine at high or low speed. Is the extra expense of lower speed more than offset by the increased precision and hence the lower defect rate? You would like to choose the strategy with the lower expected cost. Your decision tree is shown in Figure 9.8.

To decide, we need to know the probability of a defective unit under both machine settings. Because the width of the part follows a normal distribution in each case, we can get these probabilities by calculating z values and using Appendix E:

$$
\begin{aligned}
\text{P(Defective} \mid \text{Low Speed)} &= 1 - \text{P(Not Defective} \mid \text{Low Speed)} \\
&= 1 - \text{P}_\text{N}(3.995 \leq Y \leq 4.005 \mid \mu = 4, \sigma = 0.0019) \\
&= 1 - \text{P}\left(\frac{3.995 - 4}{0.0019} \leq Z \leq \frac{4.005 - 4}{0.0019}\right) \\
&= 1 - \text{P}(-2.63 \leq Z \leq 2.63) \\
&= 1 - [\text{P}(Z \leq 2.63) - \text{P}(Z \leq -2.63)] \\
&= 1 - (0.9957 - 0.0043) \\
&= 1 - 0.9914 \\
&= 0.0086
\end{aligned}
$$

Likewise, we can calculate P(Defective | High Speed):

$$
\begin{aligned}
\text{P(Defective} \mid \text{High Speed)} &= 1 - \text{P(Not Defective} \mid \text{High Speed)} \\
&= 1 - \text{P}_\text{N}(3.995 \leq Y \leq 4.005 \mid \mu = 4, \sigma = 0.0026) \\
&= 1 - \text{P}\left(\frac{3.995 - 4}{0.0026} \leq Z \leq \frac{4.005 - 4}{0.0026}\right) \\
&= 1 - \text{P}(-1.92 \leq Z \leq 1.92) \\
&= 1 - [\text{P}(Z \leq 1.92) - \text{P}(Z \leq -1.92)] \\
&= 1 - (0.9726 - 0.0274) \\
&= 1 - 0.9452 \\
&= 0.0548
\end{aligned}
$$

Figure 9.8

Decision tree for a quality-control problem.

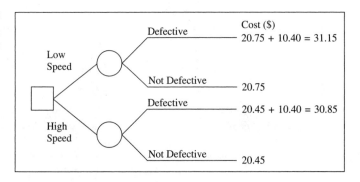

With these probabilities, it now is possible to calculate the expected cost for each alternative:

$$E(\text{Cost} \mid \text{Low Speed}) = 0.9914(\$20.75) + 0.0086(\$31.15)$$

$$= \$20.84$$

$$E(\text{Cost} \mid \text{High Speed}) = 0.9452(\$20.45) + 0.0548(\$30.85)$$

$$= \$21.02$$

Thus, in this case the increased cost from the slower speed is more than offset by the increased precision and lower defect rate, and you would definitely choose the slower speed.

The Beta Distribution

Suppose you are interested in the proportion of voters who will vote for the Republican candidate in the next presidential election. If you are uncertain about this proportion, you may want to encode your uncertainty as a continuous probability distribution. Because the proportion can take only values between 0 and 1, neither the exponential nor the normal distribution can adequately reflect the uncertainty you face. The *beta* distribution, however, may be appropriate. Let Q denote an uncertain quantity that can take any value between 0 and 1. Then the beta density function is

$$f_\beta(q \mid r, n) = \frac{(n-1)!}{(r-1)! \; (n-r-1)!} q^{r-1}(1-q)^{n-r-1} \qquad \textbf{(9.5)}$$

As before, q represents the possible values between 0 and 1 that Q can take.

The numbers r and n are parameters that determine the shape of the density function. If n is large, the distribution is fairly "tight," whereas if n is small, the distribution is more "spread out" (Figure 9.9). If $r = n/2$, the density function is symmetric around 0.5. If this is not the case, however, the distribution is skewed to the right or left depending on whether $r < n/2$ or $r > n/2$ as in Figure 9.10. As usual with density functions, the area under the curve represents probability. Thus, in Figure 9.10 the shaded area represents $P_\beta(0.2 \leq Q \leq 0.4 \mid n = 6, r = 4)$. A table in Appendix F provides cumulative probabilities for a wide variety of different beta distributions. We will demonstrate this table's use shortly.

Formula (9.5) for the density function looks much like the binomial distribution [Formula (9.1)]. Keep in mind that the beta distribution is a distribution for Q, a continuous random variable, whereas the binomial distribution is a distribution for R, the number of successes in n trials. Thus, we are considering two entirely different uncertain quantities. The two distributions, however, are closely related.

The expected value of a beta random variable is $E(Q) = r/n$, and the variance is $\text{Var}(Q) = r(n-r)/[n^2(n+1)]$. Looking at the formula for the expected value, r/n, the

Figure 9.9

Some symmetric beta distributions.

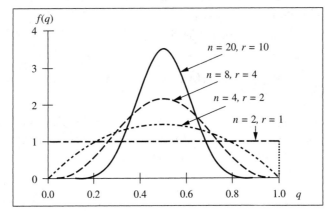

Figure 9.10

Some asymmetric beta distributions.

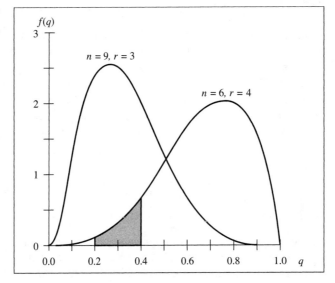

relationship between the beta and the binomial becomes apparent. Loosely speaking, r and n still can be interpreted as r successes in n trials. For example, if you had observed 4 successes in 10 trials from a binomial distribution with unknown proportion Q, your best guess for Q would be $\frac{4}{10}$, or 0.40. Moreover, you might concede that Q might not be exactly 0.40, although it should be close to 0.40.

One way to use a beta distribution to model your subjective beliefs about an uncertain Q is to imagine a sample that would be roughly equivalent to your information. In the case of the proportion of voters who will vote Republican, you may have a feeling that the proportion is "around 0.30," and perhaps you feel as if all of your information (reading newspapers, talking to friends) is roughly equivalent to having polled a random sample of 20 people. These beliefs about Q might be represented by a beta distribution with $r = 6$ and $n = 20$.

Another way to fit a beta distribution is to use the cumulative probability table for the beta distribution (Appendix F). This table works a lot like the others. First, find the appropriate r and n along the left margin; for example, look at a distribution with $r = 6$ and $n = 20$. Read across the row to find the cumulative probabilities for values of q from 0 to 1 (column headings). For example, reading across, you find the number ".11" in the $q = 0.18$ column (Figure 9.11). This means that for a beta distribution with $n = 20$ and $r = 6$, $P_\beta(Q \leq 0.18 \mid n = 20, r = 6) = 0.11$, or there is an 11% chance that Q is less than 0.18. Likewise, you can see that $P_\beta(Q \leq 0.28 \mid n = 20, r = 6) = 0.45$, and $P_\beta(Q \leq 0.30 \mid n = 20, r = 6) = 0.53$. From these two values, we can deduce that the median for this distribution is approximately 0.29. That is, $P_\beta(Q \leq 0.29 \mid n = 20, r = 6)$ is approximately 0.50. We also can see that $P_\beta(Q \leq 0.48 \mid n = 20, r = 6) = 0.95$. This statement means there is a 95% chance that Q is less than 0.48 in this distribution, and hence only a 5% chance that Q is greater than 0.48.

To use the table to fit a beta distribution to your beliefs, you might assess the median and upper and lower quartiles of your subjective distribution for Q using the techniques discussed in Chapter 8 for assessing a continuous subjective probability distribution. Then look in the table for a combination of r and n that gives a beta distribution with the same (or nearly the same) median and quartiles. For example, suppose you assess the median of your subjective distribution to be 0.40, the lower quartile to be 0.29, and the upper quartile to be 0.52. That is,

$$P(Q \leq 0.29) = 0.25$$
$$P(Q \leq 0.40) = 0.50$$
$$P(Q \leq 0.52) = 0.75$$

Looking in the table, a beta distribution with $r = 4$ and $n = 10$ has approximately the same median and quartiles:

$$P_\beta(Q \leq 0.29 \mid n = 10, r = 4) = 0.25$$
$$P_\beta(Q \leq 0.40 \mid n = 10, r = 4) = 0.52$$
$$P_\beta(Q \leq 0.52 \mid n = 10, r = 4) = 0.78$$

Given the closeness between this beta distribution and your subjective assessments, you might find it useful to represent your subjective beliefs with the distribution.

Figure 9.11

Using the beta distribution table to find
$P_\beta(Q \leq 0.18 \mid n = 20, r = 6) = 0.11$.

n	r	Values for Q	0.16	0.18	0.20
	.		.	.	.
	.		.	.	.
20	2		.83	.88	.92
	4		.36	.45	.54
	6	. . . .	.07	.11	.16
	8		.01	.01	.02
	.		.	.	.
	.		.	.	.

Suppose, however, that you had assessed

$$P(Q \leq 0.36) = 0.25$$
$$P(Q \leq 0.40) = 0.50$$
$$P(Q \leq 0.52) = 0.75$$

No beta distributions have characteristics that are very close to these. The one with $r = 2$ and $n = 5$ has (roughly) the correct median and upper quartile: $P_\beta(Q \leq 0.40 \mid r = 2, n = 5) = 0.52$ and $P_\beta(Q \leq 0.52 \mid r = 2, n = 5) = 0.72$. On the lower end, however, the approximation is poor: $P_\beta(Q \leq 0.36 \mid r = 2, n = 5) = 0.45$. In this case, we would have three choices. We could be satisfied with a not-so-great approximation, use our subjective assessment directly (perhaps with a discrete approximation), or find another way to approximate our beliefs.

Soft Pretzels One Last Time   Let us return to the issue of how much market share your pretzel will capture. Denote the market share by $Q;$ it is an uncertain quantity that must be between 0 and 1. You might want to model your beliefs about Q with a beta distribution. You know that Q is not likely to be close to 1. Suppose also that you think that the median is approximately 0.20 and the upper quartile is 0.38. To represent these assessments, you might choose a beta distribution with $r = 1$ and $n = 4$. From the table, you can see that this distribution for Q has a median around 0.21 and an upper quartile around 0.37:

$$P_\beta(Q \leq 0.20 \mid r = 1, n = 4) = 0.49$$
$$P_\beta(Q \leq 0.38 \mid r = 1, n = 4) = 0.76$$

The expected value of Q is $r/n = \frac{1}{4} = 0.25$.

Should you become a pretzel entrepreneur? Recall that the decision was to price the pretzels at 50 cents each and that the total market is estimated to be 100,000 pretzels. You figure that the total fixed cost of the project amounts to $8000 for marketing, finance costs, and overhead. The variable cost of each pretzel is 10 cents.

You only have one year to prove yourself, and then you will go on to graduate school no matter what happens. It also is apparent that this venture alone will not provide enough income to live on, but embarking on it will not interfere with other plans for gainful employment. Your rich uncle has agreed to help you financially, and you have savings of your own. The only question is whether the net contribution of the project will be positive or negative.

Write the equation for net contribution as the difference between total revenue and total costs:

$$\text{Net Contribution} = 100{,}000\ (0.50)Q - 100{,}000\ (0.10)Q - 8000$$
$$= 100{,}000\ (0.40)Q - 8000$$

The marginal contribution per pretzel sold is 40 cents, or 0.40. The expected net contribution is

$$E(\text{Net Contribution}) = 40{,}000E(Q) - 8000$$
$$= 40{,}000(0.25) - 8000$$
$$= 10{,}000 - 8000$$
$$= 2000$$

On the basis of expected value, becoming a pretzel entrepreneur is a good idea! But the project also is a pretty risky proposition. What is the probability, for instance, that the pretzels could result in a loss? To answer this, we can find the specific value for Q (call it $q*$) that would make the return equal to zero:

$$0 = 40,000q* - 8000$$

$$q* = \frac{8000}{40,000} = 0.20$$

If Q turns out to be less than 0.20, you would have been better off with the savings account. We assessed the median of Q to be approximately 0.20. This means that there is just about a 50% chance that the pretzel project will result in a loss. Are you willing to go ahead with the pretzels?

Viewing Theoretical Distributions with RISKview

In this chapter we learned to model uncertainties using five theoretical distributions (binomial, Poisson, exponential, normal, and beta). We also learned how to compute probabilities for these distributions. We can expand our repertoire beyond these to the 35 theoretical distributions in RISKview's library. With this array of distributions, we will have considerable flexibility for matching a distribution to a specific situation, allowing us to model a wide variety of uncertainties.

RISKview's graphical interface can help choose and mold a distribution to fit specific situations because it displays changes in both the graph and the numerical summaries as the parameter values change. Seeing these changes graphically helps demystify the complex formulas that describe many distributions. For example, few people can envision the wide variety of shapes the beta distribution can take on by merely examining Formula (9.5). With RISKview, however, you can actually watch as the beta distribution changes from a bell-shape, to a J-shape, and to a U-shape as the parameter values change. The graphical interface provides a quick and easy check of whether a particular distribution makes sense for modeling a specific uncertainty. We will demonstrate how to select a distribution and how to modify its parameters for a particular situation.

STEP 1

1.1   Open Excel and then @RISK, enabling any macros if prompted. RISKview is both a stand-alone program and is embedded within @RISK. We demonstrate RISKview within @RISK to highlight the integration between the two programs.

1.2   There are four on-line help options available. Pull down the @**RISK** menu and choose the **Help** command. *"How Do I"* answers frequently asked questions about @RISK, *Online Manual* is an electronic copy (pdf file) of the user's guide, *@RISK Help* is a searchable database of the

> user's guide, and *PDF Help* lists various facts about the probability distribution functions available in @RISK and RISKview. The user's guide describes how to operate RISKview, whereas *PDF Help* lists particular facts about each distribution.

Discrete Distributions

STEP 2

2.1   Highlight cell **A1** in Excel.

2.2   Open RISKview by either clicking on the **Define Distributions** button (third from the left in the *@RISK* toolbar) or **right-clicking** the mouse and choosing **@RISK** and then **Define Distributions**. This opens an *@RISK Definition for A1* window showing the standard normal distribution. We demonstrate how to construct the binomial distribution used in the soft-pretzel example and how to compute probabilities for this distribution. Similar procedures hold for other discrete distributions.

2.3   Unfurl the distribution list by clicking on the **black triangle** to the right of the *Dist* name box that currently contains the name *Normal*.

2.4   Scroll until you see binomial. Click on **binomial**. The binomial density that appears has $n = 5$ and $p = 0.5$.

2.5   Change n to **20** and change p to **0.30,** as shown in Figure 9.12. You can either delete the current entry and type in your parameter value or you can use the up/down buttons. The up/down buttons allow you to watch changes in the distribution's graph and summary statistics as n increases to 20 and p decreases to 0.30.

2.6   To the right of the graph, various summary statistics are given for this binomial distribution; for example, you see that the mean, mode, and median are all equal to 6. Note that the minimum is 0 and the maximum is 20, but the x-axis scale goes from –5 to 25. To change the x-axis scale, click on one of the **clear triangles** in the upper corners of the graph, and slide the vertical bar left or right. Fix your scale to a minimum of **0** and a maximum of **20**. If you wish to return to the original scale settings, click on *AutoScaling Off.*

STEP 3

3.1   The two interior vertical bars overlaying the graph indicate x values and the x value's corresponding cumulative probabilities. You can easily compute any probability value you wish by using these bars, which are called delimiters by RISKview. For example, the leftmost delimiter in Figure 9.12 is at the x value of 3 (3.000 is shown under the scrollbar), and 10%

Figure 9.12

The RISKview window in @RISK showing the binomial distribution with n = 20 and p = 0.3.

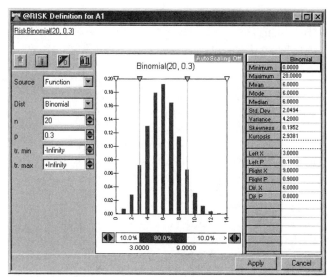

is shown in the scrollbar. Thus, the probability that $R \leq 3$ is 10%; that is, $P_B(R \leq 3 \mid n = 20, p = 0.30) = 0.10$.

3.2   Similarly, $P_B(R \leq 9 \mid n = 20, p = 0.30) = 0.90$ because the rightmost delimiter is at 9.000 and a total of 90% (10% + 80%) is shown to the left of 9 on the scrollbar. Figure 9.12 also shows that $P_B(3 \leq R \leq 9 \mid n = 20, p = 0.30) = 0.80$.

3.3   The delimiters can be moved in three ways: (1) clicking on the **gray triangle** at top and moving the mouse left or right, (2) repeatedly clicking on the **black triangle** on the scrollbar, and (3) typing in the exact x value or p value in the cells to the right of the graph. If you type a value in the **Left X** cell, then RISKview calculates the cumulative probability and reports it in the *Left P* cell. If you type in the cumulative probability, RISKview reports the corresponding x value. Experiment with these three methods.

3.4   In the soft pretzel example, we found $P_B(R = 5 \mid n = 20, p = 0.30) = 0.179$ using Appendixes A and B. To find this probability using RISKview, we need to type in a *Left X* value of less than 5, but greater than 4, and a *Right X* value of 5 or more but less than 6. Why? Click in the cell to the right of *Left X*, and enter **4.5**. Click in the cell to the right of *Right X*, and enter **5**.

3.4   The *Left P* cell below *Left X* now reads 0.2375, which in our notation is $P_B(R \leq 4.5 \mid n = 20, p = 0.30) = 0.2375$. The *Right P* cell below *Right X* reads 0.4165, which in our notation is $P_B(R \leq 5 \mid n = 20, p = 0.30) = 0.4164$. Thus,

$$P_B(R = 5 \mid n = 20, p = 0.30) = 0.4164 - 0.2375 = 0.1789.$$

3.6   The difference between the left and right probabilities is computed for you in the last cell on the right, labeled *Dif. P.*

There are eight discrete distributions available in RISKview; five are standard (Binomial, Geometric, HyperGeometric, NegativeBinomial, and Poisson) and the other three (DUniform, Discrete, and IntUniform) are designed for constructing custom discrete distributions. Using Steps 2 and 3, you can view each of these standard distributions and calculate any probabilities you wish. The custom distributions allow you to build completely general discrete distributions. The most general option is called Discrete, in which you enter all the values and their corresponding probabilities. For example, you would use Discrete to model the defective rates in the quality-control problem in this chapter. The DUniform and IntUniform options are more specialized in that they assume that all the entered values have equal probabilities. For example, you could use these options in modeling a fair die because the outcomes 1, 2, 3, 4, 5, and 6 are equally likely. The difference between DUniform and IntUniform is the x values that you input. DUniform lets you enter any set of x values, whereas IntUniform requires a consecutive sequence of integers. Thus, for the fair die you would need to enter 1, 2, 3, 4, 5, and 6 for DUniform, but need only enter the minimum (1) and the maximum (6) for IntUniform.

STEP 4

4.1 If you wish to change the appearance of the graph (view the cumulative distribution, add titles, etc.), **right-click** the mouse when the cursor is over the graph and choose **Format Graph**. See page 331 in Chapter 8 for a description of the formatting options available. Click **OK** when you have made your changes.

4.2 If you wish to place the distribution in cell A1, click **Apply**.

Continuous Distributions

STEP 5

5.1 Highlight cell **A2** in the Excel worksheet.

5.2 Open the @*RISK Definition* window by either clicking on the **Define Distribution** button (third button from the left on the @RISK toolbar) or **right-clicking** the mouse and choosing **@RISK,** then **Define Distributions**.

5.3 Choose the beta distribution from the distribution list by clicking on the **black triangle** to the right of *Normal*, scrolling up and selecting **Beta**. The beta(2, 2) distribution appears.

5.4 To see a few of the possible shapes that the beta distribution can assume, click on the **downward** arrow button for $\alpha 1$ until you reach **1.0**. Watch as the distribution's shape and the summary statistics change as the parameter decreases. The shape should now be a right triangle.

> 5.5 Now click on the **downward** arrow button for α2 until you reach **1.0**. The distribution is now flat, which is the uniform distribution between 0 and 1. Continue to decrease α2 until you reach **0.6**.
>
> 5.6 Return to α1 and decrease it until **α1 = 0.6**. The distribution is now U-shaped.

You may have noticed that RISKview's beta distribution behaves differently from the beta distribution described in the text. For example, the uniform distribution occurs when α1 = α2 = 1 in RISKview, but in the text it occurs when $n = 2$ and $r = 1$. The discrepancy is because RISKview uses a different parameterization or formula from what we used in the text. To reconcile the formulas, we need to find the correspondence between RISKview's choice of parameters (α1 and α2) and our choice (n and r). The correspondence is α1 = r and α2 = $n - r$. For example, when $n = 2$ and $r = 1$, then α1 = 1 and α2 = 2 − 1 = 1.

Let's view the particular beta distribution we used to model market share for the pretzels. Referring back to the example, we used the parameters $r = 1$ and $n = 4$, which for RISKview correspond to α1 = 1 and α2 = 4 − 1 = 3.

> 5.7 Change the RISKview parameters for the beta distribution to **1** for α1 and **3** for α2. The distribution is shown in Figure 9.13.

Figure 9.13

RISKview window showing the beta(1, 3) distribution.

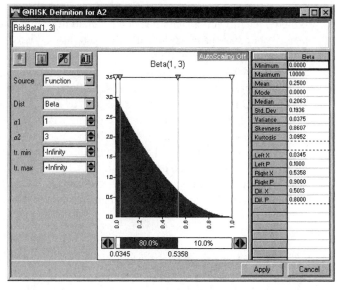

STEP 6

6.1   In the column to the right, various summary statistics are listed along with the *x* values and corresponding cumulative probabilities of the delimiters. The expected value (mean) of a beta(1, 3) distribution is 0.25, the median is 0.2063, and the probability is 0.80 that market share will fall between 3.45% and 53.58%. Make sure you know where to find these numbers.

6.2   To calculate the upper quartile, enter **0.75** for the *Right P* value. The corresponding *Right X* value is the upper quartile value of 0.3700.

The only other discrepancy between the formula RISKview uses and what we used in the text is for the exponential distribution. Whereas we used the formula $f_E(t \mid m) = me^{-mt}$, RISKview uses $f_E(t \mid \beta) = (1/\beta)e^{-(1/\beta)t}$. The two formulas are easily reconciled by letting $m = 1/\beta$, which establishes the correspondence between the parameters β and m. Thus, you would use $\beta = 1/m = 0.5$ in RISKview to generate the same exponential distribution we did with $m = 2$ in the text.

The last feature of RISKview we discuss in this section is truncating a distribution. Truncation means that within specified bounds the distribution is defined in its standard form and outside those bounds it is defined to equal zero. To set the lower and upper bounds for truncating, use the *tr. min* and *tr. max* entry cells, which are located to the left of the graph (Figure 9.13). The idea behind truncating is that a particular distribution may model the uncertainty well within a range, but the variable cannot take on values outside that range. For example, the normal distribution is often used to model heights of individuals, but height cannot be negative. Thus, we might use the truncated normal distribution with a lower bound of zero. Truncation may seem like a good idea, but it is often not necessary if the boundary limits are sufficiently far away from the mean. For example, if you are using a normal distribution to model height with mean 5 feet and standard deviation 1 foot, the probability of a negative number (negative height) with this distribution is less than 0.000000287 (less than one out of a million). This is a small enough probability that we might choose to go ahead with the untruncated distribution for height.

We can build custom continuous distributions in a couple of different ways with RISKview. One possibility is to use the general button, which is similar to the discrete button. Be cautious, however, when using RISKview's general distribution because you must enter the density height that corresponds to each *x* value. Unlike discrete distributions, this is not the probability that corresponds to the *x* value. Another possibility is to start with a theoretical distribution and modify it in the Distribution Artist window using techniques described Chapter 8. Whatever method you choose, be sure to verify that your distribution matches the uncertainty you are modeling.

SUMMARY   In this chapter we examined ways to use theoretical probability distributions in decision-analysis problems. The distributions we considered were the binomial, Poisson, exponential, normal, and beta. These are only a few of the simpler ones; many other theoretical distributions are available. Each is appropriate in different situations, and a decision analyst must develop expertise in recognizing which distribution provides the best model of the uncertainty in a decision situation.

EXERCISES

9.1 In the binomial example with Melissa Bailey (page 356), what is the probability that she will be gone six or more weekends?

9.2 Suppose you are interested in an investment with an uncertain return. You think that the return could be modeled as a normal random variable with mean $2000 and standard deviation $1500. What is the probability that the investment will end up with a loss? What is the probability that the return will be greater than $4000?

9.3 If there are, on average, 3.6 chocolate chips per cookie, what is the probability of finding no chocolate chips in a given cookie? Fewer than 5 chocolate chips? More than 10 chips?

9.4 Refer to the discussion of the pretzel problem in the section on the beta distribution (pages 372–373). Find the probability that the net contribution of the pretzel project would be greater than $4500.

9.5 Table look-up and calculation practice (or use RISKview)

    a Binomial distribution. Find the following probabilities:

$$P_B(R = 5 \mid n = 10, p = 0.22) \qquad P_B(R = 10 \mid n = 15, p = 0.63)$$
$$P_B(R \le 1 \mid n = 7, p = 0.04) \qquad P_B(R \le 3 \mid n = 10, p = 0.35)$$
$$P_B(7 < R \le 10 \mid n = 12, p = 0.50) \qquad P_B(R > 1 \mid n = 19, p = 0.06)$$
$$P_B(R = 4 \mid n = 6, p = 0.42) \qquad P_B(R < 2 \mid n = 4, p = 0.67)$$

    b Poisson distribution. Find the following probabilities:

$$P_P(X = 3 \mid m = 2.0) \qquad P_P(X \ge 17 \mid m = 15)$$
$$P_P(X > 4 \mid m = 5.2) \qquad P_P(X < 7 \mid m = 10.0)$$
$$P_P(X \le 1 \mid m = 3.9) \qquad P_P(3 \le X < 7 \mid m = 1.5)$$
$$P_P(X = 4 \mid m = 1.75) \qquad P_P(X \le 2 \mid m = 3.56)$$

    c Exponential distribution. Find the following probabilities:

$$P_E(T \le 5 \mid m = 1) \qquad P_E(T \ge 4 \mid m = 0.25)$$
$$P_E(0.25 < T < 1 \mid m = 2) \qquad P_E(T \ge 3.2 \mid m = 2)$$
$$P_E(T = 3.2 \mid m = 2) \qquad P_E(T \ge k \mid m = \tfrac{1}{k})$$

    d Normal distribution. Find the following probabilities:

$$P_N(Y \le 12 \mid \mu = 10, \sigma = 4) \qquad P_N(Y > 50 \mid \mu = 100, \sigma = 30)$$
$$P_N(100 < Y \le 124 \mid \mu = 133, \sigma = 15) \qquad P_N(Y > 20 \mid \mu = 15, \sigma = 4)$$
$$P_N(Y > 0 \mid \mu = -10, \sigma = 4) \qquad P_N(-44 \le Y \le 12 \mid \mu = 10, \sigma = 20)$$
$$P_N(Y = 12 \mid \mu = 10, \sigma = 4) \qquad P_N(Y < 12 \mid \mu = 10, \sigma = 4)$$

    e Beta distribution. Find the following probabilities:

$$P_\beta(Q \le 0.9 \mid n = 10, r = 9) \qquad\qquad P_\beta(Q \ge 0.5 \mid n = 5, r = 2)$$
$$P_\beta(Q \le 0.44 \mid n = 30, r = 20) \qquad\quad P_\beta(0.04 < Q \le 0.38 \mid n = 18, r = 8)$$
$$P_\beta(0.12 < Q \le 0.25 \mid n = 14, r = 4) \qquad P_\beta(Q = 0.76 \mid n = 18, r = 14)$$

9.6   Find values for z such that

$$P_N(Z \le z \mid \mu = 0, \sigma = 1) = 0.05 \qquad\qquad P_N(Z > z \mid \mu = 0, \sigma = 1) = 0.25$$
$$P_N(Z \le z \mid \mu = 0, \sigma = 1) = 0.50 \qquad\qquad P_N(Z > z \mid \mu = 0, \sigma = 1) = 0.10$$

9.7   Find the parameters (μ and σ) for a normal distribution whose first and third quartiles are 125 and 275. That is, $P_N(Y \le 125 \mid \mu, \sigma) = 0.25$ and $P_N(Y \le 275 \mid \mu, \sigma) = 0.75$. What are μ and σ?

9.8   An exponential distribution has $P_E(T \ge 5 \mid m) = 0.24$. Find m.

9.9   A Poisson distribution has $P_P(X \ge 12 \mid m) = 0.01$ and $P_P(X \le 2 \mid m) = 0.095$. Find m.

9.10 A Poisson distribution has $P_P(X = 0 \mid m) = 0.175$. Calculate m.

9.11 Use Appendix E or RISKview to verify the Empirical Rule (page 365) for normal distributions, which states that the probability is approximately 0.68 that a normal random variable is within one standard deviation of the mean μ and the probability is approximately 0.95 that the random variable is within two standard deviations of the mean.

QUESTIONS AND PROBLEMS

9.12 The amount of time that a union stays on strike is judged to follow an exponential distribution with a mean of 10 days.

    a Find the probability that a strike lasts less than one day.

    b Find the probability that a strike lasts less than six days.

    c Find the probability that a strike lasts between six and seven days.

    d Find the conditional probability that a strike lasts less than seven days, given that it already has lasted six days. Compare your answer to part a.

9.13 A photographer works part-time in a shopping center, two hours per day, four days per week. On the average, six customers arrive each hour, and the arrivals appear to occur independently of one another. Twenty minutes after she arrives one day, the photographer wonders what the chances are that exactly one customer will arrive during the next 15 minutes. Find this probability (i) if two customers just arrived within the first 20 minutes and (ii) if no customers have come into the shop yet on this particular day.

9.14 A consumer is contemplating the purchase of a new compact disc player. A consumer magazine reports data on the major brands. Brand A has lifetime (T_A), which is exponentially distributed with $m = 0.2$; and Brand B has lifetime (T_B), which is exponentially distributed with $m = 0.1$. (The unit of time is one year.)

    a Find the expected lifetimes for A and B. If a consumer must choose between the two on the basis of maximizing expected lifetime, which one should be chosen?

b Find the probability that A's lifetime exceeds its expected value. Do the same for B. What do you conclude?

c Suppose one consumer purchases a Brand A compact disc, and another consumer purchases a Brand B compact disc. Find the mean and variance of (i) the average lifetime of the two machines and (ii) the difference between the lifetimes of the two machines. (*Hint:* You must use the rules about means and variances of linear transformations that we discussed in Chapter 7.)

9.15 On the basis of past data, the owner of an automobile dealership finds that, on average, 8.5 cars are sold per day on Saturdays and Sundays during the months of January and February, with the sales rate relatively stable throughout the day. Moreover, purchases appear to be independent of one another. The dealership is open for 10 hours per day on each of these days. There is no reason to believe that the sales for the upcoming year will be any different from in the past.

a On the first Saturday in February, the dealership will open at 9 A.M. Find the probability that the time until the first sale is more than two hours, $P_E(T \geq 2 \text{ hours} \mid m = 8.5$ cars per 10 hours).

b Find the probability that the number of sales before 11 A.M. is equal to zero, $P_P(X = 0 \text{ in 2 hours} \mid m = 8.5$ cars per 10 hours). Compare your answer with that from part a. Can you explain why the answers are the same?

c The owner of the dealership gives her salespeople bonuses, depending on the total number of cars sold. She gives them $20 whenever exactly 13 cars are sold on a given day, $30 whenever 14 cars are sold, $50 whenever 15 cars are sold, and $70 whenever 16 or more cars are sold. On any given Saturday or Sunday in January or February, what is the expected bonus that the dealer will have to pay?

d Consider the bonus scheme presented in part c. February contains exactly four Saturdays and four Sundays. What is the probability that the owner will have to pay the $20 bonuses exactly twice in those days?

9.16 Reconsider your assessed 0.05 and 0.95 fractiles in Problem 8.13. If you are perfectly *calibrated* when judging these fractiles, then you would expect that in any given situation the actual value has a 0.90 chance of falling between your assessed fractiles (for that variable).

a Assuming you are perfectly calibrated, what is the probability that 0, 1, or 2 of the 10 actual values fall between the assessed fractiles?

b To justify using the binomial distribution to answer part a, you must assume that the chance of any particular value falling between its fractiles is the same regardless of what happens with the other variables. Do you think this is a reasonable assumption? Why or why not?

9.17 In the soft pretzels example concerning the taste test—in which we used binomial probabilities—suppose the results had been that 4 out of the 20 people sampled preferred your pretzel. Use Bayes' theorem to find your posterior probability $P(\text{Hit} \mid r = 4, n = 20)$ for the following pairs of prior probabilities:

a $P(\text{Hit}) = 0.2$, $P(\text{Flop}) = 0.8$

b $P(\text{Hit}) = 0.4$, $P(\text{Flop}) = 0.6$

c $P(\text{Hit}) = 0.5$, $P(\text{Flop}) = 0.5$

d $P(\text{Hit}) = 0.75$, $P(\text{Flop}) = 0.25$

e P(Hit) = 0.90, P(Flop) = 0.10

f P(Hit) = 1.0, P(Flop) = 0

Create a graph of the posterior probability as a function of the prior probability. (We did something similar in Chapter 7 in discussing the Hinckley trial. Refer to Figure 7.20.)

9.18 In a city, 60% of the voters are in favor of building a new park. An interviewer intends to conduct a survey.

a If the interviewer selects 20 people randomly, what is the probability that more than 15 of them will favor building the park?

b Instead of choosing 20 people as in part a, suppose that the interviewer wants to conduct the survey until he has found exactly 12 who are in favor of the park. What is the probability that the first 12 people surveyed all favor the park (in which case the interviewer can stop)? What is the probability that the interviewer can stop after interviewing the thirteenth subject? What is the probability that the interviewer can stop after interviewing the eighteenth subject?

9.19 In bottle production, bubbles that appear in the glass are considered defects. Any bottle that has more than two bubbles is classified as "nonconforming" and is sent to recycling. Suppose that a particular production line produces bottles with bubbles at a rate of 1.1 bubbles per bottle. Bubbles occur independently of one another.

a What is the probability that a randomly chosen bottle is nonconforming?

b Bottles are packed in cases of 12. An inspector chooses one bottle from each case and examines it for defects. If it is nonconforming, she inspects the entire case, replacing nonconforming bottles with good ones. This process is called rectification. If the chosen bottle conforms (has two or fewer bubbles), then she passes the case. In total, 20 cases are produced. What is the probability that at least 18 of them pass?

c What is the expected number of nonconforming bottles in the 20 cases after they have been inspected and rectified using the scheme described in part b?

9.20 In our discussion of the Poisson distribution, we used this distribution to represent the process by which customers arrive at the pretzel stand. Is it reasonable to assume that the Poisson distribution is appropriate for finding the probabilities that we need? Why or why not?

9.21 You are the mechanical engineer in charge of maintaining the machines in a factory. The plant manager has asked you to evaluate a proposal to replace the current machines with new ones. The old and new machines perform substantially the same jobs, and so the question is whether the new machines are more reliable than the old. You know from past experience that the old machines break down roughly according to a Poisson distribution, with the expected number of breakdowns at 2.5 per month. When one breaks down, $150 is required to fix it. The new machines, however, have you a bit confused. According to the distributor's brochure, the new machines are supposed to break down at a rate of 1.5 machines per month on average and should cost $170 to fix. But a friend in another plant that uses the new machines reports that they break down at a rate of approximately 3.0 per month (and do cost $170 to fix). (In either event, the number of breakdowns in any month appears to follow a Poisson distribution.) On the basis of this information, you judge that it is equally likely that the rate is 3.0 or 1.5 per month.

a Based on minimum expected repair costs, should the new machines be adopted?

b Now you learn that a third plant in a nearby town has been using these machines. They have experienced 6 breakdowns in 3.0 months. Use this information to find the posterior probability that the breakdown rate is 1.5 per month.

c Given your posterior probability, should your company adopt the new machines in order to minimize expected repair costs?

d Consider the information given in part b. If you had read it in the distributor's brochure, what would you think? If you had read it in a trade magazine as the result of an independent test, what would you think? Given your answers, what do you think about using sample information and Bayes' theorem to find posterior probabilities? Should the source of the information be taken into consideration somehow? Could this be done in some way in the application of Bayes' theorem?

9.22 In the soft pretzels example for the beta distribution, suppose you assessed the median of Q to be approximately 0.30 and the upper quartile 0.50. In this case you could use a beta distribution with parameters $r = 1$ and $n = 3$ to represent your beliefs.

a Plug the values $r = 1$ and $n = 3$ into the formula for the beta distribution. The expression simplifies considerably. (Recall that $0! = 1$ by definition.)

b Draw a graph of the distribution of Q.

c Use what you know about areas of triangles to show that this distribution is a reasonably good representation of your subjective beliefs in the sense that the median and upper quartile are fairly close to your subjectively assessed median and upper quartile.

d Find the expected profit for the pretzel project with this new distribution.

e Find the probability that the pretzel project results in a loss under this distribution.

9.23 Sometimes we use probability distributions that are not exact representations of the physical processes that they are meant to represent. (For example, we might use a normal distribution for a distribution of individuals' weights, even though no one can weigh less than zero pounds.) Why do we do this?

9.24 You are an executive at Procter and Gamble and are about to introduce a new product. Your boss has asked you to predict the market share (Q, a proportion between 0 and 1) that the new product will capture. You are unsure of Q, and you would like to communicate your uncertainty to the boss. You have made the following assessments:

There is a 1-in-10 chance that Q will be greater than 0.22, and also a 1-in-10 chance that Q will be less than 0.08.

The value for Q is just as likely to be greater than 0.14 as less than 0.14.

a What should your subjective probabilities $P(0.08 < Q < 0.14)$ and $P(0.14 < Q < 0.22)$ be in order to guarantee coherence?

b Use Appendix F to find a beta distribution for Q that closely approximates your subjective beliefs.

c The boss tells you that if you expect that the market share will be less than 0.15, the product should not be introduced. Write the boss a memo that gives an expected value and also explains how risky you think it would be to introduce the product. Use your beta approximation.

9.25 Suppose you are considering two investments, and the critical issues are the rates of return (R_1 and R_2). For Investment 1, the expected rate of return (μ_1) is 10%, and the standard deviation (σ_1) is 3%. For the second investment, the expected rate of return (μ_2) is 20%, and the standard deviation (σ_2) is 12%.

a Does it make sense to decide between these two investments on the basis of expected value alone? Why or why not?

b Does it make sense to represent the uncertainty surrounding the rates of return with normal distributions? What conditions do we need for the normal distribution to provide a good fit?

c Suppose you have decided to use normal distributions (either because of or in spite of your answer to part b). Use Appendix E to find the following probabilities:

$$P(R_1 < 0\%)$$
$$P(R_2 < 0\%)$$
$$P(R_1 > 20\%)$$
$$P(R_2 < 10\%)$$

d How can you find the probability that $R_1 > R_2$? Suppose R_1 and R_2 are correlated (as they would be if, say, both of the investments were stocks). Then the random variable $\Delta R = R_1 - R_2$ is normal with mean $\mu_1 - \mu_2$ and variance $\sigma_1^2 + \sigma_2^2 - 2\rho\sigma_1\sigma_2$ where ρ is the correlation between R_1 and R_2. If $\rho = 0.5$, find $P(R_1 > R_2)$. [*Hint:* Think about it in terms of ΔR, and find $P(\Delta R > 0)$.]

e How could you use the information from the various probabilities developed in this problem to choose between the two investments?

9.26 Your inheritance, which is in a blind trust, is invested entirely in McDonald's or in U.S. Steel. Because the trustee owns several McDonald's franchises, you believe the probability that the investment is in McDonald's is 0.8. In any one year, the return from an investment in McDonald's is approximately normally distributed with mean 14% and standard deviation 4%, while the investment in U.S. Steel is approximately normally distributed with mean 12% and standard deviation 3%. Assume that the two returns are independent.

a What is the probability that the investment earns between 6% and 18% (i) if the trust is invested entirely in McDonald's, and (ii) if the trust is invested entirely in U.S. Steel?

b Without knowing how the trust is invested, what is the probability that the investment earns between 6% and 18%?

c Suppose you learn that the investment earned more than 12%. Given this new information, find your posterior probability that the investment is in McDonald's.

d Suppose that the trustee decided to split the investment and put one-half into each of the two securities. Find the expected value and the variance of this portfolio.

9.27 A continuous random variable X has the following density function:

$$f(x) = \begin{cases} 0.5 & \text{for } 3 \le x \le 5 \\ 0 & \text{otherwise} \end{cases}$$

a Draw a graph of this density. Verify that the area under the density function equals 1.

b A density function such as this one is called a *uniform* density, or sometimes a rectangular density. It is extremely easy to work with because probabilities for intervals can be found as areas of rectangles. For example, find $P_U(X \le 4.5 \mid a = 3, b = 5)$. (The parameters a and b are used to denote the lower and upper extremes, respectively.)

c Find the following uniform probabilities:

$P_U(X \le 4.3 \mid a = 3, b = 5)$ $\qquad\qquad$ $P_U(X > 3.4 \mid a = 0, b = 10)$

$P_U(0.25 \le X \le 0.75 \mid a = 0, b = 1)$ $\qquad$ $P_U(X < 0 \mid a = -1, b = 4)$

d Plot the CDF for the uniform distribution where $a = 0$, $b = 1$.

e The expected value of a uniform distribution is $E(X) = (b + a)/2$, and the variance is $Var(X) = (b - a)^2/12$. Calculate the expected value and variance of the uniform density with $a = 3$, $b = 5$.

9.28 The length of time until a strike is settled is distributed uniformly from 0 to 10.5 days. (See the previous problem for an introduction to the uniform density.)

a Find the probability that a strike lasts less than one day.

b Find the probability that a strike lasts less than six days.

c Find the probability that a strike lasts between six and seven days.

d Find the conditional probability that a strike lasts less than seven days, given that it already has lasted six days.

e Compare your answers with those from Problem 9.12.

9.29 In a survey in a shopping center, the interviewer asks customers how long their shopping trips have lasted so far. The response (T) given by a randomly chosen customer is uniformly distributed from 0 to 1.5 hours.

a Find the probability that a customer has been shopping for 36 minutes or less.

b The interviewer surveys 18 customers at different times. Find the probability that more than one-half of these customers say that they have been shopping for 36 minutes or less.

9.30 A continuous random variable X has the following density function:

$$f(x) = \begin{cases} \frac{1}{3}x - \frac{2}{3} & \text{for } 2 \le x \le 4 \\ -\frac{2}{3}x + \frac{10}{3} & \text{for } 4 \le x \le 5 \\ 0 & \text{otherwise} \end{cases}$$

a Draw a graph of this density. Verify that the area under the density function equals 1.

b A density function such as this one is called a *triangular* density. It is almost as easy to work with as the uniform density; probabilities for intervals can be calculated easily by calculating the areas of triangles and quadrilaterals. For example, find $P(3 \le X \le 4.5)$ for this distribution.

c Find the value $x_{0.50}$ such that $P(X \le x_{0.50}) = 0.5$. (That is, find the median of this distribution.)

d Find the upper and lower quartiles of this distribution. That is, find $x_{0.25}$ such that $P(X \le x_{0.25}) = 0.25$, and find $x_{0.75}$ such that $P(X \le x_{0.75}) = 0.75$.

9.31 A greeting card shop makes cards that are supposed to fit into 6-inch (in.) envelopes. The paper cutter, however, is not perfect. The length of a cut card is normally distributed with mean 5.9 in. and standard deviation 0.0365 in. If a card is longer than 5.975 in., it will not fit into a 6-in. envelope.

a Find the probability that a card will not fit into a 6-in. envelope.

b The cards are sold in boxes of 20. What is the probability that in one box there will be two or more cards that do not fit in 6-in. envelopes?

9.32 You are the maintenance engineer for a plant that manufactures consumer electronic goods. You are just about to leave on your vacation for two weeks, and the boss is concerned about certain machines that have been somewhat unreliable, requiring your expertise to keep them running. The boss has asked you how many of these machines you expect to fail while you are out of town, and you have decided to give him your subjective probability distribution. You have made the following assessments:

1 There is a 0.5 chance that none of the machines will fail.

2 There is an approximate 0.15 chance that two or more will fail.

3 There is virtually no chance that four or more will fail.

Being impatient with this slow assessment procedure, you decide to try to fit a theoretical distribution.

a Many operations researchers would use a Poisson distribution in this case. Why might the Poisson be appropriate? Why might it not be appropriate?

b Find a Poisson distribution that provides a good representation of your assessed beliefs. Give a specific value for the parameter m.

c Given your answer to b, what is the expected number of machines that will break down during your absence?

9.33 After you have given your boss your information (Problem 9.32), he considers how accurate you have been in the past when you have made such assessments. In fact, he decides you are somewhat optimistic (and he believes in Murphy's Law), so *he assigns a Poisson distribution with $m = 1$* to the occurrence of machine breakdowns during your two-week vacation. Now the boss has a decision to make. He either can close the part of the plant involving the machines in question, at a cost of $10,000, or he can leave that part up and running. Of course, if there are no machine failures, there is no cost. If there is only one failure, he can work with the remaining equipment until you return, so the cost is effectively zero. If there are two or more failures, however, there will be assembly time lost, and he will have to call in experts to repair the machines immediately. The cost would be $15,000. What should he do?

9.34 Regarding the soft pretzels again, suppose you decide to conduct a taste test of a new recipe at your stand. On average, one person comes to the stand every four minutes, and the arrivals seem to follow a Poisson distribution fairly closely. You decide to check for 30 minutes to see how many customers during that time prefer the new recipe. Suppose the probability is 0.4 that any arriving customer prefers the new recipe over the old one. What is the probability that you will find four or more customers who prefer the new recipe during your 30-minute test period?

9.35 A factory manager must decide whether to stock a particular spare part. The part is absolutely essential to the operation of certain machines in the plant. Stocking the part costs $10 per day in storage and cost of capital. If the part is in stock, a broken machine can be repaired immediately, but if the part is not in stock, it takes one day to get the part from the distributor, during which time the broken machine sits idle. The cost of idling one machine for a day is $65. There are 50 machines in the plant that require this particular part. The probability that any one of them will break and require the part to be replaced on any one day is only 0.004 (regardless of how long since the part was previously replaced). The machines break down independently of one another.

a If you wanted to use a probability distribution for the number of machines that break down on a given day, would you use the binomial or Poisson distribution? Why?

b Whichever theoretical distribution you chose in part a, what are appropriate parameters? That is, if you chose the binomial, what are the values for p and n? If you chose the Poisson, what is the value for m?

c If the plant manager wants to minimize his expected cost, should he keep zero, one, or two parts in stock? Draw a decision tree and solve the manager's problem. (Do not forget that more than one machine can fail in one day!)

9.36 Another useful distribution that is based on the normal is the *lognormal* distribution. Among other applications, this distribution is used by environmental engineers to represent the distribution of pollutant levels, by economists to represent the distribution of returns on investments, and by actuaries to represent the distribution of insurance claims.

Finding probabilities from a lognormal distribution is "as easy as falling off a log"! If X is lognormally distributed with parameters μ and σ, then $Y = \ln(X)$ is normally distributed and has mean μ and σ^2. Thus, the simplest way to work with a lognormal random variable X is to work in terms of $Y = \ln(X)$. It is easy to obtain probabilities for Y from the normal table. The expected value and variance of X are given by the following formulas:

$$E(X) = e^{\mu + 0.5\sigma^2} \qquad \text{Var}(X) = (e^{2\mu})(e^{\sigma^2} - 1)(e^{\sigma^2})$$

For example, if X is lognormally distributed with parameters $\mu = 0.3$ and $\sigma = 0.2$, then Y is normal with mean 0.3 and standard deviation 0.2. Finding probabilities just means taking logs:

$$P_L(X \geq 1.4 \mid \mu = 0.3, \sigma = 0.2) = P_N(Y \geq \ln(1.4) \mid \mu = 0.3, \sigma = 0.2)$$

$$= P_N(Y \geq 0.336 \mid \mu = 0.3, \sigma = 0.2)$$

$$= P_N(Z \geq 0.18)$$

$$= 0.4286$$

The mean and expected value of X are

$$E(X) = e^{0.3 + 0.5(0.2)^2}$$

$$= 1.38$$

$$\text{Var}(X) = (e^{2(0.3)})(e^{(0.2)^2} - 1)(e^{(0.2)^2})$$

$$= 0.077$$

After all that, here is a problem to work. After a hurricane, claims for property damage pour into the insurance offices. Suppose that an insurance actuary models noncommercial property damage claims (X, in dollars) as being lognormally distributed with parameters $\mu = 10$ and $\sigma = 0.3$. Claims on different properties are assumed to be independent.

a   Find the mean and standard deviation of these claims.

b   Find the probability that a claim will be greater than $50,000.

c   The company anticipates 200 claims. If the state insurance commission requires the company to have enough cash on hand to be able to satisfy all claims with probability 0.95, how much money should be in the company's reserve? [*Hint:* The total claims can be represented by the variable $Q = \sum_1^{200} X_i$ and Q will be approximately normally distributed with mean 200 E(X) and variance 200 Var(X).]

C A S E   S T U D I E S

OVERBOOKING

Most airlines practice *overbooking*. That is, they are willing to make more reservations than they have seats on an airplane. Why would they do this? The basic reason is simple; on any given flight a few passengers are likely to be "no-shows." If the airline overbooks slightly, then it still may be able to fill the airplane. Of course, this policy has its risks. If more passengers arrive to claim their reservations than there are seats available, the airline must "bump" some of its passengers. Often this is done by asking for volunteers. If a passenger with a reserved seat is willing to give up his or her seat, the airline typically will give a refund as well as provide a free ticket to the same or another destination. The fundamental trade-off is whether the additional expected revenue gained by flying an airplane that is nearer to capacity on average is worth the additional expected cost of refunds and free tickets.

To study the overbooking policy, let us look at a hypothetical situation. Mockingbird Airlines has a small commuter airplane with places for 16 passengers. The airline uses this jet on a route for which it charges $225 for a one-way fare. Every flight has a fixed cost of $900 (for pilot's salary, fuel, airport fees, and so on). Each passenger costs Mockingbird an additional $100. Finally, the no-show rate is 4%. That is, on average approximately 4% of those passengers holding confirmed reservations do not show up. Refunds for unused tickets are made only if the reservation is canceled at least 24 hours before scheduled departure.

How many reservations should Mockingbird be willing to sell on this airplane? The strategy will be to calculate the expected profit for a given number of reservations. For example, suppose that the Mockingbird manager decides to sell 18 reservations. The revenue is $225 times the number of reservations:

$$R = \$225(18)$$
$$= \$4050$$

The cost consists of two components. The first is the cost of flying the plane and hauling the passengers who arrive (but not more than the airplane's capacity of 16):

$$C_1 = \$900 + \$100 \times \text{Min(Arrivals, 16)}$$

The second component is the cost of refunds and free tickets that must be issued if 17 or 18 passengers arrive:

$$C_2 = (\$225 + \$100) \times \text{Max(0, Arrivals} - 16)$$

In this expression for C_2, the \$225 represents the refund for the purchased ticket, and the \$100 represents the cost of the free ticket. The Max ( ) expression calculates the number of excess passengers who show up (zero if the number of arrivals is fewer than 16).

Questions

1   Find the probability that more than 16 passengers will arrive if Mockingbird sells 17 reservations (Res = 17). Do the same for 18 and 19.

2   Find:

$$E(R \mid \text{Res} = 16)$$
$$E(C_1 \mid \text{Res} = 16)$$
$$E(C_2 \mid \text{Res} = 16)$$

Finally, calculate

$$E(\text{Profit} \mid \text{Res} = 16) = E(R \mid \text{Res} = 16) - E(C_1 \mid \text{Res} = 16) - E(C_2 \mid \text{Res} = 16)$$

3   Repeat Question 2 for 17, 18, and 19 reservations. What is your conclusion? Should Mockingbird overbook? By how much?

4   Since the airlines were deregulated in the 1970s, pricing has become more competitive. One of the promotional schemes is the "supersaver" fare that requires early payment and restrictions on refunds. For example, to receive the special fare, a customer may be required to purchase a nonrefundable ticket two weeks in advance, after which changes can be made only by paying a penalty (e.g., \$75). How do you think this policy has affected the airlines' overbooking policy?

EARTHQUAKE PREDICTION

Because of the potential damage and destruction that earthquakes can cause, geologists and geophysicists have put considerable effort into understanding when and where earthquakes occur. The ultimate aim is the accurate prediction of earthquakes on the basis of movements in the earth's crust, although this goal appears to be some way off. In the meantime, it is possible to examine past data and model earthquakes probabilistically.

Fortunately, considerable data exist on the basis of which to model earthquakes as a probabilistic phenomenon. Gere and Shah provide the information shown in Table 9.1. Richter magnitude refers to the severity of the earthquake. For example, if an earthquake is in the 8.0–8.9 category, by definition the ground would shake strongly for 30 to 90 seconds over an area with a diameter of 160 to 320 kilometers. Earthquakes of magnitude less than 4.0 are not dangerous and, for the most part, are not noticed by laypeople.

Table 9.1

Earthquake frequency data for California.

Richter Magnitude	Average Number of Earthquakes per 100 Years in California
8.0–8.9	1
7.0–7.9	12
6.0–6.9	80
5.0–5.9	400
4.0–4.9	2000

Source: J. M. Gere and H. C. Shah (1984) *Terra Non Firma: Understanding and Preparing for Earthquakes.* Stanford, CA: Stanford Alumni Association. Reprinted by permission.

An earthquake of magnitude 8.0 or greater could cause substantial damage and a large number of deaths if it were to occur in a highly populated part of the world. In fact, the San Francisco earthquake of April 6, 1906, was calculated later as measuring 8.3 on the Richter scale. The resulting fire burned much of the city, and some 700 people died. California is particularly susceptible to earthquakes because the state straddles two portions of the earth's crust that are slipping past each other, primarily along the San Andreas Fault. For this reason, we will consider the probability of a severe earthquake happening again in California in the near future.

Questions

1   We can model the occurrence of earthquakes using a Poisson distribution. Strictly speaking, the independence requirement for the Poisson is not met for two reasons. First, the geologic processes at work in California suggest that the probability of a large earthquake increases as time elapses following an earlier large quake. Second, large earthquakes often are followed by aftershocks. Our model will ignore these issues and hence can be viewed only as a first-cut approximation at constructing a probabilistic model for earthquakes.

The data from Gere and Shah indicate that, on average, 2493 earthquakes with magnitude 4.0 or greater will occur in California over a 100-year period. Thus, we might consider using a Poisson distribution with $m = 24.93$ to represent the probability distribution for the number of earthquakes (all magnitudes greater than 4.0) that will hit California during the next year. Use this distribution to find the following probabilities:

$P_P(X \leq 10$ in Next Year $\mid m = 24.93$ Earthquakes per Year)

$P_P(X \leq 7$ in Six Months $\mid m = 24.93$ Earthquakes per Year)

$P_P(X > 3$ in Next Month $\mid m = 24.93$ Earthquakes per Year)

2   We also can model the probability distribution for the magnitude of an earthquake. For example, the data suggest that the probability of an earthquake in California of magnitude 8.0 or greater is 1/2493, or approximately 0.0004. If we use an exponential distribution to model the distribution of magnitudes, assuming that 4.0 is the least possible, then we might use the following model. Let M denote the magnitude, and let $M' = M - 4$. Then, using the exponential formula, we have $P(M \geq 8) = P(M' \geq 4) = e^{-4m} = 0.0004$. Now we can solve for m:

$$e^{-4m} = 0.0004$$

$$\ln(e^{-4m}) = \ln(0.0004)$$

$$-4m = -7.824$$

$$m = 1.96$$

Thus, the density function for M is given by

$$f(M) = 1.96e^{-1.96(M - 4)}$$

Plot this density function.

We now can find the probability that any given earthquake will have a magnitude within a specified range on the Richter scale. For example, use this model to find

$$P_E(M \leq 6.0 \mid m = 1.96)$$

$$P_E(5.0 \leq M \leq 7.5 \mid m = 1.96)$$

$$P_E(M \geq 6.4 \mid m = 1.96)$$

You may find it instructive to use this distribution to calculate the probability that an earthquake's magnitude falls within the five ranges of magnitude shown in Table 9.1. Here is a sensitivity-analysis issue: How might you find other reasonable values for m? What about a range of possible values for m?

3   We now have all of the pieces in the puzzle to find the probability of at least one severe (8.0 magnitude or more) earthquake occurring in California in the near future, say, within the next six months. Our approach will be to find the probability of the complement

$$P(X_{8+} \geq 1) = 1 - P(X_{8+} = 0)$$

where X_{8+} is used to denote the number of earthquakes having magnitude 8.0 or greater. Now expand $P(X_{8+} = 0)$ using total probability:

$$P(X_{8+} = 0) = P(X_{8+} = 0 \mid X = 0) \; P(X = 0) + P(X_{8+} = 0 \mid X = 1) \; P(X = 1)$$

$$+ P(X_{8+} = 0 \mid X = 2) \; P(X = 2)$$

$$+ \cdots + P(X_{8+} = 0 \mid X = k) \; P(X = k) + \cdots$$

$$= \sum_{k=0}^{\infty} P(X_{8+} = 0 \mid X = k) \; P(X = k)$$

The probabilities $P(X = k)$ are just the Poisson probabilities from Question 1:

$$P(X = k) = P_P(X = k \mid m = 12.5)$$

where $m = 12.5$ because we are interested in a 6-month period. The probability of no earthquakes of magnitude 8.0 out of the k that occur is easy to find. If $k = 0$, then $P(X_{8+} = 0 \mid X = 0) = 1$. If $k = 1$, then

$$P(X_{8+} = 0 \mid X = 1) = P_E(M < 8.0 \mid m = 1.96)$$
$$= 1 - e^{-1.96(8 - 4)}$$
$$= 0.9996$$

Likewise, if $k = 2$ then

$$P(X_{8+} = 0 \mid X = 2) = (0.9996)^2 = 0.9992$$

because this is just the probability of two independent earthquakes each having magnitude less than 8.0. Generalizing,

$$P(X_{8+} = 0 \mid X = k) = (0.9996)^k$$

Now we can substitute these probabilities into the formula:

$$P(X_{8+} \geq 0) = 1 - P(X_{8+} = 0)$$
$$= 1 - \sum_{k=0}^{\infty} P(X_{8+} = 0 \mid X = k) \ P(X = k)$$
$$= 1 - \sum_{k=0}^{\infty} (0.9996)^k P_P(X = k \mid m = 12.5)$$

To calculate this, you must calculate with k until the Poisson probability is so small that the remaining probabilities do not matter. It turns out that the probability of at least one earthquake of magnitude 8.0 or more within six months is approximately 0.005.

Now that you have seen how to do this, try calculating the probability of at least one earthquake of magnitude 8.0 or more (i) within the next year and (ii) within the next five years. For these, you may want to use a computer program to calculate the Poisson probabilities. How does the probability of at least one severe earthquake vary as you use the different reasonable values for m from your exponential model in Question 2? [*Hint:* Calculating the Poisson probabilities may be difficult, even on an electronic spreadsheet, because of the large exponential and factorial terms. An easy way to calculate these probabilities is to use the recursive equation

$$P_P(X = k + 1 \mid m) = \frac{m}{k+1} P_P(X = k \mid m)$$

[This equation can be used easily in an electronic spreadsheet or calculator without having to calculate large factorial and exponential terms.]

4   Using the probability model described above, it turns out that the probability of at least one earthquake of magnitude 8.0 or more within the next 20 years in California is approximately 0.2 (or higher, depending on the value used for m in the exponential distribution for the magnitude, M). That is a 1-in-5 chance. Now imagine that you are a policy maker in California's state government charged with making recommendations regarding earthquake preparedness. How would this analysis affect

Table 9.2

Probabilities for major earthquakes in California from two different probability models.

Time	USGS Probability	Poisson Model Probability
Next 5 years	0.27	0.29
Next 10 years	0.49	0.50
Next 20 years	0.71	0.75
Next 30 years	0.90	0.87

Source for USGS probabilities: U.S. Geological Survey (1988) "Probabilities of Large Earthquakes Occurring in California on the San Andreas Fault," by the Working Group on California Earthquake Probabilities. USGS Open-File Report No. 88-398, Menlo Park, CA.

your recommendations? What kinds of issues do you think should be considered? What about the need for more research regarding precise earthquake prediction at a specific location? What about regulations regarding building design and construction? What other issues are important?

The probabilistic model that we have developed using the information from Gere and Shah is based on a very simplistic model and does not account for geologic processes. Geologists do, however, use probability models in some cases as a basis for earthquake predictions. For example, as mentioned at the beginning of Chapter 8, a recent U.S. Geological Survey report concluded that the probability of an earthquake of 7.5–8.0 magnitude along the southern portion of the San Andreas Fault within the next 30 years is approximately 60%. The authors of the report actually constructed separate probability models for the occurrence of large quakes in different segments of major faults using data from the individual segments. Rather than a Poisson model, they used a lognormal distribution to model the uncertainty about the time between large earthquakes. Although their approach permits them to make probability statements regarding specific areas, their results can be aggregated to give probabilities for at least one major earthquake in the San Francisco Bay Area, along the southern San Andreas Fault in Southern California, or along the San Jacinto Fault. Table 9.2 compares their probabilities, which were developed for "large" earthquakes with expected magnitudes of 6.5–8.0, with our Poisson model probabilities of at least one earthquake having magnitude of 7.0 or greater. It is comforting to know that our model, even with its imperfections, provides probabilities that are not radically different from the geologists' estimates.

MUNICIPAL SOLID WASTE

Linda Butner considered her task. As the risk analysis expert on the city's Incineration Task Force (ITF), she was charged with reporting back to the ITF and to the city regarding the risks posed by constructing an incinerator for disposal of the city's solid waste. It was not a question of whether such an incinerator would be constructed. The city landfill site would be full within three years, and no alternative sites were available at a reasonable cost.

In particular, the state Department of Environmental Quality (DEQ) required information regarding levels of pollutants the incinerator was expected to produce. DEQ was concerned about organic compounds, metals, and acid gases. It was assumed that the plant would incorporate appropriate technology and that good combustion practices would be followed. Residual emissions were expected, however, and the officials were interested in obtaining close estimates of these. Linda's task was to provide any analysis of anticipated emissions of dioxins and furans (organic compounds), particulate matter (PM, representing metals), and sulfur dioxide (SO_2, representing the acid gases). She figured that a thorough analysis of these substances would enable her to answer questions about others.

The current specifications called for a plant capable of burning approximately 250 tons of waste per day. This placed it at the borderline between small- and medium-sized plants according to the Environmental Protection Agency's (EPA) guidelines. In part, this size was chosen because the EPA had proposed slightly different permission levels for these two plant sizes, and the city would be able to choose the plant size that was most advantageous. A smaller (less than 250 tons/day) plant would be expected to have an electrostatic precipitator for reducing particulate matter but would not have a specified SO_2 emission level. A larger plant would have a fabric filter instead of an electrostatic precipitator and would also use dry sorbent injection—the injection of chemicals into the flue—to control the SO_2 level. A summary of the EPA's proposed emission levels is shown in Table 9.3.

Standard practice in environmental risk analysis called for assessment and analysis of "worst case" scenarios. But to Linda's way of thinking, this kind of approach did not adequately portray the uncertainty that might exist. Incineration of municipal solid waste (MSW) was particularly delicate in this regard, because the levels of various pollutants could vary dramatically with the content of the waste being burned. Moreover, different burning conditions within the incineration chamber (more or less oxygen, presence of other gasses, different temperatures, and so on) could radically affect the emissions. To capture the variety of possible emission levels for the

Table 9.3
Proposed pollutant emission levels.

Pollutant	Plant Capacity (Tons of Waste per Day)	
	Small (Less than 250)	Medium (250 or More)
Dioxins/furans (ng/Nm^3)	500	125
PM (mg/dscm)	69	69
SO_2 (ppmdv)	—	30

Notes: ng/Nm^3 = nanograms per normal cubic meter; mg/dscm = milligrams per dry standard cubic meter; ppmdv = parts per million, by dry volume.

Table 9.4

Lognormal distribution
parameters μ and σ for
pollutants.

Pollutant	μ	σ
Dioxins/furans	3.13	1.20
PM	3.43	0.44
SO_2	3.20	0.39

Figure 9.14

*Lognormal density
function for SO_2 emis-
sions from incineration
plant.*

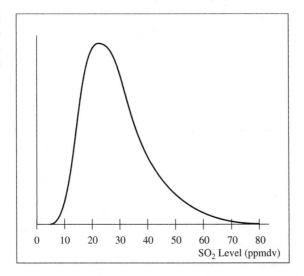

pollutants, Linda decided to represent the uncertainty about a pollutant-emission level with a probability distribution.

The lognormal distribution makes sense as a distribution for pollutant-emission levels. (See Problem 9.36 for an introduction to the lognormal.) After consulting the available data for the content of pollutants in MSW and the pollutant-emission levels for other incinerators, Linda constructed a table (Table 9.4) to show the parameters for the lognormal distributions for the three pollutants in question. Figure 9.14 illustrates the lognormal distribution for the SO_2 emissions.

As Linda looked at this information, she realized that she could make certain basic calculations. For example, it would be relatively straightforward to calculate the probability that the plant's emissions would exceed the proposed levels in Table 9.3. Having these figures in hand, she felt she would be able to make a useful presentation to the task force.

Questions

1   The plant will be required to meet established emission levels for dioxins/furans and PM on an annual basis. Find the probabilities for exceeding the small-plant levels specified in Table 9.3 for these two pollutants. Repeat the calculations for the medium-plant emission levels.

2 If the plant is subject to SO_2 certification, its emissions of this pollutant will be monitored on a continual basis and the average daily emission level must remain below the specified level in Table 9.3. The numbers in Table 9.4, however, refer to the probability distribution of a single observation. That is, we could use the specified lognormal distribution to find the probability that a single reading of the SO_2 level exceeds the specified level. Finding the probability that an average daily emission exceeds the specified level, though, requires more analysis.

Let us assume that the average emission level will be calculated by taking n observations and then calculating the *geometric mean*. To do this, we multiply the n observations together and then take the nth root of the product. In symbols, let G denote the geometric mean:

$$G = \left(\prod_{i=1}^{n} X_i \right)^{1/n}$$

It turns out that if each X_i is drawn independently from a distribution that is lognormal with parameters μ and σ, then G has a lognormal distribution with parameters μ and $\sigma/\sqrt{n}$. In our case we will take the 24th root of a product of 24 hourly observations of emission levels.

Find the probability that the geometric mean of the 24 hourly emission observations exceeds the SO_2 limit specified in Table 9.3. Compare this with the probability that a single observation will exceed the same limit.

3 Discuss the issues that the city should consider in deciding whether to build a small- or medium-sized plant.

Sources: J. Marcus and R. Mills (1988) "Emissions from Mass Burn Resource Recovery Facilities," *Risk Analysis*, No. 8, 315–327; and (1989) "Emission Guidelines: Municipal Waste Combustors," *Federal Register*, 54 (243), 52209.

REFERENCES

In this chapter we have only scratched the surface of theoretical probability distributions, although we have discussed most of the truly useful probability models. Theoretical distributions are widely used in operations research and in the construction of formal models of dynamic and uncertain systems. For additional study as well as many illustrative examples, consult the texts by DeGroot (1970); Feller (1968); Olkin, Gleser, and Derman (1980); or Winkler (1972). Johnson and Kotz (1969, 1970a, 1970b, and 1972) have compiled encyclopedic information on a great variety of probability distributions.

DeGroot, M. (1970) *Optimal Statistical Decisions.* New York: McGraw-Hill.

Feller, W. (1968) *An Introduction to Probability Theory and Its Applications, Vol. 1*, 3rd ed. New York: Wiley.

Johnson, N. L., and S. Kotz (1969) *Distributions in Statistics: Discrete Distributions.* New York: Houghton Mifflin.

Johnson, N. L., and S. Kotz (1970a) *Distributions in Statistics: Continuous Univariate Distributions I.* Boston: Houghton Mifflin.

Johnson, N. L., and S. Kotz (1970b) *Distributions in Statistics: Continuous Univariate Distributions II.* Boston: Houghton Mifflin.

Johnson, N. L., and S. Kotz (1972) *Distributions in Statistics: Continuous Multivariate Distributions.* New York: Wiley.

Olkin, I., L. J. Gleser, and C. Derman (1980) *Probability Models and Applications.* New York: Macmillan.

Winkler, R. L. (1972) *Introduction to Bayesian Inference and Decision.* New York: Holt.

E P I L O G U E   The case study on earthquake prediction was written in early October 1989, just two weeks before the Loma Prieta earthquake of magnitude 7.1 occurred near Santa Cruz, California, on the San Andreas Fault. Strong ground shaking lasted for approximately 15 seconds. The results were 67 deaths, collapsed buildings around the Bay Area, damage to the Bay Bridge between San Francisco and Oakland, and the destruction of a one-mile stretch of freeway in Oakland. This was a small earthquake, however, relative to the 8.3 magnitude quake in 1906. The "big one" is still to come and may cause even more damage.

CHAPTER 10

Using Data

We have discussed subjective judgments and theoretical distributions as sources for probabilities when modeling uncertainty in a decision problem. In this chapter we consider an obvious source for information about probabilities—historical data. It is possible to use data alone to develop probability distributions; we cover the development of discrete and continuous distributions in the first part of the chapter. We also consider the use of data to understand and model relationships among variables. It is possible to use data in conjunction with theoretical probability models; that is, the data can provide information to help refine assessments regarding the parameters of a theoretical distribution. In the last part of the chapter, we discuss the use of data in conjunction with some of the theoretical models discussed in Chapter 9.

Using Data to Construct Probability Distributions

Using past data when it is available is a straightforward idea, and most likely you have done something like this in the past, at least informally. Suppose, for example, that you are interested in planning a picnic at the Portland Zoo on an as-yet-undetermined day during February. Obviously, the weather is a concern in this case, and you want to assess the probability of rain. If you were to ask the National Weather Service for advice in this regard, forecasters would report that the probability of rain on any given day in February is approximately 0.47. They base this estimate on analysis of weather during past years; on 47% of the days in February over the past several years rain has fallen in Portland.

We can think about developing both discrete and continuous probability distributions on the basis of empirical data. For the discrete situation, the problem really becomes one of creating a relative frequency histogram from the data. In the case of a continuous situation, we can use the data to draw an empirically based CDF. We will look briefly at each of these.

Histograms

Imagine that you are in charge of a manufacturing plant, and you are trying to develop a maintenance policy for your machines. An integral part of the analysis leading to policy selection most likely would involve an examination of the frequency of machine failures. For example, you might collect the following data over 260 days:

- No Failures   217 days
- One Failure   32 days
- Two Failures   11 days

These data lead to the following relative frequencies, which could be used as estimates of probabilities of machine failures in your analysis:

No Failures   0.835 = 217/260
One Failure   0.123 = 32/260
Two Failures   0.042 = 11/260

Thus, we would have a histogram that looks like Figure 10.1, and in a decision tree, we would have a chance node with three branches like that in Figure 10.2.

Our treatment of histograms and estimation of discrete probabilities is brief precisely because the task is simply a matter of common sense. The only serious consideration to keep in mind is that you should have enough data to make a reliable estimate of the probabilities. (One nuclear power plant accident is not enough to

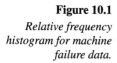

Figure 10.1

Relative frequency histogram for machine failure data.

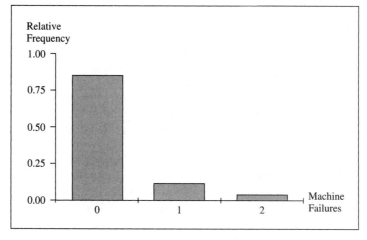

Figure 10.2

The decision-tree representation of uncertainty regarding machine failures.

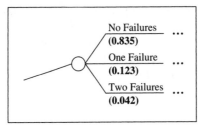

develop a probability distribution!) The data requirements depend on the particular problem, but the minimum should be approximately five observations in the least likely category. The other categories, of course, will have more observations.

Finally, keep in mind that your probability estimates are just that—estimates. The goal is to model uncertainty. Always ask yourself whether the probabilities estimated on the basis of the data truly reflect the uncertainty that you face. If you are not satisfied with the representation based on the data, you may need to model your uncertainty using subjective assessment methods. In particular, this might be the case if you think that the past may not indicate what the future holds.

Empirical CDFs

Continuous probability distributions are common in decision-making situations. We have seen how to assess a continuous probability distribution subjectively through assessment of a CDF. Our use of data to estimate a continuous distribution will follow the same basic strategy. To illustrate the principles, we will use an example and data concerning the costs related to correctional halfway houses.

HALFWAY HOUSES

A halfway house is a facility that provides a residence and some supervision and support for individuals who have been recently released from prison. The purpose of these halfway houses is to ease the transition from prison life to normal civilian life, with the ultimate goal being to improve an ex-convict's chance of successful integration into society. The National Advisory Committee on Criminal Justice Standards and Goals is responsible for providing information pertaining to the costs and resource implications of correctional standards that are related to halfway houses.

The main purpose of the reports made by the advisory committee is to provide cost information to state and local decision makers regarding the many services that halfway houses perform. One important variable is yearly per-bed rental costs. Denote this variable by C. Table 10.1 shows yearly per-bed rental costs (in dollars)—values of C—for a random sample of 35 halfway houses. The data are arranged in ascending order.

Table 10.1

Yearly bed-rental costs
for 35 halfway houses.

Rental Costs ($)				
52	205	303	400	643
76	250	313	402	693
100	257	317	408	732
136	264	325	417	749
137	280	345	422	750
186	282	373	472	791
196	283	384	480	891

Source: T. Sincich (1989) *Business Statistics by Example,* 3rd ed. San Francisco: Dellen.

To create a smooth CDF from these data, recall the idea of a cumulative probability. Suppose we look at the middle value, 325. Eighteen of the 35 values are less than or equal to 325. It also is true that 18 values are less than or equal to 326, 327, . . . , up to 344.99. So $P(C \leq 325) = P(C \leq 326) = \cdots = P(C \leq 344.99) = 18/35$, or 0.514. So how can we estimate the 0.514 fractile? Take 335 as the best estimate of the 0.514 fractile; 335 is the value that is halfway between 325 and 345. Figure 10.3 shows what we are doing in terms of a CDF graph. The data points define the ends of the flat steps on the graph. Our estimate, the smooth curve, will go through the centers of the flat steps.

To get a CDF, we do the same thing for all data points. The procedure first rank orders the data and then calculates the centers of the flats; this is just a matter of calculating the halfway points between adjacent data points. For example, consider the halfway point between the first and second data points. The halfway point is

Figure 10.3

Building the CDF by drawing a smooth curve through the centers of the flat steps.

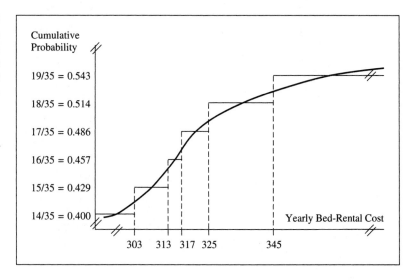

$(52 + 76)/2 = 64$. Call this x_1. Now calculate $x_2, \ldots, x_{n-1}$, where n is the number of data points. Now associate with each x_m its approximate cumulative probability. The value x_m is in the mth position, and so it has cumulative probability estimated as m/n. In symbols, $P(X \le x_m)$ is approximately m/n. In our example, $P(C \le 64)$ is approximately 1/35, or 0.029. Likewise, $x_{15} = 308$, so $P(C \le 308)$ is approximately 15/35, or 0.429. Table 10.2 shows the calculations for all 35 data points. The final step is to plot these points as in Figure 10.4. The tails of the CDF are sketched as smooth extrapolations of the curve.

Although all of the calculations that must be done to obtain a CDF look complicated, these are the kinds of calculations that can be done easily on an electronic spreadsheet and then plotted on a graph. Alternatively, RISKview will perform these calculations for you. RISKview provides a graph of the distribution and also exports the equation for the graph to Excel.

Once we have the CDF, we can use it in the same way as before. For example, we could use it to make probability statements about the uncertain quantity. With the halfway house, for example, we could say that there is a 50% chance that the yearly bed-rental cost will fall between $240 and $450. Likewise, we could say that there is

Table 10.2

Estimated cumulative probabilities for the halfway-house data.

Obs. No.	Cost	x_m	Cumulative Probability	Obs. No.	Cost	x_m	Cumulative Probability
1	52	64.0	0.029	19	345	359.0	0.543
2	76	88.0	0.057	20	373	378.5	0.571
3	100	118.0	0.086	21	384	392.0	0.600
4	136	136.5	0.114	22	400	401.0	0.629
5	137	161.5	0.143	23	402	405.0	0.657
6	186	191.0	0.171	24	408	412.5	0.686
7	196	200.5	0.200	25	417	419.5	0.714
8	205	227.5	0.229	26	422	447.0	0.743
9	250	253.5	0.257	27	472	476.0	0.771
10	257	260.5	0.286	28	480	561.5	0.800
11	264	272.0	0.314	29	643	668.0	0.829
12	280	281.0	0.343	30	693	712.5	0.857
13	282	282.5	0.371	31	732	740.5	0.886
14	283	293.0	0.400	32	749	749.5	0.914
15	303	308.0	0.429	33	750	770.5	0.943
16	313	315.0	0.457	34	791	841.0	0.971
17	317	321.0	0.486	35	891		
18	325	335.0	0.514				

Figure 10.4

Estimated CDF for the halfway-house data.

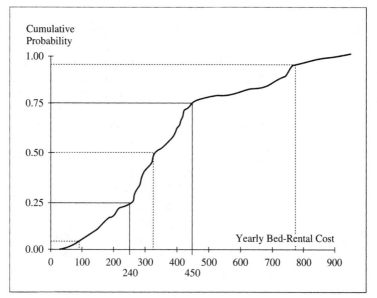

a 25% chance that the cost would fall below $240 and a 25% chance that it would fall above $450.

Another way we could use the empirical CDF would be to derive a discrete approximation (Chapter 8) for inclusion in a decision tree. In the case of the halfway-house data, Figure 10.5 shows a three-point discrete approximation that has been constructed using the extended Pearson-Tukey method described in Chapter 8. Recall that the 0.05 and 0.95 fractiles each are given probability 0.185 and the median is given probability 0.63. These fractiles are indicated in Figure 10.4.

There are other alternatives for using data to approximate a continuous distribution. One way that may be familiar to you is to split the data into reasonable groups and create a relative frequency histogram of the grouped data. This procedure is straightforward and similar to the discussion above regarding discrete probabilities. Again, no category should contain fewer than five observations, and typically the widths of the intervals that define the categories should be the same. If you need more direction on the use of histograms for continuous distributions, consult any introductory statistics textbook. The text by Sincich, from which the halfway-house data were taken, is an excellent source.

Figure 10.5

Three-point approximation for the halfway-house data.

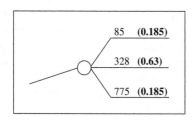

Using Data to Fit Theoretical Probability Models

Another way to deal with data is simply to fit a theoretical distribution to it. Typically, this involves two steps. First, we must decide what kind of distribution is appropriate (binomial, Poisson, normal, and so on). The choice should be based both on an understanding of the situation (for example, whether the uncertain quantity must be between 0 and 1, which suggests a beta distribution, or whether a discrete distribution is appropriate) as well as an inspection of the distribution of the data. For example, it makes little sense to fit a normal distribution, which is symmetric, to data whose distribution is highly skewed.

Having chosen the kind of distribution, the next step is to choose the values of the distribution parameters. In the case of the binomial, for instance, we need values for n and p. For the normal, μ and σ are required. Parameter values can be chosen by calculating some summary statistics (mean, standard deviation, and so on) for a sample and then simply using those as the parameter values. For example, it is possible to calculate the sample mean ($\bar{x}$) and sample variance (s^2) for the 35 halfway-house observations:

$$n = 35$$

$$\bar{x} = \frac{\sum_{i=1}^{n} x_i}{n} = 380.4$$

$$s^2 = \frac{\sum_{i=1}^{n} (x_i - \bar{x})^2}{n-1} = 47,344.3$$

$$s = \sqrt{47,344.3} = 217.6$$

Thus, we might choose a normal distribution with mean $\mu = 380.4$ and standard deviation $\sigma = 217.6$ to represent the distribution of the yearly bed-rental costs.

Another possibility is to fit a theoretical distribution using fractiles. That is, find a theoretical distribution whose fractiles match as well as possible with the fractiles of the empirical data. This is the data-based counterpart of the procedure discussed in Chapter 9 for fitting a theoretical distribution to a subjectively assessed distribution. In this case we would be fitting a theoretical distribution to a data-based distribution.

For most initial attempts to model uncertainty in a decision analysis, it may be adequate to use the sample mean and variance as estimates of the mean and variance of the theoretical distribution and to establish parameter values in this way. Refinement of the probability model may require more careful judgment about the kind of distribution as well as more care in fitting the parameters. Statisticians have devised many clever parameter-estimation methods. A discussion of these techniques is beyond the scope of this treatment but may be found in advanced textbooks in statistics.

Fitting Distributions to Data

In this section we will learn how to use the program BestFit to fit theoretical distributions to input data. BestFit is a program that uses sophisticated parameter-estimation methods in an environment that is easy to use and understand. For our part, we need only enter the data. BestFit will then run a fully automated fitting procedure and generate a report that ranks a set of distributions according to how well they fit the input data. BestFit also provides statistical measures and graphs so that we can numerically and visually compare the fit of the theoretical distributions to the data. We will use BestFit primarily to identify particular theoretical distributions that could be used to model specific uncertainties.

Before running BestFit, let's take a moment to understand how the program works. Various statistics are computed based on the input data, such as the sample mean and the sample standard deviation. Then, one by one, a distribution family is chosen and its parameters are manipulated until a best-fitting distribution in that family is found. For example, to fit the normal distribution, BestFit manipulates the normal distribution's parameters (the mean and the standard deviation) until the normal distribution fits the data as well as possible. For the normal distribution, this is straightforward because statisticians have shown that the best-fitting normal distribution occurs when the mean equals the sample mean and the standard deviation equals the sample standard deviation. After finding the best-fitting distribution within each family, BestFit ranks the distributions according to a goodness-of-fit (GOF) measure. The output consists of the following: (1) a list of the distributions with the best-fitting distribution ranked first, the second best ranked second, and so on; (2) a variety of graphs so that we can visually compare fits; and (3) a set of elaborate statistics on how well each distribution fits the input data. We demonstrate BestFit using the halfway-house data.

STEP 1

1.1   BestFit is like RISKview in that it is both a stand-alone program and also embedded within @RISK. We demonstrate BestFit within @RISK to highlight the integration between the two programs. Open Excel and then @RISK, enabling any macros if prompted.

1.2   There are four on-line help options available. Pull down the **@RISK** menu and choose the **Help** command. *"How Do I"* answers frequently asked questions about @RISK, *Online Manual* is an electronic copy of the user's guide, *@RISK Help* is a searchable database of the user's guide, and *PDF Help* lists various facts about the probability distribution functions available in @RISK and RISKview. The user's guide describes how to operate BestFit, whereas *PDF Help* lists particular facts about each distribution.

STEP 2

2.1   Click on the **Show @RISK—Model** button on the @RISK toolbar (the second button from the right). A new window pops up labeled *@RISK—Model*.

2.2   BestFit is opened by clicking on the **Insert** heading and choosing **FitTab**. A new worksheet titled *Fit1* is added to the *@RISK—Model* window, which contains a single column titled *Samples*. (You can also **right-click** on the existing tab to insert the **FitTab**.)

2.3   Enter the halfway-house data from Table 10.1 into the *Samples* column, as shown in Figure 10.6. For your convenience, the data sets from this chapter are on the Palisade CD: Examples\Chapter10\Data.xls. You can copy and paste cells from Excel to BestFit. Do not use the open command (leftmost button) in BestFit. This opens BestFit files only and not Excel files.

STEP 3

3.1   Pull down the **Fitting** menu and choose the **Run Fit** command. When the program finishes fitting the distributions, a *Fit Results* window pops up in the *@RISK—Model* window (see Figure 10.6). You have just run the automated feature of BestFit.

Figure 10.6

BestFit results of fitting the halfway-house data.

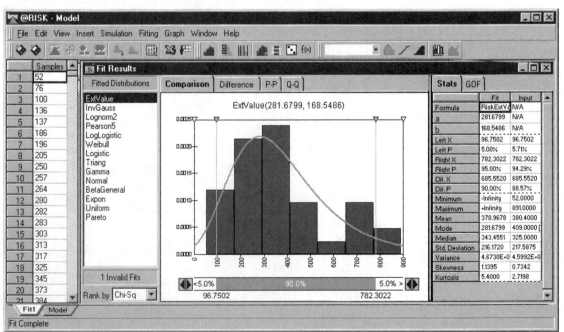

3.2   The *Fit Results* window shows that the extreme value (*ExtValue*) distribution fits the data most closely (it is first in the list), followed by the inverse Gaussian (*InvGauss*), and so on until the worst-fitting distribution in our list is the *Pareto* distribution. There was one invalid fit as stated below the fitted distribution list. An invalid fit merely means that the data are incompatible with a distribution family, and so this family has been excluded from the fitting procedures. Holding the cursor over **1 Invalid Fits** (do not click) identifies that the Pearson 6 distribution was invalid.

3.3   BestFit ranks the fit of the distributions to the data using three different methods: chi-square (Chi-Sq), Anderson-Darling (A-D), and Kolmogorov-Smirnov (K-S). (See the help manual for their descriptions.) Currently, the distributions are ranked using the chi-square measure, as shown by the *Chi-Sq* in the lower left corner of the *Fit Results* window. Click on the **triangle** to the right of *Chi-Sq* and choose **A-D**. Now according to the Anderson-Darling measure, the LogLogistic distribution is the best fitting and the extreme value distribution is fourth best. Highlight the **LogLogistic** in the list to view its fit.

3.4   Click on the **ExtValue** distribution. The graph in the middle shows the density of the extreme value distribution overlaying the histogram of the input data. Steps 5.1 and 5.2 will describe how to compare CDFs with CDFs or densities with densities. Click on a few of the fitted distribution names to view their densities compared with the input histogram.

3.5   To see how well the extreme value distribution fits according to all three measures (Chi-Sq, A-D, and K-S), highlight **ExtValue** in the list, and click on the **GOF** tab in the right-hand corner. Reading across the third row, labeled *Rank*, you see that its Chi-Sq rank is 1, its A-D rank is 4, and its K-S rank is 3. Click on **LogLogistic** in the list and its Chi-Sq rank is 4, but its A-D and K-S ranks are both 1.

STEP 4

4.1   This step describes how to read and manipulate the graph. Return to the original window by clicking on the **Stats** tab to the left of *GOF*, highlight the **ExtValue** distribution on the right, and choose *Rank by* **Chi-Sq**.

4.2   There are four vertical lines in the graph: The two outermost determine the width of the *x* axis, and the two innermost, called delimiters, help you determine cumulative probability values. Any one of these lines can be moved by clicking on the triangles at their top, holding down the mouse button, and sliding the cursor left or right. Click on a **white triangle** of an outermost line and move the cursor. To undo this, click on **AutoScaling Off**.

4.3   The left delimiter is currently at the *x* value 96.7502, as shown under the scrollbar at the very bottom (Figure 10.6). Move the left delimiter to a value of **200** by clicking on the **gray triangle**. The scrollbar now reads 19.7%, which means 19.7% of the area under this extreme value distribution lies to the left of 200. The smallest increment the delimiter will move is one unit of *x*.

4.4   Click one of the **black triangles** on the scrollbar and the corresponding delimiter moves one percentile per click.

4.5   You can also move the delimiter to an exact *x* value by typing that value into the *Fit* column to the right of the graph. Click in the **text box** immediately to the right of *Left X* and enter **230**. Thus, the area to the left of 230 for the extreme value distribution is 25.7%, which can be seen either in the scrollbar or in the *Left P* cell immediately below 230.

4.6   For comparison purposes, the value of 230 is automatically entered in the text box in the *Input* column. The *Left P* value in the *Input* column reports the area of the *input distribution* that is to the left of the *Left X* entry. This allows you to compare the percentiles of the theoretical distribution with the percentiles of the input distribution. In this case, the extreme value is reporting slightly more area to the left of 230 (25.7%) than the input distribution (22.9%).

4.7   Conversely, you can enter the exact percentile for the delimiter in the *Left P* or *Right P* text box, and BestFit will report the corresponding *x* value. Click in the **text box** immediately to the right of *Left P* and enter **15**. The corresponding *x* value is 173.75 for the extreme value. Again, for comparison purposes, the same *x* value (173.75) was automatically entered in the input column. Thus, the 15th percentile of the theoretical distribution is compared with the 14.29th percentile of the input distribution.

4.8   In Steps 4.4 to 4.7, you learned how to move the delimiters in four different ways: gray triangles at the top, scrollbar at the bottom, *x* value and *p*-value at the right. Use these methods to determine the 5th, 50th, and 95th percentiles to the halfway-house data using the extreme value distribution. (Answer: The 5th percentile is 96.75, the 50th percentile is 343.46, and the 95th percentile is 782.30.)

STEP 5

5.1   The graph can be formatted in many ways. For example, if you wish to compare cumulative distributions, click on the **Format Graph as Cumulative** button (squiggly upward-line button, fourth from the left in the *@RISK—Model* window). To return to the histogram, click on **Format Graph as Density** button (fifth from the left in the *@RISK—Model* window).

5.2   To access all the formatting options, click on the **Graph Formatting Options** button (second from the left in the *@RISK—Model* window). See page 331 of Chapter 8 for a full description of the formatting options.

Now that we know how to read the graphs and modify them to fit our needs, how do we determine which distribution we should use in a decision model? While there is no simple answer, we can provide some general guidelines. An important criterion for choosing a distribution is that it match the properties of the uncertainty that is being modeled. For example, if we were modeling the uncertain time between customer arrivals, then we would naturally want to use the exponential distribution as we did in Chapter 9. If we had data on arrival times, we could use BestFit to see how well the exponential distribution fits the data and then find the specific parameter value for the exponential. The PDF Help mentioned in Step 1.2 describes some of the common uses of the different distributions. For example, from the PDF Help file we see that the inverse Gaussian is used to model the distribution of an average of sample numbers. If the yearly bed-rental costs were averages for each halfway house, then we would use the inverse Gaussian distribution. This is because the inverse Gaussian fits the data very well and the type of data (sample averages) fit the distribution well.

If there are no clear properties of the uncertainty, then the choice must be based on the three measures: Chi-Sq, A-D, and K-S. The chi-square measure is a good overall measure of fit because it indicates which distributions match the input distribution along segments of the axis. The other two measures report the maximum distance between the input and fitted distribution. The difference between these measures is that the A-D measure penalizes distributions that do not closely match the tails of the input distribution. Thus, the A-D measure indicates the distributions that fit the tails (the extremes) of the distribution, which can be an important consideration.

The decision of which distribution to use can also be based on comparing the mean, median, mode, standard deviation, skewness, and kurtosis of the input distribution with the theoretical distribution. If the distribution is symmetric, then the normal distribution is a good choice because it will always have the same mean and standard deviation as the sample. If the distribution is skewed, though, the normal distribution is inappropriate.

Don't let the choice of which distribution to use worry you too much. Certainly, your choice can influence the results, but often most of the best-fitting distributions are very close to one another (no surprise!). It is a relatively easy exercise to determine the impact on your decision model of different distribution choices by simply replacing one distribution with another.

BestFit uses a different procedure for fitting cumulative distributions from what it does for sample data. For cumulative distributions, BestFit constructs an empirical CDF by joining the assessed values with straight-line segments. Then summary statistics, such as the mean and standard deviation (more precisely, the first four moments), are calculated based on the piecewise linear CDF. These estimates are used as the initial inputs for each distribution. BestFit then tries to improve the fit between the piecewise linear CDF and the theoretical distribution by modifying the initial input values of the theoretical distribution. The quality of fit is measured by the square root of the average squared errors ("root mean square," or RMS, error) with the best-fitting distribution having the smallest RMS error.

STEP 6   We demonstrate fitting theoretical distributions to assessed values using the pretzel problem in Chapter 8.

6.1   Insert a new *FitTab* by clicking on the **Insert** heading and choosing **FitTab**. This keeps the previous output and inserts a new worksheet titled *Fit2*.

6.2 Click on the **Fitting** heading in the *@RISK—Model* toolbar and highlight the **Input Data Options** button.

6.3 In the pop-up dialog box, click the **Cumulative Curve** button. If you were fitting a discrete distribution to the data, you would choose *Discrete* under *Domain* in this window.

6.4 Clicking **OK** changes the *@RISK—Model* window by inserting two columns (*X* and *P*) along with *X Min* and *X Max* entry boxes, as shown in Figure 10.7.

6.5 Enter **0** in the *X min* text box and **60000** in the *X max* text box.

6.6 Input the value-probability pairs as shown in Figure 10.7. You can copy and paste these values from the Data.xls worksheet supplied with the text.

6.7 Click on the **Fit Distributions to Input Data** button.

6.8 When the program finishes fitting the distributions, a *Fit Results* window pops up in the *@RISK—Model* window, showing that the inverse Gaussian (*InvGauss*) distribution fits the assessed distribution the best.

6.9 You can read and modify the output as described in Steps 4 and 5.

Recall that in Chapter 8 we used RISKview to construct an empirical distribution that fit the assessed values from the pretzel problem and learned how to modify the

Figure 10.7

Fitting theoretical distributions to the assessed values for the pretzel problem.

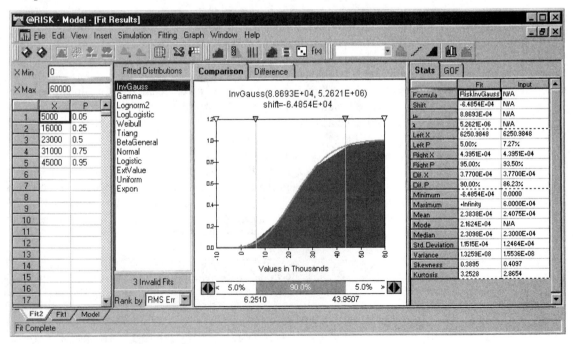

distribution in the Distribution Artist window. In this chapter, we used BestFit to see which theoretical distributions best fit the same assessed values from the pretzel problem. Thus, we now have a choice between using a theoretical distribution or an empirical distribution. Notice that in both programs the starting point for constructing the distributions is the piecewise linear CDF.

Step 7 shows how to modify the settings of BestFit.

STEP 7

7.1   Click on the **Fitting** heading in the *@RISK—Model* toolbar and highlight **Specify Distributions to Fit**.

7.2   You can see the distributions in the list to the right that BestFit has chosen to fit to your data. You can override any of its choices by clicking on a distribution.

7.3   Choosing the lower and upper bounds of the fitted distributions can substantially affect the fit. BestFit provides four choices. First, you can fix the bound (lower or upper) at a specified value. For example, if you were modeling grade point average (GPA), then 0 and 4 are the usual bounds. Second, if there are bounds but you do not know their values, choose **Bounded, but unknown**. This changes the distribution list to distributions that have a lower bound. Third, if there are no bounds, then choose **Open (extends to – infinity)**. Finally, if you do not know whether there are bounds, choose **Unsure**. Click **OK** to close the dialog box.

7.4   You can also test whether specific distributions fit the data. Under the heading Parameter Estimation, click on Fit to Predefined Distributions. If, for example, you hypothesize that the bed-rental costs in the halfway-house data set follow a normal distribution, then name your hypothesis Normal Test, enter the mean and standard deviation, and click Add →. When you click OK and run the fit, BestFit will fit only the distributions you have specified.

7.5   To return to the default distributions, open the *Specify Distributions to Fit* window in Step 7.1 and click **Find BestFit Parameters**.

There are three additional modifications that we encourage you to explore on your own. The one that could have the most influence on how well distributions fit the input data is called *Chi-Square Binning*. After the data are entered, BestFit divides the data into bins, similar to the division of the data into bars in a histogram. Because the fitting procedures do not work with the individual data, but with the bins, the definition of the bins can highly influence the ranking of the distributions. The *Chi-Square Binning* option allows the user to specify the bins used by BestFit and is found under the *Fitting* heading on the menu. The other two options are *Generate Random Data* and *Transform Data*. *Generate Random Data* creates a random sample from any of BestFit's distributions, and *Transform Data* applies a transformation (such as taking the logarithm) to the input data.

Using Data to Model Relationships

One of the most important things that we do with data is to try to understand the relationships that exist among different phenomena in our world. The following list includes just a few examples:

Causes of Cancer. Our smoking habits, diet, exercise, stress level, and many other aspects of our lives have an impact on our overall health as well as the risk of having cancer. Scientists collect data from elaborate experiments to understand the relationships among these variables.

Sales Revenue. For a business, perhaps the most crucial issue is understanding how various economic conditions, including its own decisions, can impact demand for products and, hence, revenue. Firms try to understand the relationships among sales revenue, prices, advertising, and other variables.

Economic Conditions. Economists develop statistical models to study the complex relationships among macroeconomic variables like disposable income, gross domestic product, unemployment, and inflation.

Natural Processes. Much scientific work is aimed at understanding the relationships among variables in the real world. A few examples include understanding the relationships among weather phenomena, movements in the earth's crust, changes in animal and plant populations due to ecological changes, and causes of violence in society.

In most cases, the motivation for studying relationships among phenomena that we observe is to gain some degree of control over our world. In many cases we hope to make changes in those areas where we have direct control in order to accomplish a change in another area. For example, on occasion the U.S. Federal Reserve Board buys or sells securities in order to have an impact on interest rates in the U.S. money market. Doing so requires an understanding of how their operations in the market can affect interest rates. Firms want to set prices and advertising budgets so that they result in an optimal level of profits, but doing so requires an understanding of what affects demand for their products and services. Scientific studies of natural processes often are used as inputs in government policy and regulation decisions.

In this section, we will focus on the problem of using data on a number of auxiliary variables (which we will denote as $X_1, \ldots, X_k$) to determine the distribution of some other variable of interest (Y) that is related to the X's. Y is sometimes called a *response variable,* because its probability distribution changes in response to changes in the X's. Likewise, the X's sometimes are called *explanatory variables,* because they can be used to help explain the changes in Y. (Y and the X's are also sometimes called *dependent* and *independent* variables, although this terminology can be misleading. Y does not necessarily *depend* on the X's in a causal sense, and the X's can be highly dependent—in a probabilistic sense—among themselves.)

In a business context, a common example is forecasting sales, and to come up with a conditional distribution for sales (Y), we might want to use explanatory variables such as price (X_1), advertising (X_2), a competitor's price (X_3), or other important microeconomic variables. These variables could be either quantities that are uncertain (like the competitor's price) or variables that are under the control of the decision maker (e.g., price and advertising). Often it is possible to obtain data indicating the amount of sales that was associated with a particular price, amount of advertising, and so on. How to use such data to create a model of the relationships among the variables is our topic.

The use of data to understand relationships is not trivial. Consider the influence diagram in Figure 10.8. The brute force approach would require obtaining enough data to estimate the conditional distribution for the particular variable of interest (Y) *for every possible combination of values for its conditioning or predecessor variables* (X_1 and X_2)! In addition, of course, we would need to know what are feasible values for the decision variable (X_1), and we would have to assess a distribution for the possible values for the uncertain variable (X_2). That would require a lot of data, and even in simple problems this could be a tedious or infeasible task.

One possible simplification, of course, is to use approximations. For example, in Figure 10.8 we could reduce the continuous distribution for X_2 to an extended Pearson-Tukey three-point approximation, and we could consider three possible values for X_1 (low, medium, and high, say). Even in this case, we would need to assess nine different conditional probability distributions for Y based on the possible scenarios. An approach like this may be possible in a simple situation, but the data requirements again become unwieldy as soon as there are more than just a few conditioning arcs. For example, suppose we were to add two more variables, X_3 and X_4, as in Figure 10.9. Using three-point approximations for the uncertain variables (X_2 and X_3) and considering only three possible values for each of the two decision variables (X_1 and X_4), we would still need to come up with 81 conditional distributions. (There are 81 conditional distributions because there are $3^4 = 81$ different combinations of the values for the X's.) Clearly, we need a way to streamline this process!

Figure 10.8

An influence diagram for modeling relationships among X_1, X_2, and Y.

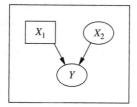

Figure 10.9

An influence diagram relating two uncertain quantities and two decision variables to Y.

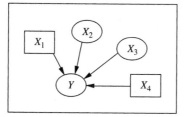

The Regression Approach

One way to approach this problem is to split it up into two pieces. First is determining the conditional expected value of Y given the X's, $E(Y \mid X_1, \ldots, X_k)$. The second step is to consider the conditional probability distribution around that expected value. Statisticians refer to this general modeling approach as *regression*.

We focus here on the simplest regression model, whose assumptions naturally lead it to be called *linear regression*. Beginning with the expected value:

1   *The conditional expected value of Y is linear in the X's.* In symbols,

$$E(Y \mid X_1, \ldots, X_k) = \beta_0 + \beta_1 X_1 + \cdots + \beta_k X_k$$

The β's are coefficients, and they serve the purpose of combining the X values to obtain a conditional expected value for Y. Moreover, every unit increase in X_i leads to a β_i change (positive or negative, as indicated by the sign of β_i) in the conditional expected value of Y. For example, suppose we have an equation relating regional expected sales (in \$1000s) of a high-volume consumer electronics item (say, a portable stereo) to the amount of money spent on advertising in that region (in \$1000s), the price set forth for that item in the ads (\$), and the current price of a competitive product (\$):

E(Sales | Advertising, Price, Competition Price)

= 2000 + 14.8(Advertising) − 500(Price) + 500(Competitive Price)

An influence diagram portraying this relationship is shown in Figure 10.10. Note that we have placed the coefficients on the arrows, indicating the effect that each variable has on the expected value of Sales. For example, every additional \$1000 spent on Advertising leads to an increase in expected Sales of \$14,800 (14.8 times \$1000, the units for Sales), given that Price and Competition Price stay the same. The opposite signs on the coefficients for Price and Competition Price indicate that these variables are expected to have opposite effects; an increase in our price should decrease our expected sales (everything else being equal), as indicated by the negative coefficient (−500). But an increase in Competition Price (again, with everything

Figure 10.10

Relating sales to three explanatory variables.

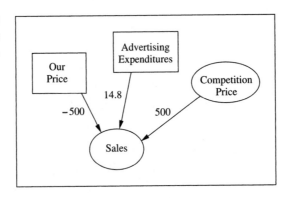

else held constant) is good news for us and should increase our expected sales, as indicated by the positive coefficient (+500) on Competition Price. Moreover, the price effects are quite strong, reflecting the competitive nature of the market; an increase in Price of $1 leads to a $500,000 decrease in expected Sales. Likewise, a $1 increase in the Competition Price leads to an increase in expected Sales of $500,000.

It is important to remember that the equation defines a relationship between the explanatory variables and the *expected Y* (Sales). The *actual Y* value will be above or below this expected value to some extent; this is where the uncertainty and the conditional probability distribution of *Y* come into play. The regression approach makes the following assumption about this distribution:

2 *The distribution around the conditional expected value has the same shape regardless of the particular X values.*

A convenient way to think about this is in terms of random errors (or "noise") added to the conditional expected value. Denote a random error as ϵ and let us say that these errors are, on average, equal to zero: $E(\epsilon) = 0$. Now we can write an expression for an actual *Y* value as follows:

$$Y = E(Y \mid X_1, \ldots, X_k) + \epsilon$$

This implies that the conditional distribution (and the corresponding density) of *Y*, given the *X*'s, has the same shape as the distribution (or density) of the errors, but it is just shifted so that the distribution is centered on the expected value $E(Y \mid X_1, \ldots, X_k)$.

Take the Sales example. Suppose that the errors—the unexplainable factors that make Sales fall above or below its conditional expected value—have a CDF as shown in Figure 10.11. Now, suppose we decide to spend $40,000 on Advertising ($X_1$) and to set Price ($X_2$) at $97.95, and Competition Price (X_3) turns out to be $94.99. Then expected Sales (*Y*), conditional on these values, would be

$$E(Y \mid X_1, X_2, X_3) = 2000 + 14.8(40) - 500(97.95) + 500(94.99)$$

$$= 1112(\$1000s)$$

Figure 10.11

A CDF for error in the sales example.

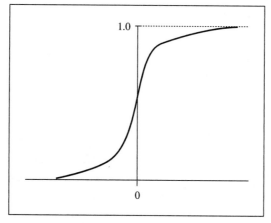

Suppose in an entirely different scenario we had Advertising equal to $70,000, Price equal to $93.95, and Competition Price equal to $98.99. Then conditional expected Sales would be

$$E(Y \mid X_1, X_2, X_3) = 2000 + 14.8(70) - 500(93.95) + 500(98.99)$$
$$= 5556(\$1000s)$$

The conditional CDFs for Sales, given the three conditioning variables in each scenario and the distribution for errors, are displayed in Figure 10.12. As you can see, the distributions have the same shape as the distribution for the errors shown in Figure 10.11. In Figure 10.12, though, the location of the distributions are different, and this is due entirely to differences in the explanatory variables.

Our two assumptions—a linear expression for conditional expected value and the constant shape for the conditional distribution—take us quite a long way toward being able to use data to study relationships among variables. You should keep in mind, however, that this is only one way to model the relationships and that it is quite possible for the expected-value relationships to be nonlinear or for the distribution to change shape. An example of a nonlinear relationship would be the degree of acidity in a lake and the number of microorganisms. If the lake is highly acidic, then reducing the acidity can lead to an increase in microorganisms up to a point, beyond which further changes are detrimental to the microorganisms' population. Changes in the distribution can be somewhat more subtle. Consider a retailer who is interested in entering a new market segment. Although the retailer may be fairly sure of the amount of sales resulting from ads aimed at her traditional market, the same advertising dollars spent in an effort to attract the new customers might lead to a great deal of uncertainty.

Although our approach has some limitations, it will provide an excellent base that can be extended via additional modeling techniques. We will mention some of these techniques later. For now, though, let us turn to the problem of using data within the regression framework we have established.

Figure 10.12
Conditional CDFs
for sales in
two scenarios.

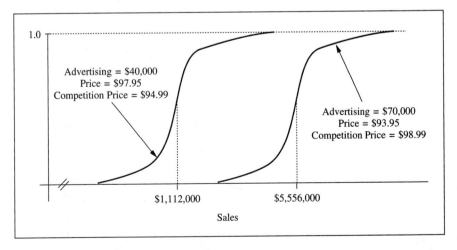

Estimation: The Basics

The discussion above has assumed that we know the β coefficients and the distribution of the errors. Of course, we may be able to make subjective judgments of these quantities, but if data are available, we may want to use that information in a systematic way to estimate the β's and to construct a data-based distribution of errors. To understand the process, let us simplify the sales example that we have been using. Suppose that all we have are data for Sales and Advertising, and we want to study the relationship between these two variables alone. The data, observations of Advertising and Sales during 36 different promotions in the eastern United States, are shown in Table 10.3. Figure 10.13 shows a *scatterplot* of these data; each point on the graph represents one of the observations.

Using the linear-regression assumptions from the previous section, we establish the equation relating expected Sales (Y) and Advertising (X_1):

$$E(Y \mid X_1) = \beta_0 + \beta_1 X_1$$

We use our data to estimate β_0 and β_1. Estimating β_0 and β_1 amounts to finding a line that passes through the cloud of points in the scatterplot; Figure 10.14 shows two different candidates with their corresponding expressions. The expressions show what those particular lines use as estimates of β_0 and β_1. The dashed line, for example, estimates β_0 to be 1900 and β_1 to be 15. Clearly, no single line can pass precisely through all of the points at once, but we would like to find one that in some sense is the "best fitting" line. And, as you can see from the graph, there are many reasonable estimates for β_0 and β_1.

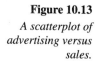

Figure 10.13

A scatterplot of advertising versus sales.

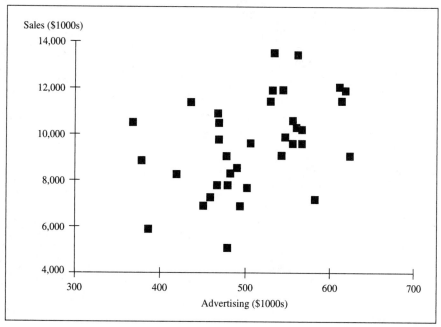

Table 10.3
Data for sales
example.

Promotion	Advertising ($1000s)	Sales ($1000s)
1	366	10541
2	377	8891
3	387	5905
4	418	8251
5	434	11461
6	450	6924
7	457	7347
8	466	10972
9	467	7811
10	468	10559
11	468	9825
12	475	9130
13	479	5116
14	479	7830
15	481	8388
16	490	8588
17	494	6945
18	502	7697
19	505	9655
20	529	11516
21	532	11952
22	533	13547
23	542	9168
24	544	11942
25	547	9917
26	554	10666
27	556	9717
28	560	13457
29	561	10319
30	566	9731
31	566	10279
32	582	7202
33	609	12103
34	612	11482
35	617	11944
36	623	9188

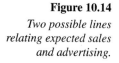

*Two possible lines
relating expected sales
and advertising.*

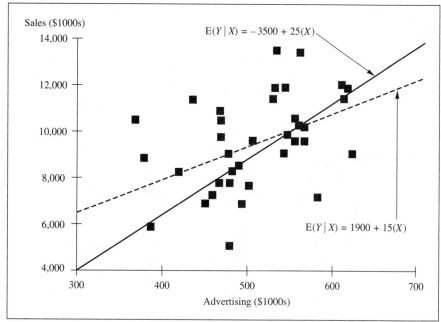

How will we choose the best-fitting line through the data points? Although there are many possible answers to this question, we will choose a way that many statisticians have found convenient and useful. In simple terms, we will choose the line that minimizes the sum of the squared vertical distances between the line and each point. Figure 10.15 shows how this will work for two sample points, A and B. Between the line and any given point (x_i) there is a certain vertical distance, which we will call the *residual* (e_i). For any given line, we can calculate the residuals for all of the points. Then we square each residual and add up the squares. And the problem is to find the line that minimizes this total of squared residuals. (The decision to minimize squared residuals is somewhat arbitrary. We could just as well choose estimates that minimize the sum of the absolute values of residuals, for example. Any error measure that penalizes large errors above or below the line will work. Mathematicians have focused on the sum of squared residuals because this sum is easy to work with mathematically.)

In symbols, let b_0 and b_1 represent the estimates of β_0 and β_1. With this we can calculate the ith residual as

$$e_i = y_i - (b_0 + b_1 x_i)$$

Squaring and summing over the n residuals gives the total (which we denote by SSE for the sum of squared errors):

$$SSE = \sum_{i=1}^{n} [y_i - (b_0 + b_1 x_i)]^2$$

Figure 10.15

Calculating residuals.

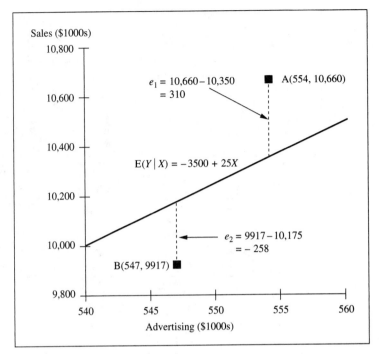

We want to choose b_0 and b_1 so that SSE is minimized. While this sounds like a complicated problem, it turns out to be a rather straightforward calculus problem. And even more good news is that you will never need to do the calculations yourself! Many computer programs are available that do the calculations for you automatically.

Figure 10.16 shows the spreadsheet in Microsoft Excel that results from doing the calculations using Excel's built-in "Regression" procedure in the Sales-Advertising example. The figure highlights Excel's estimates $b_0 = 3028.18$ and $b_1 = 12.95$. These are the coefficients that define the minimum-SSE line. (In fact, SSE = 117,853,513 and is shown in the column labeled "Sum of Squares.") Excel does many other calculations that are useful in drawing inferences from the data. For our purposes now, that of modeling relationships among variables, we will stick to the basics. All we require at this point are b_0 and b_1, the estimates of β_0 and β_1. At the end of the chapter we will briefly discuss the idea of statistical inference and how it relates to decision-analysis modeling.

We can write out the expression for the line defined by the estimates b_0 and b_1:

$$E(Y \mid X_1) = 3028.18 + 12.95X_1$$

This equation makes sense. In general, we expect Sales to increase when we spend more on Advertising, as indicated by the positive coefficient 12.95. In particular, this coefficient can be interpreted as follows: For every additional $1000 spent on advertising, Sales are expected to increase by $12,950. (Recall that Advertising and Sales are measured in $1000s.)

Figure 10.16

Using Excel to estimate regression coefficients.

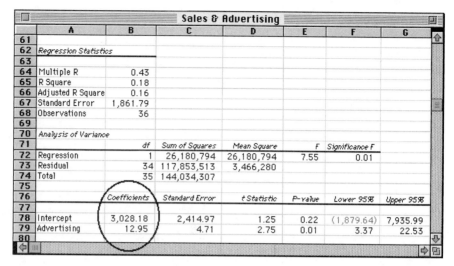

	A	B	C	D	E	F	G
61							
62	Regression Statistics						
63							
64	Multiple R	0.43					
65	R Square	0.18					
66	Adjusted R Square	0.16					
67	Standard Error	1,861.79					
68	Observations	36					
69							
70	Analysis of Variance						
71			df	Sum of Squares	Mean Square	F	Significance F
72	Regression		1	26,180,794	26,180,794	7.55	0.01
73	Residual		34	117,853,513	3,466,280		
74	Total		35	144,034,307			
75							
76			Coefficients	Standard Error	t Statistic	P-value	Lower 95% Upper 95%
77							
78	Intercept		3,028.18	2,414.97	1.25	0.22	(1,879.64) 7,935.99
79	Advertising		12.95	4.71	2.75	0.01	3.37 22.53
80							

For our simple model, we now have an indication of how expected Sales change as Advertising changes. How can we use the data to construct a model of the distribution of errors? This step is straightforward; for every data point we have a residual, which can be thought of as an estimate of the error associated with that particular point. Thus, the distribution of the residuals should be a good approximation to the distribution of the errors. Table 10.4 shows the calculation of the residuals, and Figure 10.17a displays a cumulative distribution constructed from those residuals in the same way that we constructed the distribution of halfway-house bed costs earlier (Figure 10.4). Figure 10.17b is a histogram of the residuals.

Figure 10.17

CDF and histogram of residuals.

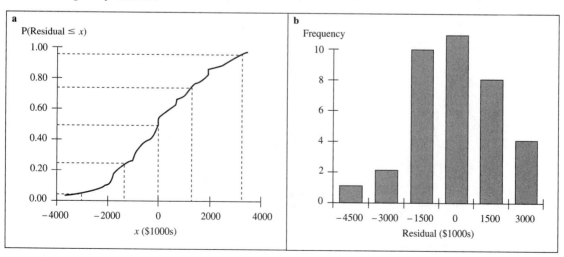

Table 10.4

Residuals for the sales example.

Observation	Sales (Y)	Advertising (X)	$E(Y \mid X)$ = 3028.18 + 12.95X	Residual = $Y - E(Y \mid X)$
1	10541	366	7768	2773
2	8891	377	7911	980
3	5905	387	8040	−2135
4	8251	418	8442	−191
5	11461	434	8649	2812
6	6924	450	8856	−1932
7	7347	457	8947	−1600
8	10972	466	9063	1909
9	7811	467	9076	−1265
10	10559	468	9089	1470
11	9825	468	9089	736
12	9130	475	9180	−50
13	5116	479	9232	−4116
14	7830	479	9232	−1402
15	8388	481	9258	−870
16	8588	490	9374	−786
17	6945	494	9426	−2481
18	7697	502	9530	−1833
19	9655	505	9568	87
20	11516	529	9879	1637
21	11952	532	9918	2034
22	13547	533	9931	3616
23	9168	542	10048	−880
24	11942	544	10074	1868
25	9917	547	10112	−195
26	10666	554	10203	463
27	9717	556	10229	−512
28	13457	560	10281	3176
29	10319	561	10294	25
30	9731	566	10358	−627
31	10279	566	10358	−79
32	7202	582	10566	−3364
33	12103	609	10915	1188
34	11482	612	10954	528
35	11944	617	11019	925
36	9188	623	11097	−1909

Figure 10.18

Extended Pearson-Tukey approximation for error distribution.

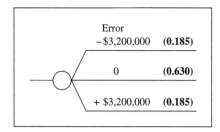

The CDF can be used to make probability statements about the errors. For example, we can see that the quartiles are approximately –1300 and +1300, so there is approximately a 50% chance that the error falls between these values. Or we can use the CDF to construct a discrete approximation. The dashed lines in Figure 10.17a for the 0.05, 0.50, and 0.95 fractiles represent the required numbers for a three-point extended Pearson-Tukey approximation as shown in Figure 10.18.

The CDF can also be used as a basis for representing the uncertainty in Sales given Advertising. To create a conditional distribution for Sales given Advertising, simply add the conditional expected Sales to the horizontal scale. For example, if Advertising = $110,000, then E(Sales | Advertising) = $4,452,680; add this amount to the horizontal scale to get the distribution for Sales given Advertising = $110,000 as shown in Figure 10.19. This distribution can in turn be used to generate conditional probability statements about Sales. Or we could create a discrete approximation of the distribution for use in an influence diagram or decision tree. Finally, we

Figure 10.19

CDF for sales given advertising = $110,000.

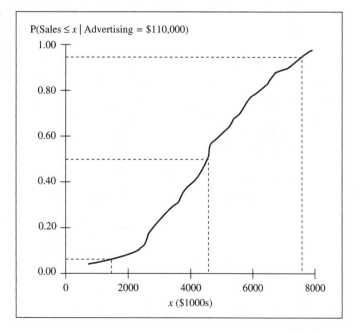

Figure 10.20

Influence diagram model showing relationship between sales and advertising.

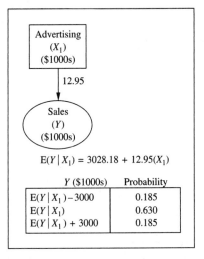

Figure 10.20

Influence diagram model showing relationship between sales and advertising.

could also fit a theoretical distribution. Typically in regression analysis, a normal distribution is fit to the residuals, but virtually any continuous distribution that makes sense for the situation could be used. (For example, the residuals could be used as data in BestFit to find the best-fitting distribution.)

All of our effort has paid off in modeling terms. With the coefficient estimates and the distribution of errors, we have a complete, if simplified, model of the uncertainty in Sales given its conditioning variable, Advertising. Figure 10.20 shows an influence diagram that represents the simple model we have created, including a three-point approximation for Sales given Advertising.

Estimation: More than One Conditioning Variable

Our advertising-and-sales example above demonstrated the basics. The model that we came up with, though, really is quite simplistic. In fact, with Advertising as the only explanatory variable, we implicitly assume that nothing else matters. But that clearly is too simple, and we have already argued that both the price of our product and our competitor's price also matter.

An augmented data set is shown in Table 10.5. For each of the 36 occasions, we now also have information on Price (X_2) and Competition Price (X_3). Can we incorporate these data into our analysis? The answer is yes, and the procedure is essentially the same as it was before.

We want to come up with estimates for β_0, β_1, β_2, and β_3 in the following expression:

$$E(Y \mid X_1, X_2, X_3) = \beta_0 + \beta_1 X_1 + \beta_2 X_2 + \beta_3 X_3$$

As before, we will let b_0, b_1, b_2, and b_3 denote the estimates of the corresponding β coefficients. Analogous to the one-variable version of the problem, we can calculate expected Sales, given the values of the conditioning variables. In turn, we can

Table 10.5
Augmented data set for sales example.

Promotion	Advertising ($1000s)	Price ($)	Competition Price ($)	Sales ($1000s)
1	366	90.99	96.95	10541
2	377	90.99	93.99	8891
3	387	94.99	90.99	5905
4	418	96.99	97.95	8251
5	434	92.99	97.95	11461
6	450	95.95	93.95	6924
7	457	93.95	90.99	7347
8	466	91.95	96.95	10972
9	467	96.95	94.99	7811
10	468	92.95	96.95	10559
11	468	97.99	98.95	9825
12	475	91.95	90.99	9130
13	479	99.95	91.95	5116
14	479	96.99	95.95	7830
15	481	91.95	90.95	8388
16	490	96.99	96.99	8588
17	494	96.95	91.95	6945
18	502	98.95	95.95	7697
19	505	94.99	96.99	9655
20	529	93.99	97.95	11516
21	532	91.99	95.99	11952
22	533	92.99	97.99	13547
23	542	93.99	92.95	9168
24	544	90.95	95.95	11942
25	547	94.99	93.95	9917
26	554	89.95	90.95	10666
27	556	96.95	95.95	9717
28	560	91.99	97.95	13457
29	561	98.99	97.95	10319
30	566	93.95	91.99	9731
31	566	94.99	94.99	10279
32	582	98.99	91.99	7202
33	609	89.95	92.99	12103
34	612	92.95	92.99	11482
35	617	92.95	94.95	11944
36	623	94.99	91.99	9188

Figure 10.21

Regression analysis for full model.

	A	B	C	D	E	F	G
	Regression Analysis						
128							
129	Regression Statistics						
130							
131	Multiple R	0.98					
132	R Square	0.95					
133	Adjusted R Square	0.95					
134	Standard Error	459.10					
135	Observations	36					
136							
137	Analysis of Variance						
138		df	Sum of Squares	Mean Square	F	Significance F	
139	Regression	3	137,289,638	45,763,213	217.12	0.00	
140	Residual	32	6,744,669	210,771			
141	Total	35	144,034,307				
142							
143		Coefficients	Standard Error	t Statistic	P-value	Lower 95%	Upper 95%
144							
145	Intercept	2199.34	3839.74	0.57	0.57	-5621.94	10020.62
146	Advertising	15.05	1.17	12.83	0.00	12.66	17.44
147	Price	-503.76	28.34	-17.77	0.00	-561.50	-446.03
148	Comp Price	499.67	30.56	16.35	0.00	437.42	561.92
149							

calculate the residual by subtracting the expected sales calculation from the actual sales for that observation. And, as before, we want to specify the b's so that the sum of the squared residuals is minimized. Accomplishing this is just a slightly more general version of the one-variable problem described above.

The same computer program used for the one-variable problem can be used here. Figure 10.21 shows the Excel screen with the results of the regression analysis with all three conditioning variables. As before, b_0, b_1, b_2, and b_3 are in the column labeled "Coefficients." With these values, we can now write out the formula for expected Sales, given the three conditioning variables:

$$E(Y \mid X_1, X_2, X_3) = 2199.34 + 15.05X_1 - 503.76X_2 + 499.67X_3$$

This expression makes sense. As before, every additional $1000 spent on Advertising increases expected Sales, but now in a more complete model this effect is estimated to be $15,050. The coefficients on Price and Competition Price are interpreted similarly; a $1 Price increase would lead to a decrease in expected Sales of $503,760, whereas a $1 increase in the Competition Price would bring expected Sales up by $499,670.

Also as before, we can calculate the residuals and use them to construct a CDF for the errors. Table 10.6 gives the residuals for our full model, and Figure 10.22 displays the CDF of the residuals. From the graph (especially the horizontal scale), you can see that the distribution of errors is much tighter in the full model than it was in the model above in which we used only Advertising. In fact, an estimate of the standard deviation (often called the *standard error* and usually abbreviated as s_e) of the current distribution is $459,100 and is given in Cell B134 in Figure 10.21. For the single-variable model, the standard deviation was over $1.8 million, which is shown in the corresponding cell in Figure 10.16. The interpretation is straightforward; Price

Table 10.6
Residuals for the full model.

Observation	Sales(Y)	$E(Y \mid X_1, X_2, X_3)$	Residual
1	10541	10312	229
2	8891	8999	−108
3	5905	5635	270
4	8251	8572	−321
5	11461	10827	634
6	6924	7578	−654
7	7347	7212	135
8	10972	11333	−361
9	7811	7850	−39
10	10559	10859	−300
11	9825	9320	505
12	9130	8490	640
13	5116	5000	116
14	7830	8490	−660
15	8388	8561	−173
16	8588	9175	−587
17	6945	6737	208
18	7697	7849	−152
19	9655	10408	−753
20	11516	11753	−237
21	11952	11826	126
22	13547	12337	1210
23	9168	9450	−282
24	11942	12511	−569
25	9917	9521	396
26	10666	10667	−1
27	9717	9669	48
28	13457	13227	230
29	10319	9716	603
30	9731	9352	379
31	10279	10327	−48
32	7202	7054	148
33	12103	12514	−411
34	11482	11047	435
35	11944	12102	−158
36	9188	9686	−498

Figure 10.22

CDF of residuals for the full model.

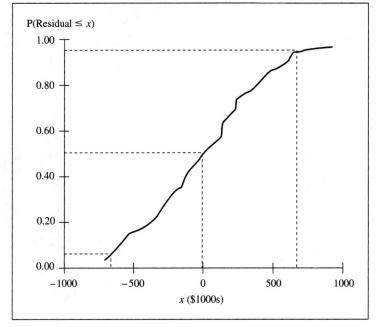

Figure 10.23

CDF of sales given advertising = $110,000, price = $94.50, and competition price = $96.69.

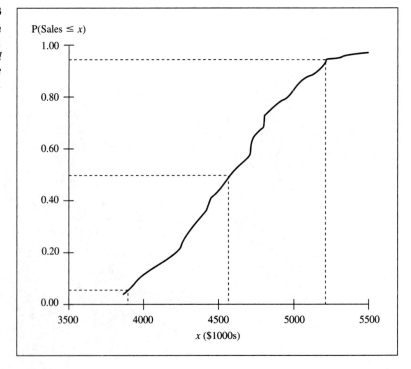

and Competition Price are variables that are relevant (and useful!) for understanding the uncertainty about Sales. Without them in the previous model, we were ignoring important information. With them in the current model, the result ultimately is less uncertainty about Sales. For example, with Advertising set at $110,000, Price at $94.50, and Competition Price at $96.69, conditional expected Sales would be $4,562,600. This is almost the same value that we had before when we used only Advertising. Now, though, you can see in Figure 10.23 that the uncertainty about Sales is much less, as reflected in the tighter distribution.

The tightness of the distribution, as indicated by the CDF of the errors and the standard deviation of that distribution, is the best available indication of the value of the model. For example, a product manager might look at the $1.8 million standard deviation for the previous single-variable model and scoff. Such a broad distribution makes planning difficult at best. On the other hand, the $459,100 standard deviation of the residuals in the full model may represent enough accuracy that this model could be used for planning production and distribution for a major promotion. And further model enhancements (adding other variables or improved modeling through advanced techniques) might reduce the uncertainty even further, thereby making the model that much more useful as a forecasting and planning tool.

The influence diagram in Figure 10.24 shows the entire model, including a three-point approximation of the error distribution to represent the conditional uncertainty about Sales.

Regression Analysis and Modeling: Some Do's and Don't's

The examples and discussion above give you a good idea of how regression analysis can be used to create models in decision analysis. However, you should be aware that we have barely scratched the surface of this powerful statistical

Figure 10.24

Influence diagram model of sales given advertising, price, and competition price.

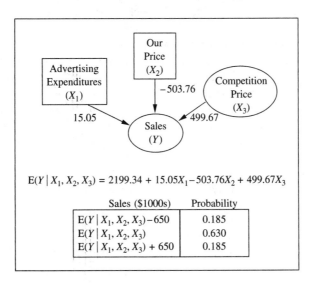

$$E(Y \mid X_1, X_2, X_3) = 2199.34 + 15.05X_1 - 503.76X_2 + 499.67X_3$$

Sales ($1000s)	Probability
$E(Y \mid X_1, X_2, X_3) - 650$	0.185
$E(Y \mid X_1, X_2, X_3)$	0.630
$E(Y \mid X_1, X_2, X_3) + 650$	0.185

technique. Complete coverage of regression is beyond the scope of this text, but many excellent statistical textbooks are available. This section provides a brief survey of some of the issues in regression modeling, especially as they relate to decision analysis.

We have seen a number of approaches to creating uncertainty models, including subjective assessment, the use of theoretical distributions, and data. The use of data can be very powerful; a statistical model and analysis, when based on an appropriate data set of adequate size, is very persuasive. The drawback is that data collection can take time and resources and may not always be feasible. In such cases, an analyst might be better off relying on theoretical distributions and expert subjective judgments. Still, when data are available, they can be valuable.

Although you should by now have an understanding of regression basics (which will improve with practice on some problems), there is much more to learn. Many courses in statistics are available, as are innumerable textbooks, some of which are listed in this chapter's reference section. Also, become familiar with at least one computer program for doing regression. The examples have used the simple (in some ways primitive) regression facility that is available with Excel. Packages developed specifically for statistical analysis may provide a more complete set of analysis options, but the flexible spreadsheet environment can be very useful to a decision analyst.

Creating an uncertainty model with regression can be quite powerful. It does have some important limitations, however. Two in particular warrant consideration in the context of decision analysis. First, the data set must have an adequate number of observations. Just what "adequate" means is debatable; many rules of thumb have been proposed, and some work better than others. The approach we have described above for decision-analysis modeling will perform poorly without enough data to estimate the coefficients and the error distribution. Although "enough data" may mean different things in different contexts, a conservative rule of thumb would be to have at least 10 observations for each explanatory variable and never less than 30 observations total.

The main reason for having adequate data is that many data points may be necessary to fully represent all of the different ways in which the variables can occur together. You would like to have as good a representation as possible of all the possible combinations. In the sales example, we would like to have occasions when prices are high, low, and in between, and for each different price level, advertising expenditures should range from high to low. Note that there is no guarantee that a large data set will automatically provide adequate coverage of all possibilities. (Inadequate coverage can result in nonsensical coefficient estimates. Advanced techniques can diagnose this situation and provide ways to get around it to some extent.) On the other hand, very small data sets almost by definition will not provide adequate coverage.

Another way to view this problem is to realize that a small number of observations means that we cannot be very sure that the b's are accurate or reliable estimates of the β's. When we use the b's to calculate the conditional expected value, however, we are in essence assuming that the b's are good estimates and that virtually all of the uncertainty in the response variable comes from the error term. In fact, with a small data set, we may also be very unsure about what the conditional expected value itself should be. In some cases you might want to model the uncertainty about the estimates explicitly and incorporate this additional uncertainty by an appropriate broadening of Y's conditional probability distribution. Statistical procedures for generating appropriate probability intervals under these circumstances are included in most computer regression programs (although not in Excel).

Even with an adequate data set and a satisfactory model and analysis, there remains an important limitation. Recall that the regression model has a particular form; it is a linear combination of the explanatory variables. And the coefficient estimates are based on the particular set of observations in the data set. The upshot is that your model may be a terrific approximation of the relationship for the variables in the neighborhood of the data that you have. But if you try to predict the response variable outside of the range of your data, you may find that your model performs poorly. For example, our sales example above was based on prices that ranged from about $89 to $100. If the product manager was contemplating reducing the price to $74.95, it is unlikely that an extrapolation of the model would give an accurate answer; there is no guarantee that the linear approximation used in the first place would extend so far outside of the range of the data. The result can be an inaccurate conditional expected value and a poor representation of the uncertainty, usually in the form of a too-narrow distribution, resulting in turn in poor planning. In practical terms, this means that the company could be taken by surprise by how many (or how few) units are sold during the promotion.

The limitation with the range of the data goes somewhat further yet when we use multiple explanatory variables, and this is related to the discussion of the data requirement above. You may try to predict the response variable for a combination of the explanatory variables that is poorly represented in the data. Even though the value of each explanatory variable falls within its own range, the combination for which you want to create a forecast could be very unusual. In the sales example, it would not be unreasonable to find in such a competitive situation that both Price and Competition Price move together. Thus, we might be unlikely to observe these prices more than a few dollars apart. Using Price = $89.99 and Competition Price = $99.95 (or visa versa) could turn out to be a situation poorly represented by the data, and in this case the analysis may provide a very misleading representation of the expected Sales and the uncertainty. Although there is no simple way to avoid this problem, one helpful step is to prepare scatterplots of all pairs of explanatory variables and to ensure that the point for which you wish to predict the response variable lies within the data cloud in each of the scatterplots. Thus, in the sales example, we would be sure that the combination of prices and advertising that we wanted to analyze fell within the data clouds for each of three different scatterplots: Price versus Competition Price, Price versus Advertising, and Competition Price versus Advertising.

Regression Analysis: Some Bells and Whistles

As mentioned, our coverage of regression is necessarily limited. In this section, we look briefly at some of the ways in which regression analysis can be extended.

Nonlinear Models   First, we can easily get past the notion that the conditional expected value is a linear function of the explanatory variables. For example, suppose an analyst encounters the scatterplot in Figure 10.25, which relates average temperature (X) and energy consumption (Y), measured in kilowatt-hours, for small residences. A linear relationship between these two quantities clearly would not work well, and the reason is obvious: When the temperature deviates from a comfortable average temperature (around 20° C), we consume additional energy for either heating or air conditioning. To model this, a quadratic expression of the form $E(Y \mid X) = \beta_0 + \beta_1 X + \beta_2 X^2$ might work well. Implementation is easy: For each observation calculate X^2, thereby creating a second explanatory variable. Now run the regression with two explanatory variables, X and X^2. The coefficient estimates obtained define a parabolic curve that fits the data best in a least-squares sense. A similar approach works for fitting exponential or logarithmic functional forms, some of which have very compelling and useful economic interpretations. A textbook on regression modeling such as Neter, Wasserman, and Kutner (1989) discusses such modeling techniques in depth.

Figure 10.25

A nonlinear relationship between temperature and energy consumption.

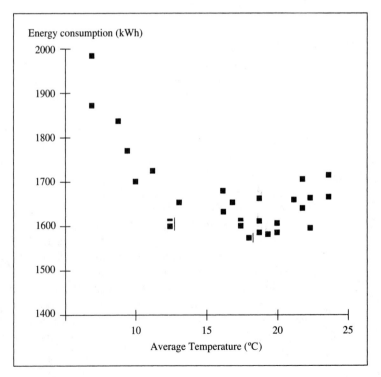

Categorical Explanatory Variables   Up to this point, all of the explanatory variables have been continuous. But not all useful measurements occur along a continuum. We might want to treat sales regions or sexes differently, for example. Or it might be useful to classify people into different age categories (child, adolescent, young adult, middle age, senior citizen) for marketing purposes. Is it possible to deal with such situations in a regression model?

The answer is yes, and the technique is straightforward. Suppose we have marketing data $(X_1, \ldots, X_k)$ for both men and women, but we would like to treat them separately. We can create a categorical variable (X_{k+1}) that equals 1 for men and 0 for women:

$$X_{k+1} = \begin{cases} 1 & \text{if male} \\ 0 & \text{if female} \end{cases}$$

This adds an extra column to the data matrix, and in the regression equation we include the new categorical variable:

$$E(Y \mid X_1, ..., X_k, X_{k+1}) = \beta_0 + \beta_1 X_1 + \cdots + \beta_k X_k + \beta_{k+1} X_{k+1}$$

Now consider how this equation works for men and women separately. Women have X_{k+1} coded as a 0, so the last term on the right-hand side drops out. Thus, for women the expression becomes

Women: $E(Y \mid X_1, ..., X_k, X_{k+1} = 0) = \beta_0 + \beta_1 X_1 + \cdots + \beta_k X_k$

For men, on the other hand, $X_{k+1} = 1$, and so the last term becomes β_{k+1}:

Men: $E(Y \mid X_1, ..., X_k, X_{k+1} = 1) = \beta_0 + \beta_1 X_1 + \cdots + \beta_k X_k + \beta_{k+1}$
$$= (\beta_0 + \beta_{k+1}) + \beta_1 X_1 + \cdots + \beta_k X_k$$

In other words, the coefficients for X_1 through X_k are exactly the same, but the expressions have different constant terms to represent the different genders. The interpretation of β_{k+1} is that this is the incremental difference in expected Y for a man as compared to a woman. For example, if Y represents mail-order clothing purchases per year, then $\beta_{k+1} = -\$50$ means that, everything else (age, income, and so on) being equal, a man is expected to spend $50 less per year on mail-order clothing than a woman.

Actually doing the analysis is easy. Create the categorical variable, include it as one of the explanatory variables in the regression analysis, and obtain the coefficient for the categorical variable from the computer printout. The rest is the same as before: To generate a prediction for a particular person, plug in the appropriate demographic data for all of the variables (including gender), calculate $E(Y \mid X_1, \ldots, X_k, X_{k+1})$, and use the error distribution to construct the conditional probability distribution for Y.

Note that we had two categories (male and female), and we were able to distinguish between them with just one categorical variable. What if you have more than one category? The rule is that you should use *one less categorical variable than there are categories.* For example, take the age example that we mentioned. There are five categories: child, adolescent, young adult, middle age, and senior citizen. (These categories must, of course, be defined clearly enough so that any individual can be placed unambiguously into one of the categories.) Now define four categorical variables:

$$X_{k+1} = \begin{cases} 1 & \text{if child} \\ 0 & \text{otherwise} \end{cases}$$

$$X_{k+2} = \begin{cases} 1 & \text{if adolescent} \\ 0 & \text{otherwise} \end{cases}$$

$$X_{k+3} = \begin{cases} 1 & \text{if young adult} \\ 0 & \text{otherwise} \end{cases}$$

$$X_{k+4} = \begin{cases} 1 & \text{if adult} \\ 0 & \text{otherwise} \end{cases}$$

Each individual will have a 1 or a 0 for each of these variables. Imagine seeing the values of these variables for some particular person. If you see that $X_{k+3} = 1$ and the others are zero, you know that this person is a young adult, and likewise for the other variables. In fact, there will never be more than one nonzero value, and that nonzero value will be a 1, indicating the person's category. But what if you see zeros for all four of these variables? Simple: This person falls into none of the four categories associated with the variables and therefore must be a senior citizen.

The rule of having one less categorical variable than you have categories is easy to forget. It may seem natural to set up one variable for each category. There is a fail-safe mechanism, however. Your computer program will not run with the extra categorical variable! It will most likely produce some cryptic error message like ANALYSIS FAILED—MATRIX SINGULAR, which is its way of indicating that it is unable to solve the required equations. Although this type of message can indicate other problems, if you have categorical variables, it almost surely means that you have one too many.

Categorical Response Variable (Y)   The regression procedure we have presented is designed especially for situations in which Y is a continuous variable. Often, though, you want to use data to predict whether an observation falls into a particular category (e.g., whether a particular customer will place another order or whether a particular individual will develop cancer). In decision-analysis terms, you would want to specify the conditional probability of falling into that category, given the values of the conditioning variables.

Statistical procedures are available to use data for this very purpose. However, these procedures have special requirements, and it is beyond the scope of this short introduction to describe how to use them. Suffice it to say that the standard approach laid out above for continuous variables is *not* appropriate for categorical response variables. If you would like to use data to model conditional probabilities for such a variable, consult one of the advanced regression texts mentioned at the end of this chapter.

Regression Modeling: Decision Analysis versus Statistical Inference

Our approach in this section has very much reflected the decision-analysis paradigm. We have used data as a basis for constructing a model of the uncertainty surrounding a response variable. Our focus has been particularly on coming up with a conditional probability distribution for Y, given values of conditioning variables. The motivation has been that understanding and modeling the relationships among the explanatory and response variables can be of use in making decisions.

If you have been exposed to regression analysis before, perhaps in a statistics course, you no doubt realize that the perspective taken here is somewhat different from the conventional approach. Although our approach is not entirely inconsistent with conventional thinking, there are some key differences:

- Statistical inference often focuses on various types of *hypothesis tests* to answer questions like, "Is a particular coefficient equal to zero?" Or, "Is this set of explanatory variables able to explain any of the variation in Y?" As decision analysts, such questions would more likely be framed as, "Is it reasonable to include X_i as an explanatory variable, or should it be left out?" A decision analyst might answer such a question with standard sensitivity-analysis techniques, which would permit consideration of the context of the current decision, including costs, benefits, and other risks. The conventional hypothesis-testing approach ignores these issues.

- Conventional regression modeling and the inference that is typically done (including hypothesis tests and the calculation of confidence and prediction intervals for Y) rely on the assumption that the errors follow a normal distribution. While the analysis can be quite powerful when this condition is met, it can be limiting or even misleading for situations in which normality is inappropriate. As decision analysts, we have focused on using the data to create a useful model of the uncertainty in the system for forecasting and decision-making purposes, which includes coming up with a suitable error distribution.

- In the standard regression approach, a question left unaddressed is how to model uncertainty in the explanatory variables. For those variables that are themselves uncertain (like Competition Price in our example), it may be important from a decision-analysis point of view to model that uncertainty. The influence-diagram approach that we have developed in this chapter provides the basis for such modeling. Either historical data or subjective judgments can be used to assign a probability distribution to an uncertain explanatory variable. Thus included in the model, that uncertainty becomes an integral part of any further analysis that is done.

An Admonition: Use with Care

With a little knowledge of regression, you have a very powerful data-analysis tool. A temptation is to apply it indiscriminately to a wide variety of situations. Part of its appeal is that it is using "objective" data, and hence the analysis and results are free of subjective judgments and bias. This is far from true! In fact, you should appreciate the broad range of judgments and assumptions that we made: linear expected values, the error probability distribution that does not change shape, the inclusion of particular explanatory variables in the model. More fundamentally, we implicitly make the judgment that the past data we use are appropriate for understanding current and future relationships among the variables. When we use a conditional probability distribution based on regression analysis, all of these judgments are implicit. Thus, the admonition is to use regression with care (especially with regard to its data requirements) and to be cognizant of its limitations and implicit assumptions. Regression is a rich and complex statistical tool. This section has provided only a brief introduction: further study of regression analysis will be especially helpful for developing your skill in using data for modeling in decision analysis.

Natural Conjugate Distributions (Optional)

So far we have considered modeling uncertainty using subjective judgments, theoretical models, and data. It is clear that we can mix these techniques to varying degrees. We have seen examples in which we have used a theoretical model to represent a subjective probability distribution (the use of the beta distribution in the soft pretzel problem), others in which data were used to estimate parameters for a theoretical distribution (the halfway-house data), and still others in which data were used to modify subjective probabilities regarding the parameter of a theoretical distribution (the soft pretzel examples using the binomial and Poisson distributions). Our topic now continues in this vein. We will see how to use Bayes' theorem in a systematic way to use data to update beliefs in the context of theoretical models.

Recall the Poisson example in Chapter 9 (page 360) in which we wondered about the quality of a new location for the soft pretzel kiosk. The new location could have been good, bad, or dismal, with P(Good) = 0.7, P(Bad) = 0.2, and P(Dismal) = 0.1. Each characterization is associated with a specific value of m, the expected number of pretzels sold per hour; a good location implies $m = 20$, bad implies $m = 10$, and dismal implies $m = 6$. The uncertainty about the nature of the location also can be thought of as uncertainty about the parameter m: $P(m = 20) = 0.7$, $P(m = 10) = 0.2$, and $P(m = 6) = 0.1$. Furthermore, we might have had more than three possible values for m; in fact, the location might have been anything between dismal and wonderful, or might have had m values anywhere from 0 to, say, 50 (or even higher).

Uncertainty about a parameter (such as m) often can be modeled with a continuous probability distribution. As before, this would be the prior probability distribution. Then, as in the Poisson example, we may obtain data that provide information about the parameter of the process. Bayes' theorem provides the mechanism for using the data to update the prior probability distribution in order to arrive at a posterior probability distribution.

In this section, we will explore this process in some detail. It turns out that the mathematical details are such that performing the calculations can be quite difficult unless the prior distribution and the distribution for the data match in a particular way. We will look at two such situations. The first is when the decision maker is uncertain about the parameter p in a binomial distribution and represents that uncertainty with a beta prior distribution. The decision maker then observes data, the outcome of a binomial random variable, and uses these data to update the prior distribution. In the second situation, the decision maker is uncertain about the mean μ of a normal distribution and represents that uncertainty with another normal distribution for μ. Now the decision maker takes a sample, calculates a sample mean, and uses this information to update the prior distribution for μ.

Our plan of attack is as follows. First, we will discuss the overall process. The process is not particularly complicated, but the concepts involved are somewhat demanding. A thorough understanding of the process requires attention to several conceptual issues. Next, we will show how the process works in the two specific situations that were mentioned above, and we will see examples. We will end with a discussion of how the uncertainty about a distribution parameter influences one's prediction or forecast.

Uncertainty About Parameters and Bayesian Updating

The flowchart (*not* an influence diagram) in Figure 10.26 shows the process that we will go through. The symbol θ is used to represent a parameter of a theoretical distribution in which we may be interested. The process goes as follows.

Figure 10.26

Using data to update uncertainty about parameter θ via Bayes' theorem.

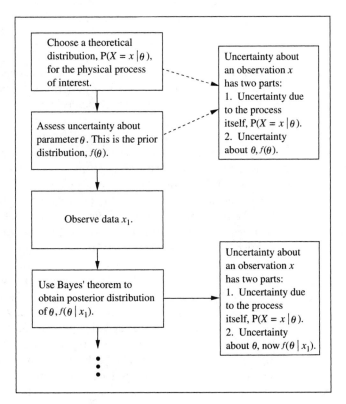

First, the decision maker chooses some theoretical distribution to model his or her beliefs regarding an uncertain quantity X (for example, a normal distribution to model the scores of students on a standardized exam or a Poisson distribution to model the number of customers who arrive at a soft pretzel kiosk). Let $P(X = x \mid \theta)$ denote this theoretical distribution, which has some parameter θ about which the decision maker is uncertain. Being uncertain about θ, the decision maker assesses a *prior probability distribution* $f(\theta)$ for the parameter. At this point, $P(X = x \mid \theta)$ and $f(\theta)$ together embody all of the decision maker's beliefs about the possible outcomes of the uncertain quantity X. If necessary, the decision maker can use this model of the uncertainty in solving a decision problem. Our concern here, however, is how to use data. Thus, the next step in the process is the acquisition of data and the incorporation of the new information into the model.

Suppose that the decision maker, having assessed $P(X = x \mid \theta)$ and $f(\theta)$, now observes an x, an outcome from the physical system (for example, a sample of exam scores or actual arrivals at the soft pretzel kiosk over a period of time). Label the data as x_1. The data contain information that the decision maker can use to refine his or her probabilities. In particular, Bayes' theorem allows for the updating of the prior distribution $f(\theta)$ so that it becomes a posterior distribution $f(\theta \mid x_1)$:

$$f(\theta \mid x_1) = \frac{P(x_1 \mid \theta)f(\theta)}{\int_{-\infty}^{+\infty} P(x_1 \mid \theta)f(\theta)d\theta} \tag{10.1}$$

Equation (10.1) is simply Bayes' theorem when $f(\theta)$ is a continuous distribution. The output of Bayes' theorem in this case is a posterior distribution of θ after having observed x_1. That is, $f(\theta \mid x_1)$ represents the decision maker's uncertainty about θ after having seen data x_1.

The term $P(x_1 \mid \theta)$ in Equation (10.1) is called the *likelihood function*. We calculated likelihoods for the binomial and Poisson soft pretzel examples in Chapter 9: We did it when we calculated the probability of observing the data for each possible parameter value. In the binomial case, we calculated binomial probabilities for the observed data depending on whether the pretzels were a hit ($p = 0.3$) or a flop ($p = 0.1$). In the Poisson case, we calculated Poisson probabilities for the observed number of arrivals depending on whether the new location was good ($m = 20$ per Hr), bad ($m = 10$ per Hr), or dismal ($m = 6$ per Hr).

The flowchart in Figure 10.26 is shown with three dots at the bottom, indicating that the process can continue. That is, the decision maker now has uncertainty about θ, as represented by $f(\theta \mid x_1)$. New data (x_2) might be observed, and the decision maker again can incorporate this new information via Bayes' theorem. The result would be $f(\theta \mid x_1, x_2)$. This sequential updating process can continue indefinitely.

In principle, Equation (10.1) can be used regardless of the form of the likelihood function or the proper distribution $f(\theta)$. With specific mathematical expressions for these functions, it always is possible to have a computer crunch the numbers to generate a posterior distribution $f(\theta \mid x_1)$. It may not be possible to write down a mathematical expression for $f(\theta \mid x_1)$, but the computer could grind out the calculations, evaluate $f(\theta \mid x_1)$ for many different values of θ, and finally plot the distribution. But an easier way to do this would be nice.

Fortunately, there is an easier way in certain situations. Recall that we are moving from a prior distribution to a posterior distribution. In certain cases, the prior and posterior distributions have the same form. This form is called the *natural conjugate distribution* for the theoretical distribution that represents the physical process. Moreover, in these cases relatively simple rules guide the use of observed data in adjusting the prior distribution to obtain the posterior. In other words, this elegant result reduces the seemingly complicated Equation (10.1) to a few simple formulas in these cases. Two such cases involve the binomial and normal distributions. We will discuss each one in turn, using examples.

Binomial Distributions: Natural Conjugate Priors for *p*

In this case, the chosen theoretical distribution to represent the physical process is binomial with parameters n and p. That is, the decision maker knows that the process generates many observations that are either successes or failures, and that in a group of n observations there will be some number X of successes. Thus, $P(X = x \mid \theta)$ in this case is equal to the binomial probability $P_B(X = r \mid n, p)$, which is given in Equation (9.1).

Of course, $P_B(X = r \mid n, p)$ is easy to calculate for any value of r as long as p is known. In this case, however, we want to incorporate uncertainty about p. The natural conjugate prior distribution for p is a beta distribution. That is, if the decision maker uses the beta distribution to model the prior uncertainty about p and then updates that beta prior on the basis of an observation from the process, then the posterior distribution of p also will be a beta distribution. Note that the beta distribution is a reasonable model, because p must be between 0 and 1. Here is how the updating process works:

1   The decision maker considers a physical process and concludes that its uncertainty can be represented with a binomial distribution. That is, out of n trials, the probability of r successes is $P_B(X = r \mid n, p)$.

2   The decision maker assesses a beta prior distribution for p, $f_\beta(p \mid r_0, n_0)$. That is, the decision maker carefully considers the uncertainty about p and concludes that it can be adequately represented with a beta distribution having parameters r_0 and n_0. Based on this prior distribution for p, a good estimate for p would be the mean of this distribution, r_0/n_0.

3   The decision maker observes n_1 independent trials, and r_1 of these are successes. Based on these data alone, of course, a good estimate for p would be r_1/n_1. The issue is how to combine this information with the prior distribution.

4   Combining the data with the prior distribution via Bayes' theorem gives a posterior distribution for p that is also a beta distribution, $f_\beta(p \mid r^*, n^*)$, where

$$r^* = r_0 + r_1 \qquad\qquad\qquad (10.2a)$$

and

$$n^* = n_0 + n_1 \qquad\qquad\qquad (10.2b)$$

That is, the decision maker only has to add the r's and n's to find the parameters of the posterior distribution for p.

In addition, we can see how this process could continue as new data are observed. The posterior $f_\beta(p \mid r^*, n^*)$ becomes the prior distribution in the next round. Data r_2 and n_2 are observed. Thus, the parameters for the posterior distribution after observing r_2 and n_2 are $r^{**} = r^* + r_2 = r_0 + r_1 + r_2$ and $n^{**} = n^* + n_2 = n_0 + n_1 + n_2$.

In Chapter 9 we discussed briefly that in subjectively assessing a beta distribution, the decision maker may think of r and n as representing "equivalent information." If the decision maker believes that the available information is equivalent to having observed 5 successes in 10 trials, then appropriate beta parameters would be $r = 5$ and $n = 10$. In performing the updating of a prior distribution via Bayes' theorem here, we can see just how appropriate this analogy is. Essentially, the r's and n's are added together as if they had all resulted from sample observations.

As an example, let us return to the soft pretzel example in Chapter 9, specifically pages 372–373. There we considered whether to proceed with the marketing of your pretzels. Of course, if you are convinced that your new pretzel will be a great seller, you certainly would undertake the project. The problem is that the proportion (Q) of the market that your pretzel might capture is uncertain. In the example, this uncertainty about Q is represented with a beta distribution having parameters $r = 1$ and $n = 4$. Let us take this as a prior distribution, $f_\beta(Q \mid r_0 = 1, n_0 = 4)$. The expected value of this distribution is $E(Q) = \frac{1}{4} = 0.25$, which would be a good estimate for Q based on the prior distribution.

Now suppose you run a taste test. You bake 20 pretzels, and 7 of 20 tasters prefer your pretzel over the competition's. Thus, $r_1 = 7$ and $n_1 = 20$. Combining these data with the prior distribution, the posterior probability distribution for Q is $f_\beta(Q \mid r^* = 1 + 7 = 8, n^* = 4 + 20 = 24)$. The posterior expected value of Q is $8/24 = 0.33$. The data have considerably affected your beliefs about Q. Figure 10.27 shows both the prior and posterior distributions. The additional data have greatly reduced the uncertainty about Q.

Normal Distributions: Natural Conjugate Priors for μ

Suppose the decision maker chooses a normal distribution to represent the physical process. That is, the decision maker knows that the outcomes (the x's) that arise from the process are numbers that can take on any value in a continuum and believes that a normal distribution provides a good representation for the x's distribution. In this case, $P(x \mid \theta)$ is the expression $f_N(x \mid \mu, \sigma)$ from Equation (9.4).

We will assume that the standard deviation of the process, σ, is known. The decision maker, however, does now know μ but is willing to assess a probability distribution for it. The natural conjugate prior distribution for μ also turns out to be a

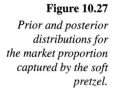

Figure 10.27

Prior and posterior distributions for the market proportion captured by the soft pretzel.

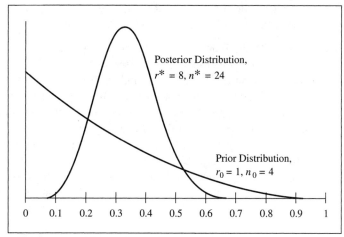

Posterior Distribution,
$r^* = 8, n^* = 24$

Prior Distribution,
$r_0 = 1, n_0 = 4$

normal distribution. That is, if the decision maker chooses a normal prior distribution to represent the uncertainty about μ and then updates that prior based on observations from the process, the resulting posterior distribution for μ also will be normal. Let us suppose that the decision maker assesses a normal distribution for μ, $f_N(\mu \mid m_0, \sigma_0)$.

This situation can be a bit confusing because so many different normal distributions are floating around. First, the original distribution is normal, $f_N(x \mid \mu, \sigma)$. But we also have normal distributions for the parameter μ itself. The prior is $f_N(\mu \mid m_0, \sigma_0)$, and the posterior also will be normal, $f_N(\mu \mid m^*, \sigma^*)$. Here are the steps in the updating process:

1 The decision maker considers a physical process and concludes that the uncertainty of the process can be represented with a normal distribution. That is, the distribution of observations from the process is $f_N(x \mid \mu, \sigma)$.

2 The decision maker assesses a normal prior distribution for μ, $f_N(\mu \mid m_0, \sigma_0)$. That is, the decision maker carefully considers the uncertainty about μ and concludes that it can be modeled with a normal distribution having parameters m_0 and σ_0. Based on this prior distribution for μ, a good estimate for μ would be the mean m_0.

3 The decision maker actually observes n_1 independent observations from the process, and the average of these sample observations is $\bar{x}_1$. Based on these data alone, of course, a good estimate for μ would be $\bar{x}_1$.

4 Combining the data with the prior distribution via Bayes' theorem gives a posterior distribution for μ that also is a normal distribution, $f_N(\mu \mid m^*, \sigma^*)$, where

$$m^* = \frac{m_0 \sigma^2 / n_1 + \bar{x}_1 \sigma_0^2}{\sigma^2 / n_1 + \sigma_0^2} \qquad (10.3a)$$

and

$$\sigma^* = \sqrt{\frac{\sigma_0^2 \sigma^2 / n_1}{\sigma^2 / n_1 + \sigma_0^2}} \qquad (10.3b)$$

That is, the decision maker combines the prior information and the data via Equations (10.3a) and (10.3b) to obtain the posterior distribution of μ.

As with the binomial model, one can see how the sequential incorporation of data would be done in this case. Observing the first sample would lead to the posterior as given above. In the next round, m^* and σ^* become the parameters for a new prior distribution that is updated on the basis of a second set of observations, using Equations (10.3a) and (10.3b) with m^* and σ^* in place of m_0 and σ_0. Further sequential updating would follow the same pattern.

In the binomial situation above, it was easy to see how the prior information could be interpreted as "equivalent" sample information. The same is true here, although it is not evident from (10.3a) and (10.3b). Suppose, however, that we write $\sigma_0^2 = \sigma^2 / n_0$. Then n_0 is interpreted as our "equivalent sample size." Having made this transformation, Equations (10.3a) and (10.3b) become

$$m^* = \frac{n_0 m_0 + n_1 \bar{x}_1}{n_0 + n_1} \qquad \text{(10.4a)}$$

and

$$\sigma^* = \sqrt{\frac{\sigma^2}{n_0 + n_1}} \qquad \text{(10.4b)}$$

From these expressions, it is clear how n_0 and n_1 are interacting. The posterior mean m^* is simply a weighted average of m_0 and $\bar{x}_1$, where the weights are based on the "sample sizes" n_0 and n_1. Posterior variance σ^{*2} has the same form as σ_0^2, with the cumulative sample size $n_0 + n_1$ used as the divisor of the process variance σ^2.

As an example of the updating process for the normal distribution, let us again look at the halfway-house data. Suppose that a decision maker believes that yearly bed-rental costs are normally distributed with mean μ and standard deviation $\sigma = \$220$. The expected value μ is unknown, but the decision maker assesses a normal distribution with mean $m_0 = \$345$ and standard deviation $\sigma_0 = \$50$, $f_N(\mu \mid m_0 = \$345, \sigma_0 = \$50)$. For example, this means that the decision maker believes there is roughly a 68% chance that μ is between \$295 and \$395, and roughly a 95% chance that μ is between \$245 and \$445.

Now suppose that the decision maker obtains the 35 observations listed in Table 10.1. The sample mean of these observations is \$380.40. We have $n_1 = 35$ and $\bar{x}_1 = \$380.40$. Applying Equations (10.3a) and (10.3b), we obtain the posterior distribution of μ, a normal distribution with parameters $m^* = \$367.80$ and $\sigma^* = \$29.80$. The prior and posterior distributions are illustrated in Figure 10.28. Notice that, with the added information, the distribution for μ is tighter, reflecting less uncertainty about μ.

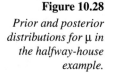

Figure 10.28

Prior and posterior distributions for μ in the halfway-house example.

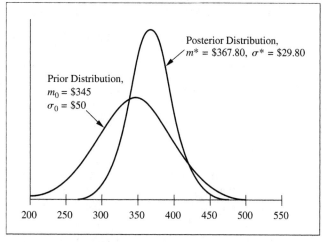

Prior Distribution,
m_0 = $345
σ_0 = $50

Posterior Distribution,
m^* = $367.80, σ^* = $29.80

Predictive Distributions

We have represented the uncertainty about the physical process under investigation with the distribution $P(X = x \mid \theta)$. That is, if we only knew θ, we would have everything we need for the distribution of the x's. We do not know θ, however, but we do have a probability distribution for it, $f(\theta)$. To give a distribution for the x's, we must combine $P(X = x \mid \theta)$ and $f(\theta)$. As previously mentioned, these two kinds of uncertainty, taken together, provide a complete model of the uncertainty that the decision maker faces regarding the x's. We simply must figure out how to put these pieces together.

To see this, imagine the binomial process. The probability of r successes in n trials is $P_B(X = r \mid n, p)$. But we need to know p in order to calculate probabilities using Equation (9.1). What we can do, however, is take something like an average over all possible values for p. This is exactly the approach we will take.

If we want a distribution for the x's that does not depend on θ, we want $P(X = x)$, not $P(X = x \mid \theta)$. We will call $P(X = x)$ the *predictive distribution* because it describes our uncertainty about X—which x values are most likely, the probability that X will fall into different intervals, and so on. To get $P(X = x)$, we must "integrate out" θ:

$$P(X = x) = \int_{-\infty}^{+\infty} P(X = x \mid \theta) f(\theta) d\theta$$

You might recognize this expression as the denominator of Bayes' theorem as it was written in Equation (10.1). Figuring out this integral will not be any easier than it was in Equation (10.1). It will always be possible to use a computer to calculate this integral numerically, but numerical integration techniques are beyond the scope of this text. In cases in which we have a natural conjugate prior distribution for q, it may be possible to find a simple expression for this distribution. We will take a look at the normal and binomial processes.

Predictive Distributions: The Normal Case

If the physical process follows a normal distribution, the integral becomes

$$P(X = x) = \int_{-\infty}^{+\infty} f_N(\mu, \sigma) f(\mu \mid m_0, \sigma_0) d\mu$$

$$= f_N\left(x \mid m_0, \sigma_p = \sqrt{\sigma^2 + \sigma_0^2}\right)$$

We will express this in words. Taking into account the prior uncertainty about μ, the distribution of the next sampled observation from the process is normal with mean m_0 and variance $\sigma_p^2 = \sigma^2 + \sigma_0^2$. The fact that the mean for x is m_0 is reasonable; our best guess for x would be μ, and our best guess for μ is m_0. The fact that the variance is $\sigma^2 + \sigma_0^2$ shows that indeed the overall uncertainty about x is composed of uncertainty about the process that generates the x's (σ^2) and uncertainty about $\mu(\sigma_0^2)$. If we knew exactly what μ was, then the variance of the distribution simply would be σ^2.

What we have just done can be generalized to any normal distribution for μ. If some data already have been collected and the decision maker has a posterior distribution for μ, $f_N(\mu \mid m^*, \sigma^*)$, then the predictive distribution for x is normal with mean m^* and variance $\sigma_p^2 = \sigma^2 + \sigma^{*2}$.

Continuing our halfway-house example, suppose that, having collected the data and updated the prior distribution for μ, the decision maker would like to express the uncertainty about the next yearly bed rental (x). The distribution desired is the predictive distribution. We know that $\sigma = \$220$ and that the posterior distribution for μ was normal with mean $m^* = \$367.80$ and standard deviation $\sigma^* = \$29.80$. Thus, the predictive distribution will be normal with mean $\$367.80$ and variance $\sigma_p^2 = (220)^2 + (29.8^2)$. Performing the calculations, we find that $\sigma_p = \$222$, which is just barely more than the process standard deviation of $\$220$. Thus, as we might have suspected, most of the uncertainty about the bed-rental cost is simply the result of fluctuation in these costs. A little bit of uncertainty results from the uncertainty about μ.

Predictive Distributions: The Binomial Case

If the physical process is one that can be represented with a binomial distribution, then the matter is slightly more complicated. We really want to know how likely we would be to get different numbers of successes in n trials. The integral to find the predictive distribution $P(X = r)$ becomes

$$P(X = r) = \int_0^1 P_B(r \mid n, p) f_\beta(p \mid r_0, n_0) dp$$

$$= \frac{(r + r_0 - 1)! \, (n + n_0 - r - r_0 - 1)! \, n! \, (n_0 - 1)!}{r! \, (r_0 - 1)! \, (n - r)! \, (n_0 - r_0 - 1)! \, (n + n_0 - 1)!}$$

This rather unusual looking expression is called a *beta-binomial* probability distribution. The number of successes r can take on only discrete values of $0, 1, \ldots, n$. The probability for any of these outcomes can be calculated by plugging values for r, n, r_0, and n_0 into the beta-binomial expression.

As with the normal distribution, this result generalizes as long as the distribution of p is a beta distribution. For example, in the case of our soft pretzels, the posterior distribution for p was a beta distribution with parameters $r^* = 8$ and $n^* = 24$. The predictive distribution of r (the number of tasters out of n who would prefer the new pretzel) would be the same as given above, but with r^* and n^* replacing r_0 and n_0. For example, the probability that 6 out of the next 12 tasters will prefer the new pretzel would be

$$P(X = 6) = \frac{(r + r^* - 1)! \ (n + n^* - r - r^* - 1)! \ n! \ (n^* - 1)!}{r! \ (r^* - 1)! \ (n - r)! \ (n^* - r^* - 1)! \ (n + n^* - 1)!}$$

$$= \frac{(6 + 8 - 1)! \ (12 + 24 - 6 - 8 - 1)! \ 12! \ (24 - 1)!}{6! \ (8 - 1)! \ (12 - 6)! \ (24 - 8 - 1)! \ (12 + 24 - 1)!}$$

$$= \frac{13! \ 21! \ 12! \ 23!}{6! \ 7! \ 6! \ 15! \ 35!}$$

$$= 0.1116$$

Probabilities for other possible values of r can be calculated in the same way.

The use of data to update information in a systematic way via Bayes' theorem can be very useful in decision analysis. We have spent much time in discussing natural conjugate distributions here, and they can be quite helpful. What is perhaps more critical, however, is that you understand the concept that one can be uncertain about a parameter and encode that uncertainty as a probability distribution. This uncertainty then has an impact on the decision maker's probability of future outcomes. Thus, the underlying ideas of distributions on parameters and predictive distributions for future outcomes are quite basic.

In the next chapter, we will take this process a step further. In simulation we will generate random variables from theoretical probability distributions with specific parameters. This is a standard technique in decision analysis and risk analysis. But we also will be able to specify our parameters as uncertain. By specifying parameters themselves as probabilistic, we will be doing the simulation counterpart of the natural conjugate analysis that we have discussed here. Furthermore, we will not be limited to certain families of distributions. We will be able to specify any kind of probability distribution that seems reasonable; the computer will do the calculations for us.

A Bayesian Approach to Regression Analysis (Optional)

Can the Bayesian approach described above be applied to regression analysis as well as the normal and binomial models described in the last section? The answer is yes, although certain assumptions must be made. The most important is that some sort of theoretical distribution for the errors must be assumed; this provides the basis for defining the likelihood function for the data. A typical assumption is that the errors follow a normal distribution with variance σ_e^2.

The general Bayesian approach is to model uncertainty about the model parameters. In a regression model, the parameters would be the error variance σ_e^2 and the coefficients $\beta_0, \ldots, \beta_k$. It turns out that there are natural conjugate prior distributions for these parameters and that data can be used to update these priors. Moreover, once the updating has been done, one can generate predictive distributions for future observations of the Y variable, given the X's. Thus, the Bayesian approach provides a systematic way to incorporate uncertainty about the coefficients as well as the error term, thereby giving a more complete accounting of uncertainty in the system. For more discussion of Bayesian regression, see Zellner (1971) or Berry (1995).

SUMMARY

In this chapter we have seen some ways in which data can be used in the development of probabilities and probability distributions for decision analysis. We began with the basics of constructing histograms and empirically based CDFs. A short discussion concerned the use of data to estimate parameters for theoretical distributions. The last part of the chapter discussed the use of data to update prior beliefs about parameters of a distribution. This process is based on Bayes' theorem. We saw how natural conjugate distributions worked for both normal and binomial cases. Finally, we discussed predictive distributions, or the distribution for future outcomes of a process when the parameters are uncertain.

EXERCISES

10.1 Explain in your own words the role that data can play in the development of models of uncertainty in decision analysis.

10.2 Why might a decision maker be reluctant to make subjective probability judgments when historical data are available? In what sense does the use of data still involve subjective judgments?

10.3 Suppose that an analyst for an insurance company is interested in using regression analysis to model the damage caused by hurricanes when they come ashore. The response variable is Property Damage, measured in millions of dollars, and the explanatory variables are Diameter of Storm, Barometric Pressure, Wind Speed, and Time of Year.

a What subjective judgments, both explicit and implicit, must the analyst make in creating and using this model?

b If X_1, Diameter of Storm, is measured in miles, what is the interpretation of the coefficient β_1?

c   Suppose the analyst decides to introduce a categorical variable X_5, which equals 1 if the eye of the hurricane comes ashore near a large city (defined as a city with population $\geq 500,000$) and 0 otherwise. How would you interpret β_5?

10.4 Estimate the 0.65 and 0.35 fractiles of the distribution of yearly bed-rental costs, based on the CDF in Figure 10.4.

10.5 Choose appropriate intervals and create a relative frequency histogram based on the halfway-house data in Table 10.1.

10.6 Estimate the 0.25 and 0.75 fractiles of the conditional distribution for Sales in Figure 10.23.

QUESTIONS AND PROBLEMS

10.7 It was suggested that five is the minimum number of data points for the least likely category in constructing histograms. In many cases, however, we must estimate probabilities that are extremely small or for which relatively few data are available. In the example about machine failures at the beginning of the chapter (pages 399–400), suppose that we had one day out of 260 in which three failures occurred. What do you think we should do in such a situation?

10.8 As discussed in the text, it often is possible to use a theoretical distribution as an approximation of the distribution of some sample data. It is always important, however, to check to make sure that your data really do fit the distribution you propose to use.

Consider the halfway-house data again. We calculated the sample mean $\bar{x} = 380.4$ and the sample standard deviation $s = 217.6$. Taking these as approximately equal to μ and σ, use BestFit or the Z table in Appendix E to find normal cumulative probabilities for dollar costs of 200, 300, 400, 500, 600, and 700. How do these theoretical normal probabilities compare with the data-based probabilities from the CDF in Figure 10.4? Do you think that the normal distribution is an appropriate distribution to use in this case? Compare the fit of the normal distribution CDF to the empirical CDF by superimposing them onto one graph.

10.9 A scientist collected the following weights (in grams) of laboratory animals:

9.79	9.23	9.11	9.62
8.73	11.93	10.39	8.68
9.76	9.59	11.49	9.86
11.41	9.60	7.24	

a   Use these data to create a data-based CDF for the distribution of weights for lab animals. Estimate the probability that an animal's weight will be less than 9.5 grams.

b   Fit a normal distribution to these data on the basis of the sample mean and sample variance. Use this normal distribution to estimate the probability that an animal's weight will be less than 9.5 grams.

c   Do you think the normal distribution used in part b is a good choice for a theoretical distribution to fit these data? Why or why not?

10.10 A plant manager is interested in developing a quality-control program for an assembly line that produces light bulbs. To do so, the manager considers the quality of the products that come from the line. The light bulbs are packed in boxes of 12, and the line produces several thousand boxes of bulbs per day. To develop baseline data, some of the workers test all the bulbs in 100 boxes. They obtain the following results:

No. of Defective Bulbs/Box	No. of Boxes
0	68
1	27
2	3
3	2

a Plot a histogram of these results. To use BestFit, you must open the @RISK—Model window, click on Fitting, and click on Input Data Options. Choose Discrete and Data in counted format.

b What kind of theoretical distribution might fit this situation well? Explain.

c Estimate parameters for the distribution you chose in part b by calculating the sample mean and variance. Use the theoretical distribution with its estimated parameters to estimate P(0 Defective), P(1 Defective), P(2 Defective), and P(3 Defective). How do these estimates compare to the relative frequency of these events in the data?

10.11 A retail manager in a discount store wants to establish a policy of the number of cashiers to have on hand and also when to open a new cash register. The first step in this process is to determine the rate at which customers arrive at the cash register. One day, the manager observes the following times (in minutes) between arrivals of customers at the cash registers:

0.1	2.6	2.9	0.5
1.2	1.8	4.8	3.3
1.7	0.2	1.5	2.0
4.2	0.6	1.0	2.6
0.9	3.4	1.7	0.4

a Plot a CDF based on these data.

b What kind of theoretical distribution do you think would be appropriate for these data? Why?

c Calculate the sample mean and sample standard deviation of the data, and use these to estimate parameters for the theoretical distribution chosen in part b.

d Plot a few points from the CDF of your theoretical distribution, and draw a smooth curve through them. How does this theoretical curve compare to the data-based CDF?

10.12 An ecologist studying the breeding habits of birds sent volunteers from the local chapter of the Audubon Society into the field to count nesting sites for a particular species. Each team was to survey five acres of land carefully. Because she was interested in studying the distribution of nesting sites within a particular kind of ecological system, the ecologist was careful to choose survey sites that were as similar as possible. In total, 24 teams surveyed five acres each and reported the following numbers of nesting sites in each of the five-acre parcels:

7	12	6	9
5	2	9	9
7	3	9	9
5	1	7	10
1	8	6	3
4	5	3	13

a Plot a histogram for these data.

b What kind of theoretical distribution might you select to represent the distribution of the number of nesting sites? Why? (Because the data are discrete, do not forget to change the input data options in BestFit to discrete. See Steps 6.2 and 6.3.)

c Estimate the parameters of the theoretical distribution by calculating the sample mean and sample variance.

d Plot the probability distribution based on your theoretical model from parts b and c. How does it compare to your histogram from part a?

10.13 Decision analyst Sandy Baron has taken a job with an up-and-coming consulting firm in San Francisco. As part of the move, Sandy will purchase a house in the area. There are two houses that are especially attractive, but their prices seem high. To study the situation a bit more, Sandy obtained data on 30 recent real-estate transactions involving properties roughly similar to the two candidates. The data are shown in Table 10.7. The Attractiveness Index measure is a score based on several aspects of a property's character, including overall condition and features (e.g., swimming pool, view, seclusion).

a Run a regression analysis on the data in Table 10.7 with Sale Price as the response variable and House Size, Lot Size, and Attractiveness as explanatory variables. What are the coefficients for the three explanatory variables? Write out the expression for the conditional expected Sale Price given House Size, Lot Size, and Attractiveness.

b Create a CDF based on the residuals from the regression in part a.

c The two properties that Sandy is considering are as follows:

House	List Price	House Size	Lot Size	Attractiveness
1	575	2700	1.6	75
2	480	2000	2.0	80

What is the expected Sale Price for each of these houses according to the regression analysis from part a?

d Create a probability distribution of the Sale Price for each house in part c. Where does the List Price for each house fall in its respective distribution? What advice can you give Sandy for negotiating on these houses?

Property	House Size (Sq. Ft.)	Lot Size (Acres)	Attractiveness Index	Sale Price ($1000s)
1	3000	3.6	64	550
2	2300	1.2	69	461
3	3300	1.3	72	501
4	2100	3.2	71	455
5	3900	1.1	40	503
6	3100	2.0	74	529
7	3600	1.6	69	478
8	2900	2.5	85	562
9	2000	2.6	70	417
10	3500	1.3	74	566
11	3100	2.3	79	494
12	3200	1.5	75	515
13	2800	1.3	62	490
14	3300	3.3	62	537
15	3000	3.9	70	527
16	3400	2.4	81	577
17	2800	1.7	77	490
18	2000	3.4	67	486
19	2400	2.9	68	450
20	3600	2.9	84	674
21	2400	1.9	75	454
22	3000	2.8	63	523
23	2200	3.6	78	469
24	3600	2.4	73	628
25	2900	1.1	85	570
26	3000	4.4	69	564
27	3100	1.8	54	444
28	2200	2.1	75	494
29	2500	3.9	61	479
30	2900	1.1	74	477

10.14 Ransom Global Communications (RGC), Inc., sells communications systems to companies that have worldwide operations. Over the next year, RGC will be instituting important improvements in its manufacturing operations. The CEO and majority stockholder, Thomas Ransom, has been extremely optimistic about the outlook for the firm and made the statement that sales will certainly exceed $6 million in the first half of 1996.

Semiannual data for RGC over the past 19 years are presented in Table 10.8. The Spending Index is related to the amount of disposable income that consumers have,

Table 10.8

Twenty years of data for Ransom Global Communications, Inc.

Year	Spending Index	System Price ($1000s)	Capital Investment ($1000s)	Advertising and Marketing ($1000s)	Sales ($1000s)
1977	39.8	56.2	49.9	76.9	5540
	36.9	59.0	16.6	88.8	5439
1978	26.8	56.7	89.2	51.3	4290
	48.4	57.8	106.7	39.6	5502
1979	39.4	59.1	142.6	51.7	4872
	33.2	60.1	61.3	20.5	4708
1980	33.6	59.8	−30.4	40.2	4628
	38.3	60.1	−44.6	31.6	4110
1981	28.5	63.1	−28.4	12.5	4123
	27.7	62.3	75.7	68.3	4842
1982	45.6	64.9	144.0	52.5	5740
	35.5	64.9	112.9	76.7	5094
1983	36.4	63.4	128.3	96.1	5383
	32.0	65.6	10.1	48.0	4888
1984	31.1	67.0	−24.8	27.2	4033
	36.2	66.9	116.7	72.7	4942
1985	40.8	66.2	120.4	62.3	5313
	43.3	67.9	121.8	24.7	5140
1986	35.9	68.9	71.1	73.9	4397
	47.6	71.4	−4.2	63.3	5149
1987	41.5	69.3	−46.9	28.7	5151
	42.0	69.7	7.6	91.4	4989
1988	53.6	73.2	127.5	74.0	5927
	43.2	73.4	−49.6	16.2	4704
1989	43.6	73.1	100.1	43.0	5366
	41.5	74.9	−40.2	41.1	4630
1990	46.2	73.2	68.2	92.5	5712
	42.9	74.2	88.0	83.3	5095
1991	51.7	74.3	27.1	74.9	6124
	32.8	77.1	59.3	87.5	4787
1992	41.8	78.6	142.0	74.5	5036
	51.5	77.1	126.4	21.3	5288
1993	41.2	78.2	29.6	26.5	4647
	45.5	77.9	18.0	94.6	5316
1994	55.4	81.0	42.4	92.5	6180
	44.1	79.9	−21.6	50.0	4801
1995	41.7	80.6	148.4	83.2	5512
	46.1	82.3	−17.6	91.2	5272

adjusted for inflation and averaged over the industrial nations. System Price is the average sales price of systems that RGC sells, Capital Investment refers to capital investment in the business, and Advertising and Marketing and Sales are self-explanatory.

a Use the data in Table 10.8 to run a regression analysis with Sales as the response variable and the other four variables as explanatory variables. Write out the expression for Expected Sales, conditional on the other variables. Interpret the coefficients.

b The forecast for the Spending Index in the first half of 1996 is 45.2. The company has committed $145,000 for capital improvements during this time. In order to make his goal of $6 million in sales, Ransom has indicated that the company will offer discounts on systems (to the extent that the average system price could be as low as $70,000) and will spend up to $90,000 on advertising and various marketing programs. If the firm drops the price as much as possible and spends the full $90,000 on Advertising and Marketing, what is the expected value for Sales over the next six months? Estimate the probability that Sales will exceed $6 million.

c Given that the company spends the full $90,000 on Advertising and Marketing, how low would Price have to be in order to ensure a 90% chance that Sales will exceed $6 million? What advice would you give Thomas?

10.15 Before performing the experiment, the scientist in Problem 10.9 thought a bit about the animals that he typically used in his lab. He knew (based on information from the animals' supplier) that the standard deviation of their weights was 1.5 grams. But the supplier was unable to specify the average weight precisely. The company did say that there was a 68% chance that the average weight was between 9.0 and 9.8 grams and a 95% chance that the average weight was between 8.6 and 10.2 grams.

a Use the stated probabilities to find a natural conjugate prior distribution for the average weight of the animals. What is $P(\mu \geq 10 \text{ grams} \mid m_0, \sigma_0)$?

b Find the posterior distribution for the average weight of the animals after having seen the data in Problem 10.9. What is $P(\mu \geq 10 \text{ grams} \mid m^*, \sigma^*)$?

10.16 Continuing Problem 10.15:

a Based on the prior probability distribution for μ, find the predictive probability that a single lab animal will weigh more than 11 grams, $P(x \geq 11 \text{ grams} \mid m_0, \sigma_0)$.

b After having seen the data and updating the beliefs about μ, find the predictive probability that a lab animal weighs more than 11 grams, $P(x \geq 11 \text{ grams} \mid m^*, \sigma^*)$?

10.17 The plant manager in Problem 10.10 actually began her investigation by assessing her subjective probability distribution for Q, the proportion of defective bulbs coming off the assembly line. Her assessed distribution was a beta distribution with parameters $r_0 = 1$, $n_0 = 20$; that is, her distribution was $f_\beta(q \mid r_0 = 1, n_0 = 20)$.

a Plot this prior distribution for Q.

b What would her posterior distribution for Q be after having observed the data in Problem 10.10?

10.18 A political analyst was interested in the proportion C of individuals who would vote for a controversial ballot measure. While he thought that it would be a close call, he was unsure of the precise value for C. He assessed a beta distribution for C with parameters $r_0 = 3$, $n_0 = 6$.

a Plot this prior distribution for C.

b The analyst is about to ask four individuals about their preferences. What is the probability that more than two of these individuals will express their support for the ballot measure?

c Having questioned the four individuals, the analyst found that three would indeed vote for the ballot measure. Find the analyst's posterior distribution for C. Plot this posterior distribution and compare it with the prior distribution plotted in part a.

d The analyst is now about to survey another 10 people. What is the probability that more than five of these people will support the ballot measure in this new poll?

e Suppose that 6 of the 10 people surveyed said they would vote for the ballot measure. The analyst now must write up his results. What is his probability that the ballot measure will pass?

10.19 A comptroller was preparing to analyze the distribution of balances in the various accounts receivable for her firm. She knew from studies in previous years that the distribution would be normal with a standard deviation of $1500, but she was unsure of the mean μ. She thought carefully about her uncertainty about this parameter and assessed a normal distribution for μ with mean $m_0 = \$10,000$ and $\sigma_0 = \$800$.

Over lunch, she discussed this problem with her friend, who also worked in the accounting division. Her friend commented that she also was unsure of μ but would have placed it somewhat higher. The friend said that "better" estimates for m_0 and σ_0 would have been $12,000 and $750, respectively.

a Find $P(\mu > \$11,000)$ for both prior distributions.

b That afternoon, the comptroller randomly chose nine accounts and calculated $\bar{x} = \$11,003$. Find her posterior distribution for μ. Find the posterior distribution of μ for her friend. Calculate $P(\mu > \$11,000)$ for each case.

c A week later the analysis had been completed. Of a total of 144 accounts (including the nine reported in part b), the average was $\bar{x} = \$11,254$. Find the posterior distribution for μ for each of the two prior distributions. Calculate $P(\mu > \$11,000)$ for each case.

d Discuss your answers to parts a, b, and c. What can you conclude?

CASE STUDIES

TACO SHELLS

Martin Ortiz, purchasing manager for the True Taco fast food chain, was contacted by a salesperson for a food service company. The salesperson pointed out the high breakage rate that was common in the shipment of most taco shells. Martin was aware of this fact, and noted that the chain usually experienced a 10% to 15% breakage rate. The salesperson then explained that his company recently had designed a new shipping container that reduced the breakage rate to less than 5%, and he produced the results of an independent test to support his claim.

When Martin asked about price, the salesperson said that his company charged $25 for a case of 500 taco shells, $1.25 more than True Taco currently was paying. But the salesperson claimed that the lower breakage rate more than compensated for the higher cost, offering a lower cost per usable taco shell than the current supplier. Martin, however, felt that he should try the new product on a limited basis and develop his own evidence. He decided to order a dozen cases and compare the breakage rate in these 12 cases with the next shipment of 18 cases from the current supplier. For each case received, Martin carefully counted the number of usable shells. The results are shown below:

Usable Shells				
New Supplier		Current Supplier		
468	467	444	441	450
474	469	449	434	444
474	484	443	427	433
479	470	440	446	441
482	463	439	452	436
478	468	448	442	429

Questions

1   Martin Ortiz's problem appears to be which supplier to choose to achieve the lowest expected cost per usable taco shell. Draw a decision tree of the problem, assuming he orders one case of taco shells. Should you use continuous fans or discrete chance nodes to represent the number of usable taco shells in one case?

2   Develop CDFs for the number of usable shells in one case for each supplier. Compare these two CDFs. Which appears to have the highest expected number of usable shells? Which one is riskier?

3   Create discrete approximations of the CDFs found in Question 2. Use these approximations in your decision tree to determine which supplier should receive the contract.

4   Based on the sample data given, calculate the average number of usable tacos per case for each supplier. Use these sample means to calculate the cost per usable taco for each supplier. Are your results consistent with your answer to Question 3? Discuss the advantages of finding the CDFs as part of the solution to the decision problem.

5   Should Martin Ortiz account for anything else in deciding which supplier should receive the contract?

Source: This case was adapted from W. Mendenhall, J. Reinmuth, and R. Beaver (1989) *Statistics for Management and Economics,* 6th ed. Boston: PWS-KENT.

FORECASTING SALES

Sales documents were scattered all over Tim Hedge's desk. He had been asked to look at all of the available information and to try to forecast the number of microwave ovens that NewWave, Inc., would sell over the upcoming year. He had been with the company for only a month and had never worked for a microwave company before. In fact, he had never worked for a company that was involved in consumer electronics. He had been hired because the boss had been impressed with his ability to grasp and analyze a wide variety of different kinds of decision problems.

Tim had dug up as much information as he could, and one of the things he had found was that one of NewWave's salespeople, Al Morley, had kept detailed records over the past 14 years of his own annual forecasts of the number of microwaves that the company would sell. Over the years, Morley had been fairly accurate. (His sales performance had been pretty good, too.) Of course, Morley had been only too happy to provide his own forecast of sales for the upcoming year. Tim thought this was fine, and reported back to his boss, Bill Maught.

"Yes," said Bill after he had listened to Tim, "I am aware that Al has been making these forecasts over the years. We have never really kept track of his forecasts, though. Even though he claims to have been fairly accurate over the years, I can remember one or two when he was not very close. For example, I think last year he was off by about 6000 units. Of course, if you ask him about those cases, he can explain them in hindsight. But there is always the possibility of unusual unforeseen circumstances in any given year. Besides, I think he forecasts low. When he does that, his sales quota is set lower, and he's more likely to get a bonus. I'd say there's about a 95 percent chance that on average he underforecasts by 1000 units or more. In fact, I'd bet even money that his average forecast error [sales – forecast] is above 1700 units."

Tim asked the obvious question. "Would you like me to look into this more? Maybe I could get a handle on just how good he is."

"Fine with me," answered Bill. "Besides, if we really could get a handle on how good he is, then his forecast might be a good basis for us to work with each year."

After that discussion, Tim had visited the accounting department to find out the number of units actually sold in each year for which Morley had made a forecast. He now had assembled the data and considered the numbers:

Forecast (Units)	Actual Sales	Error (Sales – Forecast)
39,000	41,553	2553
44,000	46,223	2223
46,000	49,351	3351
54,000	55,393	1393
60,000	61,607	1607

Forecast (Units)	Actual Sales	Error (Sales – Forecast)
59,000	68,835	9835
99,000	101,647	2647
124,000	123,573	–427
149,000	156,473	7473
145,000	146,333	1333
159,000	155,668	–3332
169,000	167,168	–1832
171,000	171,477	477
179,000	185,529	6529

Questions

1   Calculate the average and standard deviation of Morley's forecasting errors. What do you think about Bill Maught's opinion of Morley's forecasting?

2   Plot a CDF based on these data. Based on these data alone, what is the probability that Morley's forecast this year will be too low by 1700 units or more?

3   Assume that Morley's forecast errors follow a normal distribution with standard deviation $\sigma = 3000$. Translate Bill Maught's probability assessments into a prior distribution for Morley's average error, μ. On the basis of these data, what would be Maught's posterior distribution? What is his posterior probability that Morley's average error is greater than 1700 units?

4   Morley has forecast sales of 187,000 units for the coming year. Based on Maught's posterior distribution for μ (Morley's average error), what is the probability distribution for sales for this coming year? What is the probability that sales will be greater than 190,000 units? Then sketch a CDF for this distribution.

OVERBOOKING, PART II

Consider again the overbooking issue as discussed in the case study at the end of Chapter 9 (pages 388–389). Suppose the Mockingbird Airlines operations manager believes that the no-show rate (N) is approximately 0.04 but is not exactly sure of this value. She assesses a probability distribution for N and finds that a beta distribution with $r_0 = 1$ and $n_0 = 15$ provides a good fit to her subjective beliefs.

Questions

1   Suppose that Mockingbird sells 17 reservations on the next flight. On the basis of this prior distribution, find the predictive probability that all 17 passengers will

show up to claim their reservations. Find the probability that 16 will show up, 15, 14, and so on.

2   Should Mockingbird have overbooked for the next flight? If so, by how much? If not, why not? Support your answer with the necessary calculations.

3   Now suppose 17 reservations are sold, and 17 passengers show up. Find the operations manager's posterior distribution for *N*.

4   On the basis of this new information and the manager's posterior distribution, should Mockingbird overbook on the next flight? If so, by how much? If not, why not? Support your answers with the necessary calculations. [*Hint:* You may have to be careful in calculating the factorial terms. Many calculators and electronic spreadsheets will not have sufficient precision to do these calculations very well. One trick is to cancel out as many terms in the numerator and denominator as possible. Another is to do the calculation by alternately multiplying and dividing. For example, $5!/6! = (5/6)(4/5)(3/4)(2/3)(1/2)$.]

REFERENCES

Using data as the basis for probability modeling is a central issue in statistics. Some techniques that we have discussed, such as creating histograms, are basic tools; any basic statistics textbook will cover these topics. The text by Vatter et al. (1978) contains an excellent discussion of the construction of a data-based CDF. Fitting a theoretical distribution using sample statistics such as the mean and standard deviation is also a basic and commonly used technique (for example, Olkin, Gleser, and Derman 1980), although it is worth noting that statisticians have many different mathematical techniques for fitting distributions to empirical data.

Virtually every introductory statistics text covers regression analysis at some level. More advanced texts include Chatterjee and Price (1977), Draper and Smith (1981), and Neter, Wasserman, and Kutner (1989). The treatment in Chapter 9 of Vatter et al. (1978) is similar to the one presented here.

Bayesian updating and the use of natural conjugate priors is a central element of Bayesian statistics. Winkler (1972) has the most readable fundamental discussion of these issues. For a more complete treatment, including discussion of many more probabilistic processes and analysis using natural conjugate priors, see Raiffa and Schlaifer (1961) or DeGroot (1970). Zellner (1971) discusses Bayesian regression analysis.

Berry, D. (1995) *Statistics: A Bayesian Perspective*. Belmont, CA: Duxbury.

Chatterjee, S., and B. Price (1977) *Regression Analysis by Example*. New York: Wiley.

DeGroot, M. (1970) *Optimal Statistical Decisions*. New York: McGraw-Hill.

Draper, N., and H. Smith (1981) *Applied Regression Analysis,* 2nd ed. New York: Wiley.

Neter, T., W. Wasserman, and M. Kutner (1989) *Applied Linear Regression Models,* 2nd ed. Homewood, IL: Irwin.

Olkin, I., L. J. Gleser, and C. Derman (1980) *Probability Models and Applications*. New York: Macmillan.

Raiffa, H., and R. Schlaifer (1961) *Applied Statistical Decision Theory*. Cambridge, MA: Harvard University Press.

Vatter, P., S. Bradley, S. Frey, and B. Jackson (1978) *Quantitative Methods in Management: Text and Cases*. Homewood, IL: Irwin.

Winkler, R. L. (1972) *Introduction to Bayesian Inference and Decision*. New York: Holt.

Zellner, A. (1971) *An Introduction to Bayesian Inference in Econometrics*. New York: Wiley.

CHAPTER 11

Monte Carlo Simulation

The problems we have dealt with so far have allowed us to calculate expected values or to find probability distributions fairly easily. In real-world situations, however, many factors may be subject to some uncertainty. You can imagine what becomes of a decision tree that involves many uncertain events. The only way to prevent it from being a bushy mess is to present a skeleton version as in Figure 11.1, in which A and B represent alternative courses of action, each affected by many different uncertain quantities. We know that consideration of the uncertainties involved is important, but how will we deal with this much uncertainty? It is not at all clear that we will be able to use the techniques we have learned so far. We could, of course, painstakingly develop discrete approximations for all of the continuous distributions and construct the decision tree or influence diagram. In many cases this will work out fine. The decision tree, however, may become extremely complex.

Figure 11.1

Decision tree representing a complex decision situation with many sources of uncertainty.

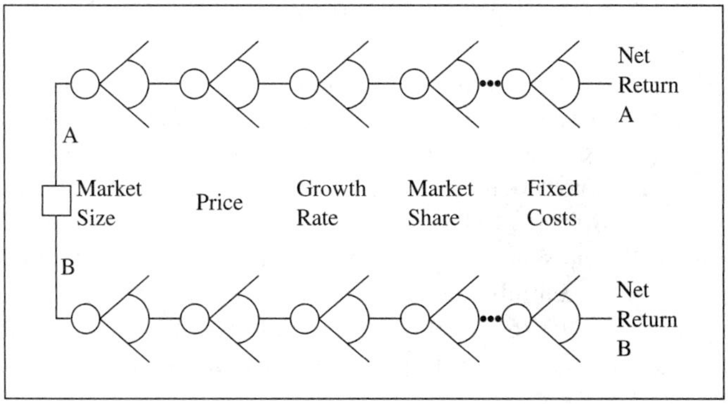

Figure 11.2

Influence diagram corresponding to decision tree in Figure 11.1. Interdependencies also may exist among the uncertainty nodes, thus complicating the influence diagram further.

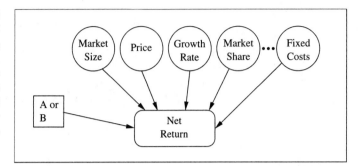

An influence diagram can help somewhat in this kind of a situation. As a more compact representation of a decision problem, an influence diagram can provide a clear picture of the way that multiple sources of uncertainty affect the decision. Figure 11.2 is an influence diagram that corresponds to the skeleton decision tree in Figure 11.1. Careful structuring can lead to a complete influence diagram, especially if the probability distributions can be represented adequately by discrete distributions in the chance nodes. But if there are multiple interrelated uncertain quantities represented by complicated continuous distributions, then even an influence-diagram approach may not be adequate.

In many decision-analysis problems, the ultimate goal is the calculation of an expected value. If a complicated uncertainty model includes several continuous distributions with expected values that are easily found, then analyzing the problem may be relatively straightforward. But this is not always true. In the following realistic example, it may not be obvious how to calculate an expected value.

FASHIONS

Janet Dawes is in a quandary. She is the purchaser for a factory that produces fashion clothes. Her current task is to choose a supplier who can furnish fabric for a new line of garments. To ensure its supply for the upcoming year, her company must sign a contract with one of several textile suppliers. After a few inquiries, she has narrowed the choice to two specific suppliers. The first will supply as much fabric as she needs during the upcoming year for $2.00 per yard and will guarantee supply through the year at this price.

The second supplier has a price schedule that depends on how much Janet orders over the next year. The first 20,000 yards will cost $2.10 per yard. The next 10,000 yards will cost $1.90 per yard. After this, the price drops to $1.70 per yard for the next 10,000 yards, and then $1.50 per yard for anything more than 40,000 yards.

After carefully considering the uncertainty about sales for the new garment line, Janet decides to model her uncertainty about the amount of fabric required over the next year as a normal distribution with mean 25,000 yards and standard deviation 5000 yards.

If the new line of garments is successful, Janet Dawes can save considerable money by going with the second supplier. On the other hand, if she signs the contract with Supplier 2 and the new line of garments does not prove successful, the cost of materials will have been higher than it might have been with Supplier 1, thus adding to the cost of an already expensive experiment. What should she do? Her decision tree is shown in Figure 11.3.

In Janet Dawes's problem, it is relatively easy to figure out the expected cost for Supplier 1: It is $2.00 per yard times 25,000 yards, or $50,000. But it is not so easy to figure out the expected cost for Supplier 2. To do this, we must know (a) the probability that X is within each interval and (b) the expected value of X for each interval. Part a is no problem; we could figure out the probabilities using the normal distribution as described in Chapter 9. It is not so clear, however, how to find the conditional expected values for X within each interval.

One approach to dealing with complicated uncertainty models such as that discussed above is through computer simulation. Think of the entire decision situation as an uncertain event, and "play the game" represented by the decision tree many times. In Janet Dawes's problem, we could imagine the computer making a random drawing from a normal distribution to find the amount of fabric required. Based on this specific amount, the computer then would calculate the total cost using the expressions given in the decision tree. After performing this procedure many times, it would be possible to draw a histogram or risk profile of the cost figures and thus to calculate the average cost. On the basis of these results, and comparing them with the distribution of costs for Supplier 1 (normal with mean $50,000 and standard deviation $10,000), a choice can be made.

For another example, imagine using a computer simulation to evaluate Alternative A in Figure 11.1. We would let the computer "flip a coin" to find the market size, then flip another to find the price, another to find the growth rate, and so on, until all uncertain quantities are chosen, each according to its own probability distribution. Of course, having the computer flip a coin means having it choose randomly the market

Figure 11.3

Decision tree for the fabric buyer. The expressions at the ends of the branches for Supplier 2 include the discounts. For example, if X is between 30,000 and 40,000 yards, the total cost is $42,000 for the first 20,000 yards, plus $19,000 for the next 10,000 yards, plus $1.70 per yard over 30,000.

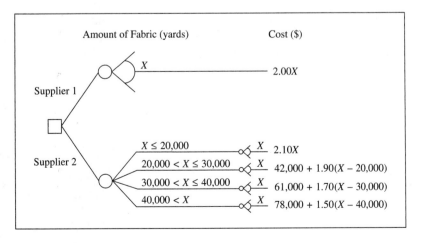

size, price, growth rate, and so on. Once all of the necessary values had been determined in this random fashion, the computer then could calculate the net return. We then would repeat the entire procedure many times and track the results. At the end, it would be possible to graph the distribution of net returns and to examine descriptive statistics such as the mean, standard deviation, and probability of a negative return.

The term *Monte Carlo simulation* often is used to refer to this process because, as in gambling, the eventual results depend on random selections of values. The basic idea is straightforward: If the computer "plays the game" long enough, we will have a very good idea of the distribution of the possible results. Although the concept is easy, its implementation requires attention to many details.

We can use Monte Carlo simulation to cope with situations in which uncertainty abounds; the objective is to represent the uncertainty surrounding the possible payoffs for the different alternatives. How does simulation fit into our perspective of building models to represent decision situations? When we put together all probability distributions for all uncertain quantities, we are building a simulation model that we believe captures the relevant aspects of the uncertainty in the problem. After running the simulation many times, we have an approximation of the probability distribution for the payoffs from the different alternatives. The more simulations we can do, the more accurate that approximation. Finally, the results, both risk profiles and average outcomes, can be used in the decision analysis to make an appropriate decision.

For the main example in this chapter, we will continue the soft pretzel problem from Chapter 9. You are thinking about manufacturing and marketing soft pretzels using a new recipe, but you face many sources of uncertainty. In the process of analyzing the decision, you might ask whether choosing to go into the soft pretzel business would have a positive expected value. By using the Monte Carlo simulation, we can examine the probability distribution of returns that would be associated with your pretzel business.

Let us suppose that the market size is unknown, but your subjective beliefs about the market's size can be represented as a normal random variable with mean 100,000 pretzels and standard deviation 10,000. The proportion (P) of the market that you will capture is unknown and, strictly speaking, is a continuous random variable that could range anywhere between 0 and 100%. You have decided, however, that your beliefs can be modeled adequately with the following discrete distribution:

Proportion (%)	Probability
16	0.15
19	0.35
25	0.35
28	0.15

Your pretzel's selling price is known to be $0.50. Variable costs are a uniform random variable between $0.08 and $0.12 per pretzel. Fixed costs also are random, with a distribution we will describe later.

Putting all of the pieces together to calculate net contribution yields

$$\text{Net Contribution} = (\text{Size} \times P/100) \times (\text{Price} - \text{Variable Cost}) - \text{Fixed Cost}$$

From this equation you can see the overall strategy; for each iteration, values for Size, P, Variable Cost, and Fixed Cost will be chosen from their respective distributions and then combined to get a figure for Net Contribution. Doing so for many iterations will yield a distribution for Net Contribution.

Using Uniform Random Numbers as Building Blocks

If we are going to construct random numbers from various different probability distributions, we will need a place to start. The starting point is a random variable that is uniformly distributed between 0 and 1. That is, all values between 0 and 1 are possible and equally likely. (See Problems 9.27–9.29.) Let x denote the number generated from such a distribution, and suppose we require a random number y^* from the distribution (CDF) $F(y) = P(Y \le y)$. An example CDF is plotted in Figure 11.4. Now generate a uniform x^* between 0 and 1. Locate that x^* on the vertical axis. Read across from x^* to the CDF and down to the horizontal axis. The number on the horizontal is the required y^*.

This approach is straightforward and intuitive; we first generate a uniform random number and then work backward through the CDF to get y. In principle, we can do this for any distribution. Unfortunately, it is not possible in some cases because of the mathematical form of $F(y)$. In these cases, other methods can be used.

Figure 11.4

Generating a random number y* *on the basis of a uniform random number* x*.

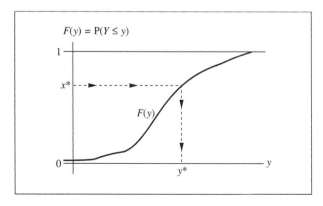

General Uniform Distributions

Suppose that we want to generate a uniform random number between 1 and 2 instead of between 0 and 1. The solution would be simply to calculate

$$y = x + 1$$

Then y would be uniformly distributed between 1 and 2. If we wanted a random number between 0 and 2, we simply multiply x by 2. This would "stretch out" the uniform distribution so that it would cover the interval from 0 to 2.

In general, suppose you want a uniform random number between a and b ($a < b$). The procedure is to generate x, then move it to the right place and stretch out the distribution to cover the desired interval. The formula is

$$y = a + x(b - a)$$

Multiplying x by $(b - a)$ stretches out the distribution, and adding a moves the distribution to the right place. If x turns out to be 0, then $y = a$, and if x is 1, then $y = b$.

In our soft pretzel example, variable cost is uniformly distributed between \$0.08 and \$0.12. Thus, the calculation is

$$\text{Variable Cost} = 0.08 + x(0.12 - 0.08)$$

This procedure for generating general uniform random numbers, which we developed intuitively, is fully compatible with the general simulation approach described above. The CDF for a uniform distribution between a and b is a straight line as shown in Figure 11.5. The line has the equation $F(y) = (y - a)/(b - a)$. (Can you verify this?) To work backward through this CDF, we set the uniform x^* equal to $F(y^*) = (y^* - a)/(b - a)$ and solve for y^*. That is, we need to know what

Figure 11.5

Generating a uniform y* between a *and* b.

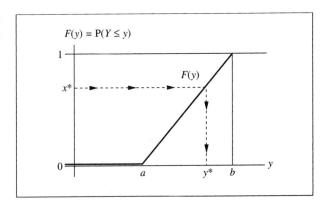

y^* corresponds to our uniform x^* between 0 and 1. Solving for y^*, we obtain $y^* = a + x^*(b - a)$.

Exponential Distributions

Another continuous random variable that is easy to simulate using our general approach is an exponential random variable T. Suppose you want to generate a random number that is exponentially distributed with rate m. The CDF for this distribution is $F(t) = 1 - e^{-tm}$. Thus, we generate a uniform x^* and set $x^* = F(t^*) = 1 - e^{-t^*m}$. Solving for t^* gives

$$t^* = \frac{-\ln(1 - x^*)}{m}$$

Figure 11.6 shows the CDF and the graphical equivalent of solving for t^*.

The formula derived above can be programmed easily into an electronic spreadsheet. But we can do a little bit better. If x^* is from a uniform distribution between 0 and 1, then so is $1 - x^*$. Thus, we can simplify the procedure by calculating

$$t' = \frac{-\ln(x^*)}{m}$$

The random number t' also has an exponential distribution with rate m. The reason for the simplification is to avoid unnecessary calculations that would slow down the computer simulation.

Figure 11.6

Generating an exponential random number.

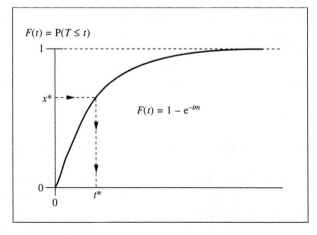

Discrete Distributions

Discrete distributions also are easy to handle in a way that is similar to using the CDF. Suppose, for example, that we want to simulate flips of a fair coin. We want the probability of a head to be 0.50. First we generate x. Then if $x \leq 0.50$, we will say that we have a head, and if $x > 0.50$, we will have a tail. Because x is from a uniform distribution, it has a 50% chance of being between 0 and 0.50 and a 50% chance of being between 0.50 and 1.

The basic strategy is to split the interval from 0 to 1 into smaller intervals corresponding to the outcomes of the discrete event. The outcome of the event then corresponds to the interval into which x falls when it is generated. The probabilities are controlled by the widths of the intervals for x; smaller intervals have smaller probabilities, and larger intervals have larger probabilities.

In our soft pretzel example, we need to generate the market proportion captured (P) using discrete probabilities. The four intervals we need are as follows:

- If $x \leq 0.15$, then $P = 16\%$.
- If $0.15 < x \leq 0.50$, then $P = 19\%$.
- If $0.50 < x \leq 0.85$, then $P = 25\%$.
- If $0.85 < x$, then $P = 28\%$.

Generating x, determining the interval in which it falls, and choosing P accordingly will generate discrete values of P with the appropriate probabilities.

Other Distributions

As you know, there are many other distributions that one might wish to use in a simulation. The procedure we have described above can be used for many of these distributions, but there are other approaches as well. For example, a normal random variable can be created by adding up a number of independent uniform random variables. Another approach to generating normal random variables uses a pair of independent uniforms and, with complex transformations, produces a pair of independent normals. For more information on simulating random variables, see Law and Kelton (1991).

Simulating Spreadsheet Models Using @RISK

The advent of high-speed personal computers has greatly reduced the cost of simulation. Managers now have at their fingertips powerful simulation programs that they can use to analyze many business problems, such as capacity planning, project management, financial planning, and option pricing. Simulation is also instrumental in designing and advancing computer chip technology, which leads to more powerful computers and, indirectly, to more powerful simulation programs.

@RISK provides the power to run full-scale simulations within a familiar and flexible spreadsheet environment. We will use the soft pretzel example to demonstrate the basics of building and running a simulation model. Then, we will move on to demonstrate some advanced features, such as running multiple simulations and simulating correlated input variables.

Step 1

1.1   Open Excel and @RISK, enabling any macros if prompted.

1.2   There are four on-line help options available. Pull down the **@RISK** menu and choose the **Help** command. *"How Do I"* answers frequently asked questions about @RISK, *Online Manual* is an electronic copy (pdf file) of the user's guide, *@RISK Help* is a searchable database of the user's guide, and *PDF Help* lists various facts about the probability distribution functions available in @RISK and RISKview.

Step 2

2.1   Build the soft pretzel spreadsheet as shown in Figure 11.7. Enter the numerical values for each assumption in the B column, and enter the formula (shown in cell C12) for *Net Contribution* in cell B12. *Net Contribution* should equal $966.67 when finished. The spreadsheet can also be found on the Palisade CD: Examples\Chapter11\Pretzel.xls.

The soft pretzel spreadsheet model constructed in Step 2 is said to be *deterministic* because the inputs are fixed numerical values that uniquely determine Net Contribution (the output). This deterministic model fails to capture the natural variability (or fluctuations) that occur in Net Contribution because it assumes the input values are static. Simulation, however, models variability by replacing the static

Figure 11.7

The soft pretzel spreadsheet.

	A	B	C	D	E
1					
2					
3					
4	ASSUMPTIONS:			Market	
5	*Market Size*	100,000		Proportion:	Probability:
6	*Market Proportion*	22%		16%	0.15
7	*Price*	$0.50		19%	0.35
8	*Variable Cost*	$0.10		25%	0.35
9	*Fixed Cost*	$7,833		28%	0.15
10					
11	FORECAST:				
12	*Net Contribution*	$966.67	=B5*B6*(B7-B8)-B9		

input values with probability distributions. When the simulation runs, the input values change as different values are sampled from the distributions, thus causing Net Contribution to change as well.

As mentioned earlier, after running the simulation, we will focus on the set of values for Net Contribution that resulted from the various sampled input values. Taken together, the Net Contribution values form a distribution, which we call the output distribution. Our goal in performing the simulation is to learn as much as we can about the output distribution. For example, we might calculate the expected Net Contribution and understand the uncertainty associated with Net Contribution. We might also perform a sensitivity analysis to pinpoint which input variables are the most important drivers of the uncertainty and which have little or no influence.

Step 3 shows the process of modeling an input variable as a distribution. Price is not included because it is a decision variable; its value is set by the decision maker. We will show later how to run a series of simulations—one for each price level that you select—in order to compare different pricing strategies.

STEP 3

3.1 Model the input variable *Market Size* with a normal distribution:

    3.1.1 **Right-click** when the cursor is over B5, select **@RISK**, and click on **Define Distributions**. A window pops up, titled *@RISK Definition for B5*, showing a normal distribution (Figure 11.8). @RISK automatically enters the current value in B5 (100,000) as the mean value; in this case, 1.0000E+05 is entered for μ (the mean).

    3.1.2 Enter **10000** for σ (standard deviation), as shown in Figure 11.8. Do not enter commas when entering numbers for the parameters. You can easily delete the current entry by double-clicking in the text box to highlight the number and typing in your number.

    3.1.3 The @RISK Definition window updates the graph of the distribution as you enter the parameter values. The formula appears in the text box above the curve. While the formula remains red and underlined, it models the distribution being displayed in the graph.

    3.1.4 To export the formula to cell B5 in the spreadsheet, click **Apply**.

3.2 Next, we model *Market Proportion* as a discrete distribution:

    3.2.1 Bring up the @RISK Definition window by **right-clicking** when the cursor is over B6, selecting **@RISK**, and clicking on **Define Distributions**.

    3.2.2 Display the list of distributions by clicking on the **black triangle** to the right of *Normal*. Slide the scrollbar up and click on **Discrete**.

    3.2.3 To enter the first *x* value, click the cell in the row labeled 1 below *X* and type in the value **0.16**. Hitting **enter** moves the cursor into the cell to the right and directly below *P*. Enter the corresponding probability **0.15**.

Figure 11.8

Defining Market Size (cell B5) as a normal distribution with mean 100,000 and standard deviation 10,000.

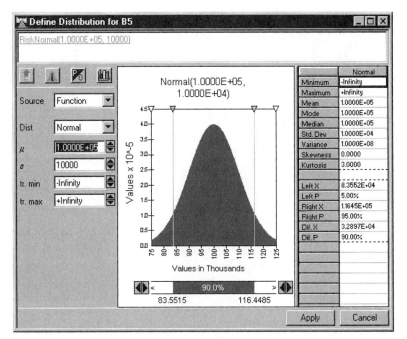

3.2.4 Hitting **enter** again places the cursor in the second row in the *x* column directly below 0.16.

3.2.5 Continue repeating Steps 3.2.3 and 3.2.4 for successive rows until you have entered the remaining three ordered pairs: **(0.19, 0.35)**, **(0.25, 0.35)**, and **(0.28, 0.15)**.

3.2.6 Click **Apply**, and cell B6 now contains this discrete distribution.

3.3 Enter the distributions for the two remaining input variables by following the above steps. Specifically, set up *Variable Cost* as a **Uniform(0.08, 0.12)** distribution in cell **B8**, where 0.08 is the *minimum* and 0.12 is the *maximum*; and *Fixed Cost* as a **Triang(6500, 8000, 9000)** distribution in **B9**, where 6500 is the *minimum,* 8000 is the *most likely,* and 9000 is the *maximum.* Be sure to click the **Apply** buttons to insert the distributions into cells B8 and B9.

3.4 An alternative to the @RISK Definition window is to use Excel's function wizard. To do this, first click the f_x button on Excel's toolbar, selcet the @*RISK Distribution* function category, then choose the particular distribution (e.g., *RiskNormal*), as shown in Figure 11.9.

@RISK provides considerable flexibility in defining an input distribution. You can select from among the 30 theoretical distributions in @RISK's library, or you can construct custom distributions, as we did with RiskDiscrete. In addition, RISKview and BestFit are specifically designed to help you construct distributions to match your uncertainty. Using RISKview, you can easily build and customize

Figure 11.9

Excel function wizard selection window showing @RISK categories.

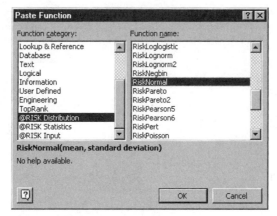

input distributions (see Chapters 8 and 9 for examples). Alternatively, if you have data available, then you can use BestFit to identify appropriate theoretical distributions (see Chapter 10 for a review of BestFit).

Recall that the point of running a simulation is to learn about the output distribution. Thus, at least one cell must be identified as an output variable. @RISK will monitor the output cell during the simulation and generate reports and graphs when the simulation finishes. Any cell or any number of cells can be identified as an output—even input cells can be identified as output cells! Our output variable is Net Contribution in cell B12.

Step 4

4.1   **Right-click** when the cursor is over B12, highlight **@RISK**, and click on **Add Output**.

4.2   Notice that *RiskOutput( )* has been added to the formula for Net Contribution in the formula bar (above the columns of the spreadsheet). The word "RiskOutput" identifies B12 as an output and the parentheses provide a way to name the output cells. Place the cursor between the parentheses in the formula bar, **click**, and type **"Net Contribution"** (including the double quotes). @RISK will now name the output from this cell Net Contribution. Hit **enter**.

Step 5

5.1   Click the **Run Simulation** button (fourth from the right in @RISK's toolbar), and @RISK will simulate the model for 100 iterations. Figure 11.10 shows the *@RISK—Results* window. The numbers in your window may be somewhat different because your particular run may have randomly chosen different input values, even though your model has exactly the same distributions.

Figure 11.10

Simulation results for the soft pretzel model.

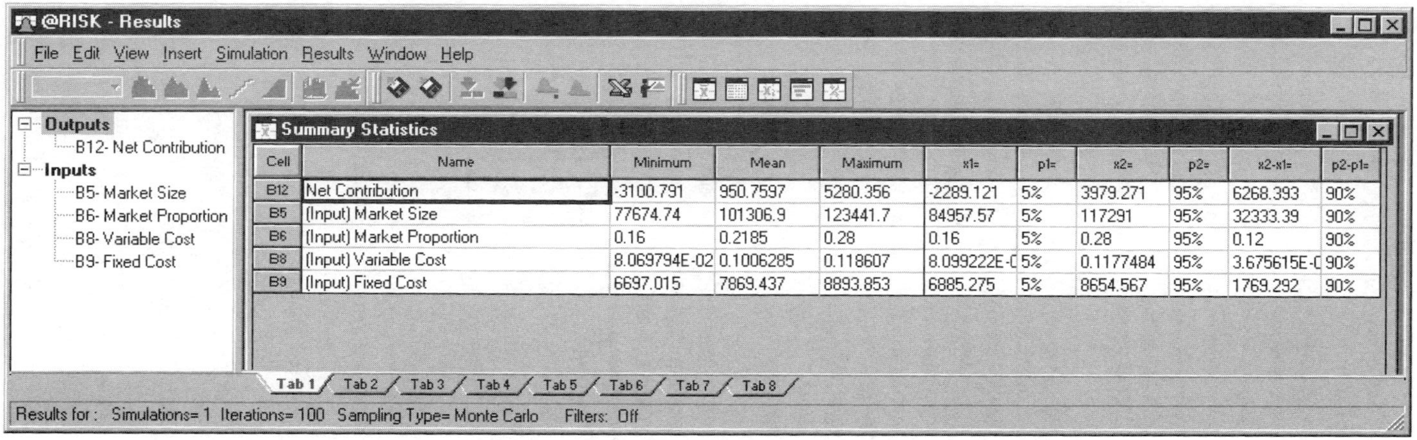

@RISK - Results

File Edit View Insert Simulation Results Window Help

- **Outputs**
  - B12- Net Contribution
- **Inputs**
  - B5- Market Size
  - B6- Market Proportion
  - B8- Variable Cost
  - B9- Fixed Cost

Summary Statistics

Cell	Name	Minimum	Mean	Maximum	x1=	p1=	x2=	p2=	x2-x1=	p2-p1=
B12	Net Contribution	-3100.791	950.7597	5280.356	-2289.121	5%	3979.271	95%	6268.393	90%
B5	[Input] Market Size	77674.74	101306.9	123441.7	84957.57	5%	117291	95%	32333.39	90%
B6	[Input] Market Proportion	0.16	0.2185	0.28	0.16	5%	0.28	95%	0.12	90%
B8	[Input] Variable Cost	8.069794E-02	0.1006285	0.118607	8.099222E-0	5%	0.1177484	95%	3.676615E-0	90%
B9	[Input] Fixed Cost	6697.015	7869.437	8893.853	6885.275	5%	8654.567	95%	1769.292	90%

Tab 1 Tab 2 Tab 3 Tab 4 Tab 5 Tab 6 Tab 7 Tab 8

Results for : Simulations= 1 Iterations=100 Sampling Type= Monte Carlo Filters: Off

Figure 11.10 shows the outputs and inputs listed along the left-hand side and the *Summary Statistics* window on the right-hand side. The *Summary Statistics* window summarizes the input and output distributions by listing the minimum, the mean, the maximum, and the 5th and 95th percentiles. For our particular simulation run, Net Contribution was between a minimum of –$3101 and a maximum of $5280 with an expected value of $951. The 95th percentile is $3979, which means that 95% of the Net Contribution values were at or below $3979.

Thus, the model predicts that the business will be profitable on average (positive expected value), but monthly Net Contribution could fluctuate by thousands of dollars. Your results will be somewhat different because the simulation chooses values randomly, but as you increase the number of iterations (see Step 7), the results should begin to converge.

Now, let's delve deeper into the results. In particular, we can create the risk profile for Net Contribution and calculate the probability that this enterprise will at least break even. That is, we will calculate P(Net Contribution ≥ 0).

STEP 6

6.1   In the @*RISK—Results* window, highlight **Net Contribution** under *Outputs* in the leftmost column, **right-click**, highlight **Histogram** in the pop-up menu, and click on **Histogram**. The histogram of Net Contribution appears in a new window, as shown in Figure 11.11. Notice that the expected Net Contribution ($951) is displayed over the histogram, and summary statistics, such as the standard deviation ($1891), are displayed in the right-hand column.

6.2   The two delimiters (gray vertical lines overlaying the graph) are markers that allow you to determine cumulative probabilities. In Figure 11.11, the leftmost delimiter is at the fifth percentile as shown by the 5% in the scrollbar below the curve. This delimiter is at the *x* value of –2289.12 as shown below the scrollbar. Therefore, according to the model, there is a 5% probability that the Net Contribution will be less than or equal to –$2289. The 90% shown in the scrollbar indicates that there is a 90% probability that demand will be greater than –$2289 but less than $3979. The rightmost delimiter shows that there is a 5% probability that the Net Contribution will be greater than or equal to $3979.

6.3   The two delimiters can be moved to display different cumulative probabilities. Click on a **delimiter bar** and drag the mouse left or right while holding the mouse button down. The bar moves correspondingly. Note that the smallest increment it will move is $1000. Find the probability that Net Contribution will be between **–2000** and **3800**. (Answer: ≈ 86%.)

6.4   Clicking on the triangles to the left and right of the scrollbar below the graph also moves the bar, but this time by one percentile. Find the *x* values so that there is a **20%** chance that Net Contribution will be below the first *x* value and a **20%** chance that Net Contribution will be above the second *x* value. (Answer: ≈ –$913 and $2633.)

6.5   You can also type exact values for either the *x* value or the percentile into the spreadsheet cells located to the right of the graph. The *Left X* and *Left*

P refer to the leftmost delimiter and the *Right X* and *Right P* refer to the rightmost. If you enter the *x* value, @RISK reports the *p* value and vice versa. *Diff. X* and *Diff. P* calculate the difference in the *x* values and *p* values; these cells cannot be typed in. Enter **25** for *Left P* and **75** for *Right P*. The corresponding *x* values are approximately –$476 and $2435. Thus, there is a 50% chance that Net Contribution will be between –$476 and $2435.

6.6   You now have three different ways to calculate the break-even probability: Click on the delimiter bar, click on the scroll bar, or type in zero for the *Left X* value. Find P(Net Contribution $\geq$ 0). (Answer: $\approx$ 40%.)

6.7   You have various options for formatting the graph. The buttons highlighted in Figure 11.11 will reformat the graph to prespecified configurations. For example, to draw the risk profile as a CDF, click on **Redraw graph as ascending cumulative line** (fourth button from the right). Alternatively, you can access all the formatting options by **right-clicking** when the cursor is over the graph, and choosing **Format Graph**. The functions of the tabs of the *Graph Format* dialog box are explained below.

The six tabs along the top of the Graph Format dialog box accessed in Step 6.7 allow you to make various alterations to the graph:

*   The *Type* tab allows you to view either a density curve or a CDF curve.

*   The *Scaling* tab allows you to change the scale (the units and the max and min) of the *x* and *y* axes.

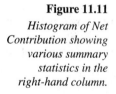

Figure 11.11

Histogram of Net Contribution showing various summary statistics in the right-hand column.

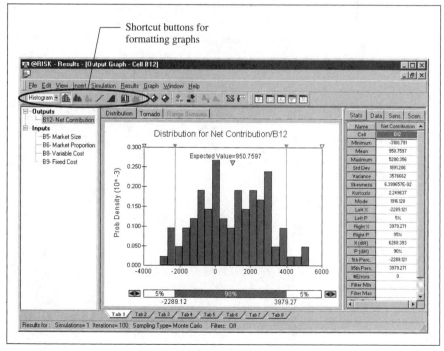

- The *Style* tab allows you to change the patterns and colors of the graph.
- The *Titles* tab allows you to label the axis, title the graph, and add a legend.
- The *Delimiters* tab allows you to remove or modify the delimiters.
- The *Variables to Graph* allows you to graph multiple output distributions on one graph.

Before moving on to some advanced simulation techniques, we explain how to alter the settings of @RISK to (1) increase the number of iterations and (2) display different sampled values in the input cells from each distribution.

STEP 7

7.1   In the *@RISK—Results* window, click on the **Simulation** heading, then click on **Settings**. (Alternatively, in Excel, pull down the **@RISK** menu, select **Simulation,** and click on **Settings**.) Figure 11.12 shows the *Simulation Settings* dialog box that appears.

7.2   To change the number of times the model is simulated, click the **Iterations** tab and type **1000** in the text box for *# Iterations*.

7.3   To have Excel display different input values, click the **Sampling** tab and choose **Monte Carlo** under the *Standard Recalc* heading. Click **OK**.

Before performing Step 7.3, Excel displayed a single number (the expected value) in each input cell. Step 7.3 causes Excel to display the values actually sampled from the distributions. Every time a new sample is drawn, Excel updates the spreadsheet with the new values and calculates the corresponding Net Contribution. A new sample is drawn each time the spreadsheet is recalculated. The F9 key manually recalculates the spreadsheet; press it a few times and watch what happens. Step 7.3 affects only the *appearance* of the spreadsheet and not the output of a simulation run.

Figure 11.12
Simulation Settings dialog box.

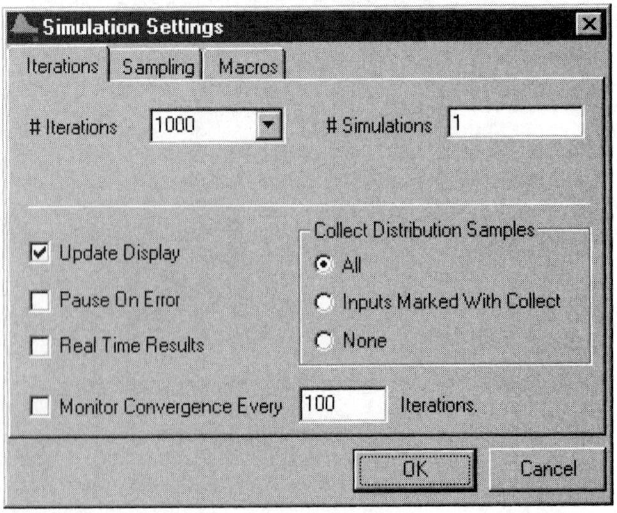

Power users may want to access the data used in the simulation:

STEP 8

8.1   In the *@RISK—Results* window, the five rightmost buttons insert new windows. The leftmost button of these five inserts a summary statistics window, which we discussed above. Click on the **Insert New Detailed Statistics Window**. This window provides a complete array of statistics on all the output and input distributions. Scrolling down, you see that you can do such things as define scenarios and filter the data. @RISK's on-line help has explanations.

8.2   Click on the **Insert New Data Window**. This window displays the actual sampled input values along with the corresponding calculated Net Contribution values.

Multiple Output Models

Price is the one variable whose influence we have not investigated. Because Price is a decision variable, it does not make sense to model it as a distribution, nor would that help determine an optimal pricing strategy. Modeling Price as a distribution would be analogous to rolling dice every morning to determine the day's price! Rather, it makes more sense to run a series of simulations, one for each specified price level. Then we will be able to compare the expected values and risk profiles for each price. @RISK has a feature specifically designed for this purpose. It will run a sequence of simulations, one for each price level, and record all the results in one spreadsheet.

STEP 9

9.1   Return to your spreadsheet and enter the prices **0.40, 0.45, 0.50, 0.55,** and **0.60** in cells C15 to C19. These are the five prices we might charge for the soft pretzels.

9.2   Highlight cell **B7** and click on the **Paste Function** button (f_x button) in the Excel toolbar.

9.3   In the *Paste Function* window, scroll down the *Function category* list (left side) until you see the @RISK categories. Select **@RISK Distribution** (Figure 11.9).

9.4   Scroll down the *Function name* list (right side) until you find **RiskSimTable**, click **OK**, move the cursor to the spreadsheet, and high-light cells **C15** to **C19**. (*Note*: Keep the mouse button depressed while highlighting the range of cells so that C15:C19 is placed in the text box.) Click **OK**. This changes Net Contribution to ($1233.33) because "Price" is $0.40.

9.5   We must now modify the settings so that @RISK knows to run five simu-lations. Click the **Settings** button (fifth button from the right) and select the **Iterations** tab shown in Figure 11.12.

9.6   Change the *# Simulations* = from 1 to **5**. @RISK will now run five sepa-
       rate simulations, the first when Price = .40, the second when Price = .45,
       and so on until Price = .60. Click **OK**.

9.7   Click the **Run Simulation** button.

A new *@RISK—Results* window appears that summarizes the five simulations
(Figure 11.13). The first row of the *Summary Statistics* window reads *B12(Sim#1)
Net Contribution*, which corresponds to the first price level of $0.40, the second row
reads *B12(Sim#2) Net Contribution*, corresponding to the second price level of
$0.45, and so on. Figure 11.13 shows that when Price = $0.40 the expected Net
Contribution is –$1228 and can range from –$4798 to $3566. Step 10 describes how
to investigate the effects of the different price levels by (1) comparing the means and
the standard deviations of the output distributions, (2) comparing the break-even
probabilities, and (3) graphing all five risk profiles. The risk profiles are very instructive because they clearly show that higher prices dominate lower prices.
(Technical note: Each of the five simulations uses the same input values. This allows
us to compare the outputs directly without having to factor in the variation from
using different input values across simulations.)

STEP 10

10.1   The means or expected values increase dramatically as the price in-
       creases, as shown in Figure 11.13. For example, the expected Net
       Contribution equals $3176 when Price = $0.60, which is nearly the maxi-
       mum Net Contribution of $3566 when Price = $0.40.

Figure 11.13

The simulation output from running five simultaneous simulations for five input price levels using RiskSimTable.

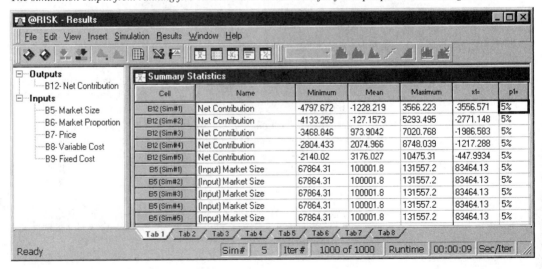

10.2   The standard deviations are found by inserting the detailed statistics window. Click on **Tab 2** in the *@RISK—Results* window and click on the **Insert New Detailed Statistics Window** button (fourth from the right). The fourth row of this window lists the standard deviation of all the input and output variables. The standard deviation increases as the price increases. Thus, along with the higher expected value there is more uncertainty.

10.3   To calculate the break-even probabilities, return to **Tab 1**. Find the column labeled $x1=$ and **double-click** on the entry in the first row (the row corresponding to *Sim #1*). Type in **0** and hit **enter**. The break-even probability is displayed in the column immediately to the right, labeled $p1=$. The break-even probability for our simulation run is 77%. Continue highlighting the entries in the $x1=$ column by double-clicking and entering 0 for the remaining price levels. The break-even probabilities will decrease as the price increases.

10.4   To graph all five risk profiles, first graph a single CDF. Click on **Tab 3**, click on **Net Contribution** under *Outputs* along the left, **right-click**, choose **Cumulative**, and click on **Ascending—Line**. In the *Graph Results* pop-up window, click **OK**. The resulting curve should be the CDF for *Net Contribution (Sim #1)*.

10.5   To overlay all five risk profiles, as shown in Figure 11.14, do the following: **Right-click** when the cursor is over the graph, choose **Format Graph,** click on the rightmost tab **Variables to Graph**, highlight **all five variables,** and click **OK**. Figure 11.14 shows that the risk profile for Price = \$0.60 dominates all of the others, as discussed in Chapter 4.

Figure 11.14

The five risk profiles for the five price levels showing that higher prices dominate lower price levels.

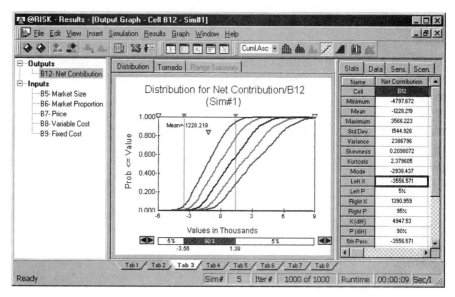

The preceding analysis shows that charging a higher price is better, all else being equal. Unfortunately, everything else is not equal; we have not included the relationship between demand and price in the model. We would expect demand to decrease as price increases.

In order to have a more realistic model, we need to understand the relationship between demand and price. In practice, we could collect data from potential customers. For this example, let's assume that the data we have at hand indicate that a $0.10 increase in price implies a five percentage point decrease in Market Proportion. Figure 11.15 shows how we have modified the original spreadsheet to reflect this simple negative dependence. We have explicitly constructed all five Net Contribution formulas in cells B19 through B23, one for each price and its corresponding market size. In each formula, we have replaced the input variable Market Proportion *(P)* with Market Proportion plus its change:

$$\text{Net Contribution} = \text{Size}*(P + \text{Change in } P)*(\text{Price} - \text{Variable Cost}) - \text{Fixed Cost}$$

You can see from Figure 11.15 that the $0.50 price appears to be the best price because it has the highest Net Contribution. But this is just for the values in the spreadsheet and does not take into account any of the uncertainty. What happens if we run a simulation?

STEP 11

11.1   Close all worksheets so that @RISK clears the inputs-by-outputs table. Open a new worksheet.

Figure 11.15

Spreadsheet that models dependence between Price and Market Proportion.

	A	B	C	D	E	F
4	ASSUMPTIONS:					
5	Market Size	100,000	=RiskNormal(100000,10000)			
6	Market Proportion	22%	=RiskDiscrete(E12:E15,F12:F15)			
7	Variable Cost	$0.10	=RiskUniform(0.08,0.12)			
8	Fixed Cost	$7,833	=RiskTriang(6500,8000,9000)			
9						
10			Change in		Market	
11			Market Proportion:		Proportion:	Probability:
12	Price Levels:	$0.30	10%		16%	0.15
13		$0.40	5%		19%	0.35
14		$0.50	0%		25%	0.35
15		$0.60	-5%		28%	0.15
16		$0.70	-10%			
17						
18	FORECAST:					
19	Net Contribution $.30	($1,433.00)	=$B$5*(C12+$B$6)*(B12-$B$7)-$B$8			
20	Net Contribution $.40	$267.00				
21	Net Contribution $.50	$967.00				
22	Net Contribution $.60	$667.00				
23	Net Contribution $.70	($633.00)				

11.2   Build the soft pretzel model as shown in Figure 11.15. The spreadsheet can also be found on the Palisade CD: Examples\Chapter11\PretzelPrice.xls.

   11.2.1   Enter the **distributions** shown in C5 to C8 for the assumptions in B5 to B8.

   11.2.2   Enter the price levels: **$.30, $.40, $.50, $.60, $.70** in B12 to B16 and the corresponding *Change in Market Proportion:* **10%, 5%, 0%, –5%, –10%** in C12 to C16.

   11.2.3   Enter the five forecast **formulas** in cells B19 to B23. The formula for cell B19 is shown in cell C19. After entering this formula into B19, copy it down to cells B20 to B23.

11.3   Collectively, highlight all five of the *Net Contribution* equations (cells **B29:B23**), **right-click**, and enter **Net Contribution** for the name of this output range.

11.4   Click the **Run Simulation** button. Figure 11.16 shows the @*RISK— Results* window that pops up when the simulation finishes.

We see the interaction effects of market share with price exhibited in the means of the Net Contribution variables (Figure 11.16). The means increase as Price moves from $0.30 to $0.40 and from $0.40 to $0.50, but then decrease as Price moves beyond $0.50. The maximum expected Net Contribution of $964 occurs when Price equals $0.50 per pretzel.

   Not only do the means increase as Price moves from $0.30 to $0.40 to $0.50, but we can show that the $0.50 price dominates both the $0.30 and $0.40 prices:

Figure 11.16

Summary statistics for the five different prices (B19 to B23) charged for soft pretzels.

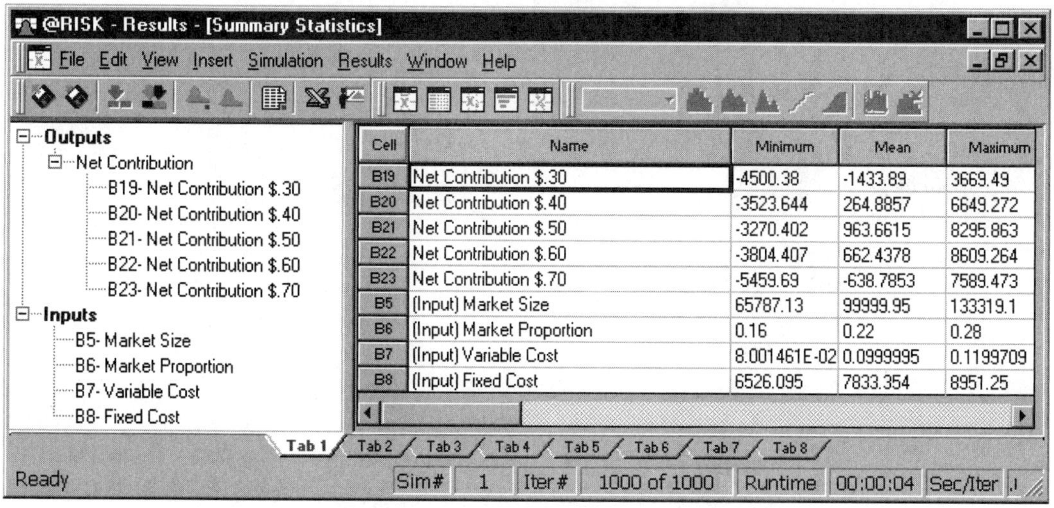

Cell	Name	Minimum	Mean	Maximum
B19	Net Contribution $.30	-4500.38	-1433.89	3669.49
B20	Net Contribution $.40	-3523.644	264.8857	6649.272
B21	Net Contribution $.50	-3270.402	963.6615	8295.863
B22	Net Contribution $.60	-3804.407	662.4378	8609.264
B23	Net Contribution $.70	-5459.69	-638.7853	7589.473
B5	(Input) Market Size	65787.13	99999.95	133319.1
B6	(Input) Market Proportion	0.16	0.22	0.28
B7	(Input) Variable Cost	8.001461E-02	0.0999995	0.1199709
B8	(Input) Fixed Cost	6526.095	7833.354	8951.25

STEP 12

12.1 Click on **Tab 2** in the *@RISK—Results* window, highlight **Net Contribution $.50** under the *Outputs* heading, **right-click**, choose **Cumulative**, and click on **Ascending—Line**.

12.2 **Right-click** when the cursor is over the graph, choose **Format Graph**, click on the rightmost tab **Variables to Graph**, highlight **Net Contribution $.30/B19** and **$.40/B20.** Clicking **OK** produces a graph of all three risk profiles showing the dominance of the $.50 price (Figure 11.17).

Summary graphs are another way to graphically represent the means and the uncertainty about Net Contribution for all simulations:

STEP 13

13.1 Click on **Tab 3** in the *@RISK—Results* window, **right-click** the **Net Contribution** heading directly below *Outputs*, and click on **Summary Graph**.

Figure 11.18 shows the summary graph with the *y* axis being the net contribution values and the *x* axis being the price levels from $0.30 (B19) to $0.70 (B23). The middle line in the graph represents the mean for the five different price levels, the narrower band represents plus/minus one standard deviation about the mean, and the wider band represents 90% of the distribution. This graph clearly shows the change in expected net contribution as price changes. The $0.30 price level shows a distribution of mainly losses. According to our model, a price of $0.30 is simply too low. As the price increases, the expected net contribution first rises then falls as the mar-

Figure 11.17

Risk profiles for the $.30, $.40, and $.50 pretzel prices incorporating the dependence between market share and price.

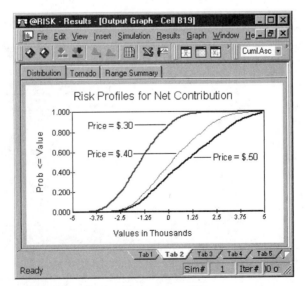

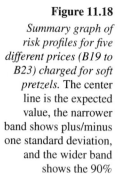

Figure 11.18

Summary graph of risk profiles for five different prices (B19 to B23) charged for soft pretzels. The center line is the expected value, the narrower band shows plus/minus one standard deviation, and the wider band shows the 90% probablility limits.

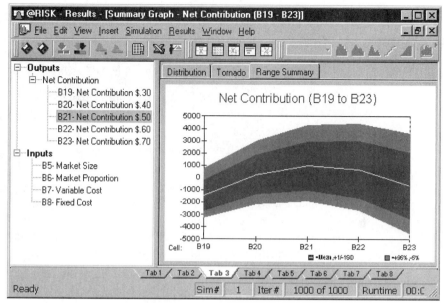

ket proportion shrinks. In addition, the distribution becomes broader as price increases, because the higher price amplifies the uncertainties in Market Size and Market Proportion. Both the risk profiles and the summary graph indicate that a price around $0.50 is the best choice, based on this model.

Distributions on Parameters (Optional)

Suppose that in the soft pretzel problem you cannot settle on a single value for the mean of the normal distribution for the market size. Your uncertainty about this parameter can also be modeled with a probability distribution. This is similar to what we did in Chapter 10 in our discussion of natural conjugate prior distributions, but here there is no need to restrict ourselves to specific classes of distributions. In a sense, we will be placing a distribution on another distribution; such models are sometimes called *hierarchical models,* and the uncertainty about the parameter is sometimes called *second-order uncertainty.*

After carefully considering your beliefs about the mean of the market-size distribution, let us suppose that you decide to model your uncertainty about the mean of market size as a uniform random variable between 90,000 and 110,000 pretzels. How would this change the model? Making the modification is straightforward:

STEP 14

14.1   Close all worksheets, so that @RISK clears the inputs-by-outputs table, and reopen the worksheet shown in Figure 11.7 in the Palisade CD: Examples\Chapter11\Pretzel.xls.

14.2   Modify the worksheet to match Figure 11.19. Specifically, insert a new fourth row, and type **Market Mean** in A4 and **=RiskUniform(90000,110000)** in B4.

14.3   Enter the distributions shown in the C column for the remaining input variables. In cell B5, be sure to indicate that the mean of market size is B4; that is, enter **=RiskNormal(B4,10000)** in B5.

14.4   Highlight cell **B12**, **right-click**, and click **Add Output**. If desired, name this cell.

14.5   Click **Run Simulation**.

For each iteration, @RISK will generate a new mean (uniform random number between 90,000 and 110,000) and then it will generate a specific market size based on the normal distribution with the new mean and standard deviation of 10,000. Why is the standard deviation for Net Contribution in this hierarchical model larger than the standard deviation of $1848 we originally found?

Dependent Input Variables (Optional)

Imagine that you have just finished constructing a complex simulation model that forecasts annual revenue streams for a large construction company. During a meeting with your staff to review the model and its performance, you become more and more convinced that, although all the formulas are correct, there is a fundamental problem with the model. "Isn't it well known," you ask, "that as interest rates rise, housing starts tend to fall? Shouldn't we incorporate this and other dependencies between the input variables into our model?" The answer is yes. If there are dependencies, then they could dramatically affect the output distribution. Let's see how @RISK incorporates dependent input variables into the Eagle Airlines example from Chapter 5.

Dick Carothers, the president of Eagle Airlines, is interested in purchasing a Piper Seneca to increase his fleet. The sensitivity-analysis results from Chapter 5

Figure 11.19

Incorporating uncertainty about the mean of the distribution for market.

	A	B	C	D	E	F	G
1							
2							
3	ASSUMPTIONS:						
4	Market Mean	100,000	=RiskUniform(90000,110000)			Market	
5	Market Size	100,000	=RiskNormal(B4,10000)			Proportion:	Probability:
6	Market Proportion	22%	=RiskDiscrete(F6:F9,G6:G9)			16%	0.15
7	Price	$0.50				19%	0.35
8	Variable Cost	$0.10	=RiskUniform(0.08,0.12)			25%	0.35
9	Fixed Cost	$7,833	=RiskTriang(6500,8000,9000)			28%	0.15
10							
11	FORECAST:						
12	Net Contribution	$966.67	=B5*B6*(B7-B8)-B9				

showed that four of the ten input variables most influenced Carothers's decision. We will describe how to model these four influential variables as distributions and simulate the profit from purchasing the plane. We will also show how to incorporate dependencies among these four variables.

The mechanics of modeling correlated random variables in @RISK is very simple—you need only type the correlation values into a matrix designed by @RISK (Step 16). Choosing appropriate correlation values is a more difficult problem. If data are available, then it is straightforward to compute an estimated correlation. In simulation models, though, suitable data may not be available. In these cases, experts might be able to provide subjective estimates based on their knowledge of the relationships.

We assume that Carothers has supplied the following correlation values. He believes that Capacity and Hours Flown are highly positively correlated (.80) because as demand increases, the number of seats filled and the number of flights will both tend to increase. He also believes that as the Ticket Price increases, both Capacity and Hours Flown will tend to decrease. He gave both these correlations a value of −.80. He believes that there is zero correlation (that is, no relationship) between Operating Cost and Hours Flown and between Operating Costs and Ticket Price because the cost is measured per hour. He does believe that there is a positive but lower correlation (.50) between Operating Cost and Capacity because as Capacity increases, there tend to be additional maintenance costs.

STEP 15

15.1   Close all worksheets and construct the worksheet shown in Figure 11.20 or open the spreadsheet Examples\Chapter11\EagleAir.xls.

15.2   Enter the distributions for **Hours Flown, Capacity, Ticket Price,** and **Operating Cost** given below:

Hours Flown	BetaGeneral(4, 2, 67, 1135)
Capacity	Beta(20, 20)
Ticket Price	BetaGeneral(9, 15, 82, 134)
Operating Cost	Normal(245, 12)

BetaGeneral(α1, α2, min, max) is an @RISK beta distribution that is shifted from the usual interval [0, 1] to the interval [*min, max*].

15.3   Highlight cell **B19**, **right-click**, highlight @**RISK**, and click on **Add Output**. Name this cell Profit by typing **"Profit"** in the formula bar between the parentheses in *RiskOutput( )* (see Step 4).

15.4   At this point we can run a simulation that assumes independence between the input variables. Before doing this, however, let's put the expected value, standard deviation, and break-even probability of the profit distribution into this spreadsheet for later comparisons. Highlight cell **B21**, click on Excel's **Paste Function** button, (the f_x button), select @**RISK Statistics** from the *Function category* list (left side), select **RiskMean** from the *Function Name* list (right side), click **OK**, enter **B19** for the

Figure 11.20

The Eagle Airlines model.

	A	B	C	D	E	F
1						
2						
3						
4	ASSUMPTIONS:					
5	*Hours Flown*	779				
6	*Capacity*	50%				
7	*Ticket Price*	$ 101.50				
8	*Charter Price*	$ 330	=3.25*B7			
9	*Charter Proportion*	50%				
10	*Operating Cost*	$ 245				
11	*Insurance*	$ 20,000				
12	*Price*	$ 87,500				
13	*Interest Rate*	11.5%				
14	*Proportion Financed*	40%				
15						
16	*Finance Cost*	$ 4,025	=B12*B13*B14			
17	*Revenue*	$ 227,322	=(B5*B8*B9)+((1-B9)*B5*B6*B7*5)			
18	*Total Cost*	$ 214,880	=B5*B10+B11+B16			
19	*Profit*	$ 12,442	=B17-B18			
20						
21	Estimated Mean					
22	Estimated St. Dev.					
23	Estimated Break-even					

Data source, and click **OK**. When the simulation finishes, cell B21 will contain the mean or expected value of the profit distribution.

15.5 Repeat (Step 15.4) for cells **B22** and **B23**, but select **RiskStDev** and **RiskTarget** instead of *RiskMean*. Cell B22 will contain the standard deviation of the profit distribution, and B23 will contain the break-even probability.

15.6 Click **Run Simulation**.

The simulation results show that the expected profit is $12,545, with a standard deviation of $23,870 and a break-even probability of 32%. These values are also in the spreadsheet in cells B21, B22, and B23. Now, let's see what happens when we correlate the four input variables.

STEP 16

16.1 Open the @*RISK—Model* window by clicking on the **Display List of Outputs and Inputs** button (double-headed red and blue arrow button).

16.2 Highlight all four of the input variables: **Hours Flown, Capacity, Ticket Price,** and **Operating Cost**. (Hold down the **Shift** key when highlighting multiple selections.)

Figure 11.21

The correlation matrix for the four variables (Hours Flown, Capacity, Ticket Price, and Operating Cost) in the Eagle Airlines model.

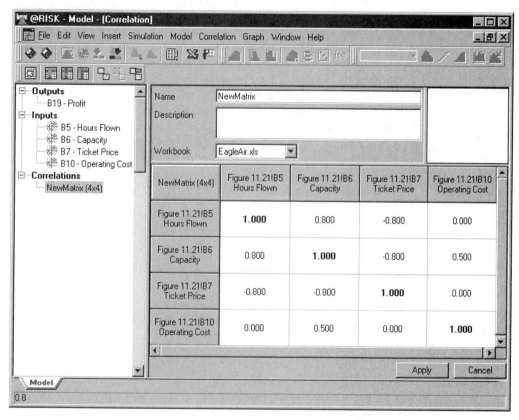

16.3 Click on the **Correlate Distributions** button. Figure 11.21 shows the *Correlation* window that opens, which contains a 4 × 4 matrix for entering correlation values. Enter the values as shown. Because correlation matrices are symmetric, you need only enter the lower diagonal values.

16.4 Click **Apply** to enter the correlation matrix into the simulation. You will notice that variables are now marked with the correlation symbol. Also, the correlation matrix has been added to the spreadsheet and can be modified either in the spreadsheet or in the *@RISK—Model* window.

16.5 Click **Run Simulation**.

The simulation results now report that the expected profit is $12,662, with a standard deviation of $12,870 and a break-even probability of 17%. Although the expected value is not very different, the standard deviation and break-even probability are nearly halved! Thus, there is much less uncertainty in the profit of purchasing the plane when the correlations are incorporated into the model. The difference between the results of these two models is significant. Carothers could decide not to purchase the plane, given the great deal of uncertainty and high probability of losing money (32%), based on the results from the first model but might choose to purchase the plane based on the results of the second model.

We conclude on a technical note. Individually, each correlation value must be between −1 and +1. Collectively, however, there are additional constraints that are too complex to state here. If correlation values are entered that do not satisfy these more subtle constraints, @RISK will warn you when you attempt to apply the matrix. A message will appear stating that the matrix is not "self-consistent." @RISK will modify the matrix so that it is self-consistent if you select the *OK* button, in which case most of the correlations may change. Selecting *Cancel* lets you go back to the original matrix to change only those values you want to change. For example, your reasoning about the relationships may include strong arguments for leaving some of the correlations equal to zero.

Simulation, Decision Trees, and Influence Diagrams

Clearly, Monte Carlo simulation provides another modeling tool for decision analysis. Indeed, simulation is an important tool, and probably will become more widely used because of the ease with which small simulations can be performed in a spreadsheet environment. Because of this, it is worthwhile to spend a bit of time thinking about how simulation relates to the other modeling tools in decision analysis.

Simulation is an excellent tool for developing a model of uncertainty. We have used it here to develop risk profiles for decision alternatives; with risk profiles for different alternatives in hand, a decision maker would have some basis for choosing one alternative over the other. As mentioned above, however, simulation also could be used as a subsidiary modeling tool to construct a probability model for a particularly messy part of a problem. For example, if a policy maker is attempting to evaluate alternatives regarding chemical spills, he or she may ask an analyst to develop a probability model for accidents that lead to spills. Such a model can be developed in the context of a simulation, and once an appropriate probability distribution is constructed, it can be used within a larger analysis.

The ease with which simulation can be performed, along with the flexibility of the simulation environment, makes it an attractive analytical tool. But this ease of use and flexibility does not mean that the decision maker can get away with less effort. In fact, subtle issues in simulation require careful thought. For example, we typically build the simulation in such a way that many of the random numbers are independent draws from their respective distributions. We saw in the previous section that incorporating dependence in the Eagle Airlines problem can have a dramatic effect on the results. The analyst may even model uncertainty about parameters through the specification of distributions on those parameters, but still have the parameters independent from one another. But would they be? If the analyst has been optimistic in assessing one parameter, perhaps other estimates have been subject to the same optimism.

A somewhat less subtle issue in simulation modeling is that an analyst may be tempted to include all possible sources of uncertainty in the model. This is relatively easy to do, after all, and somewhat more effort is required to do the sensitivity analysis to determine whether an uncertain quantity really matters in the outcome of the model. Hence, there may be a tendency with simulation not to take certain analysis steps that can lead to real insights as to what matters in a model, what issues should be considered more fully, or what uncertainties may demand more attention, in either more careful assessment or information acquisition.

Consider the Eagle Airlines case and suppose that the sensitivity-analysis step was skipped. Then we would be oblivious to the fact that only 4 of the 10 input variables contributed significantly to the output distribution and analysis of the decision. Thus, instead of attentively modeling the 4 influential variables and their corresponding 6 correlation values, we would needlessly spend our time modeling all 10 variables along with the 45 corresponding correlation values. The results from this massive modeling effort would not differ substantially from modeling only the four influential variables (see Problem 11.11).

In short, constructing a Monte Carlo simulation model requires the same careful thought that is required in any decision modeling. Simulation does have its own advantages (flexibility and ease of use) as well as disadvantages (rampant independence assumptions and a tendency to solve problems with brute force) and hence, to some extent, leads to some special problems. But the same is true of simulation that is true of decision modeling in general. The decision maker and the decision analyst still are required to think clearly about the problem at hand and to be sure that the decision model addresses the important issues appropriately. Clear thinking is the key, not fancy quantitative modeling. The objective with any decision-analysis tool is to arrive at a requisite model of the decision, one that appropriately addresses all essential elements of the decision problem. It is through the process of constructing a requisite model, which includes careful thought about the issues, that the decision maker will gain insight and understanding about the problem.

SUMMARY

As we have seen in this chapter, Monte Carlo simulation is another approach to dealing with uncertainty in a decision situation. The basic approach is to construct a model that captures all of the relevant aspects of the uncertainty, and then to translate this model into a form that a computer can use; we focused on the development of such models within the environment of electronic spreadsheets using @RISK to perform the simulation and analyze the results. We discussed how sensitivity analysis and simulation can be used together, the possibility of including second-order uncertainty for distribution parameters, and the role of simulation in creating a requisite decision-analysis model.

EXERCISES

11.1   Explain in your own words how Monte Carlo simulation may be useful to a decision maker.

11.2   Explain how the simulation process works to produce results that are helpful to a decision maker.

11.3   A simulation model has produced the three cumulative risk profiles displayed in Figure 11.22. What advice would you give a decision maker on the basis of this output?

11.4   A friend of yours has just learned about Monte Carlo simulation methods and has asked you to do a simulation of a complicated decision problem to help her make a choice. She would be happy to have you solve the problem and then recommend what action she should take. Explain why she needs to be involved in the simulation modeling process and what kind of information you need from her.

QUESTIONS AND PROBLEMS

11.5   Find the expected cost for Supplier 2 in Janet Dawes's purchasing problem as diagrammed in Figure 11.3.

11.6   Simulation is one way to find an expected value for Janet Dawes's problem as diagrammed in Figure 11.3. How could you construct a discrete approximation that would at least provide an approximate expected cost for Supplier 2?

11.7   What other real-world situations involve step functions like the one that Janet Dawes faces?

11.8   An investor has purchased a call option for 100 shares of Alligator stock and intends to hold it until the day the option expires, at which time he will sell it if he can. The option is worth nothing on its expiration date unless the price of Alligator stock is more than $45 per share. For values of the stock greater than $45, the option will be worth

Figure 11.22

Three cumulative risk profiles from a simulation.

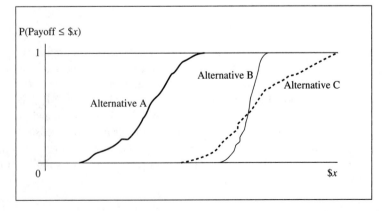

100(Share Price − $45). The reasoning behind this value is that the call option permits the option's owner to purchase 100 shares at $45 per share. Thus, if the share price is greater than $45, then the option owner could buy the shares and immediately resell them at the market price, pocketing the difference. Of course, the investor is uncertain about Alligator's eventual share price on the exercise date, but his uncertainty can be modeled using a normal distribution having mean $45.50 and standard deviation $5.00. Construct a simulation to estimate the expected value of the option on the exercise date.

11.9   Your boss has asked you to work up a simulation model to examine the uncertainty regarding the success or failure of five different investment projects. He provides probabilities for the success of each project individually: $p_1 = 0.50$, $p_2 = 0.35$, $p_3 = 0.65$, $p_4 = 0.58$, $p_5 = 0.45$. Because the projects are run by different people in different segments of the investment market, you both agree that it is reasonable to believe that, given these probabilities, the outcomes of the projects are independent. He points out, however, that he really is not fully confident in these probabilities and that he could be off by as much as 0.05 in either direction on any given probability.

   a   How can you incorporate his uncertainty about the probabilities into your simulation?

   b   Now suppose he says that if he is optimistic about the success of one project, he is likely to be optimistic about the others as well. For your simulation, this means that if one of the probabilities increases, the others also are likely to increase. How might you incorporate this information into your simulation?

11.10   A decision maker is working on a problem that requires her to study the uncertainty surrounding the payoff of an investment. There are three possible levels of payoff—$1000, $5000, and $10,000. As a rough approximation, the decision maker believes that each possible payoff is equally likely. But she is not fully comfortable with the assessment that each probability is exactly $\frac{1}{3}$, and so would like to conduct a sensitivity analysis. In fact, she believes that each probability could range from 0 to $\frac{1}{2}$.

   a   Show how a Monte Carlo simulation could facilitate a sensitivity analysis of the probabilities of the payoffs.

   b   Suppose the decision maker is willing to say that each of the three probabilities could be chosen from a uniform distribution between 0 and 1. Could you incorporate this information into your simulation? If so, how? If not, explain why not, or what additional information you would need.

11.11   Perform a simulation of the Eagle Airlines problem:

   a   Treat all of the variables in Table 5.1 as uncertain, and create triangular distributions for each of them in @RISK. For each variable, use the base value as the most likely value, and the high and low values as the upper and lower bounds of the triangular distribution. Run the simulation to obtain a risk profile for Profit. What are the 0.05, 0.25, 0.50, 0.75, and 0.95 fractiles of this risk profile?

   b   Now eliminate the uncertainty on all variables *except* Operating Cost, Hours Flown, Capacity, and Charter Price, leaving all of the other variables at their base values. Now rerun the simulation. Obtain the risk profile for Profit, being sure to note the 0.05, 0.25, 0.50, 0.75, and 0.95 fractiles. Compare the risk profile obtained from this restricted model with the results from part a.

   c   Rerun the model incorporating the correlations given in Figure 11.21 and compare the results with those from parts a and b.

CASE STUDIES

CHOOSING A MANUFACTURING PROCESS

AJS, Ltd., is a manufacturing company that performs contract work for a wide variety of firms. It primarily manufactures and assembles metal items, and so most of its equipment is designed for precision machining tasks. The executives of AJS currently are trying to decide between two processes for manufacturing a product. Their main criterion for measuring the value of a manufacturing process is net present value (NPV). The contractor will pay AJS $8 per unit. AJS is using a three-year horizon for its evaluation (the current year and the next two years).

Process 1

Under the first process, AJS's current machinery is used to make the product. The following inputs are used:

Demand   Demand for each of the three years is unknown. These three quantities are modeled as discrete random variables denoted D_0, D_1, and D_2 with the following probability distributions:

D_0	$P(D_0)$	D_1	$P(D_1)$	D_2	$P(D_2)$
11K	0.2	8K	0.2	4K	0.1
16K	0.6	19K	0.4	21K	0.5
21K	0.2	27K	0.4	37K	0.4

Variable Cost   Variable cost per unit changes each year, depending on the costs for materials and labor. Let V_0, V_1, and V_2 represent the three variable costs. The uncertainty surrounding each variable is represented by a normal distribution with mean $4 and standard deviation $0.40.

Machine Failure   Each year, AJS's machines fail occasionally, but obviously it is impossible to predict when or how many failures will occur during the year. Each time a machine fails, it costs the firm $8000. Let Z_0, Z_1, and Z_2 represent the number of machine failures in each of the three years, and assume that each is a Poisson random variable with parameter $\lambda = 4$.

Fixed Cost   Each year a fixed cost of $12,000 is incurred.

Process 2

The second process involves scrapping the current equipment (it has no salvage value) and purchasing new equipment to make the product at a cost of $60,000. Assume that the firm pays cash for the new machine, and ignore tax effects.

Demand   Because of the new machine, the final product is slightly altered and improved, and consequently the demands are likely to be higher than before, although more uncertain. The new demand distributions are:

D_0	$P(D_0)$	D_1	$P(D_1)$	D_2	$P(D_2)$
14K	0.3	12K	0.36	9K	0.4
19K	0.4	23K	0.36	26K	0.1
24K	0.3	31K	0.28	42K	0.5

Variable Cost   Variable cost still changes each year, but this time V_0, V_1, and V_2 are each judged to be normal with mean $3.50 and standard deviation $1.00.

Machine Failures   Equipment failures are less likely with the new equipment, occurring each year according to a Poisson distribution with parameter $\lambda = 3$. They also tend to be less serious, costing only $6000.

Fixed Cost   The fixed cost of $12,000 is unchanged.

Questions

1   Draw an influence diagram for this decision problem. Do you think it would be feasible to solve this problem with an influence diagram? Explain.

2   Write out the formula for the NPV for both processes described above. Use the variable names as specified, and assume a 10% interest rate.

3   For Process 1, construct a model and perform 1000 simulation trials. Estimate the mean and standard deviation of NPV for this process. Print a histogram of the results, and estimate the probability of a negative NPV occurring.

4   Repeat Question 3 for Process 2.

5   Compare the distribution of NPV for each of the two alternatives. Which process would be better for AJS? Why?

Source: This case was provided by Tom McWilliams.

ORGANIC FARMING

Jane Keller surveyed the freshly plowed field on her farm. She and her husband, Tim, had taken over the farm from her parents 10 years ago. Since that time, she and Tim had worked hard to improve the profitability of the farm. Even though the work was hard, the lifestyle was rewarding. And she found that common sense combined with basics from some business courses had helped her in making difficult decisions.

She faced one such decision now. Over the years, she had noticed more and more of her neighbors adopting a variety of organic farming methods. Many of the techniques were easy to adopt and made good sense. For example, companion planting and promoting a balanced ecology on the farm helped to create an environment in which plants were less susceptible to insect and disease damage. Last year she had grown some produce using only organic methods and had sold it locally. She learned a lot on that small-scale project, and in particular she learned that she did not know much about natural methods for preventing specific diseases and controlling insects.

Still, Jane was intrigued by the possibility of organic farming. Growers who could label their produce as "Organically Grown" commanded premium prices from specialty stores and at the local farmer's market. The Organic Farmers Association (OFA) provided the necessary certification. Meeting OFA's requirements involved (1) documenting the use of the land over a period of years to ensure little or no contamination from nonorganic pesticides, herbicides, and fertilizers; and (2) adhering to the farming methods that OFA deemed "organic."

This year Tim wanted to expand the operation by planting in a new field that had not been farmed in 20 years. Because the new field would be some distance from the other planted areas, it would be an ideal location to grow organic produce. There would be no problem getting the new field certified by the OFA. When she made this suggestion to Tim, he agreed with the idea in principle but wanted to think seriously about the project. They had several questions to answer. Would they really be able to make money from such a project? What kinds of prices could they anticipate for organic produce? What risks were involved?

Jane agreed to do some research. After visiting with her neighbors and spending a lot of time at the local organic gardening store, she was beginning to develop a plan. Although she could plant a large variety of different vegetables and herbs, most of the area would be devoted to tomatoes, green beans, and potatoes. She and Tim had plenty of experience with these crops, and she felt most comfortable experimenting with them. They would plant enough area so that, if the weather cooperated, they could expect to harvest 250 bushels of each crop. She was very uncertain about the exact yield, however, because of the variety of possible diseases and insects that could cause trouble and because of her own lack of experience with organic methods. For the tomatoes, she judged that there was a 68% chance that the yield would be between 235 and 265 bushels, and she was "almost sure" (95% chance) that the yield would be between 220 and 280 bushels. The green beans and potatoes she judged were somewhat less sensitive; for both of these crops, she estimated a 68% chance that the yield would fall between 242 and 258 bushels, and she was 95% sure that the yield would fall between 234 and 266 bushels.

The uncertain yield was complicated further by the effects of weather. Jane knew that the weather could be either too dry or too rainy for the crops, and that this would reduce her yield somewhat. At the same time, adverse weather would affect all growers in her region, leading to a smaller supply of produce and hence higher prices. The worst possible scenario for the Kellers would be for the weather to be perfect, resulting in a bumper crop around the region, while the Keller's venture into organics resulted in a low yield because of insects, disease, and their own inexperience. Prices would be low and they would have relatively little to sell.

Jane realized that she was facing a difficult judgmental task; ideally she would have to assess her uncertainty for the weather, the yield for each crop under different weather conditions, and the possible prices under the various conditions. To simplify matters, she decided to think about two scenarios. Under the first scenario, the weather would not affect the yield adversely. Under these conditions, she judged that the bushel price of tomatoes could range from $5.00 to $5.80. For potatoes, the range was $4.15 to $4.60; and for green beans, between $5.90 and $6.80. In each case, she figured that all prices in the specified range were equally likely.

Under the second scenario, the weather would have adverse effects on the region's crops. Jane estimated a probability of 0.15 for adverse weather. Because tomatoes were the most sensitive in this regard, the crop size would be reduced by some 20%. With smaller yields in the region, prices would be higher, ranging from $5.50 to $6.00 per bushel. Potatoes were the least sensitive, with only a 7% crop-size reduction and prices between $4.50 and $4.80. Green beans were not terribly sensitive to the weather in terms of quantity, so Jane estimated a 4% reduction. Under adverse weather conditions, however, the quality of the beans could be highly variable, and so she estimated their price range under this scenario to be between $5.50 and $7.00. Again, she judged that all prices in the specified ranges were equally likely.

Costs also had to be factored in. Organic methods were more labor-intensive than conventional methods, but because Jane and Tim did the work themselves, this did not really affect profits. Even so, Jane estimated that costs for the field would be approximately $800. With the uncertainty about the methods, however, she decided to represent costs with a normal distribution having mean $800 and standard deviation $50.

When she discussed this with Tim, he asked what this meant in terms of a "bottom line." What could they expect their gross sales to be? Could she give him some idea of how uncertain their profit would be? If they went with conventional farming methods, they could expect profits to be approximately $2700, with a standard deviation of some $100. If Jane could develop a probability distribution for profits under organic methods, perhaps the two distributions could be compared.

Tim added another wrinkle. He showed her an advertisement and a related story in the newspaper about a newly developed, genetically engineered bacterial pesticide named VegeTech. The ad claimed (and the story confirmed) that tests with VegeTech in their own geographical area had resulted in an average 10% increase in yield for all kinds of produce. The cost for this increase was approximately 20 cents per bushel. Given anticipated produce prices, this would appear to be cost-effective. The complication was that, while VegeTech had been fully approved by the appropriate federal agencies, OFA had not yet decided whether to approve its use as an organic substance. Proponents argued that it consisted of bacteria that attacked insects, while opponents argued that the bacteria had been synthesized and were not "organic" in the classic sense of the word. OFA had guaranteed that it would run tests and make a decision late this summer about the use of the substance; unfortunately, farmers would have to decide whether or not to use VegeTech before learning of OFA's decision. If the Kellers decided to use VegeTech, and if the OFA failed to certify it as an accepted organic substance, then the Kellers would not be able to label their produce

"Organically Grown." The net effect would be a 15% reduction in the prices they could charge. By all indications, there was a 50–50 chance that the OFA would certify VegeTech.

Question

1   Construct a simulation model to address the issues that the Kellers face. Do you think that they should stick with the conventional methods or try organic agriculture? If they go organic, should they try VegeTech this year? Support your conclusions with appropriate simulation outputs (graphs, expected values, and so on).

OVERBOOKING, PART III

Consider again Mockingbird Airlines' problem as described in the overbooking case study in Chapter 9 (pages 388–389).

Questions

1   Construct a simulation model of the system, and use it to find Mockingbird's optimal policy regarding overbooking. Compare this answer with the one based on the analysis done in Chapter 9.

2   Suppose that you are uncertain about the no-show rate. It could be as low as 0.02 or it could be as high as 0.06, and all values in between are equally likely. Furthermore, the cost of satisfying the bumped passengers may not be constant. That is, the airline may in some cases be able to entice a passenger or two to relinquish their seats in exchange for compensation that would be less than a refund and another free ticket. Alternatively, in some cases the total cost, including loss of goodwill, might be construed as considerably higher. Suppose, for example, that the cost of satisfying an excess customer is normally distributed with mean $300 and standard deviation $40.

Modify the simulation model constructed in Question 1 to include the uncertainties about the no-show rate and the cost. Do these sources of uncertainty affect the optimal overbooking policy?

3   How else might Mockingbird's analysts address the uncertainty about the no-show rate and the cost?

REFERENCES

Hertz's (1964) article in *Harvard Business Review* extolled the virtues of simulation for the decision-analysis community early on. Hertz and Thomas (1983, 1984) provide discussion and examples of the use of simulation for decision analysis. Other texts that include introductory material on simulation are Holloway (1979) and Samson (1988). Vatter et al. (1978) contains several interesting simulation case studies in decision making. More technical introductions to Monte Carlo simulation at a moderate level are provided by Law and Kelton (1991) and Watson (1989).

Hertz, D. B. (1964) "Risk Analysis in Capital Investment." *Harvard Business Review.* Reprinted in *Harvard Business Review,* September–October 1979, 169–181.

Hertz, D. B., and H. Thomas (1983) *Risk Analysis and Its Applications.* New York: Wiley.

Hertz, D. B., and H. Thomas (1984) *Practical Risk Analysis.* New York: Wiley.

Holloway, C. A. (1979) *Decision Making under Uncertainty: Models and Choices.* Englewood Cliffs, NJ: Prentice Hall.

Law, A. M., and D. Kelton (1991) *Simulation Modeling and Analysis,* 2nd ed. New York: McGraw-Hill.

Samson, D. (1988) *Managerial Decision Analysis.* Homewood, IL: Irwin.

Vatter, P., S. Bradley, S. Frey, and B. Jackson (1978) *Quantitative Methods in Management: Text and Cases.* Homewood, IL: Irwin.

Watson, G. (1989) *Computer Simulation,* 2nd ed. New York: Wiley.

CHAPTER 12

Value of Information

Decision makers who face uncertain prospects often gather information with the intention of reducing uncertainty. Information gathering includes consulting experts, conducting surveys, performing mathematical or statistical analyses, doing research, or simply reading books, journals, and newspapers. The intuitive reason for gathering information is straightforward; to the extent that we can reduce uncertainty about future outcomes, we can make choices that give us a better chance at a good outcome.

In this chapter, we will work a few examples that should help you understand the principles behind information valuation. Naturally, the examples also will demonstrate the techniques used to calculate information value. At the end of the chapter we will consider a variety of issues, including information in complex decisions, the use of information evaluation as an integral part of the decision-analysis process, what to do in the case of multiple nonmonetary objectives, and the problem of evaluating and selecting experts for the information they can provide.

The main example for this chapter is the stock market example that we introduced in Chapter 5 in our discussion of sensitivity analysis (pages 190–193). For convenience, the details are repeated here.

INVESTING IN THE STOCK MARKET

An investor has some funds available to invest in one of three choices: a high-risk stock, a low-risk stock, or a savings account that pays a sure $500. If he invests in the stocks, he must pay a brokerage fee of $200.

Figure 12.1

(a) Influence-diagram and (b) decision-tree representations of the investor's problem.

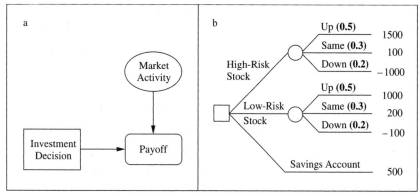

Figure 12.1

(a) Influence-diagram and (b) decision-tree representations of the investor's problem.

His payoff for the two stocks depends on what happens to the market. If the market goes up, he will earn $1700 from the high-risk stock and $1200 from the low-risk stock. If the market stays at the same level, his payoffs for the high- and low-risk stocks will be $300 and $400, respectively. Finally, if the stock market goes down, he will lose $800 with the high-risk stock but still gain $100 with the low-risk stock.

The investor's problem can be modeled with either an influence diagram or a decision tree. These two representations are shown in Figure 12.1.

Value of Information: Some Basic Ideas

Before we begin an in-depth study of information from a decision-analysis perspective, let us consider certain fundamental notions. What does it mean for an expert to provide perfect information? How does probability relate to the idea of information? What is an appropriate basis on which to evaluate information in a decision situation? This section addresses these questions and thus sets the stage for a complete development of the value of information in the rest of the chapter.

Probability and Perfect Information

An expert's information is said to be perfect if it is always correct. We can use conditional probabilities to model perfect information. Suppose that when state S will occur, the expert always says so (and never says that some other state will occur). In our stock market example, imagine an expert who always correctly identifies a situation in which the market will increase:

P(Expert Says "Market Up" | Market Really Does Go Up) = 1

Because the probabilities must add to 1, we also must have

P(Expert Says "Market Will Stay the Same or Fall"

| Market Really Will Go Up) = 0

But this is only half the story. The expert also must never say that state S will occur if any other state $(\overline{S})$ will occur. There must be no chance of our expert saying that the market will rise when it really will not:

P(Expert Says "Market Will Go Up"

| Market Really Will Stay the Same or Fall) = 0

Notice the difference between this probability statement and the preceding. Both are conditional probabilities, but the conditions are different.

If the expert's information is perfect, then upon hearing the expert's report no doubt about the future remains; if the expert says the market will rise, then we know that the market really will rise. Having used conditional probabilities to model the expert's perfect information, we can use Bayes' theorem to "flip" the probabilities as we did in Chapter 7 and show that there is no uncertainty after we have heard the expert. We want to know P(Market Really Will Go Up | Expert Says "Market Will Go Up"). Some notation will make our lives easier:

Market Up = The market really goes up

Market Down = The market really stays flat or goes down

Exp Says "Up" = The expert says the market will go up

Exp Says "Down" = The expert says the market will stay flat or go down

Now we can apply Bayes' theorem:

P(Market Up | Exp Says "Up")

$$= \frac{\text{P(Exp Says "Up" | Market Up)P(Market Up)}}{[\text{P(Exp Says "Up" | Market Up) P(Market Up)} + \text{P(Exp Says "Up" | Market Down) P(Market Down)}]}$$

$$= \frac{1 \text{ P(Market Up)}}{1 \text{ P(Market Up)} + 0 \text{ P(Market Down)}}$$

$$= 1$$

Observe that the posterior probability P(Market Up | Expert Says "Up") is equal to 1 regardless of the prior probability P(Market Up). This is because of the conditional probabilities that we used to represent the expert's perfect performance. Of course, this situation is not typical of the real world. In real problems we rarely can eliminate uncertainty altogether. If the expert sometimes makes mistakes, these conditional probabilities would not be 1's and 0's and the posterior probability would not be 1 or 0; there still would be some uncertainty about what would actually happen.

This exercise may seem a bit arcane. Its purpose is to introduce the idea of thinking about information in a probabilistic way. We can use conditional probabilities and Bayes' theorem to evaluate all kinds of information in virtually any decision setting.

The Expected Value of Information

How can we place a value on information in a decision problem? For example, how could we decide whether to hire the expert described in the last section? Does it depend on what the expert says? In the investment decision, the optimal choice is to invest in the high-risk stock. Now imagine what could happen. If the expert says that the market will rise, the investor still would choose the high-risk stock. In this case, the information appears to have no value in the sense that the investor would have taken the same action regardless of the expert's information. On the other hand, the expert might say that the market will fall or remain the same, in which case the investor would be better off with the savings account. In this second case, the information has value because it leads to a different action, one with a higher expected value than what would have been experienced without the expert's information.

We can think about information value after the fact, as we have done in the preceding paragraph, but it is much more useful to consider it before the fact—that is, before we actually get the information or before we hire the expert. What effects do we anticipate the information will have on our decision? We will talk about the *expected value of information*. By considering the expected value, we can decide whether an expert is worth consulting, whether a test is worth performing, or which of several information sources would be the best to consult.

The worst possible case would be that, regardless of the information we hear, we still would make the same choice that we would have made in the first place. In this case, the information has zero expected value! If we would take the same action regardless of what an expert tells us, then why hire the expert in the first place? We are just as well off as we would have been without the expert. Thus, at the worst, the expected value of information is zero. But if there are certain cases—things an expert might say or outcomes of an experiment—on the basis of which we would change our minds and make a different choice, then the expected value of the information must be positive; in those cases, the information leads to a greater expected value. The expected value of information can be zero or positive, but never negative.

At the other extreme, perfect information is the best possible situation. Nothing could be better than resolving all of the uncertainty in a problem. When all uncertainty is resolved, we no longer have to worry about unlucky outcomes; for every choice, we know exactly what the outcome will be. Thus, the expected value of perfect information provides an upper bound for the expected value of information in general. Putting this together with the argument in the previous paragraph, the expected value of any information source must be somewhere between zero and the expected value of perfect information.

Finally, you might have noticed that we continue to consider the expected value of information in terms of the particular choices faced. Indeed, the expected value of information is critically dependent on the particular decision problem at hand. For this reason, different people in different situations may place different values on the same information. For example, General Motors may find that economic forecasts from an expensive forecaster may be a bargain in helping the company refine its production plans. The same economic forecasts may be an extravagant waste of money for a restaurateur in a tourist town.

Expected Value of Perfect Information

Now we will see how to calculate the expected value of perfect information (EVPI) in the investment problem. For an expected value-maximizing investor, the optimal choice is the high-risk stock because it has the highest EMV ($580); however, this is partly because the investor is optimistic about what the market will do. How much would he be willing to pay for information about whether the market will move up, down, or sideways?

Suppose he could consult an expert with perfect information—a clairvoyant—who could reveal exactly what the market would do. By including an arrow from "Market Activity" to "Investment Decision," the influence diagram in Figure 12.2 represents the decision situation in which the investor has access to perfect information. Remember, an arrow leading from an uncertainty node to a decision node means that the decision is made knowing the outcome of the uncertainty node. This is exactly what we want to represent in the case of perfect information; the investor knows what the market will do before he invests his money.

With a representation of the decision problem including access to perfect information, how can we find the EVPI? Easy. Solve each influence diagram, Figures 12.1a and 12.2. Find the EMV of each situation. Now subtract the EMV for Figure 12.1a ($580) from the EMV for Figure 12.2 ($1000). The difference ($420) is the EVPI. We can interpret this quantity as the maximum amount that the investor should be willing to pay the clairvoyant for perfect information. PrecisionTree further simplifies EVPI calculations because it solves influence diagrams autometically. To calculate the EVPI from Figure 12.1a, we need only add an arc from the chance node to the decision node, and PrecisionTree reports the new EMV of $1000.

It also is useful to look at the decision-tree representation. To do this, draw a decision tree that includes the opportunity to obtain perfect information (Figure 12.3). As in the influence-diagram representation, the EMV for consulting the clairvoyant is $1000. This is $420 better than the EMV obtained by acting without the information. As before, EVPI is the difference, $420.

Recall that in a decision tree the order of the nodes conforms to a chronological ordering of the events. Is this what happens in the perfect-information branch in

Figure 12.2

Perfect information in the investor's problem.

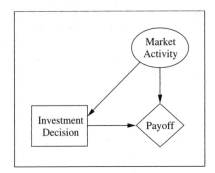

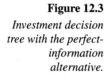

Figure 12.3

Investment decision tree with the perfect-information alternative.

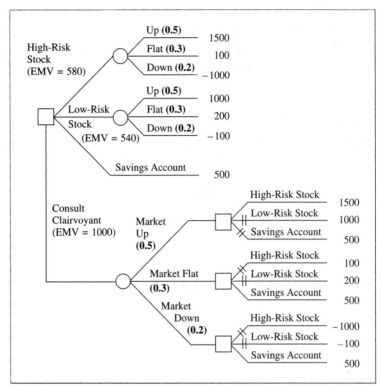

Figure 12.3? Yes and no. Yes in the sense that the uncertainty regarding the stock market's activity is resolved before the investment decision is made. This is the important part. But once the decision is made, the market still must go through its performance. It is simply that the investor knows exactly what that performance will be.

This points out a useful aspect of expected-value-of-information analysis with decision trees. If a decision maker faces some uncertainty in a decision, which is represented by those uncertainty nodes that come after his decision in a decision tree, redrawing the tree to capture the idea of perfect information is easy. Simply reorder the decision and uncertainty nodes! That is, redraw the tree so that the uncertainty nodes for which perfect information is available come before the decision node. This is exactly what we did in the perfect-information branch in Figure 12.3; in this branch, the "Market Activity" and "Investment Decision" nodes are reversed relative to their original positions. Reordering the nodes of a decision tree is simple in PrecisionTree because of its copy and paste features. At the end of this chapter, we show that a few clicks of the mouse are all that is needed to calculate EVPI in PrecisionTree.

It is worth reiterating that the way we are thinking about the value of information is in a strictly *a priori* sense. The decision-tree representation reinforces this notion because we actually include the decision branch that represents the possibility of consulting the clairvoyant. The investor has not yet consulted the clairvoyant; rather, he is considering whether to consult the clairvoyant in the first place. That action increases the expected value of the decision. Specifically, in this case there is a 50% chance that the clairvoyant will say that the market is not going up, in which case the appropriate decision would be the savings account rather than the high-risk stock.

Expected Value of Imperfect Information

We rarely have access to perfect information. In fact, our information sources usually are subject to considerable error. Thus, we must extend our analysis to deal with imperfect information.

The analysis of imperfect information parallels that of perfect information. We still consider the expected value of the information before obtaining it, and we will call it the *expected value of imperfect information* (EVII) to express the notion of collecting some information from a sample.

In the investment example, suppose that the investor hires an economist who specializes in forecasting stock market trends. Because he can make mistakes, however, he is not a clairvoyant, and his information is imperfect. For example, suppose his track record shows that if the market actually will rise, he says "up" 80% of the time, "flat" 10%, and "down" 10%. We construct a table (Table 12.1) to characterize his performance in probabilistic terms. The probabilities therein are conditional; for example, P(Economist Says "Flat" | Flat) = 0.70. The table shows that he is better when times are good (market up) and worse when times are bad (market down); he is somewhat more likely to make mistakes when times are bad.

How should the investor use the economist's information? Figure 12.4 shows an influence diagram that includes an uncertainty node representing the economist's forecast. The structure of this influence diagram should be familiar from Chapter 3; the economist's information is an example of imperfect information. The arrow from "Market Activity" to "Economic Forecast" means that the probability distribution for the particular forecast is conditioned on what the market will do. This is reflected in the distributions in Table 12.1. In fact, the distributions contained in the "Economic Forecast" node are simply the conditional probabilities from that table.

Table 12.1

Conditional probabilities characterizing economist's forecasting ability.

| | True Market State | | |
Economist's Prediction	Up	Flat	Down
"Up"	0.80	0.15	0.20
"Flat"	0.10	0.70	0.20
"Down"	0.10	0.15	0.60
	1.00	1.00	1.00

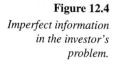

Figure 12.4

Imperfect information in the investor's problem.

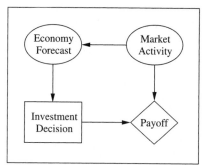

Solving the influence diagram in Figure 12.4 gives the EMV associated with obtaining the economist's imperfect information before action is taken. The EMV turns out to be $822. As we did in the case of perfect information, we calculate EVII as the difference between the EMVs from Figures 12.4 and 12.1a, or the situation with no information. Thus, EVII equals $822 − $580 = $242.

The influence-diagram approach is easy to discuss because we actually do not see the detail calculations. On the other hand, the decision-tree approach shows the calculation of EVII in its full glory. Figure 12.5 shows the decision-tree representation of the situation, with a branch that represents the alternative of consulting the economist. Look at the way in which the nodes are ordered in the "Consult Economist" alternative. The first event is the economist's forecast. Thus, we need probabilities P(Economist Says "Up"), P(Economist Says "Flat"), and P(Economist Says "Down"). Then the investor decides what to do with his money. Finally, the market goes up, down, or sideways. Because the "Market Activity" node follows the "Economists Forecast" node in the decision tree, we must have conditional probabilities for the market such as P(Market Up | Economist Says "Up") or P(Market Flat | Economist Says "Down"). What we have, however, is the opposite. We have probabilities such as P(Market Up) and conditional probabilities such as P(Economist Says "Up" | Market Up).

As we did when we first introduced the notion of the value of an expert's information at the beginning of this chapter, we must use Bayes' theorem to find the posterior probabilities for the actual market outcome. For example, what is P(Market Up | Economist Says "Up")? It stands to reason that after we hear him say "up," we should think it more likely that the market actually will go up than we might have thought before.

We used Bayes' theorem to "flip" probabilities in Chapter 7. There are several ways to think about this situation. First, applying Bayes' theorem is tantamount to reversing the arrow between the nodes "Market Activity" and "Economic Forecast" in Figure 12.4. In fact, reversing this arrow is the first thing that must be done when solving the influence diagram (Figure 12.6). Or we can think in terms of flipping a probability tree as we did in Chapter 7. Figure 12.7a represents the situation we have, and Figure 12.7b represents what we need.

Figure 12.5

Incomplete decision tree for the investment example, including the alternative for consulting the economist.

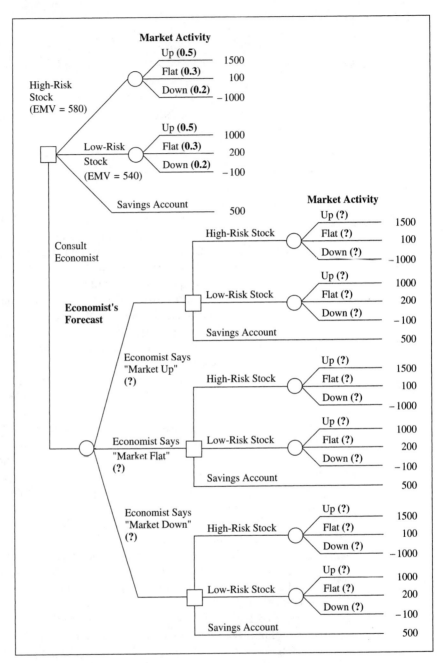

Figure 12.6

First step in solving the influence diagram. We reverse the arrow between "Economic Forecast" and "Market Activity."

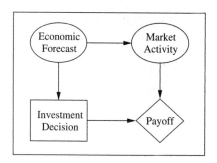

Figure 12.7

Flipping the probability tree to find posterior probabilities required for value-of-information analysis. In (a) we see what we have; in (b) we see what we need.

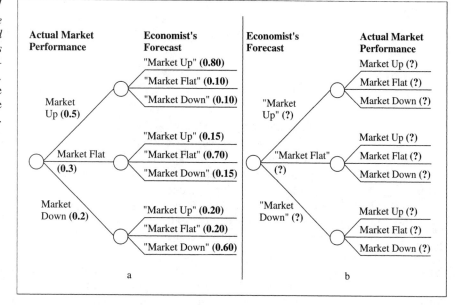

Whether we think of the task as flipping a probability tree or reversing an arrow in an influence diagram, we still must use Bayes' theorem to find the probabilities we need. For example,

P(Market Up | Economist Says "Up")

= P(Up | "Up")

$$= \frac{P(\text{``Up''} \mid \text{Up}) \, P(\text{Up})}{P(\text{``Up''} \mid \text{Up}) \, P(\text{Up}) + P(\text{``Up''} \mid \text{Flat}) \, P(\text{Flat}) + P(\text{``Up''} \mid \text{Down}) \, P(\text{Down})}$$

P(Up), P(Flat), and P(Down) are the investor's prior probabilities, while P(Economist Says "Up" | Up), and so on, are the conditional probabilities shown in Table 12.1. From the principle of total probability, the denominator is P(Economist Says "Up").

	Posterior Probability for:		
Economist's Prediction	**Market Up**	**Market Flat**	**Market Down**
"Up"	0.8247	0.0928	0.0825
"Flat"	0.1667	0.7000	0.1333
"Down"	0.2325	0.2093	0.5581

Substituting in values for the conditional probabilities and priors,

$$P(\text{Market Up} \mid \text{Economist Says "Up"}) = \frac{0.8(0.5)}{0.8(0.5) + 0.15(0.3) + 0.2(0.2)}$$

$$= \frac{0.400}{0.485}$$

$$= 0.8247$$

P(Economist Says "Up") is given by the denominator and is equal to 0.485.

Of course, we need to use Bayes' theorem to calculate nine different posterior probabilities to fill in the gaps in the decision tree in Figure 12.5. Table 12.2 shows the results of these calculations; these probabilities are included on the appropriate branches in the completed decision tree (Figure 12.8).

We also noted that we needed the marginal probabilities P("Up"), P("Flat"), and P("Down"). These probabilities are P("Up") = 0.485, P("Flat") = 0.300, and P("Down") = 0.215; they also are included in Figure 12.8 to represent our uncertainty about what the economist will say. As usual, the marginal probabilities can be found in the process of calculating the posterior probabilities because they simply come from the denominator in Bayes' theorem.

From the completed decision tree in Figure 12.8 we can tell that the EMV for consulting the economist is $822, while the EMV for acting without consulting him is (as before) only $580. The EVII is the difference between the two EMVs. Thus, EVSI is $242 in this example, just as it was when we solved the problem using influence diagrams. Given this particular decision situation, the investor would never want to pay more than $242 for the economic forecast.

As with perfect information, $242 is the value of the information only in an expected-value sense. If the economist says that the market will go up, then we would invest in the high-risk stock, just as we would if we did not consult him. Thus, if he does tell us that the market will go up, the information turns out to do us no good. But if he tells us that the market will be flat or go down, we would put our money in the savings account and avoid the relatively low expected value associated with the high-risk stock. In those two cases, we would "save" 500 − 187 = 313 and 500 − (−188) = 688, respectively, with the savings in terms of expected value. Thus, EVSI also can be calculated as the "expected incremental savings," which is 0(0.485) + 313(0.300) + 688(0.215) = 242.

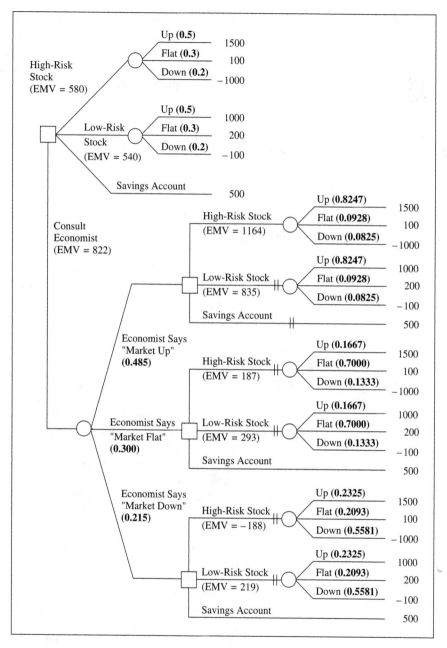

Figure 12.8

Completed decision tree for the investment example.

Such probability calculations can make value-of-information analysis tedious and time-consuming. This is where computers can play a role. PrecisionTree can perform all of the necessary probability calculations and thus make finding EVSI a simple matter. See instructions at the end of the chapter. It is also possible to construct an electronic spreadsheet model that will perform the calculations.

One last note regarding value-of-information calculations. If you have used your calculator to work through the examples in this chapter, you may have found that your answers differed slightly from those in this book. This is because calculations involving Bayes' theorem and value of information tend to be highly sensitive to rounding error. The appropriate strategy is to carry calculations out to many decimal places throughout a given problem, rounding off to dollars or cents only at the end of the problem. Rounding off intermediate results that are used in later calculations sometimes can make a large difference in the final answer.

Value of Information in Complex Problems

The example we have looked at in this section has been a fairly simple one. There was only one uncertain event, the behavior of the market, and we modeled the uncertainty with a simple discrete distribution. As we know, however, most real-world problems involve considerably more complex uncertainty models. In particular, we need to consider two specific situations. First, how can we handle continuous probability distributions? Second, what happens when there are many uncertain events and information is available about some or all of them?

The answer to the first question is straightforward conceptually, but in practice the calculation of EVPI or EVII may be difficult when dealing with continuous probability distributions. The principle is the same. Evaluate decision options with and without the information, and find the difference in the EMVs, as we have done in the discrete case. The problem, of course, is calculating the EMVs. Obviously, it always is possible to construct a discrete approximation as discussed in Chapter 8. Another possibility is to construct a Monte Carlo simulation model. Finally, for some theoretical probability models, analytical results are possible. The mathematics for such analysis, however, tend to be somewhat complicated and are beyond the scope of this introductory textbook. References for interested readers are included at the end of the chapter.

The second question asks how we handle value-of-information problems when there are many uncertain events. Again, in principle, the answer is easy, and is most transparent if we think in terms of influence diagrams. Perfect information about any particular event simply implies the presence of an informational arc from the event to the decision node. Naturally, it is possible to include such arcs for a subset of events. The only requirement is that the event not be downstream in the diagram from the decision node, because the inclusion of the information arc would lead to a cycle in the influence diagram. Solving the influence diagram with the informational arcs in place provides the EMV with the information, and this then may be compared to the EMV without the information to obtain the EVPI for the particular information sought.

Consider the same problem when the model is in decision-tree form. In the decision tree in Figure 12.3, the information branch was constructed by reversing the event node and the decision node. The same principle can apply if there are many sources of uncertainty; simply move those chance nodes for which information is to be obtained so that they precede the decision node. Now calculating EMV for the information branch will give the EMV in the case of perfect information for those events that precede the decision node in the decision tree.

For imperfect information, the same general principles apply. For influence diagrams, include an imperfect-information node that provides information to the decision maker. An excellent example is the toxic-chemicals influence diagram in Figure 3.20. Two sources of imperfect information are included in that model, the exposure survey and the lab test. In a decision-tree model, it would be a matter of constructing a tree having the appropriate informational chance nodes preceding the decision node. Unfortunately, if there are more than one or two such chance nodes, the decision tree can become extremely unwieldy. Moreover, it may be necessary to calculate and track the marginal and posterior probabilities for the decision tree; these calculations can be done automatically in the influence diagram.

Value of Information, Sensitivity Analysis, and Structuring

Our motivation for studying value of information has been to examine situations in which information is available and to show how decisions can be made systematically regarding what source of information to select and how much an expert's information might be worth. Strictly speaking, this is precisely what value-of-information analysis can do. But it also can play an elegant and subtle role in the structuring of decisions and in the entire decision-analysis process of developing a requisite decision model. Recall the ideas of sensitivity analysis from Chapter 5. In that chapter we talked about a process of building a decision structure. The first step is to find out, using a tornado diagram, those variables to which the decision was sensitive; these variables require probabilistic modeling. The second step, after constructing a probabilistic model, may be to perform sensitivity analysis on the probabilities.

A third step in the structuring of a probabilistic model would be to calculate the EVPI for each uncertain event. This analysis would indicate where the analyst or decision maker should focus subsequent efforts in the decision-modeling process. That is, if EVPI is very low for an event, then there is little sense in spending a lot of effort in reducing the uncertainty by collecting information. But if EVPI for an event is relatively high, it may indeed be worthwhile to put considerable effort into the collection of information that relates to the event. Such information can have a relatively large payoff by reducing uncertainty and improving the decision maker's EMV. In this way, EVPI analysis can provide guidance to the decision analyst as to what issues should be tackled next in the development of a requisite decision model.

We will end this chapter with a short description of a rather unusual application of value-of-information analysis. Although few applications are as elaborate as this one, this example does provide an idea of how the idea of information value, as developed in decision analysis, can be used to address real-world concerns.

SEEDING HURRICANES

Hurricanes pack tremendous power in their high winds and tides. Recent storms such as Camille (1969) and Hugo (1989) caused damage in excess of a billion dollars and amply demonstrated the destructive potential of large hurricanes. In the 1960s, the U.S. government experimented with the seeding of hurricanes, or the practice of dropping silver iodide into the storm to reduce peak winds by forcing precipitation. After early limited experiments, Hurricane Debbie was seeded in 1969 with massive amounts of silver iodide on two separate occasions. Each seeding was followed by substantial drops in peak wind speed. Given these results, should hurricanes that threaten highly populated areas be seeded on the basis of current knowledge? Should the government pursue a serious research program on the effects of hurricane seeding?

Howard, Matheson, and North addressed these specific questions about hurricane seeding. They asked whether it would be appropriate to seed hurricanes, or would a research program be appropriate, or should the federal government simply not pursue this type of weather modification? To answer these questions, they adopted a decision-analysis framework. On the basis of a relatively simple probabilistic model of hurricane winds, along with the relationships among wind speed, damage, and the effect of seeding, they were able to calculate expected dollar losses for two decision alternatives—seeding and not seeding a typical threatening hurricane. On the basis of their model, they concluded that seeding would be the preferred alternative if the federal government wanted to reduce expected damage.

The authors realized that the government might be interested in more than the matter of reducing property damage. What would happen, for example, if the decision was made to seed a hurricane, the wind speed subsequently increased, and increased damage resulted? The government most likely would become the target of many lawsuits for having taken action that appeared to have adverse effects. Thus, the government also faced an issue of responsibility if seeding were undertaken. On the basis of the authors' analysis, however, the government's "responsibility cost" would have to have been relatively high in order for the optimal decision to change.

To address the issue of whether further research on seeding would be appropriate, the authors used a value-of-information approach. They considered the possibility of repeating the seeding experiment that had been performed on Hurricane Debbie and the potential effects of such an experiment. By modeling the possible outcomes of this experiment and considering the effect on future seeding decisions, the authors were able to calculate the expected value of the research. Including reasonable costs for government responsibility and anticipated damage over a single hurricane season, the expected value of the experiment was determined to be approximately $10.2 million. The authors also extended their analysis to include all future hurricane seasons, discounting future costs by 7% per year. In this case, the expected value of the research was $146 million. To put these numbers in perspective, the cost of the seeding experiment would have been approximately $500,000. [*Source:* R. A. Howard, J. E. Matheson, and D. W. North (1972) "The Decision to Seed Hurricanes." *Science,* 176, 1191–1202.]

Value of Information and Nonmonetary Objectives

Throughout this chapter we have calculated the expected value of information on the basis of EMVs, implicitly assuming that the only objective that matters is making money. This has been a convenient fiction; as you know, in many decision situations there are multiple objectives. For example, consider the Federal Aviation Administration (FAA) bomb-detection example again (Chapter 3, Figure 3.8). Recall that the FAA was interested in maximizing the detection effectiveness and passenger acceptance of the system while at the same time minimizing the cost and time to implementation. For any given system, there may be considerable uncertainty about the level of passenger acceptance, for example, but this uncertainty could be reduced in many ways. A less expensive, but highly imperfect, option would be to run a survey. A more expensive test would be to install some of the systems at a few airports and try them out. But how much should be spent on such efforts?

With the FAA example, the answer is relatively straightforward, because minimizing cost happens to be one of the objectives. The answer would be to find the additional cost that makes the net expected value of getting the information equal to the expected value without the information (and, naturally, without the cost). If there is a clear trade-off rate that can be established for, say, dollars of cost per additional point on the passenger-acceptance scale, then the increase in expected passenger-acceptance level that results from additional information can be translated directly into dollars of cost. (Any increase in the expected value of the other objectives—which may happen due to choosing different options under different information scenarios—would also have to be included in the calculations. Doing so can also be accomplished through the specification of similar trade-offs between cost and the other objectives. Understanding such trade-offs is not complex and is treated in more detail in Chapter 15.)

When a decision situation does not involve a monetary objective, the same techniques that we have developed in this chapter can still be used. Suppose one objective is to minimize the decision maker's time; different choices and different outcomes require different amounts of time from the decision maker. Information can be valued in terms of time; the expected value of perfect information might turn out to be, say, five days of work. If resolving some uncertainty would take more time than that, it would not be worth doing!

Value of Information and Experts

In Chapter 8 we discussed the role of experts in decision analysis. One of the issues that analysts face in using experts is how to value them and how to decide how many and which experts to consult. In general, the valuation part is not a practical concern; the high stakes involved in typical public-policy risk analyses typically warrant the use of experts who may charge several thousand dollars per day in consulting fees. What is less obvious, however, is that experts typically provide information that is somewhat interrelated. Because experts tend to read the same journals, go to the same conferences, use the same techniques in their studies, and even communicate with each other, it comes as no surprise that the information they provide can be highly redundant. The real challenge in expert use is to recruit experts who look at the same problem from very different perspectives. Recruiting experts from different fields, for example, can be worthwhile if the information provided is less redundant. It can even be the case that a highly diverse set of less knowledgeable (and less expensive) experts can be much more valuable than the same number of experts who are more knowledgeable (and cost more) but give redundant information!

Calculating EVPI and EVII with PrecisionTree

We have seen in this chapter that calculating EVPI is relatively simple and involves nothing more than adding an arc to an influence diagram or, equivalently, reordering the nodes in a decision tree. Calculating EVPI is not difficult, but it can be tedious if done by hand. We will demonstrate how PrecisionTree simplifies the process with its one-click copying and pasting of decision trees. EVII calculations are more complex in that they require additional nodes and a Bayesian revision of the probabilities. PrecisionTree greatly simplifies EVII calculations for influence diagrams because the probabilities are automatically revised using Bayes' theorem. We will demonstrate how to perform both EVPI and EVII calculations in PrecisionTree using the investment problem.

EVPI

EVPI is usually calculated after the decision problem has been structured, so our starting point is a constructed decision tree and influence diagram of the investor's problem.

STEP 1

1.1   Open Excel and PrecisionTree.

1.2   Either construct both the influence diagram and decision tree for the investor's problem, as shown in Figure 12.1, or open the workbook in Examples\Chapter12\Investment.xls. The decision tree is in the worksheet titled "Base Tree" and the influence diagram is in the worksheet titled "Base ID."

Influence Diagrams

It is very simple to compute EVPI with an influence diagram; merely add an arc from the specified chance node to the appropriate decision node. PrecisionTree updates the expected value and displays it in the upper left-hand corner of the spreadsheet. EVPI is the difference between the updated expected value and the original expected value. We demonstrate for the investor's problem.

STEP 2

2.1   Open the influence-diagram worksheet titled **Base ID**.

2.2   To add an arc, click on the **New Influence Arc** button, click inside the chance node **Market Activity**, and while keeping the mouse button depressed, move the cursor to the decision node **Investment Decision**. An arc is drawn when the button is released.

2.3   Choose **Value** and **Timing** (Implied by Value) for the arc's influence type, and click **OK**.

PrecisionTree solves the new influence diagram and reports the summary information in the upper left-hand corner of the spreadsheet. Table 12.3 lists the summary statistics for the two influence diagrams: the diagram shown in Figure 12.1 without perfect information and the diagram shown in Figure 12.2 with perfect information.

Table 12.3
Comparing summary statistics for the influence diagrams in Figures 12.1 and 12.2.

	No Information	Knowing Market Activity
Expected Value	$580	$1000
Standard Deviation	$996	$500
Minimum	–$1000	$500
Maximum	$1500	$1500

Let's examine what we can learn from this table. First, EVPI equals $420, which is found by subtracting the expected values: $1000 – $580. Second, the standard deviation is cut nearly in half (compare $996 to $500) and downside risks are reduced (compare the two minimums, –$1000 and $500) when you know the market activity before choosing your investment.

Decision Trees

Calculating the expected value of perfect information is more cumbersome for decision trees because the nodes must be reordered. In our investment problem tree (Figure 12.1b), we need to move the "Market Activity" chance node in front of the "Investment Decision" node.

STEP 3

3.1   Open the Base Tree worksheet as described in Step 1.2. In a new worksheet start a decision tree by clicking the **New Tree** button and clicking in cell **A1**.

3.2   Name the tree **EVPI(Market Activity)**.

3.3   The first node in the tree will be the chance node "Market Activity." Go back to the Base Tree and click directly on the chance node **Market Activity** (not the name, but the red circle). Choose **Copy** in the *Node Settings* dialog box.

3.4   Return to the new tree, click on the only end node, and choose **Paste** in the *Node Settings* dialog box. The chance node, its three outcome branches, and their associated values and probabilities are added to the first position of the new tree.

3.5   Change the values for each branch to **0**, but do not change the probabilities.

Forgetting to set the branch values to zero is a common mistake for students. Because we have not yet chosen an investment (stocks or savings), we have no cash flows at this point.

STEP 4

4.1   We now paste copies of the Base Tree at the end nodes of the three outcomes. Return to the Base Tree and click directly on the decision node **Investment Decision** (not the name, but the green square). Choose **Copy** in the *Node Settings* dialog box.

4.2   Return to the new tree, click on an end node, and choose **Paste** in the *Node Settings* dialog box. Repeat for the remaining two end nodes.

STEP 5

5.1   Complete the tree by updating the probabilities of the second "Market Activity" chance node to reflect that the market activity outcome is known. For example, at first, there is a 50% chance that the market will go up, but once we find out for sure that it *will* go up, the probability is 100%. To incorporate this change in probability, delete the branches that do not occur in the second "Market Activity" chance node and give the one remaining branch a probability of 100%. Use Figure 12.9 as a guide. Branches are deleted by clicking directly on the branch name and choosing **Delete** in the *Branch Settings* window.

Figure 12.9

Investment decision tree knowing the market states before making the decision.

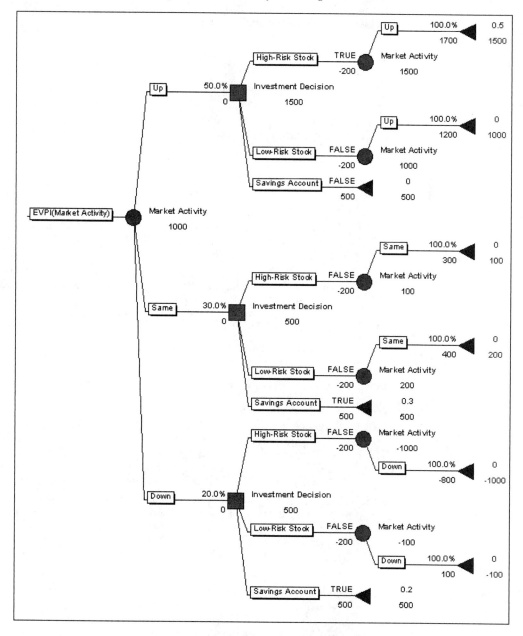

Your completed diagram should look exactly like Figure 12.9 and should have an expected value of $1000. Hence, the EVPI is $1000 – $580 = $420, which agrees with the influence diagram. Running a decision analysis on this tree will produce the summary statistics listed in Table 12.3.

EVII

We show here how to use PrecisionTree to calculate EVII in an influence diagram, including the automatic application of Bayes' theorem.

STEP 6

6.1 Open the **Base ID** worksheet. Our goal is to create the influence diagram found in Figure 12.4. (If necessary, remove the arc between Market Activity and Investment Decision that was added in Step 2.2. Delete an arc by clicking on the arc and then clicking on the *Delete Arc* button.)

STEP 7

7.1 We start by adding the expert's opinion by inserting a chance node. Click on the **Add New Influence Diagram/Node** button (second from the left, PrecisionTree toolbar), then click in cell **D6**.

7.2 Choose **Chance** node, change the node name to **Economic Forecast**, and name the outcomes **Says Up**, **Says Down**, and **Says Same**. Outcomes are added by clicking the *Add* button.

7.3 Click **OK**. We will return to this node to add the values and probabilities after adding the arcs.

STEP 8

8.1 Add two arcs. Add the first by clicking on the **New Influence Arc** button (third from left) and dragging the crosshair cursor from **Market Activity** to **Economic Forecast**. In the pop-up dialog box, choose both the **Value** and **Timing** checkboxes and click **OK**. Value was chosen because we believe that the future market outcomes and the expert's forecast are related.

8.2 Add the second arc from **Economic Forecast** to **Investment Decision** by following the same procedure. Choose both the **Value** and **Timing** (Implied by Value checkboxes). We now return to *Economic Forecast* and enter the numbers.

STEP 9

9.1 Click on the **Economic Forecast** name, and then the **Values** button in the *Settings* dialog box.

9.2 Enter the values and probabilities into the *Influence Value Editor,* as shown in Figure 12.10. Click **OK**.

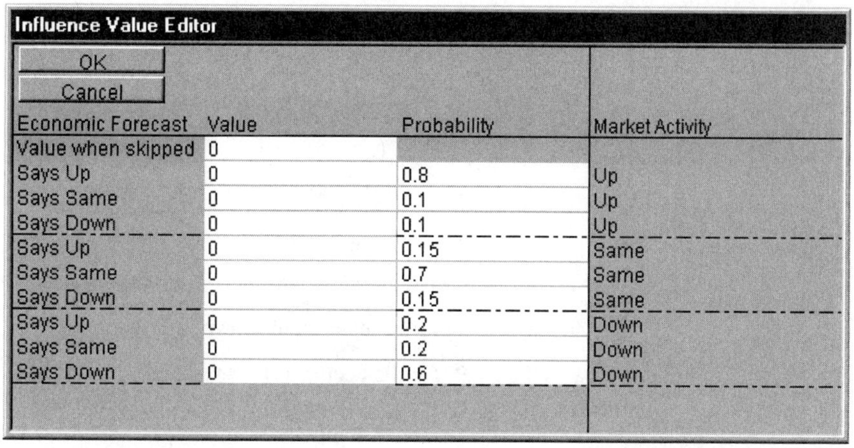

Figure 12.10

Probabilities and values for the "Economic Forecast" chance node of the investment problem.

Economic Forecast	Value	Probability	Market Activity
Value when skipped	0		
Says Up	0	0.8	Up
Says Same	0	0.1	Up
Says Down	0	0.1	Up
Says Up	0	0.15	Same
Says Same	0	0.7	Same
Says Down	0	0.15	Same
Says Up	0	0.2	Down
Says Same	0	0.2	Down
Says Down	0	0.6	Down

Influence Value Editor — OK / Cancel

9.3   Click on the **Investment Decision** name, and then the **Values** button in the *Settings* dialog box. Enter **0** for each value and click **OK**.

The updated summary statistics show that the expected value of the influence diagram is $822, so, EVII = $822 – $580, or $242. By converting this influence diagram into a decision tree (Convert to Tree is on the PrecisionTree menu), you will see that the chance node "Economic Forecast" comes first and the revised probabilities match those in Figure 12.8. We leave it to the reader to construct a decision tree for computing EVII for the investor's problem. Use Figure 12.8 as a guide and the copy/paste feature to expedite structuring.

SUMMARY

By considering the expected value of information, we can make better decisions about whether to obtain information or which information source to consult. We saw that the expected value of any bit of information must be zero or greater, and it cannot be more than the expected value of perfect information. Both influence diagrams and decision trees can be used as frameworks for calculating expected values. Influence diagrams provide the neatest representation because information available for a decision can be represented through appropriate use of arcs and, if necessary, additional uncertainty nodes representing imperfect information. In contrast, representing the expected value of imperfect information with decision trees is more complicated, requiring the calculation of posterior and marginal probabilities. The expected value of information is simply the difference between the EMV calculated both with and without the information.

The final sections in the chapter discussed generally how to solve value-of-information problems in more complex situations. We concluded with discussions of the role that value-of-information analysis can play in the decision-analysis process of developing a requisite decision model, how to value information related to non-monetary objectives, and evaluation and selection of experts.

EXERCISES

12.1 Explain why in decision analysis we are concerned with the *expected* value of information.

12.2 Calculate the EVPI for the decision shown in Figure 12.11.

12.3 What is the EVPI for the decision shown in Figure 12.12? Must you perform any calculations? Can you draw any conclusions regarding the relationship between value of information and deterministic dominance?

12.4 For the decision tree in Figure 12.13, assume Chance Events **E** and **F** are independent.

   a Draw the appropriate decision tree and calculate the EVPI for Chance Event **E** only.

   b Draw the appropriate decision tree and calculate the EVPI for Chance Event **F** only.

   c Draw the appropriate decision tree and calculate the EVPI for both Chance Events **E** and **F**: that is, perfect information for both **E** and **F** is available before a decision is made.

12.5 Draw the influence diagram that corresponds to the decision tree for Exercise 12.4. How would this influence diagram be changed in order to answer parts a, b, and c in Exercise 12.4?

QUESTIONS AND PROBLEMS

12.6 The claim was made in the chapter that information always has positive value. What do you think of this? Can you imagine any situation in which you would prefer *not* to have some unknown information revealed?

   a Suppose you have just visited your physician because of a pain in your abdomen. The doctor has indicated some concern and has ordered some tests whose results the two of you are expecting in a few days. A positive test result will suggest that you may have a life-threatening disease, but even if the test is positive, the doctor would want to confirm it with further tests. Would you want the doctor to tell you the outcome of the test? Why or why not?

   b Suppose you are selling your house. Real-estate transaction laws require that you disclose what you know about significant structural defects. Although you know of no such defects, a buyer insists on having a qualified engineer inspect the house. Would you want to know the outcome of the inspection? Why or why not?

   c Suppose you are negotiating to purchase an office building. You have a lot of experience negotiating commercial real-estate deals, and your agent has explained this to the seller, who is relatively new to the business. As a result, you expect to do very

Figure 12.11

Generic decision tree for Exercise 12.2.

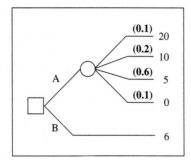

Figure 12.12

Generic decision tree for Exercise 12.3.

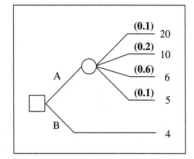

Figure 12.13

Generic decision tree for Exercise 12.4.

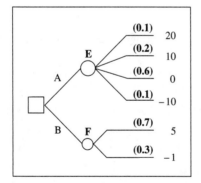

well in this negotiation. Because of the unique circumstance of the building, your agent has suggested obtaining an appraisal of the property by a local expert. You know exactly how you would use this information; it would provide an upper bound on what you are willing to pay. This fact is also clear to the seller, who will know whether you obtained the appraisal but will only find out the appraised value if you elect to reveal it. Would you obtain the appraisal? Why or why not?

12.7 Consider another oil-wildcatting problem. You have mineral rights on a piece of land that you believe may have oil underground. There is only a 10% chance that you will strike oil if you drill, but the payoff is $200,000. It costs $10,000 to drill. The alternative is not to drill at all, in which case your profit is zero.

   a Draw a decision tree to represent your problem. Should you drill?

b Draw an influence diagram to represent your problem. How could you use the influence diagram to find EVPI?

c Calculate EVPI. Use either the decision tree or the influence diagram.

d Before you drill you might consult a geologist who can assess the promise of the piece of land. She can tell you whether your prospects are "good" or "poor." But she is not a perfect predictor. If there is oil, the conditional probability is 0.95 that she will say prospects are good. If there is no oil, the conditional probability is 0.85 that she will say poor. Draw a decision tree that includes the "Consult Geologist" alternative. Be careful to calculate the appropriate probabilities to include in the decision tree. Finally, calculate the EVII for this geologist. If she charges $7000, what should you do?

12.8 In Problem 4.16, find:

a EVPI for the cost of making the processor.

b EVPI for the cost of subcontracting the processor.

c EVPI for both uncertain events.

12.9 Look again at Problem 8.11, which concerns whether you should drop your decision-analysis course. Estimate EVPI for your score in the course.

12.10 In Problem 9.33, the issue is whether or not to close the plant, and if your boss knew exactly how many machines would fail during your absence, he would be able to decide what to do without the fear of making a mistake.

a Find his EVPI concerning the number of machine failures during your absence over the next two weeks.

b Suppose that the cost of repairing the broken machines is $30,000. Then what is EVPI?

c Suppose that the cost of repairing the machines is $15,000 but the cost of closing the plant is $20,000. Now calculate EVPI.

12.11 Consider the Texaco-Pennzoil example from Chapter 4 (pages 111–112).

a What is EVPI to Hugh Liedtke regarding Texaco's reaction to a counteroffer of $5 billion? Can you explain this result intuitively?

b The timing of information acquisition may make a difference.

(i) For example, suppose that Liedtke could obtain information about the final court decision before making his current decision (take the $2 billion or counteroffer $5 billion). What would be EVPI of this information?

(ii) Suppose that Liedtke knew he would be able to obtain perfect information only after he has made his current decision but before he would have to respond to a potential Texaco counteroffer of $3 billion. What would be EVPI in this case?

c In part b, EVPI for (ii) should be less than EVPI calculated in (i). Can you explain why? (Incidentally, if your results disagree with this, you should check your calculations!)

12.12 In the Texaco-Pennzoil case, what is EVPI if Liedtke can learn both Texaco's reaction and the final court decision before he makes up his mind about the current $2 billion offer? (*Hint:* Your answer should be more than the sum of the EVPIs for Texaco's reaction and the court decision calculated separately in Problem 12.11.) Can you explain why the interaction of the two bits of information should have this effect?

12.13   In Problem 5.9, assume that the grower's loss incurred with the burners would be $17,500 and that the loss incurred with the sprinklers would be $27,500.

    a   Find EVPI for the weather conditions (freeze or not).

    b   Now assume that the loss incurred with the burners is uniformly distributed between $15,000 and $20,000. Also assume that the loss incurred with the sprinklers is uniformly distributed between $25,000 and $30,000. Now estimate EVPI regarding these losses, under the assumption that a better weather forecast cannot be obtained.

    c   Do you think the farmer should put more effort into learning his costs more precisely or should he concentrate on obtaining better weather forecasts?

12.14   Reconsider the Eagle Airlines example from Chapter 5, particularly as diagrammed in Figure 5.7. Assume that q and r are both 0.5. Calculate EVPI for each of the three uncertain events individually. What can you conclude from your analysis?

12.15   In Exercise 12.4, is it necessary to assume that the events are independent? What other assumption could be made?

CASE STUDIES

TEXACO-PENNZOIL REVISITED

Often when we face uncertainty, we would like more than simply to know the outcome; it would be nice to control the outcome and bring about the best possible result! A king might consult his wizard as well as his clairvoyant, asking the wizard to cast a spell to cause the desired outcome to occur. How much should such a wizard be paid for these services? The expected value of his wizardry naturally depends on the decision problem at hand, just as the expected value of information does. But the way to calculate the "expected value of wizardry" (to use Ron Howard's term) is very similar to solving the calculations for the expected value of perfect information.

To demonstrate this idea, we again will examine Hugh Liedtke's decision situation as diagrammed in Figure 4.2. Now consider the entirely hypothetical possibility that Liedtke could pay someone to influence Texaco's CEO, James Kinnear.

Questions

1   What would be the most desirable outcome from the "Texaco Reaction" chance node?

2   Construct the decision tree now with three alternatives: "Accept $2 Billion," "Counteroffer $5 Billion," and "Counteroffer $5 Billion and Influence Kinnear."

3   Solve your decision tree from Question 2. What is the maximum amount that Liedtke could afford to pay in order to influence Kinnear?

MEDICAL TESTS

One of the principles that arises from a decision-analysis approach to valuing information is that information is worthless if no possible informational outcome will change the decision. For example, suppose that you are considering whether to make a particular investment. You are tempted to hire a consultant recommended by your Uncle Jake (who just went bankrupt last year) to help you analyze the decision. If, however, you think carefully about the things that the consultant might say and conclude that you would (or would not) make the investment regardless of the consultant's recommendation, then you should not hire the consultant. This principle makes perfectly good sense in the light of our approach; do not pay for information that cannot possibly change your mind.

In medicine, however, it is standard practice for physicians to order extensive batteries of tests for patients. Although different kinds of patients may be subjected to different overall sets of tests, it is nevertheless the case that many of these tests provide information that is worthless in a decision-analysis sense; the doctor's prescription would be the same regardless of the outcome of a particular test.

Questions

1   As a patient, would you be willing to pay for such tests? Why or why not?

2   What incentives do you think the doctor might have for ordering such tests, assuming he realizes that his prescription would not change?

3   How do his incentives compare to yours?

DUMOND INTERNATIONAL, PART II

[Refer back to the DuMond International case study at the end of Chapter 5 (pages 211–213).] Nancy Milnor had returned to her office, still concerned about the decision. Yes, she had persuaded the directors that their disagreements did not affect her analysis; her analysis still showed the new product to be the appropriate choice. The members of the board, however, had not been entirely satisfied. The major complaint was that there was still too much uncertainty. Could she find out more about the likelihood of a ban, or could she get a better assessment from engineering regarding the delay? What about a more accurate sales forecast for the new product?

Nancy gazed at her decision tree (Figure 5.28). Yes, she could address each of those questions, but where should she start?

Question

1   Calculate EVPI for the three uncertain events in DuMond's decision as diagrammed in Figure 5.28. Where should Nancy Milnor begin her investigation?

REFERENCES

This chapter has focused primarily on the technical details of calculating the value of information. The most complete, and most highly technical, reference for this kind of analysis is Raiffa and Schlaifer (1961). Winkler (1972) provides an easily readable discussion of value of information. Both texts contain considerably more material than what is included here, including discussion of EVPI and EVII for continuous distributions. In particular, they delve into questions of the value of sample information for updating natural conjugate prior distributions for model parameters (Chapter 10 in this text) and the selection of appropriate sample sizes, a decision problem intimately related to value of information. Another succinct and complete reference is Chapter 8 in LaValle (1978).

Three articles by LaValle (1968a, b, c) contain many results regarding the expected value of information, including the result that the expected value of information about multiple events need not equal the sum of the values for individual events. Ponssard (1976) and LaValle (1980) show that expected value of information can be negative in some competitive situations. Clemen and Winkler (1985) discuss the value of information from experts, especially focusing on the effect of probabilistic dependence among experts.

At the end of the chapter, we discussed the way in which the value of information can be related to sensitivity analysis and the decision-analysis process. These ideas are Ron Howard's and have been part of his description of decision analysis since the early 1960s. Many of the articles in Howard and Matheson (1983) explain this process, and illustrative applications show how the process has been applied in a variety of real-world decision problems.

Clemen, R., and R. Winkler (1985) "Limits for the Precision and Value of Information from Dependent Sources." *Operations Research,* 33, 427–442.

Howard, R. A., and J. E. Matheson (eds.) (1983) *The Principles and Applications of Decision Analysis* (2 volumes). Palo Alto, CA: Strategic Decisions Group.

LaValle, I. (1968a) "On Cash Equivalents and Information Evaluation in Decisions under Uncertainty; Part I. Basic Theory." *Journal of the American Statistical Association,* 63, 252–276.

LaValle, I. (1968b) "On Cash Equivalents and Information Evaluation in Decisions under Uncertainty; Part II. Incremental Information Decisions." *Journal of the American Statistical Association,* 63, 277–284.

LaValle, I. (1968c) "On Cash Equivalents and Information Evaluation in Decisions under Uncertainty; Part III. Exchanging Partition-*J* for Partition-*K* Information." *Journal of the American Statistical Association,* 63, 285–290.

LaValle, I. (1978) *Fundamentals of Decision Analysis.* New York: Holt.

LaValle, I. (1980) "On Value and Strategic Role of Information in Semi-Normalized Decisions." *Operations Research,* 28, 129–138.

Ponssard, J. (1976) "On the Concept of the Value of Information in Competitive Situations." *Management Science,* 22, 737–749.

Raiffa, H., and R. Schlaifer (1961) *Applied Statistical Decision Theory.* Cambridge, MA: Harvard University Press.

Winkler, R. L. (1972) *Introduction to Bayesian Inference and Decision.* New York: Holt.

3

Modeling Preferences

We have come a long way since the first chapters. The first part of the book talked about structuring problems, and the second discussed modeling uncertainty through the use of probability. Now we turn to the problem of modeling preferences.

Why should we worry about modeling preferences? Because virtually every decision involves some kind of trade-off. In decision making under uncertainty, the fundamental trade-off question often is, How much risk is a decision maker willing to assume? After all, expected monetary value is not everything! Often the alternative that has the greatest EMV also involves the greatest risk.

Chapters 13 and 14 look at the role of risk attitudes in decision making. In Chapter 13, basic concepts are presented, and you will learn how to model your own risk attitude. We will develop the concept of a utility function. Modeling your preferences by assessing your utility function is a subjective procedure much like assessing subjective probabilities. Because a utility function incorporates a decision maker's attitude toward risk, the decision maker may decide to choose the alternative that maximizes his or her expected utility rather than expected monetary value.

Chapter 14 discusses some of the foundations that underlie the use of utility functions. The essential reason for choosing alternatives to maximize expected utility is that such behavior is consistent with some fundamental choice and behavior patterns that we call axioms. The paradox is that, even though most of us agree that intuitively the axioms are reasonable, there are cases for all of us when our actual choices are not consistent with the axioms. In many situations these inconsistencies have little effect on a decision maker's choices. But occasionally they can cause trouble, and we will discuss some of these difficulties and their implications.

Dealing with risk attitudes is an important aspect of decision making under uncertainty, but it is only part of the picture. As we discussed in Section 1, many problems involve conflicting objectives. Decision makers must balance many different aspects of the problem, and try to accomplish many things at once. Even a simple decision such as deciding where to go for dinner involves trade-offs: How far are you willing to drive? How much should you spend? How badly do you want Chinese food?

Chapters 15 and 16 deal with modeling preferences in situations in which the decision maker has multiple and conflicting objectives. In both chapters, one of the fundamental subjective assessments that the decision maker must make is how to trade off achievement in one dimension against achievement in another. Chapter 15 presents a relatively straightforward approach that is easy and intuitive, extending the introductory approach presented in Chapters 3 and 4. Chapter 16 extends the discussion to include interaction among utility attributes.

CHAPTER 13

Risk Attitudes

T his chapter marks the beginning of our in-depth study of preferences. Before we begin, let us review where we have been and think about where we are going. The first six chapters provided an introduction to the process of structuring decision problems for decision analysis and an overview of the role that probability and utility theory play in making choices. Chapters 7 through 12 have focused on probability concerns: using probability in a variety of ways to model uncertainty in decision problems, including the modeling of information sources in value-of-information problems.

At this point, we change directions and look at the preference side of decision analysis. How can we model a decision maker's preferences? This chapter looks at the problems associated with risk and return trade-offs. Chapter 14 briefly explores the axiomatic foundations of utility theory and discusses certain paradoxes from cognitive psychology. These paradoxes generally indicate that people do not make choices that are perfectly consistent with the axioms, even though they may agree that the axioms are reasonable! Although such inconsistencies generally do not have serious implications for most decisions, there are certain occasions when they can cause difficulty.

The primary motivating example for this chapter comes from the history of railways in the United States. Imagine what might have gone through E. H. Harriman's mind as he considered his strategy for acquiring the Northern Pacific Railroad in March 1901.

E. H. HARRIMAN FIGHTS FOR THE NORTHERN PACIFIC RAILROAD

"How could they do it?" E. H. Harriman asked, still angry over the fact that James Hill and J. P. Morgan had bought the Burlington Railroad out from under his nose. "Every U.S. industrialist knows I control the railroads in the West. I have the Illinois Central, the Union Pacific, the Central and Southern Pacific, not to mention the Oregon Railroad and Navigation Company. Isn't that true?"

"Yes, sir," replied his assistant.

"Well, we will put the pressure on Messrs. Hill and Morgan. They will be surprised indeed to find out that I have acquired a controlling interest in their own railroad, the Northern Pacific. I may even be able to persuade them to let me have the Burlington. By the way, how are the stock purchases going?"

"Sir, we have completed all of the purchases that you authorized so far. You may have noticed that our transactions have driven the price of Northern Pacific stock up to more than $100 per share."

Harriman considered this information. If he bought too fast, he could force the stock price up high enough and fast enough that Hill might begin to suspect that Harriman was up to something. Of course, if Harriman could acquire the shares quickly enough, there would be no problem. On the other hand, if he bought the shares slowly, he would pay lower prices, and Hill might not notice the acquisition until it was too late. His assistant's information, however, suggested that his situation was somewhat risky. If Harriman's plan were discovered, Hill could persuade Morgan to purchase enough additional Northern Pacific shares to enable them to retain control. In that case, Harriman would have paid premium prices for the stock for nothing! On the other hand, if Hill did not make the discovery immediately, the triumph would be that much sweeter.

"How many more shares do we need to have control?" asked Harriman.

"If you could purchase another 40,000 shares, sir, you would own 51 percent of the company."

Another 40,000 shares. Harriman thought about giving Hill and Morgan orders on how to run their own railroad. How enjoyable that would be! Yes, he would gladly increase his investment by that much.

"Of course," his assistant continued, "if we try to purchase these shares immediately, the price will rise very quickly. You will probably end up paying an additional $15 per share above what you would pay if we were to proceed more slowly."

"Well, $600,000 is a lot of money, and I certainly would not want to pay more. But it would be worth the money to be sure that we would be able to watch Hill and Morgan squirm! Send a telegram to my broker in New York right away to place the order. And be quick! It's already Friday. If we are going to do this, we need to do it today. I don't want Hill to have the chance to think about this over the weekend."

Risk

Basing decisions on expected monetary values (EMVs) is convenient, but it can lead to decisions that may not seem intuitively appealing. For example, consider the following two games. Imagine that you have the opportunity to play one game or the other, but only one time. Which one would you prefer to play? Your choice also is drawn in decision-tree form in Figure 13.1.

Game 1	Win $30 with probability 0.5
	Lose $1 with probability 0.5
Game 2	Win $2000 with probability 0.5
	Lose $1900 with probability 0.5

Game 1 has an expected value of $14.50. Game 2, on the other hand, has an expected value of $50.00. If you were to make your choice on the basis of expected value, then you would choose Game 2. Most of us, however, would consider Game 2 to be riskier than Game 1, and it seems reasonable to suspect that most people actually would prefer Game 1.

Using expected values to make decisions means that the decision maker is considering only the average or expected payoff. If we take a long-run frequency approach, the expected value is the average amount we would be likely to win over many plays of the game. But this ignores the range of possible values. After all, if we play each game 10 times, the worst we could do in Game 1 is to lose $10. On the other hand, the worst we could do in Game 2 is lose $19,000!

Many of the examples and problems that we have considered so far have been analyzed in terms of expected monetary value (EMV). EMV, however, does not capture risk attitudes. For example, consider the Texaco-Pennzoil example in Chapter 4 (pages 111–112). If Hugh Liedtke were afraid of the prospect that Pennzoil could end

Figure 13.1

Two lottery games. Which game would you choose?

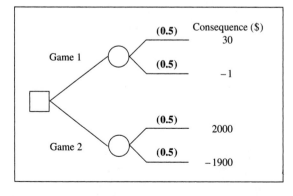

up with nothing at the end of the court case, he might be willing to take the $2 billion that Texaco offered. To consider the Eagle Airlines case (Chapter 5, pages 174–175), purchasing the airplane is a much riskier alternative than leaving the money in the bank. If Carothers were sensitive to risk, he might prefer to leave the money in the bank. Even someone like E. H. Harriman considered the riskiness of the situations in which he found himself. In our example, Harriman weighed the value of a riskless alternative (immediately purchasing the 40,000 shares that were required to gain control) against the risky alternative of not purchasing the shares and the possible outcomes that might then follow. Even though all of the dollar amounts were not specified, it is clear that Harriman was not thinking in terms of EMV.

Individuals who are afraid of risk or are sensitive to risk are called *risk-averse*. We could explain risk aversion if we think in terms of a *utility function* (Figure 13.2) that is curved and opening downward (the technical term for a curve with this shape is *concave*). This utility function represents a way to translate dollars into "utility units." That is, if we take some dollar amount *(x)*, we can locate that amount on the horizontal axis. Read up to the curve and then horizontally across to the vertical axis. From that point we can read off the utility value U(x) for the dollars we started with.

A utility function might be specified in terms of a graph, as in Figure 13.2, or given as a table, as in Table 13.1. A third form is a mathematical expression. If graphed, for example, all of the following expressions would have the same general concave shape (opening downward) as the utility function graphed in Figure 13.2:

$$U(x) = \log(x)$$
$$U(x) = 1 - e^{-x/R}$$
$$U(x) = +\sqrt{x} \qquad [\text{or } U(x) = x^{0.5}]$$

Of course, the utility and dollar values in Table 13.1 also could be graphed, as could the functional forms shown above. Likewise, the graph in Figure 13.2 could be con-

Figure 13.2

A utility function that displays risk aversion.

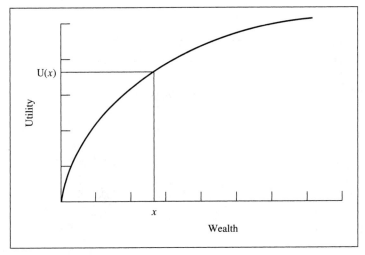

Table 13.1

A utility function in
tabular form.

Wealth	Utility Value
2500	1.50
1500	1.24
1000	0.93
600	0.65
400	0.47
0	0.15

verted into a table of values. The point is that the utility function makes the translation from dollars to utility regardless of its displayed form.

Risk Attitudes

We think of a typical utility curve as (1) upward sloping and (2) concave (the curve opens downward). An upward-sloping utility curve makes fine sense; it means that more wealth is better than less wealth, everything else being equal. Few people will argue with this. Concavity in a utility curve implies that an individual is risk-averse.

Imagine that you are forced to play the following game:

Win $500 with probability 0.5

Lose $500 with probability 0.5

Would you pay to get out of this situation? How much? The game has a zero expected value, so if you would pay something to get out, you are avoiding a risky situation with zero expected value. Generally, if you would trade a gamble for a sure amount that is less than the expected value of the gamble, you are risk-averse. Purchasing insurance is an example of risk-averse behavior. Insurance companies analyze a lot of data in order to understand the probability distributions associated with claims for different kinds of policies. Of course, this work is costly. To make up these costs and still have an expected profit, an insurance company must charge more for its insurance policy than the policy can be expected to produce in claims. Thus, unless you have some reason to believe that you are more likely than others in your risk group to make a claim, you probably are paying more in insurance premiums than the expected amount you would claim.

Not everyone displays risk-averse behavior all the time, and so utility curves need not be concave. A convex (opening upward) utility curve indicates risk-seeking behavior (Figure 13.3). The risk seeker might be eager to enter into a gamble; for example, he or she might pay to play the game just described. An individual who plays a state lottery exhibits risk-seeking behavior. State lottery tickets typically cost $1.00 and have an expected value of approximately 50 cents.

Figure 13.3

Three different shapes for utility functions.

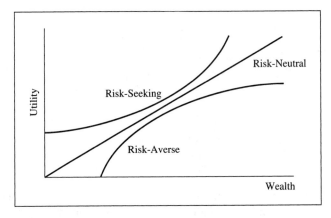

Finally, an individual can be risk-neutral. Risk neutrality is reflected by a utility curve that is simply a straight line. For this type of person, maximizing EMV is the same as maximizing expected utility. This makes sense; someone who is risk-neutral does not care about risk and can ignore risk aspects of the alternatives that he or she faces. Thus, EMV is a fine criterion for choosing among alternatives, because it also ignores risk.

Although most of us are not risk-neutral, it often is reasonable for a decision maker to assume that his or her utility curve is nearly linear in the range of dollar amounts for a particular decision. This is especially true for large corporations that make decisions involving amounts which are small relative to their total assets. In many cases, it may be worthwhile to use EMV in a first-cut analysis, and then check to see whether the decision would be sensitive to changes in risk attitude. If the decision turns out to be fairly sensitive (that is, if the decision would change for a slightly risk-averse or slightly risk-seeking person), then the decision maker may want to consider modeling his or her risk attitude carefully.

This discussion makes it sound as though individuals can be reduced to their utility functions, and those utility functions can reveal whether the individual is risk-averse or risk-seeking. Keep in mind, however, that the utility function is only a model of an individual's attitude toward risk. Moreover, our development of utility functions in this chapter is intended to help with the modeling of risk attitudes at a fundamental level, and our model may not be able to capture certain complicated psychological aspects. For example, some individuals may be extremely frightened by risk. Others may find that small wagers greatly increase their enjoyment in watching a sporting event, for example. Still others may find that waiting for the uncertainty to be resolved is a source of excitement and exhilaration, although concern about losing money is a source of anxiety. For some people, figuring out exactly what their feelings are toward risky alternatives may be extremely complicated and may depend on the amount at stake, the context of the risk, and the time horizon.

Investing in the Stock Market, Revisited

If we have a utility function that translates from dollars to utility, how should we use it? The whole idea of a utility function is that it should help to choose from among alternatives that have uncertain payoffs. Instead of maximizing expected value, the decision maker should maximize expected utility. In a decision tree or influence-diagram payoff table, the net dollar payoffs would be replaced by the corresponding utility values and the analysis performed using those values. The best choice then should be the action with the highest expected utility.

As an example, let us reconsider the stock market–investment example from Chapters 5 and 12. You will recall that an investor has funds that he wishes to invest. He has three choices: a high-risk stock, a low-risk stock, or a savings account that would pay $500. If he invests in the stocks, he must pay a $200 brokerage fee.

With the two stocks his payoff depends on what happens to the market. If the market goes up, he will earn $1700 from the high-risk stock and $1200 from the low-risk stock. If the market stays at the same level, his payoffs for the high- and low-risk stocks will be $300 and $400, respectively. Finally, if the stock market goes down, he will lose $800 with the high-risk stock but still earn $100 from the low-risk stock. The probabilities that the market will go up, stay the same, or go down are 0.5, 0.3, and 0.2, respectively.

Figure 13.4 shows his decision tree, including the brokerage fee and the payoffs for the two stocks under different market conditions. Note that the values at the ends of the branches are the *net* payoffs, taking into account both the brokerage fee and the investment payoff. Table 13.2 gives his utility function.

We already calculated the expected values of the three investments in Chapter 12. They are

$$\text{EMV (High-Risk Stock)} = 580$$

$$\text{EMV (Low-Risk Stock)} = 540$$

$$\text{EMV (Savings Account)} = 500$$

Figure 13.4

Decision tree for the stock market investor.

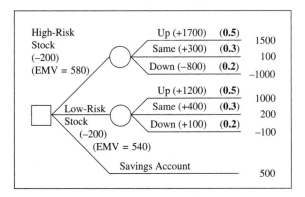

Table 13.2

Utility function for the
investment problem.

Dollar Value	Utility Value
1500	1.00
1000	0.86
500	0.65
200	0.52
100	0.46
-100	0.33
-1000	0.00

As a result, an expected-value maximizer would choose the high-risk stock.

Figure 13.5 shows the investor's decision tree with the utility values instead of the payoffs. Solving this decision tree, we calculate the expected utility (EU) for the three investments:

$$EU \text{ (High-Risk Stock)} = 0.638$$

$$EU \text{ (Low-Risk Stock)} = 0.652$$

$$EU \text{ (Savings Account)} = 0.650$$

Now the preferred action is to invest in the low-risk stock because it provides the highest expected utility, although it does not differ much from that for the savings account. You can see how the expected utilities make it possible to rank these investments in order of preference. According to the utility function we are using, this investor dislikes risk enough to find the high-risk stock the least preferred of his three alternatives.

Figure 13.5

Decision tree for stock market investor—utility values instead of dollars.

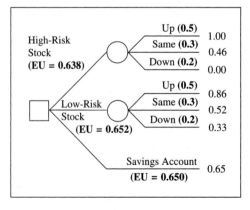

Expected Utility, Certainty Equivalents, and Risk Premiums

Two concepts are closely linked to the idea of expected utility. One is that of a *certainty equivalent,* or the amount of money that is equivalent in your mind to a given situation that involves uncertainty. For example, suppose you face the following gamble:

Win $2000 with probability 0.50

Lose $20 with probability 0.50

Now imagine that one of your friends is interested in taking your place. "Sure," you reply, "I'll sell it to you." After thought and discussion, you conclude that the least you would sell your position for is $300. If your friend cannot pay that much, then you would rather keep the gamble. (Of course, if your friend were to offer more, you would take it!)

Your certainty equivalent for the gamble is $300. This is a sure thing; no risk is involved. From this, the meaning of certainty equivalent becomes clear. If $300 is the least that you would accept for the gamble, then the gamble must be equivalent in your mind to a sure $300.

In the example at the beginning of the chapter, Harriman decided that he would pay the additional $600,000 simply to avoid the riskiness of the situation. His thinking at the time was that committing the additional money would ensure his control of the Northern Pacific Railroad. He indicated that he did not want to pay more, and so we can think of $600,000 as his certainty equivalent for the gamble of purchasing the shares more slowly and risking detection.

Let us again consider the stock market investor. We can make certain inferences about his certainty equivalent for the gambles represented by the low-risk and high-risk stocks because we have information about his utility function. For example, his expected utility for the low-risk stock is 0.652, which is just a shade more than U($500) = 0.650. Thus, his certainty equivalent for the low-risk stock must be only a little more than $500. Likewise, his expected utility for the high-risk stock is 0.638, which is somewhat less than 0.650. Therefore, his certainty equivalent for the high-risk stock must be less than $500 but not as little as $200, which has a utility of 0.520.

You can see also that we can rank the investments by their certainty equivalents. The high-risk stock, having the lowest certainty equivalent, is the least preferred. The low-risk stock, on the other hand, has the highest certainty equivalent, and so is the most preferred. Ranking alternatives by their certainty equivalents is the same as ranking them by their expected utilities. If two alternatives have the same certainty equivalent, then they must have the same expected utility, and the decision maker would be indifferent to a choice between the two.

Closely related to the idea of a certainty equivalent is the notion of *risk premium.* The risk premium is defined as the difference between the EMV and the certainty equivalent:

Risk Premium = EMV – Certainty Equivalent

Consider the gamble between winning $2000 and losing $20, each with probability 0.50. The EMV of this gamble is $990. On reflection, you assessed your certainty equivalent to be $300, and so your risk premium is

$$\text{Risk Premium} = \$990 - \$300$$
$$= \$690$$

Because you were willing to trade the gamble for $300, you were willing to "give up" $690 in expected value in order to avoid the risk inherent in the gamble. You can think of the risk premium as the premium you pay (in the sense of a lost opportunity) to avoid the risk.

Figure 13.6 graphically ties together utility functions, certainty equivalents, and risk premiums. Notice that the certainty equivalent and the expected utility of a gamble are points that are "matched up" by the utility function. That is,

$$\text{EU(Gamble)} = \text{U(Certainty Equivalent)}$$

In words, the utility of the certainty equivalent is equal to the expected utility of the gamble. Because these two quantities are equal, the decision maker must be indifferent to the choice between them. After all, that is the meaning of certainty equivalent.

Now we can put all of the pieces together in Figure 13.6. Imagine a gamble has expected utility Y. The value Y is in utility units, so we must first locate Y on the vertical axis. Trace a horizontal line from the expected utility point until the line intersects the utility curve. Now drop down to the horizontal axis to find the certainty equivalent. The difference between the expected value and the certainty equivalent is the risk premium.

Figure 13.6

Graphical representation of risk premium.

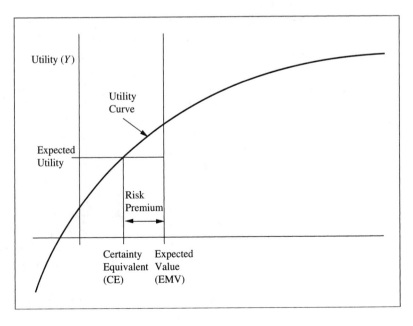

For a risk-averse individual, the horizontal EU line reaches the concave utility curve before it reaches the vertical line that corresponds to the expected value. Thus, for a risk-averse individual the risk premium must be positive. If the utility function were convex, the horizontal EU line would reach the expected value before the utility curve. The certainty equivalent would be greater than the expected value, and so the risk premium would be negative. This would imply that the decision maker would have to be paid to give up an opportunity to gamble.

In any given situation, the certainty equivalent, expected value, and risk premium all depend on two factors: the decision maker's utility function and the probability distribution for the payoffs. The values that the payoff can take combine with the probabilities to determine the EMV. The utility function, coupled with the probability distribution, determines the expected utility and hence the certainty equivalent. The degree to which the utility curve is nonlinear determines the distance between the certainty equivalent and the expected payoff.

If the certainty equivalent for a gamble is assessed directly, then finding the risk premium is straightforward—simply calculate the EMV of the gamble and subtract the assessed certainty equivalent. In other cases, the decision maker may have assessed a utility function and now faces a particular gamble that he or she wishes to analyze. If so, there are four steps in finding the gamble's risk premium:

1 Find the EU for the gamble.

2 Find the certainty equivalent, or the sure amount that has the utility value equal to the EU that was found in Step 1.

3 Calculate the EMV for the gamble.

4 Subtract the certainty equivalent from the expected payoff to find the risk premium. This is the difference between the expected value of the risky situation and the sure amount for which the risky situation would be traded.

Here is a simple example. Using the hypothetical utility function given in Figure 13.7, we will find the risk premium for the following gamble:

Win $4000 with probability	0.40
Win $2000 with probability	0.20
Win $0 with probability	0.15
Lose $2000 with probability	0.25

The first step is to find the expected utility:

$$EU = 0.40 \ U(\$4000) + 0.20 \ U(\$2000) + 0.15 \ U(\$0) + 0.25 \ U(-\$2000)$$

$$= 0.40(0.90) + 0.20(0.82) + 0.15(0.67) + 0.25(0.38)$$

$$= 0.72$$

The second line is simply a matter of estimating the utilities from Figure 13.7 and substituting them into the equation.

Figure 13.7

Utility function for a risk-averse decision maker.

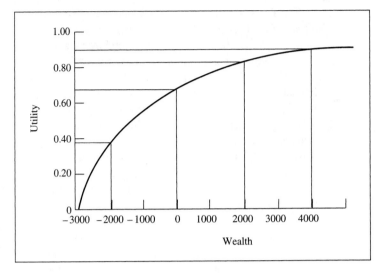

Figure 13.8

Finding a certainty equivalent.

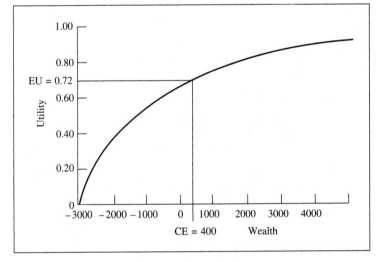

For Step 2, the certainty equivalent is the sure amount that gives the same utility as the expected utility of the gamble. Figure 13.8 shows the process of finding the certainty equivalent for the gamble that has EU = 0.72. We start at the vertical axis with the utility value of 0.72, read across to the utility curve, and then drop down to the horizontal axis. From Figure 13.8, we can see that the certainty equivalent is approximately $400.

Step 3 calculates the expected payoff or EMV:

$$EMV = 0.40(\$4000) + 0.20(\$2000) + 0.15(\$0) + 0.25(-\$2000)$$
$$= \$1500$$

Finally, in Step 4, we calculate the risk premium by subtracting the certainty equivalent from the EMV:

$$\text{Risk Premium} = \$1500 - \$400 = \$1100$$

Keeping Terms Straight

One problem that students often have is that of confusion about the terms we use in utility theory. The basic idea, remember, is to use a utility function to translate dollars into utility units. If we compare two risky projects on the basis of expected utility (EU), we are working in utility units. When we calculate certainty equivalents or risk premiums, however, we are working in dollars. Thus, a certainty equivalent is not the same as the expected utility of a gamble. The two measurements provide equivalent information, but only in the sense that the certainty equivalent for a gamble is the sure amount that gives the same utility as the expected utility of the gamble. The translation from certainty equivalent to expected utility and back again is through the utility function, as depicted in Figures 13.6 and 13.8. Again, a certainty equivalent is a dollar amount, whereas expected utility is in utility units. Be careful to use these terms consistently.

Utility Function Assessment

Different people have different risk attitudes and thus are willing to accept different levels of risk. Some are more prone to taking risks, while others are more conservative and avoid risk. Thus, assessing a utility function is a matter of subjective judgment, just like assessing subjective probabilities. In this section we will look at two utility-assessment approaches that are based on the idea of certainty equivalents. The following section introduces an alternative approach.

It is worth repeating at this point our credo about modeling and decision making. Remember that the objective of the decision-analysis exercise is to help you make a better decision. To do this, we construct a model, or representation, of the decision. When we assess a utility function, we are constructing a mathematical model or representation of preferences. This representation then is included in the overall model of the decision problem and is used to analyze the situation at hand. The objective is to find a way to represent preferences that incorporates risk attitudes. A perfect representation is not necessary. All that is required is a model that represents feelings about risk well enough to understand and analyze the current decision.

Assessment Using Certainty Equivalents

The first assessment method requires the decision maker to assess several certainty equivalents. Suppose you face an uncertain situation in which you may have $10 in the worst case, $100 in the best case, or possibly something in between. You have a variety of options, each of which leads to some uncertain payoff between $10 and $100. To evaluate the alternatives, you must assess your utility for payoffs from $10 to $100.

We can get the first two points of your utility function by arbitrarily setting U(100) = 1 and U(10) = 0. This may seem a bit strange but is easily explained. The idea of the utility function, remember, is to rank-order risky situations. We can always take any utility function and rescale it—add a constant and multiply by a positive constant—so that the best outcome has a utility of 1 and the worst has a utility of 0. The rank ordering of risky situations in terms of expected utility will be the same for both the original and rescaled utility functions. What we are doing here is taking advantage of this ability to rescale. We are beginning the assessment process by setting two utility points. The remaining assessments then will be consistent with the scale set by these points. (We could just as well set the endpoints at 100 and 0, or 100 and –50, say. We are using 1 and 0 because this choice of endpoints turns out to be particularly convenient. But we will have to be careful not to confuse these utilities with probabilities!)

Now imagine that you have the opportunity to play the following lottery, which we will call a *reference lottery:*

Win $100 with probability 0.5

Win $10 with probability 0.5

What is the minimum amount for which you would be willing to sell your opportunity to play this game? $25? $30? Your job is to find your certainty equivalent (CE) for this reference gamble. A decision tree for your choice is shown in Figure 13.9.

Finding your certainty equivalent is where your subjective judgment comes into play. The CE undoubtedly will vary from person to person. Suppose that for this reference gamble your certainty equivalent is $30. That is, for $31 you would take the money, but for $29 you would rather play the lottery; $30 must be your true indifference point.

Figure 13.9

A "reference gamble" for assessing a utility function. Your job is to find the certainty equivalent (CE) so that you are indifferent to options A and B.

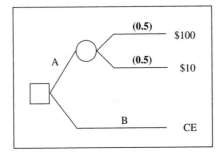

The key to the rest of the analysis is this: *Because you are indifferent between $30 and the risky gamble, the utility of $30 must equal the expected utility of the gamble.* We know the utilities of $10 and $100, so we can figure out the expected utility of the gamble:

$$U(30) = 0.5\ U(100) + 0.5\ U(10)$$
$$= 0.5(1) + 0.5(0)$$
$$= 0.5$$

We have found a third point on your utility curve. To find another, take a different reference lottery:

Win $100 with probability 0.5

Win $30 with probability 0.5

Now find your certainty equivalent for this new gamble. Again, the certainty equivalent will vary from person to person, but suppose that you settle on $50. We can do exactly what we did before, but with the new gamble. In this case, we can find the utility of $50 because we know U(100) and U(30) from the previous assessment.

$$U(50) = 0.5\ U(100) + 0.5\ U(30)$$
$$= 0.5(1) + 0.5(0.5)$$
$$= 0.75$$

This is the fourth point on your utility curve.

Now consider the reference lottery:

Win $30 with probability 0.5

Win $10 with probability 0.5

Again you must assess your CE for this gamble. Suppose it turns out to be $18. Now we can do the familiar calculations:

$$U(18) = 0.5\ U(30) + 0.5\ U(10)$$
$$= 0.5(0.5) + 0.5(0)$$
$$= 0.25$$

We now have five points on your utility curve, and we can graph and draw a curve through them. The graph is shown in Figure 13.10. A smooth curve drawn through the assessed points should be an adequate representation of your utility function for use in solving your decision problem.

Assessment Using Probabilities

The CE approach requires that you find a dollar amount that makes you indifferent between the gamble and the sure thing in Figure 13.9. Another approach involves setting the sure amount in Alternative B and adjusting the probability in the reference gamble to achieve indifference. We will call this the *probability-equivalent* (PE) assessment technique.

For example, suppose you want to know your utility for $65. This is not one of the certainty equivalents that you assessed, and thus U(65) is unknown. You could make an educated guess. Based on the previous assessments and the graph in Figure 13.10, U(65) must be between 0.75 and 1.00; it probably is around 0.85. But rather than guess, you can assess the value directly. Consider the reference lottery:

Win $100 with probability p

Win $10 with probability $(1 - p)$

This gamble is shown in Figure 13.11.

To find your utility value for $65, adjust p until you are indifferent between the sure $65 and the reference gamble. That is, think about various probabilities that make the chance of winning $100 greater or less until you are indifferent between Alternatives C and D in Figure 13.11. Now you can find U(65) because you know that U(100) = 1 and U(10) = 0:

$$U(65) = p \, U(100) + (1 - p) \, U(10)$$
$$= p(1) + (1 - p)(0)$$
$$= p$$

Figure 13.10

Graph of the utility function assessed using the certainty-equivalent approach.

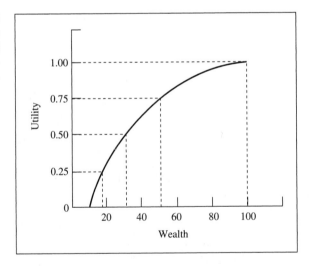

Figure 13.11

A reference gamble for assessing the utility of $65 using the probability-equivalent method.

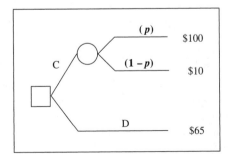

The probability that makes you indifferent just happens to be your utility value for $65. For example, if you chose $p = 0.87$ to achieve indifference, then U(65) = 0.87.

Gambles, Lotteries, and Investments

As with the assessment of subjective probabilities, we have framed utility assessment in terms of gambles or lotteries. For many individuals this evokes images of carnival games or gambling in a casino, images that may seem irrelevant to the decision at hand or even distasteful. An alternative is to think in terms of investments that are risky. Instead of considering if you would accept or reject a particular gamble, think about whether you would make a particular investment. Would you agree to invest in a security with a p chance of yielding $1000 and a $1 - p$ chance of yielding nothing? How much would you be willing to pay for such a security? Framing your utility-assessment questions in this way may help you to think more clearly about your utility, especially for investment decisions.

Risk Tolerance and the Exponential Utility Function

The assessment process just described works well for assessing a utility function subjectively, and it can be used in any situation, although it can involve a fair number of assessments. An alternative approach is to base the assessment on a particular mathematical function, such as one of those that we introduced early in the chapter. In particular, let us consider the *exponential utility function:*

$$U(x) = 1 - e^{-x/R}$$

This utility function is based on the constant e = 2.71828 . . . , the base of natural logarithms. This function is concave and thus can be used to represent risk-averse preferences. As x becomes large, U(x) approaches 1. The utility of zero, U(0), is equal to 0, and the utility for negative x (being in debt) is negative.

In the exponential utility function, R is a parameter that determines how risk-averse the utility function is. In particular, R is called the *risk tolerance.* Larger values of R make the exponential utility function flatter, while smaller values make it more concave or more risk-averse. Thus, if you are less risk-averse—if you can tolerate more risk—you would assess a larger value for R to obtain a flatter utility function. If you are less tolerant of risk, then you would assess a smaller R and have a more curved utility function.

How can R be determined? A variety of ways exist, but it turns out that R has a very intuitive interpretation that makes its assessment relatively easy. Consider the gamble

Win $$Y$ with probability 0.5

Lose $$Y$/2 with probability 0.5

Figure 13.12

Assessing your risk tolerance. Find the largest value of Y for which you would prefer Alternative E.

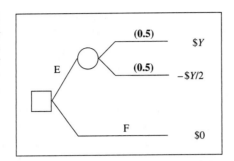

Would you be willing to take this gamble if Y were $100? $2000? $35,000? Or, framing it as an investment, how much would you be willing to risk ($Y/2) in order to have a 50% chance of tripling your money (winning $Y and keeping your $Y/2)? At what point would the risk become intolerable? The decision tree is shown in Figure 13.12.

The largest value of Y for which you would prefer to take the gamble rather than not take it is approximately equal to your risk tolerance. This is the value that you can use for R in your exponential utility function. For example, suppose that after considering the decision tree in Figure 13.12 you conclude that the largest Y for which you would take the gamble is $Y = \$900$. Hence, $R = \$900$. Using this assessment in the exponential utility function would result in the utility function

$$U(x) = 1 - e^{-x/900}$$

This exponential utility function provides the translation from dollars to utility units.

Once you have your R value and your exponential utility function, it is fairly easy to find certainty equivalents. For example, suppose that you face the following gamble:

Win $2000 with probability 0.4

Win $1000 with probability 0.4

Win $500 with probability 0.2

The expected utility for this gamble is

$$EU = 0.4\ U(\$2000) + 0.4\ U(\$1000) + 0.2\ U(\$500)$$
$$= 0.4(0.8916) + 0.4(0.6708) + 0.2(0.4262)$$
$$= 0.7102$$

To find the CE we must work backward through the utility function. We want to find the value x such that $U(x) = 0.7102$. Set up the equation

$$0.7102 = 1 - e^{-x/900}$$

Subtract 1 from each side to get

$$-0.2898 = -e^{-x/900}$$

Multiply through to eliminate the minus signs:

$$0.2898 = e^{-x/900}$$

Now we can take natural logs of both sides to eliminate the exponential term:

$$\ln(0.2898) = \ln(e^{-x/900}) = \frac{-x}{900}$$

The rule (from algebra) is that $\ln(e^y) = y$. Now we simply solve for x:

$$\ln(0.2898) = \frac{-x}{900}$$
$$x = -900[\ln(0.2898)]$$
$$= \$1114.71$$

The procedure above requires that you use the exponential utility function to translate the dollar outcomes into utilities, find the expected utility, and finally convert to dollars to find the exact certainty equivalent. That can be a lot of work, especially if there are many outcomes to consider. Fortunately, an approximation is available from Pratt (1964) and also discussed in McNamee and Celona (1987). Suppose you can figure out the expected value and variance of the payoffs. Then the CE is approximately

$$\text{Certainty Equivalent} \approx \text{Expected Value} - \frac{0.5(\text{Variance})}{\text{Risk Tolerance}}$$

In symbols,

$$CE \approx \mu - \frac{0.5\sigma^2}{R}$$

where μ and σ^2 are the expected value and variance, respectively. For example, in the gamble above, the expected value (EMV or μ) equals \$1300, and the standard deviation (σ) equals \$600. Thus, the approximation gives

$$CE \approx \$1300 - \frac{0.5(\$600)^2}{900}$$
$$\approx \$1100$$

The approximation is within \$15. That's pretty good! This approximation is especially useful for continuous random variables or problems where the expected value and variance are relatively easy to estimate or assess compared to assessing the entire probability distribution. The approximation will be closest to the actual value when the outcome's probability distribution is a symmetric, bell-shaped curve.

What are reasonable R values? For an individual's utility function, the appropriate value for R clearly depends on the individual's risk attitude. As indicated, the less risk-averse a person is, the larger R is. Suppose, however, that an individual or a group (a board of directors, say) has to make a decision on behalf of a corporation. It is important that these decision makers adopt a decision-making attitude based on corporate goals and acceptable risk levels for the corporation. This can be quite different from an individual's personal risk attitude; the individual director may be unwilling to risk $10 million, even though the corporation can afford such a loss. Howard (1988) suggests certain guidelines for determining a corporation's risk tolerance in terms of total sales, net income, or equity. Reasonable values of R appear to be approximately 6.4% of total sales, 1.24 times net income, or 15.7% of equity. These figures are based on observations that Howard has made in the course of consulting with various companies. More research may refine these figures, and it may turn out that different industries have different ratios for determining reasonable R's.

Using the exponential utility function seems like magic, doesn't it? One assessment, and we are finished! Why bother with all of those certainty equivalents that we discussed above? You know, however, that you never get something for nothing, and that definitely is the case here. The exponential utility function has a specific kind of curvature and implies a certain kind of risk attitude. This risk attitude is called *constant risk aversion.* Essentially it means that no matter how much wealth you have—how much money in your pocket or bank account—you would view a particular gamble in the same way. The gamble's risk premium would be the same no matter how much money you have. Is constant risk aversion reasonable? Maybe it is for some people. Many individuals might be less risk-averse if they had more wealth.

In later sections of this chapter we will study the exponential utility function in more detail, especially with regard to constant risk aversion. The message here is that the exponential utility function is most appropriate for people who really believe that they would view gambles the same way regardless of their wealth level. But even if this is not true for you, the exponential utility function can be a useful tool for modeling preferences. The next section shows how to model risk attitudes in PrecisionTree and how sensitivity analysis can be performed using risk tolerance.

Modeling Preferences Using PrecisionTree

It is straightforward to model a decision maker's risk preferences in PrecisionTree. You need only choose the utility curve, and PrecisionTree does the rest. It converts all end-node monetary values into utility values and calculates both expected utilities and certainty equivalents. You can choose one of the two built-in utility functions (the exponential or logarithmic) or you can define a customized utility curve. The logarithmic function is the easiest to use because it has no parameters, but it is also the least flexible because using it implies the same risk preferences for all decision makers. The exponential function is more flexible because the risk-tolerance parameter (R) allows different risk preferences to be modeled. Defining your own utility curve provides the greatest flexibility, but also requires some programming. We will demonstrate the exponential function using the Eagle Airlines example from Chapter 5.

Recall that Dick Carothers, the president of Eagle Airlines, was deciding whether he should purchase an additional aircraft to expand his operations. After considerable modeling, we worked the problem down to the consideration of three uncertain variables: the capacity of the scheduled flights (proportion of seats sold), the operating cost, and the total number of hours flown during the year. Figure 13.13 shows the decision tree with specific probabilities on the branches. Using these probabilities, we calculate the EMV for purchasing the plane as $9644, which is considerably more than the $4200 that he can earn by investing in the money market. It is also true, however, that purchasing the plane is substantially riskier than investing in the money market. The question is whether Carothers is willing to take on the added risk that accompanies the increased expected value.

Let's assume that Carothers is risk-averse with a risk tolerance of 10,000. Later we will see what happens when we vary the risk-tolerance parameter in a sensitivity analysis.

Step 1

1.1 Either build the decision tree shown in Figure 13.13 or open Examples\Chapter13\EagleAirlines.xls. The supplied worksheet (EagleAirlines.xls) is a linked decision tree in which the values at the end nodes are calculated by an Excel table. You can either construct a linked tree or simply enter the profit levels as the end-node values, as shown in Figure 13.13.

Step 2

2.1 Click on the name **Eagle Airlines** and the *Tree Settings* dialog box opens (Figure 13.14).

2.2 Choose the **Use Utility Function** checkbox.

2.3 For the *Function* option, choose the default **Exponential** and type in **10000** for the *R value*.

2.4 Choose **Expected Utility** for your *Display,* and click **OK**.

We now see the utility values and the expected utilities of each chance node (Figure 13.15). Purchasing the plane has an expected utility of 0.220, which is lower than the expected utility of 0.343 for investing in the money market. Thus, for a risk-tolerance level of 10,000, the risk associated with purchasing the plane outweighs the expected profit of $9644 when compared to a guaranteed gain of $4200.

From the above discussion, we know that the expected profit of $9644 together with the associated risks is less valuable to Carothers than the for-sure gain of $4200. How much less valuable is it? In utility values it is 0.220 compared to 0.343. In dollar values it is the difference between the certainty equivalents. Remember that the certainty equivalent is the for-sure amount that Carothers would equate with the risky project of purchasing the plane. Computing certainty equivalents in PrecisionTree is simple:

Figure 13.13

Eagle Airlines' decision tree.

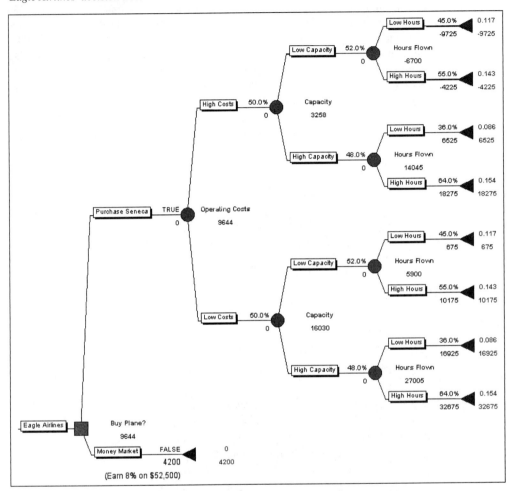

Figure 13.14

Tree Settings dialog box implementing the exponential utility function with risk tolerance equal to 10,000.

Figure 13.15

Eagle Airlines tree using the exponential utility function with risk tolerance equal to 10,000.

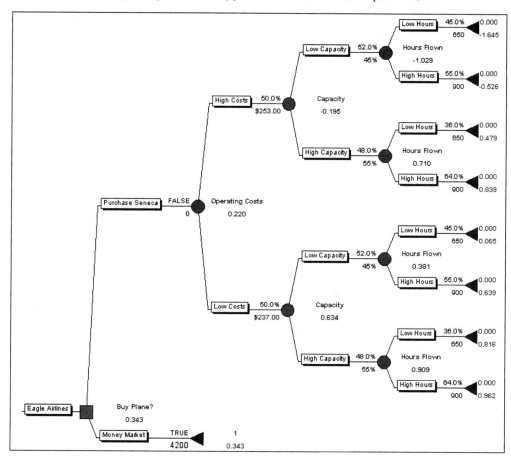

STEP 3

3.1   Click on the tree's name **Eagle Airlines**, choose **Certainty Equivalent**
       for the *Display*, and click **OK**.

The decision tree now shows the certainty equivalents for each node. We see that
the certainty equivalent of purchasing the plane is $2483. As expected, this is less
than $4200. Carothers equates the guaranteed amount of $2483 with the risky alternative of purchasing the airplane. Theoretically, he would be indifferent between
purchasing the plane (and accepting the associated gamble) or taking a guaranteed

$2483. For Carothers, the key is that he would have to give up a for-sure $4200 for a certainty equivalent of $2483, which is not a reasonable thing to do!

We know that for risk-averse people, the certainty equivalent is less than the expected value because these people view risks as an added cost. We can calculate that Carothers believes these particular risks are worth $7161 because he would be willing to forgo $7161 = $9644 – $2483 in expected profit in order to avoid the risks surrounding the plane purchase. Put another way, $7161 is the risk premium.

Sensitivity Analysis of Risk Tolerance

In Chapter 5 we learned how to use sensitivity analysis. Here we can use sensitivity analysis to study risk preferences. One approach is to vary the risk tolerance in the exponential utility function to find the point where the decision changes. In our example, as Carothers becomes less risk-averse (as R increases), the airplane purchase becomes more attractive relative to the money market. What is the risk tolerance R where the plane purchase has the same expected utility as the money market?

Answering this question is conceptually simple. All we need to do is find the critical value of R such that the expected utility of purchasing the plane equals the expected utility of investing in the money market. It is a simple matter in PrecisionTree to try different values for R until we find the critical value. Once found, the question is whether Carothers's risk tolerance is above or below the critical value. If it is above the critical value, then the risk is sufficiently low that he should buy the airplane; if it is below the critical value, then he should go with the less risky money market.

Finding the critical value is just a matter of trying different values of R until the expected utilities are equal:

STEP 4

4.1   Click on the tree's name **Eagle Airlines**, enter an R value, choose **Expected Utility** for the *Display*, and click **OK**.

4.2   If the expected utility of purchasing the airplane is more than the expected utility of the money market, then repeat Step 4.1 with a smaller R value.

4.3   If the expected utility of purchasing the airplane is less than the expected utility of the money market, then repeat Step 4.1 with a larger R value.

4.4   Continue until the expected utilities are equal.

We found the critical value to the nearest dollar to be $14,241. Now Carothers must ask whether he would be willing to accept an investment in which he would have an equal chance of winning $14,241 or losing $7,120.50. If he concludes that he would not take this gamble, then his risk tolerance must be smaller than $14,241, and he should not purchase the airplane. But if he would gladly participate in this gamble, then his risk tolerance must be greater than $14,241, and he should buy the airplane.

Decreasing and Constant Risk Aversion (Optional)

In this section we will consider how individuals might deal with risky investments. Suppose you had $1000 to invest. How would you feel about investing $500 in an extremely risky venture in which you might lose the entire $500? Now suppose you have saved more money and have $20,000. Now how would you feel about that extremely risky venture? Is it more or less attractive to you? How do you think a person's degree of risk aversion changes with wealth?

Decreasing Risk Aversion

If an individual's preferences show *decreasing risk aversion,* then the risk premium decreases if a constant amount is added to all payoffs in a gamble. Expressed informally, decreasing risk aversion means the more money you have, the less nervous you are about a particular bet.

For example, suppose an individual's utility curve can be described by a logarithmic function:

$$U(x) = \ln(x)$$

where x is the wealth or payoff and $\ln(x)$ is the natural logarithm of x. Using this logarithmic utility function, consider the gamble

Win $10 with probability 0.5

Win $40 with probability 0.5

To find the certainty equivalent, we first find the expected utility. The utility values for $10 and $40 are

$$U(\$10) = \ln(10) = 2.3026$$
$$U(\$40) = \ln(40) = 3.6889$$

Calculating expected utility:

$$EU = 0.5(2.3026) + 0.5(3.6889) = 2.9957$$

To find the certainty equivalent, you must find the certain value x that has $U(x) = 2.9957$; thus, set the utility function equal to 2.9957:

$$2.9957 = \ln(x)$$

Now solve for x. To remove the logarithm, we take antilogs:

$$e^{2.9957} = e^{\ln(x)} = x$$

The rule here corresponds to what we did with the exponential function. Here we have $e^{\ln(y)} = y$. Finally, we simply calculate $e^{2.9957}$:

$$x = e^{2.9957} = \$20 = CE$$

To find the risk premium, we need the expected payoff, which is

$$\text{EMV} = 0.5(\$10) + 0.5(\$40) = \$25$$

Thus, the risk premium is EMV − CE = $25 − $20 = $5.

Using the same procedure, we can find risk premiums for the lotteries as shown in Table 13.3. Notice that the sequence of lotteries is constructed so that each is like having the previous one plus $10. For example, the $20–$50 lottery is like having the $10–$40 lottery plus a $10 bill. The risk premium decreases with each $10 addition. The decreasing risk premium reflects decreasing risk aversion, which is a property of the logarithmic utility function.

An Investment Example

For another example, suppose that an entrepreneur is considering a new business investment. To participate, the entrepreneur must invest $5000. There is a 25% chance that the investment will earn back the $5000, leaving her just as well off as if she had not made the investment. But there is also a 45% chance that she will lose the $5000 altogether, although this is counterbalanced by a 30% chance that the investment will return the original $5000 plus an additional $10,000. Figure 13.16 shows the entrepreneur's decision tree.

We will assume that this entrepreneur's preferences can be modeled with the logarithmic utility function $U(x) = \ln(x)$, where x is interpreted as total wealth. Suppose that the investor now has $10,000. Should she make the investment or avoid it?

The easiest way to solve this problem is to calculate the expected utility of the investment and compare it with the expected utility of the alternative, which is to do nothing. The expected utility of doing nothing simply is the utility of the current wealth, or U(10,000), which is

$$U(10,000) = \ln(10,000) = 9.2103$$

The expected utility of the investment is easy to calculate:

$$\text{EU} = 0.30\, U(20,000) + 0.25\, U(10,000) + 0.45\, U(5000)$$
$$= 0.30(9.9035) + 0.25(9.2103) + 0.45(8.5172)$$
$$= 9.1064$$

Table 13.3
Risk premiums from logarithmic utility function.

50–50 Gamble Between ($)	Expected Value ($)	Certainty Equivalent ($)	Risk Premium ($)
10, 40	25	20.00	5.00
20, 50	35	31.62	3.38
30, 60	45	42.43	2.57
40, 70	55	52.92	2.08

Figure 13.16

*Entrepreneur's invest-
ment decision.* Current
wealth is denoted by x.

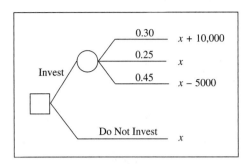

Because the expected utility of the investment is less than the utility of not investing, the investment should not be made.

Now, suppose that several years have passed. The utility function has not changed, but other investments have paid off handsomely, and she currently has $70,000. Should she undertake the project now? Recalculating with a base wealth of $70,000 rather than $10,000, we find that the utility of doing nothing is U(70,000) = 11.1563, and the EU for the investment is 11.1630. Now the expected utility of the investment is greater than the utility of doing nothing, and so she should invest.

The point of these examples is to show how decreasing risk aversion determines the way in which a decision maker views risky prospects. As indicated, the wealthier a decreasingly risk-averse decision maker is, the less anxious he or she will be about taking a particular gamble. Generally speaking, decreasing risk aversion makes sense when we think about risk attitudes and the way that many people appear to deal with risky situations. Many would feel better about investing money in the stock market if they were wealthier to begin with. For such reasons, the logarithmic utility function is commonly used by economists and decision theorists as a model of typical risk attitudes.

Constant Risk Aversion

An individual displays constant risk aversion if the risk premium for a gamble does not depend on the initial amount of wealth held by the decision maker. Intuitively, the idea is that a constantly risk-averse person would be just as anxious about taking a bet regardless of the amount of money available.

If an individual is constantly risk-averse, the utility function is exponential. It would have the following form:

$$U(x) = 1 - e^{-x/R}$$

For example, suppose that the decision maker has assessed a risk tolerance of $35:

$$U(x) = 1 - e^{-x/35}$$

We can perform the same kind of analysis that we did with the logarithmic utility function above. Consider the gamble

Win $10 with probability 0.5

Win $40 with probability 0.5

As before, the expected payoff is $25. To find the CE, we must find the expected utility, which requires plugging the amounts $10 and $40 into the utility function:

$$U(10) = 1 - e^{-10/35} = 0.2485$$

$$U(40) = 1 - e^{-40/35} = 0.6811$$

Thus, EU = 0.5 (0.2485) + 0.5 (0.6811) = 0.4648. To find the certainty equivalent, set the utility function to 0.4648. The value for x that gives the utility of 0.4648 is the gamble's CE:

$$0.4648 = 1 - e^{-x/35}$$

Now we can solve for x as we did earlier when working with the exponential utility function:

$$0.5352 = e^{-x/35}$$

$$\ln(0.5352) = -0.6251 = -x/35$$

$$x = 0.6251(35) = \$21.88 = CE$$

Finally, the expected payoff (EMV) is $25, and so the risk premium is

$$\text{Risk Premium} = \text{EMV} - \text{CE} = \$25 - \$21.88 = \$3.12$$

Using the same procedure, we can find the risk premium for each gamble in Table 13.4. The risk premium stays the same as long as the difference between the payoffs does not change. Adding a constant amount to both sides of the gamble does not change the decision maker's attitude toward the gamble.

Alternatively, you can think about this as a situation where you have a bet in which you may win $15 or lose $15. In the first gamble above, you face this bet with $25 in your pocket. In the constant-risk-aversion situation, the way you feel about the bet (as reflected in the risk premium) is the same regardless of how much money is added to your pocket. In the decreasing-risk-aversion situation, adding something to your pocket made you less risk-averse toward the bet, thus resulting in a lower risk premium.

Table 13.4

Risk premiums from exponential utility function.

50–50 Gamble Between ($)	Expected Value ($)	Certainty Equivalent ($)	Risk Premium ($)
10, 40	25	21.88	3.12
20, 50	35	31.88	3.12
30, 60	45	41.88	3.12
40, 70	55	51.88	3.12

Figure 13.17

Logarithmic and exponential utility functions plotted over the range of $10 to $100.

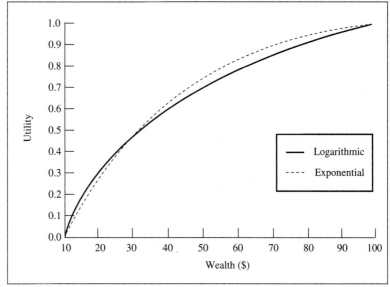

Figure 13.17 plots the two utility functions on the same graph. They have been rescaled so that U(10) = 0 and U(100) = 1 in each case. Note their similarity. It does not take a large change in the utility curve's shape to alter the nature of the individual's risk attitude.

Is constant risk aversion appropriate? Consider the kinds of risks that railroad barons such as E. H. Harriman undertook. Would a person like Harriman have been willing to risk millions of dollars on the takeover of another railroad had he not already been fairly wealthy? The argument easily can be made that the more wealth one has, the easier it is to take larger risks. Thus, decreasing risk aversion appears to provide a more appropriate model of preferences than does constant risk aversion. This is an important point to keep in mind if you decide to use the exponential utility function and the risk-tolerance parameter; this utility function displays constant risk aversion.

After all is said and done, while the concepts of decreasing or constant risk aversion may be intriguing from the point of view of a decision maker who is interested in modeling his or her risk attitude, precise determination of a decision maker's utility function is not yet possible. Decision theorists still are learning how to elicit and measure utility functions. Many unusual effects arise from human nature; we will study some of these in the next chapter. It would be an overstatement to suggest that it is possible to determine precisely the degree of an individual's risk aversion or whether he or she is decreasingly risk-averse. It is a difficult enough problem just to determine whether someone is risk-averse or risk-seeking!

Thus, it may be reasonable to use the exponential utility function as an approximation in modeling preferences and risk attitudes. A quick assessment of risk tolerance, and you are on your way. And if even that seems a bit strained, then it is always possible to use the sensitivity-analysis approach; it may be that a precise assessment of the risk tolerance is not necessary.

Some Caveats

A few things remain to be said about utilities. These are thoughts to keep in mind as you work through utility assessments and use utilities in decision problems.

1   Utilities do not add up. That is, $U(A + B) \neq U(A) + U(B)$. This actually is the whole point of having a nonlinear utility function. Thus, when using utilities in a decision analysis, you must calculate net payoffs or net contributions at the endpoints of the decision tree before transforming to utility values.

2   Utility differences do not express strength of preferences. Suppose that $U(A_1) - U(A_2) > U(A_3) - U(A_4)$. This does not necessarily mean that you would rather go from A_2 to A_1 instead of from A_4 to A_3. Utility only provides a numerical scale for ordering preferences, not a measure of their strengths. Whether this is reasonable is a matter of some debate. For example, von Winterfeldt and Edwards (1986) give the following example: You are first told that you will receive $100, and then told that you actually will receive $500, and then finally told that the actual payment will be $10,000. It would indeed be a pleasant surprise to go from $100 to $500, but for most of us, the delight we would experience from $500 to $10,000 would eclipse the difference between $100 and $500. Von Winterfeldt and Edwards argue that we can make judgments of just this sort. Whether one agrees or not, it is necessary to interpret utility carefully in this regard.

3   Utilities are not comparable from person to person. A utility function is a subjective personal statement of an individual's preferences and so provides no basis for comparing utilities among individuals.

SUMMARY   In this chapter we have explored some basic concepts that underlie risk and return trade-offs, with the aim of being able to understand how to model a decision maker's risk preferences. We discussed the notion of a risk premium (EMV – CE), which can be thought of as a measure of how risk-averse a decision maker is in regard to a particular risky situation. The basic procedure for assessing a utility function requires comparison of lotteries with riskless payoffs. Once a utility function has been determined, the procedure is to replace dollar payoffs in a decision tree or influence diagram with utility values and solve the problem to find the option with the greatest expected utility. We also studied the exponential utility function and the notion of risk tolerance. Because of its nature, the exponential utility function is particularly useful for modeling preferences in decision analysis. The concepts of decreasing and constant risk aversion also were discussed.

EXERCISES

13.1 Why is it important for decision makers to consider their attitudes toward risk?

13.2 We have not given a specific definition of risk. How would you define it? Give examples of lotteries that vary in riskiness in terms of your definition of risk.

13.3 Explain in your own words the idea of a certainty equivalent.

13.4 Explain what is meant by the term *risk premium*.

13.5 Explain in your own words the idea of risk tolerance. How would it apply to utility functions other than the exponential utility function?

13.6 Suppose a decision maker has the utility function shown in Table 13.1. An investment opportunity has EMV $1236 and expected utility 0.93. Find the certainty equivalent for this investment and the risk premium.

13.7 A decision maker's assessed risk tolerance is $1210. Assume that this individual's preferences can be modeled with an exponential utility function.

   a Find U($1000), U($800), U($0), and U(−$1250).

   b Find the expected utility for an investment that has the following payoff distribution:

$$P(\$1000) = 0.33$$

$$P(\$800) = 0.21$$

$$P(\$0) = 0.33$$

$$P(-\$1250) = 0.13$$

   c Find the exact certainty equivalent for the investment and the risk premium.

   d Find the approximate certainty equivalent using the expected value and variance of the payoffs.

   e Another investment possibility has expected value $2400 and standard deviation $300. Find the approximate certainty equivalent for this investment.

13.8 Many firms evaluate investment projects individually on the basis of expected value and at the same time maintain diversified holdings in order to reduce risk. Does this make sense in light of our discussion of risk attitudes in this chapter?

13.9 A friend of yours, who lives in Reno, has life insurance, homeowner's insurance, and automobile insurance and also regularly plays the quarter slot machines in the casinos. What kind of a utility function might explain this kind of behavior? How else might you explain such behavior?

13.10 Two risky gambles were proposed at the beginning of the chapter:

Game 1   Win $30 with probability 0.5

Lose $1 with probability 0.5

Game 2   Win $2000 with probability 0.5

Lose $1900 with probability 0.5

Many of us would probably pay to play Game 1 but would have to be paid to participate in Game 2. Is this true for you? How much would you pay (or have to be paid) to take part in either game?

QUESTIONS AND PROBLEMS

13.11 **St. Petersburg Paradox.** Consider the following game that you have been invited to play by an acquaintance who always pays his debts. Your acquaintance will flip a fair coin. If it comes up heads, you win $2. If it comes up tails, he flips the coin again. If heads occurs on the second toss, you win $4. If tails, he flips again. If heads occurs on the third toss, you win $8, and if tails, he flips again, and so on. Your payoff is an uncertain amount with the following probabilities:

Payoff	Probability	
2	0.50	
4	0.25	
8	0.125	
.	.	
.	.	
.	.	(where n is the number
2^n	0.5^n	of the toss when the
.	.	first head occurs)
.	.	
.	.	

This is a good game to play because you are bound to come out ahead. There is no possible outcome from which you can lose. How much would you pay to play this game? $10? $20? What is the expected value of the game? Would you be indifferent between playing the game and having the expected value for sure?

13.12 Assess your utility function in two different ways.

  a Use the certainty-equivalent approach to assess your utility function for wealth over a range of $100 to $20,000.

b Use the probability-equivalent approach to assess U($1500), U($5600), U($9050), and U($13,700). Are these assessments consistent with the assessments made in part a? Plot these assessments and those from part a on the same graph and compare them.

13.13 Assess your risk tolerance (R). Now rescale your exponential utility function—the one you obtain by substituting your R value into the exponential utility function—so that U($100) = 0 and U($20,000) = 1. [That is, find constants a and b so that $a + b(1 - e^{-100/R}) = 0$ and $a + b(1 - e^{-20,000/R}) = 1$.] Now plot the rescaled utility function on the same graph with the utility assessments from Problem 13.12. How do your assessments compare?

13.14 Let us return to the Texaco-Pennzoil example from Chapter 4 and think about Liedtke's risk attitude. Suppose that Liedtke's utility function is given by the utility function in Table 13.5.

    a Graph this utility function. Based on this graph, how would you classify Liedtke's attitude toward risk?

    b Use the utility function in conjunction with the decision tree sketched in Figure 4.2 to solve Liedtke's problem. With these utilities, what strategy should he pursue? Should he still counteroffer $5 billion? What if Texaco counteroffers $3 billion? Is your answer consistent with your response to part a?

    c Based on this utility function, what is the least amount (approximately) that Liedtke should agree to in a settlement? (*Hint:* Find a sure amount that gives him the same expected utility that he gets for going to court.) What does this suggest regarding plausible counteroffers that Liedtke might make?

13.15 Of course, Liedtke is not operating by himself in the Texaco-Pennzoil case; he must report to a board of directors. Table 13.6 gives utility functions for three different directors. Draw graphs of these. How would you classify each director in terms of his or her attitude toward risk? What would be the strategies of each? (That is, what would each one do with respect to Texaco's current offer, and how would each react to a Texaco counteroffer of $3 billion? To answer this question, you must solve the decision tree—calculate expected utilities—for *each* director.)

13.16 How do you think Liedtke (Problem 13.14) and the directors in Problem 13.15 will be able to reconcile their differences?

13.17 Rescale the utility function for Director A in Problem 13.15 so that it ranges between 0 and 1. That is, find constants a and b so that when you multiply the utility function by a

Table 13.5
Utility function for Liedtke.

Payoff (Billions)	Utility
10.3	1.00
5.0	0.75
3.0	0.60
2.0	0.45
0.0	0.00

Table 13.6

Utility functions for three Pennzoil directors.

Payoff	Utility		
	Director A	Director B	Director C
10.3	3.0	100	42.05
5.0	2.9	30	23.50
3.0	2.8	15	16.50
2.0	2.6	8	13.00
0.0	1.0	0	6.00

and then add b, the utility for $10.30 is 1 and the utility for $0 is 0. Graph the rescaled utility function and compare it to the graph of the original utility function. Use the rescaled utility function to solve the Texaco-Pennzoil decision tree. Is the optimal choice consistent with the one you found in Problem 13.15?

13.18 What if Hugh Liedtke were risk-averse? Based on Figure 4.2, find a critical value for Hugh Liedtke's risk tolerance. If his risk tolerance is low enough (very risk-averse), he would accept the $2 billion offer. How small would his risk tolerance have to be for EU(Accept $2 Billion) to be greater than EU(Counteroffer $5 Billion)?

13.19 The idea of dominance criteria and risk aversion come together in an interesting way, leading to a different kind of dominance. If two risky gambles have the same expected payoff, on what basis might a risk-averse individual choose between them without performing a complete utility analysis?

13.20 This problem is related to the ideas of dominance that we discussed in Chapters 4 and 8. Investment D below is said to show "second-order stochastic dominance" over Investment C. In this problem, it is up to you to explain why D dominates C.

You are contemplating two alternative uncertain investments, whose distributions for payoffs are below.

Payoff	Probabilities	
	Investment C	Investment D
50	1/3	1/4
100	1/3	1/2
150	1/3	1/4

a If your preference function is given by $U(x) = 1 - e^{-x/100}$, calculate EU for both C and D. Which would you choose?

b Plot the CDFs for C and D on the same graph. How do they compare? Use the graph to explain intuitively why any risk-averse decision maker would prefer D. (*Hint:* Think about the concave shape of a risk-averse utility function.)

13.21 Utility functions need not relate to dollar values. Here is a problem in which we know little about five abstract outcomes. What is important, however, is that a person who does know what A to E represent should be able to compare the outcomes using the lottery procedures we have studied.

A decision maker faces a risky gamble in which she may obtain one of five outcomes. Label the outcomes A, B, C, D, and E. A is the most preferred, and E is least preferred. She has made the following three assessments.

- She is indifferent between having C for sure or a lottery in which she wins A with probability 0.5 or E with probability 0.5.

- She is indifferent between having B for sure or a lottery in which she wins A with probability 0.4 or C with probability 0.6.

- She is indifferent between these two lotteries:

  1: A 50% chance at B and a 50% chance at D
  2: A 50% chance at A and a 50% chance at E

What are $U(A)$, $U(B)$, $U(C)$, $U(D)$, and $U(E)$?

13.22 You have considered insuring a particular item of property (such as an expensive camera, your computer, or your Stradivarius violin), but after considering the risks and the insurance premium quoted, you have no clear preference for either purchasing the insurance or taking the risk. The insurance company then tells you about a new scheme called "probabilistic insurance." You pay half the above premium but have coverage only in the sense that in the case of a claim there is a probability of one-half that you will be asked to pay the other half of the premium and will be completely covered, or that you will not be covered and will have your premium returned. The insurance company can be relied on to be fair in flipping the coin to determine whether or not you are covered.

a Do you consider yourself to be risk-averse?

b Would you purchase probabilistic insurance?

c Draw a decision tree for this problem.

d Show that a risk-averse individual always should prefer the probabilistic insurance.

(*Hint:* This is a difficult problem. To solve it you must be sure to consider that you are indifferent between the regular insurance and no insurance. Write out the equation relating these two alternatives and see what it implies. Another strategy is to select a specific utility function—the log utility function $U(x) = \log(x)$, say—and then find values for the probability of a claim, your wealth, the insurance premium, and the value of your piece of property so that the utility of paying the insurance premium is equal to the expected utility of no insurance. Now use these values to calculate the expected utility of the probabilistic insurance. What is the result?)

13.23 An investor with assets of $10,000 has an opportunity to invest $5000 in a venture that is equally likely to pay either $15,000 or nothing. The investor's utility function can be described by the log utility function $U(x) = \ln(x)$, where x is his total wealth.

a What should the investor do?

b Suppose the investor places a bet with a friend before making the investment decision. The bet is for $1000; if a fair coin lands heads up, the investor wins $1000, but if it lands tails up, the investor pays $1000 to his friend. Only after the bet has been resolved will the investor decide whether or not to invest in the venture. What is an appropriate strategy for the investor? If he wins the bet, should he invest? What if he loses the bet?

c   Describe a real-life situation in which an individual might find it appropriate to gamble before deciding on a course of action.

Source: D. E. Bell (1988) "Value of Pre-Decision Side Bets for Utility Maximizers." *Management Science, 34,* 797–800.

13.24   A bettor with utility function $U(x) = \ln(x)$, where x is total wealth, has a choice between the following two alternatives:

> **A**   Win $10,000 with probability 0.2
> Win $1000 with probability 0.8

> **B**   Win $3000 with probability 0.9
> Lose $2000 with probability 0.1

a   If the bettor currently has $2500, should he choose A or B?

b   Repeat a, assuming the bettor has $5000.

c   Repeat a, assuming the bettor has $10,000.

d   Do you think that this pattern of choices between A and B is reasonable? Why or why not?

Source: D. E. Bell (1988) "One-Switch Utility Functions and a Measure of Risk." *Management Science,* 34, 1416–1424.

13.25   Repeat Problem 13.24 with $U(x) = 0.0003x - 8.48e^{-x/2775}$. A utility function of this form is called *linear-plus-exponential,* because it contains both linear ($0.0003x$) and exponential terms. It has a number of interesting and useful properties, including the fact that it only switches once among any pair of lotteries (such as those in Problem 13.24) as wealth increases (see Bell 1995a, b).

13.26   Show that the linear-plus-exponential utility function in Problem 13.25 has decreasing risk aversion. (*Hint:* You can use this utility function to evaluate a series of lotteries like those described in the text and analyzed in Tables 13.3 and 13.4. Show that the risk premium for a gamble decreases as wealth increases.)

13.27   Buying and selling prices for risky investments obviously are related to certainty equivalents. This problem, however, shows that the prices depend on exactly what is owned in the first place!

Suppose that Peter Brown's utility for total wealth (A) can be represented by the utility function $U(A) = \ln(A)$. He currently has $1000 in cash. A business deal of interest to him yields a reward of $100 with probability 0.5 and $0 with probability 0.5.

a   If he owns this business deal in addition to the $1000, what is the smallest amount for which he would sell the deal?

b   Suppose he does not own the deal. What equation must be solved to find the largest amount he would be willing to pay for the deal?

c For part b, it turns out that the most he would pay is $48.75, which is not exactly the same as the answer in part a. Can you explain why the amounts are different?

d (*Extra credit for algebra hotshots.*) Solve your equation in part b to verify the answer ($48.75) given in part c.

Source: This problem was suggested by R. L. Winkler.

13.28   We discussed decreasing and constant risk aversion. Are there other possibilities? Think about this as you work through this problem.

Suppose that a person's utility function for total wealth is

$$U(A) = 200A - A^2 \quad \text{for } 0 \le A \le 100$$

where A represents total wealth in thousands of dollars.

a Graph this preference function. How would you classify this person with regard to her attitude toward risk?

b If the person's total assets are currently $10K, should she take a bet in which she will win $10K with probability 0.6 and lose $10K with probability 0.4?

c If the person's total assets are currently $90K, should she take the bet given in part b?

d Compare your answers to parts b and c. Does the person's betting behavior seem reasonable to you? How could you intuitively explain such behavior?

Source: R. L. Winkler (1972).

13.29   Suppose that a decision maker has the following utility function:

$$U(x) = -0.000156x^2 + 0.028125x - 0.265625$$

Use this utility function to calculate risk premiums for the gambles shown in Tables 13.3 and 13.4; create a similar table but based on this quadratic utility function. How would you classify the risk attitude of a decision maker with this utility function? Does such a risk attitude seem reasonable to you?

13.30   The CEO of a chemicals firm must decide whether to develop a new process that has been suggested by the research division. His decision tree is shown in Figure 13.18. There are two sources of uncertainty. The production cost is viewed as a continuous random variable, uniformly distributed between $1.75 and $2.25, and the size of the market

Figure 13.18

Decision tree for Problem 13.30.

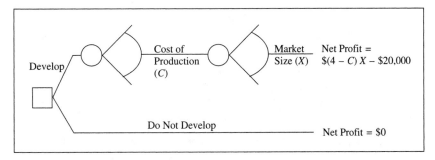

(units sold) for the product is normally distributed with mean 10,300 units and standard deviation 2200 units.

The firm's CEO is slightly risk-averse. His utility function is given by

$$U(Z) = 1 - e^{-Z/20,000} \qquad \text{where } Z \text{ is the net profit}$$

Should the CEO develop the new process? Answer this question by running a computer simulation, using 200 trials. Should the decision maker be concerned about the fact that if he develops the new process, the utility could be less than or greater than the utility for $0? On what basis should he make his decision?

13.31   The year is 2020, and you are in the supercomputer business. Your firm currently produces a machine that is relatively expensive (list price $6 million) and relatively slow (for supercomputers in the twenty-first century). Speed of supercomputers is measured in gigaflops per second (gps), where one "flop" is one calculation. Thus, one 1 gps = 1 billion calculations per second. Your current machine is capable of 150 gps. If you could do it, you would prefer to develop a supercomputer that costs less (to beat the competition) and is faster.

You have a research-and-development (R&D) decision to make based on two alternatives. You can choose one or the other of the following projects, or neither, but budget constraints prevent you from engaging in both projects.

A   The super-supercomputer. This project involves the development of a machine that is extremely fast (800 gps) and relatively inexpensive ($5 million). But this is a fairly risky project. The engineers who have been involved in the early stages estimate that there is only a 50% chance that this project would succeed. If it fails, you will be stuck with your current machine.

B   The better supercomputer. This project would pursue the development of an $8 million machine capable of 500 gps. This project also is somewhat risky. The engineers believe that there is only a 40% chance that this project will achieve its goal. They quickly point out, however, that even if the $8 million, 500 gps machine does not materialize, the technology involved is such that they would at least be able to produce a $5 million machine capable of 350 gps.

The decision tree is shown in Figure 13.19. To decide between the two alternatives, you have made the following assessments:

I   The best possible outcome is the $5 million, 800 gps machine, and the worst outcome is the status quo $6 million, 150 gps machine.

II   If you had the choice, you would be indifferent between Alternatives X and Y shown in Figure 13.20a.

III If you had the choice, you would be indifferent between Alternatives X' and Y' in Figure 13.20b.

   a   Using assessments I, II, and III, decide between Projects A and B. Justify your decision.

   b   Explain why Project A appears to be riskier than Project B. Given that A is riskier than B, would you change your answer to part a? Why or why not?

13.32   Show that the value of Y that yields indifference between the two alternatives in Figure 13.12 is within about 4% of the risk tolerance R.

Figure 13.19

Choosing between two risky supercomputer development projects.

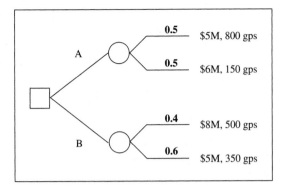

Figure 13.20

Assessments to assist your choice from among the supercomputer R&D projects in Problem 13.31.

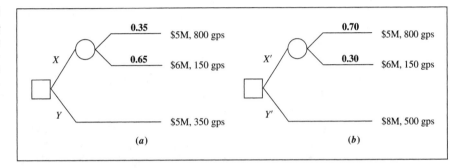

C A S E   S T U D I E S

INTERPLANTS, INC.

Don Newcomb was perplexed. It had been five years since he had founded Interplants, Inc., a research-and-development firm that developed genetically engineered plants for interplanetary space flight. During that five years, he and his scientists had made dramatic advances in developing special plants that could be used for food and air-purification systems in space stations and transports. In fact, he mused, their scientific success had been far greater than he had ever expected.

Five years ago, after the world superpowers had agreed to share space-travel technology, the outlook had been quite rosy. Everyone had been optimistic. Indeed, he was one of many investors who had jumped at the chance to be involved in the development of such technology. But now, after five tumultuous years, the prospects were less exciting.

First, there had been the disappointing realization that none of the superpowers had made substantial progress on an ion engine to power space vehicles. Such an engine was absolutely crucial to the success of interplanetary space flight, because—theoretically, at least—it would make travel 10 times as fast as conventionally powered ships. When the importance of such an engine became obvious, the superpowers had generously funded a huge multinational research project. The project had made substantial progress, but many hurdles remained. Don's risk assessors estimated that there was still a 15% chance that the ion engine would prove an infeasible power source. If this were to happen, of course, Don and the many other investors in space-travel technology would lose out.

Then there was the problem with the settlement policy. The superpowers could not agree on a joint policy for the settlement of interplanetary space, including the deployment of space stations as well as settlements on planets and their satellites. The United American Alliance urged discretion and long-range planning in this matter, suggesting that a multinational commission be established to approve individual settlement projects. Pacificasia and the Allied Slavic Economic Community were demanding that space be divided now. By immediately establishing property rights, they claimed, the superpowers would be able to develop the optimum space economy in which the superpowers could establish their own economic policies within their "colonies" as well as determine trade policies with the other superpowers. Europa favored the idea of a commission, but also was eager to explore other available economic possibilities.

The discussion among the superpowers had been going on since long before the founding of Interplants. Five years ago, progress was being made, and it appeared that an agreement was imminent. But 18 months ago the process stalled. The participants in the negotiations had established positions from which they would not budge. Don had followed the discussions closely and had even provided expert advice to the negotiators regarding the potential for interplanetary agricultural developments. He guessed that there was only a 68% chance that the superpowers would eventually arrive at an agreement. Naturally, until an agreement was reached there would be little demand for space-traveling plants.

Aside from these external matters, Don still faced the difficult issue of developing a full-scale production process for his plants. He and his engineers had some ideas about the costs of the process, but at this point, all they could do was come up with a probability distribution for the costs. In thinking about the distribution, Don had decided to approximate it with a three-point discrete distribution. Thus, he characterized the three branches as "inexpensive," "moderate," and "costly," with probabilities of 0.185, 0.63, and 0.185, respectively. Of course, his eventual profit (or loss) depended on the costs of the final process.

Don also had thought long and hard about the profits that he could anticipate under each of the various scenarios. Essentially, he thought about the uncertainty in two stages. First was the determination of costs, and second was the outcome of the external factors (the ion-engine research and the negotiations regarding settlement policy). If costs turned out to be "inexpensive," then, in the event that the superpowers agreed and the ion engine was successful, he could expect a profit of 125 billion credits. He would lose 15 billion credits if either the engine or the negotiations failed. Likewise, if costs were "moderate," he could anticipate either a profit of 100 billion credits if both of the external factors resulted in a positive outcome, or a loss of 18 billion if either of the external factors were negative. Finally, the corresponding figures in the case of a "costly" production process were profits of 75 billion credits or a loss of 23 billion.

"This is so confusing," complained Don to Paul Fiester, his chief engineer. "I really never expected to be in this position. Five years ago none of these risks were apparent to me, and I guess I just don't tolerate risk well."

After a pause, Paul quietly suggested "Well, maybe you should sell the business."

Don considered that. "Well, that's a possibility. I have no idea how many crazy people out there would want it."

"Some of the other engineers and I might be crazy enough," Paul replied. "Depending on the price, of course. At least we'd be going in with our eyes open. We know what the business is about and what the risks are."

Don gave the matter a lot of thought that night. "What should I sell the company for? I hate to give up the possibility of earning up to 125 billion credits. But I don't like the possibility of losing 23 billion either—no one would!" As he lay awake, he finally decided that he would let the business go—with all its risks—for 20 billion credits. If he could get that much for it, he'd sell. If not, he'd just as soon stick with it, in spite of his frustrations with the risks.

Questions

1   Draw a decision tree for Don Newcomb's problem.

2   What is the significance of his statement that he would sell the business for 20 billion credits?

3   Suppose that Don's risk attitude can be modeled with an exponential utility function. If his certainty equivalent were 15 billion credits, find his risk tolerance. What would his risk tolerance be if his CE were 20 billion?

TEXACO-PENNZOIL ONE MORE TIME

In Problem 12.11 (page 520), we made EVPI calculations for Liedtke's decision in the Texaco-Pennzoil case. Calculating EVPI is somewhat different when the decision maker is risk-averse; that is, we cannot simply fold a tree back using expected values and then compare the expected values with and without information. And we cannot fold back the tree in terms of expected utility and compare the expected utilities. In fact, in general we must find a value C such that when we subtract that value from each endpoint of the decision tree on the "Acquire Information" branch, the expected utility of this branch is equal to the expected utility of the "No Information" branch. Figure 13.21 illustrates the principle by including a branch for acquiring information about Texaco's response.

Figure 13.21

Finding EVPI in the Texaco-Pennzoil decision when the decision maker is risk-averse. Find C so that the expected utility of the "Acquire Information" branch has the same expected utility as the next-best alternative.

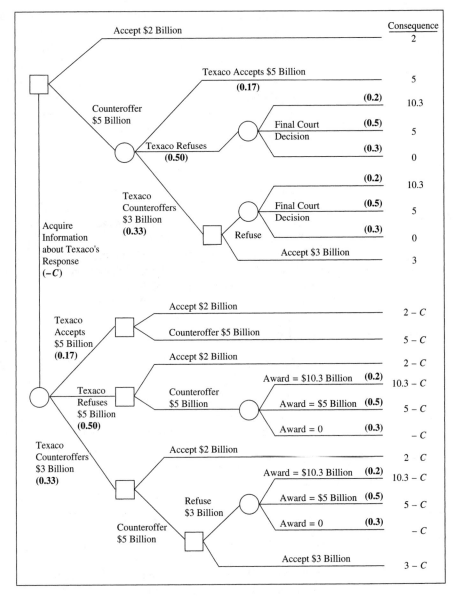

When using the exponential utility function, however, there is a simple shortcut. Because the exponential utility function displays constant risk aversion, we can work in terms of certainty equivalents. That is, we can calculate CEs for the "Acquire Information" and "No Information" branches and compare them. The difference between the two is the expected value of the information! Consider Liedtke's decision as diagrammed in Figure 13.21. Suppose that his decision analyst has modeled Pennzoil's corporate preferences with an exponential utility function and a risk tolerance of $1 billion. An acquaintance of Liedtke's knows Texaco CEO James Kinnear quite well and has offered to find out how Kinnear would react to a $5 billion counteroffer. Thus, the third alternative available to Liedtke is to acquire information about Texaco.

Question

1   Find the expected value of perfect information regarding the Texaco reaction. Compare your answer to the answer from Problem 12.11a.

STRENLAR, PART III

Consider once again Fred Wallace's decision in the Strenlar case study at the end of Chapter 4 (pages 167–169). What if Fred is risk-averse? Assume that Fred's attitude toward risk in this case can be adequately modeled using an exponential utility function in which the utility is calculated for net present value. Thus,

$$U(NPV) = 1 - e^{-NPV/R}$$

Question

1   Check the sensitivity of Fred's decision to his risk tolerance, R. What is the critical R value for which his optimal decision changes? What advice can you give to Fred?

REFERENCES

For the most part, the material presented in this chapter has been my own version of familiar material. Good basic treatments of expected utility and risk aversion are available in most decision-analysis textbooks, including Bunn (1984), Holloway (1979), Raiffa (1968), Vatter et al. (1978), von Winterfeldt and Edwards (1986), and Winkler (1972). Keeney and Raiffa (1976) offer a somewhat more advanced treatment that focuses on multiattribute utility models, although it does contain an excellent exposition of the basic material at a somewhat higher mathematical level. The text by von Winterfeldt and Edwards (1986) points out that we tend to think about different kinds of money (pocket money, monthly income, investment capital) in different ways; thus, how utility-assessment questions are framed in terms of these different funds can have a strong effect on the utility function.

The material on the exponential utility function and risk tolerance is based primarily on Holloway (1979) and McNamee and Celona (1987). Both books contain excellent discussion and problems on this material.

For students who wish more detail on decreasing and constant risk aversion, look in the financial economics literature. In financial models, utility functions for market participants are important for modeling the economic system. An excellent starting point is Copeland

and Weston (1979). A classic article that develops the idea of a risk-aversion coefficient (the reciprocal of risk tolerance) is Pratt (1964). Bell (1995a, b) discusses the "linear-plus-exponential" utility function $aw - be^{-cw}$, where w is total wealth. This utility function has many desirable properties; see Problems 13.25 and 13.26.

Bell, D. (1995a) "Risk, Return, and Utility." *Management Science,* 41, 23–30.

Bell D. (1995b) "A Contextual Uncertainty Condition for Behavior under Risk." *Management Science*, 41, 1145–1150.

Bunn, D. (1984) *Applied Decision Analysis.* New York: McGraw-Hill.

Copeland, T. E., and J. F. Weston (1979) *Financial Theory and Corporate Policy.* Reading, MA: Addison-Wesley.

Holloway, C. A. (1979) *Decision Making under Uncertainty: Models and Choices.* Englewood Cliffs, NJ: Prentice Hall.

Howard, R. A. (1988) "Decision Analysis: Practice and Promise." *Management Science,* 34, 679–695.

Keeney, R., and H. Raiffa (1976) *Decisions with Multiple Objectives.* New York: Wiley.

McNamee, P., and J. Celona (1987) *Decision Analysis for the Professional with Supertree.* Redwood City, CA: Scientific Press.

Pratt, J. (1964) "Risk Aversion in the Small and in the Large." *Econometrica,* 32, 122–136.

Raiffa, H. (1968) *Decision Analysis.* Reading, MA: Addison-Wesley.

Vatter, P. A., S. P. Bradley, S. C. Frey, Jr., and B. B. Jackson (1978) *Quantitative Methods in Management.* Homewood, IL: Irwin.

Von Winterfeldt, D., and W. Edwards (1986) *Decision Analysis and Behavioral Research.* Cambridge: Cambridge University Press.

Winkler, R. L. (1972) *Introduction to Bayesian Inference and Decision.* New York: Holt.

EPILOGUE   Harriman's broker, Jacob H. Schiff, was at his synagogue when Harriman's order for 40,000 shares was placed; the shares were never purchased. By the following Monday, Hill had cabled Morgan in France, and they had decided to buy as many Northern Pacific shares as they could. The share price went from $114 on Monday to $1000 on Thursday. In the aftermath, Hill and Morgan agreed that Harriman should be represented on the board of directors. Harriman, however, had little if any influence; James Hill continued to run the Great Northern, Northern Pacific, and Burlington railroads as he saw fit. [*Source:* S. H. Holbrook (1958) "The Legend of Jim Hill." *American Heritage,* IX(4), 10–13, 98–101.]

CHAPTER 14

Utility Axioms, Paradoxes, and Implications

In this chapter we will look at several issues. First, we will consider some of the foundations of utility theory. From the basis of a few behavioral axioms, it is possible to establish logically that people who behave according to the axioms should make choices consistent with the maximization of expected utility. But since the early 1950s, cognitive psychologists have noted that people do not always behave according to expected utility theory, and a large literature now covers these behavioral paradoxes. We review part of that literature here. Because decision analysis depends on foundational axioms, it is worthwhile to consider some implications of these behavioral paradoxes, particularly with regard to the assessment of utility functions.

The following example previews some of the issues we will consider. This one is a participatory example. You should think hard about the choices you are asked to make before reading on.

PREPARING FOR AN INFLUENZA OUTBREAK

The United States is preparing for an outbreak of an unusual Asian strain of influenza. Experts expect 600 people to die from the disease. Two programs are available that could be used to combat the disease, but because of limited resources only one can be implemented.

Program A (Tried and True) 400 people will be saved.

Program B (Experimental) There is an 80% chance that 600 people will be saved and a 20% chance that no one will be saved.

Which of these two programs do you prefer?

Now consider the following two programs:

Program C   200 people will die.

Program D   There is a 20% chance that 600 people will die and an 80% chance that no one will die.

Would you prefer C or D?

Source: Reprinted with permission from Tversky, A., and D. Kahneman, "The Framing of Decisions and the Psychology of Choice," *Science*, 211, 453–458. Copyright 1981 by the American Association for the Advancement of Science.

Axioms for Expected Utility

Our first step in this chapter is to look at the behavioral assumptions that form the basis of expected utility. These assumptions, or *axioms,* relate to the consistency with which an individual expresses preferences from among a series of risky prospects. Instead of axioms, we might call these *rules for clear thinking.* In the following discussion, the axioms are presented at a fairly abstract level. Simple examples are given to clarify their meaning. As we put them to work in the development of the main argument, the importance and intuition behind the axioms should become clearer.

1   *Ordering and transitivity.* A decision maker can order (establish preference or indifference) any two alternatives, and the ordering is transitive. For example, given any alternatives A_1, A_2, and A_3, either A_1 is preferred to A_2 (which is sometimes written as "$A_1 \succ A_2$"), A_2 is preferred to A_1, or the decision maker is indifferent between A_1 and A_2 ($A_1 \sim A_2$). Transitivity means that if A_1 is preferred to A_2 and A_2 is preferred to A_3, then A_1 is preferred to A_3. For example, this axiom says that an individual could express his or her preferences regarding, say, cities in which to reside. If that person preferred Amsterdam to London and London to Paris, then he or she would prefer Amsterdam to Paris.

2   *Reduction of compound uncertain events.* A decision maker is indifferent between a compound uncertain event (a complicated mixture of gambles or lotteries) and a simple uncertain event as determined by reduction using standard probability manipulations. This comes into play when we reduce compound events into reference gambles. The assumption says that we can perform the reduction without affecting the decision maker's preferences. We made use of this axiom in Chapter 4 in our discussion of risk profiles. The progression from Figure 4.20 through Figure 4.22 is a matter of reducing to simpler terms the compound uncertain event that is associated with the counteroffer.

3   *Continuity.* A decision maker is indifferent between a consequence A (for example, win 100) and some uncertain event involving only two basic consequences A_1 and A_2, where $A_1 \succ A \succ A_2$. This simply says that we can construct a reference gamble

with some probability p, $0 < p < 1$, for which the decision maker will be indifferent between the reference gamble and A. For example, suppose you find yourself as the plaintiff in a court case. You believe that the court will award you either $5000 or nothing. Now imagine that the defendant offers to pay you $1500 to drop the charges. According to the continuity axiom, there must be some probability p of winning $5000 (and the corresponding $1 - p$ probability of winning nothing) for which you would be indifferent between taking or rejecting the settlement offer. Of course, if your subjective probability of winning happens to be lower than p, then you would accept the proposal.

4   *Substitutability.* A decision maker is indifferent between any original uncertain event that includes outcome A and one formed by substituting for A an uncertain event that is judged to be its equivalent. Figure 14.1 shows how this works. This axiom allows the substitution of uncertain reference gambles into a decision for their certainty equivalents and is just the reverse of the reduction axiom already stated. For example, suppose you are interested in playing the lottery, and you are just barely willing to pay 50 cents for a ticket. If I owe you 50 cents, then you should be just as willing to accept a lottery ticket as the 50 cents in cash.

5   *Monotonicity.* Given two reference gambles with the same possible outcomes, a decision maker prefers the one with the higher probability of winning the preferred outcome. This one is easy to see. Imagine that two different car dealerships each can order the new car that you want. Both dealers offer the same price, delivery, warranty, and financing, but one is more likely to provide good service than the other. To which one would you go? The one that has the better chance of providing good service, of course.

6   *Invariance.* All that is needed to determine a decision maker's preferences among uncertain events are the payoffs (or consequences) and the associated probabilities.

7   *Finiteness.* No consequences are considered infinitely bad or infinitely good.

Most of us agree that these assumptions are reasonable under almost all circumstances. It is worth noting, however, that many decision theorists find some of the axioms controversial! The reasons for the controversy range from introspection regarding particular decision situations to formal psychological experiments in which human subjects make choices that clearly violate one or more of the axioms. We will discuss some of these experiments in the next section.

Figure 14.1

Two decision trees. If A is equivalent to a lottery with a p chance at C and $1 - p$ chance at D, then Decision Tree I is equivalent to Decision Tree II.

Decision Tree I

Decision Tree II

For example, the substitutability axiom is a particular point of debate. For some decision makers, the fact of having to deal with two uncertain events in Decision Tree II of Figure 14.1 can be worse than facing the single one in Decision Tree I. Moreover, individuals might make this judgment and at the same time agree that in a single-stage lottery, A is indeed equivalent to the risky prospect with a p chance at C and a $1 - p$ chance at D.

As another example, we can pick on the apparently innocuous transitivity axiom. In Figure 14.2, you have two lotteries from which to choose. Each of the six outcomes has probability $\frac{1}{6}$. One way to look at the situation is that the prize in Game B is better than Game A's in five of the six outcomes, and thus it may be reasonable to prefer B, even though the structure of the lotteries is essentially the same. Now consider the games in Figure 14.3. If B was preferred to A in Figure 14.2, then by the same argument C would be preferred to B, D to C, E to D, F to E, and, finally, A would be preferred to F. Thus, these preferences do not obey the transitivity axiom, because transitivity would never permit A to be preferred to something else that is in turn preferred to A.

The controversy about individual axioms notwithstanding, if you accept axioms 1 through 7, then logically you also must accept the following proposition.

Figure 14.2

A pair of lotteries. Outcomes 1 through 6 each occurs with probability $\frac{1}{6}$. Would you prefer to play Game A or Game B?

		Outcome					
		1	2	3	4	5	6
Prize	Game A	100	200	300	400	500	600
	Game B	200	300	400	500	600	100

Figure 14.3

More games to consider. Outcomes 1 through 6 each still has probability $\frac{1}{6}$.

		Outcome					
		1	2	3	4	5	6
Prize	Game B	200	300	400	500	600	100
	Game C	300	400	500	600	100	200
	Game D	400	500	600	100	200	300
	Game E	500	600	100	200	300	400
	Game F	600	100	200	300	400	500
	Game A	100	200	300	400	500	600

Proposition: Given any two uncertain outcomes B_1 and B_2, if assumptions 1 through 7 hold, there are numbers U_1, U_2, . . . , U_n representing preferences (or utilities) associated with the consequences such that the overall preference between the uncertain events is reflected by the expected values of the U's for each consequence. In other words, if you accept the axioms, (1) it is possible to find a utility function for you to evaluate the consequences, and (2) you should be making your decisions in a way that is consistent with maximizing expected utility.

In the following pages we will demonstrate how the axioms permit the transformation of uncertain alternatives into reference gambles (gambles between the best and worst alternatives) with different probabilities. It is on the basis of this kind of transformation that the proposition can be proved.

Suppose you face the simple decision problem shown in Figure 14.4. For convenience, assume the payoffs are in dollars. The continuity axiom says that we can find reference gambles that are equivalent to the outcomes at the ends of the branches. Suppose (hypothetically) that you are indifferent between $15 and the following reference gamble:

> Win $40 with probability 0.36
>
> Win $10 with probability 0.64

Likewise, suppose you are indifferent between $20 and the next reference gamble:

> Win $40 with probability 0.60
>
> Win $10 with probability 0.40

The substitutability axiom says that we can replace the original outcomes with their corresponding reference gambles, as in Figure 14.5. (We have replaced the outcomes in Lottery B with "trivial" lotteries. The reason for doing so will become apparent.) The substitutability axiom says that you are indifferent between A and A' and also between B and B'. Thus, the problem has not changed.

Now use the reduction-of-compound-events axiom to reduce the decision tree (Figure 14.6). In performing this step, the overall probability of winning 40 in A" is 0.5(0.60) + 0.5(0.36), or 0.48; it is similarly calculated for winning 10. For the lower half of the tree, we just retrieve B again. The monotonicity axiom means we prefer A" to B", and so by transitivity (which says that $A'' \sim A' \sim A$ and $B'' \sim B' \sim B$) we must prefer A to B in the original decision.

To finish the demonstration, we must show that it is possible to come up with numbers that represent utilities so that a higher expected utility implies a preferred alternative, and vice versa. In this case, we need utilities that result in a higher expected utility for Alternative A. Use the probabilities assessed above in the reference gambles as those utilities. (Now you can see the reason for the extension of B in Figure 14.5; we need the probabilities.) We can redraw the original decision tree, as in Figure 14.7, with the utilities in place of the original monetary payoffs.

Figure 14.4

A simple decision problem under uncertainty.

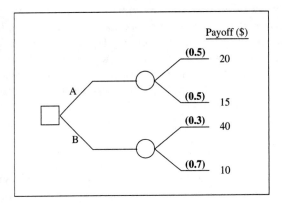

Figure 14.5

Decision tree after substituting reference gambles for outcomes.

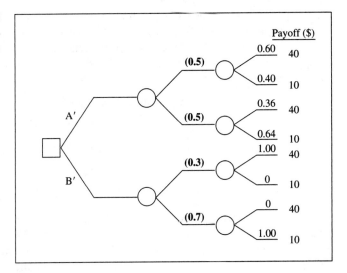

Figure 14.6

Reducing the decision tree.

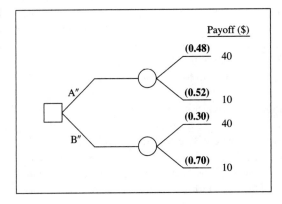

Figure 14.7

Original decision tree with utility values replacing monetary values.

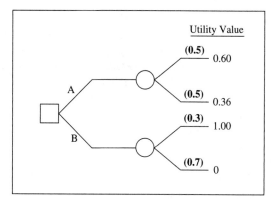

Calculating expected utilities shows that A has the higher expected utility and hence should be preferred:

$$EU(A) = 0.5(0.60) + 0.5(0.36) = 0.48$$
$$EU(B) = 0.3(1.00) + 0.7(0) = 0.30$$

This exercise may seem both arcane and academic. But put simply, if you think it is reasonable to behave according to the axioms at the beginning of the section, then (1) it is possible to find your utility values, and (2) you should make decisions that would be consistent with the maximization of expected utility.

In our initial Chapter 13 discussions of utility assessment, the claim was made that a utility function could be scaled (multiplied by a positive constant and added to a constant) without changing anything. Now you should see clearly that the whole purpose of a utility function is to rank-order risky prospects. Take Alternatives A and B. We have concluded that A is preferred to B because EU(A) = 0.48, which is greater than EU(B) = 0.30. Suppose that we take the utility numbers, as shown in Figure 14.7, and scale them. That is, let

$$U'(x) = a + bU(x)$$

where $b > 0$ and $U(x)$ is our original utility function (from Figure 14.7). We can calculate the expected utilities of A and B on the basis of the following new utility function:

$$EU'(A) = 0.5[a + bU(20)] + 0.5[a + bU(15)]$$
$$= a + b[0.5U(20) + 0.5U(15)]$$
$$= a + bEU(A)$$
$$= a + b(0.48)$$

$$EU'(B) = 0.3[a + bU(40)] + 0.7[a + bU(10)]$$
$$= a + b[0.3U(40) + 0.7U(10)]$$
$$= a + bEU(B)$$
$$= a + b(0.30)$$

It should be clear that EU'(A) will be greater than EU'(B) as long as $b > 0$. As indicated, the implication is that we can scale our utility functions linearly with the constants a and b without changing the rankings of the risky alternatives in terms of expected utility. Specifically, this means that no matter what the scale of a utility function, we always can rescale it so that the largest value is 1 and the smallest is 0. (For that matter, you can rescale it so that the largest and smallest values are whatever you want!)

If you look carefully at our transformations of the original decision problem above, you will see that we never explicitly invoked either the invariance or finiteness axioms. Why are these axioms important? The invariance axiom says that we need nothing but the payoffs and the probabilities; nothing else matters. (In the context of multiattribute utility as discussed in the next two chapters, we would need consequences on all relevant dimensions.)

The finiteness axiom assures us that expected utility will never be infinite, and so we always will be able to make meaningful comparisons. To see the problem with unbounded utility, suppose that you have been approached by an evangelist who has told you that unless you accept his religion, you will burn in Hell for eternity. If you attach an infinitely negative utility to an eternity in Hell (and it is difficult to imagine a worse fate), then no matter how small your subjective probability that the evangelist is right, as long as it is even slightly positive you must be compelled to convert. This is simply because any small positive probability multiplied by the infinitely negative utility will result in an infinitely negative expected utility. Similar problems are encountered if an outcome is accorded infinite positive utility; you would do anything at all if doing so gave you even the slightest chance at achieving some wonderful outcome. Thus, if an outcome in your decision problem has unbounded utility, then the expected utility approach does not help much when it comes to making the decision.

Paradoxes

Even though the axioms of expected utility theory appear to be compelling when we discuss them, people do not necessarily make choices in accordance with them. Research into these behavioral paradoxes began almost as early as the original research into utility theory itself, and now a large literature exists for many aspects of human behavior under uncertainty. Much of this literature is reviewed in von Winterfeldt and Edwards (1986) and Hogarth (1987). We will cover a few high points to indicate the nature of the results.

Framing effects are among the most pervasive paradoxes in choice behavior. Tversky and Kahneman (1981) show how an individual's risk attitude can change depending on the way the decision problem is posed—that is, on the "frame" in which a problem is presented. The difficulty is that the same decision problem usually can be expressed in different frames. A good example is the influenza-outbreak problem at the beginning of the chapter. You may have noticed that Program A is the same as C and that B is the same as D. It all depends on whether you think in terms of deaths or lives saved. Many people prefer A on one hand, but D on the other.

To a great extent, the reason for the inconsistent choices appears to be that different points of reference are used to frame the problem in two different ways. That is, in Programs A and B the reference point is that 600 people are expected to die, but some may be saved. Thus, we think about gains in terms of numbers of lives saved. On the other hand, in Programs C and D, the reference point is that no people would be expected to die without the disease. In this case, we tend to think about lives lost. One of the important general principles that psychologists Kahneman and Tversky and others have discovered is that people tend to be risk-averse in dealing with gains but risk-seeking in deciding about losses. A typical assessed utility function for changes in wealth is shown in Figure 14.8. These results have been obtained in many different behavioral experiments (for example, see Swalm 1966).

More fundamental than the risk-averse–risk-seeking dichotomy is that the reference point or status quo can be quite flexible in some situations and inflexible in others. For example, the influenza-outbreak example can be viewed with relative ease from either frame. For many people, the financial status quo changes as soon as they file their income-tax return in anticipation of a refund; they "spend" their refund, usually in the form of credit, long before the check arrives in the mail. In other cases, individuals may maintain a particular reference point far longer than they should. A case in point is that gamblers often try to make up their losses; they throw good money after bad. Here "gamblers" can refer to individuals in casinos as well as to stock market investors or even managers who maintain a commitment to a project that has obviously gone sour. Typically, such a gambler will argue that backing out of a failed project amounts to a waste of the resources already spent.

Is a specific axiom being violated when a decision maker's choices exhibit a framing effect? The answer to this question is not exactly clear. Although many possibilities exist, the invariance axiom may be the weak link in this case. It may be that

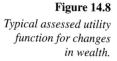

Figure 14.8

Typical assessed utility function for changes in wealth.

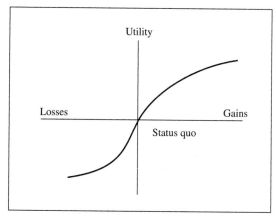

payoffs (or utilities) and probabilities are not sufficient to determine a decision maker's preferences. Some understanding of the decision maker's frame of reference also may be required.

For another example, consider the problem:

Allais Paradox

You have two decisions to make.

Decision 1:   **A**   Win $1 million with probability 1
                  B   Win $5 million with probability 0.10
                          Win $1 million with probability 0.89
                          Win $0 with probability 0.01

Before proceeding, choose A or B. Would you give up a sure $1 million for a small chance at $5 million and possibly nothing?

Decision 2:   **C**   Win $1 million with probability 0.11
                          Win $0 with probability 0.89
                  D   Win $5 million with probability 0.10
                          Win $0 million with probability 0.90

Now choose C or D in Decision 2.

This is the well-known Allais Paradox (Allais 1953; Allais and Hagen 1979). The decisions are shown in decision-tree form in Figure 14.9. Experimentally, as many as 82% of subjects prefer A over B and 83% prefer D over C. But we can easily show that choosing A on the one hand and D on the other is contrary to expected utility maximization. Let $U(\$0) = 0$ and $U(\$5,000,000) = 1$; they are the best and worst outcomes. Then,

$$EU(A) = U(\$1 \text{ million})$$
$$EU(B) = 0.10 + 0.89U(\$1 \text{ million})$$

Thus, A is preferred to B if and only if

$$U(\$1 \text{ million}) > 0.10 + 0.89 \ U(\$1 \text{ million})$$

or

$$U(\$1 \text{ million}) > 0.91$$

Now for Decision 2,

$$EU(C) = 0.11 \ U(\$1 \text{ million})$$
$$EU(D) = 0.10$$

so D is preferred to C if and only if $U(\$1 \text{ million}) < 0.91$.

Figure 14.9

Choices in the Allais Paradox.

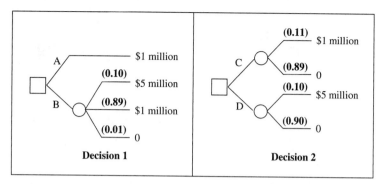

U($1 million) cannot be both greater than and less than 0.91 at the same time, so choosing A and D is not consistent with expected utility. Consistent choices are A and C or B and D. Kahneman and Tversky (1981) have attributed this common inconsistency to the *certainty effect,* whereby individuals tend to place too much weight on a certain outcome relative to uncertain outcomes. In the Allais Paradox, the certainty effect would tend to make individuals overvalue A in Decision 1, possibly leading to an inconsistent choice. When confronted with their inconsistency, some individuals revise their choices. Would you revise yours in light of this discussion?

Another way to look at the Allais Paradox is to structure the decision problem using lottery tickets in a hat. Imagine that 100 tickets are numbered sequentially from 1 to 100 and placed in a hat. One ticket will be drawn at random, and you will receive a prize depending on the option you choose and the number on the ticket. Prizes for A and B are described in Table 14.1. For A, you would win $1 million regardless of the ticket chosen, but for B you would win nothing if Ticket 1 is chosen, $5 million for Tickets 2 through 11, and $1 million for Tickets 12 through 100. Note that you win the same ($1 million) in the two lotteries for Tickets 12 through 100. Thus, your choice should depend only on your preferences regarding the outcomes for Tickets 1 through 11.

The same kind of thing can be done for Options C and D, which also are shown in Table 14.1. If you choose C, you win $1 million for Tickets 1–11 and nothing for Tickets 12–100. In D you would win nothing for Ticket 1, $5 million for Tickets 2–11, and nothing for Tickets 12–100. Again, you win exactly the same thing (nothing) in both C and D for Tickets 12–100, and so your preferences between C and D should depend only on your preferences regarding Tickets 1–11. As you can see, the prizes associated with Tickets 1–11 are the same for Options A and C on the one hand and B and D on the other. Thus, if you prefer A to B you also should prefer C to D, and vice versa.

It is intuitively reasonable that your preferences should not depend on Tickets 12–100, because the outcome is the same regardless of the decision made. This is an example of the *sure-thing principle,* which says that our preferences over lotteries or risky prospects should depend only on the parts of the lotteries that can lead to different outcomes. The idea of the sure thing is that, in the choice between A and B,

Table 14.1

Prizes in the Allais Paradox. Tickets 1–100 are placed in a hat, and one ticket is drawn randomly. The dollar amounts shown in the table are the prizes for the four options.

	Tickets		
Option	1	2–11	12–100
A	$1 million	$1 million	$1 million
B	0	$5 million	$1 million
C	$1 million	$1 million	0
D	0	$5 million	0

winning $1 million is a sure thing if one of the tickets from 12 to 100 is drawn, regardless of your choice. If, as in the Allais choices, there are possible outcomes that have the same value to you regardless of the option you choose, and these outcomes occur with the same probability regardless of the option chosen, then you can ignore this part of the lottery. The sure-thing principle can be derived logically from our axioms. For our purposes, it is best to think of it as a behavioral principle that is consistent with our axioms; if you agree to behave in accordance with the axioms, then you also should obey the sure-thing principle. If your choices in the Allais Paradox were inconsistent, then your preferences violated the sure-thing principle.

Implications

As we learned above, people do not always behave according to the behavioral axioms. This fact has distinct implications for the practice of decision analysis. First, there are implications regarding how utility assessments should be made. Second, and perhaps more intriguing, are the implications for managers and policy makers whose jobs depend on how people actually do make decisions. It may be important for such managers to consider some of the above-described behavioral phenomena. In this section, we will look at the issues involved in each of these areas.

Implications for Utility Assessment

We rely on assessments of certainty equivalents and other comparisons to find a utility function. Given the discussion above, it is clear that, in the assessment process, we ask decision makers to perform tasks that we know they do not always perform consistently according to the axioms! Thus, it is no surprise that the behavioral paradoxes discussed above may have some impact on the way that utility functions should be assessed. Our discussion here will be brief, not because the literature is particularly large, but because many research questions remain to be answered.

There are several approaches to the assessment of utilities. In Chapter 13 we introduced the certainty-equivalent approach (find a CE for a specified gamble) and the probability-equivalent approach (find a probability that makes a reference lottery equivalent to a specific certain amount). If we think of the general case (Figure 14.10) as being indifferent between CE for sure and the reference lottery:

Win G with probability p

Win L with probability $1 - p$

then it is clear that we must preset three out of the four variables p, CE, G, and L. The selection of the fourth, which makes the decision maker indifferent between the two prospects, allows the specification of a utility value. Most practioners use either the CE or PE approaches. Assessing a certainty equivalent involves fixing G, L, and p and assessing CE; the probability-equivalent method involves fixing CE, G, and L and assessing p.

The relative merits of the possible assessment techniques have been discussed to some degree. Hershey, Kunreuther, and Shoemaker (1982) report that the use of the CE approach tends to result in more risk-averse responses than does the PE approach when the consequences are gains. On the other hand, when the consequences are losses, the CE approach results in more risk-seeking behavior. When using the PE approach, many people appear to exhibit certain forms of probability distortion. Although the evidence is far from conclusive, it indicates that people deal best with 50–50 chances. This empirical result appears to be related to the certainty effect discussed above.

Clearly, these results have an impact on the assessment of utility functions; the nature of the decision maker's responses and hence the deduced risk attitude can depend on the way that questions have been posed in the assessment procedure. McCord and De Neufville (1986) have suggested, in light of the distortion from the certainty effect, that utilities should not be assessed using the CE approach, which requires a decision maker to compare a lottery with a certain quantity. The CE approach, they argue, contains a built-in bias. They suggest that it would be more appropriate to assess utilities by comparing lotteries. For example, suppose that A is the best outcome and C is the worst, and we would like to assess U(B), which is somewhere between A and C. McCord and De Neufville's technique would have the decision maker assess the probability p that produces indifference between the two

Figure 14.10

A general framework for assessing utilities.

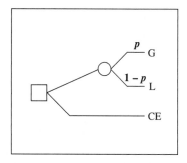

If three of the four variables (G, L, CE, and p) are fixed, assessment of the fourth such that the decision maker is indifferent between the two decision branches permits establishment of utility values.

lotteries in the decision tree in Figure 14.11. Because p makes the decision maker indifferent, we can set up the equation

$$0.5\ U(B) + 0.5\ U(C) = p\ U(A) + (1-p)\ U(C)$$

Substituting $U(A) = 1$ and $U(C) = 0$ and rearranging, this becomes

$$U(B) = 2p$$

Thus, the utility is relatively easy to find once the assessment has been made.

The McCord–De Neufville approach does indeed lead to utility assessments that are less risk-averse than those made with certainty equivalents. Figure 14.12 compares the two methods in terms of hypothetical utility functions that might be assessed. It may be the case, however, that decision makers have a harder time thinking about comparable lotteries than about certainty equivalents.

Managerial and Policy Implications

The idea that people actually make decisions that are sometimes inconsistent with decision-analysis principles is not new. In fact, the premise of a book such as this one

Figure 14.11

The McCord–De Neufville utility-assessment procedure.

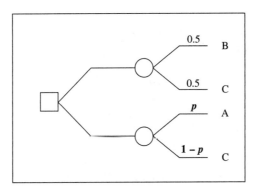

Figure 14.12

Comparing utility functions assessed with different approaches.

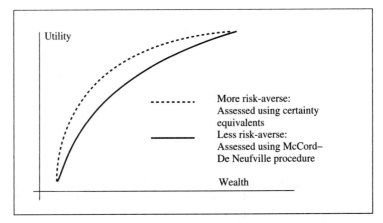

is that it is possible to improve one's decision-making skills. But now we have seen that individuals behave in certain specific and predictable ways. What implications does this have for managers and policy makers?

The most fundamental issue has to do with the reference point or status quo. What is the status quo in a particular decision situation? How do people establish a status quo for decision-making purposes? Can the perception of the status quo be manipulated, and, if so, how? Complete answers to these questions are certainly beyond the scope of our discussion here, but we can discuss important examples in which the status quo plays an important role.

First is the problem of "sunk costs." As briefly mentioned earlier, managers frequently remain committed to a project that obviously has gone bad. In decision-analysis terms, the money that already has been spent on the project no longer should influence current and future decisions. Any decisions, particularly whether to continue or abandon a project, must be forward-looking—that is, consider only future cash flows. To account for the sunk costs, they would have to be accounted for on every branch and hence could not affect the relative ordering of the alternatives.

What do we say to an individual who seems unable to ignore sunk costs? One piece of advice is to make sure that the individual understands exactly what the status quo is. If the individual wants to "throw good money after bad," then he or she may be operating on the basis of a previous status quo; abandoning the project may look like accepting a sure loss of the money already invested. From this perspective, it might seem quite reasonable instead to seek a risky gain by remaining committed to the project. But the real status quo is that the project is unlikely to yield the anticipated benefits. From this new perspective, abandoning the project amounts to the avoidance of a losing venture; funds would be better invested elsewhere.

In other cases, a problem's frame may be specified with equal validity in several ways. Consider the case of seat belts. If people view seat belts as inconvenient and uncomfortable, then they may refuse to wear them; the status quo is the level of comfort when unbuckled. Suppose, however, the status quo were established as how well off people are in general. Who would want to risk the loss of a healthy family, a productive life, and possibly much of one's savings? The use of a seat belt is a form of inexpensive insurance to avoid a possible dramatic loss relative to this new status quo.

A similar argument can be made in the area of environmental protection. People may view environmental programs as anything from inconveniences (being forced to separate recyclable materials from trash) to major economic impediments (adhering to EPA's complex regulations). The implied status quo is the current level of direct costs; the programs represent a sure loss relative to this status quo. If, however, emphasis is given to the current overall condition of the community, nation, or world, then the increased cost induced by the programs can be viewed as a small amount of insurance necessary to avoid a large loss relative to our current overall welfare.

A third situation in which the perception of the status quo can be productively manipulated is in the area of creativity enhancement. In Chapter 6 we discussed removing blocks to creativity. In organizations, one key approach is to make the environment as nonthreatening as possible so that individuals will not be afraid to present new ideas. In the terminology of this section, the goal is to establish a status quo in the organizational environment such that individuals will view the presentation of a new idea as one in which the loss is small if the idea fails but otherwise the potential gain is large.

In the discussion so far we have focused our attention on issues concerning the status quo. We turn now to the certainty effect, or the fact that we tend to overvalue sure outcomes relative to uncertain ones. This psychological phenomenon can have interesting effects on our decision making, especially under conditions of uncertainty.

In our western culture, we abhor uncertainty. Indeed, it is appropriate to expend resources for information in order to reduce uncertainty. This is true only up to a point, however, as we discussed in Chapter 12. But where may the certainty effect lead? We may try too hard to eliminate uncertainty altogether. For example, in organizations, there may be a tendency to spend far too much effort and far too many resources on tracking down elusive information in an effort to know the "truth." More insidious is a tendency to ignore uncertainty altogether. We often tend to view forecasts, for example, as perfect indicators of the future, ignoring their inherent uncertainty. The use of carefully assessed probabilistic forecasts is an important way to avoid this problem.

As a society, the certainty effect may be a factor in our preoccupation with certain kinds of health and environmental risks. Many individuals—activists in various causes—would prefer to eliminate risk altogether and thus they call for stringent regulations on projects that appear to pose risks to the population or environment. Certainly these individuals play a valuable role in drawing our attention to important issues, but achieving zero risk in our society is not only impractical, but also impossible. A sound approach to the management of risk in our society requires a consideration of both the benefits and costs of reducing risk to various levels.

In this brief section we have considered certain implications of the behavioral paradoxes for the application of decision analysis (especially utility assessment), as well as for organizational decisions and policy making. But we have only scratched the surface here. As research progresses in this fertile area, many other behavioral paradoxes and their implications will be studied. The examples we have considered illustrate the pervasiveness of the effects as well as their importance.

A Final Perspective

We have discussed a variety of inconsistencies in human behavior. Do these inconsistencies invalidate expected utility theory? The argument all along has been that people do not seem to make coherent decisions without some guidance. If a decision maker does wish to make coherent decisions, then a careful decision-analysis approach, including a careful assessment of personal preferences, can help the decision maker in looking for better decisions.

It is easy to get the impression that utility and probability numbers reside in a decision maker's mind and that the assessment procedure simply elicits those numbers.

But the process is more than that. Just as the process of structuring a decision problem helps the decision maker understand the problem better and possibly leads to the recognition of new alternatives for action, so may the assessment process provide a medium in which the decision maker actually can develop his or her preferences or beliefs about uncertainty. The assessment process helps to mold the decision maker's subjective preferences and beliefs. How many of us, for example, have given much thought to assessing a utility function for a cup of coffee (Problem 14.16) or considered the probability distribution for a grade in a decision-analysis course (Problem 8.11)? Certainly the assessed numbers do not already exist inside our heads; they come into being through the assessment procedure. This is exactly why assessment often requires hard thinking and why decisions are hard to make. Thus, perhaps the best way to think about the assessment process is in constructive terms. Reflecting on the decision problem faced and considering our preferences and beliefs about uncertainty provides a basis not only for building a model of the problem and necessary beliefs, but also for constructing those beliefs in the first place. This view may explain some of the behavioral paradoxes that we have discussed; individuals who have not thought long and hard in thoroughly developing their preferences and beliefs might have a tendency to make inconsistent judgments.

This constructive view of the assessment process is a fundamental matter. You may recall that we introduced these ideas way back in Chapter 1 when we suggested that the decision-analysis process actually helps a decision maker develop his or her thoughts about the structure of the problem, beliefs about uncertainty, and preferences. Again, the idea of a requisite model is appropriate. A decision model is requisite in terms of preferences if it captures the essential preference issues that matter to the decision maker for the problem at hand. Thus, this constructive view suggests that decision analysis should provide an environment in which the decision maker can systematically develop his or her understanding of the problem, including preferences and beliefs about uncertainty. But how can we be sure that decision analysis does provide such an environment? More research is required to answer this question completely. The decision-analysis approach, however, does encourage decision makers to think about the issues in a systematic way. Working through the decision-analysis cycle (modeling, analyzing, performing sensitivity analysis, and then modeling again) should help the decision maker to identify and think clearly about the appropriate issues. Thus, the argument is that good initial decision-analysis structuring of a problem will lead to appropriate assessments and careful thought about important issues.

SUMMARY

We have covered a lot of ground in this chapter. We started with the axioms that underlie utility theory and showed how those axioms imply that an individual should make decisions that are consistent with the maximization of expected utility. Then we examined some of the behavioral paradoxes that have been documented. These are situations in which intelligent people make decisions that violate one or more of the axioms, and thus make decisions that are inconsistent with expected utility.

These paradoxes do not invalidate the idea that we should still make decisions according to expected utility; recall that the basic goal of decision analysis is to help people improve their decision-making skills. But the paradoxes do have certain implications. We explored some of these implications, including the possibility of assessing equivalent lotteries rather than certainty equivalents and implications for policy makers. We ended with a constructive perspective of assessment, whereby it provides a medium within which the decision maker can explore and develop preferences and beliefs.

EXERCISES

14.1 In your own words explain why the axioms that underlie expected utility are important to decision analysis.

14.2 From a decision-analysis perspective, why is it worthwhile to spend time studying the kinds of paradoxes described in this chapter?

14.3 Most people are learning constantly about themselves and their environment. Our tastes develop and change as our environment changes. As new technologies arise, new risks are discovered. What are the implications of this dynamic environment for decision analysis?

14.4 **a** Find the value for p that makes you indifferent between

| **Lottery 1** | Win $1000 with probability p |
| | Win $0 with probability $1 - p$ |

and

| **Lottery 2** | Win $400 for sure |

b Now find the q that makes you indifferent between

| **Lottery 3** | Win $400 with probability 0.50 |
| | Win $0 with probability 0.50 |

and

| **Lottery 4** | Win $1000 with probability q |
| | Win $0 with probability $1 - q$ |

c According to your assessment in part a, U($400) = p. In part b, U($400) = $2q$. Explain why this is the case.

d To be consistent, your assessments should be such that $p = 2q$. Were your assessments consistent? Would you change them? Which assessment do you feel most confident about? Why?

QUESTIONS AND PROBLEMS

14.5 We used the phrase "rules for clear thinking" to refer to the axioms described in this chapter. However, these "rules" do not cover every aspect of decision making as we have discussed it in this book. In particular, the axioms do not tell us in detail how to structure a problem, how to generate creative alternatives, or how to perform a sensitivity analysis. Can you give some "rules for clear thinking" that would help a decision maker in these aspects of decision making? (See Frisch and Clemen 1994.)

14.6 Imagine that you collect bottles of wine as a hobby. How would you react to these situations?

a You have just learned that one of the bottles (Wine A) that you purchased five years ago has appreciated considerably. A friend has offered to buy the bottle from you for $100. His offer is a fair price. Would you sell the bottle of wine?

b A second bottle of wine (Wine B) is viewed by experts as being equivalent in value to Wine A. You never purchased a bottle of Wine B, but a casual acquaintance did. In conversation at a party, he offers to sell you his bottle of Wine B for $100. Would you buy it?

Many people would neither sell Wine A nor buy Wine B. Explain why this pattern of choices is inconsistent with expected utility. What other considerations might be taken into account?

14.7 Consider the following two scenarios:

a You have decided to see a show for which tickets cost $20. You bought your ticket in advance, but as you enter the theater, you find that you have lost the ticket. You did not make a note of your seat number, and the ticket is not recoverable or refundable in any way. Would you pay another $20 for a second ticket?

b You have decided to see a show whose tickets cost $20. As you open your wallet to pay for the ticket, you discover that you have lost a $20 bill. Would you still buy a ticket to see the show?

Many individuals would not purchase a second ticket under the first scenario, but they would under the second. Explain why this is inconsistent with expected utility. How would you explain this kind of behavior?

14.8 Imagine yourself in the following two situations:

a You are about to go shopping to purchase a stereo system at a local store. Upon reading the newspaper, you find that another stereo store across town has the system you are interested in for $1089.99. You had been planning to spend $1099.95, the best price you had found after considerable shopping. Would you drive across town to purchase the stereo system at the other store?

b You are about to go shopping to purchase a popcorn popper at a local hardware store for $19.95, the best price you have seen yet. Upon reading the paper, you discover that a department store across town has the same popper for $9.99. Would you drive across town to purchase the popper at the department store?

Many people would drive across town to save money on the popcorn popper, but not on the stereo system. Why is this inconsistent with expected utility? What explanation can you give for this behavior?

14.9 A classified advertisement, placed by a car dealer, reads as follows:

> ***NEW***
> 1991 **BMW** 525i
> Retail $38,838.50
> **WAS** $32,998
> **NOW**
> $32,498

a What is your reaction to this advertisement?

b In the same paper, a computer dealer advertised a $500 discount on a computer system, marked down from $3495 to $2995. Which do you think is a better deal, the special offer on the car or the computer? Why?

14.10 Consider these two scenarios:

a You have made a reservation to spend the weekend at the coast. To get the reservation, you had to make a nonrefundable $50 deposit. As the weekend approaches, you feel a bit out of sorts. On balance, you decide that you would be happier at home than at the coast. Of course, if you stay home, you forfeit the deposit. Would you stay home or go to the coast? What arguments would you use to support your position?

b You have decided to spend the weekend at the coast and have made a reservation at your favorite resort. While driving over, you discover a new resort. After looking it over, you realize that you would rather spend your weekend here. Your reservation at the coast can be canceled easily at no charge. Staying at the new resort, however, would cost $50 more. Would you stay at the new resort or continue to the coast?

Are your decisions consistent in parts a and b? Explain why or why not.

14.11 Even without a formal assessment process, it often is possible to learn something about an individual's utility function just through the preferences revealed by choice behavior. Two persons, A and B, make the following bet: A wins $40 if it rains tomorrow and B wins $10 if it does not rain tomorrow.

a If they both agree that the probability of rain tomorrow is 0.10, what can you say about their utility functions?

b If they both agree that the probability of rain tomorrow is 0.30, what can you say about their utility functions?

c Given no information about their probabilities, is it possible that their utility functions could be identical?

d If they both agree that the probability of rain tomorrow is 0.20, could both individuals be risk-averse? Is it possible that their utility functions could be identical? Explain.

Source: R. L. Winkler (1972) *Introduction to Bayesian Inference and Decision.* New York: Holt.

14.12   Assess your utility function for money in the bank over a range from $100 to $20,000 using the McCord–De Neufville procedure described in this chapter. Plot the results of your assessments on the same graph as the assessments made for Problem 13.12. Discuss the differences in your assessments from one method to the other. Can you explain any inconsistencies among them?

14.13   Assume that you are interested in purchasing a new model of a personal computer whose reliability has not yet been perfectly established. Measure reliability in terms of the number of days in the shop over the first three years that you own the machine. (Does this definition of reliability pass the clarity test?) Now assess your utility function for computer reliability over the range from 0 days (best) to 50 days (worst). Use whatever assessment technique with which you feel most comfortable, and use computer assessment aids if they are available.

14.14   You are in the market for a new car. An important characteristic is the life span of the car. (Define lifespan as the number of miles driven until the car breaks down, requiring such extensive repairs that it would be cheaper to buy an equivalent depreciated machine.) Assess your utility function for automobile life span over the range from 40,000 to 200,000 miles.

14.15   Being a student, you probably have well-developed feelings about homework. Given the same amount of material learned, the less the better, right? (I thought so!) Define homework as the number of hours spent outside of class on various assignments that enter into your final grade. Now, assuming that the amount of material learned is the same in all instances, assess your utility function for homework over the range from 0 hours per week (best) to 20 hours per week (worst). (*Hint:* You may have to narrow the definition of homework. For example, does it make a difference what kind of course the homework is for? Does it matter whether the homework is term papers, case studies, short written assignments, oral presentations, or something else?)

14.16   We usually think of utility functions as always sloping upward (more is better) or downward (less is better; fewer nuclear power plant disasters, for example). But this is not always the case. In this problem, you must think about your utility function for coffee versus milk.

Imagine that you are about to buy a cup of coffee. Let c ($0 \leq c \leq 1$) represent the proportion of the contents of the cup accounted for by coffee, and $1 - c$ the proportion accounted for by milk.

a   Assess your utility function for c for $0 \leq c \leq 1$. Note that if you like a little milk in your coffee, the high point on your utility function may be at a value of c somewhere between 0 and 1.

b   Compare (A) the mixture consisting of proportions c of coffee and $1 - c$ of milk in a cup and (B) the lottery yielding a cup of coffee with probability c and a cup of milk with probability $1 - c$. (The decision tree is shown in Figure 14.13.) Are the expected amounts of milk and coffee the same in A and B? [That is, if you calculate E(c), is it the same in A and B?] Is there any value of c for which you are indifferent between A and B? (How about when c is 0 or 1?) Are you indifferent between A and B for the value of c at the high point of your utility function?

c   How would you describe your risk attitude with respect to c? Are you risk-averse or risk-prone, or would some other term be more appropriate?

Figure 14.13

Decision tree for Problem 14.16b.

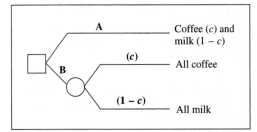

14.17   In a court case, a plaintiff claimed that the defendant should pay her $3 million for damages. She did not expect the judge to award her this amount; her expected value actually was $1 million. The defendant also did not believe that the court would award the full $3 million, and shared the plaintiff's expected value of $1 million.

    a Assuming that the plaintiff is thinking about the award in terms of gains, explain why you might expect her to be risk-averse in this situation. If she is risk-averse, what kind of settlement offer might she accept from the defendant?

    b Assuming that the defendant is thinking about the situation in terms of losses, explain why you might expect him or her to be risk-seeking. What would this imply about settlement offers to which the defendant might agree? (*Hint:* Draw an example of a risk-seeking utility curve over the possible negative payoffs for the defendant. Now find the certainty equivalent.)

    c Discuss your answers to parts a and b. What are the implications for settlements in real-world court cases? What would happen if the defendant's expected value were less than the plaintiff's?

CASE STUDIES

THE LIFE INSURANCE GAME

Peggy Ewen sat back in her chair and listened as Tom Pitman tried to explain. "I don't know what's going on," Tom said. "I have no trouble making the phone calls, and I seem to be able to set up in-home visits. I am making at least as many visits as anyone else in the office. For some reason, though, I cannot talk them into buying the product. I seem to be unlucky enough to have run into a lot of people who just are not interested in life insurance."

Peggy thought about this. Tom had been with the company for five months now. He was bright and energetic. He had gone through the training program easily, and had appeared to hit the ground running. His paperwork was always perfect. For some reason, though, his career selling life insurance was virtually stalled. His sales rate was only one-third that of the next best salesperson. Why?

Peggy asked, "How do you feel about going to the in-home visits?"

"Fine," Tom replied. "Well, I'm starting to feel a little apprehensive about it just because I'm becoming less sure of myself."

"Well, that's something we'll have to work on. But how do the visits go? What do you talk about? Tell me what a typical visit would be like."

"Let's see. Usually I'll come in, sit down, and we'll chat briefly. They'll offer me a cup of coffee. After a short visit, we get right down to business. I go through the presentation material provided by the company. Eventually we get around to talking about the reasons for purchasing life insurance. I really stress the importance of being able to make up for the loss of income. The presentation material stresses the idea of building up savings for sending kids to school or for retirement. You know, the idea of being sure that the extra money will be there down the road. But I really don't think that's why most people buy life insurance. I think they buy it to be sure that their family will be able to make up for a loss. For just a small premium, they can be sure that the loss won't happen, or at least they can minimize the loss."

Peggy seemed interested in Tom's account. "So you really stress the idea that for a little bit of money they can insure against the loss of income."

"Yes," Tom answered. "I'd rather have them look at life insurance as protection against a potential loss, rather than as a savings mechanism that would provide some sure amount in the future. Most of them know that there are better ways to save, anyway."

"And how would you classify your typical client? What kind of income bracket?"

"Mostly young couples just starting out," said Tom. "Maybe they've just had their first child. Not much income yet. Not much savings, either. We usually discuss this early on in the conversation. In general they seem to be quite aware of their financial situation. Occasionally they are even quite sensitive about it."

Peggy looked at Tom and grinned. "Tom, I do believe that there's something you can do right now to improve your sales rate."

Questions

1   About what issue is Peggy thinking?

2   What are the implications for people interested in selling financial securities such as insurance, annuities, and so on?

3   What are the implications for their customers?

NUCLEAR POWER PARANOIA

Ray Kaplan was disgusted. The rally against the power plant had gone poorly. The idea had been to get a lot of people out to show the utility company that the community did not support the idea of nuclear power. It was just too dangerous! Too much chance of an accident. Sure, the officials always pointed out that there had never been a serious nuclear power plant accident in the United States. They always pointed to the safeguards in the plants, and to the safety features required by federal regulations. "You're safer in our plant than on the freeway," they claimed. Well, the same officials met with the protesters again today and said the same old things. Ray was getting tired of it. He hopped on his motorcycle and rode home. Fast.

He was still ruminating about the rally while he broiled his steak over the charcoal that evening. As he snacked on a bowl of ice cream after dinner, he asked himself, "Can't they see the potential for disaster?"

The next day, Ray decided to mow his lawn and then ride out to the beach. About a mile from his house, he realized that he wasn't wearing his motorcycle helmet. Well, he thought, the beach was only about 20 miles away. Besides, it would be nice to feel the wind in his hair. The ride was nice, even though the traffic was heavier than he had expected. The fumes from one of the trucks really got to him briefly, but fortunately the exit for the beach was right there. As he lay down on his towel to soak up the sunshine, his mind went back to the rally. "Darn," he thought. "I'll probably develop an ulcer just worrying about that silly power plant."

Question

1   What do you think about Ray Kaplan's behavior?

THE MANAGER'S PERSPECTIVE

Ed Freeman just couldn't understand it. Why were the activists so blind? For years the information had been available showing just how safe nuclear power plants were. Study after study had concluded that the risk of fatalities from a nuclear power plant accident was far less than driving a car, flying in a commercial airliner, and many other commonplace activities in which people freely chose to engage. Sure, there had been some close calls, such as Three Mile Island, and there had been the terrible accident at Chernobyl in the Soviet Union. Still, the overall record of the nuclear power industry in the United States was excellent. No one could deny it if they would only compare the industry to others.

His risk assessors had gone through their own paces, documenting the safety features of his plant for the Nuclear Regulatory Commission; it was up to date and, in fact, one of the safest plants in the country. The experts had estimated the probability of an accident at the plant as nearly zero. Furthermore, even if an accident were to occur, the safety systems that had been built in would minimize the public's exposure.

Given all this, he just could not understand the public opposition to the plant. He knew that these were bright people. They were articulate, well read, and able to marshal their supporters with great skill. But they seemed to ignore all of the data as well as the experts' reports and conclusions.

"I guess it takes all kinds," he sighed as he prepared to go back to work.

Questions

1   This case and "Nuclear Power Paranoia" go together. People often are willing to engage voluntarily in activities that are far more risky (in the sense of the probability of a serious injury or death) than living near a nuclear power plant. Why do you think this is the case?

2   What makes new technologies seem risky to you?

REFERENCES

The axioms of expected utility, along with the notion of subjective probability, were first discussed by Ramsey (1931), but the world appears to have ignored him. Most economists refer to "von Neumann-Morgenstern utility functions" because in 1947 von Neumann and Morgenstern published their celebrated *Theory of Games and Economic Behavior* in which they also set forth a set of axioms for choice behavior that leads to maximization of expected utility. The axioms subsequently appeared in a wide variety of forms and in many textbooks. Some examples are Luce and Raiffa (1957), Savage (1954), DeGroot (1970), and, more recently, French (1986). French's text is excellent for those interested in the axiomatic mathematics that underlie decision theory.

The various axioms have been debated widely. Our discussion of the transitivity axiom, for example, was suggested by Dr. Peter Fishburn as part of his acceptance speech for the Ramsey Medal, an award for distinguished contributions to decision analysis. If Peter Fishburn, one of the foremost scholars of decision theory, is willing to concede that intransitive preferences might not be unreasonable, perhaps we should pay attention! Fishburn (1989) summarizes many of the recent developments in the axioms and theory.

The text by von Winterfeldt and Edwards (1986) also contains much intriguing discussion of the axioms from the point of view of behavioral researchers and a discussion of the many paradoxes found in behavioral decision theory. Hogarth (1987) also covers this topic. Tversky and Kahneman (1981) provide an excellent and readable treatment of framing effects, and Kahneman and Tversky (1979) present a theory of behavioral decision making that accounts for many anomalies in individual decision behavior.

The fact that people do not normally follow the axioms perfectly has been the source of much debate about expected utility theory. Many theorists have attempted to relax the axioms in ways that are consistent with the observed patterns of choices that people make (e.g., Fishburn 1988). For the purpose of decision analysis, though, the question is whether we should model what people actually do or whether we should help them to adhere to axioms that are compelling but in some instances difficult to follow because of our own frailties. This debate is taken up by Rex Brown (1992) and Ron Howard (1992) and is a central theme of a collection of articles in Edwards (1992). Luce and von Winterfeldt (1994) make an argument for decision analysis in which the status quo is explicitly used as an essential element in decision models, and subjective assessments are made relative to the status quo.

The debate about axioms is fine as far as it goes, but it is important to realize that many aspects of decision making are not covered by the axioms. In particular, the axioms tell you what to do once you have your problem structured and probabilities and utilities assessed. The axioms do indicate that decisions problems can be decomposed into issues of value (utility) and uncertainty (probability), but no guidance is provided regarding how to determine the important dimensions of value or the important uncertainties in a decision situation. The axioms provide little help in terms of creativity, sensitivity analysis, or even how to recognize a decision situation. Frisch and Clemen (1994) discuss this issue from the point of view of psychological research on decision making and the prescriptive goals of decision analysis.

The constructionist view of decision analysis that is presented in the last section of this chapter does not appear to be widely discussed, although such a view has substantial and fundamental implications for research in decision theory as well as in decision-analysis practice. For more discussion of this topic with regard to utility assessment, see von Winterfeldt and Edwards (1986, p. 356), Fischer (1979), and Payne, Bettman, and Johnson (1993). On the probability side, Shafer and Tversky (1986) view probability and structuring of inference problems in an interesting "constructive" way. Phillips's notion of the development of a requisite model (1982, 1984) and Watson and Buede's (1987) approach to decision analysis contain many of the elements of the constructionist view.

For many people, risk refers primarily to dangerous circumstances that may cause bodily injury, disease, or even death. The two cases, "Nuclear Power Paranoia" and "The Manager's Perspective," introduce the idea of risk to life and limb instead of risk in monetary gambles. Although one might think that measuring such risks would be a straightforward matter of assessing probability distributions for injuries or deaths—for example, with respect to options associated with building or siting a power plant—it turns out that people are very sensitive to certain kinds of risk. How individuals perceive risk has an important impact on how risks should be managed and how to communicate risk information to the public. For good introductory material, see Slovic (1987) and Morgan (1993).

Allais, M. (1953) "Le Comportement de l'Homme Rationnel Devant le Risque: Critique des Postulats et Axiomes de l'École Americaine." *Econometrica,* 21, 503–546.

Allais, M., and J. Hagen (eds.) (1979) *Expected Utility Hypotheses and the Allais Paradox.* Dordrecht, The Netherlands: Reidel.

Brown, R. (1992) "The State of the Art of Decision Analysis: A Personal Perspective." *Interfaces,* 22, 5–14.

DeGroot, M. (1970) *Optimal Statistical Decisions.* New York: McGraw-Hill.

Edwards, W. (ed.) (1992) *Utility Theories: Measurements and Applications.* Boston, MA: Kluwer.

Fischer, G. (1979) "Utility Models for Multiple Objective Decisions: Do They Accurately Represent Human Preferences?" *Decision Sciences,* 10, 451–479.

Fishburn, P. C. (1988) *Nonlinear Preference and Utility Theory.* Baltimore: Johns Hopkins.

Fishburn, P. C. (1989) "Foundations of Decision Analysis: Along the Way." *Management Science,* 35, 387–405.

French, S. (1986) *Decision Theory: An Introduction to the Mathematics of Rationality.* London: Wiley.

Frisch, D., and R. Clemen (1994) "Beyond Expected Utility: Rethinking Behavioral Decision Research." *Psychological Bulletin,* 116, 46–54.

Hershey, J. C., H. C. Kunreuther, and P. J. H. Shoemaker (1982) "Sources of Bias in Assessment Procedures for Utility Functions." *Management Science,* 28, 936–954.

Hogarth, R. (1987) *Judgment and Choice,* 2nd ed. New York: Wiley.

Howard, R. (1992) "Heathens, Heretics, and Cults: The Religious Spectrum of Decision Aiding." *Interfaces,* 22, 15–27.

Kahneman, D., and A. Tversky (1979) "Prospect Theory: An Analysis of Decision under Risk." *Econometrica,* 47, 263–291.

Luce, R. D., and H. Raiffa (1957) *Games and Decisions: Introduction and Critical Survey.* New York: Wiley.

Luce, R. D., and D. von Winterfeldt (1994) "What Common Ground Exists for Descriptive, Prescriptive, and Normative Utility Theories?" *Management Science,* 40, 263–279.

McCord, M., and R. De Neufville (1986) " 'Lottery Equivalents': Reduction of the Certainty Effect Problem in Utility Assessment." *Management Science,* 32, 56–60.

Morgan, G. (1993) "Risk Analysis and Management." *Scientific American,* July, 32–41.

Payne, J., J. Bettman, and E. Johnson (1993) *The Adaptive Decision Maker.* Cambridge: Cambridge University Press.

Phillips, L. D. (1982) "Requisite Decision Modelling." *Journal of the Operational Research Society,* 33, 303–312.

Phillips, L. D. (1984) "A Theory of Requisite Decision Models." *Acta Psychologica,* 56, 29–48.

Ramsey, F. P. (1931) "Truth and Probability." In R. B. Braithwaite (ed.) *The Foundations of Mathematics and Other Logical Essays.* New York: Harcourt, Brace.

Savage, L. J. (1954) *The Foundations of Statistics.* New York: Wiley.

Shafer, G., and A. Tversky (1986) "Languages and Designs for Probability Judgment." *Cognitive Science,* 9, 309–339.

Slovic, P. (1987) "Perception of Risk." *Science,* 236, 280–285.

Swalm, R. (1966) "Utility Theory—Insights into Risk Taking." *Harvard Business Review,* 123–136.

Tversky, A., and D. Kahneman (1981) "The Framing of Decisions and the Psychology of Choice." *Science,* 211, 453–458.

von Neumann, J., and O. Morgenstern (1947) *Theory of Games and Economic Behavior.* Princeton, NJ: Princeton University Press.

von Winterfeldt, D., and W. Edwards (1986) *Decision Analysis and Behavioral Research.* Cambridge: Cambridge University Press.

Watson, S. R., and D. M. Buede (1987) *Decision Synthesis: The Principles and Practice of Decision Analysis.* Cambridge: Cambridge University Press.

EPILOGUE   We began the chapter with the Asian influenza example. This kind of study has been done repeatedly by many different experimenters, and the results are always the same; many of the subjects make inconsistent choices that depend on the framing of the problem. Of course, many of these experiments have been done using college students and other individuals who are not used to making this kind of decision. It would be nice to think that individuals who make difficult decisions often would not be susceptible to such inconsistencies. Unfortunately, such is not the case. Tversky and Kahneman (1981) report the same kinds of inconsistencies among decisions made by university faculty and physicians.

CHAPTER 15

Conflicting Objectives I: Fundamental Objectives and the Additive Utility Function

The utility functions for money that we have considered have embodied an important fundamental trade-off: monetary return versus riskiness. We have argued all along that the basic reason for using a utility function as a preference model in decision making is to capture our attitudes about risk and return. Accomplishing high returns and minimizing exposure to risk are two conflicting objectives, and we already have learned how to model our preference trade-offs between these objectives using utility functions. Thus, we already have addressed two of the fundamental conflicting objectives that decision makers face.

There are many other trade-offs that we make in our decisions, however. Some of the examples from Chapters 3 and 4 involved cost versus safety or fun versus salary. The FAA bomb-detection example balanced detection effectiveness, passenger acceptance, time to implement, and cost. Other examples are familiar: When purchasing cars or computers, we consider not only reliability and life span but also price, ease of use, maintenance costs, operating expenses, and so on. When deciding among school courses, you might be interested in factors such as the complementarity of material covered, importance in relation to your major and career goals, time schedule, and quality of the instructor. Still other examples and possible objectives appear in Figure 15.1 (abstracted from Keeney and Raiffa 1976).

As individuals, we usually can do a fair job of assimilating enough information so that we feel comfortable with a decision. In many cases, we end up saying things like, "Well, I can save some money now and buy a new car sooner," "You get what you pay for," and "You can't have everything." These are obvious intuitive statements that reflect the informal trade-offs that we make. Understanding trade-offs in detail, however, may be critical for a company executive who is interested in acquiring hundreds of personal computers or a large fleet of automobiles for the firm.

Figure 15.1

Four examples of decisions involving complicated preference trade-offs. Source: Keeney, R., and Raiffa, H. (1976) *Decisions with Multiple Objectives: Preference and Value Tradeoffs,* pp. 1–4. Reprinted with the permission of Cambridge University Press.

1 A mayor must decide whether to approve a major new electric power generating station. The city needs more power capacity, but the new plant would worsen the city's air quality. The mayor might consider the following issues:

- The health of residents
- The economic conditions of the residents
- The psychological state of the residents
- The economy of the city and the state
- Businesses
- Local politics

2 Imagine the issues involved in the treatment of heroin addicts. A policy maker might like to:

- Reduce the size of the addict pool
- Reduce costs to the city and its residents
- Reduce crimes against property and persons
- Improve the "quality of life" of addicts
- Improve the "quality of life" of nonaddicts
- Curb organized crime
- Live up to the high ideals of civil rights and civil liberties
- Decrease the alienation of youth
- Get elected to higher political office

3 In choosing a site for a new airport near Mexico City, the head of the Ministry of Public Works had to balance such objectives as:

- Minimize the costs to the federal government
- Raise the capacity of airport facilities
- Improve the safety of the system
- Reduce noise levels
- Reduce access time to users
- Minimize displacement of people for expansion
- Improve regional development (roads, for instance)
- Achieve political aims

4 A doctor prescribing medical treatment must consider a variety of issues:

- Potential health complications for the patient (perhaps death)
- Money cost to the patient
- Patient's time spent being treated
- Cost to insurance companies
- Payments to the doctor
- Utilization of resources (nurses, hospital space, equipment)
- Information gained in treating this patient (may be helpful in treating others)

In this chapter we will present a relatively straightforward way of dealing with conflicting objectives. Essentially, we will create an *additive* preference model; that is, we will calculate a utility score for each objective and then add the scores, weighting them appropriately according to the relative importance of the various objectives. The procedure is easy to use and intuitive. Computer programs are available that make the required assessment process fairly simple. But with the simple additive form comes limitations. Some of those limitations will be exposed in the problems at the end of the chapter. In Chapter 16, we will see how to construct more complicated preference models that are less limiting.

Where are we going? The first part of the chapter reviews some of the issues regarding identifying objectives, constructing objective hierarchies, and creating useful attribute scales. With objectives and attribute scales specified, we move on to the matter of understanding trade-offs. In the section titled "Trading Off Conflicting Objectives: The Basics," we will look at an example that offers a relatively simple choice involving three automobiles and two objectives. In this initial discussion, we will develop intuitive ways to trade off two conflicting objectives. The purpose of this discussion is to introduce ideas, help you focus on the primary issues involved, and provide a framework for thinking clearly.

The next section formally introduces the additive utility function and explores some of its properties. In particular, we introduce indifference curves and the marginal rate of substitution between objectives. We also show that the simple multiobjective approach described in Chapters 3 and 4 is consistent with the additive utility function.

The additive utility function is composed of two different kinds of elements, scores on individual attribute scales and weights for the corresponding objectives. Many different methods exist for assessing the scores and the weights, and these are the topics of the following sections.

In the last section of the chapter, we consider an actual example in which the city of Eugene, Oregon, evaluates four candidate sites for a new public library. The example shows the process of defining objectives and attribute scales, rating the alternatives on each attribute scale, assessing weights, and then putting all of the assessments together to obtain an overall comparison of the four sites.

Objectives and Attributes

In Chapter 3 we discussed at length the notion of conflicting objectives in decision making. Understanding objectives is an important part of the structuring process. We stressed the importance of identifying fundamental objectives, the essential reasons that matter in any given decision context. Fundamental objectives are organized into a hierarchy in which the lower levels of the hierarchy explain what is meant by the higher (more general) levels. Figure 15.2 shows a fundamental-objectives hierarchy for evaluating alternative-energy technologies. We also discussed the notion of means objectives, which are not important in and of themselves but help, directly or indirectly, to accomplish the fundamental objectives. Distinguishing means and fundamental objectives is important because we would like to measure the available alternatives relative to the fundamental objectives, the things we really care about.

The discussion in Chapter 3 also introduced attribute scales. Attribute scales provide the means to measure accomplishment of the fundamental objectives. Some attribute scales are easily defined; if minimizing cost is an objective, then measuring cost in dollars is an appropriate attribute scale. Others are more difficult. In Figure 15.2, for example, there is no obvious way to measure risks related to aesthetic aspects of the environment. In cases like these, we discussed the use of constructed scales and related measurements as proxies.

For convenience, we use the term attribute to refer to the quantity measured on an attribute scale. For example, if an objective is to minimize cost, then the attribute scale might be defined in terms of dollars, and we would refer to dollar cost as the attribute for this objective. Use of the term "attribute" is common in the decision-analysis literature, and many authors use *multiattribute utility theory* (MAUT) to refer to topics covered in Chapters 15 and 16.

Figure 15.2

Objectives hierarchy for evaluating alternative-energy technologies. [Source: von Winterfeldt and Edwards (1986).]

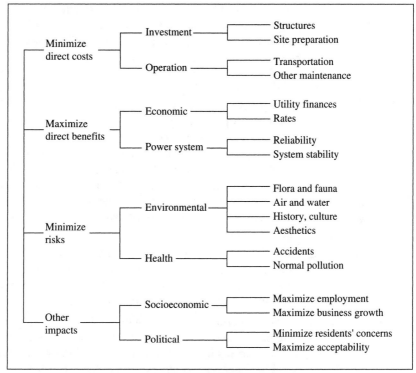

There are a number of essential criteria for fundamental objectives and their attributes. Many of these we discussed in Chapters 3 and 4, and you are encouraged to review those sections. Here is an encapsulation:

1 The set of objectives, as represented by the fundamental-objectives hierarchy, should be complete; it should include all relevant aspects of a decision. The fact that important objectives are missing can be indicated by reluctance to accept the results of an analysis or simply the gnawing feeling that something is missing. If the results "just don't feel right," ask yourself what is wrong with the alternatives that the analysis suggests should be preferred. Careful thought and an honest answer should reveal the missing objectives.

2 At the same time, the set of objectives should be as small as possible. Too many objectives can be cumbersome and hard to grasp. Keep in mind that the objectives hierarchy is meant to be a useful representation of objectives that are important to the decision maker. Furthermore, each objective should differentiate the available alternatives. If all of the alternatives are equivalent with regard to a particular objective (as measured by its corresponding attribute), then that objective will not be of any help in making the decision.

3 The set of fundamental objectives should not be redundant. That is, the same objectives should not be repeated in the hierarchy, and the objectives should not be closely related.

4   As far as possible, the set of objectives should be *decomposable.* That is, the decision maker should be able to think about each objective easily without having to consider others. For example, in evaluating construction bids, the cost of the project and the amount of time required may be important attributes. In most cases, we can think about these attributes separately; regardless of the cost, it always would be preferable to complete the project sooner, and vice versa. Thus, the objectives would be decomposable into these two attributes, which can be considered independently. On the other hand, if you are deciding which course to take, you may want to choose the most interesting topic and at the same time minimize the amount of effort required. These attributes, however, are related in a way that does not permit decomposition; whether you want to put in a lot of effort may depend on how interested you are in the material. Hence, you may have to alter your set of objectives. For example, you might construct a scale that measures something like the extent to which the course inspires you to work harder and learn more.

5   Means and fundamental objectives should be distinguished. Even if fundamental objectives are difficult to measure and means objectives are used as proxies, it is important to remain clear on why the decision matters in the first place. Doing so can help avoid choosing an inappropriate alternative for the wrong reason.

6   Attribute scales must be operational. They should provide an easy way to measure performance of the alternatives or the outcomes on the fundamental objectives. Another way to put it is that the attribute scales should make it straightforward (if not easy) to fill in the cells in the consequence matrix.

Trading Off Conflicting Objectives: The Basics

The essential problem in multiobjective decision making is deciding how best to trade off increased value on one objective for lower value on another. Making these trade-offs is a subjective matter and requires the decision maker's judgment. In this section, we will look at a simple approach that captures the essence of trade-offs. We will begin with an example that involves only two objectives.

Choosing an Automobile: An Example

Suppose you are buying a car, and you are interested in both price and life span. You would like a long expected life span—that is, the length of time until you must replace the car—and a low price. (These assumptions are made for the purpose of this example; some people might enjoy purchasing a new car every three years, and for them a long life span may be meaningless.) Let us further suppose that you have narrowed your choices down to three alternatives: the Portalo (a relatively expensive sedan with a reputation for longevity), the Norushi (renowned for its reliability), and the Standard Motors car (a relatively inexpensive domestic automobile). You have done some research and have evaluated all three cars on both attributes as shown in Table 15.1. Plotting these three alternatives on a graph with expected life span on the horizontal axis and price on the vertical axis yields Figure 15.3. The Portalo, Norushi, and Standard show up on the graph as three points arranged on an upward-sloping curve. That the three points are ordered in this way reflects the notion, "You get what you pay for." If you want a longer expected life span, you have to pay more money.

Occasionally alternatives may be ruled out immediately by means of a dominance argument. For example, consider a hypothetical car that costs $15,000 and has an expected life span of seven years (Point A in Figure 15.3). Such a car would be a poor choice relative to the Norushi, which gives a longer expected life for less money. Thus, A would be dominated by the Norushi.

On the other hand, none of the cars under consideration is dominated. With this being the case, how can you choose? The question clearly is, "How much are you willing to pay to increase the life span of your car?" To answer this question, we will start with the Standard and assume that you will purchase it if the others are not better. Is it worthwhile to switch from the Standard to the Norushi? Note that the slope of the line connecting the Norushi and the Standard is $666.67 per year. The switch would be worthwhile if you were willing to pay at least $666.67 for each additional year of life span, or $2000 to increase the expected life span by three years. Would you be willing to pay more than $2000 to increase the expected life of your car by

Table 15.1

Automobile purchase alternatives.

	Portalo	Norushi	Standard Motors
Price ($1000s)	17	10	8
Life Span (Years)	12	9	6

Figure 15.3

Graph of three cars, comparing price and expected life span.

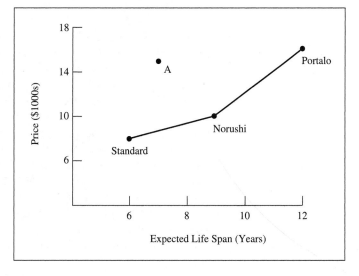

three years? This is a subjective assessment! If you would, then it is worthwhile to switch to the Norushi from the Standard. If not, do not switch.

For the sake of continuing the example, assume that you make the switch, and that you have decided that the Norushi is better for you than the Standard; you would pay at least $2000 for the additional three years of life span. Now, should you switch to the Portalo? Notice that the slope of the line connecting the Norushi and the Portalo is $2333.33 per year, or $7000 for an additional three years. You were willing to pay at least $666.67 for an extra year. Now, what about an extra $7000 for another three years? Are the extra years of expected life span worth this much to you? If so, make the switch to the Portalo. If not, stick with the Norushi.

This simple procedure permits you to move systematically through the alternatives. The idea is to start at one corner of the graph (for example, the Standard) and then consider switching with each of the other alternatives, always moving in the same direction. Once a switch is made, there is never a need to reconsider the previous alternative. (After all, the one to which you changed must be better.) If there were many cars to choose from, the procedure would be to plot them all on a graph, eliminate the dominated alternatives, and then systematically move through the nondominated set, beginning at the lower left-hand side of the graph and considering the alternatives that are to the upper right-hand side.

This procedure works well in the context of two conflicting objectives, and you can see how we are trading off the two. At each step we ask whether the additional benefit from switching (the increase in expected life span) is worth the cost (the increase in price). The same kind of procedure can be used with three or more attributes. The trade-offs become more complicated, however, and graphical interpretation is difficult.

The Additive Utility Function

The automobile example above is intended to give you an intuitive idea of how trade-offs work. This particular example is easy because it seems natural to many of us to think in terms of dollars, and we often can reduce nonmonetary attributes to dollars. But what if we wanted to trade off life span and reliability? We would like to have a systematic procedure that we can apply in any situation fairly easily. To do this, we must find satisfactory ways to answer two questions. The first question has to do with comparing the attribute levels of the available alternatives. We are comparing three different automobiles. How do they compare on the two attributes? Is the Portalo "twice as good" on life span as the Norushi? How does the Standard compare (quantitatively) to the Portalo on price? In the energy example in Figure 15.2, alternative technologies must be ranked on each of their attributes. For example, substantial differences may exist among these technologies on all of the detail attributes. To get anywhere with the construction of a quantitative model of preferences, we must assess numerical scores for each alternative that reflect the comparisons.

The second question asks how the attributes compare in terms of importance. In the automobile example, is life span twice as important as price, and exactly what does "twice as important" mean? In the alternative-energy example, how do the environmental and health risks compare in terms of importance within the "Minimize Risks" fundamental objective? As with the scores, numerical weights must be assessed for each attribute.

The model that we will adopt in this chapter is called the *additive utility function.* We assume that we have individual utility functions $U_1(x_1), \ldots, U_m(x_m)$ for m different attributes x_1 through x_m. These are utility functions just like those that we discussed in Chapters 13 and 14. In particular, we assume that each utility function assigns values of 0 and 1 to the worst and best levels on that particular objective. The additive utility function is simply a weighted average of these different utility functions. For an outcome that has levels $x_1, \ldots, x_m$ on the m objectives, we would calculate the utility of this outcome as

$$U(x_1, \ldots, x_m) = k_1\, U_1(x_1) + \cdots + k_m\, U_m(x_m)$$

$$= \sum_{i=1}^{m} k_i\, U_i(x_i) \tag{15.1}$$

where the weights are $k_1, \ldots, k_m$. All of the weights are positive, and they add up to 1.

First, you can see this additive utility function also assigns values of 0 and 1 to the worst and best conceivable outcomes, respectively. To see this, look at what happens when we plug in the worst level (x_i^-) for each objective. The individual utility functions then assign 0 for each objective $[U(x_i^-) = 0]$, and so the overall utility is also 0. If we plug in the best possible value for each objective (x_i^+), the individual utility functions are equal to 1 $[U(x_i^+) = 1]$, and so the overall utility becomes

$$U(x_1^+, \ldots, x_m^+) = k_1 U_1(x_1^+) + \cdots + k_m U_m(x_m^+)$$

$$= k_1 + \cdots + k_m$$

$$= 1$$

You can see that this utility function is consistent with what we developed in Chapters 3 and 4. In Chapter 3 we discussed how to assess "scores" by using attribute scales; you can see that the individual utility functions U_i in Equation (15.1) define the attribute scales. In Chapters 3 and 4 we had the attribute scales ranging from 0 to 100, whereas here we have defined the utility functions to range from 0 to 1. The transformation is simple enough; think of the 0 and 100 in Chapters 3 and 4 as 0% and 100%. In the case of quantitative natural scales we showed in Chapter 4 how to standardize those to a scale from 0 to 100, and in the case of constructed scales we had to make a subjective assessment of the value of the intermediate levels on the scale from 0 to 100. As we proceed through this chapter, we will examine some alternative ways of coming up with the individual utility functions.

In Chapter 4 we turned to the assessment of weights. We used a simple technique of comparing the importance of the range of one attribute with another. Later in this chapter we will discuss this and several other approaches for assessing weights.

Choosing an Automobile: Proportional Scores

To understand the additive utility function better, let us work through a simple example with the automobile choice. We will begin with the determination of the individual utility values, following the same proportional-scoring technique discussed in Chapter 4 (except that we will have the utility functions range from 0 to 1 rather than 0 to 100).

The first step is easy. The Standard is best on price and worst on life span, so assign it a 1 for price and a 0 for life span: $U_P(\text{Standard}) = 1$ and $U_L(\text{Standard}) = 0$, where the subscripts P and L represent price and life span, respectively. Do the opposite for the Portalo, which is worst on price and best on life span: $U_P(\text{Portalo}) = 0$ and $U_L(\text{Portalo}) = 1$. Now, how do we derive the corresponding scores for the Norushi? Because we have natural numerical measures for the objectives (dollars and years), we can simply scale the Norushi's price and life span. A general formula is handy. Calculate

$$U_i(x) = \frac{x - \text{Worst Value}}{\text{Best Value} - \text{Worst Value}} \qquad \textbf{(15.2)}$$

$$= \frac{x - x_i^-}{x_i^+ - x_i^-}$$

For the Norushi's price,

$$U_P(\$10,000) = \frac{10,000 - 17,000}{8000 - 17,000}$$

$$= 0.78$$

Likewise, the Norushi's utility for life span is

$$U_L(9 \text{ years}) = \frac{9 - 6}{12 - 6}$$

$$= 0.50$$

The intuition behind these calculations is that 9 years is exactly halfway between 6 and 12 years [thus $U_L(\text{Norushi}) = 0.50$], whereas $10,000 is 78% of the way from $17,000 to $8000. Moreover, you can see that the values of 0 and 1, which we assigned to the Standard and the Portalo, are consistent with the general formula; plugging in the worst value for x yields $U_i(\text{Worst}) = 0$, and plugging in the best value gives $U_i(\text{Best}) = 1$. The utilities for the three cars are summarized in Table 15.2. As long as the objectives have natural numerical attributes, it is a straightforward matter to scale those attributes so that the utility of the best is 1, the utility of the worst is 0, and the intermediate alternatives have scores that reflect the relative distance between the best and worst.

Table 15.2
Utilities for three cars on two attributes.

	Portalo	Norushi	Standard Motors
Price (U_P)	0.00	0.78	1.00
Life Span (U_L)	1.00	0.50	0.00

Assessing Weights: Pricing Out the Objectives

Now we must assess the weights for price and life span. But before we decide on the weights once and for all, let us look at the implications of various weights. For the automobile example, we must assess k_P and k_L, which represent the weights for price and life span, respectively.

Suppose you were to decide that price and expected life span should be weighted equally, or $k_P = k_L = 0.5$. In general, we are going to calculate

$$U(\text{Price, Life Span}) = k_P\, U_P(\text{Price}) + k_L\, U_L(\text{Life Span})$$

Thus, the weighted utilities would be

$$U(\text{Portalo}) = 0.5(0.00) + 0.5(1.00) = 0.50$$
$$U(\text{Norushi}) = 0.5(0.78) + 0.5(0.50) = 0.64$$
$$U(\text{Standard}) = 0.5(1.00) + 0.5(0.00) = 0.50$$

The Standard and the Portalo come out with exactly the same overall utility because of the way that price and life span are traded off against each other. Because the difference between 1 and 0 amounts to $9000 in price versus six years in life span, the equal weight in this case says that one additional year of life span is worth $1500. The Norushi comes out on top because you pay less than $1500 per year for the three additional years in expected life span as compared to the Standard.

Suppose that you have little money to spend on a car. Then you might think that price should be twice as important as life span. To model this, let $k_P = 0.67$ and $k_L = 0.33$. Now the overall utilities for the cars are Portalo, 0.33; Norushi, 0.69; and Standard, 0.67. In this case the weights imply that an increase in life span of one year is only worth an increase in price of $750. (You can verify this by calculating the utility for a car that costs $8750 and is expected to last seven years; such a car will have the same weighted score as the Standard.) Again the Norushi comes out as being preferred to the Standard, because its three-year increase in life span (relative to the Standard) is accompanied by only a $2000 increase in price, whereas the weights indicate that the additional three years would be worth $2250.

You may not be happy with either scheme. Perhaps you have thought carefully about the relative importance of expected life span and price, and you have decided that you would be willing to pay up to $600 for an extra year of expected life span. You have thus *priced out* the value of an additional year of expected life span. How can you translate this price into the appropriate weights? Take the Standard as your base case (although any of the three automobiles could be used for this). Essentially, you are saying that you would be indifferent between paying $8000 for six years of expected life span and $8600 for seven years of expected life span. Using Equation (15.2), we can find that such a hypothetical car (Car B) would score $\frac{1}{6} = 0.167$ on expected life span (which is one-sixth of the way from the worst to the best case) and 0.933 on price ($8600 is 0.933 of the way from the worst to the best case). Because you would be indifferent between the Standard and the hypothetical Car B, the weights must satisfy

$$U(\text{Standard}) = U(\text{Car B})$$
$$k_P(1.00) + k_L(0) = k_P(0.933) + k_L(0.167)$$

Simplify this equation to find that

$$k_P(1.00 - 0.933) = k_L(0.167)$$

$$k_P = k_L \frac{0.167}{0.067}$$

or that

$$k_P = 2.50 k_L$$

Including the condition that the weights must sum to 1, we have

$$k_P = 2.50 \, (1 - k_P)$$

or

$$k_P = 0.714 \quad \text{and} \quad k_L = 0.286$$

Note that these weights are consistent with what we did above. The weight $k_P = 0.667$ implied a price of $750 per additional year of expected life span. With a still lower price ($600), we obtained a higher weight for k_P.

The final objective, of course, is to compare the cars in terms of their overall utilities:

$$U(\text{Portalo}) = 0.714(0.00) + 0.286(1.00) = 0.286$$
$$U(\text{Norushi}) = 0.714(0.78) + 0.286(0.50) = 0.700$$
$$U(\text{Standard}) = 0.714(1.00) + 0.286(0.00) = 0.714$$

The Standard comes out only slightly better than the Norushi. This is consistent with the switching approach described earlier. The weights here came from the assessment that one year of life span was worth only $600, not the $666.67 or more required to switch from the Standard to the Norushi.

Indifference Curves

The assessment that you would trade $600 for an additional year of life span can be used to construct *indifference curves*, which can be thought of as a set of alternatives (some perhaps being hypothetical) among which the decision maker is indifferent. For example, we already have established that you would be indifferent between the Standard and hypothetical Car B, which costs $8600 and lasts seven years. Thus, in Figure 15.4 we have a line that passes through the points for the Standard and the point for the hypothetical Car B (Point B). All of the points along this line represent cars that would be equivalent to the Standard; all would have the same utility, 0.714. Other indifference curves also are shown with their corresponding utilities. Note that the indifference curves have higher utilities as one moves down and to the right, because you would rather pay less money and have a longer life span. You can see that the Norushi and the Portalo are not preferred to the Standard because they lie above the 0.714 indifference curve.

Figure 15.4

Indifference curves for the automobile decision.

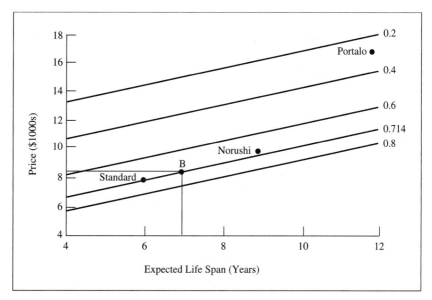

The slope of the indifference curves in Figure 15.4 is related to the trade-off rate that was assessed. Specifically, the slope is $600 per year, the price that was assessed for each year of expected life span. This also is sometimes called the *marginal rate of substitution,* or the rate at which one attribute can be used to replace another.

When using the additive utility function, it is straightforward to calculate how much a utility point in objective i is worth in terms of objective j. Let us say that you want to know how much one utility unit of attribute i is worth in terms of utility units of attribute j. Then the marginal rate of substitution between i and j is simply k_i/k_j. Thus, in the automobile case, the marginal rate of substitution in terms of the utility scales is $0.286/0.714 = 0.40$. In other words, the increase of one utility point on the life-span scale is worth 40% of the increase of one point on the price scale.

Unfortunately, knowing the marginal rate of substitution in utility terms is not so useful. If you are using the additive utility function and proportional scores as we have done here, then the marginal rate of substitution in terms of the original attributes is easily calculated. Let M_{ij} denote the marginal rate of substitution between attributes i and j. Then

$$M_{ij} = \frac{k_i/|x_i^+ - x_i^-|}{k_j/|x_j^+ - x_j^-|} \tag{15.3}$$

Thus, for the automobiles M_{LP} can be calculated as

$$M_{LP} = \frac{0.286/|12 \text{ yr} - 6 \text{ yr}|}{0.714/|\$8000 - \$17{,}000|}$$
$$= \$600 \text{ per year}$$

In general, without proportional scores or with a nonadditive overall utility function, the marginal rate of substitution can vary depending on the values of the attributes

x_i and x_j, and the reason is that the indifference curves may be just that: curves instead of straight lines. Moreover, calculating M_{ij} can be difficult, requiring calculus to determine the slope of an indifference curve at a particular point. It is often possible to graph approximate indifference curves, however, and doing so can provide insight into one's assessed trade-offs. We will return to indifference curves in Chapter 16.

Assessing Individual Utility Functions

As you can see, the essence of using the additive utility function is to be able to assess the individual utility functions and the weights. In this section we look at some issues regarding the assessment of the individual utility functions.

Proportional Scores

In the example here and in Chapter 4, we have used the proportional scoring approach. In fact, we have just taken the original values and scaled them so that they now range from 0 (worst) to 1 (best). For the automobile example, Figure 15.5 graphs the price relative to a car's score on price. The graph shows a straight line, and we know about straight lines in this context: They imply risk neutrality and all of the unusual behavior associated with risk neutrality.

Let us think about what risk neutrality implies here with another simple example. Imagine that you face two career choices. You could decide to invest your life sav-

Figure 15.5

Proportional scores for prices of three cars.

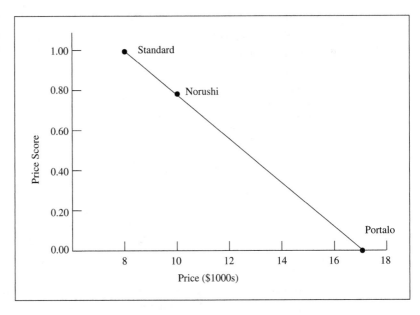

ings in an entrepreneurial venture, or you could take a job as a government bureaucrat. After considering the situation carefully, you conclude that your objectives are purely monetary, and you can think in terms of income and savings. (Actually, more than just monetary outcomes should influence your career choice!) The bureaucratic job has a well-defined career path and considerable security. After 10 years, you know with virtual certainty that you will make $30,000 per year and have $60,000 in the bank toward retirement. On the other hand, becoming an entrepreneur is a risky proposition. The income and savings outcomes could range anywhere from zero dollars in the case of failure to some large amount in the event of success. To solve the problem, you decide to assess a continuous distribution and fit a discrete approximation. The decision tree in Figure 15.6 shows your probability assessments and alternatives. You can see that for the entrepreneurial alternative, the expected values are $30,000 in income and $60,000 in savings, the same as the certain values associated with the bureaucratic job.

Now we assign scores to the outcomes. For income, the best is $60,000, the worst is 0, and $30,000 is halfway between. These outcomes receive scores of 1, 0, and $\frac{1}{2}$, respectively. Performing the same analysis for the savings dimension results in similar scores, and the decision tree with these scores appears in Figure 15.7.

We now calculate expected scores for the two alternatives; this requires calculating scores for each of the three possible outcomes. Let k_I and k_S represent the weights that we would assign to income and savings. The scores (U) for the possible outcomes are

Figure 15.6

A career decision.

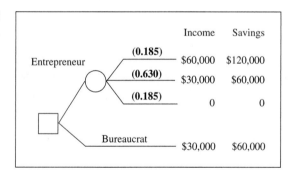

Figure 15.7

A career decision with proportional scores.

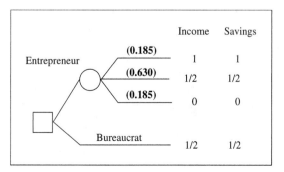

$$U(\$60,000, \$120,000) = k_I(1) + k_S(1) = k_I + k_S$$

$$U(\$30,000, \$60,000) = k_I\left(\tfrac{1}{2}\right) + k_S\left(\tfrac{1}{2}\right) = \tfrac{1}{2}(k_I + k_S)$$

$$U(0,0) = k_I(0) + k_S(0) = 0$$

Now we can calculate the expected utility (EU) for the entrepreneurial venture. That expected score is obtained by averaging the scores over the three branches:

$$EU(Entrepreneur) = 0.185(k_I + k_S) + 0.63\left[\tfrac{1}{2}(k_I + k_S)\right] + 0.185(0)$$

$$= 0.50(k_I + k_S)$$

$$= 0.50$$

because $k_I + k_S = 1$. Of course, the expected score for the bureaucratic option also is 0.50, and so, on the basis of expected scores, the two alternatives are equivalent. (In fact, it is important to note that the two alternatives have the same score regardless of the specific values of k_I and k_S.) It is obvious, however, that being an entrepreneur is riskier. If each job really has the same expected scores in all relevant dimensions, would you be indifferent? (Probably not, but we will not guess about whether you would choose the riskier job or the more secure one!)

Ratios

Another way to assess utilities—and a particularly appropriate one for attributes that are not naturally quantitative—is to assess them on the basis of some ratio comparison. For example, let us return to the automobile example. Suppose that color is an important attribute in your automobile purchase decision. Clearly, this is not something that is readily measurable on a meaningful numerical scale. Using a ratio approach, you might conclude that to you blue is twice as good as red and that yellow is $2\tfrac{1}{2}$ times as good as red. We could accomplish the same by assigning some number of points between 0 and 100 to each possible alternative on the basis of performance on the attribute. In this way, for example, you might assign 30 points to red, 60 points to blue, and 75 points to yellow. This could be represented graphically as in Figure 15.8.

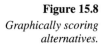

Figure 15.8

Graphically scoring alternatives.

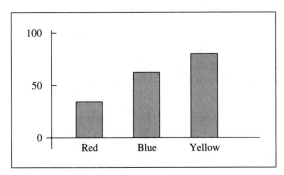

Now, however, we must scale these assessments so that they range from 0 to 1. We need to find constants a and b so that

$$0 = a + b(30)$$
$$1 = a + b(75)$$

Solving these two equations simultaneously gives

$$a = -\frac{2}{3}$$

$$b = \frac{1}{45}$$

Applying these scaling constants, we can calculate U_C, the utilities for the three colors:

$$U_C(\text{Red}) \quad = -\frac{2}{3} + \frac{30}{45} = 0$$

$$U_C(\text{Blue}) \quad = -\frac{2}{3} + \frac{60}{45} = \frac{2}{3}$$

$$U_C(\text{Yellow}) = -\frac{2}{3} + \frac{75}{45} = 1$$

Figure 15.9 shows the scaled scores, which now represent your relative preference for the different colors. They may be used to calculate weighted scores for different cars in a decision problem in which color is one attribute to consider. For example, with appropriate trade-off weights for price, color, and life span, the weighted score for a blue Portalo would be

$$U(\$17{,}000, 12 \text{ Years}, \text{Blue}) = k_P(0) + k_L(1) + k_C(0.667)$$

You can see how the ratio approach can be used to compare virtually any set of alternatives whether or not they are quantitatively measured.

Figure 15.9

Scaled scores for colors.

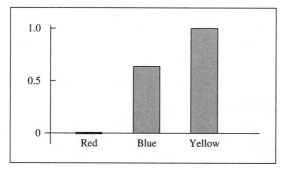

Standard Utility-Function Assessment

In principle, assessing the individual utility functions need be no more complicated than the assessment procedures described in Chapters 13 and 14. For the automobile example, we would need to assess utility functions for price, life span, and color. If you face a decision that is complicated by both uncertainty and trade-offs, this is the best solution. The utility functions, having been assessed in terms of your preferences over uncertain situations, are models of your preferences in which your risk attitude is built in.

To see the effect of building in a risk attitude, let us return to the example above about the two jobs. Suppose we assess utility functions for income and savings and find that

$$U_I(\$60,000) = 1$$
$$U_I(\$30,000) = 0.75$$
$$U_I(0) = 0$$

and

$$U_S(\$120,000) = 1$$
$$U_S(\$60,000) = 0.68$$
$$U_S(0) = 0$$

With these utilities, the expected score for becoming an entrepreneur is

$$EU(\text{Entrepreneur}) = 0.185(1k_I + 1k_S) + 0.63(0.75k_I + 0.68k_S)$$
$$+0.185(0k_I + 0k_S)$$
$$= 0.658k_I + 0.613k_S$$

Now compare EU(Entrepreneur) with U(Bureaucrat):

$$U(\text{Bureaucrat}) = 0.750k_I + 0.680k_S$$

Regardless of the specific values for the weights k_I and k_S, the expected score for being a bureaucrat always will be greater than the expected score for being an entrepreneur because the coefficients for k_I and k_S are larger for the bureaucrat than for the entrepreneur. This makes sense; the expected values (EMVs) for the two alternatives are equal for each attribute, but being an entrepreneur clearly is riskier. When we use a utility function for money that incorporates risk aversion, the riskier alternative ends up with a lower expected utility.

Assessing Weights

In the examples here and in Chapter 4, we have seen two different methods of assessing weights for the additive utility function. The approach in the automobile example is called *pricing out* because it involves determining the value of one objective in terms of another (usually dollars). The summer-job example in Chapter 4, though, used an approach that we will call *swing weighting*. We will discuss both of these methods here along with a third method based on a comparison of lotteries.

Pricing Out

The pricing-out method for assessing weights is no more complicated than what was described above in the automobile example. The essence of the procedure is to determine the marginal rate of substitution between one particular attribute (usually monetary) and any other attribute. Thus, we might say that a year of life span is worth $600. Or, in a decision regarding whether to accept a settlement offer or pursue a lawsuit, one might assess that it would be worth $350 in time and frustration to avoid an hour consulting with lawyers, giving depositions, or sitting in a courtroom. (And that would be on top of the cost of the lawyers' time!)

Assessing the marginal rate of substitution is straightforward in concept. The idea is to find your indifference point: the most you would pay for an incremental unit of benefit, or the least you would accept for an incremental unit of something undesirable. Although this idea seems straightforward, it can be a difficult assessment to make, especially for attributes with which we have little buying or selling experience.

As you can see, pricing out is especially appropriate for direct determination of the marginal rate of substitution from one attribute scale to another. Because the additive utility function implies a constant marginal rate of substitution, pricing out is consistent with the notion of proportional scores. However, when the individual utility functions are nonlinear or not readily interpretable in terms of an interval scale (as they might not be with a constructed scale), then pricing out makes less sense. One would have to ask, "How much utility on Attribute Scale 1 would I be willing to give up for a unit increase on Attribute Scale 2?" This is a more difficult assessment, and in situations like this, one of the following weight-assessment procedures might be easier.

Swing Weighting

The swing-weighting approach can be used in virtually any weight-assessment situation. It requires a thought experiment in which the decision maker compares individual attributes directly by imagining (typically) hypothetical outcomes. We demonstrate the procedure in this section with the automobile example.

To assess swing weights, the first step is to create a table like the one in Table 15.3 for the automobiles. The first row indicates the worst possible outcome, or the outcome that is at the worst level on each of the attributes. In the case of the automobiles, this would be a red car that lasts only six years and costs $17,000. This "worst case" provides a benchmark. Each of the succeeding rows "swings" one of the attributes from worst to best. For example, the second row in Table 15.3 is for a red car that lasts 12 years and costs $17,000. The last row is for a car that is worst on price and life span ($17,000 and six years, respectively) but is your favorite color, yellow.

With the table constructed, the assessment can begin. The first step is to rank-order the outcomes. You can see in Table 15.3, a "4" has been placed in the "Rank"

Table 15.3

Swing-weight assessment table for automobile example.

Attribute Swung from Worst to Best	Consequence to Compare	Rank	Rate	Weight
(Benchmark)	6 years, $17,000, red	4	___	___
Life span	12 years, $17,000, red	___	___	___
Price	6 years, $8000, red	___	___	___
Color	6 years, $17,000, yellow	___	___	___

column for the first row. There are four hypothetical cars to compare, and it is safe to assume that the benchmark car—the one that is worst on all the objectives—will rank fourth (worst) overall. The others must be compared to determine which ranks first (best), second, and third. Suppose that after some thought you conclude that the low-price car is the best, then the long life-span car, and finally the yellow car. Table 15.4 shows the partially completed table.

The next step is to fill in the "Rate" column in the table. You can see in Table 15.4 that two of the ratings are predetermined; the rating for the benchmark car is 0 and the rating for the top-ranked car is 100. The ratings for the other two cars must fall between 0 and 100. The comparison is relatively straightforward to make; how much less satisfaction do you get by swinging life span from 6 to 12 years as compared to swinging price from $17,000 to $8000? What about swinging color from red to yellow as compared to swinging price? You can even think about it in percentage terms; considering the increase in satisfaction that results from swinging price as 100%, what percentage of that increase do you get by swinging life span from worst to best?

Suppose that after careful thought, your conclusion is to assign 75 points to life span and 10 points to color. Essentially, this means that you think improving life span from worst to best is worth 75% of the value you get by improving the price from $17,000 to $8000. Likewise, changing the color from yellow to red is worth only 10% of the improvement in price. With these assessments, the table can be completed and weights calculated. Table 15.5 shows the completed table. The weights are the normalized ratings; recall that by convention we have the weights add up to 1. For example, k_P is calculated as $100/(100 + 75 + 10) = 0.541$. Likewise, we have $k_L = 75/(100 + 75 + 10) = 0.405$, and $k_C = 10/(100 + 75 + 10) = 0.054$.

Table 15.4

Swing-weight assessment table with ranks assessed.

Attribute Swung from Worst to Best	Consequence to Compare	Rank	Rate	Weight
(Benchmark)	6 years, $17,000, red	4	0	___
Life span	12 years, $17,000, red	2	___	___
Price	6 years, $8000, red	1	100	___
Color	6 years, $17,000, yellow	3	___	___

Table 15.5

Completed swing-weight assessment table.

Attribute Swung from Worst to Best	Consequence to Compare	Rank	Rate	Weight
(Benchmark)	6 years, $17,000, red	4	0	_____
Life span	12 years, $17,000, red	2	75	0.405 = 75/185
Price	6 years, $8000, red	1	100	0.541 = 100/185
Color	6 years, $17,000, yellow	3	10	0.054 = 10/185
		Total	185	1.000

With the weights determined, we can calculate the overall utility for different alternatives or outcomes. For example, we now can finish calculating the utility for a blue Portalo:

$$U(\$17,000, 12 \text{ Years, Blue}) = k_P(0) + k_L(1) + k_C(0.667)$$

$$= 0.541(0) + 0.405(1) + 0.054(0.667)$$

$$= 0.441$$

Why do swing weights work? The argument is straightforward. Here are the utilities for the hypothetical cars that you have considered:

$$U(\text{Worst Conceivable Outcome}) = U(\$17,000, 6 \text{ Years, Red})$$

$$= k_P(0) + k_L(0) + k_C(0)$$

$$= 0$$

$$U(\$8000, 6 \text{ Years, Red}) \quad = k_P(1) + k_L(0) + k_C(0) = k_P$$

$$U(\$17,000, 12 \text{ Years, Red}) \quad = k_P(0) + k_L(1) + k_C(0) = k_L$$

$$U(\$17,000, 6 \text{ Years, Yellow}) = k_P(0) + k_L(0) + k_C(1) = k_C$$

From the first two equations, you can see that the increase in satisfaction from swinging price from worst to best is just k_P. Likewise, the improvement from swinging any attribute from worst to best is simply the value of the corresponding weight. When you compare the relative improvements in utility by swinging the attributes one at a time, you are assessing the ratios k_L/k_P and k_C/k_P. These assessments, along with the constraint that the weights add to 1, allow us to calculate the weights. Figure 15.10 graphically shows how swing weights work.

Swing weights have a built-in advantage in that they are sensitive to the range of values that an attribute takes on. For example, suppose you are comparing two personal computers, and price is an attribute. One computer costs $3500 and the other $3600. When you work through the swing-weight assessment procedure, you probably will conclude that the increase in utility from swinging the price is pretty small. This would result, appropriately, in a small weight for price. But if the difference in price is $1000 rather than $100, the increase in utility experienced by swinging from worst to best would be much larger, resulting in a larger weight for price.

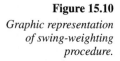
Figure 15.10

Graphic representation of swing-weighting procedure.

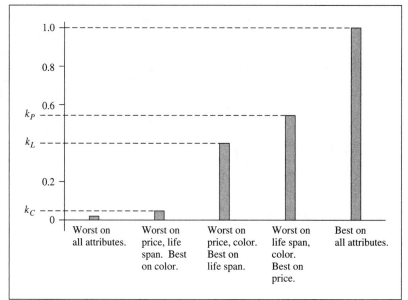

If you have a hard time thinking about the "Worst Conceivable Outcome," you might try reversing this procedure. That is, use as a benchmark the "Best Conceivable Outcome," best on all attributes, and consider decreases in satisfaction from swinging attributes from high to low. Assess relative decreases in satisfaction, and use those assessments in exactly the same way that we used the relative utility increases.

Lottery Weights

It should come as no surprise that we also can use lottery-comparison techniques to assess weights. In fact, the technique we will use is a version of the probability-equivalent assessment technique introduced in Chapter 13. The general assessment setup is shown in Figure 15.11.

The assessment of the probability p that makes you indifferent between the lottery and the sure thing turns out to be the weight for the one odd attribute in the sure thing. We will see how this works in the case of the automobiles. Figure 15.12 shows the assessment decision for determining the weight associated with price.

Suppose that the indifference probability in Figure 15.12 turns out to be 0.55. Write down the equation that is implied by the indifference:

$$k_P U_P(\$8000) + k_L U_L(6\text{ Years}) + k_C U_C(\text{Red})$$
$$= .055[k_P U_P(\$8000) + k_L U_L(12\text{ Years}) + k_C U_C(\text{Yellow})]$$
$$+0.45[k_P U_P(\$17{,}000 + k_L U_L(6\text{ Years}) + k_C U_C(\text{Red})]$$

As before, the individual utilities range from 0 to 1. This means that

Figure 15.11

Assessing weights using a lottery technique. The task is to assess the probability *p* that makes you indifferent between the lottery (A) and the sure thing (B).

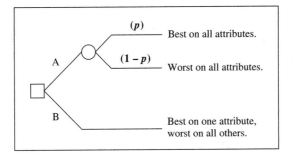

Figure 15.12

Assessing the weight for price.

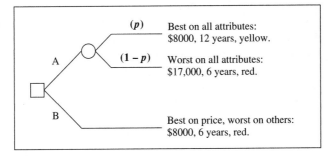

$$U_P(\$8000) = 1.00 \qquad U_P(\$17{,}000) = 0$$

$$U_L(12 \text{ Years}) = 1.00 \qquad U_L(6 \text{ Years}) = 0$$

$$U_C(\text{Yellow}) = 1.00 \qquad U_C(\text{Red}) = 0$$

Substitute these values into the equation to obtain

$$k_P = 0.55(k_P + k_L + k_C)$$

Because $k_P + k_L + k_C = 1$, we have $k_P = 0.55$, which is simply the indifference probability that was assessed. Thus, we have a direct way to find the trade-off weight for the price attribute. Repeating this procedure one more time for the life-span attribute gives k_L, and then k_C follows because the weights must add to 1. Of course, a simple way to check for consistency is to repeat the procedure a third time for the color attribute. If assessed weights do not add to 1 and are not even close, then the additive model that we are using in this chapter is not appropriate. Chapter 16 discusses more complicated multiattribute utility models that may be able to accommodate such preferences.

Finally, you may have wondered why we always scale the individual utility functions from 0 to 1. The answer is contained in the equations above that underlie the weight-assessment techniques. In each case, we needed the values of 0 and 1 for best and worst cases in order to solve for the weights; the idea that scores are scaled from 0 to 1 is built into these assessment techniques. The result is that weights have very specific meanings. In particular, the swing-weight approach implies that we can interpret the weights in terms of improvements in utility that result from changing one attribute from low to high, and those low and high values may be specific to the alternatives being considered. The lottery-assessment method suggests that the weights can be interpreted as an indifference probability in a comparison of lotteries. The specific low and high values among the alternatives are important anchors for interpreting these indifference probabilities.

Keeping Concepts Straight: Certainty versus Uncertainty

Up to this point we have been doing something that is, strictly speaking, not correct. We have been mixing up decision making under conditions of certainty with decision making under conditions of uncertainty. For example, the decision regarding which automobile to buy, as we have framed it, is one that is made under certainty regarding the outcomes; the Norushi, Portalo, and Standard Motors cars have particular characteristics (price, color, and advertised life span) that you know for sure when you buy them. On the other hand, we could reframe the problem and consider life span to be uncertain. In this case, we would assess a probability distribution over possible life spans for each car. The decision model we would create would require us to assess a utility function that covers all of the possible life-span outcomes so that we could appropriately value the probability distributions or lotteries for life span that come with each car. As we also saw, the entrepreneur-bureaucrat example involves uncertainty, and we needed an appropriate utility function in order to evaluate those two options in a way that made sense.

Why does it matter whether we are talking about certainty or uncertainty? Some of the utility-assessment methods we have talked about are appropriate for decisions under uncertainty, and some are not. When making decisions under certainty, we can use what are called *value functions* (also known as *ordinal utility functions*). Value functions are only required to rank-order sure outcomes in a way that is consistent with the decision maker's preferences for those outcomes. There is no concern with lotteries or uncertain outcomes. In fact, different value functions that rank sure outcomes in the same way may rank a set of lotteries in different ways.

If you face a decision under uncertainty, you should use what is called a *cardinal utility function*. A cardinal utility function appropriately incorporates your risk attitude so that lotteries are rank-ordered in a way that is consistent with your risk attitude. All of the discussion in Chapters 13 and 14 concerned cardinal utility; we constructed preference models to incorporate risk attitudes.

The good news is that most of the assessment techniques we have examined are appropriate for conditions of both certainty and uncertainty. Of the methods for assessing individual utility functions, only the ratio approach is, strictly speaking, limited only to decisions under certainty; nothing about this assessment method encodes the decision maker's attitude about risk. The proportional-scores technique is a special case; it may be used under conditions of uncertainty by a decision maker who is risk-neutral for the specified attributes. Of course, all of the lottery-based utility-assessment methods from Chapters 13 and 14 are appropriate for decisions under uncertainty. And all of the weight-assessment methods described in this chapter can be used regardless of the presence or absence of uncertainty; once a specific multiattribute preference model is established, the weight-assessment methods amount to various ways of establishing indifference among specific lotteries or consequences. The weights can then be derived on the basis of these indifference judgments.

How much does the distinction between ordinal and cardinal utility models really matter? Although the distinction can be important in theoretical models of economic behavior, practical decision-analysis applications rarely distinguish between the two. Human decision makers need help understanding objectives and trade-offs, and all of the assessment techniques described in this chapter are useful in this regard. As always, we adopt a modeling view. The objective of assessing a multiattribute utility function is to gain insight and understanding of the decision maker's trade-offs among multiple objectives. Any assessment method or modeling technique that helps to do this—which includes all that are mentioned in this chapter—should result in a reasonable preference model that can lead the decision maker to a clear choice among available alternatives.

In a way, the situation is similar to the discussion on decreasing and constant risk aversion in Chapter 13, where we argued that it is difficult enough to determine the extent of a given individual's risk aversion, let alone whether it is decreasing or not. When facing multiple objectives, any approach that will help us to understand trade-offs among objectives is welcome. Attention to preference ordering of lotteries rather than sure outcomes can be important in some situations but would come after obtaining a good basic understanding of the objectives and approximate trade-off weights.

A similar argument can be made for the additive utility function itself. Although strictly speaking it has some limitations and special requirements that will be explained in Chapter 16, it is an exceptionally useful and easy way to model preferences in many situations. Even if used only as an approximation, the additive utility function takes us a long way toward understanding our preferences and resolving a difficult decision.

An Example: Library Choices

With the discussion of assessing utility functions and weights behind us, we now turn to a realistic problem. This example will demonstrate the development of an additive utility function when there are many objectives. The issue is site selection for a new public library.

THE EUGENE PUBLIC LIBRARY

In 1986 a solution was sought for the overcrowded and inadequate conditions at the public library in Eugene, Oregon. The current building, with approximately 38,000 square feet of space, had been built in 1959 with the anticipation that it would serve satisfactorily for some 30 years. In the intervening years, Eugene's population had grown to the point that, on a per capita basis, the Eugene Public Library was one of the most heavily used public libraries in the western United States. All available space had been used. Even the basement, which had not been designed originally for patron use, had been converted to a periodicals reading room. Low-circulation materials had to be removed from the stacks and placed in storage to make room for patrons and books. Expansion was imperative; consultants estimated that 115,000 square feet of space were needed. The current building could be expanded, but because of the original design it could not be operated as efficiently as a brand new building. Other potential sites were available, but all had their own benefits and drawbacks. After much thought, the possibilities were reduced to four sites in or near downtown Eugene, one of which was the current site. Some were less expensive for one reason or another, others had better opportunities for future expansion, and so on. How could the city choose from among them?

In evaluating the four proposed library sites, the first task was to create a fundamental-objectives hierarchy. What aspects of site location were important? The committee in charge of the study created the hierarchy shown in Figure 15.13 with seven root-level fundamental objectives and a second level of detail objectives. Parking, for example, was broken down into patron parking, off-site parking, night staff parking, and bookmobile space. (Without knowledge of Eugene's financial system, certain attributes in the "Related Cost" category may not make sense.) An important objective that is conspicuous by its absence is minimizing the cost of building a library. The committee's strategy was to compare the sites on the seven fundamental criteria first, and then consider the price tags for each. We will see later how this comparison can be made.

Comparing the four candidate sites (alternatives) required these four steps:

1 Evaluate the alternatives on each attribute.

2 Weight the attributes. This can be done with any of the three weight-assessment methods (although pricing out may be difficult because the overall cost is not included in this part of the analysis).

3 Calculate overall utility with the additive utility function.

4 Choose the alternative with the greatest overall utility.

Application of this procedure in the analysis of the four library sites produced the matrix shown in Table 15.6. The relative weights and the individual utilities are all shown in this table. (Actually, the committee's scores did not range from 0 to 1. Their scores have been thusly rescaled here, and the weights have been adjusted to be consistent with the new numbers. Six attributes on which all four sites were scored the same also have been eliminated.)

The overall utility is calculated and shown on a scale from 0 to 100, rather than from 0 to 1; this just makes the table easier to read. Under each fundamental objective is a subtotal for each site. For example, the subtotal for Site 1 under "Parking" is

Figure 15.13

Fundamental objectives for the Eugene Public Library site-evaluation study.

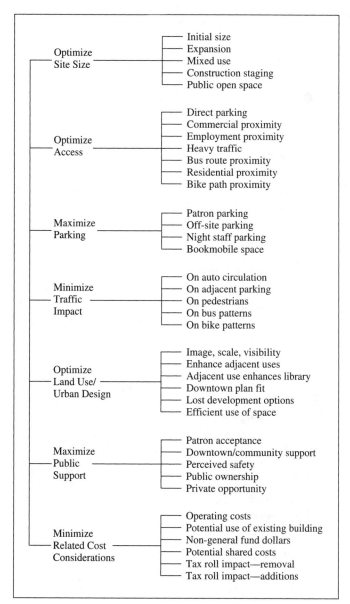

$$\text{Subtotal (Parking}_1) = [(0.053 \times 0.20 \times 1.00) + (0.053 \times 0.60 \times 0.00)$$
$$+ (0.053 \times 0.20 \times 1.00)] \times 100$$
$$= 2.12$$

(The factor of 100 at the end simply changes the scale so that it ranges from 0 to 100.) Once all individual utilities have been calculated for each fundamental objective, they are simply added to find the overall utility for each site. For example,

Table 15.6

Matrix of weights and utilities for four library sites.

Attributes	%	Utilities			
		Site 1	Site 2	Site 3	Site 4
Site Size (21.1%)					
Initial	38	1.00	0.00	1.00	1.00
Expansion (Horizontal)	13	0.00	0.00	0.00	1.00
Mixed Use	25	0.00	1.00	1.00	1.00
Construction Staging	12	1.00	0.00	0.00	1.00
Public Open Space	12	1.00	0.00	0.00	0.00
Subtotals		*13.08*	*5.28*	*13.29*	*18.57*
Access (20.6%)					
Direct Parking	8	0.00	1.00	0.00	0.00
Commercial Proximity	23	0.00	1.00	0.67	1.00
Employment Proximity	15	0.50	1.00	0.00	1.00
Heavy Traffic	23	0.33	0.33	1.00	0.00
Bus Route Proximity	15	0.00	0.50	0.50	1.00
Residential Proximity	16	1.00	0.00	1.00	0.50
Subtotals		*6.40*	*12.58*	*12.75*	*12.57*
Parking (5.3%)					
Patron Parking	20	1.00	0.00	1.00	1.00
Off-Site Parking	60	0.00	1.00	0.33	0.33
Bookmobile Parking	20	1.00	0.00	1.00	1.00
Subtotals		*2.12*	*3.18*	*3.17*	*3.17*
Traffic Impacts (4.5%)					
Auto Circulation	47	0.00	0.75	1.00	0.00
Adjacent Parking	29	0.00	0.00	1.00	0.00
Bus Patterns	24	1.00	1.00	1.00	0.00
Subtotals		*1.08*	*2.67*	*4.50*	*0.00*
Land Use/Design (8.4%)					
Image/Scale/Visibility	13	0.00	1.00	0.00	0.00
Enhance Adjacent Uses	13	0.00	1.00	1.00	1.00
Adj. Uses Enhance Lib.	38	0.00	1.00	1.00	0.00
Downtown Plan Fit	13	1.00	0.00	1.00	1.00
Lost Devel. Options	23	1.00	0.00	0.00	0.00
Subtotals		*3.02*	*5.38*	*5.38*	*2.18*

Table 15.6
(Continued)

Attributes	%	Site 1	Site 2	Site 3	Site 4
Public Support (19.0%)					
Patron Acceptance	25	1.00	0.33	0.67	0.00
DT/Community Support	25	1.00	0.67	0.33	0.00
Perceived Safety	25	1.00	0.33	1.00	0.00
Public Ownerhip	17	0.00	1.00	1.00	0.00
Private Opportunity	8	1.00	0.00	1.00	1.00
Subtotals		*15.77*	*9.55*	*14.25*	*1.52*
Related Costs (21.1%)					
Operating Costs	20	0.00	1.00	1.00	1.00
Use of Existing Building	20	1.00	0.00	0.00	0.00
No General Fund $	30	0.00	1.00	1.00	1.00
Tax Roll Impact, Removal	10	0.00	1.00	1.00	0.00
Tax Roll Impact, Added	20	0.00	1.00	1.00	1.00
Subtotals		*4.22*	*16.88*	*16.88*	*14.77*
Weighted Score		*45.70*	*55.51*	*70.22*	*52.78*

Source: Adapted from Robertson/Sherwood/Architects (1987) "Preliminary Draft Report: Eugene Public Library Selection Study. Executive Summary." Eugene, OR: Robertson/Sherwood.

$$U(\text{Site } 1) = 13.08 + 6.40 + 2.12 + 1.08 + 3.02 + 15.77 + 4.22$$
$$= 45.70$$

You can see that the overall weight given to a specific attribute is the product of the specific weight at the lower level and the overall weight for the fundamental objective. For example, Site 1 has U(Construction Staging) = 1.00. This utility then is multiplied by 0.12, the weight for construction staging, and then multiplied by 0.211, the weight for the fundamental attribute of "Site Size." Thus, in this grand scheme, the utilities are weighted by a product of the weights at the two levels in the value tree. To express this more formally, let k_i represent the weight of the ith fundamental objective, and k_{ij} and U_{ij} the weight and utility, respectively, for the jth attribute under fundamental objective i. If there are m fundamental objectives and m_i detail attributes under fundamental objective i, then the overall utility for a site is

$$
\begin{aligned}
U(\text{Site}) &= k_1(k_{11}U_{11} + k_{12}U_{12} + k_{13}U_{13} + \cdots + k_{1m_1}U_{1m_1}) \\
&\quad + k_2(k_{21}U_{21} + k_{22}U_{22} + k_{23}U_{23} + \cdots + k_{2m_2}U_{2m_2}) \\
&\quad + \cdots + k_m(k_{m1}U_{m1} + k_{m2}U_{m2} + k_{m3}U_{m3} + \cdots + k_{mm_k}U_{mm_k}) \\
&= k_1k_{11}U_{11} + k_1k_{12}U_{12} + \cdots + k_mk_{mm_k}U_{mm_k} \\
&= \sum_{i=1}^{m} k_i \left[\sum_{j=1}^{m_i} k_{ij}U_{ij} \right]
\end{aligned}
$$

From these expressions, we can see that the utilities on the individual attributes are being weighted by the product of the appropriate weights and then added. Thus, we still have an additive score that is a weighted combination of the individual utilities, just as we did in the simpler two- and three-attribute examples above. Moreover, it also should be clear that as the hierarchy grows to have more levels, the formula also grows, multiplying the individual utilities by all of the appropriate weights in the hierarchy.

The result of all of the calculations for the library example? Site 3 ranked the best with 70.22 points, Site 2 was second with 55.51 points, and Sites 4 and 1 (the current location) were ranked third and fourth overall with 52.78 and 45.70 points, respectively.

There is another interesting and intuitive way to interpret this kind of analysis. Imagine that 100 points are available to be awarded for each alternative, depending on how a given alternative ranks on each attribute. In the library case, 21.1 of the 100 points are awarded on the basis of "Site Size," 20.6 on the basis of "Access," 5.3 for "Parking," and so on. Within the "Site Size" category, the weights on the detail objectives determine how the 21.1 points for "Site Size" will be allocated; 38% of the 21.1 points (or 8.02 points) will be awarded on the basis of "Initial Size," 13% of the 21.1 points (2.74 points) will be awarded on the basis of "Expansion," and so on. We can see how this subdivision could continue through many layers in a hierarchy. Finally, when the ends of the branches are reached, we must determine utilities for the alternatives. If the utilities range from 0 to 1, then the utility indicates what proportion of the available points are awarded to the alternative for the particular detail attribute being considered. For example, Site 3 has a utility of 0.67 on "Commercial Proximity," so it receives 67% of the points available for this detail attribute. How many points are available? The weight of 23% tells us that 23% of the total points for "Access" are allocated to "Commercial Proximity," and the 20.6% weight for "Access" says that 20.6 points total are available for "Access." Thus, Site 3 earns $0.67 \times 0.23 \times 20.6 = 3.17$ points for "Commercial Proximity." For each detail attribute, calculate the points awarded to Site 3. Now add those points; the total is Site 3's overall utility (70.22) on a scale from 0 to 100.

Recall that cost was an important attribute that was not included in the analysis of the library sites. The committee studying the problem decided to ignore construction costs until the sites were well understood in terms of their other attributes. But now that we have ranked the sites on the basis of the attributes, we must consider money. Table 15.7 shows the costs associated with each site, along with the overall utilities from Table 15.6, and Figure 15.14 shows the same information, graphically plotting cost against overall utility.

Table 15.7 and Figure 15.14 show clearly that Sites 1 and 4 are dominated. That is, if you like Site 4, then you should like Sites 2 or 3 even better, because each has a greater utility for less money. Likewise, Site 2 clearly dominates Site 1. Thus, Sites 1 and 4 can be eliminated from the analysis altogether on the basis of dominance. This leaves Sites 2 and 3. Is it worthwhile to pay the additional $2.72 million to gain 14.71 additional points in terms of overall utility? Alternatively, is an increase of one point in the utility worth $184,908?

Table 15.7

Cost and overall utility for four library sites.

Site	Cost ($Million)	Overall Utility
1	21.74	45.70
2	18.76	55.51
3	24.48	70.22
4	24.80	52.78

Figure 15.14

Library site costs plotted against overall utility.

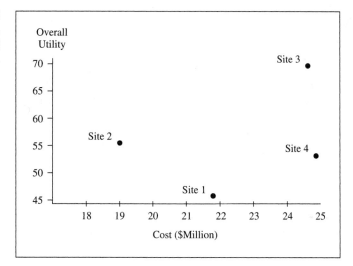

Obviously, we are trying to price out the value of a single point on our 1-to-100 scale, just as we previously priced out the value of changes in attributes. But answering this question now is difficult because one point in the overall utility may have many components. One possible approach is to return to the detail attributes, look for specific attributes on which Site 3 is ranked higher than Site 2, and consider how much we would be willing to pay (in dollars) to bring Site 2 up to Site 3's level on this attribute. For example, Site 3 scored much higher than Site 2 for access during heavy traffic periods. The difference is that Site 2 is in a relatively congested area of downtown and on one of the city's main thoroughfares. In contrast, Site 3 is located at the edge of downtown, and access would be through relatively low-volume streets. How much would altering the traffic patterns be worth so that Site 2 would be just as good as Site 3 for this attribute? One million dollars? More?

Let us suppose that we have assessed that the difference between the sites in terms of access during heavy traffic is indeed worth $1 million. It turns out that Site 3's advantage increases its weighted score by 3.17 points. (This is the difference between the sites in points awarded for this attribute.) Thus, the assessment would indicate that 3.17 points of overall utility are worth $1 million, or $315,457 per point of utility. This is greater than the $184,908 per point required to switch from Site 2 to Site 3.

A pricing-out approach such as this can be used to assess the dollar value of one point of overall utility. Rather than making only one assessment, however, we should make several on different attributes where the two alternatives differ. For some of these assessments it may be possible to make estimates of true costs in terms of the market price of adjacent land, redesigning traffic patterns, constructing parking lots, and so on, and these cost estimates then might be helpful in assessing the dollar value of specific differences between sites. If the assessments in terms of prices per point of overall utility come out fairly close, then the average of these assessments can be used as a reasonable measure of the value of one point of overall utility. In the case of the library sites, if the final price per point is less than $184,908, then Site 2 is preferred; otherwise Site 3 is preferred. If the price per point turns out to be very close to $184,908, then the two sites are approximately equivalent.

The analysis probably would have been more complete had the committee elected to include minimizing cost as one of its fundamental objectives. In fact, all of the other attributes could have been priced out in terms of dollars, thus making the comparisons and trade-offs more intuitive. But it is not unusual for groups to ignore costs in an initial stage. The motivation often is political; it may be easier to gain constituent support by playing up the benefits of a project such as a new library early before talking about the "bottom line."

Using Software for Multiple-Objective Decisions

As you can imagine, multiple-objective decisions can become very complex. The library example involved only four alternatives and a relatively simple objectives hierarchy. Still, there were many assessments to make, and the calculations involved, though straightforward, are tedious. In Chapter 4 we demonstrated two methods that relieve you of the tedious calculations when modeling multiple-objective decisions. Both methods take advantage of the fact that PrecisionTree runs within a spreadsheet environment, providing easy access to side calculations. In the first method, you enter the attribute scores and weights into the spreadsheet along with the corresponding utility functions. Then Excel will calculate the overall utility of each alternative and PrecisionTree will calculate the expected utility. In the second method, you build a linked decision tree that takes its input values directly from the tree and outputs the expected utility value.

S U M M A R Y   This chapter has introduced the basics of making decisions that involve trade-offs. The basic problem is to create a model of the decision maker's preferences in terms of the various objectives. The first step in doing so, as discussed in detail in Chapter 3, is to understand which objectives are most important in making your decision; this requires introspection as to the fundamental objectives. The goal of this step is to create a fundamental-objectives hierarchy and a set of operational attribute scales to measure the performance of each alternative or outcome on each of the fundamental objectives.

To evaluate the alternatives, we introduced the additive utility function, which calculates an overall utility for an alternative or outcome as a weighted sum of individual utility functions for each fundamental objective. Assessing the individual utility functions is best done through a consistent method of comparison that results in utilities ranging from 0 to 1. We discussed three methods: calculation of proportional scores, assessment of ratios, and the conventional lottery-based assessment discussed in Chapters 13 and 14. Once these utility functions are established, weights must also be assessed. Several assessment procedures are possible here, all providing mechanisms for determining the rate at which the attributes can be traded off against one another. Sometimes trade-off rates can be assessed in terms of dollars by pricing out the other attributes. The swing-weighting and lottery-based assessment techniques can be used even if it is difficult to think of the attributes in dollar terms, and these techniques lead to clear interpretations of the assessed weights. With weights and individual utility functions determined, overall utilities are calculated by applying the additive utility formula. We demonstrated the procedure with an example in which potential sites for a public library were evaluated.

EXERCISES

15.1   Why is it important to think carefully about decision situations that involve multiple conflicting objectives?

15.2   Explain the general decision-analysis approach to dealing with multiple objectives.

15.3   Explain what is meant by the term *indifference curve*.

15.4   Explain the idea of dominance in the context of multiobjective decision making.

15.5   Imagine that you are working on a committee to develop an employment-conditions policy for your firm. During the discussion of safety, one committee member states that employee safety is twice as important as benefits such as flexible work time, employer-sponsored day care, and so on.

    a   How might you respond to such a statement?

    b   Suppose the statement was that employee safety was twice as important as insurance benefits. How might you translate this statement into a form (perhaps based on dollar values) so that it would be possible to evaluate various policy alternatives in terms of safety and insurance in a way that would be consistent with the statement?

15.6   Explain why proportional scores represent a risk attitude that is risk-neutral.

15.7   An MBA student evaluating weather outcomes for an upcoming party has concluded that a sunny day would be twice as good as a cloudy day, and a cloudy day would be three times as good as a rainy day. Use these assessments to calculate utilities that range from 0 to 1 for sunny, rainy, and cloudy days.

15.8   Explain in your own words why swing weights produce meaningful weights for evaluating alternatives.

15.9   A decision maker is assessing weights for two attributes using the swing-weight method. When he imagines swinging the attributes individually from worst to best, he concludes that his improvement in satisfaction from Attribute A is 70% of the improvement from swinging Attribute B. Calculate k_A and k_B.

15.10 Explain in your own words why the lottery method works to produce meaningful weights for evaluating alternatives.

15.11 A decision maker is assessing weights for three attributes using the lottery-assessment method. In considering the lotteries, she concludes that she is indifferent between:

> **A**   Win the best possible combination with probability 0.34
> Win the worst possible combination with probability 0.66

and

> **B**   Win a combination that is worst on Attributes 1 and 3 and best on 2

She also has concluded that she is indifferent between

> **C**   Win the best possible combination with probability 0.25
> Win the worst possible combination with probability 0.75

and

> **D**   Win a combination that is worst on Attributes 2 and 3 and best on 1

Find weights k_1, k_2, and k_3.

QUESTIONS AND PROBLEMS

15.12 Suppose that you are searching for an apartment in which to live while you go to school. Apartments near campus generally cost more than equivalent apartments farther away. Five apartments are available. One is right next to campus, and another is one mile away. The remaining apartments are two, three, and four miles away.

a   Suppose you have a tentative agreement to rent the apartment that is one mile from campus. How much more would you be willing to pay in monthly rent to obtain the one next to campus? (Answer this question on the basis of your own personal experience. Other than rent and distance from campus, the two apartments are equivalent.)

b   Now suppose you have a tentative agreement to rent the apartment that is four miles away. How much more would you be willing to pay in monthly rent to move to the apartment that is only three miles from campus?

c   What are the implications of your answers to parts a and b? Would it be appropriate to rank the apartments in terms of distance using the proportional-scoring technique?

d Sketch an indifference curve that reflects the way you would trade off rent versus proximity to campus. Is your indifference curve a straight line?

15.13 A friend of yours is in the market for a new computer. Four different machines are under consideration. The four computers are essentially the same, but they vary in price and reliability. The least expensive model is also the least reliable, the most expensive is the most reliable, and the other two are in between.

a Describe to your friend how you would approach the decision.

b Define reliability in a way that would be appropriate for the decision. Do you need to consider risk?

c How might your friend go about establishing a marginal rate of substitution between reliability and price?

15.14 Continuing Problem 15.13, the computers are described as follows:

A Price: $998.95   Expected number of days in the shop per year: 4

B Price: $1300.00 Expected number of days in the shop per year: 2

C Price: $1350.00 Expected number of days in the shop per year: 2.5

D Price: $1750.00 Expected number of days in the shop per year: 0.5

The computer will be an important part of your friend's livelihood for the next two years. (After two years, the computer will have a negligible salvage value.) In fact, your friend can foresee that there will be specific losses if the computer is in the shop for repairs. The magnitude of the losses are uncertain but are estimated to be approximately $180 per day that the computer is down.

a Can you give your friend any advice without doing any calculations?

b Use the information given to determine weights k_P and k_R, where R stands for reliability. What assumptions are you making?

c Calculate overall utilities for the computers. What do you conclude?

d Sketch three indifference curves that reflect your friend's trade-off rate between reliability and price.

e What considerations other than losses might be important in determining the trade-off rate between cost and reliability?

15.15 Throughout the chapter we have assessed individual utility functions that range from 0 to 1. What is the advantage of doing this?

15.16 You are an up-and-coming developer in downtown Seattle and are interested in constructing a building on a site that you own. You have collected four bids from prospective contractors. The bids include both a cost (millions of dollars) and a time to completion (months):

Contractor	Cost	Time
A	100	20
B	80	25
C	79	28
D	82	26

The problem now is to decide which contractor to choose. B has indicated that for another $20 million he could do the job in 18 months, and you have said that you would be indifferent between that and the original proposal. In talking with C, you have indicated that you would just as soon pay her an extra $4 million if she could get the job done in 26 months. Who gets the job? Explain your reasoning. (It may be convenient to plot the four alternatives on a graph.)

15.17 Once you decide that you are in the market for a personal computer, you have many different considerations. You should think about how you will use the computer, and so you need to know whether appropriate software is available and at what price. The nature of the available peripheral equipment (printers, disk drives, and so on) can be important. The "feel" of the computer, which is in some sense determined by the operating system and the user interface, can be critical. Are you an experienced user? Do you want to be able to program the machine, or will you (like most of us) rely on existing or over-the-counter software? If you intend to use the machine for a lot of number crunching, processor speed may be important. Reliability and service are other matters. For many students, an important question is whether the computer will be compatible with other systems in any job they might eventually have. Finally, of course, price and operating costs are important.

Create a fundamental-objectives hierarchy to compare your options. Take care in doing this; be sure that you establish the fundamental objectives and operational attributes that will allow you to make the necessary comparisons. (Note that the attributes suggested above are *not* exhaustive, and some may not apply to you!)

Use your model to evaluate at least three different computers (preferably from different manufacturers). You will have to specify precisely the packages that you compare. It also might be worthwhile to include appropriate software and peripheral equipment. (Exactly what you compare is up to you, but make the packages meaningful.) Be sure that your utilities are such that the best alternative gets a 1 and the worst a 0 for each attribute. Assess weights using pricing out, swing weighting, or lottery weights. Calculate overall utilities for your alternatives.

Try using the utility functions for money and computer reliability that you assessed in Problems 14.12 and 14.13. You may have to rescale the utility functions to obtain scores so that the best alternative in this problem scores 1 and the worst scores 0.

If possible, use a computer-based multiattribute decision program to do this problem.

15.18 When you choose a place to live, what objectives are you trying to accomplish? What makes some apartments better than others? Would you rather live close to campus or farther away and spend less money? What about the quality of the neighborhood? How about amenities such as a swimming pool?

Create a fundamental-objectives hierarchy that allows you to compare apartment options. Take care in doing this; be sure to establish the fundamental objectives and operational attributes that will allow you to make the necessary comparisons.

Once you are satisfied with your hierarchy, use it to compare available housing alternatives. Try ranking different apartments that are advertised in the classified section of the newspaper, for example. Be sure that your individual utilities follow the rules: Best takes a 1 and worst takes 0. (Try using the utility function for money that you assessed in Problem 14.12. You may have to rescale it so that your best alternative gets a 1 and worst gets a 0.) Assess weights using pricing out, swing weighting, or lottery weights. Evaluate your alternatives with the additive utility function.

If possible, use a computer-based multiattribute decision program to do this problem.

15.19 What is important to you in choosing a job? Certainly salary is important, and for many people location matters a lot. Other considerations might involve promotion potential, the nature of the work, the organization itself, benefits, and so on.

Create a fundamental-objectives hierarchy that allows you to compare job offers. Be sure to establish the fundamental objectives and operational attributes that will allow you to make the necessary comparisons.

Once you are satisfied with your hierarchy, use it to compare your job offers. You also may want to think about your "ideal" job in order to compare your current offers with your ideal. You also might consider your imaginary worst possible job in all respects. (For the salary attribute, try using the utility function for money that you assessed in Problem 14.12, or some variation of it. You may have to rescale it so that your best alternative is 1 and worst 0.) Assess weights for the attributes using pricing out, swing weighting, or lottery weights, being careful to anchor your judgments in terms of both ideal and worst imaginable jobs. Evaluate your various job offers with the additive utility function.

If possible, use a computer-based multiattribute decision program to do this problem.

15.20 How can you compare your courses? When you consider those that you have taken, it should be clear that some were better than others and that the good ones were, perhaps, good for different reasons. What are the important dimensions that affect the quality of a course? Some are obvious, such as the enthusiasm of an instructor, topic, and amount and type of work involved. Other aspects may not be quite so obvious; for example, how you perceive one course may depend on other courses you have had.

In this problem, the objective is to create a "template" that will permit consistent evaluation of your courses. The procedure is essentially the same as it is for any multiattribute decision, except that you will be able to use the template to evaluate future courses. Thus, we do not have a set of alternatives available to use for the determination of scores and weights. (You want to think, however, about current and recent courses in making your assessments.)

First, create a fundamental-objectives hierarchy that allows you to compare courses. Be sure to establish the fundamental objectives and operational attributes that will allow for the necessary comparisons. Constructing a set of objectives for comparing courses is considerably more difficult than comparing computers, apartments, or jobs. You may find that many of the attributes you consider initially will overlap with others, leading to a confusing array of attributes that are interdependent. It may take considerable thought to reduce the degree of redundancy in your hierarchy and to arrive at one that is complete, decomposable, small enough to be manageable, and involves attributes that are easy to think about.

Once you are satisfied with your objectives and attributes, imagine the best and worst courses for each attribute. Create attribute scales (constructed scales where appropriate) for each objective. The idea is to be able to return to these scales with any new course and determine utilities for each objective with relative ease. (Try using the homework utility function that you developed in Problem 14.15. You may have to rescale it so that your best alternative gets a 1 and the worst a 0.)

Once you have created the attribute scales, you are ready to assess the weights. Try the swing-weighting or lottery approach for assessing the weights. (Pricing out may be difficult to do in this particular example. Can you place a dollar value on your attributes?)

Finally, with scales and weights established, you are ready to evaluate courses. Try comparing three or four of your most recent courses. (Try evaluating one that you took more than a year ago. Can you remember enough about the course to assess the individual utilities with some degree of confidence?)

If possible, implement your course-evaluation template using a computer-based multiattribute decision program. Alternatively, you might create a spreadsheet template that you could use to evaluate courses.

15.21 Refer to the discussion of the automobiles in the section on "Trading Off Conflicting Objectives: The Basics." We discussed switching first from the Standard to the Norushi, and then from the Norushi to the Portalo. Would it make sense to consider a direct switch from the Standard to the Portalo? Why or why not?

15.22 In Chapter 2 we discussed net present value (NPV) as a procedure for evaluating consequences that yield cash flows at different points in time. If x_i is the cash flow at year i and r is the discount rate, then the NPV is given by

$$\text{NPV} = \sum \frac{x_i}{(1+r)^i}$$

where the summation is over all future cash flows including the current x_0.

a   Explain how the NPV criterion is similar to the additive utility function that was discussed in this chapter. What are the attributes? What are the weights? Describe the way cash at time period i is traded off against cash at time period $i + 1$. (*Hint:* Review Chapter 2!)

b   Suppose that you can invest in one of two different projects. Each costs $20,000. The first project is riskless and will pay you $10,000 each year for the next three years. The second one is risky. There is a 50% chance that it will pay $15,000 each year for the next three years and a 50% chance that it will pay only $5000 per year for the next three years. Your discount rate is 9%. Calculate the NPV for both the riskless and risky project. Compare them. What can you conclude about the use of NPV for deciding among risky projects?

c   How might your NPV analysis in part b be modified to take risk into account? Could you use a utility function? How does the idea of a risk-adjusted discount rate fit into the picture? How could the interest rate be adjusted to account for risk? Would this be the same as using a utility function for money?

15.23 Following up Problem 15.22, even though we take riskiness into account, there still is difficulty with NPV as a decision criterion. Suppose that you are facing the two risky projects shown in Figure 15.15.

Project A pays either $10,000 for each of two years or $100 for those two years. Project B pays $10,000 either in the first year or the second year and $100 in the other year. Assume that the cash flows are annual new profits.

a   Which of these two risky investments would you prefer? Why?

b   Calculate the expected NPV for both projects, using the same 9% interest rate from Problem 15.22. Based on expected NPV, in which project would you invest?

Figure 15.15

Decision tree for Problem 15.23. Which of these two risky investments would you prefer?

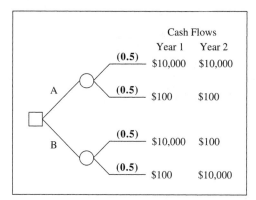

c   After careful assessment, you have concluded that you are risk-averse and that your utility function can be adequately represented by $U(X_i) = \ln(X_i)$, where X_i represents cash flow during year i. Calculate the expected net present utility for each project. Net present utility is given by

$$NPU = \sum \frac{U(X_i)}{(1+r)^i}$$

d   NPU in part c should incorporate your attitude toward risk. Are your NPU calculations in part c consistent with your preferences in part a? What is there about these two projects that is not captured by your utility function? Can you think of any other way to model your preferences?

15.24   Instead of calculating a "discounted" utility as we did in Problem 15.23, let us consider calculating U(NPV). That is, calculate NPV first, using an appropriate interest rate, and then calculate a utility value for the NPV. For your utility function, use the exponential utility function $U(NPV) = 1 - e^{-NPV/5000}$. Use this approach to calculate the expected utility of Projects A and B in Figure 15.15. Which would you choose? Are there any problems with using this procedure for evaluating projects?

15.25   A policy maker in the Occupational Safety and Health Administration is under pressure from industry to permit the use of certain chemicals in a newly developed industrial process. Two different versions of the process use two different chemicals, A and B. The risks associated with these chemicals are not known with certainty, but the available information indicates that they may affect two groups of people in the following ways:

- *Chemical A*   There is a 50% chance that Group 1 will be adversely affected, while Group 2 is unaffected; and a 50% chance that Group 2 is adversely affected, while Group 1 is unaffected.

- *Chemical B*   There is a 50% chance that both groups will be adversely affected, and a 50% chance that neither group will be affected.

Assume that "adversely affected" means the same in every case—an expected increase of one death in the affected group over the next two years. The decision maker's problem looks like the decision tree in Figure 15.16.

a   Calculate the expected number of deaths for each chemical.

Figure 15.16

Deciding between alternative chemicals in Problem 15.25.

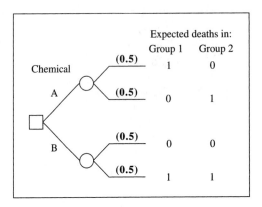

b A decision maker who values consequences using an overall utility might calculate the utility for each consequence as

$$U(\text{Chemical}) = k_1 U_1(\text{Group 1 Deaths}) + k_2 U_2(\text{Group 2 Deaths})$$

For both U_1 and U_2, the best and worst possible outcomes are 0 deaths and 1 death, respectively. Thus,

$$U_1(1 \text{ Death}) = 0 \qquad U_1(1 \text{ Death}) = 1$$
$$U_2(1 \text{ Death}) = 0 \qquad U_2(1 \text{ Death}) = 1$$

Explain why k_1 and $1 - k_1$ may not be equal.

c Assume that $k_1 = 0.4$. Show that the decision maker who evaluates the two chemicals in terms of their expected overall utilities (as defined above) would be indifferent between them. Does the value of k_1 matter?

d Why might the decision maker *not* be indifferent between the two programs? (Most people think about the decision maker's risk attitude toward the number of deaths or lives saved. Besides this, think about the following: Suppose you are a member of Group 1, and the decision maker has chosen Chemical A. It turned out that Group 1 was affected. How would you feel? What would you do? What does this imply for the decision maker?)

15.26 Refer to the discussion of the three automobiles in the section "Trading Off Conflicting Objectives: The Basics." Suppose we had the following individual utility functions for price and life span:

Life span	Price
$U_L(6 \text{ years}) = 0.00$	$U_P(17{,}000) = 0.00$
$U_L(9 \text{ Years}) = 0.75$	$U_P(10{,}000) = 0.50$
$U_L(12 \text{ Years}) = 1.00$	$U_P(8000) = 1.00$

The additive utility model discussed in this chapter would give us the following:

$$U(\text{Price, Life Span}) = k_L U_L(\text{Life Span}) + (1 - k_L) U_P(\text{Price})$$

a With $k_L = 0.45$, calculate the utility for the three cars. Which would be chosen?

b Suppose that you are not completely comfortable with the assessment of $k_L = 0.45$. How *large* could k_L be before the decision changes, and what would be the new choice? How *small* could k_L be before the decision changes, and what would be the new choice? Would you say that the choice among these three cars is very sensitive to the assessment of k_L?

15.27 Refer to Table 15.6. Your boss is a member of the library-site selection committee. She is not perfectly satisfied with the assessments shown in the table. Specifically, she wonders to what extent the assessed weights and individual utilities on the detail attributes could change without affecting the overall ranking of the four sites. Use an electronic spreadsheet to answer this question, and write a memo that discusses your findings. (*Hint:* There is no specific way to attack a sensitivity analysis like this. One possibility is to establish reasonable ranges for the weights and then create a tornado diagram. Be sure that your weights add to 1 in each category!)

15.28 Refer to Problem 15.11. Suppose that the decision maker has made a third assessment, concluding that she is indifferent between

> **E**   Win the best possible combination with probability 0.18
> Win the worst possible combination with probability 0.82

and

> **F**   Win a combination that is worst on Attributes 1 and 2, and best on 3

What does this assessment imply for the analysis? Is it consistent with your answer for k_3 in Problem 15.11? What should you do now?

CASE STUDIES

THE SATANIC VERSES

In early 1989, the Ayatollah Khomeini of Iran decreed that Salman Rushdie, the British author of *The Satanic Verses,* should be put to death. In many ways, Rushdie's novel satirized Islam and the Prophet Muhammed. Khomeini declared that Rushdie should die and that whoever killed him would go to Heaven.

Many bookstores in both Europe and the United States that carried *The Satanic Verses* found themselves in a bind. Some Muslims threatened violence unless the bookstores stopped selling the book. Some booksellers removed the book from their shelves and sold it only to customers who specifically asked for it. Others refused to sell it altogether on the grounds that it was too risky. Still others defied the threats.

One bookseller in Berkeley, California, continued selling the book on the grounds that he would not allow anyone to interfere with the principle of freedom of the press. His store was bombed, and damage was substantial. His reaction? He increased security.

Questions

1   Imagine that you are the owner of a bookstore faced with the decision of what to do about *The Satanic Verses*. In deciding on a course of action, there are several conflicting objectives. Develop an objectives hierarchy and operational attribute scales.

2   What alternatives do you have?

3   What risks do you face? Do the risks differ depending on the alternative you choose? Sketch a simple influence diagram or decision tree for your problem.

DILEMMAS IN MEDICINE

In *Alpha and Omega: Ethics at the Frontiers of Life and Death,* author Ernlé Young specifies four fundamental principles that must be considered in making medical decisions: beneficence, nonmaleficence, justice, and autonomy. The following descriptions of these principles have been abstracted from the book (pp. 21–23):

- *Beneficence* implies that the physician's most important duty is to provide services that are beneficial to the patient. In many cases, this can mean taking measures that are intended to preserve the patient's life.

- *Nonmaleficence* is the duty not to cause harm to the patient. A medical aphorism of uncertain origin proclaims *primum non nocere*—above all, do no harm. Harm can mean different things in different situations and for different patients, and can include death, disability, separation from loved ones, or deprivation of pleasure or freedom. The difficulty is that many medical procedures entail at least some harm or the potential for harm. This always must be weighed against the potential benefits.

- *Justice* in this context refers to the fair use of resources. It is, after all, impossible to do absolutely everything that would be medically justifiable for all patients. Thus, decisions must be made regarding the allocation of scarce resources. The issue is how to make these decisions fairly or equitably. For example, how should we decide what patients have priority for receiving donated organs? Is it appropriate to admit a terminally ill patient to an intensive care unit?

- *Autonomy* requires allowing a patient to make his or her own decisions regarding medical treatment as far as is possible. The patient, operating as an independent, self-determining agent, should be able to obtain appropriate information and participate fully in the decisions regarding the course of treatment and, ultimately, the patient's life.

In most medical situations, these principles do not conflict. That is, the physician can provide beneficial care for the patient without causing harm, the treatment can be provided equitably, and the patient can easily make his or her own decisions. In a few cases, however, the principles are in conflict and it is impossible to accomplish all of them at once. For example, consider the case of a terminally ill patient who insists that everything possible be done to extend his or her life. Doing so may violate both nonmaleficence and justice while at the same time providing limited benefit. But not providing the requested services violates autonomy. Thus, the physician would be in a very difficult dilemma.

Questions

1   Discuss the relationship between the medical ethics here and decision making in the face of conflicting objectives. Sketch an objectives hierarchy for a physician who must cope with difficult problems such as those described above. Can you explain or expand on the four fundamental objectives by developing lower-level objectives?

2   Neonatology is the study and treatment of newborn infants. Of particular concern is the treatment of low-birth-weight infants who are born prematurely. Often these babies are the victims of poor prenatal care and may be burdened with severe deformities. Millions of dollars are spent annually to save the lives of such infants. Discuss the ways in which the four principles conflict in this situation. Could you give any guidelines to a panel of doctors and hospital administrators grappling with such problems?

3   Terminally ill patients face the prospect of death within a relatively short period of time. In the case of cancer victims, their last months can be extremely painful. Increasing numbers of such patients consider taking their own lives, and much controversy has developed concerning euthanasia, or mercy killing. Imagine that a patient is terminally ill and mentions that he or she is considering suicide. If everything reasonable has been done to arrest the disease without success, this may be a reasonable option. Furthermore, the principle of autonomy should be respected here, provided that the patient is mentally stable and sound and understands fully the implications. But how deeply should the physician or loved one be involved? There are varying degrees of involvement. First, the physician might simply provide counseling and emotional support. The next step would be encouraging the patient by removing obstacles. Third, the physician might provide information about how to end one's life effectively and without trauma. The next step would be to assist in the procurement of the means to commit suicide. Helping the patient to end his or her life represents still another step, and actually killing the patient—by lethal injection or removal of a life support system, for example—would represent full involvement.

   Suppose that one of your loved ones were terminally ill and considering suicide. What issues would you want him or her to consider carefully? Draw an objectives hierarchy for the patient's decision.

Now suppose that the patient has asked you to assist in his or her suicide. What issues would you want to consider when deciding on your level of involvement? Sketch an objectives hierarchy for your own decision. Compare this hierarchy with the patient's.

Source: E. Young (1989) *Alpha and Omega: Ethics at the Frontiers of Life and Death*, Palo Alto, CA: Stanford Alumni Association.

A MATTER OF ETHICS

Paul Lambert was in a difficult situation. When he started his current job five years ago, he understood clearly that he would be working on sensitive defense contracts for the government. In fact, his firm was a subcontractor for some major defense contractors. What he did not realize at the time—indeed, he only discovered this gradually over the past two years—was that the firm was overcharging. And it was not just a matter of a few dollars. In some cases, the government was overcharged by as much as a factor of 10.

Three weeks ago, he inadvertently came across an internal accounting memo that documented one particularly flagrant violation. He quietly made a copy and locked it in his desk. At the time, he was amazed, then righteously indignant. He resolved to take the evidence immediately to the appropriate authorities. But the more he thought about it, the more confused he became. Finally, he called his brother-in-law, Jim Grillich. Jim worked for another defense-related firm and agreed to have lunch with Paul. After exchanging stories about their families and comments on recent sporting events, Paul laid his cards on the table.

"Looks as though you could really make some waves," Jim commented, after listening to Paul's story.

"I guess I could. But I just don't know. If I blow the whistle, I'd feel like I'd have to resign. And then it would be tough to find another job. Nancy and I don't have a lot of savings, you know." The thought of dipping into their savings made Paul shake his head. "I just don't know."

The two men were silent for a long time. Then Paul continued, "To make matters worse, I really believe that the work that the company is doing, especially in the research labs, is important. It may have a substantial impact on our society over the next 20 years. The CEO is behind the research 100 percent, and I gather from the few comments I've overheard that he's essentially funding the research by overcharging on the subcontracts. So if I call foul, the research program goes down the drain."

"I know what you mean." Jim went on to recount a similar dilemma that he faced a few years before.

"So what did you do?"

"The papers are still in my desk. I always wanted to talk to someone about it. I even thought about calling you up, but I never did. After a while, it seemed like it was pretty easy just to leave the papers in there, locked up, safe and sound."

Questions

1   What trade-offs is Paul trying to make? What appear to be his fundamental objectives?

2   Suppose that Paul's take-home pay is currently $2400 per month. In talking to an employment company, he is told that it will probably take two months to find a similar job if he leaves his current one, and he had better expect three months if he wants a better job. In looking at his savings account of $10,500, he decides that he cannot justify leaving his job, even though this means keeping quiet about the overcharging incident. Can you say anything about an implicit trade-off rate between the fundamental objectives that you identified above?

3   Have you ever been in a situation in which it was difficult for you to decide whether to take an ethically appropriate action? Describe the situation. What made the decision difficult? What trade-offs did you have to make? What did you finally do?

FDA AND THE TESTING OF EXPERIMENTAL DRUGS

The Food and Drug Administration (FDA) of the federal government is one of the largest consumer-protection agencies in the world. One of the FDA's charges is to ensure that drugs sold to consumers do not pose health threats. As a result, the testing procedure that leads to a new drug's approval is rigorous and demanding. So much so, in fact, that some policy makers are calling for less stringent standards. Below are some of the dilemmas that FDA faces:

- If an experimental drug shows promise in the treatment of a dangerous disease such as AIDS, should the testing procedure be abbreviated in order to get the drug to market more quickly?

- The FDA already is a large and costly bureaucracy. By easing testing standards, substantial dollars could be saved. But would it be more likely that a dangerous drug would be approved? What are the costs of such a mistake?

A fundamental trade-off is involved here. What are we gaining in the way of assurance of safe drugs, and what are we giving up by keeping the drugs away from the general public for an additional year or two?

Questions

1   What are the consequences (both good and bad) of keeping a drug from consumers for some required period of rigorous testing?

2   What are the consequences (both good and bad) of allowing drugs to reach consumers with less stringent testing?

3   Imagine that you are the FDA's commissioner. A pharmaceutical company requests special permission to rush a new AIDS drug to market. On the basis of a first round of tests, the company estimates that the new drug will save the lives of 200 AIDS victims in the first year. Your favorite pharmacologist expresses reservations, however, claiming that without running the complete series of tests, he fears that the drug may have as-yet-undetermined but serious side effects. What decision would you make? Why?

4   Suppose that the drug in Question 3 was for arthritis. It could be used by any individual who suffers from arthritis and, according to the preliminary tests, would be able to cure up to 80% of rheumatoid arthritis cases. But your pharmacologist expresses the same reservations as for the AIDS drug. Now what decision would you make? Why?

REFERENCES

The additive utility function has been described by many authors. The most comprehensive discussion, and the only one that covers swing weights, is that by von Winterfeldt and Edwards (1986). Keeney and Raiffa (1976) and Keeney (1980) also devote a lot of material to this preference model. Edwards and Barron (1994) discuss some heuristic approaches to assessing weights, including the use of only rank-order information about the objectives.

The basic idea of creating an additive utility function is fairly common and has been applied in a variety of settings. Moreover, this basic approach also has earned several different names. For example, a cost-benefit analysis typically prices out nonmonetary costs and benefits and then aggregates them. For an interesting critique of a cost-benefit analysis, see Bunn (1984, Chapter 5).

Other decision-aiding techniques also use the additive utility function implicitly or explicitly, including the Analytic Hierarchy Process (Saaty 1980) and goal programming with nonpreemptive weights (see Winston 1987). Conjoint analysis, a statistical technique used in market research to determine preference patterns of consumers on the basis of survey data, often is used to create additive utility functions. In all of these, some kind of subjective judgment forms the basis for the weights, and yet the interpretation of the weights is not always clear. For all of these alternative models, extreme care must be exercised in making the judgments on which the additive utility function is based. There is no substitute for thinking hard about trade-off issues. This text's view is that the decision-analysis approach discussed in this chapter provides the best systematic framework for making those judgments.

Bunn, D. (1984) *Applied Decision Analysis.* New York: McGraw-Hill.

Edwards, W., and F. H. Barron (1994) "SMARTS and SMARTER: Improved Simple Methods for Multiattribute Utility Measurement." Organizational Behavior and Human Decision Processes, 60, 306–325.

Keeney, R. (1980) *Siting Energy Facilities.* New York: Academic Press.

Keeney, R., and H. Raiffa (1976) *Decisions with Multiple Objectives.* New York: Wiley.

Saaty, T. (1980) *The Analytic Hierarchy Process.* New York: McGraw-Hill.

von Winterfeldt, D., and W. Edwards (1986) *Decision Analysis and Behavioral Research.* Cambridge: Cambridge University Press.

Winston, W. (1987) *Operations Research: Applications and Algorithms.* Boston: PWS-KENT.

EPILOGUE

What happened with the Eugene Public Library? After much public discussion, an alternative emerged that had not been anticipated. An out-of-state developer expressed a desire to build a multistory office building in downtown Eugene (at Site 2, in fact) and proposed that the library could occupy the lower two floors of the building. By entering into a partnership with the developer, the city could save a lot in construction costs. Many citizens voiced concerns about the prospect of the city's alliance with a private developer, others were concerned about the complicated financing arrangement, and still others disapproved of the proposed location for a variety of reasons. On the other hand, the supporters pointed out that this might be the only way that Eugene would ever get a new library. In March 1989, the proposal to accept the developer's offer was submitted to the voters. The result? They turned down the offer.

Eugene had still more chances to get a new library. In 1991, a study commissioned by the city concluded that a building vacated by Sears in downtown Eugene—only a few blocks from the current library—would be a suitable site for the new facility. In a March election of that year, nearly three-quarters of those who voted agreed with this assessment. A subsequent vote in May of 1994 to authorize issuance of bonds to pay for refurbishing the building and moving the library, however, was defeated by a handful of absentee ballots. The argument was made that the ballot measure was too complex; it proposed using the funds for other city facilities as well as the library. A measure that focused on funding only the library would surely fare better. In November of 1994, such a measure was placed on the ballot. And defeated by fewer than 100 absentee ballots.

CHAPTER 16

Conflicting Objectives II: Multiattribute Utility Models with Interactions

The additive utility function described in Chapter 15 is an easy-to-use technique. It is incomplete, however, because it ignores certain fundamental characteristics of choices among multiattribute alternatives. We discussed one problem with proportional scores—they assume risk neutrality. Problem 15.22 demonstrated this in the important context of the common NPV choice criterion. More subtle are situations in which attributes interact. For example, two attributes may be substitutes for one another to some extent. Imagine a CEO who oversees several divisions. The simultaneous success of every division may not be terribly important; as long as some divisions perform well, cash flows and profits will be adequate. On the other hand, attributes can be complementary. An example might be the success of various phases of a research-and-development project. The success of each individual project is valuable in its own right. But the success of all phases might make possible an altogether new technology or process, thus leading to substantial synergistic gains in many ways. In this case, high achievement on all attributes (success in the various R&D phases) is worth more than the sum of the value obtained from the individual successes.

Such interactions cannot be captured by the additive utility function. That model is essentially an additive combination of preferences for individual attributes. To capture the kinds of interactions that we are talking about here, as well as risk attitudes, we must think more generally. Let us think in terms of a *utility surface,* such as the one depicted in Figure 16.1 for two attributes. Although it is possible to think about many attributes at once, we will develop multiattribute utility theory concepts using only two attributes. The ideas are readily extended to more attributes, and at the end of the chapter we will say a bit about multiattribute utility functions for three or more attributes.

Much of this chapter is fairly abstract and technical. To do a good job with the material, the mathematics are necessary. After theoretical development, which is

Figure 16.1

A utility surface for two attributes.

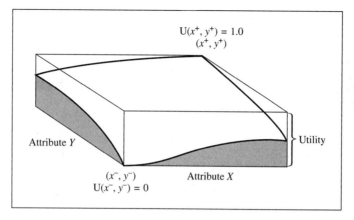

$U(x^+, y^+) = 1.0$
(x^+, y^+)

Attribute Y

Utility

(x^-, y^-)
$U(x^-, y^-) = 0$

Attribute X

sprinkled with illustrative examples, we will process a complete example that involves the assessment of a two-attribute utility function for managing a blood bank. We also discuss briefly how to deal with three or more attributes and demonstrate such a model in a large-scale electric-utility example.

Multiattribute Utility Functions: Direct Assessment

To assess a utility function like the one in Figure 16.1, we can use the same basic approach that we already have used. For example, consider the reference-gamble method. The appropriate reference gamble has the worst pair (x^-, y^-) and the best pair (x^+, y^+) as the two possible outcomes:

Win (x^+, y^+) with probability p

Win (x^-, y^-) with probability $1 - p$

Now for any pair (x, y), where $x^- \leq x \leq x^+$ and $y^- \leq y \leq y^+$ find the probability p to use in the reference gamble that will make you indifferent between (x, y) and the reference gamble. As before, you can use p as your utility $U(x, y)$ because $U(x^+, y^+) = 1$, and $U(x^-, y^-) = 0$. Figure 16.2 shows the decision tree that represents the assessment

Figure 16.2

Directly assessing a multiattribute utility. The probability that makes you indifferent between the lottery and the sure thing is your utility value for (x, y).

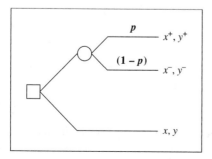

p
x^+, y^+

$(1 - p)$
x^-, y^-

x, y

Figure 16.3

*Sketching indifferent
curves.* The point val-
ues are the assessed
utility values for the
corresponding
(*x, y*) pair.

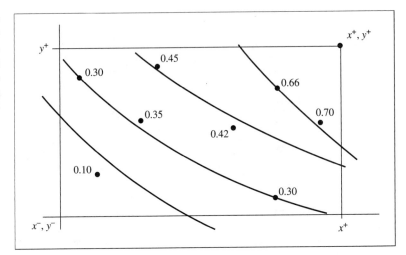

Figure 16.3

Sketching indifferent curves. The point values are the assessed utility values for the corresponding (*x, y*) pair.

situation. This is simply the standard probability-equivalent utility assessment technique that we have seen before.

You can see that we will wind up with many utility numbers after making this assessment for a reasonable number of (*x, y*) pairs. There may be several pairs with the same utility, and you should be indifferent among such pairs. Thus, (*x, y*) pairs with the same utilities must fall on an indifference curve. One approach to understanding your multiattribute preferences is simply to plot the assessed points on a graph, as in Figure 16.3, and sketch rough indifference curves.

To find a good representation of preferences through direct assessment, however, there is a drawback: You must assess utilities for a substantial number of points. And even though it is straightforward to see how this approach might be extended to three or more attributes, the more being considered, the more points you must assess, and the more complicated graphical representations become. An easier way would be convenient.

Another approach that would ease the assessment burden would be to think about a multiattribute utility function that is made up of the individual utility functions. Mathematically, we might represent the most general case as

$$U(x, y) = f[U_X(x), U_Y(y)]$$

The $f[\cdot\,, \cdot]$ notation means that $U(x, y)$ is a function of the individual utility functions $U_X(x)$ and $U_Y(y)$. In Chapter 15, the form we used was

$$U(x, y) = k_X U_X(x) + k_Y U_Y(y)$$

In this chapter we will consider

$$U(x, y) = c_1 + c_2 U_X(x) + c_3 U_Y(y) + c_4 U_X(x)U_Y(y)$$

The importance of any such formulation is that it greatly eases the assessment burden; as in Chapter 15, we only require the individual utility functions and enough information to put them together. Being able to break down the multiattribute utility function this way sometimes is called *separability;* the overall utility function can be "separated" into chunks that represent different attributes.

Is such an arrangement possible? Yes, but it requires some interesting conditions for the combined utility function. These conditions concern how the preferences interact among the attributes, a point suggested in the beginning of this chapter. We will digress briefly to discuss these conditions.

Independence Conditions

Preferential Independence

One thing we need in order to have the kind of separability mentioned above is *mutual preferential independence.* An attribute Y is said to preferentially independent of X if preferences for specific outcomes of Y do not depend on the level of attribute X. As an example, let Y be the time to completion of a project and X its cost. If we prefer a project time of 5 days to one of 10 days, assuming that the cost is 100 in each case, and if we also prefer a project time of 5 days to one of 10 days if the cost is 200 in both cases, then Y is preferentially independent of $X;$ it does not matter what the cost is—we still prefer the shorter completion time.

We need mutual preferential independence, so we also need the cost to be preferentially independent of the completion time. If we prefer lower cost no matter what the completion time, then X is preferentially independent of Y. Then we can say that the two attributes are mutually preferentially independent.

Preferential independence seems to be a pretty reasonable condition to assume, especially in cases like the one involving costs and time to completion. But it is easy to imagine situations in which preferential independence might not hold. For example, in Chapter 15 it was suggested that your preference for amount of homework effort might depend on course topic. Bunn (1984) relates a nice hypothetical example in which preferential independence might not hold. Consider a decision with outcomes that affect both the place where you live and the automobile that you drive. Let X be an outcome variable that could denote either Los Angeles or an African farm, and Y an outcome variable denoting either a Cadillac or a Land Rover. The value of X (whether you live in Los Angeles or on an African farm) may well affect your preference for a Cadillac or a Land Rover. Therefore, Y would not be preferentially independent of X. Consider the reverse: You may prefer Los Angeles to an African farm (or vice versa) regardless of the car you own. Thus, one attribute would be preferentially independent of the other, but the two are not mutually preferentially independent.

It probably is fair to say that mutual preferential independence holds for many people and many situations, or that at least it is a reasonable approximation. Mutual preferential independence is like the decomposability property for an objectives hierarchy. If a decision maker has done a good job of building a decomposable hierarchy, mutual preferential independence probably is a reasonable assumption. But it should never be taken for granted.

If you recall the discussion about value functions and utility functions from Chapter 15, you will realize that mutual preferential independence, being about sure outcomes, is a condition that applies to value functions. In fact, mutual preferential independence is exactly the condition that is needed for the additive (ordinal) utility function to be appropriate when making a decision with no uncertainty. That is, when a decision maker's preferences display mutual preferential independence, then the additive utility function, assessed using any of the techniques described in Chapter 15, is appropriate for decisions under certainty. Once we move to decision under uncertainty, however, mutual preferential independence is not quite strong enough. Although it is a necessary condition for obtaining separability of a (cardinal) multiattribute utility function, it is not sufficient. Thus, we must look at stronger conditions.

Utility Independence

Utility independence is slightly stronger than preferential independence. An attribute Y is considered utility independent of attribute X if preferences for *uncertain choices* involving different levels of Y are independent of the value of X. Imagine assessing a certainty equivalent for a lottery involving only outcomes in Y. If our certainty equivalent amount for the Y lottery is the same no matter what the level of X, then Y is utility independent of X. If X also is utility independent of Y, then the two attributes are mutually utility independent.

Utility independence clearly is analogous to preferential independence, except that the assessments are made under conditions of uncertainty. For the project evaluation example above, suppose we assess that the certainty equivalent for an option giving, say, a 50% chance of $Y = 5$ and a 50% chance of $Y = 10$ does not depend on the level at which the cost X is fixed. As long as our preferences for lotteries in the completion-time attribute are the same (as, say, measured by their certainty equivalents) regardless of the fixed level of cost, then completion time is utility independent of cost.

Keeney and Raiffa (1976) discuss an example in which utility independence might not hold. Suppose that X and Y are the rates of serious crime in two precincts of a metropolitan police division. In determining the joint utility function for this region's police chief, the issue of utility independence for X and Y would be faced. With Y fixed at 5, a relatively low rate of crime, he may be quite risk-averse to rates of crime in region X. He may not want to appear as though he is neglecting a particular precinct. Thus, his certainty equivalent for an option giving a 50% chance of $X = 0$ and a 50% chance of $X = 30$ may be 22 when Y is fixed at 5. If Y were fixed at the higher rate of 15, however, his certainty equivalent may be less risk-aversely assessed at 17. Thus, one must not assume that utility independence will hold in all cases. Even so, almost all reported multiattribute applications assume utility independence and thus are able to use a decomposable utility function.

Determining Whether Independence Exists

How can you determine whether your preferences are preferentially independent? The simplest approach is to imagine a series of paired comparisons that involve one of the attributes. With the other attribute fixed at its lowest level, decide which outcome in each pair is preferred. Once this is done, imagine changing the level of the fixed attribute. Would your comparisons be the same? Would the comparisons be the same regardless of the fixed level of the other attribute? If so, then preferential independence holds.

Determining whether independence holds is a rather delicate matter. The following sample dialogue is taken from Keeney and Raiffa (1976). The notation has been changed to correspond to our notation here.

ANALYST. I would now like to investigate how you feel about various Y values when we hold fixed a particular value of X. For example, on the first page of this questionnaire [this is shown to the assessor] there is a list of 25 paired comparisons between Y evaluations; each element of the pair describes levels on the Y attributes alone. On this first page it is assumed that, throughout, the X evaluations are all the same, that is, x_1 [the fixed value for X is shown to the assessor]. Is this clear?

ASSESSOR. Crystal clear, but you are asking me for a lot of work.

ANALYST. Well, I have a devious purpose in mind, and it will not take as much time as you think to find out what I want. Now on the second page of the questionnaire [this is shown to the assessor] the identical set of 25 paired comparisons are repeated, but now the fixed, common level on the X attribute is changed from x_1 to x_2 [value is shown to the assessor]. Are you with me?

ASSESSOR. All the way.

ANALYST. On page 3, we have the same 25 paired comparisons but now the common value of the X value is x_3 [shown to the assessor].

ASSESSOR. You said this would not take long.

ANALYST. Well now, here comes the punchline. Suppose that you painstakingly respond to all of the paired comparisons on page 1 where x_1 is fixed. Now when you go to the next page would your responses change to these same 25 paired comparisons?

ASSESSOR. Let's see. In the second page all paired comparisons are the same except x_1 is replaced with x_2. What difference should that make?

ANALYST. Well, you tell me. If we consider this first comparison [pointed to on the questionnaire] does it make any difference if the X values are fixed at x_1 or x_2? There could be some interaction concerning how you view the paired comparison depending on the common value of X.

ASSESSOR. I suppose that might be the case in some other situation, but in the first comparison I prefer the left alternative to the right no matter what the X value is . . . as long as they are the same.

ANALYST. Okay. Would you now feel the same if you consider the second paired comparison?

ASSESSOR. Yes. And the third and so on. Am I being naive? Is there some trick here?

ANALYST. No, not at all. I am just checking to see if the X values have any influence on your responses to the paired comparisons. So I gather that you are telling me that your responses on page 1 carry over to page 2.

ASSESSOR. That's right.

ANALYST. And to page 3, where the X value is held fixed at x_3 [shown to assessor]?

ASSESSOR. Yes.

ANALYST. Well, on the basis of this information I now pronounce that for you attribute Y is preferentially independent of attribute X.

ASSESSOR. That's nice to know.

ANALYST. That's all I wanted to find out.

ASSESSOR. Aren't you going to ask me to fill out page 1?

ANALYST. No. That's too much work. There are less painful ways of getting that information.

Source: Keeney, R., and H. Raiffa (1976) *Decisions with Multiple Objectives: Preferences and Value Tradeoffs.* New York: Wiley. Reprinted in 1993 by Cambridge University Press. Copyright Cambridge University Press.

The dialogue describes how to check for preferential independence. Checking for utility independence would be much the same, except that the paired comparisons would be comparisons between lotteries involving attribute Y rather than sure outcomes. As long as the comparisons remain the same regardless of the fixed value for X, then Y can be considered utility independent of X. Of course, to establish mutual preferential or utility independence, the roles of X and Y would have to be reversed to determine whether paired comparisons of outcomes or lotteries in X depended on fixed values for Y. If each attribute turns out to be independent of the other, then mutual utility or preferential (whichever is appropriate) independence holds.

Using Independence

If a decision maker's preferences show mutual utility independence, then a two-attribute utility function can be written as a composition of the individual utility functions. As usual, the least preferred outcome (x^-, y^-) is assigned the utility value 0, and the most preferred pair (x^+, y^+) is assigned the utility value 1.

Under mutual utility independent preferences, the two-attribute utility function can be written as

$$U(x, y) = k_X U_X(x) + k_Y U_Y(y) + (1 - k_X - k_Y) U_X(x)U_Y(y),$$

where

$U_X(x)$ = utility function on X scaled so that $U_X(x^-) = 0$ and $U_X(x^+) = 1$

$U_Y(y)$ = utility function on Y scaled so that $U_Y(y^-) = 0$ and $U_Y(y^+) = 1$

$k_X = U(x^+, y^-)$

$k_Y = U(x^-, y^+)$

The product term $U_X(x) U_Y(y)$ in this utility function is what permits the modeling of interactions among attributes. The utility functions U_X and U_Y are *conditional utility functions,* and each must be assessed with the other attribute fixed at a particular level. (For example, in assessing U_Y, imagine that X is fixed at a specific level.) To understand the last two conditions, all we must do is plug the individual utilities into the equation. For example,

$$U(x^+, y^-) = k_X U_X(x^+) + k_Y U_Y(y^-) + (1 - k_X - k_Y) U_X(x^+) U_Y(y^-)$$

$$= k_X(1) + k_Y(0) + (1 - k_X - k_Y)(1)(0)$$

$$= k_X$$

This multiattribute utility function, called a *multilinear* expression, is not as bad as it looks! Look at it from the point of view of the X attribute. Think about fixing Y at a value (say, y_a); you get a conditional utility function for X, given that Y is fixed at y_a:

$$U(x, y_a) = k_Y U_Y(y_a) + [k_X + (1 - k_X - k_Y) U_Y(y_a)] U_X(x)$$

Because Y is fixed at y_a, the terms $k_Y U_Y(y_a)$ and $[k_X + (1 + k_X - k_Y)U_Y(y_a)]$ are just constants. Thus, $U(x, y_a)$ is simply a scaled version of $U_X(x)$. Now change to another y (y_b). What happens to the utility function for X? The expression now looks like

$$U(x, y_b) = k_Y U_Y(y_b) + [k_X + (1 - k_X - k_Y) U_Y(y_b)] U_X(x)$$

This is just another linear transformation of $U_X(x)$, and so $U(x, y_b)$ and $U(x, y_a)$ must be identical in terms of the way that lotteries involving the X attribute would be ranked. We have scaled the utility function $U_X(x)$ in two different ways, but the scaling does not change the ordering of preferences. Now, notice that we can do exactly the same thing with the Y attribute; for different fixed values of X (x_a and x_b), the conditional utility functions are simply linear transformations of each other:

$$U(x_a, y) = k_X U_X(x_a) + [k_Y + (1 - k_X - k_Y) U_X(x_a)] U_Y(y)$$
$$U(x_b, y) = k_X U_X(x_b) + [k_Y + (1 - k_X - k_Y) U_X(x_b)] U_Y(y)$$

No matter what the level of one attribute, preferences over lotteries in the second attribute (Y) stay the same. This was the definition of utility independence in the last section. We have mutual utility independence because the conditional utility function for one attribute stays essentially the same no matter which attribute is held fixed.

Additive Independence

Look again at the multiattribute utility function

$$U(x, y) = k_X U_X(x) + k_Y U_Y(y) + (1 - k_X - k_Y) U_X(x) U_Y(y)$$

If $k_X + k_Y = 1$, then the utility function turns out to be simply additive:

$$U(x, y) = k_X U_X(x) + (1 - k_X) U_Y(y)$$

If this is the case, we only have to assess the two individual utility functions $U_X(x)$ and $U_Y(y)$ and the weighting constant k_X. This would be convenient: It would save having to assess k_Y. Of course, this is just the additive utility function from Chapter 15. How is this kind of multiattribute utility function related to the independence conditions? To be able to model preferences accurately with this additive utility function, we need *additive independence*, an even stronger condition than utility independence.

The statement of additive independence is the following: Suppose X and Y are mutually utility independent, and you are indifferent between Lotteries A and B:

> **A** (x^-, y^-) with probability 0.5
>
> (x^+, y^+) with probability 0.5
>
> **B** (x^-, y^+) with probability 0.5
>
> (x^+, y^-) with probability 0.5

If this is the case, then the utility function can be written as the weighted combination of the two utility functions, $U(x, y) = k_X U_X(x) + (1 - k_X) U_Y(y)$. You can see by writing out the expected utilities of the lotteries that they are equivalent:

$$EU(A) = 0.5[k_X U_X(x^-) + (1 - k_X) U_Y(y^-)] + 0.5[k_X U_X(x^+) + (1 - k_X) U_Y(y^+)]$$
$$= 0.5[k_X U_X(x^-) + (1 - k_X) U_Y(y^-) + k_X U_X(x^+) + (1 - k_X) U_Y(y^+)]$$

$$EU(B) = 0.5[k_X U_X(x^-) + (1 - k_X) U_Y(y^+)] + 0.5[k_X U_X(x^+) + (1 - k_X) U_Y(y^-)]$$
$$= 0.5[k_X U_X(x^-) + (1 - k_X) U_Y(y^+) + k_X U_X(x^+) + (1 - k_X) U_Y(y^-)]$$
$$= EU(A)$$

The intuition behind additive independence is that, in assessing uncertain outcomes over both attributes, we only have to look at one attribute at a time, and it does not matter what the other attribute's values are in the uncertain outcomes. This sounds a lot like utility independence. The difference is that, in the case of additive independence, changes in *lotteries* in one attribute do not affect preferences for lotteries in the other attribute; for utility independence, on the other hand, changes in *sure levels* of one attribute do not affect preferences for lotteries in the other attribute. Here is another way to say it: When we are considering a choice among risky prospects involving multiple attributes, if additive independence holds, then we can compare the alternatives one attribute at a time. In comparing Lotteries A and B above, we are indifferent because (1) for attribute X, each lottery gives us a 50% chance at x^- and a 50% chance at x^+; and (2) for Y, each lottery gives us a 50% chance at y^- and a 50% chance at y^+. Looking at the attributes one at a time, the two lotteries are the same.

The additive utility function from Chapter 15 requires additive independence of preferences across attributes in order to be an accurate model of a decision maker's preferences in decisions under uncertainty. Think back to some of the examples or problems. Do you think this idea of additive independence makes sense in purchasing a car? Think about reliability and quality of service, two attributes that might be important in this decision. When you purchase a new car, you do not know whether the reliability will be high or low, and you may not know the quality of the service. To some extent, however, the two attributes are substitutes for each other. Suppose you faced the hypothetical decision shown in Figure 16.4. Would you prefer Lottery A or B? Most of us probably would take A. If you have a clear preference for one or the other, then additive independence cannot hold.

Von Winterfeldt and Edwards (1986) discuss reports from behavioral decision theory that indicate additive independence usually does not hold. If this is the case, what is the justification for the use of the additive utility function? Many multiattribute decisions that we make involve little or no uncertainty, and evidence has shown that the additive model is reasonable for most situations under conditions of certainty. And in extremely complicated situations with many attributes, the additive model may be a useful rough-cut approximation. It may turn out that considering the interactions among attributes is not critical to the decision at hand.

Finally, it is possible to use simple approximation techniques to include interactions within the additive utility framework. For example, suppose that we have a decision problem with many attributes that are, for the most part, additively independent of one another. The additive representation of the utility function does not allow for any interaction among the attributes. If this is appropriate for almost all of the possible outcomes, then we may use the additive representation while including a specific "bonus" or "penalty" (depending on which is appropriate) for those outcomes with noticeable interaction effects.

Raiffa (1982) has an interesting example. Suppose a city is negotiating a new contract with its police force. Two attributes are (1) increase in vacation for officers who have less than five years of service and (2) increase in vacation for officers who have more than five years of service. The city loses points in an additive value function for increases in vacation for either group. If either group is held to no increase, then the city loses no points for that particular group. But the city would be happy if all its officers could be held to no increase in vacation time; thus, no precedent is set for the other group. To capture this interaction, the city gets a "bonus" of some points in its overall utility if there is no increase in vacation for either group.

Figure 16.4

An assessment lottery for a car purchase. If you have a clear preference for A or B, then additive independence cannot hold.

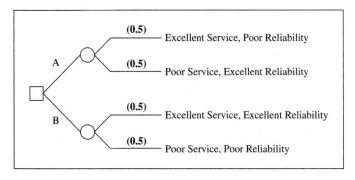

Substitutes and Complements

In the multiattribute utility function, the interaction between the attributes is captured by the term $(1 - k_X - k_Y)\, U_X(x)\, U_Y(y)$. How can we interpret this? Keeney and Raiffa (1976) give an interesting interpretation of the coefficient $(1 - k_X - k_Y)$. The sign of $(1 - k_X - k_Y)$ can be interpreted in terms of whether x and y are complements or substitutes for each other. Suppose $(1 - k_X - k_Y)$ is positive. Now examine the multiattribute utility function:

$$U(x, y) = k_X U_X(x) + k_Y U_Y(y) + (1 - k_X - k_Y)\, U_X(x)\, U_Y(y)$$

Preferred values of X and Y will give high values to the conditional utility functions, and the positive $(1 - k_X - k_Y)$ will drive up the overall utility for the pair even higher. Thus, if $(1 - k_X - k_Y)$ is positive, the two attributes complement each other. On the other hand, if $(1 - k_X - k_Y)$ is negative, high values on each scale will result in a high product term, which must be subtracted in the multiattribute preference value. In this sense, preferred values of each attribute work against each other. But if one attribute is high and the other low, the subtraction effect is not as strong. Thus, if $(1 - k_X - k_Y)$ is negative, the two attributes are substitutes.

Keeney and Raiffa (1976) offer two examples. In one, imagine a corporation with two divisions that operate in different markets altogether, and let profits in each division represent two attributes of concern to the president. To a great extent, success by the two divisions could be viewed as substitutes. That is, if profit from one division was down while the other was up, the firm would get along fine. Financial success by one division would most likely ensure the overall success of the firm.

For an example of the complementary case, Keeney and Raiffa consider the problem a general would face in a battle being fought on two fronts. If we let the consequences on the two fronts represent two distinct attributes, then these two attributes may be complementary. That is, defeat on one front may be almost as bad as defeat on both fronts, and a completely successful outcome may be guaranteed only by victory on both.

Assessing a Two-Attribute Utility Function

Now that we have seen the basics of two-attribute utility functions, we are ready to assess one. The procedure is relatively straightforward. First, we determine whether mutual utility independence holds. Provided that it does, we then assess the individual utility functions. Finally, the scaling constants are determined in order to put the individual utility functions together.

THE BLOOD BANK

In a hospital blood bank it is important to have a policy for deciding how much of each type of blood should be kept on hand. For any particular year, there is a "shortage rate," the percentage of units of blood demanded but not filled from stock because of shortages. Whenever there is a shortage, a special order must be placed to locate the required blood elsewhere or to locate donors. An operation may be postponed, but only rarely will a blood shortage result in a death. Naturally, keeping a lot of blood stocked means that a shortage is less likely. But there is also a rate at which blood is "outdated," or kept on the shelf the maximum amount of time, after which it must be discarded. Although having a lot of blood on hand means a low shortage rate, it probably also would mean a high outdating rate. Of course, the eventual outcome is unknown because it is impossible to predict exactly how much blood will be demanded. Should the hospital try to keep as much blood on hand as possible so as to avoid shortages? Or should the hospital try to keep a fairly low inventory in order to minimize the amount of outdated blood discarded? How should the hospital blood bank balance these two objectives? [*Source:* Keeney and Raiffa (1976).]

The outcome at the blood bank depends on uncertain demand over the year as well as the specific inventory policy (stock level) chosen. Thus, we can think of each possible inventory policy as a lottery over uncertain outcomes having two attributes, shortage and outdating. Shortage is measured as the annual percentage of units demanded but not in stock, while outdating is the percentage of units that are discarded due to aging. A high stock level probably will lead to less shortage but more outdating, and a low stock level will lead to more shortage and less outdating. To choose an appropriate stock level, we need to assess both the probability distribution over shortage and outdating outcomes for each possible stock level and the decision maker's utility function over these outcomes. Because each outcome has two attributes, we need a two-attribute utility function. Here we focus on the assessment of the utility function through the following steps:

1 The first step was to explain the problem to the nurse in charge of ordering blood. Maintaining an appropriate stock level was her responsibility, so it made sense to base an analysis of the problem on her personal preferences. She understood the importance of the problem and was motivated to think hard about her assessments. Without such understanding and motivation on her part, the entire project probably would have failed.

2 It was established that the annual outdating and shortage rates might range from 10% (worst case) to 0% (best case).

3 Did mutual utility independence hold? The nurse assessed a certainty equivalent for uncertain shortage rates (Attribute X), given a fixed outdating rate (Attribute Y). The certainty equivalent did not change for different outdating rates. Thus,

shortage was found to be utility independent of outdating. Similar procedures showed the reverse to be true as well. Thus, shortage and outdating were mutually utility independent, implying the multilinear form for the utility function.

4   The next step was to assess the conditional utility functions $U_X(x)$ and $U_Y(y)$. In each case, the utility function was assessed conditional on the other attribute being held constant at 0. To assess $U_X(x)$, it was first established that preferences decreased as x increased. Using the lotteries that had been assessed earlier in the utility independence step, an exponential utility function was determined. Setting $U_X(0) = 1$ (best case) and $U_X(10) = 0$ (worst case), the utility function was

$$U_X(x) = 1 + 0.375(1 - e^{x/7.692})$$

Likewise, the second utility function was determined using the previously assessed certainty equivalents, and again an exponential form was used. The utility function was

$$U_Y(y) = 1 + 2.033(1 - e^{y/25})$$

This utility function also has $U_Y(0) = 1$ and $U_Y(10) = 0$.

5   Assessing the weights k_X and k_Y is the key to finding the two-attribute utility function. The trick is to use as much information as possible to set up equations based on indifferent outcomes and lotteries, and then to solve the equations for the weights. Because we have two unknowns, k_X and k_Y, we will be solving two equations in two unknowns. To set up two equations, we will need two utility assessments.

Recall that the multilinear form can be written as

$$U(x, y) = k_X U_X(x) + k_Y U_Y(y) + (1 - k_X - k_Y)\, U_X(x)\, U_Y(y)$$

We also know that

$$U(10, 0) = k_Y$$
$$U(0, 10) = k_X$$

These follow from steps 3 and 4 above, substituting $x^- = 10$, $x^+ = 0$, $y^- = 10$, and $y^+ = 0$.

The nurse determined that she was indifferent between the two outcomes ($x = 4.75$, $y = 0$) and ($x = 0$, $y = 10$). This first assessment indicates that, for her, avoiding shortages is more important than avoiding outdating. We can substitute each one of these points into the expression for the utility function, establishing the first equation relating k_X and k_Y:

$$U(4.75, 0) = k_X\, U_X(4.75) + k_Y\, U_Y(0) + (1 - k_X - k_Y)\, U_X(4.75)\, U_Y(0)$$
$$= k_X\, U_X(4.75) + k_Y(1) + (1 - k_X - k_Y)\, U_X(4.75) \quad (1)$$

Because she was indifferent between (4.75, 0) and (0, 10), we have

$$U(4.75, 0) = U(0, 10) \quad \text{(Because she is indifferent)}$$
$$= k_X \quad \text{[from } U(0, 10) = k_X \text{, above]}$$

Substituting, we obtain

$$k_X = k_X U_X(4.75) + k_Y + (1 - k_X - k_Y) U_X(4.75) \qquad \textbf{(16.1)}$$

$$= k_Y + (1 - k_Y) U_X(4.75)$$

$$= k_Y + (1 - k_Y)[1 + 0.375(1 - e^{4.75/7.692})]$$

$$= k_Y + (1 - k_Y)0.68$$

$$= 0.68 + 0.32k_Y$$

In the second assessment, the decision maker concluded that she was indifferent between the outcome (6, 6) and a 50–50 lottery between the outcomes (0, 0) and (10, 10). Using this assessment, we can find U(6, 6):

$$U(6, 6) = 0.5\ U(0, 0) + 0.5\ U(10, 10)$$

$$= 0.5(1) + 0.5(0)$$

$$= 0.5$$

This is just a standard assessment of a certainty equivalent for a 50–50 gamble between the best and worst outcomes. Now substitute U(6, 6) = 0.5 into the two-attribute utility function to find a second equation in terms of k_X and k_Y:

$$0.5 = U(6, 6)$$

$$= k_X U_X(6) + k_Y U_Y(6) + (1 - k_X - k_Y) U_X(6) U_Y(6)$$

Substituting the values $X = 6$ and $Y = 6$ into the formulas for the individual utility functions gives

$$U_X(6) = 0.56$$

$$U_Y(6) = 0.45$$

Now plug these into the equation for U(6, 6) to get

$$0.5 = k_X(0.56) + k_Y(0.45) + (1 - k_X - k_Y)(0.56)(0.45)$$

which simplifies to

$$0.248 = 0.308k_X + 0.198k_Y \qquad \textbf{(16.2)}$$

Now we have two linear equations in k_X and k_Y—Equations (16.1) and (16.2):

$$k_X = 0.680 + 0.320k_Y$$

$$0.248 = 0.308k_X + 0.198k_Y$$

Solving these two equations simultaneously for k_X and k_Y, we find that $k_X = 0.72$ and $k_Y = 0.13$. Thus, the two-attribute utility function can be written as

$$U(x, y) = 0.72\ U_X(x) + 0.13\ U_Y(y) + 0.15\ U_X(x)\ U_Y(y)$$

where $U_X(x)$ and $U_Y(y)$ are given by the exponential utility functions defined above. Now we can find the utility for any (x, y) pair (as long as the x's and y's are each between 0 and 10, the range of the assessments). Any policy for ordering blood can be evaluated in terms of its expected utility. Table 16.1 shows utilities for different possible outcomes, and Figure 16.5 shows the indifference curves associated with the utility function. From Figure 16.5, we can verify the conditions and assessments that were used:

$$U(0, 0) = 1 \qquad\qquad U(10, 10) = 0$$

$$U(10, 0) = 0.13 = k_Y \qquad U(0, 10) = U(4.75, 0) = 0.72 = k_X$$

$$U(6, 6) = 0.50$$

The final assessed utility function is readily interpreted. The large value for k_X relative to k_Y means that the nurse is much more concerned about the percentage of shortage than she is about the percentage of outdating. This makes sense; most of us would agree that the objective of the blood bank is primarily to save lives, and we probably would rather throw out old blood than not have enough on hand when it is

Table 16.1

Utility values for shortage and outdating in the blood bank.

x Values (Shortage)	\multicolumn{6}{c}{y Values (Outdating)}
	0	2	4	6	8	10
0	1.00	0.95	0.90	0.85	0.79	0.72
2	0.90	0.86	0.81	0.76	0.70	0.64
4	0.78	0.74	0.69	0.64	0.59	0.54
6	0.62	0.58	0.54	0.50	0.45	0.40
8	0.40	0.37	0.34	0.31	0.27	0.23
10	0.13	0.11	0.08	0.06	0.03	0.00

Figure 16.5

Indifference curves for nurse's utility function for shortage and outdating. The numbers are U(x, y) for the corresponding indifference curve.

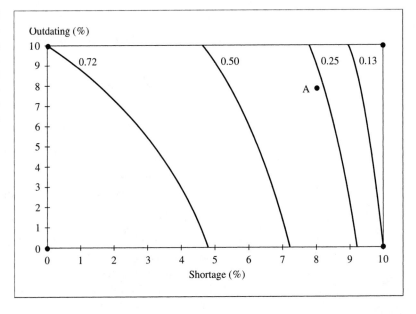

needed. The fact that $k_X + k_Y < 1$ means that the two attributes are complements rather than substitutes. We can see this in Figure 16.5. For example, imagine an outcome of (8, 8), for which the utility is 0.27 (Point A). Now imagine improving the shortage percentage to zero. This would increase the utility value to $U(0, 8) = 0.79$, for an approximate net increase of 0.52. On the other hand, improving the outdating percentage the same amount results in $U(8, 0) = 0.40$, for a net increase of 0.13. If we increased both at the same time, we would have $U(0, 0) = 1.00$, an increase of 0.73. The increase in utility from increasing both at once is greater than the sum of the individual increases $(0.73 > 0.52 + 0.13 = 0.65)$. This is the sense in which there is an interaction. This kind of phenomenon is impossible in the additive utility function.

Three or More Attributes (Optional)

When the decision problem involves three or more objectives, modeling preferences is more difficult. Building a utility function that will permit interactions across many attributes can become complex. Under certain conditions, however, the *multiplicative* utility function can be used. Let X_i denote the ith attribute, and $U_i(x_i)$ and k_i the corresponding individual utility function and scaling constant. The multiplicative utility function for n different attributes is given by the equation

$$1 + k\, U(x_1, x_2, \ldots, x_n) = \prod_{i=1}^{n} \left[kk_i U_i(x_i) + 1 \right] \tag{16.3}$$

where k is a nonzero solution to the equation

$$1 + k = \prod_{i=1}^{n} (1 + kk_i)$$

The k_i's have the same meaning they had in the two-attribute case:

$$U(x_1^-, \ldots, x_{i-1}^-, x_i^+, x_{i+1}^-, \ldots, x_n^-) = k_i$$

That is, k_i is the utility of an outcome having the best level on attribute X_i and worst on all others. Thus, we can assess the k_i's directly through our standard reference-lottery approach, which is shown in Figure 16.6. We have the reference lottery with the best possible and worst possible outcomes versus the sure thing having the best level on attribute X_i and the worst level on all others. The probability p_i that makes us indifferent between the lottery and the sure thing is our utility for the sure thing and hence is equal to k_i. The decision maker can assess each of the k_i's in this way and then find the scaling constant k that satisfies the condition given above. Finally, putting the individual utility functions together with the scaling constants gives the overall utility function.

Figure 16.6

Assessing scaling constants for the multiplicative utility function. The p_i that makes the decision maker indifferent between the lottery and the sure thing is the utility of the sure outcome and hence equals k_i.

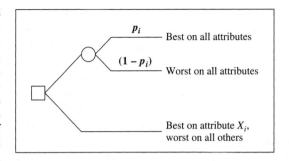

The multiplicative utility function requires a fairly strong version of utility independence. Strictly speaking, each subset of the attributes must be utility-independent of the remaining attributes. This essentially means that we should be able to partition the attributes into two subsets in any way we want, and then consider lotteries in one subset, holding the attributes in the other subset fixed. As long as preferences for the lotteries do not depend on the level of the remaining attributes, the multiplicative utility function should provide a good model of the decision maker's preferences. If the multiplicative model is not appropriate, then a more general version of the multilinear model may be a possibility. For more details, consult Keeney and Raiffa (1976).

When Independence Fails

We have just dealt in depth with situations in which the assumption of mutual utility independence results in a reasonable model of preferences. This is not always true. Suppose we are interested in assessing a two-attribute utility function over attributes X and Y but have found that neither X nor Y is utility independent of the other. Then neither the multilinear nor additive forms for the utility function are appropriate. How can we obtain a reasonable $U(x, y)$ for decision-making purposes? Several possibilities exist. One is simply to perform a direct assessment as described at the beginning of this chapter. Pick the best and worst (x, y) pairs, assign them utility values of 1 and 0, and then use reference gambles to assess the utility values for other points.

A second approach is to transform the attributes and proceed to analyze the problem with the new set. Of course, the new set of attributes still must capture the critical aspects of the problem, and they must be measurable. Take the example discussed above in which X and Y designate measures of the crime rates in two sections of a city. There may be a complicated preference structure for (x, y) pairs. For political reasons, the relative ordering of hypothetical lotteries for criminal activity in one section may be highly dependent on the level of crime in the other section. But suppose we define $s = (x + y)/2$ and $t = |x - y|$. Then s may be interpreted as an average crime index for the city and *t* as an indicator of the balance of criminal activity between the two sections. Any (x, y) outcome implies a unique (s, t) outcome, so it is easy to transform from one set of attributes to the other. Furthermore, even though x and y may not be utility independent, it may be reasonable to model s and t as being utility independent, thus simplifying the assessment procedure.

One of the most difficult and subtlest concepts in multiattribute utility theory is the notion of interaction among attributes. To use the additive utility function from Chapter 15, we must have no interaction at all. To use the multilinear utility function discussed in this chapter, we can have some interaction, but it must be of a limited form. (Any interactions must conform to the notion of utility independence.) The blood bank example demonstrated the nature of the interaction between two attributes that is possible with the multilinear utility function. In this section, we are concerned with situations in which the interactions are even more complicated than those that are possible in the multilinear case. Fortunately, evidence from behavioral research suggests that it is rarely, if ever, necessary to model extremely complex preference interactions.

Chapter 15 and 16 have presented many of the principles and decision-analysis techniques that are useful in making decisions in the face of conflicting objectives. Given the complexity of the techniques, it is important to keep in mind the modeling perspective that we have held all along. The objective of using the multiattribute decision-analysis techniques is to construct a model that is a reasonable representation of a decision maker's value structure. If minimal interactions exist among the attributes, then the additive utility function is appropriate. When attributes interact, then it may be necessary to consider multiattribute utility theory, just as we have in this chapter.

Multiattribute Utility in Action: BC Hydro

Can multiattribute utility modeling be of use? Consider the situation of the British Columbia Hydro and Power Authority (BC Hydro):

STRATEGIC DECISIONS AT BC HYDRO

In the late 1980s, BC Hydro found itself at an important juncture. Knowing that it would face many strategic decisions in the future, its directors wanted to prepare to make those decisions in the best possible way. It would have to make decisions regarding power-generation plants, placement of transmission lines, employee relations, how to communicate with public and special-interest groups, environmental impact of its activities and facilities, and policies for addressing large-scale problems such as global warming. To ensure that many different decisions would be coordinated and would all serve the organization's interests, Mr. Ken Peterson was appointed as Director of Strategic Planning. In short, Peterson's job was to make BC Hydro the leader in strategic planning among North American utility companies. To help Peterson get started, BC Hydro had a general mission statement. Like most such statements, however, BC Hydro's was broad and lacked details. It was certainly not up to the task of coordinating the myriad of diverse decisions that BC Hydro would face.

Realizing that he would need a way to think systematically about BC Hydro's strategic alternatives, Peterson enlisted Ralph Keeney and Tim McDaniels to help identify the organization's objectives and construct a multiattribute utility function. The project is described in Keeney (1992) and Keeney and McDaniels (1992). The process began with interviews of key decision makers in the organization to identify fundamental objectives. Once a satisfactory set of fundamental objectives was established, the process continued through all of the steps necessary to define and assess

Figure 16.7

Two nonlinear utility functions for BC Hydro. (a) A risk-averse utility function for government dividend. (b) A risk-prone utility function for duration of outages to large customers.

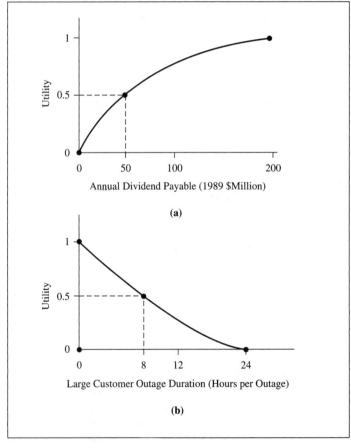

(a)

(b)

Source: Keeney (1992, p. 364).

individual utility functions for the attributes and scaling constants for the multiat-tribute utility function. The final utility function involved 18 different attributes in six major groups (economics, environment, health and safety, equity, service quality, and public-service recognition). The hierarchy of objectives with corresponding attributes and ranges is shown in Table 16.2.

Keeney reports that the component utility functions were linear (i.e., proportional scores were used) for all but two of the attributes. The two that were not linear were annual government dividend and large customer outage duration. The individual utility functions for these two attributes are shown in Figure 16.7. Note that the dividend utility function shows risk aversion, but the outage utility function is risk-seeking.

The assessment proceeded by creating multiattribute utility functions for each of the groups and subgroups in Table 16.2. For example, two-attribute utility functions were created for service quality to small customers and large customers, each one covering the number of outages and outage duration for the corresponding customer type. These two utility functions were then combined with individual utility functions for new service and inquiries to create a four-attribute utility function for service. Finally, the service utility function was combined with five similar utility functions for each of the other major groups.

All of the multiattribute utility functions (including the overall function for the six major groups) were additive except for the economic utility function, which was multiplicative as in Equation (16.3). The reason for the nonadditive economic utility function is that the economic attributes were thought to be substitutes for each other to some extent. For the additive part of the model, it is possible to collapse the hierarchy and calculate the weight that any one attribute has in the overall utility function. This is done simply by multiplying the weight for the attribute times the weight for the group (or subgroup) in which it resides, and so on, until the top level of the objectives hierarchy is reached. (You will recall that we did this in the library example in Chapter 15.) The scaling weights, thusly calculated, are shown in Table 16.3 for the additive attributes. (This table is taken from Keeney's Table 12.5 (1992). Note that scaling constants are given for public and worker health-and-safety aggregates rather than for the individual mortality and morbidity attributes. The same is true for small and large customer service attributes.)

Table 16.3 also shows in parentheses the scaling constants for the six major attribute groups, thus indicating the relative importance of these top-level objectives. The most important of the six is the economic objective, accounting for almost 40% of the total weight. The table also shows the scaling constants for the three components in the economic utility function and the constant k that defines the interactions in this multiplicative submodel. The economic utility function is given by

$$1 - 0.76 \, U(x_1, x_2, x_3) = [1 - 0.76(0.84) \, U_1(x_1)] \tag{16.4}$$

$$[1 - 0.76(0.18) \, U_2(x_2)][1 - 0.76(0.30) \, U_3(x_3)]$$

where x_1, x_2, and x_3 represent levelized energy costs, annual dividend, and resource losses, respectively.

From Table 16.3, the overall utility of any given consequence can be calculated. First, find the individual utility for each attribute. Then, for the additive components,

Table 16.2
Attributes and ranges for BC Hydro's strategic utility function.

	Worst Level	Best Level
Economics		
Levelized cost of energy from new sources (1989 $0.001/kWh)	55	35
Annualized dividend payable (1989 $ million)	0	200
Economic cost or resource losses (1989 $ million)	20	0
Environment		
Local Impacts		
Flora (hectares of mature forest lost)	10,000	0
Fauna (hectares of wildlife habitat lost)	10,000	0
Wilderness ecosystem (hectares of wilderness lost)	10,000	0
Recreation (hectares of recreational land lost)	10,000	0
Aesthetic (annual person-years of viewing high-voltage transmission lines in scenic terrain)	500,000	0
Global impact (megawatts generated from fossil fuels)	1000	0
Health and Safety		
Public		
Mortality (annual person-years of life lost)	100	0
Morbidity (annual person-years of "severe" disability)	1000	0
Employees		
Mortality (annual person-years of life lost)	100	0
Morbidity (annual person-years of lost work time)	1000	0
Equity		
Equitable pricing (constructed scale: see Keeney 1992, section 12.3)	0.5	0
Equitable compensation (annual average number of individuals who feel they are inequitably treated)	500	0
Service quality		
Small customers		
Outages (annual number per customer)	2	0
Outage duration (hours per outage)	24	0
Large customers		
Outages (annual number per customer)	2	0
Outage duration (hours per outage)	24	0
New service (installation time in workdays)	20	1
Inquiries (minutes until personal response)	1	0
Public-service recognition (constructed scale: see Keeney 1992, section 12.3)	0	4

Source: Keeney (1992, pp. 358–359).

Table 16.3

Scaling constants for BC Hydro's strategic utility function.

	Scaling Constant
Economics (0.395)	
Constants for multiplicative function:	
Levelized cost of energy	0.84
Annualized dividend	0.18
Resource cost	0.30
Scaling constant *k*	-0.76
Constants for additive function:	
Environment (0.250)	
Flora	0.023
Fauna	0.046
Wilderness ecosystem	0.093
Recreation	0.046
Aesthetic	0.023
Global impact	0.019
Health and Safety (0.089)	
Public	0.045
Worker	0.045
Equity (0.012)	
Equitable pricing	0.004
Equitable compensation	0.008
Service quality (0.250)	
Small customers	0.111
Large customers	0.125
New service	0.010
Inquiries	0.005
Public-service recognition (0.004)	0.004

multiply each utility value times its weight and add up the resulting products. For the economic attributes, calculate the utility from Equation (16.4) and multiply the result times 0.395, the weight given to the economic component. Finally, add this product to the sum of the weighted utilities from the previous step. Of course, no one would do this in such a painstaking way by hand. Fortunately, it is very straightforward to program such utility calculations either into a specialized utility program or even in an electronic spreadsheet.

Keeney (1992) shows how this utility function for BC Hydro can be used to generate important insights. For example, he shows the resource value in dollars of various attributes: Each hectare of wilderness lost is valued at $2500, but a hectare of flora lost is worth $625. Each person-year of aesthetic deterioration is worth $13. One statistical fatality, for either an employee or a member of the general public, is worth $3 million. Many other economic trade-offs are listed in Keeney's Table 12.6 (1992). These are the amounts that BC Hydro should just be willing to sacrifice (but no more) in order to save a hectare, a life, or a person-year of aesthetic damage. In addition to these trade-offs, Keeney lists a number of decision opportunities that arose from the analysis, including studying the implications of resource losses on economic activity, developing a database for fatalities, studying public environmental values, and understanding the process of responding to telephone inquiries.

Did the definition of strategic objectives and a strategic utility function have an impact on BC Hydro? Ken Peterson is quoted as saying:

> The structured set of objectives has influenced BC Hydro planning in many contexts. Two examples include our work to develop a decision framework for supply planning, and a case study of an investment to upgrade reliability, both of which have adopted a multiple objective structure. Less obvious has been an evolution in how key senior planners view planning issues. The notion of a utility function over a range of objectives (rather than a single objective, like costs) is evident in many planning contexts. The specific trade-offs in the elicitation process are less important than the understanding that trade-offs are unavoidable in electricity utility decisions and that explicit, well-structured, informed trade-offs can be highly useful. (*Interfaces,* November–December 1992, p. 109)

The use of multiattribute utility models has been a source of insight for many decision makers in diverse decision situations. For example, Keeney and Raiffa (1976) report applications involving fire department operations, strategic decision making by a technical consulting firm, evaluation of computer systems, siting of nuclear power facilities, the analysis of sites for the Mexico City airport, and many others. Other applications are described in von Winterfeldt and Edwards (1986).

SUMMARY   We have continued the discussion of making decisions in the face of conflicting objectives. Much of the chapter has been a rather technical discussion and treatment of independence conditions: preferential independence, utility independence, and additive independence. The differences among these have to do with the presence or absence of uncertainty. Preferential independence says that preferences for sure outcomes in one attribute do not depend on the level of other attributes. Utility independence requires that preferences for gambles or lotteries in an attribute do not depend on the level of other attributes. Additive independence is still stronger: Preferences over lotteries in one attribute must not depend on lotteries in the other attributes. The blood bank example showed how to apply mutual utility independence to assess a two-dimensional utility function that includes an interaction term. We saw that the attributes of shortage and outdating were complementary; they work together to increase the decision maker's utility. We briefly saw what to do when there are three or more attributes, and what to do if no independence properties hold. Finally, the BC Hydro application demonstrated multiattribute utility in a large-scale organizational setting.

EXERCISES

16.1 Explain what is meant when we speak of interaction between attributes. Why would the additive utility function from Chapter 15 be inappropriate if two attributes interact?

16.2 What are the advantages and disadvantages of directly assessing a multiattribute utility function?

16.3 Explain preferential independence in your own words. Can you cite an example from your own experience in which preferential independence holds? Can you cite an example in which it does not hold?

16.4 Explain in your own words the difference between preferential independence and utility independence.

16.5 Explain in your own words the difference between utility independence and additive independence. Why is it important to understand the concept of additive independence?

16.6 Suppose that a company would like to purchase a fairly complicated machine to use in its manufacturing operation. Several different machines are available, and their prices are more or less equivalent. But the machines vary considerably in their available technical support (Attribute X) and reliability (Attribute Y). Some machines have a high degree of reliability and relatively low support, while others are less reliable but have excellent field support. The decision maker determined what the best and worst scenarios were for both attributes, and an assessment then was made regarding the independence of the decision maker's preferences for these attributes. It was determined that utility independence held. Individual utility functions $U_X(x)$ and $U_Y(y)$ were assessed. Finally, the decision maker was found to be indifferent in comparing Lottery A with its alternative B:

    A   Best on both reliability and support with probability 0.67

                Worst on both reliability and support with probability 0.33

    B   Best on reliability and worst on support

The decision maker also was indifferent between

    C   Best on both reliability and support with probability 0.48

                Worst on both reliability and support with probability 0.52

    D   Worst on reliability and best on support

a   What are the values for k_X and k_Y? Write out the decision maker's full two-attribute utility function for reliability and support.

b   Are the two attributes substitutes or complements? Explain, both intuitively and on the basis of k_X and k_Y.

QUESTIONS AND PROBLEMS

16.7   A hospital administrator is making a decision regarding the hospital's policy of treating individuals who have no insurance coverage. The policy involves examination of a prospective patient's financial resources. The issue is what level of net worth should be required in order for the patient to be provided treatment. Clearly, there are two competing objectives in this decision. One is to maximize the hospital's revenue, and the other is to provide as much care as possible to the uninsured poor. Attributes to measure achievement toward these two objectives are (1) prospective revenue (R), and (2) percentage of uninsured poor who are treated (P).

The two attributes were examined for independence, and the administrator concluded that they were mutually utility independent. Utility functions $U_R(r)$ and $U_P(p)$ for the two attributes then were assessed. Then two more assessments were made. Lottery A and its certain alternative B were judged to be equivalent by the administrator:

   A   Best on revenue, worst on treating poor with probability 0.65

        Worst on revenue, best on treating poor with probability 0.35

   B   Levels of revenue and teatment of poor that give $U_R(r) = 0.5$ and $U_P(p) = 0.5$

In the second assessment, Lottery C and its certain alternative D were judged to be equivalent:

   C   Best on both revenue and treating poor with probability 0.46

        Worst on both revenue and treating poor with probability 0.54

   D   Worst on revenue and best on treatment of poor

a   Find values for k_R and k_P. Should the administrator consider these two attributes to be substitutes or complements? Why or why not?

b   Comment on using an additive utility model in this situation. Would such a model seriously compromise the analysis of the decision?

16.8   Suppose you face an investment decision in which you must think about cash flows in two different years. Regard these two cash flows as two different attributes, and let X represent the cash flow in Year 1, and Y the cash flow in Year 2. The maximum cash flow you could receive in any year is $20,000, and the minimum is $5000. You have assessed your individual utility functions for X and Y, and have fitted exponential utility functions to them:

$$U_X(x) = 1.05 - 2.86 \, e^{-x/5000}$$

$$U_Y(y) = 1.29 - 2.12 \, e^{-y/10,000}$$

Furthermore, you have decided that utility independence holds, and so these individual utility functions for each cash flow are appropriate regardless of the amount of the other cash flow. You also have made the following assessments:

- You would be indifferent between a sure outcome of $7500 each year for two years and a risky investment with a 50% chance at $20,000 each year, and a 50% chance at $5000 each year.

- You would be indifferent between getting (1) $18,000 the first year and $5000 the second, and (2) getting $5000 the first year and $20,000 the second.

a Use these assessments to find the scaling constants k_X and k_Y. What does the value of $(1 - k_X - k_Y)$ imply about the cash flows of the different periods?

b Use this utility function to choose between Alternatives A and B in Problem 15.23 (Figure 15.15).

c Draw indifference curves for $U(x, y) = 0.25, 0.50$, and 0.75.

16.9 Refer to Problem 15.25. A decision maker who prefers Chemical B might be said to be sensitive to equity between the two groups. The eventual outcome with Chemical A is not equitable; one group is better off than the other. On the other hand, with Chemical B, both groups are treated the same. Let X and Y denote the expected increase in the number of deaths in Groups 1 and 2, respectively, and denote a decision maker's utility function as

$$U(x, y) = k_X \, U_X(x) + k_Y \, U_Y(y) + (1 - k_X - k_Y) \, U_X(x) \, U_Y(y)$$

What can you say about the value of $(1 - k_X - k_Y)$? Are X and Y complements or substitutes?

16.10 In Problems 13.12 and 14.13 you assessed individual utility functions for money (X) and computer reliability (Y). In this problem, we will use these assessed utility functions to put together a two-attribute utility function for use in a computer-purchase decision. (If you have not already worked Problems 13.12 or 14.13, do so before continuing. Even if you already have assessed these utility functions, review them and confirm that they are good models of your preferences. When you assessed them, did you consciously think about keeping all other important attributes at the same level?)

a Are your preferences mutually utility independent? Your utility for money probably does not depend on your computer's level of reliability. But would you be less nervous about computer reliability if you had more money in the bank? Imagine that you have $5000 in the bank. Now assess a certainty equivalent (in terms of computer downtime) for a 50–50 gamble between your computer being down for 20 days next year or not breaking down at all. Now imagine that you have $30,000 in the bank, and reassess your certainty equivalent. Is it the same? Are your preferences for money and computer reliability mutually utility independent?

b Regardless of your answer to part a, let us assume that your preferences for money and computer reliability are mutually utility independent. Now make the following assessments:

    i.   Assess a certainty equivalent (in terms of both attributes: computer downtime and money) for a 50–50 gamble between the worst outcome ($1000 in the bank and 50 days of computer downtime) and the best outcome ($20,000 in the bank and no downtime).

    ii. Imagine the outcome that is $1000 and no downtime (worst level in money, and best level in computer reliability). Assess a dollar amount x so that the outcome x dollars and 50 days of downtime is equivalent to $1000 and no downtime.

c Using your individual assessed utility functions from Problems 13.12 and 14.13 and assessments i and ii from part b, calculate k_X and k_Y. Write out your two-attribute utility function.

(*Note:* This problem can be done without having fit a mathematical expression to your individual utility function. It can be done with a utility function expressed as a table or as a graph.)

16.11 Someday you probably will face a choice among job offers. Aside from the nature of the job itself, two attributes that are important for many people are salary and location. Some people prefer large cities, others prefer small towns. Some people do not have strong preferences about the size of the town in which they live; this would show up as a low weight for the population-size attribute in a multiattribute utility function.

Assess a two-attribute utility function for salary (X) and population size (Y):

a Determine whether your preferences for salary and town size are mutually utility independent.

b If your preferences display mutual utility independence, assess the two individual utility functions and the weights k_X and k_Y. Draw indifference curves for your assessed utility function. If your preferences do not display mutual utility independence, then you need to think about alternative approaches. The simplest is to assess several utility points as described at the beginning of this chapter and "eyeball" the indifference curves.

c What other attributes are important in a job decision? Would the two-attribute utility function you just assessed be useful as a first approximation if many of the other attributes were close in comparing two jobs?

16.12 Refer to Problem 15.17. For some of the attributes in your computer decision, you face uncertainty. The additive utility function assessed in Problem 15.17 essentially assumes that additive independence among attributes is reasonable.

a Check some of your attributes with formal assessments for additive independence. Follow the example in the text in setting up the two lotteries. (One lottery is a 50–50 gamble between best and worst on both attributes. The other is a 50–50 gamble between (i) best on X, worst on Y; and (ii) worst on X, best on Y.)

b How could you extend this kind of assessment for additive independence to more than two attributes?

16.13 Show that when $n = 2$, the multiplicative utility function is equivalent to the two-attribute multilinear utility function. (*Hint:* This is an algebra problem. You must show that one utility function is a scaled version of the other. Start by solving for the scaling constant k in the multiplicative function when $n = 2$. Then substitute your expression for k into the multiplicative utility function and simplify.)

16.14 In many cases a group of people must make a decision that involves multiple objectives. In fact, difficult decisions usually are dealt with by committees composed of individuals who represent different interests. For example, imagine a lumber mill owner and an environmentalist on a committee trying to decide on national forest management policy. It might make sense for the committee to try to assess a multiattribute "group utility function." But assessment of the weights would be a problem because different individuals probably would want to weight the attributes differently. Can you give any advice to a committee working on a problem that might help it to arrive at a decision? How should the discussions be structured? Can sensitivity analysis help in such a situation; if so, how?

16.15 In making a land-use policy decision, a planner had to consider three objectives: the development of the best economic mixture of industrial and residential uses, the preservation of sensitive environmental areas, and satisfying the largest industrial firm in the community. Attributes to measure achievement along these three objectives were developed. Let k_{econ} represent the scaling constant for the economy, k_{env} the scaling constant for the environmental concerns, and k_{firm} the scaling constant for satisfying the firm. After careful thought, the planner assessed $k_{econ} = 0.36$, $k_{env} = 0.25$, and $k_{firm} = 0.14$. Use Equation 16.3 for the multiplicative utility model to verify that the scaling constant k is approximately 1.303.

CASE STUDIES

A MINING-INVESTMENT DECISION

A major U.S. mining firm faced a difficult capital-investment decision. The firm had the opportunity to bid on two separate parcels of land that had valuable ore deposits. The project involved planning, exploration, and eventually production of minerals. The firm had to decide how much to bid, whether to bid alone or with a partner, and how to develop the site if the bid were successful. Overall, the company would have to commit approximately $500 million to the project if it obtained the land.

Figure 16.8 shows a skeleton version of the decision-tree model for this decision. Note that one of the immediate alternatives is not to bid at all, but to stay with and develop the firm's own property. Some of the key uncertainties are whether the bid is successful, the success of a competing venture, capital-investment requirements, operating costs, and product price.

Figure 16.9 shows cumulative distribution functions for net present value (NPV) from four possible strategies. Strategy 25—develop own property with partner—stochastically dominates all of the other strategies considered, and hence appears to be a serious candidate for the chosen alternative. The decision makers realized, however, that while they did want most to maximize the project's NPV, they also had another objective, the maximization of product output (PO). Because of this, a

Figure 16.8

Skeleton decision tree for mining-investment decision.

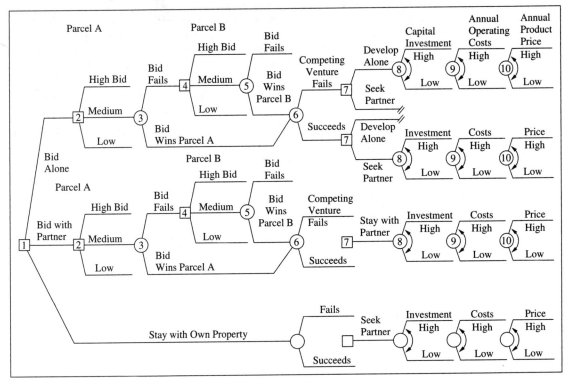

Figure 16.9

*Cumulative risk pro-
files for four strategies
in mining-investment
decision.*

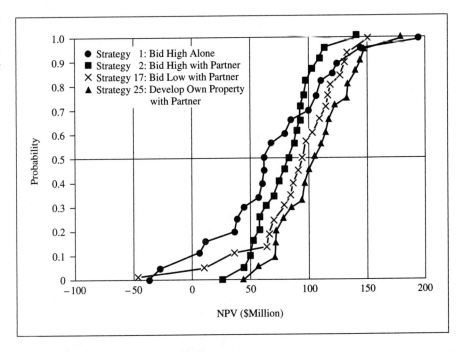

Table 16.4 Expected utilities (EU) and certainty equivalents (CE) for strategies in mining-investment decision.		Bid High Alone	Bid High with Partner	Bid Low with Partner	Develop Own Property with Partner	Do Nothing
	EU	0.574	0.577	0.564	0.567	0.308
	CE (NPV, in $Million, PO = 50)	51	52	45	47	

two-attribute utility function for NPV and PO was constructed. The individual utility functions were assessed as exponential utility functions:

$$U_{PO}(po) = 1 - e^{-po/33.33}$$
$$U_{NPV}(npv) = 1 - e^{-(npv+100)/200}$$

The scaling constants also were assessed, yielding $k_{NPV} = 0.79$ and $k_{PO} = 0.16$. Using this two-attribute model, expected utilities and certainty equivalents—a certain (NPV, PO) pair, with PO set to a specific level—were calculated. Table 16.4 shows the results for some of the strategies.

Questions

1   Which of the alternatives should be chosen? Why? Discuss the apparent conflict between the stochastic dominance results (Figure 16.9) and the results from the utility model.

2   The values of the scaling constants suggest that NPV and PO are viewed by the firm as complements. Can you explain intuitively why these two attributes would be complements rather than substitutes?

Source: A. C. Hax and K. M. Wiig (1977) "The Use of Decision Analysis in a Capital Investment Problem." In D. Bell, R. L. Keeney, and H. Raiffa (eds.) *Conflicting Objectives in Decisions,* pp. 277–297. New York: Wiley. Figures 16.8 and 16.9 reprinted by permission.

REFERENCES

Chapter 16 represents only the tip of the iceberg when it comes to general multiattribute utility modeling. If you are interested in reading more, the standard reference for multiattribute utility theory is Keeney and Raiffa (1976). Keeney (1980) covers most of the same material and is a little easier to read. Bunn (1984) provides a somewhat less technical summary of many of the techniques, and von Winterfeldt and Edwards (1986) have extensive discussion with an emphasis on behavioral issues. Further references and examples of applications can be obtained from all of these texts.

Bunn, D. (1984) *Applied Decision Analysis.* New York: McGraw-Hill.

Keeney, R. (1980) *Siting Energy Facilities.* New York: Academic Press.

Keeney, R. (1992) *Value-Focused Thinking: A Path to Creative Decisionmaking.* Cambridge, MA: Harvard University Press.

Keeney, R., and T. McDaniels (1992) "Value-Focused Thinking about Strategic Decisions at BC Hydro." *Interfaces,* 22, 94–109.

Keeney, R., and H. Raiffa (1976) *Decisions with Multiple Objectives.* New York: Wiley.

Raiffa, H. (1982) *The Art and Science of Negotiation.* Cambridge, MA: Harvard University Press.

von Winterfeldt, D., and W. Edwards (1986) *Decision Analysis and Behavioral Research.* Cambridge: Cambridge University Press.

EPILOGUE   **The Mining-Investment Decision.**   Hax and Wiig (1977) reported that the decision was to adopt Strategy 2—bid high with a partner. Just looking at NPV left the decision maker feeling uneasy. But with the product volume included in the two-attribute analysis, the decision maker was satisfied that the important objectives had been considered, so it was easy to adopt the strategy with the highest EU.

CHAPTER 17

Conclusion and Further Reading

W e all have to face hard decisions from time to time. Sometimes we must make difficult personal decisions such as how to care for an elderly loved one or which one of several job offers to accept. Many policy decisions for corporations or governmental agencies are hard to make. In fact, as time goes by, it becomes increasingly clear that as a society we must grapple with some particularly thorny problems, such as competitiveness in the world marketplace, the risks associated with new technologies, and trade-offs between short-term economic benefits and long-term environmental stability.

The argument all along has been that decision analysis can help with such hard decisions. The cycle of structuring the decision, modeling uncertainty and preferences, analyzing and then performing sensitivity analysis can lead a decision maker systematically through the issues that make the decision complicated and toward a requisite decision model, one that captures all of the essential elements of the problem. The objective is to arrive at a decision model that explicates the complex parts in a way that the decision maker can choose from the alternatives with insight and understanding.

At the same time that decision analysis provides a framework for tackling difficult decisions, it also furnishes the decision maker with a complete tool kit for constructing the necessary models of uncertainty and preferences. We have spent much time in considering probability and how to use it to model the uncertainty that a decision maker faces. Subjective assessment, theoretical models, and the use of data and simulation are all tools in the decision analyst's kit. We also discussed the tools available for modeling preferences. We considered in some depth the fundamental trade-off between risk and return. Finally, the last two chapters focused on the modeling of preferences when the decision maker must try to satisfy conflicting objectives.

Throughout the book the view of decision making has been optimistic. We all are subject to human foibles, but a person interested in making better decisions can use the principles and tools that we have discussed in order to do a better job. In day-to-day decisions, it may be simply a matter of thinking in terms of an informally decomposed problem: What is the nature of the problem? What are the objectives? Are there trade-offs to make? What uncertainties are there? Is it a risky situation? More complicated situations may warrant considerable effort and careful use of the modeling tools.

Finally, in the process of reading the text and working through the problems, you may have learned something about yourself. You may have learned how you personally feel about uncertainty in your life, how you deal with risky situations, or what kinds of trade-offs are important to you. If you have learned something about the tools of decision analysis, gained an understanding of what it means to build a model of a decision problem, and learned a little about your own decision-making personality, then your work has been worthwhile. If you feel that you are more prepared to face some of the complicated decisions that we all must face as we move into the twenty-first century, then the goal of this text has been achieved.

A Decision-Analysis Reading List

Where should you go from here? The references at the end of each chapter can lead you to more information on specific topics. You undoubtedly noticed that many of the references reappeared several times. Several textbooks cover decision analysis at a variety of levels and from different perspectives. Here are some favorites:

- Max Bazerman and Margaret Neale (1992) *Negotiating Rationally*. New York: Free Press. An introduction to behavioral aspects of negotiation, written at a slightly lower level than the companion volume by Neale and Bazerman. Taking a decision-analysis perspective, the authors also emphasize prescriptive advice for negotiators.

- Derek Bunn (1984) *Applied Decision Analysis*. New York: McGraw-Hill. An excellent book written at about the same level as this one, although more theoretical and somewhat more terse. Excellent problems.

- Robyn Dawes (1988) *Rational Choice in an Uncertain World*. San Diego, CA: Harcourt Brace. An easy-to-read introduction to the behavioral issues in decision analysis.

- Simon French (1986) *Decision Theory: An Introduction to the Mathematics of Rationality*. London: Wiley. If you liked the chapter on utility axioms, you would love this book. The text covers a lot of new material, including consensus, group decisions, and non-Bayesian approaches. The problems are good, tending to be technical and theoretical rather than applied.

- Robin M. Hogarth (1987) *Judgment and Choice,* 2nd ed. New York: Wiley. An excellent introduction to behavioral decision theory, this is decision analysis from a psychological perspective. This book covers a broad range of topics and is very easy to read.

- Ronald A. Howard and James Matheson (eds.) (1983) *The Principles and Applications of Decision Analysis* (2 volumes). Palo Alto, CA: Strategic Decisions Group. Since the early 1960s, Ron Howard has been practicing decision analysis and teaching the principles to students in the Engineering-Economic Systems Department at Stanford University. This two-volume set contains many papers presenting the principles and techniques that make up the "Stanford School" of decision analysis.

- Daniel Kahneman, Paul Slovic, and Amos Tversky (1982) *Judgment under Uncertainty: Heuristics and Biases.* Cambridge: Cambridge University Press. A reader filled with classic articles on behavioral decision theory.

- Ralph Keeney (1992) *Value-Focused Thinking: A Path to Creative Decision Making.* Cambridge, MA: Harvard University Press. As you know by now, this book provides all the details on understanding one's objectives and using them as a basis for decision-analysis models and improved decision making.

- Ralph Keeney and Howard Raiffa (1976) *Decisions with Multiple Objectives: Preferences and Value Tradeoffs.* New York: Wiley. Reprinted in 1993 by Cambridge University Press. This is *the* standard reference for multiattribute utility theory, although it also is good for decision analysis in general. Many applications are described. Much of the material is highly technical, although the mathematics are not difficult. Gems of insight and explanation are scattered through the technical material. Unfortunately, no problems are included.

- Dennis V. Lindley (1985) *Making Decisions,* 2nd ed. New York: Wiley. A classic by a founder in the field. Professor Lindley explains difficult concepts well. The problems tend to be somewhat abstract.

- M. Granger Morgan and Max Henrion (1990) *Uncertainty: A Guide to Dealing with Uncertainty in Quantitative Risk and Policy Analysis.* Cambridge: Cambridge University Press. The authors provide an in-depth treatment of the use of uncertainty for risk analysis. Especially good on the elicitation and use of expert judgment.

- Margaret Neale and Max Bazerman (1991) *Cognition and Rationality in Negotiation.* New York: Free Press. An excellent summary of behavioral issues in negotiation and group decision making. Many of the results are closely related to parallel results in individual decision making.

- Scott Plous (1993) *The Psychology of Judgment and Decision Making.* New York: McGraw-Hill. The best up-to-date summary of behavioral decision theory.

- Howard Raiffa (1968) *Decision Analysis.* Reading, MA: Addison-Wesley. Professor Raiffa also is a founder of decision analysis, and like Lindley, he explains the material well. The problems tend to be abstract. The text covers the basics (and then some) and still is worthwhile after almost 30 years.

- Howard Raiffa (1982) *The Art and Science of Negotiation.* Cambridge, MA: Belknap Press. Raiffa discusses negotiations from a decision-theoretic point of view. A well-written (and easy-to-read!) book that provides deep insights and top-flight analytical tools.

- Detlof von Winterfeldt and Ward Edwards (1986) *Decision Analysis and Behavioral Research.* Cambridge: Cambridge University Press. An up-to-date and in-depth treatment of decision analysis from a behavioral perspective. Professor Edwards has been involved in decision analysis since its beginnings, and his history (Chapter 14) is an eye-opener. The authors have a strong slant toward applications and behavioral research that provides evidence on the applicability of decision-analysis tools. Like Keeney and Raiffa, there are no problems.

- Robert L. Winkler (1972) *Introduction to Bayesian Inference and Decision.* New York: Holt. An excellent introduction to decision theory. Professor Winkler is especially interested in Bayesian models of information, and the book is slanted more toward inference and statistics than toward applied decision analysis.

Appendixes

APPENDIX A

Binomial Distribution: Individual Probabilities

$$P_B(R = r \mid n, p) = \frac{n!}{r!(n-r)!} p^r (1-p)^{n-r}$$

Figure A.1

Using Appendix A to find $P_B(R = 2 \mid n = 9, p = 0.12)$.

n	r	p	.10	.12	.14
.	.		.	.	.
.	.		.	.	.
.	.		.	.	.
9	0		.387	.316	.257
	1		.387	.388	.377
	2	. . . .	.172	.212	.245
	3		.045	.067	.093
	.		.	.	.
	.		.	.	.

n	r	.01	.02	.04	.05	.06	.08	.10	.12	.14	.15	.16	.18	.20	.22	.24	.25	.30	.35	.40	.45	.50
2	0	.980	.960	.922	.903	.884	.846	.810	.774	.740	.723	.706	.672	.640	.608	.578	.563	.490	.423	.360	.303	.250
	1	.020	.039	.077	.095	.113	.147	.180	.211	.241	.255	.269	.295	.320	.343	.365	.375	.420	.455	.480	.495	.500
	2	0	0	.002	.003	.004	.006	.010	.014	.020	.023	.026	.032	.040	.048	.058	.063	.090	.123	.160	.203	.250
3	0	.970	.941	.885	.857	.831	.779	.729	.681	.636	.614	.593	.551	.512	.475	.439	.422	.343	.275	.216	.166	.125
	1	.029	.058	.111	.135	.159	.203	.243	.279	.311	.325	.339	.363	.384	.402	.416	.422	.441	.444	.432	.408	.375
	2	0	.001	.005	.007	.010	.018	.027	.038	.051	.057	.065	.080	.096	.113	.131	.141	.189	.239	.288	.334	.375
	3	0	0	0	0	0	.001	.001	.002	.003	.003	.004	.006	.008	.011	.014	.016	.027	.043	.064	.091	.125
4	0	.961	.922	.849	.815	.781	.716	.656	.600	.547	.522	.498	.452	.410	.370	.334	.316	.240	.179	.130	.092	.063
	1	.039	.075	.142	.171	.199	.249	.292	.327	.356	.368	.379	.397	.410	.418	.421	.422	.412	.384	.346	.299	.250
	2	.001	.002	.009	.014	.019	.033	.049	.067	.087	.098	.108	.131	.154	.177	.200	.211	.265	.311	.346	.368	.375
	3	0	0	0	0	.001	.002	.004	.006	.009	.011	.014	.019	.026	.033	.042	.047	.076	.111	.154	.200	.250
	4	0	0	0	0	0	0	0	0	0	.001	.001	.001	.002	.002	.003	.004	.008	.015	.026	.041	.063
5	0	.951	.904	.815	.774	.734	.659	.590	.528	.470	.444	.418	.371	.328	.289	.254	.237	.168	.116	.078	.050	.031
	1	.048	.092	.170	.204	.234	.287	.328	.360	.383	.392	.398	.407	.410	.407	.400	.396	.360	.312	.259	.206	.156
	2	.001	.004	.014	.021	.030	.050	.073	.098	.125	.138	.152	.179	.205	.230	.253	.264	.309	.336	.346	.337	.313
	3	0	0	.001	.001	.002	.004	.008	.013	.020	.024	.029	.039	.051	.065	.080	.088	.132	.181	.230	.276	.313
	4	0	0	0	0	0	0	0	.001	.002	.002	.003	.004	.006	.009	.013	.015	.028	.049	.077	.113	.156
	5	0	0	0	0	0	0	0	0	0	0	0	0	0	.001	.001	.001	.002	.005	.010	.018	.031
6	0	.941	.886	.783	.735	.690	.606	.531	.464	.405	.377	.351	.304	.262	.225	.193	.178	.118	.075	.047	.028	.016
	1	.057	.108	.196	.232	.264	.316	.354	.380	.395	.399	.401	.400	.393	.381	.365	.356	.303	.244	.187	.136	.094
	2	.001	.006	.020	.031	.042	.069	.098	.130	.161	.176	.191	.220	.246	.269	.288	.297	.324	.328	.311	.278	.234
	3	0	0	.001	.002	.004	.008	.015	.024	.035	.041	.049	.064	.082	.101	.121	.132	.185	.235	.276	.303	.313
	4	0	0	0	0	0	.001	.001	.002	.004	.005	.007	.011	.015	.021	.029	.033	.060	.095	.138	.186	.234
	5	0	0	0	0	0	0	0	0	0	0	.001	.001	.002	.002	.004	.004	.010	.020	.037	.061	.094
	6	0	0	0	0	0	0	0	0	0	0	0	0	0	0	0	0	.001	.002	.004	.008	.016
7	0	.932	.868	.751	.698	.648	.558	.478	.409	.348	.321	.295	.249	.210	.176	.146	.133	.082	.049	.028	.015	.008
	1	.066	.124	.219	.257	.290	.340	.372	.390	.396	.396	.393	.383	.367	.347	.324	.311	.247	.185	.131	.087	.055
	2	.002	.008	.027	.041	.055	.089	.124	.160	.194	.210	.225	.252	.275	.293	.307	.311	.318	.298	.261	.214	.164
	3	0	0	.002	.004	.006	.013	.023	.036	.053	.062	.071	.092	.115	.138	.161	.173	.227	.268	.290	.292	.273
	4	0	0	0	0	0	.001	.003	.005	.009	.011	.014	.020	.029	.039	.051	.058	.097	.144	.194	.239	.273
	5	0	0	0	0	0	0	0	0	.001	.001	.002	.003	.004	.007	.010	.012	.025	.047	.077	.117	.164
	6	0	0	0	0	0	0	0	0	0	0	0	0	0	.001	.001	.001	.004	.008	.017	.032	.055
	7	0	0	0	0	0	0	0	0	0	0	0	0	0	0	0	0	0	.001	.002	.004	.008

n	r	.50	.45	.40	.35	.30	.25	.24	.22	.20	.18	.16	.15	.14	.12	.10	.08	.06	.05	.04	.02	.01
8	0	.004	.008	.017	.032	.058	.100	.111	.137	.168	.204	.248	.272	.299	.360	.430	.513	.610	.663	.721	.851	.923
	1	.031	.055	.090	.137	.198	.267	.281	.309	.336	.359	.378	.385	.390	.392	.383	.357	.311	.279	.240	.139	.075
	2	.109	.157	.209	.259	.296	.311	.311	.305	.294	.276	.252	.238	.222	.187	.149	.109	.070	.051	.035	.010	.003
	3	.219	.257	.279	.279	.254	.208	.196	.172	.147	.121	.096	.084	.072	.051	.033	.019	.009	.005	.003	0	0
	4	.273	.263	.232	.188	.136	.087	.077	.061	.046	.033	.023	.018	.015	.009	.005	.002	.001	0	0	0	0
	5	.219	.172	.124	.081	.047	.023	.020	.014	.009	.006	.003	.003	.002	.001	0	0	0	0	0	0	0
	6	.109	.070	.041	.022	.010	.004	.003	.002	.001	.001	0	0	0	0	0	0	0	0	0	0	0
	7	.031	.016	.008	.003	.001	0	0	0	0	0	0	0	0	0	0	0	0	0	0	0	0
	8	.004	.002	.001	0	0	0	0	0	0	0	0	0	0	0	0	0	0	0	0	0	0
9	0	.002	.005	.010	.021	.040	.075	.085	.107	.134	.168	.208	.232	.257	.316	.387	.472	.573	.630	.693	.834	.914
	1	.018	.034	.060	.100	.156	.225	.240	.271	.302	.331	.357	.368	.377	.388	.387	.370	.329	.299	.260	.153	.083
	2	.070	.111	.161	.216	.267	.300	.304	.306	.302	.291	.272	.260	.245	.212	.172	.129	.084	.063	.043	.013	.003
	3	.164	.212	.251	.272	.267	.234	.224	.201	.176	.149	.121	.107	.093	.067	.045	.026	.013	.008	.004	.001	0
	4	.246	.260	.251	.219	.172	.117	.106	.085	.066	.049	.035	.028	.023	.014	.007	.003	.001	.001	0	0	0
	5	.246	.213	.167	.118	.074	.039	.033	.024	.017	.011	.007	.005	.004	.002	.001	0	0	0	0	0	0
	6	.164	.116	.074	.042	.021	.009	.007	.005	.003	.002	.001	.001	0	0	0	0	0	0	0	0	0
	7	.070	.041	.021	.010	.004	.001	.001	.001	0	0	0	0	0	0	0	0	0	0	0	0	0
	8	.018	.008	.004	.001	0	0	0	0	0	0	0	0	0	0	0	0	0	0	0	0	0
	9	.002	.001	0	0	0	0	0	0	0	0	0	0	0	0	0	0	0	0	0	0	0
10	0	.001	.003	.006	.013	.028	.056	.064	.083	.107	.137	.175	.197	.221	.279	.349	.434	.539	.599	.665	.817	.904
	1	.010	.021	.040	.072	.121	.188	.203	.235	.268	.302	.333	.347	.360	.380	.387	.378	.344	.315	.277	.167	.091
	2	.044	.076	.121	.176	.233	.282	.288	.298	.302	.298	.286	.276	.264	.233	.194	.148	.099	.075	.052	.015	.004
	3	.117	.166	.215	.252	.267	.250	.243	.224	.201	.174	.145	.130	.115	.085	.057	.034	.017	.010	.006	.001	0
	4	.205	.238	.251	.238	.200	.146	.134	.111	.088	.067	.048	.040	.033	.020	.011	.005	.002	.001	0	0	0
	5	.246	.234	.201	.154	.103	.058	.051	.037	.026	.018	.011	.008	.006	.003	.001	0	0	0	0	0	0
	6	.205	.160	.111	.069	.037	.016	.013	.009	.006	.003	.002	.001	.001	0	0	0	0	0	0	0	0
	7	.117	.075	.042	.021	.009	.003	.002	.001	.001	0	0	0	0	0	0	0	0	0	0	0	0
	8	.044	.023	.011	.004	.001	0	0	0	0	0	0	0	0	0	0	0	0	0	0	0	0
	9	.010	.004	.002	.001	0	0	0	0	0	0	0	0	0	0	0	0	0	0	0	0	0
	10	.001	.001	0	0	0	0	0	0	0	0	0	0	0	0	0	0	0	0	0	0	0
11	0	.000	.001	.004	.009	.020	.042	.049	.065	.086	.113	.147	.167	.190	.245	.314	.400	.506	.569	.638	.801	.895
	1	.005	.013	.027	.052	.093	.155	.170	.202	.236	.272	.308	.325	.341	.368	.384	.382	.355	.329	.293	.180	.099
	2	.027	.051	.089	.140	.200	.258	.268	.284	.295	.299	.293	.287	.277	.251	.213	.166	.113	.087	.061	.018	.005
	3	.081	.126	.177	.225	.257	.258	.254	.241	.221	.197	.168	.152	.135	.103	.071	.043	.022	.014	.008	.001	0
	4	.161	.206	.236	.243	.220	.172	.160	.136	.111	.086	.064	.054	.044	.028	.016	.008	.003	.001	.001	0	0
	5	.226	.236	.221	.183	.132	.080	.071	.054	.039	.027	.017	.013	.010	.005	.002	.001	0	0	0	0	0

n	r	.50	.45	.40	.35	.30	.25	.24	.22	.20	.18	.16	.15	.14	.12	.10	.08	.06	.05	.04	.02	.01 (p)
11	6	.226	.193	.147	.099	.057	.027	.022	.015	.010	.006	.003	.002	.002	.001	0	0	0	0	0	0	0
	7	.161	.113	.070	.038	.017	.006	.005	.003	.002	.001	0	0	0	0	0	0	0	0	0	0	0
	8	.081	.046	.023	.010	.004	.001	.001	0	0	0	0	0	0	0	0	0	0	0	0	0	0
	9	.027	.013	.005	.002	.001	0	0	0	0	0	0	0	0	0	0	0	0	0	0	0	0
	10	.005	.002	.001	0	0	0	0	0	0	0	0	0	0	0	0	0	0	0	0	0	0
	11	0	0	0	0	0	0	0	0	0	0	0	0	0	0	0	0	0	0	0	0	0
12	0	.000	.001	.002	.006	.014	.032	.037	.051	.069	.092	.123	.142	.164	.216	.282	.368	.476	.540	.613	.785	.886
	1	.003	.008	.017	.037	.071	.127	.141	.172	.206	.243	.282	.301	.320	.353	.377	.384	.365	.341	.306	.192	.107
	2	.016	.034	.064	.109	.168	.232	.244	.266	.283	.294	.296	.292	.286	.265	.230	.183	.128	.099	.070	.022	.006
	3	.054	.092	.142	.195	.240	.258	.257	.250	.236	.215	.188	.172	.155	.120	.085	.053	.027	.017	.010	.001	0
	4	.121	.170	.213	.237	.231	.194	.183	.159	.133	.106	.080	.068	.057	.037	.021	.010	.004	.002	.001	0	0
	5	.193	.222	.227	.204	.158	.103	.092	.072	.053	.037	.025	.019	.015	.008	.004	.001	0	0	0	0	0
	6	.226	.212	.177	.128	.079	.040	.034	.024	.016	.010	.005	.004	.003	.001	0	0	0	0	0	0	0
	7	.193	.149	.101	.059	.029	.011	.009	.006	.003	.002	.001	.001	0	0	0	0	0	0	0	0	0
	8	.121	.076	.042	.020	.008	.002	.002	.001	.001	0	0	0	0	0	0	0	0	0	0	0	0
	9	.054	.028	.012	.005	.001	0	0	0	0	0	0	0	0	0	0	0	0	0	0	0	0
	10	.016	.007	.002	.001	0	0	0	0	0	0	0	0	0	0	0	0	0	0	0	0	0
	11	.003	.001	0	0	0	0	0	0	0	0	0	0	0	0	0	0	0	0	0	0	0
	12	0	0	0	0	0	0	0	0	0	0	0	0	0	0	0	0	0	0	0	0	0
13	0	.000	.000	.001	.004	.010	.024	.028	.040	.055	.076	.104	.121	.141	.190	.254	.338	.447	.513	.588	.769	.878
	1	.002	.004	.011	.026	.054	.103	.116	.145	.179	.216	.257	.277	.298	.336	.367	.382	.371	.351	.319	.204	.115
	2	.010	.022	.045	.084	.139	.206	.220	.245	.268	.285	.293	.294	.291	.275	.245	.199	.142	.111	.080	.025	.007
	3	.035	.066	.111	.165	.218	.252	.254	.254	.246	.229	.205	.190	.174	.138	.100	.064	.033	.021	.012	.002	0
	4	.087	.135	.184	.222	.234	.210	.201	.179	.154	.126	.098	.084	.071	.047	.028	.014	.005	.003	.001	0	0
	5	.157	.199	.221	.215	.180	.126	.114	.091	.069	.050	.033	.027	.021	.012	.006	.002	.001	0	0	0	0
	6	.209	.217	.197	.155	.103	.056	.048	.034	.023	.015	.008	.006	.004	.002	.001	0	0	0	0	0	0
	7	.209	.177	.131	.083	.044	.019	.015	.010	.006	.003	.002	.001	.001	0	0	0	0	0	0	0	0
	8	.157	.109	.066	.034	.014	.005	.004	.002	.001	.001	0	0	0	0	0	0	0	0	0	0	0
	9	.087	.050	.024	.010	.003	.001	.001	0	0	0	0	0	0	0	0	0	0	0	0	0	0
	10	.035	.016	.006	.002	.001	0	0	0	0	0	0	0	0	0	0	0	0	0	0	0	0
	11	.010	.004	.001	0	0	0	0	0	0	0	0	0	0	0	0	0	0	0	0	0	0
	12	.002	.001	0	0	0	0	0	0	0	0	0	0	0	0	0	0	0	0	0	0	0
	13	0	0	0	0	0	0	0	0	0	0	0	0	0	0	0	0	0	0	0	0	0
14	0	.000	.000	.001	.002	.007	.018	.021	.031	.044	.062	.087	.103	.121	.167	.229	.311	.421	.488	.565	.754	.869
	1	.001	.003	.007	.018	.041	.083	.095	.122	.154	.191	.232	.254	.276	.319	.356	.379	.376	.359	.329	.215	.123
	2	.006	.014	.032	.063	.113	.180	.195	.223	.250	.272	.287	.291	.292	.283	.257	.214	.156	.123	.089	.029	.008
	3	.022	.046	.085	.137	.194	.240	.246	.252	.250	.239	.219	.206	.190	.154	.114	.074	.040	.026	.015	.002	0

$n = 14$

r	.50	.45	.40	.35	.30	.25	.24	.22	.20	.18	.16	.15	.14	.12	.10	.08	.06	.05	.04	.02	.01
4	.061	.104	.155	.202	.229	.220	.214	.195	.172	.144	.115	.100	.085	.058	.035	.018	.007	.004	.002	0	0
5	.122	.170	.207	.218	.196	.147	.135	.110	.086	.063	.044	.035	.028	.016	.008	.003	.001	0	0	0	0
6	.183	.209	.207	.176	.126	.073	.064	.047	.032	.021	.012	.009	.007	.003	.001	0	0	0	0	0	0
7	.209	.195	.157	.108	.062	.028	.023	.015	.009	.005	.003	.002	.001	.001	0	0	0	0	0	0	0
8	.183	.140	.092	.051	.023	.008	.006	.004	.002	.001	0	0	0	0	0	0	0	0	0	0	0
9	.122	.076	.041	.018	.007	.002	.001	.001	0	0	0	0	0	0	0	0	0	0	0	0	0
10	.061	.031	.014	.005	.001	0	0	0	0	0	0	0	0	0	0	0	0	0	0	0	0
11	.022	.009	.003	.001	0	0	0	0	0	0	0	0	0	0	0	0	0	0	0	0	0
12	.006	.002	.001	0	0	0	0	0	0	0	0	0	0	0	0	0	0	0	0	0	0
13	.001	0	0	0	0	0	0	0	0	0	0	0	0	0	0	0	0	0	0	0	0
14	0	0	0	0	0	0	0	0	0	0	0	0	0	0	0	0	0	0	0	0	0

$n = 15$

r	.50	.45	.40	.35	.30	.25	.24	.22	.20	.18	.16	.15	.14	.12	.10	.08	.06	.05	.04	.02	.01
0	.000	.000	.000	.002	.005	.013	.016	.024	.035	.051	.073	.087	.104	.147	.206	.286	.395	.463	.542	.739	.860
1	.000	.002	.005	.013	.031	.067	.077	.102	.132	.168	.209	.231	.254	.301	.343	.373	.378	.366	.339	.226	.130
2	.003	.009	.022	.048	.092	.156	.171	.201	.231	.258	.279	.286	.290	.287	.267	.227	.169	.135	.099	.032	.009
3	.014	.032	.063	.111	.170	.225	.234	.246	.250	.245	.230	.218	.204	.170	.129	.086	.047	.031	.018	.003	0
4	.042	.078	.127	.179	.219	.225	.221	.208	.188	.162	.131	.116	.100	.069	.043	.022	.009	.005	.002	0	0
5	.092	.140	.186	.212	.206	.165	.154	.129	.103	.078	.055	.045	.036	.021	.010	.004	.001	.001	0	0	0
6	.153	.191	.207	.191	.147	.092	.081	.061	.043	.029	.017	.013	.010	.005	.002	.001	0	0	0	0	0
7	.196	.201	.177	.132	.081	.039	.033	.022	.014	.008	.004	.003	.002	.001	0	0	0	0	0	0	0
8	.196	.165	.118	.071	.035	.013	.010	.006	.003	.002	.001	.001	0	0	0	0	0	0	0	0	0
9	.153	.105	.061	.030	.012	.003	.003	.001	.001	0	0	0	0	0	0	0	0	0	0	0	0
10	.092	.051	.024	.010	.003	.001	0	0	0	0	0	0	0	0	0	0	0	0	0	0	0
11	.042	.019	.007	.002	.001	0	0	0	0	0	0	0	0	0	0	0	0	0	0	0	0
12	.014	.005	.002	0	0	0	0	0	0	0	0	0	0	0	0	0	0	0	0	0	0
13	.003	.001	0	0	0	0	0	0	0	0	0	0	0	0	0	0	0	0	0	0	0
14	.000	0	0	0	0	0	0	0	0	0	0	0	0	0	0	0	0	0	0	0	0
15	0	0	0	0	0	0	0	0	0	0	0	0	0	0	0	0	0	0	0	0	0

$n = 16$

r	.50	.45	.40	.35	.30	.25	.24	.22	.20	.18	.16	.15	.14	.12	.10	.08	.06	.05	.04	.02	.01
0	.000	.000	.000	.001	.003	.010	.012	.019	.028	.042	.061	.074	.090	.129	.185	.263	.372	.440	.520	.724	.851
1	.000	.001	.003	.009	.023	.053	.063	.085	.113	.147	.187	.210	.233	.282	.329	.366	.379	.371	.347	.236	.138
2	.002	.006	.015	.035	.073	.134	.148	.179	.211	.242	.268	.277	.285	.289	.275	.239	.182	.146	.108	.036	.010
3	.009	.022	.047	.089	.146	.208	.218	.236	.246	.248	.238	.229	.216	.184	.142	.097	.054	.036	.021	.003	0
4	.028	.057	.101	.155	.204	.225	.224	.216	.200	.177	.147	.131	.114	.081	.051	.027	.011	.006	.003	0	0
5	.067	.112	.162	.201	.210	.180	.170	.146	.120	.093	.067	.056	.045	.027	.014	.006	.002	.001	0	0	0
6	.122	.168	.198	.198	.165	.110	.098	.076	.055	.037	.023	.018	.013	.007	.003	.001	0	0	0	0	0
7	.175	.197	.189	.152	.101	.052	.044	.030	.020	.012	.006	.005	.003	.001	0	0	0	0	0	0	0
8	.196	.181	.142	.092	.049	.020	.016	.010	.006	.003	.001	.001	.001	0	0	0	0	0	0	0	0

n	r	.50	.45	.40	.35	.30	.25	.24	.22	.20	.18	.16	.15	.14	.12	.10	.08	.06	.05	.04	.02	.01
16	9	.175	.132	.084	.044	.019	.006	.004	.002	.001	.001	0	0	0	0	0	0	0	0	0	0	0
	10	.122	.075	.039	.017	.006	.001	.001	0	0	0	0	0	0	0	0	0	0	0	0	0	0
	11	.067	.034	.014	.005	.001	0	0	0	0	0	0	0	0	0	0	0	0	0	0	0	0
	12	.028	.011	.004	.001	0	0	0	0	0	0	0	0	0	0	0	0	0	0	0	0	0
	13	.009	.003	.001	0	0	0	0	0	0	0	0	0	0	0	0	0	0	0	0	0	0
	14	.002	.001	0	0	0	0	0	0	0	0	0	0	0	0	0	0	0	0	0	0	0
	15	0	0	0	0	0	0	0	0	0	0	0	0	0	0	0	0	0	0	0	0	0
	16	0	0	0	0	0	0	0	0	0	0	0	0	0	0	0	0	0	0	0	0	0
17	0	.000	.000	.000	.001	.002	.008	.009	.015	.023	.034	.052	.063	.077	.114	.167	.242	.349	.418	.500	.709	.843
	1	.000	.001	.002	.006	.017	.043	.051	.070	.096	.128	.167	.189	.213	.264	.315	.358	.379	.374	.354	.246	.145
	2	.001	.004	.010	.026	.058	.114	.128	.158	.191	.225	.255	.267	.278	.288	.280	.249	.194	.158	.118	.040	.012
	3	.005	.014	.034	.070	.125	.189	.202	.223	.239	.246	.243	.236	.226	.196	.156	.108	.062	.041	.025	.004	.001
	4	.018	.041	.080	.132	.187	.221	.223	.221	.209	.189	.162	.146	.129	.094	.060	.033	.014	.008	.004	0	0
	5	.047	.087	.138	.185	.208	.191	.183	.162	.136	.108	.080	.067	.054	.033	.017	.007	.002	.001	0	0	0
	6	.094	.143	.184	.199	.178	.128	.116	.091	.068	.047	.031	.024	.018	.009	.004	.001	0	0	0	0	0
	7	.148	.184	.193	.168	.120	.067	.057	.040	.027	.016	.009	.007	.005	.002	.001	0	0	0	0	0	0
	8	.185	.188	.161	.113	.064	.028	.023	.014	.008	.004	.002	.001	.001	0	0	0	0	0	0	0	0
	9	.185	.154	.107	.061	.028	.009	.007	.004	.002	.001	0	0	0	0	0	0	0	0	0	0	0
	10	.148	.101	.057	.026	.009	.002	.002	.001	0	0	0	0	0	0	0	0	0	0	0	0	0
	11	.094	.052	.024	.009	.003	.001	0	0	0	0	0	0	0	0	0	0	0	0	0	0	0
	12	.047	.021	.008	.002	.001	0	0	0	0	0	0	0	0	0	0	0	0	0	0	0	0
	13	.018	.007	.002	.001	0	0	0	0	0	0	0	0	0	0	0	0	0	0	0	0	0
	14	.005	.002	0	0	0	0	0	0	0	0	0	0	0	0	0	0	0	0	0	0	0
	15	.001	0	0	0	0	0	0	0	0	0	0	0	0	0	0	0	0	0	0	0	0
	16	0	0	0	0	0	0	0	0	0	0	0	0	0	0	0	0	0	0	0	0	0
	17	0	0	0	0	0	0	0	0	0	0	0	0	0	0	0	0	0	0	0	0	0
18	0	.000	.000	.000	.000	.002	.006	.007	.011	.018	.028	.043	.054	.066	.100	.150	.223	.328	.397	.480	.695	.835
	1	.000	.000	.001	.004	.013	.034	.041	.058	.081	.111	.149	.170	.194	.246	.300	.349	.377	.376	.360	.255	.152
	2	.001	.002	.007	.019	.046	.096	.109	.139	.172	.207	.241	.256	.268	.285	.284	.258	.205	.168	.127	.044	.013
	3	.003	.009	.025	.055	.105	.170	.184	.209	.230	.243	.244	.241	.233	.207	.168	.120	.070	.047	.028	.005	.001
	4	.012	.029	.061	.110	.168	.213	.218	.221	.215	.200	.175	.159	.142	.106	.070	.039	.017	.009	.004	0	0
	5	.033	.067	.115	.166	.202	.199	.193	.175	.151	.123	.093	.079	.065	.040	.022	.009	.003	.001	0	0	0
	6	.071	.118	.166	.194	.187	.144	.132	.107	.082	.058	.038	.030	.023	.012	.005	.002	0	0	0	0	0
	7	.121	.166	.189	.179	.138	.082	.071	.052	.035	.022	.013	.009	.006	.003	.001	0	0	0	0	0	0
	8	.167	.186	.173	.133	.081	.038	.031	.020	.012	.007	.003	.002	.001	.001	.000	0	0	0	0	0	0
	9	.185	.169	.128	.079	.039	.014	.011	.006	.003	.002	.001	0	0	0	0	0	0	0	0	0	0
	10	.167	.125	.077	.038	.015	.004	.003	.002	.001	0	0	0	0	0	0	0	0	0	0	0	0
	11	.121	.074	.037	.015	.005	.001	.001	0	0	0	0	0	0	0	0	0	0	0	0	0	0

n	r	.50	.45	.40	.35	.30	.25	.24	.22	.20	.18	.16	.15	.14	.12	.10	.08	.06	.05	.04	.02	p .01
18	12	.071	.035	.015	.005	.001	0	0	0	0	0	0	0	0	0	0	0	0	0	0	0	0
	13	.033	.013	.004	.001	0	0	0	0	0	0	0	0	0	0	0	0	0	0	0	0	0
	14	.012	.004	.001	0	0	0	0	0	0	0	0	0	0	0	0	0	0	0	0	0	0
	15	.003	.001	0	0	0	0	0	0	0	0	0	0	0	0	0	0	0	0	0	0	0
	16	.001	0	0	0	0	0	0	0	0	0	0	0	0	0	0	0	0	0	0	0	0
	17	0	0	0	0	0	0	0	0	0	0	0	0	0	0	0	0	0	0	0	0	0
	18	0	0	0	0	0	0	0	0	0	0	0	0	0	0	0	0	0	0	0	0	0
19	0	.000	.000	.000	.000	.001	.004	.005	.009	.014	.023	.036	.046	.057	.088	.135	.205	.309	.377	.460	.681	.826
	1	.000	.000	.001	.003	.009	.027	.033	.048	.068	.096	.132	.153	.176	.228	.285	.339	.374	.377	.364	.264	.159
	2	.000	.001	.005	.014	.036	.080	.093	.121	.154	.190	.226	.243	.258	.230	.285	.265	.215	.179	.137	.049	.014
	3	.002	.006	.017	.042	.087	.152	.166	.194	.218	.236	.244	.243	.238	.217	.180	.131	.078	.053	.032	.006	.001
	4	.007	.020	.047	.091	.149	.202	.210	.219	.218	.207	.186	.171	.155	.118	.080	.045	.020	.011	.005	0	0
	5	.022	.050	.093	.147	.192	.202	.199	.185	.164	.137	.106	.091	.076	.048	.027	.012	.004	.002	.001	0	0
	6	.052	.095	.145	.184	.192	.157	.146	.122	.095	.070	.047	.037	.029	.015	.007	.002	.001	0	0	0	0
	7	.096	.144	.180	.184	.153	.097	.086	.064	.044	.029	.017	.012	.009	.004	.001	0	0	0	0	0	0
	8	.144	.177	.180	.149	.098	.049	.041	.027	.017	.009	.005	.003	.002	.001	0	0	0	0	0	0	0
	9	.176	.177	.146	.098	.051	.020	.016	.009	.005	.003	.001	.001	0	0	0	0	0	0	0	0	0
	10	.176	.145	.098	.053	.022	.007	.005	.003	.001	.001	0	0	0	0	0	0	0	0	0	0	0
	11	.144	.097	.053	.023	.008	.002	.001	.001	0	0	0	0	0	0	0	0	0	0	0	0	0
	12	.096	.053	.024	.008	.002	0	0	0	0	0	0	0	0	0	0	0	0	0	0	0	0
	13	.052	.023	.008	.002	.001	0	0	0	0	0	0	0	0	0	0	0	0	0	0	0	0
	14	.022	.008	.002	.001	0	0	0	0	0	0	0	0	0	0	0	0	0	0	0	0	0
	15	.007	.002	.001	0	0	0	0	0	0	0	0	0	0	0	0	0	0	0	0	0	0
	16	.002	0	0	0	0	0	0	0	0	0	0	0	0	0	0	0	0	0	0	0	0
	17	0	0	0	0	0	0	0	0	0	0	0	0	0	0	0	0	0	0	0	0	0
	18	0	0	0	0	0	0	0	0	0	0	0	0	0	0	0	0	0	0	0	0	0
	19	0	0	0	0	0	0	0	0	0	0	0	0	0	0	0	0	0	0	0	0	0
20	0	.000	.000	.000	.000	.001	.003	.004	.007	.012	.019	.031	.039	.049	.078	.122	.189	.290	.358	.442	.668	.818
	1	.000	.000	.000	.002	.007	.021	.026	.039	.058	.083	.117	.137	.159	.212	.270	.328	.370	.377	.368	.272	.165
	2	.000	.001	.003	.010	.028	.067	.078	.105	.137	.173	.211	.229	.247	.274	.285	.271	.225	.189	.146	.053	.016
	3	.001	.004	.012	.032	.072	.134	.148	.178	.205	.228	.241	.243	.241	.224	.190	.141	.086	.060	.036	.006	.001
	4	.005	.014	.035	.074	.130	.190	.199	.213	.218	.213	.195	.182	.167	.130	.090	.052	.023	.013	.006	.001	0
	5	.015	.036	.075	.127	.179	.202	.201	.192	.175	.149	.119	.103	.087	.057	.032	.015	.005	.002	.001	0	0
	6	.037	.075	.124	.171	.192	.169	.159	.136	.109	.082	.057	.045	.035	.019	.009	.003	.001	0	0	0	0
	7	.074	.122	.166	.184	.164	.112	.100	.076	.055	.036	.022	.016	.012	.005	.002	.001	0	0	0	0	0
	8	.120	.162	.180	.161	.114	.061	.051	.035	.022	.013	.007	.005	.003	.001	0	0	0	0	0	0	0
	9	.160	.177	.160	.116	.065	.027	.022	.013	.007	.004	.002	.001	.001	0	0	0	0	0	0	0	0
	10	.176	.159	.117	.069	.031	.010	.008	.004	.002	.001	0	0	0	0	0	0	0	0	0	0	0

n	r	.50	.45	.40	.35	.30	.25	.24	.22	.20	.18	.16	.15	.14	.12	.10	.08	.06	.05	.04	.02	.01
	11	.160	.119	.071	.034	.012	.003	.002	.001	0	0	0	0	0	0	0	0	0	0	0	0	0
	12	.120	.073	.035	.014	.004	.001	.001	0	0	0	0	0	0	0	0	0	0	0	0	0	0
	13	.074	.037	.015	.004	.001	0	0	0	0	0	0	0	0	0	0	0	0	0	0	0	0
	14	.037	.015	.005	.001	0	0	0	0	0	0	0	0	0	0	0	0	0	0	0	0	0
	15	.015	.005	.001	0	0	0	0	0	0	0	0	0	0	0	0	0	0	0	0	0	0
	16	.005	.001	0	0	0	0	0	0	0	0	0	0	0	0	0	0	0	0	0	0	0
	17	.001	0	0	0	0	0	0	0	0	0	0	0	0	0	0	0	0	0	0	0	0
	18	0	0	0	0	0	0	0	0	0	0	0	0	0	0	0	0	0	0	0	0	0
	19	0	0	0	0	0	0	0	0	0	0	0	0	0	0	0	0	0	0	0	0	0
	20	0	0	0	0	0	0	0	0	0	0	0	0	0	0	0	0	0	0	0	0	0
25	0	.000	.000	.000	.000	.000	.001	.001	.002	.004	.007	.013	.017	.023	.041	.072	.124	.213	.277	.360	.603	.778
	1	.000	.000	.000	.000	.001	.006	.008	.014	.024	.038	.061	.076	.094	.140	.199	.270	.340	.365	.375	.308	.196
	2	.000	.000	.000	.002	.007	.025	.031	.048	.071	.101	.139	.161	.183	.228	.266	.282	.260	.231	.188	.075	.024
	3	.000	.000	.002	.008	.024	.064	.076	.104	.136	.170	.203	.217	.229	.239	.226	.188	.127	.093	.060	.012	.002
	4	.000	.002	.007	.022	.057	.118	.132	.161	.187	.206	.213	.211	.205	.179	.138	.090	.045	.027	.014	.001	0
	5	.002	.006	.020	.051	.103	.165	.175	.190	.196	.190	.170	.156	.140	.103	.065	.033	.012	.006	.002	0	0
	6	.005	.017	.044	.091	.147	.183	.184	.179	.163	.139	.108	.092	.076	.047	.024	.010	.003	.001	0	0	0
	7	.014	.038	.080	.133	.171	.165	.158	.137	.111	.083	.056	.044	.034	.017	.007	.002	0	0	0	0	0
	8	.032	.070	.120	.161	.165	.124	.112	.087	.062	.041	.024	.017	.012	.005	.002	0	0	0	0	0	0
	9	.061	.108	.151	.163	.134	.078	.067	.046	.029	.017	.009	.006	.004	.001	0	0	0	0	0	0	0
	10	.097	.142	.161	.141	.092	.042	.034	.021	.012	.006	.003	.002	.001	0	0	0	0	0	0	0	0
	11	.133	.158	.147	.103	.054	.019	.015	.008	.004	.002	.001	0	0	0	0	0	0	0	0	0	0
	12	.155	.151	.114	.065	.027	.007	.005	.003	.001	0	0	0	0	0	0	0	0	0	0	0	0
	13	.155	.124	.076	.035	.011	.002	.002	.001	0	0	0	0	0	0	0	0	0	0	0	0	0
	14	.133	.087	.043	.016	.004	.001	0	0	0	0	0	0	0	0	0	0	0	0	0	0	0
	15	.097	.052	.021	.006	.001	0	0	0	0	0	0	0	0	0	0	0	0	0	0	0	0
	16	.061	.027	.009	.002	0	0	0	0	0	0	0	0	0	0	0	0	0	0	0	0	0
	17	.032	.012	.003	.001	0	0	0	0	0	0	0	0	0	0	0	0	0	0	0	0	0
	18	.014	.004	.001	0	0	0	0	0	0	0	0	0	0	0	0	0	0	0	0	0	0
	19	.005	.001	0	0	0	0	0	0	0	0	0	0	0	0	0	0	0	0	0	0	0
	20	.002	0	0	0	0	0	0	0	0	0	0	0	0	0	0	0	0	0	0	0	0
	21	0	0	0	0	0	0	0	0	0	0	0	0	0	0	0	0	0	0	0	0	0
	22	0	0	0	0	0	0	0	0	0	0	0	0	0	0	0	0	0	0	0	0	0
	23	0	0	0	0	0	0	0	0	0	0	0	0	0	0	0	0	0	0	0	0	0
	24	0	0	0	0	0	0	0	0	0	0	0	0	0	0	0	0	0	0	0	0	0
	25	0	0	0	0	0	0	0	0	0	0	0	0	0	0	0	0	0	0	0	0	0

APPENDIX B

Binomial Distribution: Cumulative Probabilities

$$P_B(R \le r \mid n, p) = \sum_{i=0}^{r} \frac{n!}{i!(n-i)!} p^i (1-p)^{n-i}$$

Figure B.1
Using Appendix B to find $P_B(R \le 2 \mid n = 9, p = 0.12)$.

n	r	p	.10	.12	.14
.	.		.	.	.
.	.		.	.	.
.	.		.	.	.
9	0		.387	.316	.257
	1		.775	.705	.634
	2	. . . .	.947	.917	.880
	3		.992	.984	.973
	.		.	.	.
	.		.	.	.

n	r	.50	.45	.40	.35	.30	.25	.24	.22	.20	.18	.16	.15	.14	.12	.10	.08	.06	.05	.04	.02	.01
2	0	.250	.303	.360	.423	.490	.563	.578	.608	.640	.672	.706	.723	.740	.774	.810	.846	.884	.903	.922	.960	.980
	1	.750	.798	.840	.878	.910	.938	.942	.952	.960	.968	.974	.978	.980	.986	.990	.994	.996	.998	.998	1	1
	2	1	1	1	1	1	1	1	1	1	1	1	1	1	1	1	1	1	1	1	1	1
3	0	.125	.166	.216	.275	.343	.422	.439	.475	.512	.551	.593	.614	.636	.681	.729	.779	.831	.857	.885	.941	.970
	1	.500	.575	.648	.718	.784	.844	.855	.876	.896	.914	.931	.939	.947	.960	.972	.982	.990	.993	.995	.999	1
	2	.875	.909	.936	.957	.973	.984	.986	.989	.992	.994	.996	.997	.997	.998	.999	.999	1	1	1	1	1
	3	1	1	1	1	1	1	1	1	1	1	1	1	1	1	1	1	1	1	1	1	1
4	0	.063	.092	.130	.179	.240	.316	.334	.370	.410	.452	.498	.522	.547	.600	.656	.716	.781	.815	.849	.922	.961
	1	.313	.391	.475	.563	.652	.738	.755	.788	.819	.849	.877	.890	.903	.927	.948	.966	.980	.986	.991	.998	.999
	2	.688	.759	.821	.874	.916	.949	.955	.964	.973	.980	.986	.988	.990	.994	.996	.998	.999	.999	.999	1	1
	3	.938	.959	.974	.985	.992	.996	.997	.998	.998	.999	.999	.999	1	1	1	1	1	1	1	1	1
	4	1	1	1	1	1	1	1	1	1	1	1	1	1	1	1	1	1	1	1	1	1
5	0	.031	.050	.078	.116	.168	.237	.254	.289	.328	.371	.418	.444	.470	.528	.590	.659	.734	.774	.815	.904	.951
	1	.188	.256	.337	.428	.528	.633	.654	.696	.737	.778	.817	.835	.853	.888	.919	.946	.968	.977	.985	.996	.999
	2	.500	.593	.683	.765	.837	.896	.907	.926	.942	.956	.968	.973	.978	.986	.991	.995	.998	.999	.999	1	1
	3	.813	.869	.913	.946	.969	.984	.987	.990	.993	.996	.997	.998	.998	.999	1	1	1	1	1	1	1
	4	.969	.982	.990	.995	.998	.999	.999	.999	1	1	1	1	1	1	1	1	1	1	1	1	1
	5	1	1	1	1	1	1	1	1	1	1	1	1	1	1	1	1	1	1	1	1	1
6	0	.016	.028	.047	.075	.118	.178	.193	.225	.262	.304	.351	.377	.405	.464	.531	.606	.690	.735	.783	.886	.941
	1	.109	.164	.233	.319	.420	.534	.558	.606	.655	.704	.753	.776	.800	.844	.886	.923	.954	.967	.978	.994	.999
	2	.344	.442	.544	.647	.744	.831	.846	.875	.901	.924	.944	.953	.961	.974	.984	.991	.996	.998	.999	1	1
	3	.656	.745	.821	.883	.930	.962	.967	.976	.983	.988	.993	.994	.995	.997	.999	.999	1	1	1	1	1
	4	.891	.931	.959	.978	.989	.995	.996	.997	.998	.999	.999	1	1	1	1	1	1	1	1	1	1
	5	.984	.992	.996	.998	.999	1	1	1	1	1	1	1	1	1	1	1	1	1	1	1	1
	6	1	1	1	1	1	1	1	1	1	1	1	1	1	1	1	1	1	1	1	1	1
7	0	.008	.015	.028	.049	.082	.133	.146	.176	.210	.249	.295	.321	.348	.409	.478	.558	.648	.698	.751	.868	.932
	1	.063	.102	.159	.234	.329	.445	.470	.522	.577	.632	.689	.717	.744	.799	.850	.897	.938	.956	.971	.992	.998
	2	.227	.316	.420	.532	.647	.756	.777	.816	.852	.885	.913	.926	.938	.958	.974	.986	.994	.996	.998	1	1
	3	.500	.608	.710	.800	.874	.929	.938	.954	.967	.977	.985	.988	.991	.995	.997	.999	1	1	1	1	1
	4	.773	.847	.904	.944	.971	.987	.989	.993	.995	.997	.998	.999	.999	1	1	1	1	1	1	1	1
	5	.938	.964	.981	.991	.996	.999	.999	.999	1	1	1	1	1	1	1	1	1	1	1	1	1
	6	.992	.996	.998	.999	1	1	1	1	1	1	1	1	1	1	1	1	1	1	1	1	1
	7	1	1	1	1	1	1	1	1	1	1	1	1	1	1	1	1	1	1	1	1	1

n	r	.50	.45	.40	.35	.30	.25	.24	.22	.20	.18	.16	.15	.14	.12	.10	.08	.06	.05	.04	.02	.01
8	0	.004	.008	.017	.032	.058	.100	.111	.137	.168	.204	.248	.272	.299	.360	.430	.513	.610	.663	.721	.851	.923
	1	.035	.063	.106	.169	.255	.367	.392	.446	.503	.563	.626	.657	.689	.752	.813	.870	.921	.943	.962	.990	.997
	2	.145	.220	.315	.428	.552	.679	.703	.751	.797	.839	.877	.895	.911	.939	.962	.979	.990	.994	.997	1	1
	3	.363	.477	.594	.706	.806	.886	.900	.924	.944	.960	.973	.979	.983	.990	.995	.998	.999	.999	1	1	1
	4	.637	.740	.826	.894	.942	.973	.977	.984	.990	.993	.996	.997	.998	.999	1	1	1	1	1	1	1
	5	.855	.912	.950	.975	.989	.996	.997	.998	.999	.999	1	1	1	1	1	1	1	1	1	1	1
	6	.965	.982	.991	.996	.999	1	1	1	1	1	1	1	1	1	1	1	1	1	1	1	1
	7	.996	.998	.999	1	1	1	1	1	1	1	1	1	1	1	1	1	1	1	1	1	1
	8	1	1	1	1	1	1	1	1	1	1	1	1	1	1	1	1	1	1	1	1	1
9	0	.002	.005	.010	.021	.040	.075	.085	.107	.134	.168	.208	.232	.257	.316	.387	.472	.573	.630	.693	.834	.914
	1	.020	.039	.071	.121	.196	.300	.325	.378	.436	.499	.565	.599	.634	.705	.775	.842	.902	.929	.952	.987	.997
	2	.090	.150	.232	.337	.463	.601	.629	.684	.738	.790	.837	.859	.880	.917	.947	.970	.986	.992	.996	.999	1
	3	.254	.361	.483	.609	.730	.834	.852	.886	.914	.938	.958	.966	.973	.984	.992	.996	.999	.999	1	1	1
	4	.500	.621	.733	.828	.901	.951	.958	.971	.980	.988	.993	.994	.996	.998	.999	1	1	1	1	1	1
	5	.746	.834	.901	.946	.975	.990	.992	.995	.997	.998	.999	.999	1	1	1	1	1	1	1	1	1
	6	.910	.950	.975	.989	.996	.999	.999	.999	1	1	1	1	1	1	1	1	1	1	1	1	1
	7	.980	.991	.996	.999	1	1	1	1	1	1	1	1	1	1	1	1	1	1	1	1	1
	8	.998	.999	1	1	1	1	1	1	1	1	1	1	1	1	1	1	1	1	1	1	1
	9	1	1	1	1	1	1	1	1	1	1	1	1	1	1	1	1	1	1	1	1	1
10	0	.001	.003	.006	.013	.028	.056	.064	.083	.107	.137	.175	.197	.221	.279	.314	.400	.506	.599	.665	.817	.904
	1	.011	.023	.046	.086	.149	.244	.267	.318	.376	.439	.508	.544	.582	.658	.697	.782	.862	.914	.942	.984	.996
	2	.055	.100	.167	.262	.383	.526	.556	.617	.678	.737	.794	.820	.845	.891	.910	.948	.975	.988	.994	.999	1
	3	.172	.266	.382	.514	.650	.776	.799	.841	.879	.912	.939	.950	.960	.976	.981	.991	.997	.999	1	1	1
	4	.377	.504	.633	.751	.850	.922	.933	.952	.967	.979	.987	.990	.993	.996	.997	.999	1	1	1	1	1
	5	.623	.738	.834	.905	.953	.980	.984	.990	.994	.996	.998	.999	.999	1	1	1	1	1	1	1	1
	6	.828	.898	.945	.974	.989	.996	.997	.998	.999	1	1	1	1	1	1	1	1	1	1	1	1
	7	.945	.973	.988	.995	.998	1	1	1	1	1	1	1	1	1	1	1	1	1	1	1	1
	8	.989	.995	.998	.999	1	1	1	1	1	1	1	1	1	1	1	1	1	1	1	1	1
	9	.999	1	1	1	1	1	1	1	1	1	1	1	1	1	1	1	1	1	1	1	1
	10	1	1	1	1	1	1	1	1	1	1	1	1	1	1	1	1	1	1	1	1	1
11	0	.000	.001	.004	.009	.020	.042	.049	.065	.086	.113	.147	.167	.190	.245	.314	.400	.506	.569	.638	.801	.895
	1	.006	.014	.030	.061	.113	.197	.219	.267	.322	.385	.455	.492	.531	.613	.697	.782	.862	.898	.931	.980	.995
	2	.033	.065	.119	.200	.313	.455	.487	.551	.617	.684	.748	.779	.809	.863	.910	.948	.975	.985	.992	.999	1
	3	.113	.191	.296	.426	.570	.713	.740	.792	.839	.880	.915	.931	.944	.966	.981	.991	.997	.998	.999	1	1
	4	.274	.397	.533	.668	.790	.885	.901	.928	.950	.967	.979	.984	.988	.994	.997	.999	1	1	1	1	1
	5	.500	.633	.753	.851	.922	.966	.972	.981	.988	.993	.996	.997	.998	.999	1	1	1	1	1	1	1

n	r	.01	.02	.04	.05	.06	.08	.10	.12	.14	.15	.16	.18	.20	.22	.24	.25	.30	.35	.40	.45	.50
11	6	—	—	—	—	—	—	—	—	—	—	—	.999	.998	.996	.994	.992	.978	.950	.901	.826	.726
	7	—	—	—	—	—	—	—	—	—	—	—	—	—	—	.999	.999	.996	.988	.971	.939	.887
	8	—	—	—	—	—	—	—	—	—	—	—	—	—	—	—	—	.999	.998	.994	.985	.967
	9	—	—	—	—	—	—	—	—	—	—	—	—	—	—	—	—	—	—	.999	.998	.994
	10	—	—	—	—	—	—	—	—	—	—	—	—	—	—	—	—	—	—	—	—	—
	11	—	—	—	—	—	—	—	—	—	—	—	—	—	—	—	—	—	—	—	—	—
12	0	.886	.785	.613	.540	.476	.368	.282	.216	.164	.142	.123	.092	.069	.051	.037	.032	.014	.006	.002	.001	.000
	1	.994	.977	.919	.882	.840	.751	.659	.569	.483	.443	.405	.336	.275	.222	.178	.158	.085	.042	.020	.008	.003
	2	—	.998	.989	.980	.968	.935	.889	.833	.770	.736	.701	.630	.558	.489	.422	.391	.253	.151	.083	.042	.019
	3	—	—	.999	.998	.996	.988	.974	.954	.925	.908	.889	.845	.795	.739	.680	.649	.493	.347	.225	.134	.073
	4	—	—	—	—	—	.998	.996	.991	.982	.976	.969	.951	.927	.898	.862	.842	.724	.583	.438	.304	.194
	5	—	—	—	—	—	—	.999	.999	.997	.995	.994	.988	.981	.970	.955	.946	.882	.787	.665	.527	.387
	6	—	—	—	—	—	—	—	—	—	.999	.999	.998	.996	.993	.989	.986	.961	.915	.842	.739	.613
	7	—	—	—	—	—	—	—	—	—	—	—	—	.999	.999	.998	.997	.991	.974	.943	.888	.806
	8	—	—	—	—	—	—	—	—	—	—	—	—	—	—	—	—	.998	.994	.985	.964	.927
	9	—	—	—	—	—	—	—	—	—	—	—	—	—	—	—	—	—	.999	.997	.992	.981
	10	—	—	—	—	—	—	—	—	—	—	—	—	—	—	—	—	—	—	—	.999	.997
	11	—	—	—	—	—	—	—	—	—	—	—	—	—	—	—	—	—	—	—	—	—
	12	—	—	—	—	—	—	—	—	—	—	—	—	—	—	—	—	—	—	—	—	—
13	0	.878	.769	.588	.513	.447	.338	.254	.190	.141	.121	.104	.076	.055	.040	.028	.024	.010	.004	.001	.000	.000
	1	.993	.973	.907	.865	.819	.721	.621	.526	.439	.398	.360	.292	.234	.185	.144	.127	.064	.030	.013	.005	.002
	2	—	.998	.986	.975	.961	.920	.866	.802	.730	.692	.654	.577	.502	.430	.364	.333	.202	.113	.058	.027	.011
	3	—	—	.999	.997	.994	.984	.966	.939	.903	.882	.859	.806	.747	.684	.618	.584	.421	.278	.169	.093	.046
	4	—	—	—	—	.999	.998	.994	.986	.974	.966	.956	.932	.901	.863	.818	.794	.654	.501	.353	.228	.133
	5	—	—	—	—	—	—	.999	.998	.995	.992	.990	.982	.970	.954	.932	.920	.835	.716	.574	.427	.291
	6	—	—	—	—	—	—	—	—	.999	.999	.998	.996	.993	.988	.981	.976	.938	.871	.771	.644	.500
	7	—	—	—	—	—	—	—	—	—	—	—	.999	.999	.998	.996	.994	.982	.954	.902	.821	.709
	8	—	—	—	—	—	—	—	—	—	—	—	—	—	—	.999	.999	.996	.987	.968	.930	.867
	9	—	—	—	—	—	—	—	—	—	—	—	—	—	—	—	—	.999	.997	.992	.980	.954
	10	—	—	—	—	—	—	—	—	—	—	—	—	—	—	—	—	—	—	.999	.996	.989
	11	—	—	—	—	—	—	—	—	—	—	—	—	—	—	—	—	—	—	—	.999	.998
	12	—	—	—	—	—	—	—	—	—	—	—	—	—	—	—	—	—	—	—	—	—
	13	—	—	—	—	—	—	—	—	—	—	—	—	—	—	—	—	—	—	—	—	—
14	0	.869	.754	.565	.488	.421	.311	.229	.167	.121	.103	.087	.062	.044	.031	.021	.018	.007	.002	.001	.000	.000
	1	.992	.969	.894	.847	.796	.690	.585	.486	.397	.357	.319	.253	.198	.153	.116	.101	.047	.021	.008	.003	.001
	2	1	.998	.983	.970	.952	.904	.842	.768	.689	.648	.607	.526	.448	.376	.311	.281	.161	.084	.040	.017	.006
	3	1	1	.998	.996	.992	.979	.956	.923	.879	.853	.826	.765	.698	.628	.557	.521	.355	.220	.124	.063	.029

n = 14

r	.50	.45	.40	.35	.30	.25	.24	.22	.20	.18	.16	.15	.14	.12	.10	.08	.06	.05	.04	.02	.01
4	.090	.167	.279	.423	.584	.742	.770	.824	.870	.909	.941	.953	.964	.980	.991	.996	.999	1	1	1	1
5	.212	.337	.486	.641	.781	.888	.905	.934	.956	.973	.984	.988	.992	.996	.999	1	1	—	—	—	—
6	.395	.546	.692	.816	.907	.962	.969	.980	.988	.994	.997	.998	.999	.999	1	—	—	—	—	—	—
7	.605	.741	.850	.925	.969	.990	.992	.995	.998	.999	.999	—	—	—	—	—	—	—	—	—	—
8	.788	.881	.942	.976	.992	.998	.998	.999	—	—	—	—	—	—	—	—	—	—	—	—	—
9	.910	.957	.982	.994	.998	—	—	—	—	—	—	—	—	—	—	—	—	—	—	—	—
10	.971	.989	.996	.999	—	—	—	—	—	—	—	—	—	—	—	—	—	—	—	—	—
11	.994	.998	.999	—	—	—	—	—	—	—	—	—	—	—	—	—	—	—	—	—	—
12	.999	—	—	—	—	—	—	—	—	—	—	—	—	—	—	—	—	—	—	—	—
13	—	—	—	—	—	—	—	—	—	—	—	—	—	—	—	—	—	—	—	—	—
14	—	—	—	—	—	—	—	—	—	—	—	—	—	—	—	—	—	—	—	—	—

n = 15

r	.50	.45	.40	.35	.30	.25	.24	.22	.20	.18	.16	.15	.14	.12	.10	.08	.06	.05	.04	.02	.01
0	.000	.000	.000	.002	.005	.013	.016	.024	.035	.051	.073	.087	.104	.147	.206	.286	.395	.463	.542	.739	.860
1	.000	.002	.005	.014	.035	.080	.094	.126	.167	.219	.282	.319	.358	.448	.549	.660	.774	.829	.881	.965	.990
2	.004	.011	.027	.062	.127	.236	.264	.327	.398	.477	.561	.604	.648	.735	.816	.887	.943	.964	.980	.997	—
3	.018	.042	.091	.173	.297	.461	.498	.573	.648	.722	.791	.823	.852	.904	.944	.973	.990	.995	.998	—	—
4	.059	.120	.217	.352	.515	.686	.719	.781	.836	.883	.922	.938	.952	.974	.987	.995	.999	.999	—	—	—
5	.151	.261	.403	.564	.722	.852	.873	.910	.939	.961	.977	.983	.988	.994	.998	.999	—	—	—	—	—
6	.304	.452	.610	.755	.869	.943	.954	.970	.982	.990	.995	.996	.998	.999	—	—	—	—	—	—	—
7	.500	.654	.787	.887	.950	.983	.987	.992	.996	.998	.999	.999	—	—	—	—	—	—	—	—	—
8	.696	.818	.905	.958	.985	.996	.997	.998	.999	—	—	—	—	—	—	—	—	—	—	—	—
9	.849	.923	.966	.988	.996	.999	.999	—	—	—	—	—	—	—	—	—	—	—	—	—	—
10	.941	.975	.991	.997	.999	—	—	—	—	—	—	—	—	—	—	—	—	—	—	—	—
11	.982	.994	.998	—	—	—	—	—	—	—	—	—	—	—	—	—	—	—	—	—	—
12	.996	.999	—	—	—	—	—	—	—	—	—	—	—	—	—	—	—	—	—	—	—
13	—	—	—	—	—	—	—	—	—	—	—	—	—	—	—	—	—	—	—	—	—
14	—	—	—	—	—	—	—	—	—	—	—	—	—	—	—	—	—	—	—	—	—
15	—	—	—	—	—	—	—	—	—	—	—	—	—	—	—	—	—	—	—	—	—

n = 16

r	.50	.45	.40	.35	.30	.25	.24	.22	.20	.18	.16	.15	.14	.12	.10	.08	.06	.05	.04	.02	.01
0	.000	.000	.000	.001	.003	.010	.012	.019	.028	.042	.061	.074	.090	.129	.185	.263	.372	.440	.520	.724	.851
1	.000	.001	.003	.010	.026	.063	.075	.103	.141	.189	.249	.284	.323	.412	.515	.630	.751	.811	.867	.960	.989
2	.002	.007	.018	.045	.099	.197	.223	.283	.352	.430	.516	.561	.607	.700	.789	.866	.933	.957	.976	.996	.999
3	.011	.028	.065	.134	.246	.405	.442	.519	.598	.678	.754	.790	.824	.884	.932	.966	.987	.993	.997	—	—
4	.038	.085	.167	.289	.450	.630	.666	.735	.798	.854	.901	.921	.938	.965	.983	.993	.998	.999	—	—	—
5	.105	.198	.329	.490	.660	.810	.836	.881	.918	.947	.968	.976	.983	.992	.997	.999	—	—	—	—	—
6	.227	.366	.527	.688	.825	.920	.934	.957	.973	.985	.992	.994	.996	.998	.999	—	—	—	—	—	—
7	.402	.563	.716	.841	.926	.973	.979	.987	.993	.996	.998	.999	.999	—	—	—	—	—	—	—	—
8	.598	.744	.858	.933	.974	.993	.994	.997	.999	.999	—	—	—	—	—	—	—	—	—	—	—

n = 16

r	.01	.02	.04	.05	.06	.08	.10	.12	.14	.15	.16	.18	.20	.22	.24	.25	.30	.35	.40	.45	.50
9	—	—	—	—	—	—	—	—	—	—	—	—	—	.999	.999	.998	.993	.977	.942	.876	.773
10	—	—	—	—	—	—	—	—	—	—	—	—	—	—	—	—	.998	.994	.981	.951	.895
11	—	—	—	—	—	—	—	—	—	—	—	—	—	—	—	—	—	.999	.995	.985	.962
12	—	—	—	—	—	—	—	—	—	—	—	—	—	—	—	—	—	—	.999	.997	.989
13	—	—	—	—	—	—	—	—	—	—	—	—	—	—	—	—	—	—	—	.999	.998
14	—	—	—	—	—	—	—	—	—	—	—	—	—	—	—	—	—	—	—	—	—
15	—	—	—	—	—	—	—	—	—	—	—	—	—	—	—	—	—	—	—	—	—
16	—	—	—	—	—	—	—	—	—	—	—	—	—	—	—	—	—	—	—	—	—

n = 17

r	.01	.02	.04	.05	.06	.08	.10	.12	.14	.15	.16	.18	.20	.22	.24	.25	.30	.35	.40	.45	.50
0	.843	.709	.500	.418	.349	.242	.167	.114	.077	.063	.052	.034	.023	.015	.009	.008	.002	.001	.000	.000	.000
1	.988	.955	.853	.792	.728	.601	.482	.378	.290	.252	.219	.162	.118	.085	.060	.050	.019	.007	.002	.001	.000
2	.999	.996	.971	.950	.922	.850	.762	.665	.568	.520	.473	.387	.310	.243	.188	.164	.077	.033	.012	.004	.001
3	—	—	.996	.991	.984	.958	.917	.862	.793	.756	.716	.633	.549	.467	.389	.353	.202	.103	.046	.018	.006
4	—	—	—	.999	.997	.991	.978	.955	.922	.901	.878	.822	.758	.687	.612	.574	.389	.235	.126	.060	.025
5	—	—	—	—	—	.999	.995	.989	.977	.968	.958	.931	.894	.849	.795	.765	.597	.420	.264	.147	.072
6	—	—	—	—	—	—	.999	.998	.994	.992	.988	.978	.962	.940	.911	.893	.775	.619	.448	.290	.166
7	—	—	—	—	—	—	—	—	.999	.998	.997	.994	.989	.981	.968	.960	.895	.787	.641	.474	.315
8	—	—	—	—	—	—	—	—	—	—	—	.999	.997	.995	.991	.988	.960	.901	.801	.663	.500
9	—	—	—	—	—	—	—	—	—	—	—	—	—	.999	.998	.997	.987	.962	.908	.817	.685
10	—	—	—	—	—	—	—	—	—	—	—	—	—	—	—	.999	.997	.988	.965	.917	.834
11	—	—	—	—	—	—	—	—	—	—	—	—	—	—	—	—	.999	.997	.989	.970	.928
12	—	—	—	—	—	—	—	—	—	—	—	—	—	—	—	—	—	.999	.997	.991	.975
13	—	—	—	—	—	—	—	—	—	—	—	—	—	—	—	—	—	—	—	.998	.994
14	—	—	—	—	—	—	—	—	—	—	—	—	—	—	—	—	—	—	—	—	.999
15	—	—	—	—	—	—	—	—	—	—	—	—	—	—	—	—	—	—	—	—	—
16	—	—	—	—	—	—	—	—	—	—	—	—	—	—	—	—	—	—	—	—	—
17	—	—	—	—	—	—	—	—	—	—	—	—	—	—	—	—	—	—	—	—	—

n = 18

r	.01	.02	.04	.05	.06	.08	.10	.12	.14	.15	.16	.18	.20	.22	.24	.25	.30	.35	.40	.45	.50
0	.835	.695	.480	.397	.328	.223	.150	.100	.066	.054	.043	.028	.018	.011	.007	.006	.002	.000	.000	.000	.000
1	.986	.950	.839	.774	.706	.572	.450	.346	.260	.224	.192	.139	.099	.069	.048	.039	.014	.005	.001	.000	.000
2	.999	.995	.967	.942	.910	.830	.734	.631	.529	.480	.433	.346	.271	.208	.157	.135	.060	.024	.008	.003	.001
3	—	—	.995	.989	.980	.949	.902	.838	.762	.720	.677	.589	.501	.418	.341	.306	.165	.078	.033	.012	.004
4	—	—	.999	.998	.997	.988	.972	.944	.904	.879	.852	.788	.716	.639	.559	.519	.333	.189	.094	.041	.015
5	—	—	—	—	—	.998	.994	.985	.969	.958	.945	.911	.867	.813	.751	.717	.534	.355	.209	.108	.048
6	—	—	—	—	—	—	.999	.997	.992	.988	.983	.969	.949	.920	.883	.861	.722	.549	.374	.226	.119
7	—	—	—	—	—	—	—	.999	.998	.997	.996	.991	.984	.972	.954	.943	.859	.728	.563	.391	.240
8	—	—	—	—	—	—	—	—	—	.999	.999	.998	.996	.992	.985	.981	.940	.861	.737	.578	.407
9	—	—	—	—	—	—	—	—	—	—	—	—	.999	.998	.996	.995	.979	.940	.865	.747	.593
10	—	—	—	—	—	—	—	—	—	—	—	—	—	—	.999	.999	.994	.979	.942	.872	.760
11	—	—	—	—	—	—	—	—	—	—	—	—	—	—	—	—	.999	.994	.980	.946	.881

Values of p:

r	n	.50	.45	.40	.35	.30	.25	.24	.22	.20	.18	.16	.15	.14	.12	.10	.08	.06	.05	.04	.02	.01
12	18	.952	.982	.994	.999	—	—	—	—	—	—	—	—	—	—	—	—	—	—	—	—	—
13		.985	.995	.999	—	—	—	—	—	—	—	—	—	—	—	—	—	—	—	—	—	—
14		.996	.999	—	—	—	—	—	—	—	—	—	—	—	—	—	—	—	—	—	—	—
15		.999	—	—	—	—	—	—	—	—	—	—	—	—	—	—	—	—	—	—	—	—
16		—	—	—	—	—	—	—	—	—	—	—	—	—	—	—	—	—	—	—	—	—
17		—	—	—	—	—	—	—	—	—	—	—	—	—	—	—	—	—	—	—	—	—
18		—	—	—	—	—	—	—	—	—	—	—	—	—	—	—	—	—	—	—	—	—
0	19	.000	.000	.000	.000	.001	.004	.005	.009	.014	.023	.036	.046	.057	.088	.135	.205	.309	.377	.460	.681	.826
1		.000	.000	.001	.003	.010	.031	.038	.057	.083	.119	.168	.198	.233	.317	.420	.544	.683	.755	.825	.945	.985
2		.000	.002	.005	.017	.046	.111	.131	.178	.237	.309	.394	.441	.491	.597	.705	.809	.898	.933	.962	.994	.999
3		.002	.008	.023	.059	.133	.263	.297	.372	.455	.545	.638	.684	.729	.813	.885	.940	.976	.987	.994	—	—
4		.010	.028	.070	.150	.282	.465	.506	.590	.673	.752	.824	.856	.884	.931	.965	.985	.996	.998	.999	—	—
5		.032	.078	.163	.297	.474	.668	.705	.775	.837	.889	.930	.946	.960	.980	.991	.997	.999	—	—	—	—
6		.084	.173	.308	.481	.666	.825	.851	.897	.932	.959	.977	.984	.989	.995	.998	—	—	—	—	—	—
7		.180	.317	.488	.666	.818	.923	.937	.960	.977	.987	.994	.996	.997	.999	—	—	—	—	—	—	—
8		.324	.494	.667	.815	.916	.971	.978	.987	.993	.997	.999	.999	.999	—	—	—	—	—	—	—	—
9		.500	.671	.814	.913	.967	.991	.993	.997	.998	.999	—	—	—	—	—	—	—	—	—	—	—
10		.676	.816	.912	.965	.989	.998	.998	.999	—	—	—	—	—	—	—	—	—	—	—	—	—
11		.820	.913	.965	.989	.997	—	—	—	—	—	—	—	—	—	—	—	—	—	—	—	—
12		.916	.966	.988	.997	.999	—	—	—	—	—	—	—	—	—	—	—	—	—	—	—	—
13		.968	.989	.997	.999	—	—	—	—	—	—	—	—	—	—	—	—	—	—	—	—	—
14		.990	.997	.999	—	—	—	—	—	—	—	—	—	—	—	—	—	—	—	—	—	—
15		.998	.999	—	—	—	—	—	—	—	—	—	—	—	—	—	—	—	—	—	—	—
16		—	—	—	—	—	—	—	—	—	—	—	—	—	—	—	—	—	—	—	—	—
17		—	—	—	—	—	—	—	—	—	—	—	—	—	—	—	—	—	—	—	—	—
18		—	—	—	—	—	—	—	—	—	—	—	—	—	—	—	—	—	—	—	—	—
19		—	—	—	—	—	—	—	—	—	—	—	—	—	—	—	—	—	—	—	—	—
0	20	.000	.000	.000	.000	.001	.003	.004	.007	.012	.019	.031	.039	.049	.078	.122	.189	.290	.358	.442	.668	.818
1		.000	.000	.001	.002	.008	.024	.030	.046	.069	.102	.147	.176	.208	.289	.392	.517	.660	.736	.810	.940	.983
2		.000	.001	.004	.012	.035	.091	.109	.151	.206	.275	.358	.405	.455	.563	.677	.788	.885	.925	.956	.993	.999
3		.001	.005	.016	.044	.107	.225	.257	.329	.411	.503	.599	.648	.696	.787	.867	.929	.971	.984	.993	.999	—
4		.006	.019	.051	.118	.238	.415	.456	.542	.630	.715	.794	.830	.863	.917	.957	.982	.994	.997	.999	—	—
5		.021	.055	.126	.245	.416	.617	.657	.734	.804	.864	.913	.933	.949	.974	.989	.996	.999	—	—	—	—
6		.058	.130	.250	.417	.608	.786	.816	.870	.913	.946	.970	.978	.985	.993	.998	.999	—	—	—	—	—
7		.132	.252	.416	.601	.772	.898	.917	.946	.968	.982	.991	.994	.996	.999	—	—	—	—	—	—	—
8		.252	.414	.596	.762	.887	.959	.968	.981	.990	.995	.998	.999	.999	—	—	—	—	—	—	—	—
9		.412	.591	.755	.878	.952	.986	.990	.995	.997	.999	—	—	—	—	—	—	—	—	—	—	—
10		.588	.751	.872	.947	.983	.996	.997	.999	.999	—	—	—	—	—	—	—	—	—	—	—	—

n	r	.01	.02	.04	.05	.06	.08	.10	.12	.14	.15	.16	.18	.20	.22	.24	.25	.30	.35	.40	.45	.50
20	11	1	1	1	1	1	1	1	1	1	1	1	1	1	1	.999	.999	.995	.980	.943	.869	.748
	12	1	1	1	1	1	1	1	1	1	1	1	1	1	1	1	1	.999	.994	.979	.942	.868
	13	1	1	1	1	1	1	1	1	1	1	1	1	1	1	1	1	1	.998	.994	.979	.942
	14	1	1	1	1	1	1	1	1	1	1	1	1	1	1	1	1	1	1	.998	.994	.979
	15	1	1	1	1	1	1	1	1	1	1	1	1	1	1	1	1	1	1	1	.998	.994
	16	1	1	1	1	1	1	1	1	1	1	1	1	1	1	1	1	1	1	1	1	.999
	17	1	1	1	1	1	1	1	1	1	1	1	1	1	1	1	1	1	1	1	1	1
	18	1	1	1	1	1	1	1	1	1	1	1	1	1	1	1	1	1	1	1	1	1
	19	1	1	1	1	1	1	1	1	1	1	1	1	1	1	1	1	1	1	1	1	1
	20	1	1	1	1	1	1	1	1	1	1	1	1	1	1	1	1	1	1	1	1	1
25	0	.778	.603	.360	.277	.213	.124	.072	.041	.023	.017	.013	.007	.004	.002	.001	.001	.000	.000	.000	.000	.000
	1	.974	.911	.736	.642	.553	.395	.271	.180	.117	.093	.074	.045	.027	.016	.009	.007	.002	.000	.000	.000	.000
	2	.998	.987	.924	.873	.813	.677	.537	.409	.300	.254	.213	.147	.098	.064	.041	.032	.009	.002	.000	.000	.000
	3	1	.999	.983	.966	.940	.865	.764	.648	.529	.471	.416	.317	.234	.168	.117	.096	.033	.010	.002	.000	.000
	4	1	1	.997	.993	.985	.955	.902	.827	.733	.682	.629	.523	.421	.328	.248	.214	.090	.032	.009	.002	.000
	5	1	1	1	.999	.997	.988	.967	.929	.873	.838	.800	.712	.617	.518	.423	.378	.193	.083	.029	.009	.002
	6	1	1	1	1	.999	.997	.991	.976	.949	.930	.908	.851	.780	.697	.607	.561	.341	.173	.074	.026	.007
	7	1	1	1	1	1	.999	.998	.993	.983	.975	.964	.934	.891	.834	.765	.727	.512	.306	.154	.064	.022
	8	1	1	1	1	1	1	1	.998	.995	.992	.988	.975	.953	.921	.877	.851	.677	.467	.274	.134	.054
	9	1	1	1	1	1	1	1	1	.999	.998	.997	.992	.983	.967	.944	.929	.811	.630	.425	.242	.115
	10	1	1	1	1	1	1	1	1	1	1	.999	.998	.994	.988	.978	.970	.902	.771	.586	.384	.212
	11	1	1	1	1	1	1	1	1	1	1	1	.999	.998	.996	.992	.989	.956	.875	.732	.543	.345
	12	1	1	1	1	1	1	1	1	1	1	1	1	1	.999	.998	.997	.983	.940	.846	.694	.500
	13	1	1	1	1	1	1	1	1	1	1	1	1	1	1	.999	.999	.994	.975	.922	.817	.655
	14	1	1	1	1	1	1	1	1	1	1	1	1	1	1	1	1	.998	.991	.966	.904	.788
	15	1	1	1	1	1	1	1	1	1	1	1	1	1	1	1	1	1	.997	.987	.956	.885
	16	1	1	1	1	1	1	1	1	1	1	1	1	1	1	1	1	1	.999	.996	.983	.946
	17	1	1	1	1	1	1	1	1	1	1	1	1	1	1	1	1	1	1	.999	.994	.978
	18	1	1	1	1	1	1	1	1	1	1	1	1	1	1	1	1	1	1	1	.998	.993
	19	1	1	1	1	1	1	1	1	1	1	1	1	1	1	1	1	1	1	1	1	.998
	20	1	1	1	1	1	1	1	1	1	1	1	1	1	1	1	1	1	1	1	1	1
	21	1	1	1	1	1	1	1	1	1	1	1	1	1	1	1	1	1	1	1	1	1
	22	1	1	1	1	1	1	1	1	1	1	1	1	1	1	1	1	1	1	1	1	1
	23	1	1	1	1	1	1	1	1	1	1	1	1	1	1	1	1	1	1	1	1	1
	24	1	1	1	1	1	1	1	1	1	1	1	1	1	1	1	1	1	1	1	1	1
	25	1	1	1	1	1	1	1	1	1	1	1	1	1	1	1	1	1	1	1	1	1

APPENDIX C

Poisson Distribution: Individual Probabilities

$$P_P(X = k) = \frac{e^{-m}m^k}{k!}$$

Figure C.1

Using Appendix C to find $P_P(X = 2 \mid m = 1.5)$.

k \ m	....	1.4	1.5	1.6
0		.247	.223	.202
1		.345	.335	.323
2	....	.242	.251	.258
3		.113	.126	.138
.		.	.	.
.		.	.	.

m

x	0.00	0.00	0.00	0.00	0.00	0.00	0.01	0.01	0.01	0.01	0.02	0.03	0.04	0.05	0.06
0	.999	.998	.997	.996	.995	.994	.993	.992	.991	.990	.980	.970	.961	.951	.942
1	.001	.002	.003	.004	.005	.006	.007	.008	.009	.010	.020	.029	.038	.048	.057
2													.001	.001	.002

m

x	0.07	0.08	0.09	0.10	0.15	0.20	0.25	0.30	0.40	0.50	0.60	0.70	0.80	0.90	1.00
0	.932	.923	.914	.905	.861	.819	.779	.741	.670	.607	.549	.497	.449	.407	.368
1	.065	.074	.082	.090	.129	.164	.195	.222	.268	.303	.329	.348	.359	.366	.368
2	.002	.003	.004	.005	.010	.016	.024	.033	.054	.076	.099	.122	.144	.165	.184
3						.001	.002	.003	.007	.013	.020	.028	.038	.049	.061
4									.001	.002	.003	.005	.008	.011	.015
5												.001	.001	.002	.003
6															.001

m

x	1.1	1.2	1.3	1.4	1.5	1.6	1.7	1.8	1.9	2.0	2.1	2.2	2.3	2.4	2.5
0	.333	.301	.273	.247	.223	.202	.183	.165	.150	.135	.122	.111	.100	.091	.082
1	.366	.361	.354	.345	.335	.323	.311	.298	.284	.271	.257	.244	.231	.218	.205
2	.201	.217	.230	.242	.251	.258	.264	.268	.270	.271	.270	.268	.265	.261	.257
3	.074	.087	.100	.113	.126	.138	.150	.161	.171	.180	.189	.197	.203	.209	.214
4	.020	.026	.032	.039	.047	.055	.064	.072	.081	.090	.099	.108	.117	.125	.134
5	.004	.006	.008	.011	.014	.018	.022	.026	.031	.036	.042	.048	.054	.060	.067
6	.001	.001	.002	.003	.004	.005	.006	.008	.010	.012	.015	.017	.021	.024	.028
7				.001	.001	.001	.001	.002	.003	.003	.004	.005	.007	.008	.010
8									.001	.001	.001	.002	.002	.002	.003
9														.001	.001

x	m 4.0	3.9	3.8	3.7	3.6	3.5	3.4	3.3	3.2	3.1	3.0	2.9	2.8	2.7	2.6
0	.018	.020	.022	.025	.027	.030	.033	.037	.041	.045	.050	.055	.061	.067	.074
1	.073	.079	.085	.091	.098	.106	.113	.122	.130	.140	.149	.160	.170	.181	.193
2	.147	.154	.162	.169	.177	.185	.193	.201	.209	.216	.224	.231	.238	.245	.251
3	.195	.200	.205	.209	.212	.216	.219	.221	.223	.224	.224	.224	.222	.220	.218
4	.195	.195	.194	.193	.191	.189	.186	.182	.178	.173	.168	.162	.156	.149	.141
5	.156	.152	.148	.143	.138	.132	.126	.120	.114	.107	.101	.094	.087	.080	.074
6	.104	.099	.094	.088	.083	.077	.072	.066	.061	.056	.050	.045	.041	.036	.032
7	.060	.055	.051	.047	.042	.039	.035	.031	.028	.025	.022	.019	.016	.014	.012
8	.030	.027	.024	.022	.019	.017	.015	.013	.011	.010	.008	.007	.006	.005	.004
9	.013	.012	.010	.009	.008	.007	.006	.005	.004	.003	.003	.002	.002	.001	.001
10	.005	.005	.004	.003	.003	.002	.002	.002	.001	.001	.001	.001	0	0	0
11	.002	.002	.001	.001	.001	.001	.001	0	0	0	0	0	0	0	0
12	.001	.001	0	0	0	0	0	0	0	0	0	0	0	0	0

x	m 5.5	5.4	5.3	5.2	5.1	5.0	4.9	4.8	4.7	4.6	4.5	4.4	4.3	4.2	4.1
0	.004	.005	.005	.006	.006	.007	.007	.008	.009	.010	.011	.012	.014	.015	.017
1	.022	.024	.026	.029	.031	.034	.036	.040	.043	.046	.050	.054	.058	.063	.068
2	.062	.066	.070	.075	.079	.084	.089	.095	.100	.106	.112	.119	.125	.132	.139
3	.113	.119	.124	.129	.135	.140	.146	.152	.157	.163	.169	.174	.180	.185	.190
4	.156	.160	.164	.168	.172	.175	.179	.182	.185	.188	.190	.192	.193	.194	.195
5	.171	.173	.174	.175	.175	.175	.175	.175	.174	.173	.171	.169	.166	.163	.160
6	.157	.156	.154	.151	.149	.146	.143	.140	.136	.132	.128	.124	.119	.114	.109
7	.123	.120	.116	.113	.109	.104	.100	.096	.091	.087	.082	.078	.073	.069	.064
8	.085	.081	.077	.073	.069	.065	.061	.058	.054	.050	.046	.043	.039	.036	.033
9	.052	.049	.045	.042	.039	.036	.033	.031	.028	.026	.023	.021	.019	.017	.015
10	.029	.026	.024	.022	.020	.018	.016	.015	.013	.012	.010	.009	.008	.007	.006
11	.014	.013	.012	.010	.009	.008	.007	.006	.006	.005	.004	.004	.003	.003	.002
12	.007	.006	.005	.005	.004	.003	.003	.003	.002	.002	.002	.001	.001	.001	.001
13	.003	.002	.002	.002	.002	.001	.001	.001	.001	.001	.001	0	0	0	0
14	.001	.001	.001	.001	.001	0	0	0	0	0	0	0	0	0	0

x	m=5.6	5.7	5.8	5.9	6.0	6.1	6.2	6.3	6.4	6.5	6.6	6.7	6.8	6.9	7.0	x
0	.004	.003	.003	.003	.002	.002	.002	.002	.002	.002	.001	.001	.001	.001	.001	0
1	.021	.019	.018	.016	.015	.014	.013	.012	.011	.010	.009	.008	.008	.007	.006	1
2	.058	.054	.051	.048	.045	.042	.039	.036	.034	.032	.030	.028	.026	.024	.022	2
3	.108	.103	.098	.094	.089	.085	.081	.077	.073	.069	.065	.062	.058	.055	.052	3
4	.152	.147	.143	.138	.134	.129	.125	.121	.116	.112	.108	.103	.099	.095	.091	4
5	.170	.168	.166	.163	.161	.158	.155	.152	.149	.145	.142	.138	.135	.131	.128	5
6	.158	.159	.160	.160	.161	.160	.160	.159	.159	.157	.156	.155	.153	.151	.149	6
7	.127	.130	.133	.135	.138	.140	.142	.144	.145	.146	.147	.148	.149	.149	.149	7
8	.089	.092	.096	.100	.103	.107	.110	.113	.116	.119	.121	.124	.126	.128	.130	8
9	.055	.059	.062	.065	.069	.072	.076	.079	.082	.086	.089	.092	.095	.098	.101	9
10	.031	.033	.036	.039	.041	.044	.047	.050	.053	.056	.059	.062	.065	.068	.071	10
11	.016	.017	.019	.021	.023	.024	.026	.029	.031	.033	.035	.038	.040	.043	.045	11
12	.007	.008	.009	.010	.011	.012	.014	.015	.016	.018	.019	.021	.023	.025	.026	12
13	.003	.004	.004	.005	.005	.006	.007	.007	.008	.009	.010	.011	.012	.013	.014	13
14	.001	.001	.002	.002	.002	.003	.003	.003	.004	.004	.005	.005	.006	.006	.007	14
15	0	.001	.001	.001	.001	.001	.001	.001	.002	.002	.002	.002	.003	.003	.003	15
16	0	0	0	0	0	0	0	.001	.001	.001	.001	.001	.001	.001	.001	16
17	0	0	0	0	0	0	0	0	0	0	0	0	0	.001	.001	17

x	15.0	14.0	13.0	12.0	11.0	10.0	9.5	9.0	8.5	8.0	7.5	7.4	7.3	7.2	*m* 7.1	x
0	.000	.000	.000	.000	.000	.000	.000	.000	.000	.000	.001	.001	.001	.001	.001	0
1	.000	.000	.000	.000	.000	.000	.001	.001	.002	.003	.004	.005	.005	.005	.006	1
2	.000	.000	.000	.000	.001	.002	.003	.005	.007	.011	.016	.017	.018	.019	.021	2
3	.000	.000	.001	.002	.004	.008	.011	.015	.021	.029	.039	.041	.044	.046	.049	3
4	.001	.001	.003	.005	.010	.019	.025	.034	.044	.057	.073	.076	.080	.084	.087	4
5	.002	.004	.007	.013	.022	.038	.048	.061	.075	.092	.109	.113	.117	.120	.124	5
6	.005	.009	.015	.025	.041	.063	.076	.091	.107	.122	.137	.139	.142	.144	.147	6
7	.010	.017	.028	.044	.065	.090	.104	.117	.129	.140	.146	.147	.148	.149	.149	7
8	.019	.030	.046	.066	.089	.113	.123	.132	.138	.140	.137	.136	.135	.134	.132	8
9	.032	.047	.066	.087	.109	.125	.130	.132	.130	.124	.114	.112	.110	.107	.104	9
10	.049	.066	.086	.105	.119	.125	.124	.119	.110	.099	.086	.083	.080	.077	.074	10
11	.066	.084	.101	.114	.119	.114	.107	.097	.085	.072	.059	.056	.053	.050	.048	11
12	.083	.098	.110	.114	.109	.095	.084	.073	.060	.048	.037	.034	.032	.030	.028	12
13	.096	.106	.110	.106	.093	.073	.062	.050	.040	.030	.021	.020	.018	.017	.015	13
14	.102	.106	.102	.090	.073	.052	.042	.032	.024	.017	.011	.010	.009	.009	.008	14
15	.102	.099	.088	.072	.053	.035	.027	.019	.014	.009	.006	.005	.005	.004	.004	15
16	.096	.087	.072	.054	.037	.022	.016	.011	.007	.005	.003	.002	.002	.002	.002	16
17	.085	.071	.055	.038	.024	.013	.009	.006	.004	.002	.001	.001	.001	.001	.001	17
18	.071	.055	.040	.026	.015	.007	.005	.003	.002	.001	0	0	0	0	0	18
19	.056	.041	.027	.016	.008	.004	.002	.001	.001	0	0	0	0	0	0	19
20	.042	.029	.018	.010	.005	.002	.001	.001	0	0	0	0	0	0	0	20
21	.030	.019	.011	.006	.002	.001	0	.001	0	0	0	0	0	0	0	21
22	.020	.012	.006	.003	.001	0	0	0	0	0	0	0	0	0	0	22
23	.013	.007	.004	.002	.001	0	0	0	0	0	0	0	0	0	0	23
24	.008	.004	.002	.001	0	0	0	0	0	0	0	0	0	0	0	24
25	.005	.002	.001	0	0	0	0	0	0	0	0	0	0	0	0	25
26	.003	.001	.001	0	0	0	0	0	0	0	0	0	0	0	0	26
27	.002	.001	0	0	0	0	0	0	0	0	0	0	0	0	0	27
28	.001	0	0	0	0	0	0	0	0	0	0	0	0	0	0	28

APPENDIX D

Poisson Distribution: Cumulative Probabilities

$$P_P(X \le k) = \sum_{i=0}^{k} \frac{e^{-m} m^i}{i!}$$

Figure D.1
Using Appendix D to find $P_P(X \le 2 \mid m = 1.5)$.

k \ m	. . . .	1.4	1.5	1.6
0		.247	.223	.202
1		.592	.558	.525
2	. . . .	.833	.809	.783
3		.946	.934	.921
.		.	.	.
.		.	.	.

m

x	0.00	0.00	0.00	0.00	0.01	0.01	0.01	0.01	0.01	0.01	0.02	0.03	0.04	0.05	0.06
0	.999	.998	.997	.996	.995	.994	.993	.992	.991	.990	.980	.970	.961	.951	.942
1	1	1	1	1	1	1	1	1	1	1	1	1	.999	.999	.998
2													1	1	1

m

x	0.07	0.08	0.09	0.10	0.15	0.20	0.25	0.30	0.40	0.50	0.60	0.70	0.80	0.90	1.00
0	.932	.923	.914	.905	.861	.819	.779	.741	.670	.607	.549	.497	.449	.407	.368
1	.998	.997	.996	.995	.990	.982	.974	.963	.938	.910	.878	.844	.809	.772	.736
2	1	1	1	1	.999	.999	.998	.996	.992	.986	.977	.966	.953	.937	.920
3					1	1	1	1	.999	.998	.997	.994	.991	.987	.981
4									1	1	1	.999	.999	.998	.996
5												1	1	1	.999
6															1

m

x	1.1	1.2	1.3	1.4	1.5	1.6	1.7	1.8	1.9	2.0	2.1	2.2	2.3	2.4	2.5
0	.333	.301	.273	.247	.223	.202	.183	.165	.150	.135	.122	.111	.100	.091	.082
1	.699	.663	.627	.592	.558	.525	.493	.463	.434	.406	.380	.355	.331	.308	.287
2	.900	.879	.857	.833	.809	.783	.757	.731	.704	.677	.650	.623	.596	.570	.544
3	.974	.966	.957	.946	.934	.921	.907	.891	.875	.857	.839	.819	.799	.779	.758
4	.995	.992	.989	.986	.981	.976	.970	.964	.956	.947	.938	.928	.916	.904	.891
5	.999	.998	.998	.997	.996	.994	.992	.990	.987	.983	.980	.975	.970	.964	.958
6	1	1	1	.999	.999	.999	.998	.997	.997	.995	.994	.993	.991	.988	.986
7				1	1	1	1	.999	.999	.999	.999	.998	.997	.997	.996
8								1	1	1	1	1	.999	.999	.999
9													1	1	1

x	m 2.6	2.7	2.8	2.9	3.0	3.1	3.2	3.3	3.4	3.5	3.6	3.7	3.8	3.9	4.0
0	.074	.067	.061	.055	.050	.045	.041	.037	.033	.030	.027	.025	.022	.020	.018
1	.267	.249	.231	.215	.199	.185	.171	.159	.147	.136	.126	.116	.107	.099	.092
2	.518	.494	.469	.446	.423	.401	.380	.359	.340	.321	.303	.285	.269	.253	.238
3	.736	.714	.692	.670	.647	.625	.603	.580	.558	.537	.515	.494	.473	.453	.433
4	.877	.863	.848	.832	.815	.798	.781	.763	.744	.725	.706	.687	.668	.648	.629
5	.951	.943	.935	.926	.916	.906	.895	.883	.871	.858	.844	.830	.816	.801	.785
6	.983	.979	.976	.971	.966	.961	.955	.949	.942	.935	.927	.918	.909	.899	.889
7	.995	.993	.992	.990	.988	.986	.983	.980	.977	.973	.969	.965	.960	.955	.949
8	.999	.998	.998	.997	.996	.995	.994	.993	.992	.990	.988	.986	.984	.981	.979
9	1	.999	.999	.999	.999	.999	.998	.998	.997	.997	.996	.995	.994	.993	.992
10	1	1	1	1	1	1	1	.999	.999	.999	.999	.998	.998	.998	.997
11	1	1	1	1	1	1	1	1	1	1	1	1	.999	.999	.999
12	1	1	1	1	1	1	1	1	1	1	1	1	1	1	1

x	m 4.1	4.2	4.3	4.4	4.5	4.6	4.7	4.8	4.9	5.0	5.1	5.2	5.3	5.4	5.5
0	.017	.015	.014	.012	.011	.010	.009	.008	.007	.007	.006	.006	.005	.005	.004
1	.085	.078	.072	.066	.061	.056	.052	.048	.044	.040	.037	.034	.031	.029	.027
2	.224	.210	.197	.185	.174	.163	.152	.143	.133	.125	.116	.109	.102	.095	.088
3	.414	.395	.377	.359	.342	.326	.310	.294	.279	.265	.251	.238	.225	.213	.202
4	.609	.590	.570	.551	.532	.513	.495	.476	.458	.440	.423	.406	.390	.373	.358
5	.769	.753	.737	.720	.703	.686	.668	.651	.634	.616	.598	.581	.563	.546	.529
6	.879	.867	.856	.844	.831	.818	.805	.791	.777	.762	.747	.732	.717	.702	.686
7	.943	.936	.929	.921	.913	.905	.896	.887	.877	.867	.856	.845	.833	.822	.809
8	.976	.972	.968	.964	.960	.955	.950	.944	.938	.932	.925	.918	.911	.903	.894
9	.990	.989	.987	.985	.983	.980	.978	.975	.972	.968	.964	.960	.956	.951	.946
10	.997	.996	.995	.994	.993	.992	.991	.990	.988	.986	.984	.982	.980	.977	.975
11	.999	.999	.998	.998	.998	.997	.997	.996	.995	.995	.994	.993	.992	.990	.989
12	1	1	.999	.999	.999	.999	.999	.999	.998	.998	.998	.997	.997	.996	.996
13	1	1	1	1	1	1	1	1	.999	.999	.999	.999	.999	.999	.998
14	1	1	1	1	1	1	1	1	1	1	1	1	1	1	.999
15	1	1	1	1	1	1	1	1	1	1	1	1	1	1	1

x	m 5.6	5.7	5.8	5.9	6.0	6.1	6.2	6.3	6.4	6.5	6.6	6.7	6.8	6.9	7.0
0	.004	.003	.003	.003	.002	.002	.002	.002	.002	.002	.001	.001	.001	.001	.001
1	.024	.022	.021	.019	.017	.016	.015	.013	.012	.011	.010	.009	.009	.008	.007
2	.082	.077	.072	.067	.062	.058	.054	.050	.046	.043	.040	.037	.034	.032	.030
3	.191	.180	.170	.160	.151	.143	.134	.126	.119	.112	.105	.099	.093	.087	.082
4	.342	.327	.313	.299	.285	.272	.259	.247	.235	.224	.213	.202	.192	.182	.173
5	.512	.495	.478	.462	.446	.430	.414	.399	.384	.369	.355	.341	.327	.314	.301
6	.670	.654	.638	.622	.606	.590	.574	.558	.542	.527	.511	.495	.480	.465	.450
7	.797	.784	.771	.758	.744	.730	.716	.702	.687	.673	.658	.643	.628	.614	.599
8	.886	.877	.867	.857	.847	.837	.826	.815	.803	.792	.780	.767	.755	.742	.729
9	.941	.935	.929	.923	.916	.909	.902	.894	.886	.877	.869	.860	.850	.840	.830
10	.972	.969	.965	.961	.957	.953	.949	.944	.939	.933	.927	.921	.915	.908	.901
11	.988	.986	.984	.982	.980	.978	.975	.972	.969	.966	.963	.959	.955	.951	.947
12	.995	.994	.993	.992	.991	.990	.989	.987	.986	.984	.982	.980	.978	.976	.973
13	.998	.998	.997	.997	.996	.996	.995	.995	.994	.993	.992	.991	.990	.989	.987
14	.999	.999	.999	.999	.999	.998	.998	.998	.997	.997	.997	.996	.996	.995	.994
15	1	1	1	1	.999	.999	.999	.999	.999	.999	.999	.998	.998	.998	.998
16	1	1	1	1	1	1	1	1	1	1	.999	.999	.999	.999	.999
17	1	1	1	1	1	1	1	1	1	1	1	1	1	1	1

x	7.1	7.2	7.3	7.4	7.5	8.0	8.5	9.0	9.5	10.0	11.0	12.0	13.0	14.0	15.0	x
0	.001	.001	.001	.001	.001	.000	.000	.000	.000	.000	.000	.000	.000	.000	.000	0
1	.007	.006	.006	.005	.005	.003	.002	.001	.001	.000	.000	.000	.000	.000	.000	1
2	.027	.025	.024	.022	.020	.014	.009	.006	.004	.003	.001	.001	.000	.000	.000	2
3	.077	.072	.067	.063	.059	.042	.030	.021	.015	.010	.005	.002	.001	.000	.000	3
4	.164	.156	.147	.140	.132	.100	.074	.055	.040	.029	.015	.008	.004	.002	.001	4
5	.288	.276	.264	.253	.241	.191	.150	.116	.089	.067	.038	.020	.011	.006	.003	5
6	.435	.420	.406	.392	.378	.313	.256	.207	.165	.130	.079	.046	.026	.014	.008	6
7	.584	.569	.554	.539	.525	.453	.386	.324	.269	.220	.143	.090	.054	.032	.018	7
8	.716	.703	.689	.676	.662	.593	.523	.456	.392	.333	.232	.155	.100	.062	.037	8
9	.820	.810	.799	.788	.776	.717	.653	.587	.522	.458	.341	.242	.166	.109	.070	9
10	.894	.887	.879	.871	.862	.816	.763	.706	.645	.583	.460	.347	.252	.176	.118	10
11	.942	.937	.932	.926	.921	.888	.849	.803	.752	.697	.579	.462	.353	.260	.185	11
12	.970	.967	.964	.961	.957	.936	.909	.876	.836	.792	.689	.576	.463	.358	.268	12
13	.986	.984	.982	.980	.978	.966	.949	.926	.898	.864	.781	.682	.573	.464	.363	13
14	.994	.993	.992	.991	.990	.983	.973	.959	.940	.917	.854	.772	.675	.570	.466	14
15	.997	.997	.996	.996	.995	.992	.986	.978	.967	.951	.907	.844	.764	.669	.568	15
16	.999	.999	.999	.998	.998	.996	.993	.989	.982	.973	.944	.899	.835	.756	.664	16
17	1	1	.999	.999	.999	.998	.997	.995	.991	.986	.968	.937	.890	.827	.749	17
18	1	1	1	1	1	.999	.999	.998	.996	.993	.982	.963	.930	.883	.819	18
19	1	1	1	1	1	1	.999	.999	.998	.997	.991	.979	.957	.923	.875	19
20	1	1	1	1	1	1	1	1	.999	.998	.995	.988	.975	.952	.917	20
21	1	1	1	1	1	1	1	1	1	.999	.998	.994	.986	.971	.947	21
22	1	1	1	1	1	1	1	1	1	1	.999	.997	.992	.983	.967	22
23	1	1	1	1	1	1	1	1	1	1	1	.999	.996	.991	.981	23
24	1	1	1	1	1	1	1	1	1	1	1	.999	.998	.995	.989	24
25	1	1	1	1	1	1	1	1	1	1	1	1	.999	.997	.994	25
26	1	1	1	1	1	1	1	1	1	1	1	1	1	.999	.997	26
27	1	1	1	1	1	1	1	1	1	1	1	1	1	.999	.998	27
28	1	1	1	1	1	1	1	1	1	1	1	1	1	1	.999	28
29	1	1	1	1	1	1	1	1	1	1	1	1	1	1	1	29

m

Normal Distribution: Cumulative Probabilities

Figure E.1

Using Appendix E to find $P_N(Z \le 1.25)$.

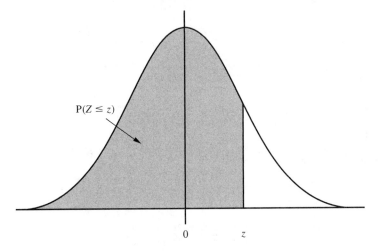

. . .	z	$P(Z \le z)$	. . .
. . .	. . .	. . .	. . .
	1.20	.8849	
	1.21	.8869	
	1.22	.8888	
	1.23	.8907	
	1.24	.8925	
. . .	1.25	.8944	. . .
	1.26	.8962	
	1.27	.8980	
	1.28	.8997	
	1.29	.9015	
. . .	. . .	. . .	. . .

z	P(Z≤z)	z	P(Z≤z)	z	P(Z≤z)	z	P(Z≤z)	z	P(Z≤z)
-3.49	.0002	-2.99	.0014	-2.49	.0064	-1.99	.0233	-1.49	.0681
-3.48	.0003	-2.98	.0014	-2.48	.0066	-1.98	.0239	-1.48	.0694
-3.47	.0003	-2.97	.0015	-2.47	.0068	-1.97	.0244	-1.47	.0708
-3.46	.0003	-2.96	.0015	-2.46	.0069	-1.96	.0250	-1.46	.0721
-3.45	.0003	-2.95	.0016	-2.45	.0071	-1.95	.0256	-1.45	.0735
-3.44	.0003	-2.94	.0016	-2.44	.0073	-1.94	.0262	-1.44	.0749
-3.43	.0003	-2.93	.0017	-2.43	.0075	-1.93	.0268	-1.43	.0764
-3.42	.0003	-2.92	.0018	-2.42	.0078	-1.92	.0274	-1.42	.0778
-3.41	.0003	-2.91	.0018	-2.41	.0080	-1.91	.0281	-1.41	.0793
-3.40	.0003	-2.90	.0019	-2.40	.0082	-1.90	.0287	-1.40	.0808
-3.39	.0003	-2.89	.0019	-2.39	.0084	-1.89	.0294	-1.39	.0823
-3.38	.0004	-2.88	.0020	-2.38	.0087	-1.88	.0301	-1.38	.0838
-3.37	.0004	-2.87	.0021	-2.37	.0089	-1.87	.0307	-1.37	.0853
-3.36	.0004	-2.86	.0021	-2.36	.0091	-1.86	.0314	-1.36	.0869
-3.35	.0004	-2.85	.0022	-2.35	.0094	-1.85	.0322	-1.35	.0885
-3.34	.0004	-2.84	.0023	-2.34	.0096	-1.84	.0329	-1.34	.0901
-3.33	.0004	-2.83	.0023	-2.33	.0099	-1.83	.0336	-1.33	.0918
-3.32	.0005	-2.82	.0024	-2.32	.0102	-1.82	.0344	-1.32	.0934
-3.31	.0005	-2.81	.0025	-2.31	.0104	-1.81	.0351	-1.31	.0951
-3.30	.0005	-2.80	.0026	-2.30	.0107	-1.80	.0359	-1.30	.0968
-3.29	.0005	-2.79	.0026	-2.29	.0110	-1.79	.0367	-1.29	.0985
-3.28	.0005	-2.78	.0027	-2.28	.0113	-1.78	.0375	-1.28	.1003
-3.27	.0005	-2.77	.0028	-2.27	.0116	-1.77	.0384	-1.27	.1020
-3.26	.0006	-2.76	.0029	-2.26	.0119	-1.76	.0392	-1.26	.1038
-3.25	.0006	-2.75	.0030	-2.25	.0122	-1.75	.0401	-1.25	.1056
-3.24	.0006	-2.74	.0031	-2.24	.0125	-1.74	.0409	-1.24	.1075
-3.23	.0006	-2.73	.0032	-2.23	.0129	-1.73	.0418	-1.23	.1093
-3.22	.0006	-2.72	.0033	-2.22	.0132	-1.72	.0427	-1.22	.1112
-3.21	.0007	-2.71	.0034	-2.21	.0136	-1.71	.0436	-1.21	.1131
-3.20	.0007	-2.70	.0035	-2.20	.0139	-1.70	.0446	-1.20	.1151
-3.19	.0007	-2.69	.0036	-2.19	.0143	-1.69	.0455	-1.19	.1170
-3.18	.0007	-2.68	.0037	-2.18	.0146	-1.68	.0465	-1.18	.1190
-3.17	.0008	-2.67	.0038	-2.17	.0150	-1.67	.0475	-1.17	.1210
-3.16	.0008	-2.66	.0039	-2.16	.0154	-1.66	.0485	-1.16	.1230
-3.15	.0008	-2.65	.0040	-2.15	.0158	-1.65	.0495	-1.15	.1251
-3.14	.0008	-2.64	.0041	-2.14	.0162	-1.64	.0505	-1.14	.1271
-3.13	.0009	-2.63	.0043	-2.13	.0166	-1.63	.0516	-1.13	.1292
-3.12	.0009	-2.62	.0044	-2.12	.0170	-1.62	.0526	-1.12	.1314
-3.11	.0009	-2.61	.0045	-2.11	.0174	-1.61	.0537	-1.11	.1335
-3.10	.0010	-2.60	.0047	-2.10	.0179	-1.60	.0548	-1.10	.1357
-3.09	.0010	-2.59	.0048	-2.09	.0183	-1.59	.0559	-1.09	.1379
-3.08	.0010	-2.58	.0049	-2.08	.0188	-1.58	.0571	-1.08	.1401
-3.07	.0011	-2.57	.0051	-2.07	.0192	-1.57	.0582	-1.07	.1423
-3.06	.0011	-2.56	.0052	-2.06	.0197	-1.56	.0594	-1.06	.1446
-3.05	.0011	-2.55	.0054	-2.05	.0202	-1.55	.0606	-1.05	.1469
-3.04	.0012	-2.54	.0055	-2.04	.0207	-1.54	.0618	-1.04	.1492
-3.03	.0012	-2.53	.0057	-2.03	.0212	-1.53	.0630	-1.03	.1515
-3.02	.0013	-2.52	.0059	-2.02	.0217	-1.52	.0643	-1.02	.1539
-3.01	.0013	-2.51	.0060	-2.01	.0222	-1.51	.0655	-1.01	.1562
-3.00	.0013	-2.50	.0062	-2.00	.0228	-1.50	.0668	-1.00	.1587

z	P(Z≤z)	z	P(Z≤z)	z	P(Z≤z)	z	P(Z≤z)	z	P(Z≤z)
-0.99	.1611	-0.49	.3121	0.01	.5040	0.51	.6950	1.01	.8438
-0.98	.1635	-0.48	.3156	0.02	.5080	0.52	.6985	1.02	.8461
-0.97	.1660	-0.47	.3192	0.03	.5120	0.53	.7019	1.03	.8485
-0.96	.1685	-0.46	.3228	0.04	.5160	0.54	.7054	1.04	.8508
-0.95	.1711	-0.45	.3264	0.05	.5199	0.55	.7088	1.05	.8531
-0.94	.1736	-0.44	.3300	0.06	.5239	0.56	.7123	1.06	.8554
-0.93	.1762	-0.43	.3336	0.07	.5279	0.57	.7157	1.07	.8577
-0.92	.1788	-0.42	.3372	0.08	.5319	0.58	.7190	1.08	.8599
-0.91	.1814	-0.41	.3409	0.09	.5359	0.59	.7224	1.09	.8621
-0.90	.1841	-0.40	.3446	0.10	.5398	0.60	.7257	1.10	.8643
-0.89	.1867	-0.39	.3483	0.11	.5438	0.61	.7291	1.11	.8665
-0.88	.1894	-0.38	.3520	0.12	.5478	0.62	.7324	1.12	.8686
-0.87	.1922	-0.37	.3557	0.13	.5517	0.63	.7357	1.13	.8708
-0.86	.1949	-0.36	.3594	0.14	.5557	0.64	.7389	1.14	.8729
-0.85	.1977	-0.35	.3632	0.15	.5596	0.65	.7422	1.15	.8749
-0.84	.2005	-0.34	.3669	0.16	.5636	0.66	.7454	1.16	.8770
-0.83	.2033	-0.33	.3707	0.17	.5675	0.67	.7486	1.17	.8790
-0.82	.2061	-0.32	.3745	0.18	.5714	0.68	.7517	1.18	.8810
-0.81	.2090	-0.31	.3783	0.19	.5753	0.69	.7549	1.19	.8830
-0.80	.2119	-0.30	.3821	0.20	.5793	0.70	.7580	1.20	.8849
-0.79	.2148	-0.29	.3859	0.21	.5832	0.71	.7611	1.21	.8869
-0.78	.2177	-0.28	.3897	0.22	.5871	0.72	.7642	1.22	.8888
-0.77	.2206	-0.27	.3936	0.23	.5910	0.73	.7673	1.23	.8907
-0.76	.2236	-0.26	.3974	0.24	.5948	0.74	.7704	1.24	.8925
-0.75	.2266	-0.25	.4013	0.25	.5987	0.75	.7734	1.25	.8944
-0.74	.2296	-0.24	.4052	0.26	.6026	0.76	.7764	1.26	.8962
-0.73	.2327	-0.23	.4090	0.27	.6064	0.77	.7794	1.27	.8980
-0.72	.2358	-0.22	.4129	0.28	.6103	0.78	.7823	1.28	.8997
-0.71	.2389	-0.21	.4168	0.29	.6141	0.79	.7852	1.29	.9015
-0.70	.2420	-0.20	.4207	0.30	.6179	0.80	.7881	1.30	.9032
-0.69	.2451	-0.19	.4247	0.31	.6217	0.81	.7910	1.31	.9049
-0.68	.2483	-0.18	.4286	0.32	.6255	0.82	.7939	1.32	.9066
-0.67	.2514	-0.17	.4325	0.33	.6293	0.83	.7967	1.33	.9082
-0.66	.2546	-0.16	.4364	0.34	.6331	0.84	.7995	1.34	.9099
-0.65	.2578	-0.15	.4404	0.35	.6368	0.85	.8023	1.35	.9115
-0.64	.2611	-0.14	.4443	0.36	.6406	0.86	.8051	1.36	.9131
-0.63	.2643	-0.13	.4483	0.37	.6443	0.87	.8078	1.37	.9147
-0.62	.2676	-0.12	.4522	0.38	.6480	0.88	.8106	1.38	.9162
-0.61	.2709	-0.11	.4562	0.39	.6517	0.89	.8133	1.39	.9177
-0.60	.2743	-0.10	.4602	0.40	.6554	0.90	.8159	1.40	.9192
-0.59	.2776	-0.09	.4641	0.41	.6591	0.91	.8186	1.41	.9207
-0.58	.2810	-0.08	.4681	0.42	.6628	0.92	.8212	1.42	.9222
-0.57	.2843	-0.07	.4721	0.43	.6664	0.93	.8238	1.43	.9236
-0.56	.2877	-0.06	.4761	0.44	.6700	0.94	.8264	1.44	.9251
-0.55	.2912	-0.05	.4801	0.45	.6736	0.95	.8289	1.45	.9265
-0.54	.2946	-0.04	.4840	0.46	.6772	0.96	.8315	1.46	.9279
-0.53	.2981	-0.03	.4880	0.47	.6808	0.97	.8340	1.47	.9292
-0.52	.3015	-0.02	.4920	0.48	.6844	0.98	.8365	1.48	.9306
-0.51	.3050	-0.01	.4960	0.49	.6879	0.99	.8389	1.49	.9319
-0.50	.3085	0.00	.5000	0.50	.6915	1.00	.8413	1.50	.9332

z	P(Z≤z)	z	P(Z≤z)	z	P(Z≤z)	z	P(Z≤z)
1.51	.9345	2.01	.9778	2.51	.9940	3.01	.9987
1.52	.9357	2.02	.9783	2.52	.9941	3.02	.9987
1.53	.9370	2.03	.9788	2.53	.9943	3.03	.9988
1.54	.9382	2.04	.9793	2.54	.9945	3.04	.9988
1.55	.9394	2.05	.9798	2.55	.9946	3.05	.9989
1.56	.9406	2.06	.9803	2.56	.9948	3.06	.9989
1.57	.9418	2.07	.9808	2.57	.9949	3.07	.9989
1.58	.9429	2.08	.9812	2.58	.9951	3.08	.9990
1.59	.9441	2.09	.9817	2.59	.9952	3.09	.9990
1.60	.9452	2.10	.9821	2.60	.9953	3.10	.9990
1.61	.9463	2.11	.9826	2.61	.9955	3.11	.9991
1.62	.9474	2.12	.9830	2.62	.9956	3.12	.9991
1.63	.9484	2.13	.9834	2.63	.9957	3.13	.9991
1.64	.9495	2.14	.9838	2.64	.9959	3.14	.9992
1.65	.9505	2.15	.9842	2.65	.9960	3.15	.9992
1.66	.9515	2.16	.9846	2.66	.9961	3.16	.9992
1.67	.9525	2.17	.9850	2.67	.9962	3.17	.9992
1.68	.9535	2.18	.9854	2.68	.9963	3.18	.9993
1.69	.9545	2.19	.9857	2.69	.9964	3.19	.9993
1.70	.9554	2.20	.9861	2.70	.9965	3.20	.9993
1.71	.9564	2.21	.9864	2.71	.9966	3.21	.9993
1.72	.9573	2.22	.9868	2.72	.9967	3.22	.9994
1.73	.9582	2.23	.9871	2.73	.9968	3.23	.9994
1.74	.9591	2.24	.9875	2.74	.9969	3.24	.9994
1.75	.9599	2.25	.9878	2.75	.9970	3.25	.9994
1.76	.9608	2.26	.9881	2.76	.9971	3.26	.9994
1.77	.9616	2.27	.9884	2.77	.9972	3.27	.9995
1.78	.9625	2.28	.9887	2.78	.9973	3.28	.9995
1.79	.9633	2.29	.9890	2.79	.9974	3.29	.9995
1.80	.9641	2.30	.9893	2.80	.9974	3.30	.9995
1.81	.9649	2.31	.9896	2.81	.9975	3.31	.9995
1.82	.9656	2.32	.9898	2.82	.9976	3.32	.9995
1.83	.9664	2.33	.9901	2.83	.9977	3.33	.9996
1.84	.9671	2.34	.9904	2.84	.9977	3.34	.9996
1.85	.9678	2.35	.9906	2.85	.9978	3.35	.9996
1.86	.9686	2.36	.9909	2.86	.9979	3.36	.9996
1.87	.9693	2.37	.9911	2.87	.9979	3.37	.9996
1.88	.9699	2.38	.9913	2.88	.9980	3.38	.9996
1.89	.9706	2.39	.9916	2.89	.9981	3.39	.9997
1.90	.9713	2.40	.9918	2.90	.9981	3.40	.9997
1.91	.9719	2.41	.9920	2.91	.9982	3.41	.9997
1.92	.9726	2.42	.9922	2.92	.9982	3.42	.9997
1.93	.9732	2.43	.9925	2.93	.9983	3.43	.9997
1.94	.9738	2.44	.9927	2.94	.9984	3.44	.9997
1.95	.9744	2.45	.9929	2.95	.9984	3.45	.9997
1.96	.9750	2.46	.9931	2.96	.9985	3.46	.9997
1.97	.9756	2.47	.9932	2.97	.9985	3.47	.9997
1.98	.9761	2.48	.9934	2.98	.9986	3.48	.9997
1.99	.9767	2.49	.9936	2.99	.9986	3.49	.9998
2.00	.9772	2.50	.9938	3.00	.9987	3.50	.9998

Beta Distribution: Cumulative Probabilities

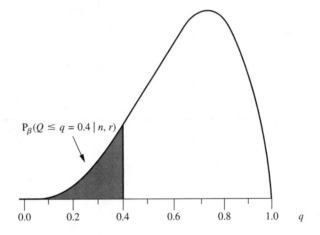

Figure F.1

Using Appendix F to find
$P_\beta(Q \le 0.18 \mid n = 20, r = 6) = 0.11.$

$P_\beta(Q \le q = 0.4 \mid n, r)$

				0.16	0.18	0.20
n	*r*		*q*			
	·			·	·	·
	·			·	·	·
	·			·	·	·
20	2			.83	.88	.92
	4			.36	.45	.54
	6	···		.07	.11	.16
	8			.01	.01	.02
	·			·	·	·
	·			·	·	·

n	r	.50	.48	.46	.44	.42	.40	.38	.36	.34	.32	.30	.28	.26	.25	.24	.22	.20	.18	.16	.14	.12	.10	.08	.06	.04	.02	0
2	1	.50	.48	.46	.44	.42	.40	.38	.36	.34	.32	.30	.28	.26	.25	.24	.22	.20	.18	.16	.14	.12	.10	.08	.06	.04	.02	0
3	1	.75	.73	.71	.69	.66	.64	.61	.59	.56	.54	.51	.48	.45	.44	.42	.39	.36	.33	.29	.26	.22	.19	.15	.12	.08	.04	0
	2	.25	.23	.21	.19	.18	.16	.14	.13	.12	.10	.09	.08	.07	.06	.06	.05	.04	.03	.03	.02	.01	.01	.01	0	0	0	0
4	1	.87	.86	.84	.82	.80	.78	.76	.74	.71	.68	.66	.63	.59	.58	.56	.52	.49	.45	.41	.36	.32	.27	.22	.17	.11	.06	0
	2	.50	.47	.44	.41	.38	.35	.32	.30	.27	.24	.22	.19	.17	.16	.15	.12	.10	.09	.07	.05	.04	.03	.02	.01	0	0	0
	3	.13	.11	.10	.09	.07	.06	.05	.05	.04	.03	.03	.02	.02	.02	.01	.01	.01	.01	0	0	0	0	0	0	0	0	0
5	1	.94	.93	.91	.90	.89	.87	.85	.83	.81	.78	.76	.73	.70	.68	.67	.63	.59	.55	.50	.45	.40	.34	.28	.22	.15	.08	0
	2	.69	.66	.63	.59	.56	.52	.49	.45	.42	.38	.35	.31	.28	.26	.24	.21	.18	.15	.12	.10	.07	.05	.03	.02	.01	0	0
	3	.31	.28	.26	.23	.20	.18	.16	.14	.12	.10	.08	.07	.06	.05	.05	.04	.03	.02	.01	.01	.01	0	0	0	0	0	0
	4	.06	.05	.04	.04	.03	.03	.02	.02	.01	.01	.01	.01	0	0	0	0	0	0	0	0	0	0	0	0	0	0	0
6	1	.97	.96	.95	.94	.93	.92	.91	.89	.87	.85	.83	.80	.78	.76	.74	.71	.67	.63	.58	.53	.47	.41	.34	.26	.18	.09	0
	2	.81	.79	.76	.73	.70	.66	.63	.59	.55	.51	.47	.43	.39	.37	.35	.30	.26	.22	.18	.15	.11	.08	.05	.03	.01	0	0
	3	.50	.46	.43	.39	.35	.32	.28	.25	.22	.19	.16	.14	.11	.10	.09	.07	.06	.04	.03	.02	.01	.01	0	0	0	0	0
	4	.19	.16	.14	.12	.10	.09	.07	.06	.05	.04	.03	.02	.02	.02	.01	.01	.01	0	0	0	0	0	0	0	0	0	0
	5	.03	.03	.02	.02	.01	.01	.01	.01	0	0	0	0	0	0	0	0	0	0	0	0	0	0	0	0	0	0	0
7	1	.98	.98	.97	.97	.96	.95	.94	.93	.92	.90	.88	.86	.83	.82	.81	.77	.74	.69	.65	.59	.53	.47	.39	.31	.22	.11	0
	2	.89	.87	.85	.82	.80	.77	.73	.70	.66	.62	.58	.54	.49	.47	.44	.39	.34	.30	.25	.20	.16	.11	.08	.05	.02	.01	0
	3	.66	.62	.58	.54	.50	.46	.41	.37	.33	.29	.26	.22	.19	.17	.15	.12	.10	.08	.06	.04	.03	.02	.01	0	0	0	0
	4	.34	.31	.27	.24	.21	.18	.15	.13	.11	.09	.07	.06	.04	.04	.03	.02	.02	.01	.01	.01	0	0	0	0	0	0	0
	5	.11	.09	.08	.06	.05	.04	.03	.03	.02	.01	.01	.01	.01	.01	0	0	0	0	0	0	0	0	0	0	0	0	0
	6	.02	.01	.01	.01	.01	0	0	0	0	0	0	0	0	0	0	0	0	0	0	0	0	0	0	0	0	0	0
8	1	.99	.99	.98	.98	.98	.97	.96	.95	.94	.93	.92	.90	.88	.86	.85	.82	.79	.75	.70	.65	.59	.52	.44	.35	.25	.13	0
	2	.94	.92	.91	.89	.87	.84	.81	.78	.75	.71	.67	.63	.58	.56	.53	.48	.42	.37	.31	.26	.20	.15	.10	.06	.03	.01	0
	3	.77	.74	.70	.66	.62	.58	.54	.49	.44	.40	.35	.31	.26	.24	.22	.18	.15	.12	.09	.06	.04	.03	.01	.01	0	0	0
	4	.50	.46	.41	.37	.33	.29	.25	.22	.18	.15	.13	.10	.08	.07	.06	.05	.03	.02	.02	.01	.01	0	0	0	0	0	0
	5	.23	.20	.17	.14	.12	.10	.08	.06	.05	.04	.03	.02	.02	.01	.01	.01	0	0	0	0	0	0	0	0	0	0	0
	6	.06	.05	.04	.03	.02	.02	.01	.01	.01	.01	0	0	0	0	0	0	0	0	0	0	0	0	0	0	0	0	0
	7	.01	.01	0	0	0	0	0	0	0	0	0	0	0	0	0	0	0	0	0	0	0	0	0	0	0	0	0

n	r	1.0	.98	.96	.94	.92	.90	.88	.86	.84	.82	.80	.78	.76	.75	.74	.72	.70	.68	.66	.64	.62	.60	.58	.56	.54	.52
2	1	1	.98	.96	.94	.92	.90	.88	.86	.84	.82	.80	.78	.76	.75	.74	.72	.70	.68	.66	.64	.62	.60	.58	.56	.54	.52
3	1	1	1	1	1	.99	.99	.99	.98	.97	.97	.96	.95	.94	.94	.93	.92	.91	.90	.88	.87	.86	.84	.82	.81	.79	.77
	2	1	.96	.92	.88	.85	.81	.77	.74	.71	.67	.64	.61	.58	.56	.55	.52	.49	.46	.44	.41	.38	.36	.34	.31	.29	.27
4	1	1	1	1	1	1	1	1	1	1	.99	.99	.99	.99	.98	.98	.98	.97	.97	.96	.95	.95	.94	.93	.91	.90	.89
	2	1	1	1	.99	.98	.97	.96	.95	.93	.91	.90	.88	.85	.84	.83	.81	.78	.76	.73	.70	.68	.65	.62	.59	.56	.53
	3	1	.94	.88	.83	.78	.73	.68	.64	.59	.55	.51	.47	.44	.42	.41	.37	.34	.31	.29	.26	.24	.22	.20	.18	.16	.14
5	1	1	1	1	1	1	1	1	1	1	1	1	1	1	1	1	.99	.99	.99	.99	.98	.98	.97	.97	.96	.96	.95
	2	1	1	1	1	1	1	.99	.99	.99	.98	.97	.96	.95	.95	.94	.93	.92	.90	.88	.86	.84	.82	.80	.77	.74	.72
	3	1	1	.99	.98	.97	.95	.93	.90	.88	.85	.82	.79	.76	.74	.72	.69	.65	.62	.58	.55	.51	.48	.44	.41	.37	.34
	4	1	.92	.85	.78	.72	.66	.60	.55	.50	.45	.41	.37	.33	.32	.30	.27	.24	.21	.19	.17	.15	.13	.11	.10	.09	.07
6	1	1	1	1	1	1	1	1	1	1	1	1	1	1	1	1	1	1	1	1	.99	.99	.99	.99	.98	.98	.97
	2	1	1	1	1	1	1	1	1	1	1	.99	.99	.99	.98	.98	.98	.97	.96	.95	.94	.93	.91	.90	.88	.86	.84
	3	1	1	1	1	1	.99	.99	.98	.97	.96	.94	.93	.91	.90	.89	.86	.84	.81	.78	.75	.72	.68	.65	.61	.57	.54
	4	1	1	.99	.97	.95	.92	.89	.85	.82	.78	.74	.70	.65	.63	.61	.57	.53	.49	.45	.41	.37	.34	.30	.27	.24	.21
	5	1	.90	.82	.73	.66	.59	.53	.47	.42	.37	.33	.29	.25	.24	.22	.19	.17	.15	.13	.11	.09	.08	.07	.06	.05	.04
7	1	1	1	1	1	1	1	1	1	1	1	1	1	1	1	1	1	1	1	1	1	1	1	.99	.99	.99	.99
	2	1	1	1	1	1	1	1	1	1	1	1	1	1	1	.99	.99	.99	.99	.98	.97	.97	.96	.95	.94	.92	.91
	3	1	1	1	1	1	1	1	1	.99	.99	.98	.98	.97	.96	.96	.94	.93	.91	.89	.87	.85	.82	.79	.76	.73	.69
	4	1	1	1	1	.99	.98	.97	.96	.94	.92	.90	.88	.85	.83	.81	.78	.74	.71	.67	.63	.59	.54	.50	.46	.42	.38
	5	1	.99	.98	.95	.92	.89	.84	.80	.75	.70	.66	.61	.56	.53	.51	.46	.42	.38	.34	.30	.27	.23	.20	.18	.15	.13
	6	1	.89	.78	.69	.61	.53	.46	.40	.35	.30	.26	.23	.19	.18	.16	.14	.12	.10	.08	.07	.06	.05	.04	.03	.02	.02
8	1	1	1	1	1	1	1	1	1	1	1	1	1	1	1	1	1	1	1	1	1	1	1	1	1	1	.99
	2	1	1	1	1	1	1	1	1	1	1	1	1	1	1	1	1	1	.99	.99	.99	.99	.98	.98	.97	.96	.95
	3	1	1	1	1	1	1	1	1	1	1	1	.99	.99	.99	.98	.98	.97	.96	.95	.94	.92	.90	.88	.86	.83	.80
	4	1	1	1	1	1	1	.99	.99	.98	.98	.97	.95	.94	.93	.92	.90	.87	.85	.82	.78	.75	.71	.67	.63	.59	.54
	5	1	1	1	.99	.99	.97	.96	.94	.91	.88	.85	.82	.78	.76	.74	.69	.65	.60	.56	.51	.46	.42	.38	.34	.30	.26
	6	1	.99	.97	.94	.90	.85	.80	.74	.69	.63	.58	.52	.47	.44	.42	.37	.33	.29	.25	.22	.19	.16	.13	.11	.09	.08
	7	1	.87	.75	.65	.56	.48	.41	.35	.30	.25	.21	.18	.15	.13	.12	.10	.08	.07	.05	.04	.04	.03	.02	.02	.01	.01

n	r	.50	.48	.46	.44	.42	.40	.38	.36	.34	.32	.30	.28	.26	.25	.24	.22	.20	.18	.16	.14	.12	.10	.08	.06	.04	.02	q 0
9	1	.99	.99	.99	.99	.98	.98	.98	.97	.96	.95	.94	.93	.91	.90	.89	.86	.83	.79	.75	.70	.64	.57	.48	.39	.28	.15	0
	2	.96	.96	.94	.93	.91	.89	.87	.85	.82	.78	.74	.70	.66	.63	.61	.55	.50	.44	.37	.31	.25	.19	.13	.08	.04	.01	0
	3	.86	.83	.80	.76	.72	.68	.64	.60	.55	.50	.45	.40	.35	.32	.30	.25	.20	.16	.12	.09	.06	.04	.02	.01	0	0	0
	4	.64	.59	.55	.50	.45	.41	.36	.32	.27	.23	.19	.16	.13	.11	.10	.08	.06	.04	.03	.02	.01	.01	0	0	0	0	0
	5	.36	.32	.28	.24	.21	.17	.14	.12	.09	.07	.06	.04	.03	.03	.02	.02	.01	.01	0	0	0	0	0	0	0	0	0
	6	.14	.12	.10	.08	.06	.05	.04	.03	.02	.02	.01	.01	.01	0	0	0	0	0	0	0	0	0	0	0	0	0	0
	7	.04	.03	.02	.02	.01	.01	.01	0	0	0	0	0	0	0	0	0	0	0	0	0	0	0	0	0	0	0	0
	8	0	0	0	0	0	0	0	0	0	0	0	0	0	0	0	0	0	0	0	0	0	0	0	0	0	0	0
10	1	1	.99	.99	.99	.99	.99	.98	.98	.97	.97	.96	.95	.93	.92	.91	.89	.86	.83	.79	.74	.68	.61	.52	.42	.30	.16	0
	2	.98	.97	.97	.96	.94	.93	.91	.89	.87	.84	.80	.77	.72	.70	.67	.62	.56	.50	.43	.37	.30	.23	.16	.10	.05	.01	0
	3	.91	.89	.86	.84	.80	.77	.73	.69	.64	.59	.54	.48	.43	.40	.37	.32	.26	.21	.16	.12	.08	.05	.03	.01	0	0	0
	4	.75	.71	.66	.62	.57	.52	.47	.42	.37	.32	.27	.23	.18	.17	.15	.11	.09	.06	.04	.03	.02	.01	0	0	0	0	0
	5	.50	.45	.40	.36	.31	.27	.23	.19	.16	.13	.10	.08	.06	.05	.04	.03	.02	.01	.01	0	0	0	0	0	0	0	0
	6	.25	.22	.18	.15	.12	.10	.08	.06	.05	.03	.03	.02	.01	.01	.01	.01	0	0	0	0	0	0	0	0	0	0	0
	7	.09	.07	.06	.04	.03	.03	.02	.01	.01	.01	0	0	0	0	0	0	0	0	0	0	0	0	0	0	0	0	0
	8	.02	.01	.02	.01	.01	0	0	0	0	0	0	0	0	0	0	0	0	0	0	0	0	0	0	0	0	0	0
	9	0	0	0	0	0	0	0	0	0	0	0	0	0	0	0	0	0	0	0	0	0	0	0	0	0	0	0
12	2	.99	.99	.99	.98	.98	.97	.96	.95	.93	.91	.89	.86	.82	.80	.78	.73	.68	.62	.55	.47	.39	.30	.22	.14	.07	.02	0
	4	.89	.86	.83	.79	.75	.70	.65	.60	.55	.49	.43	.37	.31	.29	.26	.21	.16	.12	.08	.06	.03	.02	.01	0	0	0	0
	6	.50	.45	.39	.34	.29	.25	.20	.17	.13	.10	.08	.06	.04	.03	.03	.02	.01	.01	0	0	0	0	0	0	0	0	0
	8	.11	.09	.07	.05	.04	.03	.02	.01	.01	.01	0	0	0	0	0	0	0	0	0	0	0	0	0	0	0	0	0
	10	.01	.01	0	0	0	0	0	0	0	0	0	0	0	0	0	0	0	0	0	0	0	0	0	0	0	0	0
14	2	1	1	1	.99	.99	.99	.98	.97	.97	.95	.94	.92	.89	.87	.86	.82	.77	.71	.64	.56	.47	.38	.28	.18	.09	.03	0
	4	.95	.94	.92	.89	.87	.83	.79	.75	.70	.64	.58	.52	.45	.42	.38	.32	.25	.19	.14	.10	.06	.03	.02	.01	0	0	0
	6	.71	.66	.60	.54	.48	.43	.37	.31	.26	.21	.17	.13	.09	.08	.07	.05	.03	.02	.01	.01	0	0	0	0	0	0	0
	8	.29	.24	.20	.16	.13	.10	.07	.05	.04	.03	.02	.01	.01	.01	0	0	0	0	0	0	0	0	0	0	0	0	0
	10	.05	.03	.02	.02	.01	.01	.01	0	0	0	0	0	0	0	0	0	0	0	0	0	0	0	0	0	0	0	0
	12	0	0	0	0	0	0	0	0	0	0	0	0	0	0	0	0	0	0	0	0	0	0	0	0	0	0	0

q values across top.

n	r	1.0	.98	.96	.94	.92	.90	.88	.86	.84	.82	.80	.78	.76	.75	.74	.72	.70	.68	.66	.64	.62	.60	.58	.56	.54	.52	r	n
9	1	1	1	1	1	1	1	1	1	1	1	1	1	1	1	1	1	1	1	1	1	1	1	1	1	1	.99	1	9
	2	1	1	1	1	1	1	1	1	1	1	1	1	1	1	1	1	1	1	1	1	.99	.99	.99	.98	.98	.97	2	
	3	1	1	1	1	1	1	1	1	1	1	1	1	1	1	.99	.99	.99	.98	.98	.97	.96	.95	.94	.92	.90	.88	3	
	4	1	1	1	1	1	1	1	1	1	.99	.99	.98	.98	.97	.97	.96	.94	.93	.91	.88	.86	.83	.79	.76	.72	.68	4	
	5	1	1	1	1	1	.99	.99	.98	.97	.96	.94	.92	.90	.89	.87	.84	.81	.77	.73	.68	.64	.59	.55	.50	.45	.41	5	
	6	1	1	1	.99	.98	.96	.94	.91	.88	.84	.80	.75	.70	.68	.65	.60	.55	.50	.45	.40	.36	.32	.28	.24	.20	.17	6	
	7	1	.99	.96	.92	.87	.81	.75	.69	.63	.56	.50	.45	.39	.37	.34	.30	.26	.22	.18	.15	.13	.11	.09	.07	.06	.04	7	
	8	1	.85	.72	.61	.51	.43	.36	.30	.25	.20	.17	.14	.11	.10	.09	.07	.06	.05	.04	.03	.02	.02	.01	.01	.01	.01	8	
10	1	1	1	1	1	1	1	1	1	1	1	1	1	1	1	1	1	1	1	1	1	1	1	1	1	1	1	1	10
	2	1	1	1	1	1	1	1	1	1	1	1	1	1	1	1	1	1	1	1	1	1	1	.99	.99	.99	.99	2	
	3	1	1	1	1	1	1	1	1	1	1	1	1	1	1	1	1	1	.99	.99	.99	.98	.97	.96	.95	.94	.93	3	
	4	1	1	1	1	1	1	1	1	1	1	1	.99	.99	.99	.99	.98	.97	.97	.95	.94	.92	.90	.88	.85	.82	.78	4	
	5	1	1	1	1	1	1	1	.99	.99	.99	.98	.97	.96	.95	.94	.92	.90	.87	.84	.81	.77	.73	.69	.64	.60	.55	5	
	6	1	1	1	1	1	.99	.98	.97	.96	.94	.91	.89	.85	.83	.82	.77	.73	.68	.63	.58	.53	.48	.43	.38	.34	.29	6	
	7	1	1	1	.99	.97	.95	.92	.88	.84	.79	.74	.68	.63	.60	.57	.52	.46	.41	.36	.31	.27	.23	.20	.16	.13	.11	7	
	8	1	.99	.95	.90	.84	.77	.70	.63	.57	.50	.44	.38	.33	.30	.28	.23	.20	.16	.13	.11	.09	.07	.06	.04	.03	.03	8	
	9	1	.83	.69	.57	.47	.39	.32	.26	.21	.17	.13	.11	.08	.08	.07	.05	.04	.03	.02	.02	.01	.01	.01	.01	.01	0	9	
12	2	1	1	1	1	1	1	1	1	1	1	1	1	1	1	1	1	1	1	1	1	1	1	1	1	1	1	2	12
	4	1	1	1	1	1	1	1	1	1	1	1	1	1	1	1	1	1	.99	.99	.99	.98	.97	.96	.95	.93	.91	4	
	6	1	1	1	1	1	1	1	1	1	1	.99	.98	.97	.97	.96	.94	.92	.90	.87	.83	.80	.75	.71	.66	.61	.55	6	
	8	1	1	1	.99	.99	.98	.97	.94	.92	.88	.84	.79	.74	.71	.69	.63	.57	.51	.45	.40	.35	.30	.25	.21	.17	.14	8	
	10	1	.97	.93	.86	.78	.70	.61	.53	.45	.38	.32	.27	.22	.20	.18	.14	.11	.09	.07	.05	.04	.03	.02	.02	.01	.01	10	
14	2	1	1	1	1	1	1	1	1	1	1	1	1	1	1	1	1	1	1	1	1	1	1	1	1	1	1	2	14
	4	1	1	1	1	1	1	1	1	1	1	1	1	1	1	1	1	1	1	1	1	.99	.99	.99	.98	.98	.97	4	
	6	1	1	1	1	1	1	1	1	1	1	1	1	1	.99	.99	.99	.98	.97	.96	.95	.93	.90	.87	.84	.80	.76	6	
	8	1	1	1	1	1	1	1	.99	.99	.98	.97	.95	.93	.92	.91	.87	.83	.79	.74	.69	.63	.57	.52	.46	.40	.34	8	
	10	1	1	1	.99	.98	.97	.94	.90	.86	.81	.75	.68	.62	.58	.55	.48	.42	.36	.30	.25	.21	.17	.13	.11	.08	.06	10	
	12	1	.97	.91	.82	.72	.62	.53	.44	.36	.29	.23	.18	.14	.13	.11	.08	.06	.05	.03	.03	.02	.01	.01	.01	0	0	12	

n	r	q: 0	.02	.04	.06	.08	.10	.12	.14	.16	.18	.20	.22	.24	.25	.26	.28	.30	.32	.34	.36	.38	.40	.42	.44	.46	.48	.50	n	r
16	2	0	.04	.12	.23	.34	.45	.55	.64	.72	.78	.83	.87	.91	.92	.93	.95	.96	.98	.98	.99	.99	.99	1	1	1	1	1	16	2
	4	0	0	0	.01	.03	.06	.10	.15	.21	.28	.35	.43	.50	.54	.57	.64	.70	.76	.81	.85	.88	.91	.93	.95	.96	.97	.98		4
	6	0	0	0	0	0	0	.01	.01	.02	.04	.06	.09	.13	.15	.17	.22	.28	.34	.40	.47	.53	.60	.66	.71	.76	.81	.85		6
	8	0	0	0	0	0	0	0	0	0	0	0	.01	.01	.02	.02	.03	.05	.07	.10	.13	.17	.21	.26	.32	.38	.44	.50		8
	10	0	0	0	0	0	0	0	0	0	0	0	0	0	0	0	0	0	.01	.01	.02	.02	.03	.05	.07	.09	.12	.15		10
	12	0	0	0	0	0	0	0	0	0	0	0	0	0	0	0	0	0	0	0	0	0	0	0	.01	.01	.01	.02		12
	14	0	0	0	0	0	0	0	0	0	0	0	0	0	0	0	0	0	0	0	0	0	0	0	0	0	0	0		14
18	2	0	.04	.15	.27	.40	.52	.62	.71	.78	.84	.88	.92	.94	.95	.96	.97	.98	.99	.99	.99	1	1	1	1	1	1	1	18	2
	4	0	0	0	.02	.04	.08	.14	.21	.28	.37	.45	.53	.61	.65	.68	.74	.80	.84	.88	.91	.94	.95	.97	.98	.98	.99	.99		4
	6	0	0	0	0	0	0	.01	.02	.04	.07	.11	.15	.20	.23	.27	.33	.40	.47	.55	.61	.68	.74	.79	.83	.87	.90	.93		6
	8	0	0	0	0	0	0	0	0	0	.01	.01	.02	.03	.04	.05	.07	.10	.14	.19	.24	.30	.36	.43	.49	.56	.62	.69		8
	10	0	0	0	0	0	0	0	0	0	0	0	0	0	0	0	.01	.01	.02	.03	.05	.07	.09	.12	.16	.21	.26	.31		10
	12	0	0	0	0	0	0	0	0	0	0	0	0	0	0	0	0	0	0	0	0	.01	.01	.02	.02	.04	.05	.07		12
	14	0	0	0	0	0	0	0	0	0	0	0	0	0	0	0	0	0	0	0	0	0	0	0	0	0	0	.01		14
	16	0	0	0	0	0	0	0	0	0	0	0	0	0	0	0	0	0	0	0	0	0	0	0	0	0	0	0		16
20	2	0	.05	.18	.32	.46	.58	.68	.77	.83	.88	.92	.94	.96	.97	.97	.98	.99	.99	1	1	1	1	1	1	1	1	1	20	2
	4	0	0	.01	.02	.06	.11	.19	.27	.36	.45	.54	.63	.70	.74	.77	.82	.87	.90	.93	.95	.97	.98	.98	.99	.99	1	1		4
	6	0	0	0	0	0	.01	.02	.04	.07	.11	.16	.23	.30	.33	.37	.45	.53	.60	.67	.73	.79	.84	.88	.91	.93	.95	.97		6
	8	0	0	0	0	0	0	0	0	.01	.01	.02	.04	.06	.08	.09	.13	.18	.24	.30	.37	.44	.51	.58	.65	.71	.77	.82		8
	10	0	0	0	0	0	0	0	0	0	0	0	0	.01	.01	.01	.02	.03	.05	.07	.10	.14	.19	.24	.30	.36	.43	.50		10
	12	0	0	0	0	0	0	0	0	0	0	0	0	0	0	0	0	0	.01	.01	.01	.02	.04	.05	.07	.10	.14	.18		12
	14	0	0	0	0	0	0	0	0	0	0	0	0	0	0	0	0	0	0	0	0	0	0	.01	.01	.01	.02	.03		14
	16	0	0	0	0	0	0	0	0	0	0	0	0	0	0	0	0	0	0	0	0	0	0	0	0	0	0	0		16
	18	0	0	0	0	0	0	0	0	0	0	0	0	0	0	0	0	0	0	0	0	0	0	0	0	0	0	0		18
30	4	0	0	.03	.09	.20	.33	.47	.59	.70	.79	.86	.91	.94	.95	.96	.98	.99	.99	1	1	1	1	1	1	1	1	1	30	4
	8	0	0	0	0	0	.01	.02	.04	.08	.14	.21	.30	.39	.44	.49	.59	.68	.76	.82	.87	.91	.94	.96	.98	.99	.99	1		8
	12	0	0	0	0	0	0	0	0	0	0	.01	.01	.03	.04	.05	.08	.13	.19	.26	.34	.42	.51	.60	.68	.75	.82	.87		12
	16	0	0	0	0	0	0	0	0	0	0	0	0	0	0	0	0	0	.01	.02	.03	.05	.07	.11	.15	.21	.28	.36		16
	20	0	0	0	0	0	0	0	0	0	0	0	0	0	0	0	0	0	0	0	0	0	0	0	.01	.01	.02	.03		20
	24	0	0	0	0	0	0	0	0	0	0	0	0	0	0	0	0	0	0	0	0	0	0	0	0	0	0	0		24
	28	0	0	0	0	0	0	0	0	0	0	0	0	0	0	0	0	0	0	0	0	0	0	0	0	0	0	0		28

n	r	1.0	.98	.96	.94	.92	.90	.88	.86	.84	.82	.80	.78	.76	.75	.74	.72	.70	.68	.66	.64	.62	.60	.58	.56	.54	.52 (q)
16	2	1	1	1	1	1	1	1	1	1	1	1	1	1	1	1	1	1	1	1	1	1	1	1	1	1	1
	4	1	1	1	1	1	1	1	1	1	1	1	1	1	1	1	1	1	1	1	1	1	1	1	.99	.99	.99
	6	1	1	1	1	1	1	1	1	1	1	1	1	1	1	1	1	1	.99	.99	.98	.98	.97	.95	.93	.91	.88
	8	1	1	1	1	1	1	1	1	1	1	1	.99	.99	.98	.98	.97	.95	.93	.90	.87	.83	.79	.74	.68	.62	.56
	10	1	1	1	1	1	1	.99	.99	.98	.96	.94	.91	.87	.85	.83	.78	.72	.66	.60	.53	.47	.40	.34	.29	.24	.19
	12	1	1	1	.99	.97	.94	.90	.85	.79	.72	.65	.57	.50	.46	.43	.36	.30	.24	.19	.15	.12	.09	.07	.05	.04	.03
	14	1	.96	.88	.77	.66	.55	.45	.36	.28	.22	.17	.13	.09	.08	.07	.05	.04	.02	.02	.01	.01	.01	0	0	0	0
18	2	1	1	1	1	1	1	1	1	1	1	1	1	1	1	1	1	1	1	1	1	1	1	1	1	1	1
	4	1	1	1	1	1	1	1	1	1	1	1	1	1	1	1	1	1	1	1	1	1	1	1	1	1	1
	6	1	1	1	1	1	1	1	1	1	1	1	1	1	1	1	1	1	1	1	1	.99	.99	.98	.98	.96	.95
	8	1	1	1	1	1	1	1	1	1	1	1	1	1	1	1	.99	.99	.98	.97	.95	.93	.91	.88	.84	.79	.74
	10	1	1	1	1	1	1	1	1	1	.99	.99	.98	.97	.96	.95	.93	.90	.86	.81	.76	.70	.64	.57	.51	.44	.38
	12	1	1	1	1	1	1	.99	.98	.96	.93	.89	.85	.80	.77	.73	.67	.60	.53	.45	.39	.32	.26	.21	.17	.13	.10
	14	1	1	1	.98	.96	.92	.86	.79	.72	.63	.55	.47	.39	.35	.32	.26	.20	.16	.12	.09	.06	.05	.03	.02	.02	.01
	16	1	.96	.85	.73	.60	.48	.38	.29	.22	.16	.12	.08	.06	.05	.04	.03	.02	.01	.01	.01	0	0	0	0	0	0
20	2	1	1	1	1	1	1	1	1	1	1	1	1	1	1	1	1	1	1	1	1	1	1	1	1	1	1
	4	1	1	1	1	1	1	1	1	1	1	1	1	1	1	1	1	1	1	1	1	1	1	1	1	1	1
	6	1	1	1	1	1	1	1	1	1	1	1	1	1	1	1	1	1	1	1	1	1	1	.99	.99	.99	.98
	8	1	1	1	1	1	1	1	1	1	1	1	1	1	1	1	1	1	.99	.99	.99	.98	.96	.95	.93	.90	.86
	10	1	1	1	1	1	1	1	1	1	1	1	1	.99	.99	.99	.98	.97	.95	.93	.90	.86	.81	.76	.70	.64	.57
	12	1	1	1	1	1	1	1	1	.99	.99	.98	.96	.94	.92	.91	.87	.82	.76	.70	.63	.56	.49	.42	.35	.29	.23
	14	1	1	1	1	1	.99	.98	.96	.93	.89	.84	.77	.70	.67	.63	.55	.47	.40	.33	.27	.21	.16	.12	.09	.07	.05
	16	1	1	.99	.98	.94	.89	.81	.73	.64	.55	.46	.37	.30	.26	.23	.18	.13	.10	.07	.05	.03	.02	.02	.01	.01	0
	18	1	.95	.82	.68	.54	.42	.32	.23	.17	.12	.08	.06	.04	.03	.03	.02	.01	.01	0	0	0	0	0	0	0	0
30	4	1	1	1	1	1	1	1	1	1	1	1	1	1	1	1	1	1	1	1	1	1	1	1	1	1	1
	8	1	1	1	1	1	1	1	1	1	1	1	1	1	1	1	1	1	1	1	1	1	1	1	1	1	1
	12	1	1	1	1	1	1	1	1	1	1	1	1	1	1	1	1	1	1	1	1	.99	.99	.98	.96	.94	.91
	16	1	1	1	1	1	1	1	1	1	1	1	1	1	.99	.99	.98	.97	.95	.92	.88	.83	.77	.69	.61	.53	.44
	20	1	1	1	1	1	1	1	1	.99	.97	.95	.91	.86	.83	.80	.72	.64	.54	.45	.36	.28	.21	.16	.11	.07	.05
	24	1	1	1	.99	.97	.94	.87	.79	.68	.57	.46	.36	.27	.23	.20	.14	.09	.06	.04	.02	.01	.01	0	0	0	0
	28	1	.89	.68	.47	.31	.20	.12	.07	.04	.02	.01	.01	0	0	0	0	0	0	0	0	0	0	0	0	0	0

Cumulative probabilities for given n, r, and q.

n	r	.50	.48	.46	.44	.42	.40	.38	.36	.34	.32	.30	.28	.26	.25	.24	.22	.20	.18	.16	.14	.12	.10	.08	.06	.04	.02	0
40	4	1	1	1	1	1	1	1	1	1	1	1	1	1	.99	.99	.98	.97	.94	.89	.81	.70	.56	.38	.20	.07	.01	0
	8	1	1	1	1	1	1	.99	.99	.98	.96	.93	.89	.83	.79	.75	.65	.53	.40	.28	.17	.09	.04	.01	0	0	0	0
	12	1	.99	.98	.97	.95	.91	.86	.80	.72	.62	.52	.41	.30	.25	.21	.13	.07	.04	.02	.01	0	0	0	0	0	0	0
	16	.90	.85	.78	.70	.61	.51	.41	.31	.22	.15	.09	.05	.03	.02	.01	.01	0	0	0	0	0	0	0	0	0	0	0
	20	.50	.40	.31	.22	.16	.10	.06	.04	.02	.01	0	0	0	0	0	0	0	0	0	0	0	0	0	0	0	0	0
	24	.10	.06	.04	.02	.01	.01	0	0	0	0	0	0	0	0	0	0	0	0	0	0	0	0	0	0	0	0	0
	28	0	0	0	0	0	0	0	0	0	0	0	0	0	0	0	0	0	0	0	0	0	0	0	0	0	0	0
	32	0	0	0	0	0	0	0	0	0	0	0	0	0	0	0	0	0	0	0	0	0	0	0	0	0	0	0
	36	0	0	0	0	0	0	0	0	0	0	0	0	0	0	0	0	0	0	0	0	0	0	0	0	0	0	0
60	4	1	1	1	1	1	1	1	1	1	1	1	1	1	1	1	1	1	1	.99	.97	.93	.85	.70	.48	.21	.03	0
	8	1	1	1	1	1	1	1	1	1	1	1	1	.99	.99	.98	.96	.93	.86	.75	.60	.41	.23	.10	.02	0	0	0
	12	1	1	1	1	1	1	1	1	.99	.98	.97	.93	.87	.84	.79	.67	.53	.37	.23	.11	.05	.01	0	0	0	0	0
	16	1	1	1	1	.99	.99	.97	.94	.90	.83	.73	.61	.47	.40	.33	.21	.12	.05	.02	.01	0	0	0	0	0	0	0
	20	1	.99	.98	.96	.92	.86	.78	.68	.56	.42	.30	.19	.11	.08	.06	.02	.01	0	0	0	0	0	0	0	0	0	0
	24	.94	.90	.83	.74	.63	.51	.38	.27	.17	.10	.05	.02	.01	.01	0	0	0	0	0	0	0	0	0	0	0	0	0
	28	.70	.58	.46	.34	.24	.15	.09	.05	.02	.01	0	0	0	0	0	0	0	0	0	0	0	0	0	0	0	0	0
	32	.30	.20	.13	.07	.04	.02	.01	0	0	0	0	0	0	0	0	0	0	0	0	0	0	0	0	0	0	0	0
	36	.06	.03	.01	.01	0	0	0	0	0	0	0	0	0	0	0	0	0	0	0	0	0	0	0	0	0	0	0
	40	0	0	0	0	0	0	0	0	0	0	0	0	0	0	0	0	0	0	0	0	0	0	0	0	0	0	0
	44	0	0	0	0	0	0	0	0	0	0	0	0	0	0	0	0	0	0	0	0	0	0	0	0	0	0	0
	48	0	0	0	0	0	0	0	0	0	0	0	0	0	0	0	0	0	0	0	0	0	0	0	0	0	0	0
	52	0	0	0	0	0	0	0	0	0	0	0	0	0	0	0	0	0	0	0	0	0	0	0	0	0	0	0
	56	0	0	0	0	0	0	0	0	0	0	0	0	0	0	0	0	0	0	0	0	0	0	0	0	0	0	0

(column header = q)

_q_

r	n	1.0	.98	.96	.94	.92	.90	.88	.86	.84	.82	.80	.78	.76	.75	.74	.72	.70	.68	.66	.64	.62	.60	.58	.56	.54	.52
4	40	1	1	1	1	1	1	1	1	1	1	1	1	1	1	1	1	1	1	1	1	1	1	1	1	1	1
8		1	1	1	1	1	1	1	1	1	1	1	1	1	1	1	1	1	1	1	1	1	1	1	1	1	1
12		1	1	1	1	1	1	1	1	1	1	1	1	1	1	1	1	1	1	1	1	1	1	1	1	1	1
16		1	1	1	1	1	1	1	1	1	1	1	1	1	1	1	1	1	1	1	1	1	.99	.99	.98	.96	.94
20		1	1	1	1	1	1	1	1	1	1	1	1	1	1	1	1	1	.99	.98	.96	.94	.90	.84	.78	.69	.60
24		1	1	1	1	1	1	1	1	1	1	1	.99	.99	.98	.97	.95	.91	.85	.78	.69	.59	.49	.39	.30	.22	.15
28		1	1	1	1	1	1	1	.99	.98	.96	.93	.87	.79	.75	.70	.59	.48	.38	.28	.20	.14	.09	.05	.03	.02	.01
32		1	1	1	1	.99	.96	.91	.83	.72	.60	.47	.35	.25	.21	.17	.11	.07	.04	.02	.01	.01	0	0	0	0	0
36		1	.99	.93	.80	.62	.44	.30	.19	.11	.06	.03	.02	.01	.01	0	0	0	0	0	0	0	0	0	0	0	0
4	60	1	1	1	1	1	1	1	1	1	1	1	1	1	1	1	1	1	1	1	1	1	1	1	1	1	1
8		1	1	1	1	1	1	1	1	1	1	1	1	1	1	1	1	1	1	1	1	1	1	1	1	1	1
12		1	1	1	1	1	1	1	1	1	1	1	1	1	1	1	1	1	1	1	1	1	1	1	1	1	1
16		1	1	1	1	1	1	1	1	1	1	1	1	1	1	1	1	1	1	1	1	1	1	1	1	1	1
20		1	1	1	1	1	1	1	1	1	1	1	1	1	1	1	1	1	1	1	1	1	1	1	1	1	1
24		1	1	1	1	1	1	1	1	1	1	1	1	1	1	1	1	1	1	1	1	1	1	1	.99	.99	.97
28		1	1	1	1	1	1	1	1	1	1	1	1	1	1	1	1	1	1	1	1	.99	.98	.96	.93	.87	.80
32		1	1	1	1	1	1	1	1	1	1	1	1	1	1	1	1	1	.99	.98	.95	.91	.85	.76	.66	.54	.42
36		1	1	1	1	1	1	1	1	1	1	1	1	1	.99	.99	.98	.95	.90	.83	.73	.62	.49	.37	.26	.17	.10
40		1	1	1	1	1	1	1	1	1	1	.99	.98	.94	.92	.89	.81	.70	.58	.44	.32	.22	.14	.08	.04	.02	.01
44		1	1	1	1	1	1	1	.99	.98	.95	.88	.79	.67	.60	.53	.39	.27	.17	.10	.06	.03	.01	.01	0	0	0
48		1	1	1	1	1	.99	.95	.89	.77	.63	.47	.33	.21	.16	.13	.07	.03	.02	.01	0	0	0	0	0	0	0
52		1	1	1	.98	.90	.77	.59	.40	.25	.14	.07	.04	.02	.01	.01	0	0	0	0	0	0	0	0	0	0	0
56		1	.97	.79	.52	.30	.15	.07	.03	.01	0	0	0	0	0	0	0	0	0	0	0	0	0	0	0	0	0

Answers to Selected Exercises

The following are answers to selected numerical *exercises* only. Solutions and answers to *all* exercises, questions, and problems can be found in the instructor's manual.

Chapter 2: 2.9. $158.78. **2.10.** $2003.90, 19.2%. **2.11a.** $23.71, –$28.44.
2.11b. 14.5%. **2.12a.** –$25.60. **2.12b.** $25.60. **2.12c.** –$0.08.

Chapter 4: 4.4. 3.0, 3.2. **4.6.** 8.1, 3.2. **4.8.** 5.08, 4.5.

Chapter 7: 7.3. 0.12, 0.29, 0.41, 0.65, 0.35, 0.29, 0.18, 0.17. **7.4.** 0.94.
7.6. j, c, c, c, j, c, c, c, c, c, j. **7.7.** σ_B = $3.72 million. σ_C = 0. **7.8.** 0.58, 0.34,
0.75, 0.42, 0.58, 0.66, 0.34, 0.25, 0.75. **7.9.** 0.90, 0.61, 0.61, 0.39, 0.61, 0.10, 0.90,
0.10, 0.90. **7.10.** 2.73, 0.60. **7.14.** 0.70, 0.98, 0.21, 0.79, 0.970, 0.030, 0.603,
0.397. **7.15.** 0.56. **7.16a.** 2.65, 0.728, 0.853. **7.16b.** 26.40, 2255.04, 47.49.
7.16c. 0.632, 0.233, 0.482. **7.19a.** $7000, 12250. **7.19b.** $42,895, 41871.
7.20. $1594, 430,575. **7.22.** 0.24, 0.795, 0.001. **7.24.** 0.779, 0.041, 0.656, 0.134.

Chapter 8: 8.13. (1) 39 years; (2) 4187 miles; (3) 13 countries; (4) 39 books;
(5) 2160 miles; (6) 390,000 pounds; (7) 1756; (8) 645 days; (9) 5959 miles;
(10) 36,198 feet.

Chapter 9: 9.1. 0.915. **9.2.** 0.0918, 0.0918. **9.3.** 0.027, 0.706, 0.001.
9.4. 0.33. **9.5a.** 0.037, 0.971, 0.191, 0.157; 0.205, 0.514, 0.317, 0.108.
9.5b. 0.180, 0.594, 0.099, 0.068; 0.336, 0.130, 0.190, 0.310. **9.5c.** 0.993, 0.471, 0.0;
0.368, 0.002, 0.368. **9.5d.** 0.6915, 0.2604, 0.0062, 0.0; 0.9525, 0.1056, 0.5363,
0.6915. **9.5e.** 0.39, 0.01, 0.36; 0.31, 0.30, 0. **9.6.** –1.645, 0; 0.675, 1.28.
9.7. μ = 200, σ = 111.11. **9.8.** 0.2854. **9.9.** 5.4. **9.10.** 1.743.

Chapter 10: 10.4. 275,400. **10.6.** 9300, 9850.

Chapter 12: 12.2. 1.2. **12.4.** 3.04, 1.20, 3.22.

Chapter 13: 13.6. $236. **13.7a.** 0.56, 0.48, 0, –1.81. **13.7b.** 0.052.
13.7c. $64.52, $270.98. **13.7d.** $106.18. **13.7e.** $2363.

Chapter 15: 15.7. 1, 0.4, 0. **15.9.** 0.41, 0.59. **15.11.** 0.25, 0.34, 0.41.

Chapter 16: 16.6. 0.48, 0.67.

Credits

Figure 3.1 & 3.2: Reprinted by permission of the publisher from *Value-Focused Thinking: A Path to Creative Decision Making*, by Ralph L. Keeney, Cambridge, Mass.: Harvard University Press. Copyright © 1992 by the President and Fellows of Harvard College.

Table 3.5 & 3.6: Adapted by permission of the publisher from *Value-Focused Thinking: A Path to Creative Decision Making*, by Ralph L. Keeney, Cambridge, Mass.: Harvard University Press. Copyright © 1992 by the President and Fellows of Harvard College.

Figure 5.13: From "To Catheterise or not to Catheterise?" by F. MacCartney, et al, *British Heart Journal*, 1984, *51*, 330–338. Reprinted by permission.

Table 5.2: From "Decision Analysis of a Facilities Investment and Expansion Decision," by C. S. Spetzler and R. M. Zamora. In R. A. Howard and J. E. Matheson (Eds.), *The Principles and Applications of Decision Analysis*. Copyright © 1984 Strategic Decisions Group. Reprinted by permission.

Figure 6.6 & 6.7: From *Conceptual Blockbusting: A Guide to Better Ideas* by James L. Adams. © 1974, 1976, 1979, 1986 by James L. Adams. Stanford Alumni Association, Stanford, CA and James L. Adams.

Figure 6.4: Reprinted by permission of the publisher from *Value-Focused Thinking; A Path to Creative Decision Making*, by Ralph L. Keeney, Cambridge, Mass.: Harvard University Press. Copyright © 1992 by the President and Fellows of Harvard College.

Page 229: From Synectics, by William J. J. Gordon. Copyright © 1961 Harper & Row. Reprinted by permission of the author.

Page 302: From "On the Psychology of Prediction," by D. Kahneman and A. Tversky, 1973, *Psychological Review*, *80*, 237–251. Copyright © 1973 American Psychological Association. Reprinted by permission.

Figure 8.16 & 8.17: From "Probabilistic Forecasts from Probabilistic Models Case Study: The Oil Market," by B. Abramson and A. J. Finizza, 1995, *International Journal of Forecasting*, *11*, 63–72. Reprinted by permission of Elsevier Science and the author.

Figure 16.7 & Table 16.2: Reprinted by permission of the publisher from *Value-Focused Thinking: A Path to Creative Decision Making* by Ralph L. Keeney, Cambridge, Mass.: Harvard University Press. Copyright © 1992 by the President and Fellows of Harvard College.

Author Index

Subject Index

Note: Italicized page numbers refer to problems or case studies